MICROSCALE ORGANIC LABORATORY

with Multistep and Multiscale Syntheses

SEVENTH EDITION

MICROSCALE ORGANIC LABORATORY

with Multistep and Multiscale Syntheses

SEVENTH EDITION

Dana W. Mayo

Charles Weston Pickard
Professor of Chemistry, Emeritus
Bowdoin College

Ronald M. Pike

Professor of Chemistry, Emeritus
Merrimack College

David C. Forbes

Professor of Chemistry
University of South Alabama

**With contributions by:

Dr. Nicholas E. Leadbeater
Department of Chemistry
University of Connecticut

Dr. Cynthia B. McGowan
Department of Chemistry
Merrimack College

Dr. Andrew Dicks
Department of Chemistry
University of Toronto

Dr. Elizabeth Stemmler
Department of Chemistry
Bowdoin College

WILEY

DIRECTOR	Andrea Pellerito
ASSISTANT EDITOR	Samantha Hart
SENIOR MANAGING EDITOR	Judy Howarth
PRODUCTION EDITOR	Mahalakshmi Babu
COVER PHOTO CREDIT	© Nikolay Litov/Shutterstock

This book was set in 11/13 Palatino LT Std by Straive™.

Founded in 1807, John Wiley & Sons, Inc. has been a valued source of knowledge and understanding for more than 200 years, helping people around the world meet their needs and fulfill their aspirations. Our company is built on a foundation of principles that include responsibility to the communities we serve and where we live and work. In 2008, we launched a Corporate Citizenship Initiative, a global effort to address the environmental, social, economic, and ethical challenges we face in our business. Among the issues we are addressing are carbon impact, paper specifications and procurement, ethical conduct within our business and among our vendors, and community and charitable support. For more information, please visit our website: www.wiley.com/go/citizenship.

ISBN: 978-1-119-11052-1 (PBK)
ISBN: 978-1-119-74577-8 (EVALC)

Library of Congress Cataloging-in-Publication Data

LCCN 2022046306

TO DANA AND RON, MAY THEY REST IN PEACE

When *Microscale Organic Laboratory (MOL)* was first published in 1985 as paperback Xerox copies of an unproofed manuscript, it was the only microscale organic laboratory text available. In the February 1999 *Book Buyers Guide Supplement to the Journal of Chemical Education*, however, there were seventeen laboratory manuals (of a total of thirty-nine) containing miniaturized, fully microscale, or a mixture of micro and macro experiments. Fast forward a decade and a half and without any doubt, microscale techniques have solidly established their place in chemical education. The number of lab manuals currently in print reflects the growing number of students being introduced to organic chemistry through microscale techniques. While the conversion may not yet be quite as high as the eighty percent predicted by David Brooks back in 1985, a conservative estimate would be that a solid two-thirds majority of sophomore students now work with miniaturized experiments compared with the amounts of material employed in these laboratories in the late 1970s.

The major changes that were made to *MOL* in both the fourth and fifth editions were very well received by our readers. Starting with the significant internal reorganization and rewriting with *MOL4*, *MOL5* witnessed modifications within the procedural sections to allow for inquiry-based experimentation and the inclusion of microwave heating as a tool. While *MOL6* on the surface will look very much like *MOL5* as it is nearing the fine-tuning stage in the evolution of this laboratory text, *MOL7* has undergone further internal reorganization and rewriting. Many helpful suggestions have been received from reviewers and from instructors who have used previous editions of this text. As a result, some major changes have been made for this new edition:

- A key change to the seventh edition is the addition of two dehydration labs and an Aldol condensation lab. All three build upon what was already part of *MOL6* and only makes sense to incorporate these labs, which are frequently used by many who have adopted text over the years. Our research identified each of these transformations and nicely complement the existing labs in Experiments 9 and 20 in Chapter 6. The inclusion of these experiences brings the total number of experiments in *MOL7* to 39! Furthermore, a new technique of data processing, NMR band deconvolution, is introduced both in Experiment 9c and part of an optional exercise for Experiment 17.

- Another key change throughout *MOL7*, are that sections have been added, revised, and expanded upon to illustrate current advances made in improving the "greenness" of an industrially important synthetic process and information of biological relevancy. References to green initiatives G and those of biological relevancy B are noted in the text.

- The use of microwave heating as a tool in synthetic organic chemistry continues to be a fast-growing and enabling technology. Optional instructions remain as part of *MOL7* to allow for the integration of microwave heating as a tool for performing reactions. Since reaction times are shorter than when conventional heating methods are used, students have the opportunity to supplement these activities with traditional techniques and as stated above, engage in discussions comparing the two. Optional microwave heating instructions are part of Experiments 7, 8, 15, 22, and 30. References to microwave use are noted in the text by the use of this icon $\boxed{M}$.

- The modified procedural sections allowing for question driven experimentation continues with this edition. As we highlighted in the fifth edition, this central concept is intended to develop a key skill set involving how to best monitor reactions and gauge product purity. Keeping with this format in *MOL7*, the sections that have the opportunity of a more interactive experience between groups should that be the wish of the instructor are part of Experiments 5A, 5B, 7, 19B, 24A, and 32. Experiments 11A, 16, and 28 have been modified in a way which focuses on validation of product purity. References to inquiry-based guidelines $\boxed{?}$ and validation experiences $\boxed{V}$ are noted in the text.

- And perhaps the third and most notable change you will see in *MOL7* is a complete rewrite of most of the end of chapter exercises. It is important to state that with each change, especially with the rewriting of the end of chapter exercises, no changes to the content of any question were made. The reasoning for the rewrite is as follows: Any seasoned instructor will agree that we can no longer compete with the Internet. Because what is part of *MOL7* cannot be found online, students have the opportunity to apply what is being presented and simply not copy and paste a response from the Internet. While I know this window will not remain open, it is refreshing that while it is, there is at least one complete set of problems that are "organic."

Additional Resources

Text website (www.wiley.com/go/mayo/microscaleorganiclaboratory7e)

The **Instructor's Manual** provides a list of chemicals for each experiment, setup suggestions, and anticipated outcomes. The Instructor's Manual has a separate listing for each experiment developed in the text, which often includes tips for avoiding potential trouble spots and adds considerable information and important references.

Wiley Custom Select

Wiley's custom publishing program, "Wiley Custom Select" (http://custom select.wiley.com/) gives you the freedom to build your course materials exactly the way you want them. Through a simple, on-line three step process, Wiley Custom Select allows instructors to select content from a vast database of experiments to create a customized laboratory text that meets the needs of their particular course. Each book can be fully customized—instructors can select their own output method, create a cover, arrange the sequence of content, and upload their own materials. At any time, instructors can preview a full version of what the customized book will look like, before the final order is placed.

Acknowledgments

We continue to acknowledge the outstanding contributions of the early pioneers of instructional microscale programs and techniques, such as F. Emich and F. Pregl in Austria; N. D. Cheronis (who first defined 100 mg of starting substrate in an organic reaction as a microscale transformation), L. Craig, R. C. Fuson, E. H. Huntress, T. S. Ma, A. A. Morton, F. L. Schneider, and R. L. Shriner, in the United States; and J. T. Stock in both England and the United States. These educators laid the foundation on which we were able to fashion much of the current introductory program.

In addition, we are grateful to the colleagues listed below whose careful reviews, helpful suggestions, comments, and thoughtful criticisms of the manuscript have been of such great value to us in developing the final version of this seventh edition of *MOL*.

Douglas Armstrong, *Olivet Nazarene University*

Andrew Frazer, *University of Central Florida*

Rick Heldrich, *College of Charleston*

Deborah Lieberman, *University of Cincinnati, Main Campus*

Kevin Minbiole, *Villanova University*

Evonne Rezler, *Florida Atlantic University, Boca Raton*

Kirk Voska, *Rogers State University*

We appreciate the support from Wiley and would like to extend our gratitude to Judy Howarth, Mahalakshmi Babu, Samantha Hart, and Jennifer Yee who shepherded this projected from conception to press.

We continue to applaud the widespread development of affordable glassware for use in microscale instructional laboratories. We are particularly pleased to note that the particular style of equipment (Capseal connectors) that we developed for this program at Bowdoin College has accomplished an outstanding record of survival on the battleground of the sophomore laboratory bench. Much of the credit for the granitelike character of this equipment goes to J. Ryan and Larry Riley of the ACE Glass Company. Several contributors have played long-term roles in the successful evolution of the microscale organic laboratory program, and we are happy to acknowledge them: Peter Trumper, Janet Hotham, Judy Foster, Henry Horner, Lauren Bartlett, Robert Stevens, and Samuel Butcher have all made vital contributions along the way.

We are particularly indebted to our colleagues Andrew Dicks, Nicholas Leadbeater, Cynthia McGowan, and Elizabeth Stemmler. Their willingness to contribute to this project is gratefully appreciated. Cynthia and Nicholas provided in its entirety the microwave contribution and Andy, the transition metal–catalyzed cross-coupling sequence, which was introduced in *MOL5* and continues to be a key component of *MOL7*. As it was with *MOL4*, Elizabeth's contribution of an introductory discussion on the *Application of Mass Spectrometry to Organic Chemistry* continues to offer the reader a diverse experience using this powerful technique to the introductory laboratory experience.

The development of our kinetics experiment fell on the strong shoulders of Paulette Messier, Laboratory Instructor at Bowdoin College, and adds just one more accomplishment to her unending contributions to the development of the microscale program. Paulette, more than any other person, has made this program a success in the trenches between the lab benches where it really counts.

With the publication of the seventh edition, *Microscale Organic Laboratory* might be considered to have reached a mature state. In our opinion, however, chemical education is as dynamic as the subject itself. For on our drawing boards are thoughts almost as outrageous as the idea that occurred in the early winter of 1980 to 1981—to run an introductory organic laboratory program on a milligram scale!

And finally, none of this would have been possible had it not been for the vision, drive, leadership, and effort of Dana and Ron. Starting with the first edition of *MOL*, they prioritized the reduction of costs, chemical waste, and possible hazards associated with teaching organic chemistry. Because of their efforts, we have the seventh edition of a textbook second to none. As many of you know, Dana in 2016 and Ron three years later sadly passed away. It was an honor of mine to have worked with them. Their work has been adopted by hundreds of schools across the nation reaching students in the hundreds of thousands. Today and for years to come, their work will continue to be relevant to those studying organic chemistry.

DAVID C. FORBES
December 2022

CONTENTS

Chapter 6

MICROSCALE ORGANIC LABORATORY EXPERIMENTS   119

Chapter 7

SEQUENTIAL SYNTHESES: THE TRANSITION FROM MACRO TO MICRO 440

Chapter 8

SPECTROSCOPIC IDENTIFICATION OF ORGANIC COMPOUNDS   560

Chapter 9

QUALITATIVE IDENTIFICATION OF ORGANIC COMPOUNDS 649

Chapter 10

ADVANCED MICROSCALE ORGANIC LABORATORY EXPERIMENTS 685

INTRODUCTION

You are about to embark on a challenging adventure—the microscale organic chemistry laboratory!

Your course is going to be quite different from the conventional manner in which this laboratory has been taught in past decades. You will be learning the experimental side of organic chemistry from the *microscale level*. Although you will be working with very small amounts of materials, you will be able to observe and learn more organic chemistry in one year than many of your predecessors did in nearly two years of laboratory work. You will find this laboratory an exciting and interesting place to be especially with the experiences where the outcome is not predetermined. While we cannot guarantee it for you individually, the majority of students find the microscale organic laboratory to be a surprisingly pleasant adventure.

This textbook is centered on helping you develop skills in microscale organic laboratory techniques. Its focus is twofold. For those of you in the academic environment and involved with the introductory organic laboratory, it allows the flexibility of developing your own scaling sequence without being tied to a prescribed set of quantities. For those of you working in a research environment at the advanced undergraduate or graduate level or in the industrial area, this text will provide the foundation from which you can develop a solid expertise in microscale techniques directly applicable to your work. Working at the microscale level is substantially different from using conventional operations in the organic laboratory with multigram quantities of materials.

During the last several decades, the experimental side of organic chemistry has moved ever closer to the microscale level. This conversion started in earnest nearly 30 years ago and has been spurred on by the rapidly accelerating cost of chemical waste disposal. As we have said, you will be working with very small amounts of materials, but the techniques that you will learn, and experience you will gain, will allow you to accomplish more organic chemistry in the long run than many of your predecessors.

First, we want to acquaint you with the organization and contents of the text. ***With the seventh edition, a continued effort has been made to streamline the basic reference material from the prior edition while incorporating new experiments and technologies having as our goal, a higher level of engagement for those enrolled. We hope this treatment of the laboratory will make the more important aspects of the basic text easier to access and will speed your laboratory work along.*** We then give you a few words of advice, which, if they are heeded, will allow you to avoid many of the sand traps you will find as you develop microscale laboratory techniques. Finally, we wax philosophical and attempt to describe what we think you should derive from this experience.

Chapter 1: CH₄, Methane
a substance of natural origin, known as Marsh Gas to the alchemists.

After this brief introduction, the second chapter is concerned with safety in the laboratory. This chapter supplies information that will allow you to estimate your maximum possible exposure to volatile chemicals used in the microscale laboratory. Chapter 2 also discusses general safety protocol for the laboratory. It is vitally important that you become familiar with the details of the material contained in this chapter; your health and safety depend on this knowledge.

The next three chapters are concerned primarily with the development of experimental techniques. Chapter 3 describes in detail the glassware employed in microscale organic chemistry: the logic behind its construction, tips on its usage, the common arrangements of equipment, and various other laboratory manipulations, including techniques for transferring microquantities of materials. Strange as it may sound, the number one error with any instructional laboratory experience is the proper dispensing of liquids and weighing of solids. And finally, suggestions for the organization of your laboratory notebook are presented at the end of this chapter.

Chapter 4 deals with equipment and techniques for determining a number of physical properties of microscale samples. Chapter 5 is divided into nine technique sections. Detailed discussions develop the major areas of experimental technique that are used in the microscale organic laboratory.

Chapters 6 and 7 contain the main experimental sections of this text. Chapter 6 is focused primarily on preparative organic chemistry at the microscale level and consists of 36 experiments. Six experiments (Experiments 5A, 5B, 7, 19B, 24A, and 32) in Chapter 6 have been modified in a way that replaces the posting of a reaction time with the task of monitoring the reaction by TLC until complete. The TLC technique is asked of the experimentalist in three more experiments (Experiments 11A, 16, and 28) in order to provide additional evidence of reaction purity upon recrystallization of the crude reaction mixture. And finally, five experiments (Experiments 7, 8, 15, 22, and 30) in Chapter 6 now have optional exercises that utilize microwave technologies. Additional selections of individual experiments can be drawn from those experiments presented in Chapter 7. Chapter 10 contains a series of seven experiments of a more sophisticated nature. A number of the experiments contained in Chapters 6 and Chapter 10 are of optional scale so that you may also have the opportunity to gain some experience with experimentation at larger scales. Chapter 7 consists of a set of six sequential experiments that are essentially identical to the type of problems tackled by research chemists involved in synthetic organic chemistry. A number of these multistep procedures begin the first step in the experiment with large-scale, multigram quantities of starting material, but require microscale techniques to complete the final step or two. The use of this chapter is most appropriate in the final stages of the course—for example, the latter part of the second semester of a two-semester sequence.

Chapter 8 develops the characterization of organic materials at the microscale level by spectroscopic techniques. The chapter starts with a brief discussion of the interpretation of infrared (IR) group frequencies and is followed by a more detailed treatment of nuclear magnetic resonance (NMR) spectral data, a brief discussion of ultraviolet–visible (UV–vis) spectroscopy, and a brief introduction to the theory, experimental techniques, and applications of mass spectrometry to organic chemistry. A more detailed introduction to the theoretical basis for these spectroscopic techniques is also presented on the accompanying website.

Chapter 9 develops the characterization of organic materials at the microscale level by the use of classical organic reactions to form solid derivatives. Tables of derivative data for use in compound identification by these techniques are discussed and are included on the website as Appendix A.

A list of all the experiments grouped by reaction mechanism is given on the website as Appendix B.

The organization of the experimental procedures given in Chapters 6 and 7 is arranged in the following fashion. A short opening statement describing the reaction to be studied is followed by the reaction scheme.

Generally, a brief discussion of the reaction follows, including a mechanistic interpretation. In a few cases of particularly important reactions, or where the experiment is likely to precede presentation of the topic in the classroom, a more detailed description is given. The estimated time needed to complete the work, along with a table of reactant data, comes next. For ease in organizing your laboratory time, the experimental section is divided into four subsections: *reagents and equipment, reaction conditions, isolation of product,* and *purification and characterization.*

We then introduce a series of questions and problems designed to enhance and focus your understanding of the chemistry and the experimental procedures involved in a particular laboratory exercise. Finally, a bibliography offering a list of literature references is given. Although this list comes at the end of the experimental section, we view it as a very important part of the text. The discussion of the chemistry involved in each experiment is necessarily brief. We hope that you will take time to read and expand your knowledge about the particular experiment that you are conducting. You may, in fact, find that some of these references become assigned reading.

A prompt (➡) in the text indicates that experimental apparatus involved with that stage of the experiment are shown in the margin. Important comments are italicized in the text, and **Warnings** and **Cautions** are given in boxes and also indicated in the margins.

In an effort to streamline our treatment of the laboratory we have moved a considerable quantity of material from the previous editions, MOL3, MOL4, MOL5, and MOL6, and placed it in easily accessible form on our website. An icon lets you know that supplemental material is available on the website (www.wiley.com/go/ mayo/microscaleorganiclaboratory7e). As was the case in MOL6, a detailed listing within the table of contents of all materials available online. We hope this format will make the more important aspects of the basic text easier to access and speed your laboratory work along.

GENERAL RULES FOR THE MICROSCALE LABORATORY

1. Study the experiment before you come to lab. This rule is a historical plea from all laboratory instructors. In the microscale laboratory it takes on a more important meaning. You will not survive if you do not prepare ahead of time. In microscale experiments, operations happen much more quickly than in the macroscale laboratory. Your laboratory time will be overflowing with many more events. If you are not familiar with the sequences you are to follow, you will be in deep trouble. Although the techniques employed at the microscale level are not particularly difficult to acquire, they do demand a significant amount of attention. For you to reach a successful and happy conclusion, you cannot afford to have the focus of your concentration broken by having to constantly refer to the text during the experiment. Disaster is ever present for the unprepared.

2. ALWAYS work with clean equipment. You must take the time to scrupulously clean your equipment before you start any experiment. Contaminated glassware will ultimately cost you additional time, and you will experience the frustration of inconsistent results and lower yields/recoveries. Dirty equipment is the primary cause of reaction failure at the microscale level.

3. CAREFULLY measure the quantities of materials to be used in the experiments. A little extra time at the beginning of the laboratory can speed you on your way at the end of the session. A great deal of time has been spent optimizing the conditions employed in these experiments in order to maximize yields. Many organic reactions are very sensitive to the relative quantities of substrate (the material on which the reaction is taking place) and reagent (the reactive substance or substances that bring about the change in the substrate). After equipment contamination, the second-largest cause of failed reactions is attempting to run a reaction with incorrect quantities of the reactants present. Do not be hurried or careless at the balance. Recall what was stated above, the most common error with any instructional laboratory experience is the improper dispensing of liquids or weighing of solids.

4. Clean means DRY. Water or cleaning solution can be as detrimental to the success of a reaction as dirt or sludge in the system. You often will be working with very small quantities of moisture-sensitive reagents. The glass surface areas with which these reagents come in contact, however, are relatively large. A slightly damp piece of glassware can rapidly deactivate a critical reagent and result in reaction failure. *This rule must be strictly followed.*

5. ALWAYS work on a clean laboratory bench surface, preferably glass!

6. ALWAYS protect the reaction product that you are working with from a disastrous spill by carrying out all solution or solvent transfers over a crystallizing dish.

7. ALWAYS place reaction vials or flasks in a clean beaker when standing them on the laboratory bench. Then, when a spill occurs the material is more likely to be contained in the beaker and less likely to be found on the laboratory bench or floor.

8. NEVER use cork rings to support round-bottom flasks, particularly if they contain liquids. If you do, you are inviting disaster to be a guest at your laboratory bench.

9. ALWAYS think through the next step you are going to perform before starting it. Once you have added the wrong reagent, it is back to square one.

10. ALWAYS save everything you have generated until you have ample experimental evidence that the objective(s) of experiment has been successfully completed. You can retrieve, let's say, a mislabeled chromatographic fraction from your locker, but not from the waste container!

THE ORGANIC CHEMISTRY LABORATORY

The confidence gained by mastering the microscale techniques described here will pay huge dividends as you progress into modern-day experimental chemistry. The organic laboratory has had a reputation of being smelly, long, tedious, and pockmarked with fires and explosions; but present-day organic chemistry is undergoing a revolution at the laboratory bench. New techniques are sweeping away many of the old complaints, as an increasing fraction of industrial and academic research is being carried out at the microscale level.

This book allows the interested participant to rapidly develop the skills needed to slice more deeply into organic chemistry than ever before. The attendant benefits are greater confidence and independence in acquired laboratory techniques. The happy result is that in a microscale-based organic chemistry laboratory, you are more likely to have a satisfying encounter with the experimental side of this fascinating field of knowledge.

C₂H₄apter 2

SAFETY

Research laboratories vary widely with respect to facilities and support given to safety. Large laboratories may have several hundred chemists and an extensive network of co-workers, supervisors, safety officers, and hazardous-waste managers. They also, pursuant to government regulations, have an extensive set of safety procedures and detailed practices for the storage and disposal of hazardous wastes. In small laboratories, the individual chemist may have to take care of all these aspects of safety. Some laboratories may routinely deal with very hazardous materials and may run all reactions in hoods. Others may deal mainly with relatively innocuous compounds and have very limited hood facilities.

Our approach is to raise some questions to think about and to suggest places to look for further information. In this chapter, we do not present a large list of safety precautions for use in all situations; rather, we present a list of very basic precautionary measures. A bibliography at the end of the chapter offers a list of selected references. *We urge you to consult these references concerning specific safety regulations.* Many laboratories may have safety guidelines that will supersede this very cursory treatment. This chapter is no more than a starting point.

MAKING THE LABORATORY A SAFER PLACE

Murphy's law states in brief, "If anything can go wrong, it will." Although it is often taken to be a silly law, it is not. Murphy's law means that if sparking switches are present in areas that contain flammable vapors, sooner or later there will be a fire. If the glass container can move to the edge of the shelf as items are moved around or because the building vibrates, at some time it will come crashing to the floor. If the pipet can become contaminated, then the mouth pipetter will eventually ingest a contaminant.

We cannot nullify Murphy's law, but we can do a lot to minimize the damage. We can reduce the incidence of sparks and flames and flammable vapors. We can make sure that if the accident does occur, we have the means to contain the damage and to take care of any injuries that result. All of this means thinking about the laboratory environment. Does your laboratory have or enforce regulations related to important items such as eye, face, and foot protection, safety clothing, respiratory equipment, first aid supplies, fire equipment, spill kits, hoods, and compliance regulations? *Think ahead* about what could go wrong and then *plan* and *prepare* to minimize the chance of an accident and be prepared to respond when one does occur.

NATURE OF HAZARDS

The chemistry laboratory presents a wide assortment of risks. These risks are outlined briefly here so that you can begin to think about the steps necessary to make the laboratory safer:

Chapter 2: C₂H₄, Ethylene
a substance of natural origin, released by ripening fruit.

1. **Physical hazards.** Injuries resulting from flames, explosions, and equipment (cuts from glass, electrical shock from faulty instrumentation, or improper use of instruments).

2. **External exposure to chemicals.** Injuries to skin and eyes resulting from contact with chemicals that have spilled, splashed, or been left on the bench top or on equipment.

3. **Internal exposure.** Longer-term (usually) health effects resulting from breathing hazardous vapors or ingesting chemicals.

REDUCTION OF RISKS

Many things can be done to reduce risks. The rules below may be absolute in some laboratories. In others, the nature of the materials and apparatus used may justify the relaxation of some of these rules or the addition of others.

1. **Stick to the procedures described by your supervisor.** This attention to detail is particularly important for the chemist with limited experience. In other cases, variation of the reagents and techniques may be part of the work.

2. **Wear approved safety goggles.** We can often recover quickly from injuries affecting only a few square millimeters on our bodies, unless that area happens to be in our eyes. Larger industrial laboratories often require that laboratory work clothes and safety shoes be worn. Wear them, if requested.

3. **Do not put anything in your mouth under any circumstances while in the laboratory.** This includes food, drinks, chemicals, and pipets. There are countless ways that surfaces can become contaminated in the laboratory. Since there are substances that must never be pipetted by mouth, one must get into the habit of *never* mouth pipetting anything.

4. **Be cautious with flames and flammable solvents.** Remember that the flame at one end of the bench can ignite the flammable liquid at the other end in the event of a spill or improper disposal. Flames must never be used when certain liquids are present in the laboratory, and flames must always be used with care. Check the *fire diamond* hazard symbol, if available.

5. **Be sure that you have the proper chemicals for your reaction.** Check labels carefully, and return unused chemicals to the proper place for storage. Be sure to replace caps on containers immediately after use. An open container is an invitation for a spill. Furthermore, some reagents are very sensitive to moisture, and may decompose if left open.

6. **Minimize the loss of chemicals to air or water and dispose of waste properly.** Some water-soluble materials may be safely disposed of in the water drains. Other wastes should go into special receptacles. Pay attention to the labels on these receptacles. Recent government regulations have placed stringent rules on industrial and academic laboratories for proper disposal of chemicals. *Severe penalties are levied on those who do not follow proper procedures.* We recommend that you consult general safety references nos. 3 and 4 at the end of the chapter.

7. **Minimize skin contact with any chemicals.** Use impermeable gloves when necessary, and promptly wash any chemical off your body. If you have to wash something off with water, use lots of it. Be sure that you know where the nearest water spray device is located.

NOTE. Do not use latex gloves. They are permeable to many chemicals, and some people are allergic to them. A recommended substitute are the various grades of nitrile gloves.

8. **Do not inhale vapors from volatile materials.** Severe illness or internal injury can result.

9. **Tie back or confine long hair and loose items of clothing.** You do not want them falling into a reagent or getting near flames.

10. **Do not work alone.** Too many things can happen to a person working alone that might leave him or her unable to obtain assistance. As in swimming, the "buddy system" is safest.

11. **Exercise care in assembling glass and electrical apparatus.** All operations with glass, such as separating standard taper glassware, involve the risk that the glass may break and that lacerations or punctures may result. Seek help or advice with glassware, if necessary. Special containers should be provided for the disposal of broken glass. Electrical shock can occur in many ways. When making electrical connections, make sure that your hands, the laboratory bench, and the floor are all dry and that *you* do not complete an electrical path to ground. Be sure that electrical equipment is properly grounded and insulated.

12. **Report any injury or accident to the appropriate person.** Reporting injuries and accidents is important so that medical assistance can be obtained if necessary. It also allows others to be made aware of any safety problems; these problems may be correctable.

13. **Keep things clean.** Put unused apparatus away. Immediately wipe up or care for spills on the bench top or floor. This also pertains to the balance area and to where chemicals are dispensed.

14. **Never heat a closed system.** Always provide a vent to avoid an explosion. Provide a suitable trap for any toxic gases generated in a given reaction.

15. **Learn the correct use of gas cylinders.** Even a small gas cylinder can become a lethal bomb if not properly used.

16. **Attend safety programs.** Many industrial laboratories offer excellent seminars and lectures on a wide variety of safety topics. Pay careful attention to the advice and counsel of the safety officer.

17. **Above all, use your common sense.** Think before you act.

PRECAUTIONARY MEASURES

Know the location and operation of safety equipment in the laboratory. Locate the nearest

- Fire extinguisher
- Telephone
- Exit
- First aid kit
- Emergency shower
- Fire blanket
- Eye wash

Know where to call (have the numbers posted) for

- Fire
- Medical emergency
- Spill or accidental release of corrosive or toxic chemicals

Know where to go

- In case of injury
- To evacuate the building

THINKING ABOUT THE RISKS IN USING CHEMICALS

The smaller quantities used in the microscale laboratory carry with them a reduction in hazards caused by fires and explosions; hazards associated with skin contact are also reduced. However, care must be exercised when working with even the small quantities involved.

There is great potential for reducing the exposure to chemical vapors, but these reductions will be realized only if everyone in the laboratory is careful. One characteristic of vapors emitted outside hoods is that they mix rapidly throughout the lab and will quickly reach the person on the other side of the room. In some laboratories, the majority of reactions may be carried out in hoods. When reactions are carried out in the open laboratory, each experimenter becomes a polluter whose emissions affect nearby people the most, but these emissions become added to the laboratory air and to the burden each of us must bear.

The concentration of vapor in the general laboratory air space depends on the vapor pressure of the liquids, the area of the solid or liquid exposed, the nature of air currents near the sources, and the ventilation characteristics of the laboratory. One factor over which each individual has control is evaporation, which can be reduced by the following practices:

- Certain liquids must remain in hoods.
- Reagent bottles must be recapped when not in use.
- Spills must be quickly cleaned up and the waste discarded.

Chemicals must be properly stored when not in use. Some balance must be struck between the convenience of having the compound in the laboratory where you can easily put your hands on it and the safety of having the compound in a properly ventilated and fire-safe storage room. Policies for storing chemicals will vary from place to place. There are limits to the amounts of flammable liquids that should be stored in glass containers, and fire-resistant cabinets must be used for storage of large amounts of flammable liquids. Chemicals that react with one another should not be stored in close proximity. There are plans for sorting chemicals by general reactivity classes in storerooms; for instance, Flinn Scientific Company includes a description of a storage system online.[1]

---

[1]To access their online teacher resource on how to store chemicals properly, see: https://www.flinnsci.com/flinn-chemical-storage-pattern-poster/ap6196/.

DISPOSAL OF CHEMICALS

Chemicals must also be segregated into categories for disposal. The categories used will depend on the disposal service available and upon federal, state, and local regulations. For example, some organic wastes are readily incinerated, while those containing chlorine may require much more costly treatment. Other wastes may have to be buried. For safety and economic reasons, it is important to place waste material in the appropriate container. In today's world, it often costs more to dispose of a chemical than to purchase it in the first place! The economic impact of waste generation and disposal is gigantic. Based upon the toxic release inventory from the EPA for 2019, do you realize that the chemical industry in the United States released more than 30.7 billion pounds of on-site and off-site chemical waste?[2] Each year hundreds of billions of dollars are spent per year in waste treatment, control, and disposal costs!

It is our obligation as chemists to decrease the impact that hazardous chemicals have on our environment. A movement is currently underway (referred to as "Green Chemistry" or "Benign by Design") to accomplish this goal by focusing on the design, manufacture, and use of chemicals and chemical processes that have little or no pollution potential or environmental risk.

(MATERIAL) SAFETY DATA SHEETS

Although risks are associated with the use of most chemicals, the magnitudes of these risks vary greatly. A short description of the risks is provided by a Material Safety Data Sheet, commonly referred to as an MSDS. A Safety Data Sheet or SDS, however, contains other technical and scientific information beyond Sections I–VIII and conforms to the UN Globally Harmonized System (GHS) of Classification and Labeling of Chemicals. Regardless of the format, all participants of a laboratory experience are strongly encouraged to educate themselves on the risks, large or small, of the chemicals they are scheduled to work with while in lab. The information contained on these sheets can be obtained from a number of locations. While they are normally provided by the manufacturer or vendor of the chemical, and users are required to keep on file the (M)SDS of each material stored or used, data sheets can be easily obtained online.

As an example, the 1985 MSDS for acetone is shown here. This sheet was provided by the J. T. Baker Chemical Company. Sheets from other sources, especially those online,[3] do provide a very illuminating comparison of what has transpired over 20+ years. Much of the information on these sheets is self-explanatory, but let's review the major sections of the acetone example.

Section I provides identification numbers and codes for the compound and includes a summary of the risks associated with the use of acetone. Because these sheets are available for many thousands of compounds and mixtures, there must be a means of unambiguously identifying the substance. A standard reference number for chemists is the Chemical Abstracts Service Number (CAS No.).

A quick review of the degree of risks is given by the numerical scale under Precautionary Labeling. This particular scale is a proprietary scale that ranges from 0 (very little or nonexistent risk) to 4 (extremely high risk). The National

---

[2]For online access to the EPA Toxic Release Inventory, see: www.epa.gov/tri.
[3]For an online MSDS for acetone, see: https://fscimage.fishersci.com/msds/00140.htm.

Fire Protection Association (NFPA) uses a similar scale but the risks considered are different. Other systems may use different scales, and there are some that represent low risks by the highest number! Be sure that you understand the scale being used. Perhaps some day one scale will become standard.

Section II covers risks from mixtures. Because a mixture is not considered here, the section is empty. Selected physical data are described in Section III. Section IV contains fire and explosion data, including a description of the toxic gases produced when acetone is exposed to a fire. The MSDSs are routinely made available to fire departments that may be faced with fighting a fire in a building where large amounts of chemicals are stored.

J. T. BAKER CHEMICAL CO. 222 RED SCHOOL LANE, PHILLIPSBURG, NJ 08865
MATERIAL SAFETY DATA SHEET
24-HOUR EMERGENCY TELEPHONE — (201) 859-2151
CHEMTREC # (800) 424-9300 — NATIONAL RESPONSE CENTER # (800) 424-8802

A0446 –01   ACETONE   PAGE: 1

EFFECTIVE: 10/11/85   ISSUED: 01/23/86

SECTION I – PRODUCT IDENTIFICATION

PRODUCT NAME: ACETONE
FORMULA: (CH3)2CO
FORMULA WT: 58.08
CAS NO.:   00067-64-1
NIOSH/RTECS NO.:   AL3150000
COMMON SYNONYMS: DIMETHYL KETONE; METHYL KETONE; 2-PROPANONE
PRODUCT CODES: 9010,9006,9002.9254,9009,9001,9004,5356, A134,9007,9005,9008

PRECAUTIONARY LABELLING

BAKER SAF-T-DATA(TM) SYSTEM

HEALTH   – 1
FLAMMABILITY   – 3 (FLAMMABLE)
REACTIVITY   – 2
CONTACT   – 1

LABORATORY PROTECTIVE EQUIPMENT

SAFETY GLASSES; LAB COAT; VENT HOOD; PROPER GLOVES; CLASS B EXTINGUISHER

PRECAUTIONARY LABEL STATEMENTS

DANGER
EXTREMELY FLAMMABLE
HARMFUL IF SWALLOWED OR INHALED
CAUSES IRRITATION

KEEP AWAY FROM HEAT, SPARKS, FLAME, AVOID CONTACT WITH EYES, SKIN, CLOTHING, AVOID BREATHING VAPOR, KEEP IN TIGHTLY CLOSED CONTAINER, USE WITH ADEQUATE VENTILATION, WASH THOROUGHLY AFTER HANDLING, IN CASE OF FIRE, USE WATER SPRAY, ALCOHOL FOAM, DRY CHEMICAL, OR CARBON DIOXIDE, FLUSH SPILL AREA WITH WATER SPRAY.

SECTION II – HAZARDOUS COMPONENTS

COMPONENT	%	CAS NO.
ACETONE	90-100	67-64-1

SECTION III – PHYSICAL DATA

BOILING POINT:	56 C (133 F)	VAPOR PRESSURE(MM HG):	181
MELTING POINT:	–95 C (–139 F)	VAPOR DENSITY(AIR = 1):	2
SPECIFIC GRAVITY: (H2O=1)	0.79	EVAPORATION RATE: (BUTYL ACETATE = 1)	5.6

SOLUBILITY(H2O):   COMPLETE (IN ALL PROPORTIONS) % VOLATILES BY VOLUME: 100
APPEARANCE & ODOR: CLEAR, COLORLESS LIQUID WITH FRAGRANT SWEET ODOR.

SECTION IV – FIRE AND EXPLOSION HAZARD DATA

FLASHPOINT: –18 C (0 F)   NFPA 704M RATING: 1-3-0

FLAMMABLE LIMITS: UPPER–   13%   LOWER–   2%

FIRE EXTINGUISHING MEDIA
     USE ALCOHOL FOAM, DRY CHEMICAL OR CARBON DIOXIDE.
     (WATER MAY BE INEFFECTIVE.)

SPECIAL FIRE-FIGHTING PROCEDURES
     FIREFIGHTERS SHOULD WEAR PROPER PROTECTIVE EQUIPMENT AND SELF-
     CONTAINED (POSITIVE PRESSURE IF AVAILABLE) BREATHING APPARATUS WITH
     FULL FACEPIECE.
     MOVE EXPOSED CONTAINERS FROM FIRE AREA IF IT CAN BE DONE WITHOUT RISK.
     USE WATER TO KEEP FIRE-EXPOSED CONTAINERS COOL.

UNUSUAL FIRE & EXPLOSION HAZARDS
     VAPORS MAY FLOW ALONG SURFACES TO DISTANT IGNITION SOURCES AND
     FLASH BACK, CLOSED CONTAINERS EXPOSED TO HEAT MAY EXPLODE.
     CONTACT WITH STRONG OXIDIZERS MAY CAUSE FIRE.

SECTION V – HEALTH HAZARD DATA

THRESHOLD LIMIT VALUE (TLV/TWA):   1780 MG/M3 (750 PPM)

SHORT-TERM EXPOSURE LIMIT (STEL):   2375 MG/M3 (1000 PPM)

TOXICITY: LD50 (ORAL-RAT)(MG/KG)   –   9750
          LD50 (IPR-MOUSE) (G/KG)   –   1297

EFFECTS OF OVEREXPOSURE
     CONTACT WITH SKIN HAS A DEFATTING EFFECT, CAUSING DRYING AND IRRITATION.
     OVEREXPOSURE TO VAPORS MAY CAUSE IRRITATION OF MUCOUS MEMBRANES,
     DRYNESS OF MOUTH AND THROAT, HEADACHE, NAUSEA AND DIZZINESS.

EMERGENCY AND FIRST AID PROCEDURES
     CALL A PHYSICIAN.
     IF SWALLOWED, IF CONSCIOUS, IMMEDIATELY INDUCE VOMITING.
     IF INHALED, REMOVE TO FRESH AIR. IF NOT BREATHING, GIVE ARTIFICIAL
     RESPIRATION. IF BREATHING IS DIFFICULT, GIVE OXYGEN.
     IN CASE OF CONTACT, IMMEDIATELY FLUSH EYES WITH PLENTY OF WATER FOR AT
     LEAST 15 MINUTES, FLUSH SKIN WITH WATER.

SECTION VI – REACTIVITY DATA

STABILITY: STABLE   HAZARDOUS POLYMERIZATION: WILL NOT OCCUR

CONDITIONS TO AVOID: HEAT, FLAME, SOURCES OF IGNITION

INCOMPATIBLES: SULFURIC ACID, NITRIC ACID, STRONG OXIDIZING AGENTS

SECTION VII – SPILL AND DISPOSAL PROCEDURES

STEPS TO BE TAKEN IN THE EVENT OF A SPILL OR DISCHARGE
     WEAR SUITABLE PROTECTIVE CLOTHING. SHUT OFF IGNITION SOURCES; NO
     FLARES, SMOKING, OR FLAMES IN AREA. STOP LEAK IF YOU CAN DO SO
     WITHOUT RISK, USE WATER SPRAY TO REDUCE VAPORS, TAKE UP WITH SAND
     OR OTHER NON-COMBUSTIBLE ABSORBENT MATERIAL AND PLACE INTO
     CONTAINER FOR LATER DISPOSAL. FLUSH AREA WITH WATER.

     J. T. BAKER SOLUSORB(R) SOLVENT ADSORBENT IS RECOMMENDED FOR SPILLS
     OF THIS PRODUCT.

DISPOSAL PROCEDURE
     DISPOSE IN ACCORDANCE WITH ALL APPLICABLE FEDERAL, STATE, AND LOCAL
     ENVIRONMENTAL REGULATIONS.

EPA HAZARDOUS WASTE NUMBER: U002 (TOXIC WASTE)

SECTION VIII – PROTECTIVE EQUIPMENT	
VENTILATION:	USE GENERAL OR LOCAL EXHAUST VENTILATION TO MEET TLV REQUIREMENTS.
RESPIRATORY PROTECTION:	RESPIRATORY PROTECTION REQUIRED IF AIRBORNE CONCENTRATION EXCEEDS TLV. AT CONCENTRATIONS UP TO 5000 PPM, A GAS MASK WITH ORGANIC VAPOR CANNISTER IS RECOMMENDED. ABOVE THIS LEVEL, A SELF-CONTAINED BREATHING APPARATUS WITH FULL FACE SHIELD IS ADVISED.
EYE/SKIN PROTECTION:	SAFETY GLASSES WITH SIDESHIELDS, POLYVINYL ACETATE GLOVES ARE RECOMMENDED.

SECTION IX – STORAGE AND HANDLING PRECAUTIONS

SAF-T-DATA(TM) STORAGE COLOR CODE: RED

SPECIAL PRECAUTIONS
 BOND AND GROUND CONTAINERS WHEN TRANSFERRING LIQUID. KEEP CONTAINER TIGHTLY CLOSED. STORE IN A COOL, DRY, WELL-VENTILATED, FLAMMABLE LIQUID STORAGE AREA.

SECTION X – TRANSPORTATION DATA AND ADDITIONAL INFORMATION

DOMESTIC (D.O.T.)

PROPER SHIPPING NAME	ACETONE
HAZARD CLASS	FLAMMABLE LIQUID
UN/NA	UN1090
LABELS	FLAMMABLE LIQUID

INTERNATIONAL (I.M.O.)

PROPER SHIPPING NAME	ACETONE
HAZARD CLASS	3.1
UN/NA	UN1090
LABELS	FLAMMABLE LIQUID

(TM) AND (R) DESIGNATE TRADEMARKS.
N/A = NOT APPLICABLE OR NOT AVAILABLE

THE INFORMATION PUBLISHED IN THIS MATERIAL SAFETY DATA SHEET HAS BEEN COMPILED FROM OUR EXPERIENCE AND DATA PRESENTED IN VARIOUS TECHNICAL PUBLICATIONS. IT IS THE USER'S RESPONSIBILITY TO DETERMINE THE SUITABILITY OF THIS INFORMATION FOR THE ADOPTION OF NECESSARY SAFETY PRECAUTIONS. WE RESERVE THE RIGHT TO REVISE MATERIAL SAFETY DATA SHEETS PERIODICALLY AS NEW INFORMATION BECOMES AVAILABLE.

Health hazards are described in Section V. The entries of most significance for evaluating risks from vapors are the Threshold Limit Value (or TLV) and the Short-Term Exposure Limit (STEL). The TLV is a term used by the American Conference of Governmental Industrial Hygienists (ACGIH). This organization examines the toxicity literature for a compound and establishes the TLV. This standard is designed to protect the health of workers exposed to the vapor eight hours a day, five days a week. The Occupational Safety and Health Administration (OSHA) adopts a value to protect the safety of workplaces in the United States. Their value is termed the Time-Weighted Average (TWA) and in many cases is numerically equal to the TLV. The STEL is a value not to be exceeded for even a 15-minute averaging time. TLV, TWA, and STEL values for many chemicals are summarized in a small handbook available from the ACGIH (2000); they are also collected in the *CRC Handbook of Chemistry and Physics.*

The toxicity of acetone is also described in terms of the toxic oral dose. In this case, the LD_{50} is the dose that will cause the death of 50% of the mice or rats given that dose. The dose is expressed as milligrams of chemical per kilogram of body weight of the subject animal. The figures for small animals are often used to estimate the effects on humans. If, for example, we used the mouse figure of 1297 mg/kg and applied it to a 60-kg chemist, a dose of 77,820 mg (~98.5 mL) would kill 50% of the subjects receiving that dose. As a further example, chloroform has an LD_{50} of 80 mg/kg. For our 60-kg chemist, a dose of 4800 mg (~3 mL) would be fatal for 50% of these cases. The effects of exposure of skin to the liquid and vapor are also described.

Section VI describes the reactivity of acetone and the classes of compounds with which it should not come in contact. For example, sodium metal reacts violently with a number of substances (including water) and should not come in contact with them. Strong oxidizing agents (such as nitric acid) should not be mixed with organic compounds (among other things). The final sections (Sections VII–X) are self-explanatory.

ALTERNATE SOURCES OF INFORMATION

Similar information in a more compact form can be found in the *Merck Index* (Merck). This basic reference work provides information on the toxicity of many chemicals. It often refers one to the *NIOSH Pocket Guide to Chemical Hazards* (National Institute for Occupational Safety and Health). The *Merck Index* also supplies interesting information about the common uses of the chemicals listed, particularly related to the medical area. References to the chemical literature are also provided. The *CRC Handbook of Chemistry and Physics,* which is updated each year, contains a wide range of data (located in tables) in the area of health, safety, and environmental protection. It also includes directions for the handling and disposal of laboratory chemicals. Your laboratory should have a copy of this work. Most chemical supply houses now label their containers with data showing not only the usual package size, physical properties, and chemical formula, but also pictures or codes showing hazard information. Some include a pictogram (for example, see the newer Aldrich Chemical labels on their bottles). The J. T. Baker Company uses the Baker SAF-T-DATA System.

ESTIMATING RISKS FROM VAPORS

Other things (availability, suitability) being equal, one would, of course, choose the least toxic chemical for a given reaction. Some very toxic chemicals play very important roles in synthetic organic chemistry, and the toxicity of the chemicals in common use varies greatly. Bromine and benzene have TLVs of 0.7 and 30 mg/m^3, respectively, and are at the more toxic end of the spectrum of chemicals routinely used. Acetone has a TLV of 1780 mg/m^3. These representative figures do not mean that acetone is harmless or that bromine cannot be used. In general, one should exercise care at all times (make a habit of good laboratory practice) and should take special precautions when working with highly toxic materials.

The TLV provides a simple means to evaluate the relative risk of exposure to the vapor of any substance used in the laboratory. If the quantity of the material evaporated is represented by m (milligrams/hour) and the TLV is expressed by L (milligrams per cubic meter), a measure of relative risk to the vapor is given by m/L. This quantity represents the volume of clean air required to dilute the emissions to the TLV. As an example, the emission of 1 g of bromine and 10 g of acetone in one hour leads to the values of m/L of 1400 m^3/hour (h) for the bromine and 5.6 m^3/h for acetone. These numbers provide a direct handle on the *relative* risks from these two vapors. It is difficult to assess the absolute risk to these vapors without a lot of information about the ventilation characteristics of the laboratory. If these releases occur within a properly operated hood, the threat to the worker in the laboratory is probably very small. (However, consideration must be given to the hood exhaust.)

Exposure in the general laboratory environment can be assessed if we assume that the emissions are reasonably well mixed before they are inhaled and if we know something about the room ventilation rate. The ventilation rate of the room can be measured by a number of ways.[4] Given the ventilation rate, it might be safe to assume that only 30% of that air is available for diluting the emissions. (This accounts for imperfect mixing in the room.) The effective amount of air available for dilution can then be compared with the amount of air required to dilute the chemical to the TLV.

Let us continue our example. Suppose that the laboratory has a volume of 75 m^3 and an air exchange rate of 2 air changes per hour. This value means that (75 m^3)(2/h)(0.3) = 45 m^3/h are available to dilute the pollutants. There may be enough margin for error to reduce the acetone concentration to a low level (5.6 m^3/h is required to reach the TLV), but use of bromine should be restricted to the hood. An assessment of the accumulative risk of several chemicals is obtained by adding the individual $m/L \left(\frac{\text{mg/h}}{\text{mg/m}^3} \right)$ values.

The m/L figures may also be used to assess the relative risk of performing the experiment outside a hood. Since m/L represents the volume of air for each student, this may be compared with the volume of air actually available for each student. If the ventilation rate for the entire laboratory is Q (in cubic meters per minute) for a section of n students meeting for t minutes, the volume for each student is kQt/n cubic meters. Here k is a mixing factor that allows for the fact that the ventilation air will not be perfectly mixed in the laboratory before it is exhausted. In a reasonable worst-case mixing situation a k value of 0.3 seems reasonable. Laboratories with modest ventilation rates supplied by 15–20 linear feet of hoods can be expected to provide 30–100 m^3 per student over a 3-h laboratory period if the hoods are working properly. Let us take the figure of 50 m^3 per student as an illustration. If the value of m/L for a compound (or a group of compounds in a reaction) is substantially less than 50 m^3, it may be safe to do that series of operations in the open laboratory. If m/L is comparable to or greater than 50 m^3, a number of options are available: (1) Steps using that compound may be restricted to a hood. (2) The instructional staff may satisfy themselves that much less than the assumed value is actually evaporated under conditions present in the laboratory. (3) The number of individual repetitions of this experiment may be reduced. The size of the laboratory section can be reduced or the experiment may be done in pairs or trios.

Conducting reactions in a hood does not automatically convey a stamp of safety. Hoods are designed to keep evaporating chemicals from entering the general laboratory space. For hoods to do their job, there must be an adequate flow of air into the hood, and this air flow must not be disturbed by turbulence

---

[4]Butcher, S. S.; Mayo, D. W.; Hebert, S. M.; Pike, R. M. "Laboratory Air Quality, Part I" *J. Chem. Educ.* **1985,** *62,* A238; and "Laboratory Air Quality, Part II" *J. Chem. Educ.* **1985,** *62,* A261.

at the hood face. A frequently used figure of merit for hood operation is the face velocity of 100 ft/min. This is an average velocity of air entering the hood opening. (In the event that your lab does not have monitoring systems already housed within the hoods, instruments for measuring flow rate are available and can be purchased from most major equipment suppliers.) Even with a face velocity of 100 ft/min, vapors can be drawn out of an improperly designed hood simply by people walking by the opening, or by drafts from open windows.

Hood performance should be checked at regular intervals. The face velocity will increase as the front hood opening is decreased. If an adequate face velocity cannot be maintained with a front opening height of 15 cm, use of the hood for carrying out reactions will be limited. A low face velocity may indicate that the fans and ductwork need cleaning, that the exhaust system leaks (if it operates under lower than ambient pressure), or that the supply of makeup air is not adequate. When the hood system is properly maintained, the height of the hood opening required to provide an adequate face velocity is often indicated with a sticker.

Hoods are often used for storage of volatile compounds. A danger in this practice is that the hood space can become quickly cluttered, making work in the hood difficult, and the air flow may be disturbed. Of course, hoods being used for storage must never be turned off.

MICROWAVE SAFETY*

Scientific microwave apparatus is designed for preparative chemistry and is built with safety in mind. Because of the short reaction times, this technology when compared to conventional heating is "green." *Domestic (household) microwave ovens should not be used for preparative chemistry.* When employing microwave heating, all the safety precautions that are taken when performing a reaction using conventional heating should be adhered to, particularly the fact that reaction vessels can be hot and, when sealed, residual pressure needs to be released carefully at the end of the reaction. There are also some safety precautions that are specific to reactions using microwave heating:

- Adherence to the microwave manufacturer's user manual and guidelines is essential.
- Before running reactions in sealed vessels, it is prudent to check the reaction vessels for cracks or any other signs of damage prior to use.
- Only fill the reaction vessels to manufacturer's specifications; do not over-fill the reaction vessels. An approximate "rule of thumb" is to fill the vessel to no further than half its capacity.
- Only seal a closed reaction vessel with the manufacturer's recommended cap. These caps are designed to vent and reseal in the case of an over-pressurization during a reaction.
- If the cap is a twist-on type, be sure to use the appropriate tool to tighten the cap to the manufacturer-specified torque.

*This section has been written by Dr. Nicholas E. Leadbeater from the Department of Chemistry at the University of Connecticut, and Dr. Cynthia B. McGowan from the Department of Chemistry at Merrimack College, MA.

- It is important to monitor temperature and pressure profiles during the course of a reaction and to set safety limits before starting. A reaction mixture can be heated gradually to the set temperature, or, in some cases, chemists prefer to heat up the contents as rapidly as possible. In the latter case, care needs to be taken to ensure that the temperature and pressure do not rise uncontrollably. As a result, it is best to use a low initial microwave power.

- Before performing a reaction at elevated temperatures, chemists should consider carefully the stability of the reagents and solvents they use at these temperatures.

- It is important to ensure efficient stirring, especially when using heterogeneous reaction mixtures and metal catalysts or reagents since localized heating can occur, resulting in some cases in melting of the vessel walls and, under pressure, failure of the vessel.

- The stir bars used for agitation should not be exactly 3 cm in length since this equates to ¼ wavelength of a microwave at 2450 MHz and thus acts as an antenna.

- Upon completion of a microwave run, the microwave unit will start a cooling process. In the case of the monomode microwave, the pressure sensor will release when the tube is cool enough to handle (50 °C). With multimode microwave units, the apparatus is set to cool for a period of time, but, at the end of this, the reaction vessels may still be hot. Check the temperature before removing the reaction vessels.

- When opening sealed vessels at the end of a reaction, be sure to point the vessel away from your face and any other person, preferably doing so in a hood. Any remaining pressure will release as soon as the cap is removed. If the tube is very warm, cool it in an ice bath before removing the cap.

CONCLUDING THOUGHTS

This brief chapter touches only a few of the important points concerning laboratory safety. The risk from vapor exposure is discussed in some detail, but other risks are treated briefly. Applications in some laboratories may involve reactions with a risk from radiation or infection or may involve compounds that are unstable with respect to explosion. The chemist must be aware of the potential risks and must be prepared to go to an appropriate and detailed source of information, as needed. The references cited here represent a small fraction of the safety data, texts, and journals available on this subject. It is highly recommended that the library and/or laboratory at your institution have at least this minimal selection. Of course, the selections should be kept up to date!

QUESTIONS

2-1. After bookmarking a (M)SDS URL, locate a chemical of your choice. If needing guidance, research the materials planned for one or more of the scheduled experiments for this term! Underline on the sheet the CAS No., solubility data, fire and explosion data, reactivity data, and what protective equipment is required when using this chemical. Does your laboratory meet the safety regulations required to use this chemical? Why, or why not?

2-2. Think and describe what you would do in each of the following situations which could happen in your laboratory.

    (a) You are working at your station and the 100-mL round-bottom flask in which you are running a reaction in isopropyl acetate solvent suddenly catches fire.

(b) The person working across the laboratory bench from you allows hydrogen chloride gas to escape from his or her apparatus.

(c) A reagent bottle is dropped, spilling concentrated sulfuric acid.

(d) A hot solution "bumps," splashing your face.

2-3. You are working in the laboratory using 3.0 mL of benzene in an extraction procedure. An alternative to benzene is toluene. However, three times more toluene is required to perform the extraction. The isolation of the desired product from the extraction solution requires evaporation of the solvent (benzene or toluene). This takes 0.5 h to complete. Calculate the relative risks of using these two solvents. Which solvent would you use and why?

2-4. A laboratory has four hoods; each is 39 in. wide. When the hood door is open to a height of 8 in. and the hoods are operating, the average air velocity through the hood face is 170 ft/min.

(a) Evaluate the total ventilation rate for this room, assuming that there are no other exhausts.

(b) The laboratory is designed for use by 30 students. Evaluate the air available per student if the mixing factor is 0.3 and the experiments last for 3 h.

(c) An experiment is considered in which each student would be required to evaporate 7 mL of methylene chloride (CH_2Cl_2). Estimate the average concentration of methylene chloride. Look up the TLV or the TWA for methylene chloride and consider how the evaporation might be performed.

2-5. An experiment is considered in which 1 mL of diethylamine would be used by each student. The ventilation rate for the laboratory is 5 m^3/min. Look up the TLV (or TWA) for diethylamine, $(C_2H_5)_2NH$. What restrictions might be placed on the laboratory to keep the average concentration, over a 3-h period, less than one-third of the TWA? Assume a mixing factor of 0.3.

GENERAL SAFETY REFERENCES

1. ACS Committee on Chemical Safety. *Safety in Academic Chemical Laboratories*, 8th ed.; American Chemical Society: Washington, DC, 2017.

2. *Handbook of Laboratory Safety*, 5th ed.; Furr, A. K. Jr., Ed.; CRC Press: Boca Raton, FL, 2000.

3. National Research Council. *Prudent Practices in the Laboratory: Handling and Management of Chemical Hazards, Updated Version*; National Academy Press: Washington, DC, 2011.

4. Armour, M. A. *Hazardous Laboratory Chemicals Disposal Guide*, 3rd ed.; CRC Press: Boca Raton, FL, 2003.

5. *Working Safely with Chemicals in the Laboratory*, 2nd ed.; Gorman, C. E., Ed.; Genium: Schenectady, NY, 1995.

6. Alaimo, R. J., Ed. *Handbook of Chemical Health and Safety*, 1st ed.; Oxford University Press: New York, 2001.

7. *The Sigma–Aldrich Library of Regulatory and Safety Data*, Lenga, R. E.; Votoupal, K. L., Eds.; Aldrich Chemical Co., Milwaukee, WI, 1992.

8. Young, J. A. Improving *Safety in the Chemical Laboratory: A Practical Guide*, 2nd ed.; Wiley: New York, 1991.

9. American Chemical Society. *Less Is Better (Laboratory Chemical Management for Waste Reduction)*; American Chemical Society: Washington, DC, 2008.

10. Verschueren, K. *Handbook of Environmental Data on Organic Chemicals*, 3rd ed.; Van Nostrand Reinhold: New York, 1996.

11. Lewis, R. J., Sr., *Hazardous Chemicals Desk Reference*, 6th ed.; Wiley-Interscience: New York, 2008.

12. *NIOSH Pocket Guide to Chemical Hazards*, National Institute for Occupational Safety and Health, U. S. Government Publication Office: Washington,

DC, 2008. NIOSH offers four versions of the pocket guide (https://www
.cdc.gov/niosh/npg/default.html).

13. *OSHA Regulated Hazardous Substances*, Vols. I and II, Noyes Data Corp.:
    Park Ridge, NJ, 1990.

14. Lund, G.; Sansone, E. B. "Safe Disposal of Highly Reactive Chemicals"
    J. Chem. Educ. 1994, 71, 972.

15. It is also recommended that one refer to the numerous articles on safety
    that appear regularly in the *Journal of Chemical Education* (http://jchemed
    .chem.wisc.edu) and *The Chemical Educator*.

BIBLIOGRAPHY

ACGIH. *Threshold Limit Values and Biological Exposure Indices*. Available from ACGIH, Kemper Woods Center, 1330 Kemper Meadow Drive, Cincinnati, OH 45240.

Aldrich. *Catalog Handbook of Fine Chemicals*, 1001 W. St. Paul Ave., Milwaukee, WI, 2009–2010.

Anastas, P. T.; Farris, C. A., Eds., *Benign by Design—Alternate Synthetic Design for Pollution Prevention*; ACS Symposium Series 577, American Chemical Society: Washington, DC, 1994.

Anastas, P. T.; Williamson, T. C., Eds., *Green Chemistry—Designing Chemistry for the Environment*; ACS Symposium Series 626, American Chemical Society: Washington, DC, 1996.

———, *Green Chemistry: Frontiers in Benign Chemical Synthesis and Processes;* Oxford University Press: New York, 1998.

Mayo, D. W.; Hebert, S. M.; Pike, R. M. "Laboratory Air Quality, Part I" *J. Chem. Educ.* **1985**, *62*, A238; "Laboratory Air Quality, Part II"; *J. Chem. Educ.* **1985**, 62, A261.

Flinn Scientific Company, *Chemical Catalog/Reference Manual;* Available from Flinn Scientific Co., P.O. Box 219, Batavia, IL 60510, 2011.

Handbook of Chemistry and Physics, 102nd ed.; Rumble, J. R., D. R., Ed.; CRC Press: Boca Raton, FL, 2021–2022.

The Merck Index, 15th ed.; O'Neil, M. J., Ed.; Merck Research Laboratories Publications: White-house Station, NJ, 2013. For online access, see: https://www.rsc.org/merck-index.

Molinelli, R. P.; Reale, M. J.; Freudenthal, R. I. *Material Data Safety Sheets*; Hill & Garnett: Boca Raton, FL, 1992.

C₃H₄apter 3

INTRODUCTION TO MICROSCALE ORGANIC LABORATORY EQUIPMENT AND TECHNIQUES

We begin this chapter with a description of the standard pieces of glassware that are generally employed in a microscale laboratory. Modern standard taper glassware is particularly convenient to use and gives the student a sense of the flavor of the research laboratory. It is not essential, however, for the experimental work in an instructional laboratory, and many courses use glassware with alternative connectors. We describe the standard taper glassware as just one example of microscale equipment that is available. The operations carried out in the laboratory will be very similar or identical if, for example, a plastic connector is used to assemble the experimental setup. We next consider a series of standard experimental apparatus setups that use this equipment, briefly discussing the role that they play in the laboratory. We end the chapter with a set of laws that govern how one operates in a microscale laboratory (the rules are a bit different than those for a macroscale laboratory) and a set of guidelines for recording your experimental data. The basic individual pieces of equipment are shown in Figures 3.1 to 3.7.

Figure 3.1   Reaction flasks.

Figure 3.2   Condensers, jacketed and air.

Figure 3.3   Distillation heads.

Chapter 3: **C₃H₄, Cyclopropene**
Demyanov and Doyarenko (1922).

Figure 3.4   Recrystallization tubes with Teflon plugs.

Figure 3.5   Miscellaneous items.

Figure 3.6   Gas chromatographic fraction collection items.

Figure 3.7   Hickman–Hinkle still head with side port.

Figure 3.8   Standard taper joints (⑆). (Reproduced with permission of John Wiley and Sons, Inc., from Zubrick, J. W. *The Organic Chemistry Lab Survival Manual*, 9th ed., Copyright 2013.)

MICROGLASSWARE EQUIPMENT

Standard Taper Joints

Standard taper ground-glass joints are the common mechanism for assembling all conventional research equipment in the organic laboratory. The symbol ⑆ is commonly used to indicate the presence of this type of connector. Normally, ⑆ is either followed or preceded by #/#. The first # refers to the maximum inside diameter of a female (outer) joint or the maximum outside diameter of a male (inner) joint, measured in millimeters. The second number corresponds to the total length of the ground surface of the joint (Fig. 3.8). The advantage of this type of connection is that if the joint surfaces are *lightly* greased, a vacuum seal is achieved. One of the drawbacks of using these joints is that contamination of the reacting system readily occurs if the solvents present in the reaction vessel dissolve the grease. In small-scale reactions, this contamination can be particularly troublesome.

The small joints used in the microscale experimental organic laboratory, however, have the ease of assembly and physical integrity of research-grade, standard taper, ground-glass joints along with a number of important additional features. The joint dimensions are usually ⑆ 14/10. The conical vials in which most microscale reactions are carried out use this type of connecting system. Note that in addition to being ground to a standard taper on the inside surface of the throat of the vial, these vials also have a screw thread on the outside surface (Fig. 3.9).

This arrangement allows a standard taper male joint to be sealed to the reaction flask by a septum-type (open) plastic screw cap. The screw cap applies compression to a silicone rubber retaining O-ring positioned on the shoulder of the male joint (Fig. 3.10). The compression of the O-ring thereby achieves a greaseless gas-tight seal on the joint seam, while at the same time clamping the two pieces of equipment together. The ground joint provides both protection from intimate solvent contact with the O-ring and mechanical stability to the connection. The use of this type of connector leads to a further bonus during

Figure 3.9   Threaded female joint.

Figure 3.10   Male joint with septum cap and O-ring.

construction of an experimental setup. Because the individual sections are small, light, and firmly sealed together, the entire arrangement often can be mounted on the support rack by a single clamp. In conventional systems it is often necessary to use at least two clamps. This can easily lead to strain in the glass components unless considerable care is taken in the assembly process. Clamp strain is one of the major sources of experimental glassware breakage. The ability to single-clamp most microscale setups effectively eliminates this problem.

NOTE. When ground-glass joint surfaces are grease-free, it is important to disconnect joints soon after use (particularly with basic solutions) or they may become locked or "frozen" together.

Joints of the size employed in these microscale experiments, however, are seldom a problem to separate if given proper care *(keep them clean!).*

Reaction Flasks

Both the conical vials (3 and 5 mL) and the round-bottom flasks are designed to be connected via an O-ring compression cap installed on the male joint of the adjacent part of the system (see Fig. 3.1).

Condensers

Two types of condensers (air condensers and water-jacketed condensers) are available; in most cases the water-jacketed condenser can work well as an air condenser. Condensers are usually attached to 14/10 $\overline{s}$-jointed reaction flasks. The upper female joints allow connection of the condenser to the 14/10 $\overline{s}$ drying tube and the 14/10 $\overline{s}$ capillary gas delivery tube (see Fig. 3.2).

Distillation Heads

The simple Hickman still is used with an O-ring compression cap to carry out semi-micro simple or crude fractional distillations. The Hickman–Hinkle spinning band still uses a 3-cm fractionating column and routinely develops between five and six theoretical plates. The Hickman–Hinkle still is currently available with 14/10 $\overline{s}$ joints and can be conveniently operated with the 14/10 $\overline{s}$ 3- and 5-mL conical vials (see Figs. 3.1, 3.3, and 3.7). The still head is also available with an optional side arm collection port.

Recrystallization Tubes

Craig tubes are a particularly effective method for recrystallizing small quantities of reaction products. These tubes possess a nonuniform ground joint in the outer section. The substitution of Teflon for glass in the head makes these systems quite durable and much less susceptible to breakage during centrifugation (see Fig. 3.4).

Miscellaneous Items

The Claisen head (see Fig. 3.3) is often used to facilitate the syringe addition of reagents to closed moisture-sensitive systems (such as Grignard reactions) via a septum seal in the vertical upper joint. This joint can also function to position the thermometer (using an adapter) in the well of a Hickman–Hinkle still (see Fig. 3.15). The Claisen adapter is also used to mount the drying tube in a protected position remote from the reaction chamber. The drying tube, in turn, is used to protect moisture-sensitive reaction components from atmospheric water vapor, while allowing a reacting system to remain unsealed. The capillary gas delivery tube is employed in transferring gases formed during reactions to storage containers (see Fig. 3.5).

Gas Chromatographic Fraction Collection Items

For fraction collection the gas chromatographic (GC) collection tube is connected directly to the exit port of the GC detector through a stainless steel standard taper adapter. The collected sample is then transferred to a 0.1-mL conical vial for storage. The system is conveniently employed in the resolution and isolation of two-component mixtures (see Fig. 3.6).

STANDARD EXPERIMENTAL APPARATUS

Heating and Stirring Arrangements

It is important to be able to carry out microscale experiments at accurately determined temperatures. Very often, transformations are successful, in part, because of the ability to maintain precise temperature control. In addition, many reactions require reactants to be intimately mixed to obtain a substantial yield of product. Therefore, the majority of the reactions you perform in this laboratory will be conducted with rapid stirring of the reaction mixture.

Sand Bath Technique—Hot-Plate Calibration

A most convenient piece of equipment for heating or stirring or for performing both operations simultaneously on a microscale level is the hot-plate-magnetic stirrer. Heat transfer from the hot surface to the reaction flask is generally accomplished with a crystallizing dish containing a *shallow* layer of sand that can conform to the size and shape of the particular vessel employed. The temperature (external) of the system is monitored by a thermometer embedded in the sand near the reaction vessel.

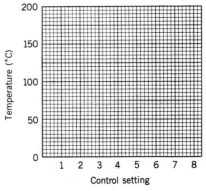

Figure 3.11 Plot your bath and/or vial temperature (°C) versus hot-plate control setting.

A successful procedure for determining the temperature inside the vial relative to the bath temperature is to mount a second thermometer in a vial containing 2 mL of high-boiling silicone oil. The vial temperature is then measured at various sand-bath temperatures and the values are entered on graphs of vial temperatures versus hot-plate settings and bath temperatures versus hot-plate settings (see Fig. 3.11) for your particular hot-plate system (see also the section on Metal Heat-Transfer Devices, p. 24). These data will save considerable time when you bring a reaction system to operating temperature. When you first enter the laboratory, it is advisable to adjust the temperature setting on the hot-plate stirrer with the heating device, or bath, in place. The setting is determined from your control setting-temperature calibration curve. This procedure will allow the heated bath to reach a relatively constant temperature by the time it is required. You will then be able to make small final adjustments more quickly, if necessary.

NOTE. Heavy layers of sand act as an insulator on the hot-plate surface, which can damage the heating element at high temperature settings. When temperatures over 150 °C are required, it is especially important to use the minimum amount of sand.

Recording the weight of sand used and the size of the crystallizing dish will help to make the graph values more reproducible.

The high sides of the crystallizing dish protect the apparatus from air drafts, and so the dish also operates somewhat as a hot-air bath. Heating can be made even more uniform by covering the crystallizing dish with aluminum foil (see Fig. 3.12). This procedure works well, but is a bit awkward and is required in only a few instances.

The insulating properties of sand provide a readily available variable heat source because the temperature of the sand is higher deeper in the bath; thus, the depth of sand used in the bath is exceedingly important. **The depth should always be kept to a minimum, in the range of 10–15 mm.** Finally, sand baths offer a significant safety advantage over oil baths. Individual grains of sand are so small that they have little heat capacity and thus are less likely to burn the chemist in the event of a spill.

Metal Heat-Transfer Devices

An alternative to the sand bath is a heat-transfer system that employs copper tube plates or aluminum metal blocks drilled to accommodate the different reaction vials and flasks.

Stirring

Stirring the reaction mixture in a conical vial is carried out with Teflon-coated magnetic spin vanes, and in round-bottom flasks with Teflon-coated magnetic stirring bars (see Fig. 3.12). It is important to put the reaction flask as close to the center and to the bottom surface of the crystallizing dish as possible when using magnetic stirring. This arrangement is a good practice in general, as it leads to using the minimum amount of sand needed in a sand bath.

If the reaction does not require elevated temperatures, but needs only to be stirred, the system can be assembled without the heat-transfer device (sand bath or metal plate). Some stirred reactions, on the other hand, require cooling.

Figure 3.12   Hot-plate-magnetic stirrer with sand bath and reaction vial.

In these cases a crystallizing dish filled with ice water, or with ice water and salt, if lower temperatures are called for, will provide the correct environment.

Reflux Apparatus

To bring about a successful reaction between two substances, it is often necessary to mix the materials together intimately and to maintain a specific temperature. The mixing operation is conveniently achieved by dissolving the materials in a solvent in which they are mutually soluble. If the reaction is carried out in solution under reflux conditions, the choice of solvent can be used to control the temperature of the reaction. Many organic reactions involve the use of a reflux apparatus in one arrangement or another.

What do we mean by *reflux?* The term means to "return," or "run back." This return is exactly how the reflux apparatus functions. When the temperature of the reaction system is raised to the solvent's boiling point (constant temperature), all vapors are condensed and returned to the reaction flask or vial; this operation is not a distillation and the liquid phase remains at a stable maximum temperature. In microscale reactions, two basic types of reflux condensers are utilized: the air-cooled condenser, or air condenser , and the water-jacketed condenser (see Fig. 3.13).The air condenser condenses solvent vapors on the cool vertical wall of an extended glass tube that dissipates the heat by contact with the laboratory room air. This arrangement functions quite effectively with liquids boiling above 150 °C. Indeed, a simple test tube can act as a reaction chamber and air condenser all in one unit, and many simple reactions can be most easily carried out in test tubes.

Air condensers can occasionally be used with lower boiling systems; however, the water-jacketed condenser is more often employed in these situations. The water-jacketed condenser employs flowing cold water to remove heat from the vertical column and thus facilitate vapor condensation. It is highly effective at condensing vapor from low-boiling liquids.

Figure 3.13 Water-jacketed condenser with conical vial, arranged for heating and magnetic stirring.

Both styles of condensers accommodate various sizes of reaction flasks and are available with 14/10 ⊺ standard taper joints. The tops of both condenser columns have a female 14/10 ⊺ joint.

In refluxing systems that do not require significant mixing or agitation, the stirrer (magnetic spin vane or bar) usually is replaced by a "boiling stone." These sharp-edged stones possess highly fractured surfaces that are very efficient at initiating bubble formation as the reacting medium approaches the boiling point. The boiling stone acts to protect the system from disastrous boilovers and also reduces "bumping." (Boiling stones should be used only once and must **never** be added to a hot solution. In the first case, the vapor cavities become filled with liquid upon cooling, and thus a boiling stone becomes less effective after its first use. In the second case, **adding the boiling stone to the hot solution may suddenly start an uncontrollable boilover**).

Distillation Apparatus

Distillation is a laboratory operation used to separate substances that have different boiling points. The mixture is heated, vaporized, and then condensed; the early fractions of condensate are enriched in the more volatile component. Unlike the reflux operation, in distillations none, or only a portion, of the condensate is returned to the flask where vaporization is taking place. Many distillation apparatus have been designed to carry out this basic operation. They differ mainly in small features that are used to solve particular types of separation problems. In several of the microscale experiments contained in Chapters 6 and 7 *semimicroscale* distillations are required. In carrying out these distillations the choice of still depends to a large degree on the difficulty of the separation required (generally, how close are the boiling points in the mixture to be separated?).

The Hickman still head (Fig. 3.14) is ideally suited for simple distillations. This system has a 14/10 ⊺ male joint for connection to conical vials or

Air condenser

14/10⊺ female

Hickman still head

Clamp

14/10⊺ male

5-or 10-mL RB flask

Boiling stone

Wire gauze on ring stand

Microburner

Gas

Figure 3.14 Hickman still head and air condenser with 5-mL round-bottom flask, arranged for microburner heating.

Figure 3.15   Hickman–Hinkle still head with side port 3- or 5-mL conical vial, Teflon spinning band, and thermometer adapter and arranged for heating and magnetic stirring.

round-bottom flasks. The still head functions as both an air condenser and a condensate trap. For a detailed discussion of this piece of equipment see Experiments [3A] and [3B]. The simple Hickman still has been modified (see Fig. 3.15) with a spinning band. The still continues to function in much the same way as the simple Hickman still, but a tiny Teflon spinning band is now mounted in a slightly extended section between the male joint and the collection collar. When the band is spun at 1500 rpm by a magnetic-stirring hot plate, this still functions as an effective short-path fractional distillation column (see Distillation, Experiment [3D]). In addition, this modified system has a built-in thermometer well that allows reasonably accurate measurement of vapor temperatures plus a sidearm port for removing distillate.

The most powerful microscale distillation system currently available is the 2.5-in. vacuum-jacketed microscale spinning-band distillation column (see Fig. 3.16 and Experiment [3C] for description and details). This still is designed for conventional downward distillate collection and nonstopcock reflux control. The column is rated at ~10 theoretical plates.

For a discussion of reduced pressure distillations, see Distillation.

Moisture-Protected Reaction Apparatus

Many organic reagents react rapidly and preferentially with water. *The success or failure of many experiments depends to a large degree on how well atmospheric moisture is excluded from the reaction system.* The "drying tube," which is lightly packed with a desiccant such as anhydrous calcium chloride and held in place using cotton plugs, is a handy way to carry out a reaction in an apparatus that

Figure 3.16 Micro spinning band distillation column (2.5 in.).

Figure 3.17 Moisture-protected water-jacketed condenser with 3- or 5-mL conical vial, arranged for heating and magnetic stirring.

is not totally closed to the atmosphere, but that is reasonably well protected from water vapor. The microscale apparatus described here is designed to be used with the 14/10 ⊺ drying tube. The reflux condensers discussed earlier are constructed with female 14/10 ⊺ joints at the top of the column, which allows convenient connection of the drying tube if the refluxing system is moisture sensitive (see Fig. 3.17).

Because many reactions are highly sensitive to moisture, successful operation at the microscale level can be rather challenging. If anhydrous reagents are to be added after an apparatus has been dried and assembled, it is important to be able to introduce these reagents without exposing the system to the atmosphere, particularly when operating in a humid atmosphere. In room-temperature reactions that do not need refluxing, adding anhydrous reagents is best accomplished by use of the microscale Claisen head adapter. The adapter has a vertical screw-threaded standard taper joint that will accept a septum cap. The septum seal allows syringe addition of reagents and avoids the necessity of opening the apparatus to the laboratory atmosphere (see Fig. 3.18).

Specialized Pieces of Equipment

Collection of Gaseous Products. Some experiments lead to gaseous products. The collection, or trapping, of gases is conveniently carried out by using the capillary gas delivery tube. This item is designed to be attached directly to a 1- or 3-mL conical vial (see Fig. 3.19), or to the female 14/10 ⊺ joint of a condenser connected to a reaction flask or vial. The tube leads to the collection system, which may be a simple, inverted, graduated cylinder; a blank-threaded

Figure 3.18   Moisture-protected Claisen head with 3- or 5-mL conical vial, arranged for syringe addition and magnetic stirring.

Figure 3.19   Conical vial (3-mL) and capillary gas delivery tube arranged for heating and magnetic stirring.

septum joint; or an air condenser filled with water (if the gaseous products are not water-soluble). The 0.1-mm capillary bore considerably reduces dead volume and increases the efficiency of product transfer.

Collection of Gas Chromatographic Effluents. The trapping and collection of gas chromatographic liquid fractions become particularly important exercises in microscale experiments. When yields of a liquid product are less than 100 µL conventional distillation, even using microscale equipment, is impractical. In this case, preparative gas chromatography replaces conventional distillation as the route of choice to product purification. A number of the reaction products in Chapters 6 and 7 depend on this approach for successful purification and isolation. The ease and efficiency of carrying out this operation is greatly facilitated by employing the 5/5 ⊺ collection tube and the 0.1-mL 5/5 ⊺ conical collection vial.

MICROWAVE HEATING AS A TOOL FOR ORGANIC CHEMISTRY*

Introduction

An appliance found in almost all homes is a microwave oven. It is possible to heat food much more quickly and easily using a microwave than using a stove top. The observation that microwave energy can be used to heat food

*This section has been written by Dr. Nicholas E. Leadbeater from the Department of Chemistry at the University of Connecticut, and Dr. Cynthia B. McGowan from the Department of Chemistry at Merrimack College, MA.

was first made by Percy Spencer, an employee of the Raytheon Corporation.[1] His company was a manufacturer of radar sets and, while working on one he noticed that the candy bar he had in his pocket had melted. Intrigued by this, the next day he brought in some popcorn from home and found that if he placed this near his radar set, it popped. In 1945, Raytheon filed a patent for the microwave cooking process, and in 1947 they built the first microwave oven called the Radarange.[2] It was almost 6 feet (1.8 m) high and weighed over 750 pounds (340 kg) and cost between $2,000 and $3,000.[3] The first popular home model was launched in 1967, and current estimates suggest that over 90% of U.S. households own a microwave oven.[4]

Just as microwave ovens prove so valuable in the kitchen, it is also possible to use similar technology in preparative chemistry. It was in 1986 that the first reports of microwave heating as a tool for organic chemistry appeared in the scientific literature.[5,6] Two research groups published results they had obtained in their laboratories using kitchen microwave ovens. They said that chemistry that usually takes hours to reach completion using conventional heating could be performed in a matter of minutes in a microwave oven. Since these first reports, the use of microwave heating in organic chemistry has grown rapidly. Today, the technology is used in industry and academic laboratories for performing a wide range of reactions. Microwave heating has opened up a range of new areas in organic chemistry, allowing chemists to perform reactions quickly and easily. As an example, in this book Experiment 7, the Cannizzaro reaction, is performed in *one hour* using conventional heating. In the microwave protocol, the reaction is complete *in just one minute.*

G The use of microwave heating addresses a number of the green chemistry principles.[7] Since it is often possible to obtain higher yields using microwave heating as opposed to conventional heating, there will be less waste and unused reagents. Also, since microwave heating is fast, there is often not enough time for products to decompose, so this makes the product purification cleaner and easier. Chemists have also used the inherent advantages of microwave heating to their advantage for developing cleaner alternatives to known reactions. Take, for example, the use of water as a solvent instead of organics such as dichloromethane and benzene. Work has shown that water is an excellent solvent for organic chemistry, especially when combined with microwave heating. It is possible to heat water well above its boiling point in a sealed reaction vessel very safely and efficiently using microwaves. At these higher temperatures, water behaves more like an organic solvent. While most organic compounds are not soluble in water at room temperature, they can be in this higher temperature water, or at least partly so. This means that reactions can take place and, when the mixture cools down at the end, the product crystallizes out and is easily removed. As well as allowing for a more environmentally friendly solvent to be used, it also makes purification easy.

---

[1]*Reader's Digest,* August 1958, page 114.
[2]Spencer, P. L. 1945. "Method of Treating Foodstuffs." US Patent 2,495,429, filed October 5, 1945, and published January 24, 1950.
[3]Gallawa, J. C. *Complete Microwave Oven Service Handbook: Operation, Maintenance, Troubleshooting, and Repair;* Prentice Hall, 2000.
[4]U.S. Bureau of Labor Statistics, see: https://www.bls.gov/cpi/quality-adjustment/microwave-ovens.htm.
[5]Gedye, R.; Smith, F.; Westaway, K.; Ali, H.; Baldisera, L.; Laberge, L.; and Rousell, J. "The Use of Microwave Ovens for Rapid Organic Synthesis," *Tetrahedron Lett.* **1986,** 27, 279.
[6]Giguere, R. J.; Bray, T. L.; Duncan, S. M.; Majetich, G. "Application of Commercial Microwave Ovens to Organic Synthesis," *Tetrahedron Lett.* **1986,** 27, 4945.
[7]Anastas, P. T.; Warner, J. C. *Green Chemistry: Theory and Practice;* Oxford University Press, New York, 1998.

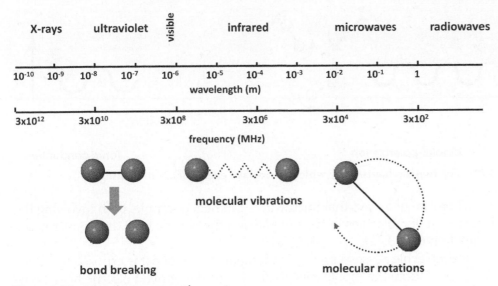

Figure 3.20   The electromagnetic spectrum.

Before looking at applications in organic chemistry, it is important to appreciate some of the physical chemistry concepts behind microwave heating. The microwave region of the electromagnetic spectrum (Fig. 3.20) is classified as that between 300 and 300,000 megahertz (MHz). Compared to ultraviolet, infrared, and visible light, microwave irradiation is of relatively low energy. As a result, microwaves are not high enough in energy to break chemical bonds. Instead they can only make molecules rotate. This is very different from the more energetic ultraviolet radiation which, when it interacts with molecules, can break bonds, giving rise to the area of chemistry known as photochemistry.

Both home and scientific microwave equipment operates at 2,450 MHz. Interestingly, the microwave region of the electromagnetic spectrum is also used for navigation, communication, and remote sensing purposes. This includes technologies such as global positioning systems, wireless Internet, and Bluetooth, as well as radar. As a result, the frequency used in microwave ovens has to be different from those used for these other applications and so is strictly regulated.

Microwaves, like all electromagnetic energy, move at the speed of light and comprise oscillating electric and magnetic fields (Fig. 3.21). These components oscillate at right angles to each other and to the direction of propagation.

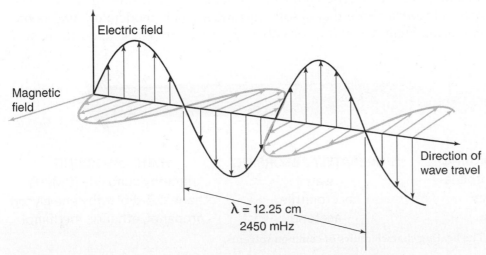

Figure 3.21   Microwave energy comprises electric and magnetic fields.

dipolar polarization   **ionic conduction**

Figure 3.22   The two mechanisms by which microwave energy leads to heating.

There are two ways that microwaves can heat a sample, both involving the interaction of molecules in the sample with the electric field of the microwave irradiation (Fig. 3.22).

If a molecule possesses a dipole moment, then when it is exposed to microwave irradiation, the dipole tries to align with the applied electric field. Since the electric field is oscillating, the dipoles constantly try to realign to follow this. Molecules have time to align with the electric field but not to follow the oscillating field exactly. This continual reorientation of the molecules results in friction and thus heat. This heating method is termed dipolar polarization. If a molecule is ionic, then the electric field component of the microwave irradiation moves the ions back and forth through the sample, also colliding them into each other. This movement again generates heat and is termed ionic conduction.

Conventionally, to heat reaction mixtures, chemists tend to use a hot plate or sand bath. These can be relatively slow and inefficient ways of transferring heat into a sample because they depend on convection currents and the thermal conductivity of the reaction mixture. Also, the walls of the reaction vessel can be hotter than the contents, thus introducing a thermal gradient. This can mean that reagents or products can be decomposed over time because they sit on the walls of the vessel. When using microwave heating, since the energy interacts with the sample directly, heating can be much more effective. In addition, microwave heating is safer than conventional heating; there are no sand baths or hot plates that can burn the chemist.

Each solvent or reagent in a reaction mixture will interact with microwave energy differently. Although not the only factor in determining the absorbance of microwave energy, the polarity of the solvent is a helpful tool for determining how well it will heat when placed into a microwave field; more polar molecules are affected more and nonpolar less. Solvents can be split into three categories: those that absorb microwaves well, moderately, and poorly (Fig. 3.23). High-absorbing solvents will heat up very fast upon microwave

LOW-ABSORBING	MODERATELY ABSORBING	HIGH-ABSORBING
dichloromethane	water	dimethylsulfoxide (DMSO)
toluene	acetonitrile	ethane-1,2-diol (ethylene glycol)
hexane	acetone	propanol, ethanol, methanol

Figure 3.23   The heating characteristics of common solvents.

Figure 3.24   Areas of high and low microwave energy are found in the cavity of a multimode microwave unit.

irradiation. Lower-absorbing solvents can still be used, but work better if one of the reagents in the reaction mixture is itself a good absorber. Interestingly, water absorbs more weakly than methanol, while it is considerably more polar. This can be attributed, at least in part, to that fact that the strong, extensive hydrogen bonding in water goes some way to restricting rotation of molecules when irradiated with microwaves.

Electric power is turned into microwave energy using a magnetron, this in essence being a high-voltage tube in which electrons generated from a heated cathode are affected by magnetic and electric fields in such a way as to produce microwave radiation. As the microwaves come into the cavity (heating chamber) of a home microwave unit, they will move around and bounce off the walls. As they do so, they will generate pockets (called modes) of high energy and low energy as the moving waves either reinforce or cancel out each other (Fig. 3.24). This means that the microwave field in the microwave cavity is not uniform. Instead, there will be hot spots and cold spots; these correspond to the pockets of high and low energy, respectively. Domestic microwave ovens are therefore called "multimode" microwave ovens.

While home microwave ovens are useful for heating food, performing chemical reactions using them presents a number of challenges. They have no accurate temperature measurement device; the microwave field inside the oven is not uniform; and they are not safe for containing hot, flammable, organic solvents. These problems have led to the need for scientific microwave apparatus specifically designed for performing chemical reactions safely and reliably. Scientific multimode microwave units have been developed for use in preparative chemistry (Fig. 3.26b). As well as being built to withstand explosions of reaction vessels inside the microwave cavity, temperature and pressure monitoring has been introduced, as well as the ability to stir reaction mixtures. It is possible to run a number of reactions at the same time in a multimode microwave oven, the samples being placed into tubes and loaded onto a turntable. As the samples move around, because they are large enough to absorb the microwave energy effectively, heating is fairly uniform.

When performing reactions on a small scale, it is sometimes difficult to heat the small volumes of reagents effectively in a multimode microwave apparatus. This is because, with the hot and cold spots that occur in the cavity of a multimode apparatus, it is hard to get constant microwave energy to irradiate the small sample. To overcome these problems, a smaller, single-mode (often called monomode) microwave apparatus has been developed. The cavity of a monomode microwave apparatus is designed for the length of only one wave (mode) (Fig. 3.25). By placing the sample in the middle of the cavity, it can be irradiated constantly with microwave energy. Using a monomode apparatus, it is possible to heat samples of as little as 0.2 mL very effectively. The upper volume limit of the monomode apparatus is determined by the size of the microwave cavity and is in the region of 100 mL (Fig. 3.26a).

Figure 3.25   The cavity of a monomode microwave unit is designed to fit just one mode.

(a)   (b)

Figure 3.26 (a) A monomode microwave unit and (b) a multimode microwave unit.

Applications

Many organic reactions require heat in order to proceed. In the lab, this is traditionally done using a hot plate, steam, oil, or sand bath, or, before that, a Bunsen burner. For those reactions that do require heat, the problem is that these heating sources are inefficient and reactions can often take a long time to reach completion. By using microwave heating, reaction times can be dramatically reduced and product yields can be higher. Shortening the time of known reactions is not the only advantage that microwave heating is having. It is impacting modern organic chemistry by opening up avenues to compounds that were previously not accessible. It is also a cleaner way to do preparative chemistry. Almost any reaction that needs heat can be performed in a microwave (Fig. 3.27). There are a few exceptions, including those that are known to be highly exothermic.

Microwave heating has proven particularly useful in the pharmaceutical industry, where compounds need to be made rapidly so they can be screened for activity as drug candidates. In an interesting experiment undertaken by Boehringer Ingelheim Pharmaceuticals, the exact amount of time saved using microwave as opposed to conventional heating was determined.[8] Two

> *oxidation*   *rearrangements*
> *reduction*   *ester and amide synthesis*
> *substitution*   *ring-forming*
> *addition*   *heterocycle synthesis*
> *cycloadditions*   *metal-catalyzed processes*

Figure 3.27 Some classes of organic reaction that can be performed using microwave heating.

---

[8]"Timesavings Associated with Microwave-assisted Synthesis: A Quantitative Approach," C. R. Sarko in *Microwave Assisted Organic Synthesis*, edited by J. P. Tierney and P. Lidstrom, Blackwell Publishing, Oxford, 2005.

(a)   *(b)*

Figure 3.28   Reactions can be performed using either a sealed tube or an open round-bottom flask.

scientists were told to make a series of compounds. One of them used microwave chemistry and the other used conventional methods. Both scientists were given the same preparative route to the molecules to follow. However, after 37 days the chemist using the conventional heating approach concluded that the molecules could not be generated using that route. The microwave chemist on the other hand optimized reaction conditions and produced the final products in two days.

Many reactions utilizing microwave heating have been performed in sealed vessels (Fig. 3.28a). These are tubes of varying sizes that can be sealed with a specially designed stopper. Reaction mixtures can then be heated to temperatures well above the boiling point of the solvent inside. This offers a very safe way to perform chemistry at high temperatures and pressures. It is much more convenient than the steel containers used traditionally for this sort of chemistry. Also, it is possible to monitor the temperature and pressure of reaction mixtures very closely, and this means it is possible to report the exact reaction conditions used so that others can use them.

Another option is to use standard laboratory glassware in a microwave. Reactions can be run in round-bottom flasks equipped with a reflux condenser (Fig. 3.28b). The flask sits inside the microwave cavity, and the reflux condenser comes out through the top of the apparatus. Often, equally good results can be obtained using an open vessel arrangement as compared to a sealed tube.

When using a monomode microwave unit, it is possible to perform reactions using sealed tubes of capacity ranging from 0.2 to 25 mL and open vessels ranging from 10 to 100 mL. Reactions are performed one at a time. When using sealed tubes, it is possible to automate the unit using robotics so that when one reaction is complete, the tube can be removed from the microwave and the next one put in. This allows for multiple reactions to be performed without the need for the operator to be present. Up to 60 reaction vessels can be lined up and run one after another.

Multimode microwave units can process multiple reaction vessels at the same time. The sealed vessels all sit in a holder (reaction carousel) that fits into the microwave cavity. Working on a scale of up to a few grams, it is possible to process up to 40 reaction vessels at a time. Up to 92 reactions can be run at a time when using microscale quantities of reagents. Another option possible when working in multimode microwave unit is to use one large reaction vessel. This can either be a larger sealed vessel (up to 1 L in volume) or an open round-bottom flask (up to 5 L in volume). This enables chemists to scale up their reactions to make more of their desired compound.

Equipment Available

There are a number of commercially available scientific microwave units. The four major microwave manufacturers are listed here.

Anton Paar is an Austrian company that manufactures a multimode microwave unit called the Synthos 3000. There are a number of reaction carousels that are available with the unit, allowing for reactions to be performed from the microliter scale up to 100 mL. On the small scale, reactions are run in specially designed silicon carbide plates with either 24 or 48 wells. Using plates made from this inert, highly microwave-absorbing material allows for equal heating of all the wells. Larger reactions are performed in glass or quartz tubes sealed with a specially designed stopper.

Biotage, a company based in Sweden, manufactures a monomode microwave unit called the Initiator. Using this instrument, it is possible to run reactions on scales from 0.2 to 20 mL in sealed tubes. It is possible to automate the unit with a robotic arm, thus allowing up to 60 reactions to be run sequentially.

CEM Corporation, a company based in North Carolina, manufactures a range of microwave units. Its monomode microwave apparatus is called the Discover platform. A number of variants are available. It is possible to run reactions from 0.5 to 60 mL in sealed tubes using this unit, as well as open round-bottom flasks of up to 125 mL. It is possible to automate the unit, allowing reactions to be run sequentially. In addition, an accessory is available for loading reaction vessels with reactive gases such as hydrogen and carbon monoxide, opening the door for performing a wide range of reactions otherwise not possible using microwave heating. CEM also manufactures a multimode microwave unit called the MARS. There are a number of reaction carousels accommodating sealed tubes that can be used with the unit. In addition, open round-bottom flasks of up to 5 L can be placed into the microwave cavity and standard reflux glassware attached. This allows for scale-up of reactions using batch processing.

Milestone, a company based in Italy, manufactures a number of microwave units. The MultiSYNTH has the capability to act as both a monomode and a multimode microwave unit. This means that conditions can be optimized in a small sealed tube using the monomode functionality and then a series of up to 12 small or 6 larger reactions can run in parallel in multimode. The unit can also accommodate a round-bottom flask of capacity up to 1 L, allowing a reaction to be performed at atmospheric pressure. The MicroSYNTH platform is a multimode microwave unit. There are a number of reaction carousels that can be used, allowing for reactions to be performed in parallel using either glass, Teflon, or quartz tubes. In addition, open round-bottom flasks can be placed into the microwave cavity and standard reflux glassware attached.

All the modern scientific microwave units have the capability to measure temperature during the course of a reaction. This can be done remotely using an infrared sensor located in the wall or the bottom of the microwave cavity. In many cases it is also possible to record the temperature inside a reaction vessel using a fiber-optic probe or thermocouple. Pressure measurement is also possible in many cases. The contents of a reaction mixture can be stirred by means of a magnetic stir plate located beneath the microwave cavity and a Teflon stir bar in the vessel.

When running a reaction, key parameters such as temperature, pressure, and microwave power can be measured throughout the run and data saved to

Figure 3.29   Example of a heating profile for a reaction performed using microwave heating.

a computer for use in reports and for reproducing the conditions at a later date (Fig. 3.29). Generally, when programming a protocol into a microwave unit, there are two important stages. The first is entering the ramp time, the time that the user wants the microwave to take to reach the target temperature. The second is entering the hold time, the time that the user wants the reaction mixture to remain at the target temperature before cooling back to room temperature. The microwave unit will use the requisite microwave power to heat the reaction mixture to temperature, and then the power will automatically fluctuate to hold the reaction at the set temperature.

Experimental Protocols

Experimental protocols using microwave heating have been added to Experiments 7, 8, 15, 22, and 30. The experiments in this book can be performed on a range of these commercially available microwave units. The procedures are split into two classes; the first is generally for use with monomode microwave apparatus (Biotage Initiator and CEM Discover) and the second for use with multimode microwave units (Anton Paar Synthos 3000, CEM MARS, and Milestone MicroSYNTH). A modified version of the monomode procedure for use with the Anton Paar Synthos 3000 in conjunction with the silicon carbide plate format for microscale chemistry is added as a footnote in the monomode protocol.

MICROSCALE LAWS

Rules of the Trade for Handling Organic Materials at the Microscale Level

Now that we have briefly looked at the equipment we will be using to carry out microscale organic reactions, let us examine the specific techniques that are used to deal with the small quantities of material involved. Microscale synthetic organic reactions, as defined by Cheronis,[9] start with 15–150 mg of the

---

[9]Cheronis, N. D. *Semimicro Experimental Organic Chemistry*; Hadrion Press: New York, 1958.

limiting reagent. These quantities sound small, and they are. Although 150 mg of a light, powdery material will fill half a 1-mL conical vial, you will have a hard time observing 15 mg of a clear liquid in the same container, even with magnification. This volume of liquid, on the other hand, is reasonably easy to observe if it is in a 0.1-mL conical vial. A vital part of the game of working with small amounts of materials is to become familiar with microscale techniques and to practice them as much as possible in the laboratory.

Rules for Working with Liquids at the Microscale Level

1. Liquids are never poured at the microscale level. Liquid substances are transferred by pipet or syringe. As we are working with small, easy-to-hold glassware, the best way to transfer liquids is to hold both containers with the fingers of one hand, with the mouths as close together as possible. The free hand is then used to operate the pipet (syringe) to withdraw the liquid and make the transfer. This approach reduces to a minimum the time that the open tip is not in or over one vessel or the other. We use three different pipets and two standard syringes to perform most experiments involving liquids. **This equipment can be a prime source of contamination.** Be very careful to thoroughly clean the pipets and syringes after each use.

    a. Pasteur pipet (often called a capillary pipet). A simple glass tube with the end drawn to a fine capillary. These pipets can hold several milliliters of liquid (Fig. 3.30*a*) and are filled using a small rubber bulb or one of the very handy, commercially available pipet pumps. Because many transfers are made with Pasteur pipets, it is suggested that several of them be calibrated for approximate delivery of 0.5, 1.0, 1.5, and 2.0 mL of liquid. This calibration is easily done by drawing the measured amount of a liquid from a graduated cylinder and marking the level of the liquid in the pipet. This mark can be made with transparent tape, or by scratching with a file. Indicate the level with a marking pen before trying to tape or file the pipet.

    b. Pasteur filter pipet. A very handy adaptation of the Pasteur pipet is a filter pipet. This pipet is constructed by taking a small cotton ball and placing it in the large open end of the standard Pasteur pipet. Hold the pipet vertically and tap it gently to position the cotton ball in the drawn section of the tube (Fig. 3.30*b*). Now form a plug in the capillary section by pushing the cotton ball down the pipet with a piece of copper wire (Fig. 3.30*c*). Finish by seating the plug flush with the end of the capillary (Fig. 3.30*d*). The optimum-size plug will allow easy movement along the capillary while it is being positioned by the copper wire. Compression of the cotton will build enough pressure against the walls of the capillary (once the plug is in position) to prevent the plug from slipping while the pipet is filled with liquid. If the ball is too big, it will wedge in the capillary before the end is reached, and wall pressure will be so great that liquid flow will be shut off. Even some plugs that are loose enough to be positioned at the end of the capillary will still have developed sufficient lateral pressure to make the filling rate unacceptably slow. If the cotton filter, however, is positioned too loosely, it may be easily dislodged from the pipet by the solvent flow. These plugs can be quickly and

Figure 3.30   **Preparation of Pasteur filter pipet.**

easily inserted with a little practice. Once in place, the plug is rinsed with 1 mL of methanol and 1 mL of hexane, and dried before use.

There are two reasons for placing the cotton plug in the pipet. First, it solves a particular problem with the transfer of volatile liquids via the standard Pasteur pipet: the rapid buildup of back pressure from solvent vapors in the rubber bulb. This pressure quickly tends to force the liquid back out of the pipet and can cause valuable product to drip on the bench top. The cotton plug tends to resist this back pressure and allows much easier control of the solution once it is in the pipet. The time-delay factor becomes particularly important when the Pasteur filter pipet is employed as a microseparatory funnel (see the discussion on extraction techniques in Technique 4, p. 71).

Second, each time a transfer of material is made, the material is automatically filtered. This process effectively removes dust and lint, which are constant problems when working at the microscale level. A second stage of filtration may be obtained by employing a disposable filter tip on the original Pasteur filter pipet as described by Rothchild.[10]

c. **Automatic pipet (considered the Mercedes-Benz of pipets).** Automatic pipets quickly, safely, and reproducibly measure and dispense specific volumes of liquids. These pipets are particularly valuable at the microscale level, because they generate the precise and accurate liquid measurements that are absolutely necessary when handling only microliters of a liquid. The automatic pipet adds considerable insurance for the success of an experiment, since any liquid can be efficiently measured, transferred, and delivered to the reaction flask.

The automatic pipet consists of a calibrated piston pipet with a specially designed disposable plastic tip. It is possible to encounter any one of three pipet styles: single volume, multirange, or continuously adjustable (see Fig. 3.31). The first type is calibrated to deliver

---

[10]Rothchild, R. *J. Chem. Educ.* 1990, 67, 425.

Filling   Emptying   Ejection

Figure 3.31   Operation of automatic delivery pipet.

only a single volume. The second type is adjustable to two or three predetermined delivery volumes. The third type is the most versatile; it can be set by the user to deliver any volume within the range of the pipet. Obviously, the price of these valuable laboratory tools goes up with increasing features. Automatic pipets are expensive, and usually must be shared in the laboratory. Treat them with respect!

The automatic pipet is designed so that the liquid comes in contact only with the disposable tip.

- Never load the pipet without the tip in place.
- Never immerse the tip completely in the liquid that is being pipetted.
- Always keep the pipet vertical when the tip is attached.
- If an air bubble forms in the tip during uptake, return the liquid, discard the tip, and repeat the sampling process.

If these three rules are followed, most automatic pipets will give many years of reliable service. A few general rules for improving reproducibility with an automatic pipet should also be followed:

- Try to use the same uptake and delivery motion for all samples. Smooth depression and release of the piston will give the most consistent results. Never allow the piston to snap back.
- *Always* depress the piston to the first stop before inserting the tip into the liquid. If the piston is depressed after submersion, formation of an air bubble in the tip becomes likely, which will result in a filling error.
- *Never* insert the tip more than 5 mm into the liquid. It is good practice not to allow the body of the pipet to contact any surface, or bottle neck, that might be wet with a chemical.
- If an air bubble forms in the tip during uptake, return the fluid, discard the tip, and repeat the sampling process.

     d. Syringes. Syringes are particularly helpful for transferring liquid reagents or solutions to sealed reaction systems from sealed reagent or solvent reservoirs. Syringe needles can be inserted through a septum, which avoids opening the apparatus to the atmosphere. Syringes are also routinely employed in the determination of ultramicro boiling points (10-µL GC syringe). It is critically important to clean the syringe needle after each use. Effective cleaning of a syringe requires as many as a dozen flushes. For many transfers, the microscale laboratory uses a low-cost glass 1-mL insulin syringe in which the rubber plunger seal is replaced with a Teflon seal (ACE Glass). For preparative GC injections, the standard 50- or 100-µL syringes are preferred (see Technique 1).

2. Liquid volumes may be converted easily to mass measures by the following relationship:

$$\text{Volume (mL)} = \frac{\text{mass (g)}}{\text{density (g/mL)}}$$

3. Work with liquids in conical vials , and work in a vial whose capacity is approximately twice the volume of the material it needs to hold. The trick here is to reduce the surface area of the flask in contact with the sample to an absolute minimum. A conical vial is thus better than the spherical surface of the conventional round-bottom flask.

Liquids may also be weighed directly. A tared container (vial) should be used. After addition of the liquid, the vial should be kept capped throughout the weighing operation. This procedure prevents loss of the liquid by evaporation. If the density of the liquid is known, the approximate volume of the liquid should be transferred to the container using an automatic delivery pipet or a calibrated Pasteur pipet. Use the above expression relating density, mass, and volume to calculate the volume required by the measured mass. Adjustment of the mass to give the desired value can then be made by adding or removing small amounts of liquid from the container by Pasteur pipet.

NOTE. Before you leave the balance area, be sure to replace all caps on reagent bottles and clean up any spills. A balance is a precision instrument that can easily be damaged by contamination.

Rules for Working with Solids at the Microscale Level

1. General considerations. Working with a crystalline solid is much easier than working with the equivalent quantity of a liquid. Unless the solid is in solution, a spill on a clean glass working surface usually can be recovered quickly and efficiently. *Be careful, however, when working with a solution. Treat solutions as you would a pure liquid.*

2. Transfer of solids. Solids are normally transferred with a microspatula, a technique that is not difficult to develop.

3. Weighing solids at the milligram level. Electronic balances can automatically tare an empty vial. Once the vial is tared, the reagent is added in small portions. The weight of each addition is instantly registered; material is added until the desired quantity has been transferred.

Solids are best weighed in glass containers (vials or beakers), in plastic or aluminum weighing trays ("boats"), or on glazed weighing paper. Filter paper or other absorbent materials are not good choices: small quantities of the weighed material will often stick to the fibers of the paper, and vice versa.

THE LABORATORY NOTEBOOK

Writing is the most important method chemists use to communicate their work. It begins with the record kept in a laboratory notebook. An experiment originally recorded in the laboratory notebook can become the source of information used to prepare scientific papers published in journals or presented at meetings. For the industrial chemist, these written records are critical in obtaining patent coverage for new discoveries.

It is important that you learn to keep a detailed account of your work. A laboratory notebook has several key components. Note how each component is incorporated into the example that follows.

Key Components of a Laboratory Experiment Write-up

1. Date experiment was conducted
2. Title of experiment
3. Purpose for running the reaction
4. Reaction scheme
5. Table of reagents and products
6. Details of procedure used and safety considerations
7. Characteristics of the product(s)
8. References to product or procedure (if any)
9. Analytical and spectral data
10. Signature of person performing the experiment and that of a witness, if required

In reference to point 6, it is the obligation of the person doing the work to list the equipment, the amounts of reagents, the experimental conditions, and the method used to isolate the product. Any color or temperature changes should be carefully noted and recorded.

Several additional points can be made about the proper maintenance of a laboratory record.

11. A hardbound, permanent notebook is essential.
12. Each page of the notebook should be numbered in consecutive order. For convenience, an index at the beginning or end of the book is recommended, and pages should be left blank for this purpose.
13. If a page is not completely filled, an "X" should be used to show that no further entry was made.
14. Data are always recorded directly into the notebook, *never* on scrap paper! Always record your data in ink. If a mistake is made, draw a neat line through the word or words so that they remain legible. Do not completely obliterate anything; you might learn from your mistakes, if you can read them later.
15. Make the record clear and unambiguous. Pay attention to grammar and spelling.
16. In industrial research laboratories, your signature, as well as that of a witness, is required, because the notebook may be used as a legal document.

17. Always write and organize your work so that someone else could come into the laboratory and follow your directions without confusion or uncertainty. *Completeness* and *legibility* are key factors.

Most of you are newcomers to the organic laboratory, and the reactions you will be performing have probably been worked out and checked in detail. Because of this, your instructor may not require you to keep your notebook in such a meticulous fashion. For example, when you describe the procedure (item 6), it may be acceptable to make a clear reference to the material in the laboratory manual and to note any modifications or deviations from the prescribed procedure. In some cases, it may be more practical to use an outline method. In any event, the following example should be studied carefully. It may be used as a reference when detailed records are important in your work. It is more important to record what you observed and what you actually did, than to record what you were supposed to observe and what you were supposed to do.

NOTE. *Because of its length, the example here is typed. Notebooks are usually handwritten. Many chemists, however, now use computers to record their data.*

The circled numbers refer to the list on p. 42.

EXAMPLE OF A LABORATORY NOTEBOOK ENTRY

19 July 2022 ①
PREPARATION OF DIPHENYL SUCCINATE ②

$$\begin{array}{l}CH_2CO_2H \\ | \\ CH_2CO_2H\end{array} + 2\ C_6H_5OH + POCl_3 \rightarrow \begin{array}{l}CH_2CO_2C_6H_5 \\ | \\ CH_2CO_2C_6H_5\end{array} + HPO_3 + 3\ HCl \quad \bigg\} ④$$

Diphenyl succinate is being prepared as one of a series of dicarboxylic acid esters that are to be investigated as growth stimulants for selected fungi species. $\bigg\}$ ③

This procedure was adapted from that reported by Daub, G. H., and Johnson, W. S. *Organic Syntheses;* Wiley: New York, 1963; Collect. Vol. IV, p. 390. $\bigg\}$ ⑧

Physical Properties of Reactants and Products					
Compound	MW[a]	Amounts	mmol	mp (°C)	bp (°C)
Succinic acid	118.09	118 mg	1.0	182	
Phenol	94.4	188 mg	2.0	40–42	182
Phosphorous oxychloride	153.33	84 µL	0.9		105.8
Diphenyl succinate	270.29			121	
[a]MW = molecular weight.					

⑤

⑥

In a 3.0-mL conical vial containing a magnetic spin vane and equipped with a reflux condenser protected by a calcium chloride drying tube were placed succinic acid (118 mg, 1.0 mmol), phenol (188 mg, 2.0 mmol), and phosphorous oxychloride (84 μL 0.9 mmol). As described in the Material Safety Data Sheets, care needs to be exercised when handling these materials. Phenol and phosphorous oxychloride are hazardous in case of skin contact. Both calcium chloride and succinic acid are known irritants. Caution should be exercised when working with the volatile organic compound toluene and a reaction which generates a severely corrosive gas (HCl). The reaction mixture was heated with stirring at 115 °C in a sand bath in the **hood** for 1.25 h. It was necessary to conduct the reaction in the **hood**, because hydrogen chloride (HCl) gas evolved during the course of the reaction. The drying tube was removed, toluene (0.5 mL) was added through the top of the condenser using a Pasteur pipet, and the drying tube was replaced. The mixture was then heated for an additional one hour at 115 °C.

The hot toluene solution was separated from the red syrupy residue of phosphoric acid using a Pasteur pipet. The toluene extract was filtered by gravity using a fast-grade filter paper and the filtrate was collected in a 10-mL Erlenmeyer flask. The phosphoric acid residue was then extracted with two additional 1.0-mL portions of hot toluene. These extracts were also separated using the Pasteur pipet and filtered, and the filtrate was collected in the same Erlenmeyer flask. The combined toluene solutions were concentrated to a volume of approximately 0.6 mL by warming them in a sand bath under a gentle stream of nitrogen (N_2) gas in the **hood**. The pale yellow liquid residue was then allowed to cool to room temperature; the diphenyl succinate precipitated as colorless crystals. The solid was collected by vacuum filtration using a Hirsch funnel, and the filter cake was washed with three 0.5-mL portions of cold diethyl ether. The product was dried in a vacuum oven at 30 °C (3 mm Hg) for 30 min.

⑦
⑧

The 181 mg (67%) of diphenyl succinate had an mp of 120–121 °C (lit. value 121 °C: *CRC Handbook of Chemistry and Physics,* 89th ed.; CRC Press: Boca Raton, FL, 2008-2009; no. 13559, p. 3–220).

⑨

The key change in the IR spectrum was a shift of the carbonyl absorption. The ester's carbonyl absorption is now approximately 50 wavenumbers to the right when compared to that of the anhydride. [At *this point, the data may be listed, or the spectrum attached to a separate page of the notebook.*]

⑩ *Marilyn C. Waris*

witnessed by
D. Jeanne d'Arc Mailhiot — 19 July 2022

CALCULATING YIELDS

Almost without exception, in each of the experiments presented in this text, you are asked to calculate the percentage yield. For any reaction, it is always important for the chemist to know how much of a product is actually produced (experimental) compared to the theoretical (maximum) amount that could have been formed. The percentage yield is calculated on the basis of the relationship

$$\% \text{ yield} = \frac{\text{actual yield (experimental)}}{\text{theoretical yield (calculated maximum)}} \times 100$$

The percentage yield is generally calculated on a weight (gram or milligram) or on a mole basis. In the present text, the calculations are made using milligrams.

Several steps are involved in calculating the percentage yield.

Step 1 Write a *balanced* equation for the reaction. For example, consider Experiment [22], the Williamson synthesis of propyl *p*-tolyl ether.

p-Cresol	Propyl iodide	Propyl *p*-tolyl ether

Physical Properties of Reactants				
Compound	MW (mg/mmol)	Amount	mmol	d(mg/μL)
p-Cresol	108.15	160 μL	1.56	1.5312
25% (by weight) NaOH soln	40.0	260 μL	~1.6	
Tetrabutylammonium bromide	322.38	18 mg	0.056	
Propyl iodide	169.99	150 μL	1.54	1.5058

Step 2 Identify the *limiting* reactant. The ratio of reactants is calculated on a millimole (or mole) basis. In the example, 1.56 mmol of *p*-cresol and ca. 1.6 mmol of sodium hydroxide are used, compared to 1.54 mmol of propyl iodide, which is therefore the limiting reagent. The tetrabutylammonium bromide is not considered because it is used as a catalyst—it is neither incorporated into the product nor consumed in the reaction. The calculation of the theoretical yield is thus based on the amount of propyl iodide, 1.54 mmol.

Step 3 Calculate the *theoretical* (maximum) amount of the product that could be obtained for the conversion, based on the limiting reactant. Here, one mole of propyl iodide produces one mole of the propyl *p*-tolyl ether. Therefore, the maximum amount of propyl *p*-tolyl ether (molecular weight = 150.2) that can be produced from 1.54 mmol of propyl iodide is 1.54 mmol, or 231 mg.

Step 4 Determine the *actual* (experimental) yield (milligrams) of product isolated in the reaction. This amount is invariably less than the theoretical quantity, unless the material is impure (one common contaminant is water). For example, student yields for the preparation of propyl *p*-tolyl ether average 140 mg.

Step 5 Calculate the *percentage yield* using the weights determined in steps 3 and 4. The percentage yield is then

$$\% \text{yield} = \frac{140 \text{ mg (actual)}}{231 \text{ mg (theoretical)}} \times 100 = 60.6\%$$

As you carry out each reaction in the laboratory, try to obtain as high a percentage yield of product as possible. The reaction conditions in this book's experiments have been carefully developed; if you master the microscale techniques for transferring reagents and isolating products, your yields will be as high as possible.

QUESTIONS

3-1. Factory A produces the wheels that are used for the frames made in Factory B. Factory C relies exclusively on the materials produced in Factories A and B. Assuming all the necessary parts minus the wheels and frames are housed in Factory C, how many bicycles can be completely assembled when Factories A and B provide 24 wheels and 13 frames, respectively? Explain.

3-2. You are provided a vial that contains 220 mg of material. This material represents a 73% isolated yield. Calculate the theoretical amount (theoretical yield) that could have been formed.

3-3. The density of 2-methyl-2-butanol is 0.806 g/mL. How many mg represent an aliquot of 513 μL? How many mmol represent an aliquot of 0.340 mL (2-methyl-2-butanol, formula weight is 88.15 g/mol)?

DETERMINATION OF PHYSICAL PROPERTIES

Determination of physical properties is important for substance identification and as an indication of material purity. Historically, the physical constants of prime interest have included boiling point, density, and refractive index in liquids and melting point in solids. In special cases, optical rotation and molecular weight determinations may be required. Today, with the widespread availability of spectroscopic instrumentation, powerful new techniques may be applied to the direct identification and characterization of materials, including the analysis of individual components of very small quantities of complex mixtures. The sequential measurement of the infrared (IR) and mass spectrometric (MS) characteristics of a substance resolved "on the fly" by capillary gas chromatography (GC) can be quickly determined and interpreted. This particular combination (GC-IR-MS), which stands out among a number of hyphenated techniques that are now available, is perhaps the most powerful system yet developed for molecular identification. The rapid development of high-field multinuclear and benchtop magnetic resonance (NMR) spectrometers has added another powerful dimension to identification techniques. NMR sensitivity, however, is still considerably lower than that of either IR or MS. The IR spectrum alone, obtained with one data point per wavenumber, can add more than 4000 measurements to the few classically determined properties. *Indeed, even compared to high-resolution MS and pulsed ^{1}H and ^{13}C NMR, the infrared spectrum of a material remains a powerful set of physical properties (transmission elements) available to the organic chemist for the identification of an unknown compound.*[1]

Simple physical constants are determined mainly to assist in establishing the purity of *known* materials. Because the boiling point or the melting point of a material can be very sensitive to small quantities of impurities, these data can be particularly helpful in determining whether a starting material needs further purification or whether a product has been isolated in acceptable purity. Gas (GC), high-performance liquid (HPLC), and thin-layer (TLC) chromatography, however, now provide more powerful purity information when such data are required. When a new composition of matter has been formed, an elemental (combustion) analysis is normally reported if sufficient material is available for this destructive analysis. For new substances we are, of course, interested in establishing not only the identity, but also the molecular structure of the materials. In this situation other modern techniques (such as ^{1}H and ^{13}C NMR spectroscopy, high-resolution MS, and single-crystal X-ray diffraction) can provide sensitive and powerful structural information.

Chapter 4: C$_4$H$_6$, Bicyclo[1.1.0]butane
Lemal, Menger, and Clark (1963).

[1]Griffiths, P. R.; de Haseth, J. A. *Fourier Transform Infrared Spectrometry*, 2nd ed.; Wiley: New York, 2007.

When comparisons are made between experimental data and values obtained from the literature, it is essential that the latter information be obtained from the most reliable sources available. This is especially true when considering the volume of misinformation found online. Certainly, judgment, which improves with experience, must be exercised in accepting any value as a standard. In addition to the wealth of data found online (e.g., SciFinder Scholar), the known classical properties of a large number of compounds can be obtained from the *CRC Handbook of Chemistry and Physics* and the *Merck Index*. The *Aldrich Catalog Handbook of Fine Chemicals* is also a readily available, inexpensive source. These reference works list physical properties for inorganic, organic, and organometallic compounds. The *Aldrich Catalog* also references IR and NMR data for a large number of substances. If recent editions of the *CRC Handbook* and the *Aldrich Catalog* are not readily available, be sure to check online through your department or library.

LIQUIDS

Ultramicro Boiling Point

Upon heating, the vapor pressure of a liquid increases, though in a nonlinear fashion. When the pressure reaches the point where it matches the local pressure, the liquid boils. That is, it spontaneously begins to form vapor bubbles, which rapidly rise to the surface. If heating is continued, both the vapor pressure and the temperature of the liquid will remain constant until the substance has been completely vaporized (Fig. 4.1).

Because microscale preparations generally yield about 30–70 μL of liquid products, using only 5 μL or less of material for boiling-point measurements is highly desirable. The modification of the earlier Wiegand ultramicro boiling-point procedure[2] to the ultramicro procedure described here has established that reproducible and reasonably accurate (±1 °C) boiling points can be observed on 3–4 μL of most liquids thermally stable at the required temperatures.

Procedure. Ultramicro boiling points can be conveniently determined in standard (90-mm-long) Pyrex glass capillary melting-point tubes. The melting-point tube replaces the conventional 3- to 4-mm (o.d.) tubing used in the Siwoloboff procedure.[3] The sample (3–4 μL) is loaded into the melting-point capillary via a 10-μL syringe and centrifuged to the bottom if necessary. A small glass bell replaces the conventional melting-point tube as the bubble generator in micro boiling-point determinations. It is formed by heating 3-mm (o.d.) Pyrex tubing with a microburner and drawing it out to a diameter small enough to be readily fit inside the melting-point capillary. A section of the drawn capillary is fused and then cut to yield two small glass bells approximately 5 mm long (Fig. 4.2*a*). It is important that the fused section be reasonably large, because it is more than just a seal. The fused glass must add enough weight to the bell that it will firmly seat itself in the bottom of the melting-point tube.

An alternative technique for preparing the glass bells follows: heat the mid-section of an open-ended melting-point capillary tube and then draw the glass to form a smaller capillary section. This section is then broken approximately in the middle and each open end is sealed. The appropriate length for the bell is

[2]Wiegand, C. *Angew. Chem.* **1955**, *67*, 77. Mayo, D. W.; Pike, R. M.; Butcher, S. S.; Meredith, M. L. *J. Chem. Educ.* **1985**, *62*, 1114.
[3]Siwoloboff, A. *Berichte* **1886**, *19*, 795.

Figure 4.1   Vapor pressure curves. (Reproduced with permission of John Wiley and Sons, Inc., from Brady, J. E.; Humiston, G. E. *General Chemistry*, 3d ed., Copyright 1982.)

Figure 4.2   (a) Preparation of small glass bell for ultramicro boiling-point determination. (b) Ultramicro boiling-point assembly. (Reproduced with permission from the American Chemical Society, from Mayo, D. W.; Pike, R. M.; Butcher, S. S.; Meredith, M. L. *J. Chem. Educ.* **1985,** *62,* 1114. Copyright 1985, American Chemical Society.)

then broken off. Thus, two bells are obtained, one from each section. The sealing process (be sure that a significant section of glass is fused during the tube closure to give the bell enough weight) can be repeated on each remaining glass section, and thus a series of bells can be prepared in a relatively short period.

A glass bell is now inserted into the loaded melting-point capillary, open end first (down), and allowed to fall (centrifuged if necessary) to the bottom. The assembled system (Fig. 4.2*b*) is then inserted onto the stage of a Thomas–Hoover Uni-Melt Capillary Melting Point Apparatus (Fig. 4.3)[4] or a similar system (such as a Mel-Temp or DigiMelt).

The temperature is rapidly raised to 15–20 °C below the expected boiling point (the temperature should be monitored carefully in the case of unknown substances), and then adjusted to a maximum rise rate of 2 °C/min and heated until a *fine stream* of bubbles is emitted from the glass bell. The heat control is then adjusted to drop the temperature. The boiling point is recorded at the point where the last escaping bubble collapses (i.e., when the vapor pressure of the substance equals the atmospheric pressure). The heater is then rapidly adjusted to again raise the temperature at 2 °C/min and induce a second stream of bubbles. This procedure may then be repeated several times. A *precise and sensitive temperature control system is essential to the successful application of this cycling technique, but it is not essential for obtaining satisfactory boiling-point data.*

Utilization of the conventional melting-point capillary as the "boiler" tube has the particular advantage that the boiling point of a liquid can readily be determined using a conventional melting-point apparatus. The illumination and magnification available make the observation of rate changes in the bubble stream easily seen. Inexpensive 10-µL GC injection syringes appear to be the most successful instrument to use for transferring the small quantities of

---

[4]Thomas Scientific, P.O. Box 99, Swedesboro, NJ 08085.

Figure 4.3   Thomas–Hoover melting-point determination device. (Reproduced with permission of John Wiley and Sons, Inc., from Zubrick, J. W. *The Organic Chemistry Lab Survival Manual*, 9th ed., Copyright 2013.)

liquids involved. The 3-in. needles normally supplied with the 10-μL barrels will not reach the bottom of the capillary; liquid samples deposited on the walls of the tube, however, are easily and efficiently moved to the bottom by centrifugation. After the sample is packed in the bottom of the capillary tube, the glass bell is introduced. The glass bell is necessary because a conventional Siwoloboff fused-capillary insert would extend beyond the top of the melting-point tube; thus, capillary action between the "boiler" tube wall and the capillary insert would draw most of the sample from the bottom of the tube up onto the walls. This effect often precludes the formation of the requisite bubble stream.

Little loss of low-boiling liquids occurs (see Table 4.1). Furthermore, if the boiling point is overrun and the sample is suddenly evaporated from the bottom of the "boiler" capillary, it will rapidly condense on the upper (cooler) sections of the tube. These sections extend above the heat-transfer fluid or metal block. The sample can easily be recentrifuged to the bottom of the tube and a new determination of the boiling point begun. Note that if the bell cavity fills completely during the cooling point of a cycle, it is often difficult to reinitiate the bubble stream without first emptying the entire cavity by overrunning the boiling point.

Observed boiling points for a series of compounds, which boil over a wide range of temperatures, are summarized in Table 4.1.

Materials that are thermally stable at their boiling point will give identical values on repeat determinations. Substances that begin to decompose will

Table 4.1 Observed Boiling Points (°C)

Compound	Observed	Literature Value	Reference
Methyl iodide	42.5	42–43	*a*
Isopropyl alcohol	82.3	82.3	*b*
2,2-Dimethoxypropane	80.0	83.0	*c*
2-Heptanone	149–159	151.1	*d*
Cumene	151–153	152.4	*e*
Mesitylene	163	164.7	*f*
p-Cymene	175–178	177.1	*g*
Benzyl alcohol	203	205.3	*h*
Diphenylmethane	263–265	265	*i*

Note. (Observed values are uncorrected for changes in atmospheric pressure [corrections all estimated to be less than ± 0.5 °C].)

Source. Adapted from CRC Handbook of Chemistry and Physics, 89th ed.; CRC Press: Boca Raton, FL, 2008–2009: [a]no. 6307, p. 3–306; [b]no. 9167, p. 3–442; [c]no. 3883, p. 3–190; [d]*no.* 5689; p. 3–274; [e]no. 6478, p. 3–314; [f]no. 10509, p. 3–540; [g]no. 6509, p. 3–316; [h]no. 780, p. 3–42; [i]no. 4498, p. 3–218.

give values that slowly drift after the first few measurements. The observation of color and/or viscosity changes, together with a variable boiling point, signal the need for caution in making extended repeat measurements.

Comparison of the boiling points obtained experimentally at various atmospheric pressures with reference boiling points at 760 torr is greatly facilitated by the use of pressure–temperature nomographs such as that shown in Figure 4.4. A straight line from the observed boiling point to the observed

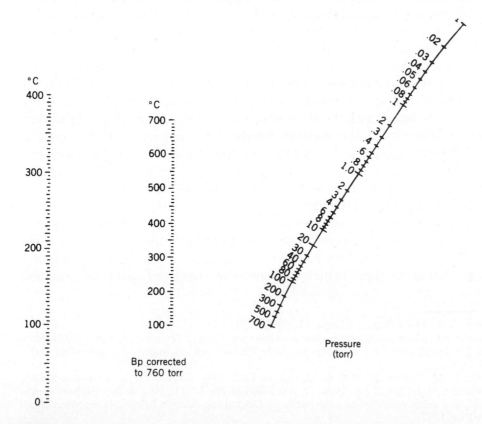

Figure 4.4 Pressure–temperature nomograph.

pressure will pass through the corrected boiling-point value. These values can be of practical importance when carrying out reduced pressure distillations.

DENSITY

Density, defined as mass per unit volume, is generally expressed as grams per milliliter (g/mL) or grams per cubic centimeter (g/cm^3) for liquids. Accurate nondestructive procedures have been developed for the measurement of this physical constant at the microscale level. A micropycnometer (density meter), developed by Clemo and McQuillen, requires approximately 2 µL (Fig. 4.5).[5]

This very accurate device gives the density to three significant figures. The system is self-filling, and the fine capillary ends do not need to be capped while temperature equilibrium is reached or during weighing (the measured values tend to degrade for substances boiling under 100 °C and when room temperatures rise much above 20 °C). In addition, the apparatus must first be tared, filled, and then reweighed on an *analytical* balance. A technique that results in less precise densities (good to about two significant figures), but which is far easier to use, is simply to substitute a 50- or 100-µL syringe for the pycnometer. The method simply requires weighing the syringe before and after filling it to a measured volume as in the conventional technique. With the volume and the weight of the liquid known, the density can be calculated. A further advantage of the syringe technique is that the pycnometer is not limited to a fixed volume. Although much larger samples are required, it is not inconvenient to utilize the entire sample obtained in the reaction for this measurement, since the material can be efficiently recovered from the syringe for additional characterization studies. Because density changes with temperature, these measurements should be obtained at a constant temperature.

An alternative to the syringe method is to use *Drummond Disposable Microcaps* as pycnometers. These precision-bore capillary tubes, calibrated to contain the stated volume from end to end (accuracy ±1%), are available from a number of supply houses.[6] These tubes are filled by capillary action or by suction using a vented rubber bulb (provided). The pipets can be obtained in various sizes, but as with the syringe, volumes of 50, 75, or 100 µL are recommended. When using this method, handle the micropipet with forceps and not with your fingers. The empty tube is first *tared*, and then filled and weighed again. The difference in these values is the weight of liquid in the pipet. For convenience, the pipet may be placed in a small container (10-mL beaker or Erlenmeyer flask) when the weighing procedure is carried out.

Two inexpensive micropycnometers can also be easily prepared: The first can be made from a Pasteur pipet as reported by Singh et al.[7] The volume of each individual pycnometer can be varied from 20 to 100 µL, or larger if desired. Values to three significant figures are obtained using an *analytical*

4 µm   0.4 mm

Figure 4.5   Pycnometer of Clemo and McQuillen. (Adapted from pycnometer from Schneider, F. L. Monographien aus dem Gebiete der qualitativen Mikroanalyse, *Qualitative Organic Microanalysis*, Vol. II; Benedetti-Pichler, A. A., Ed.; Springer: Vienna, 1964.)

[5]Clemo, G. R.; McQuillen, A. *J. Chem. Soc.* **1935**, 1220.

[6]Drummond Disposable Microcaps are available from Thomas Scientific, P.O. Box 99, Swedesboro, NJ 08085; and Sargent-Welch Scientific Co., a VWR company, P.O. Box 1026, Skokie, IL 60097.

[7]Singh, M. M.; Szafran, Z.; Pike, R. M. *J. Chem. Educ.* **1993**, 70, A36; see also Ellefson-Kuehn, J., Wilcox, C. J. *J. Chem. Educ.* **1994**, 71, A150; and Singh, M. M.; Pike, R. M.; Szafran, Z. *Microscale and Selected Macroscale Experiments for General and Advanced General Chemistry*; Wiley: New York, 1995.

balance, because evaporation is generally negligible, and if the pycnometer mouth is small.

The second pycnometer, by Pasto and co-workers, is made from a melting-point capillary tube.[8] In both of these techniques, the volume of the pycnometer must be determined. The procedure to determine the density involves the following steps. The empty micropycnometer is tared on an analytical balance, filled with the liquid in question, and reweighed (the difference in weights is the weight of the liquid). The sample is removed and the pycnometer is rinsed with acetone and dried. It is then filled with distilled water and reweighed. From the known[9] density of water at the given temperature, the volume of water can be determined and thus the volume of the pycnometer. The volume of the original liquid sample also equals this value. The weight and volume of the sample are used to calculate its density.

SOLIDS

Melting Points

In general, the crystalline lattice forces holding organic solids together are distributed over a relatively narrow energy range. The melting points of organic compounds, therefore, are usually relatively sharp—that is, less than 2 °C. The range and maximum temperature of the melting point, however, are very sensitive to impurities. Small amounts of sample contamination by soluble impurities nearly always result in melting-point depressions.

The drop in melting point is usually accompanied by an expansion of the melting-point range. Thus, in addition to the melting point acting as a useful guide in identification, it also can be a particularly effective indication of sample purity.

Procedure.   In the microscale laboratory, two different types of melting-point determinations are carried out: (1) simple capillary melting points and (2) evacuated melting points.

Simple Capillary Melting Point. Because the microscale laboratory utilizes the Thomas–Hoover Uni-Melt apparatus or a similar system for determining boiling points, melting points are conveniently obtained on the same apparatus. The Uni-Melt system utilizes an electrically heated and stirred silicone oil bath. The temperature readings require no correction in this case because the depth of immersion is held constant. (This assumes, of course, that the thermometer is calibrated to the operational immersion depth.) Melting points are determined in the same capillaries as boiling points. The capillary is loaded by introducing about 1 mg of material into the open end. The sample is then tightly packed (~2 mm) into the closed end by dropping the capillary down a length of glass tubing held vertically to the bench top. The melting-point tube is then ready for mounting on the metal stage, which is immersed in the silicone oil bath of the apparatus. If the melting point of the substance is expected to occur in a certain range, the temperature can be rapidly raised to ~2 °C below the expected value. At that point, the temperature rise should

---

[8]Pasto, D.; Johnson, C.; Miller, M. *Experiments and Techniques in Organic Chemistry;* Prentice Hall: Englewood Cliffs, NJ, 1992.

[9]Values for the density of water at various temperatures can be found in the *CRC Handbook of Chemistry and Physics*.

Figure 4.6 Melting-point composition diagram for the binary mixture, $A + B$. In this diagram, a is the melting point of solid A, b of solid B, e of eutectic mixture E, and m of the 80% A:20% B mixture, M.

be adjusted to a maximum of 2 °C/min, which is the standard rate of change at which the reference determinations are obtained. The melting-point range is recorded from the temperature at which the first drop of liquid forms (point e in Fig. 4.6) to that at which the last crystal melts (point m in Fig. 4.6).

Evacuated Melting Points. Many organic compounds begin to decompose at their melting points. This decomposition often begins as the melting point is approached and may adversely affect the values measured. The decomposition can be invariably traced to reaction with oxygen at elevated temperatures. If the melting point is obtained in an evacuated tube, therefore, much more accurate melting points can be obtained. These more reliable values arise not only from increased sample stability, but also because several repeat determinations can often be made on the same sample. The multiple measurements then may be averaged to provide more accurate data.

Evacuated melting points are quickly and easily obtained with a little practice. The procedure is as follows: Shorten the capillary portion of a Pasteur pipet to approximately the same length as a normal melting-point tube (Fig. 4.7a). Seal the capillary end by rotating in a microburner flame. Touch the pipet only to the very edge of the flame, and keep the large end at an angle below the end being sealed (Fig. 4.7b). This technique will prevent water from the flame being carried into the tube, where it will condense in the cooler sections. Then load 1–2 mg of sample into the drawn section of the pipet with a microspatula (Fig. 4.7c). Tap the pipet gently to seat the solid powder as far down the capillary as it can be worked (Fig. 4.7d). Then push the majority of the sample part way down the capillary with the same diameter copper wire that you used to seat the cotton plug in constructing the Pasteur filter pipet (Fig. 4.7e). Next, connect the pipet with a piece of vacuum tubing to a mechanical high-vacuum pump. Turn on the vacuum and evacuate the pipet for 30 s (Fig. 4.7f). With a microburner, gently warm the surface of the capillary tubing just below the drawn section. On warming, the remaining fragments of the sample (the majority of which has been forced farther down in the tube) will sublime in either direction away from the hot section. Once the traces of sample have been "chased" away, the heating is increased, and the capillary tube is

(a)

(b) rotate

(c) 1-2 mg of sample

microspatula sealed

copper wire

(d) (e) (f) to vacuum

(g) to vacuum separate

(h) to vacuum tamp to bottom drop down tubing

(i)

Figure 4.7   Procedure for obtaining evacuated melting-point capillaries.

collapsed, fused, and separated from the shank. The shank remains connected
to the vacuum system (Fig. 4.7g). The vacuum system is then vented and the
shank is discarded. The sample is tightly packed into the initially sealed end
of the evacuated capillary by dropping it down a section of glass tubing, as in
the case of packing open melting-point samples. After the sample is packed
(~2 mm in length, see Fig. 4.7h), a section of the evacuated capillary about
10–15 mm above the sample is once more gently heated and collapsed by the
microburner flame (Fig. 4.7i).

This procedure is required to trap the sample below the surface of the
heated silicone oil in the melting-point bath and thus avoid sublimation
up the tube to cooler sections during measurement of the melting point.

The operation is a little tricky and should be practiced a few times. It is very important that the tubing completely fuse. Now the sample is ready to be placed in the melting-point apparatus. The procedure beyond this point is the same as in the open capillary case, except that after the sample melts, it can be cooled, allowed to crystallize, and remelted several times, and the average value of the range reported. If these values begin to drift downward, the sample can be considered to be decomposing even under evacuated, deoxygenated conditions. In this case the first value observed should be recorded as the melting point and decomposition noted (mp xx dec, where dec = decompose).

Mixture Melting Point. Additional information can often be extracted from the sensitivity of the melting point to the presence of impurities. Where two different substances possess identical melting points (not uncommon), it would be impossible to identify an unknown sample as either material based on the melting point alone. If reference standards of the two compounds are available, however, then mixtures of the unknown and the two standards can be prepared. It is important to prepare several mixtures of varying concentrations for melting-point comparisons, since the point of maximum depression need not occur on the phase diagram at the 50:50 ratio (see Fig. 4.6). In samples that do not exhibit any decomposition at the melting point, the prepared mixtures should be first heated until a homogeneous melt is obtained. Each is then cooled and ground to a fine powder, and the definitive melting point is obtained on the ground sample. The melting points of the unknown and the mixed samples should be obtained simultaneously (the Uni-Melt stage will accept up to seven capillaries at one time). This is desirable because all the samples will then be heated at the same rate. The unknown sample and the mixture of the unknown with the correct reference will have identical values, but the mixture of the reference with a different substance will give a depressed melting point. This procedure is the classical step to positive identification of crystalline solids.

Mixtures of two different compounds only rarely fail to exhibit mixture melting-point depression, but it can happen. Some mixtures may not show a depression or show only a very small one, due to eutectic or compound formation. Elevation of the melting point has also been observed. Therefore, if mixture melting-point data are used for identification purposes, comparison of other physical constants or spectroscopic data is advocated to establish identity beyond any reasonable doubt.

QUESTIONS

4-1. Room temperature is recorded when a density determination for a given substance is performed in the laboratory. Why?

4-2. Describe how you would determine the melting point of a substance that sublimes before it melts.

4-3. You are presented with three vials, each containing a white crystalline solid. While you are provided with the fact that each vial contains pure sample, only two samples, trans-cinnamic acid and urea (both melt at 132.5–133 °C) represent the contents of all three vials. Outline a detailed plan using the melting-point technique that identifies the contents of each vial.

4-4. In the microscale method of determining boiling points, one heats the liquid until a steady stream of bubbles is observed coming out of the bell. The temperature is then lowered and the boiling point is read just as the bubbles stop. Why is this technique preferable to measuring the boiling point when the bubbles first start to appear?

4-5. What would you expect the observed boiling point to be at 2 torr of a liquid which has a boiling point of 250 °C at 760 torr?

REFRACTIVE INDEX

A beam of light appears to bend as it passes from one medium to another. For example, when viewed from above an oar looks bent at the point where it enters the water. This effect is a consequence of the refraction of light. It results from the change in velocity of the radiation at the interface of the media, and the angle of refraction, ϕ'. It is related to the velocity change as follows (see Fig. 4A.1):

$$\frac{\sin \phi}{\sin \phi'} = \frac{\text{velocity in vacuum}}{\text{velocity in sample}} = n \,(\text{refractive index})$$

where ϕ is the angle of incidence between the beam of light and the interface. Because the velocity of light in a medium must be less than that in a vacuum, the index of refraction n will always be greater than 1. In practice, n is taken as the ratio of the velocity of light in air relative to the medium being measured.

The refractive index is also wavelength dependent. The wavelength dependence gives rise to the effect of dispersion or the spreading of white light into its component colors. When we measure n, therefore, we must specify the wavelength at which the measurement is made. The standard wavelength for refractive index determinations has become the (yellow) sodium 589-nm emission, the sodium D line. Sodium, unfortunately, is a poor choice of wavelength for these measurements with organic substances. In the past, however, the sodium lamp represented one of the easiest-to-obtain monochromatic sources of light and thus it became widely used. Because the density of the medium is sensitive to temperature, the velocity of radiation also changes with temperature, so refractive index measurements must be made at constant temperatures. Many values in the literature are reported at 20 °C. The refractive index can be measured optically quite accurately to four decimal places. Because this measurement is particularly sensitive to the presence of impurities, the refractive index can be a valuable physical constant for tracking the purification of liquid samples.

For example, the measurement is reported as

$$n_{\mathrm{D}}^{20} = 1.4628$$

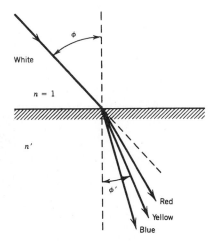

Figure 4A.1 Upon refraction, white light is spread out into a spectrum. This is called dispersion.

PROCEDURE

In the Abbe-3L refractometer (Fig. 4A.2), white light is used as the source, but compensating prisms give indexes for the D line. This refractometer is commonly used in many organic laboratories. Samples (~10 µL) are applied between the horizontal surfaces of a pair of hinged prisms (Fig. 4A.3). A sampling procedure recently developed by Ronald[10] significantly reduces the amount of sample required and allows accurate measurements of highly volatile materials. The technique involves placing a small precut 6-mm disk of good-quality lens paper at the center of the bottom prism. The sample is loaded onto the disk with a micro Pasteur pipet or microliter syringe (see Table 4A.1).

CAUTION: Do not touch the refractometer prisms with the Pasteur pipet or syringe tip. The prisms can be easily and permanently marred or scratched and the refractometer will then give erroneous results.

Figure 4A.2   Abbe-3L refractometer. (*Courtesy of Bausch-Lomb, Rochester, NY.*)

Figure 4A.3   Diagram of a typical refractometer. (*Courtesy of Milton Roy Co., Rochester, NY.*)

The refractometer is adjusted so that the field of view has a well-defined light and dark split image (see your instrument manual for the correct routine for making adjustments on your particular refractometer).

When using the refractometer, always clean the prisms with alcohol and lens paper before and after use. Record the temperature at which the reading is taken. A reasonably good extrapolation of temperature effects can be obtained by assuming that the index of refraction changes 0.0004 unit per degree Celsius, and that it varies inversely with temperature.

Table 4A.1 Refractive Index Measurements Utilizing Lens Paper Disk Technique			
Substance	T (°C)	n^t (normal) 100 μL	n^t (microdisk) 2–4 μL
Water	24.5	1.3224	1.3226
Diethyl ether	24	1.3508	1.3505
Chlorobenzene	24.5	1.5225	1.5219
Iodobenzene	24.5	1.6151	1.6151

[10]Ronald, B. P., Department of Chemistry, Idaho State University, Pocatello, ID (personal communication).

MICROSCALE LABORATORY TECHNIQUES

This chapter introduces the microscale organic laboratory techniques used throughout the experimental sections of the textbook. These must be mastered to be successful when working at this scale. Detailed discussions are given for each individual experimental technique. At the end of each discussion there is a list of the experiments in Chapters 6 and 7 that use the technique. These lists should prove useful to instructors compiling experiments to be covered in the laboratory. The lists will also be handy for students who wish to examine the application of a particular technique to other experiments not covered in their laboratory sequence.

As was the case with earlier editions, a continued effort has been made to streamline the basic reference material from the text, which now includes material available on the book's website.. We hope this treatment of the laboratory will make the more important aspects of the basic text easier to access and will speed your laboratory work along.

One of the principal hurdles in dealing with experimental chemistry is the isolation of pure materials. Characterization (identification) of a substance almost always requires a pure sample of the material. This is a particularly difficult demand of organic chemistry because most organic reactions generate several products. We are generally satisfied if the desired product is the major component of the mixture obtained. This chapter places a heavy emphasis on separation techniques.

Gas Chromatography

Technique 1 begins the discussion of the resolution (separation) of microliter quantities of liquid mixtures via preparative gas chromatography. Techniques 2 and 3 deal with semi-micro adaptations of classical distillation routines that focus on the separation of liquid mixtures involving one to several milliliters of material.

Chromatography methods revolutionized experimental organic chemistry. These methods are by far the most powerful of the techniques for separating mixtures and isolating pure substances, either solids or liquids. Chromatography is the resolution (separation) of a multicomponent mixture (several hundred components in some cases) by distribution between two phases, one stationary and one mobile. The various methods of chromatography are categorized by the phases involved: column, thin-layer, and paper (all solid–liquid chromatography); partition (liquid–liquid chromatography); and vapor-phase

---

Chapter 5: C$_5$H$_6$, Propellane
Wiberg and Walker (1982).

(gas–liquid chromatography, or simply gas chromatography). The principal mechanism these separations depend on is differential solubility, or adsorptivity, of the mixture components in the two phases involved. That is, the components must exhibit different partition coefficients (see also Technique 4 for a detailed discussion of partition coefficients).

Gas chromatography (GC, sometimes called vapor-phase chromatography) is an extraordinarily powerful technique for separating mixtures of organic compounds. The stationary phase in GC is a high-boiling liquid and the mobile phase is a gas (the carrier gas). Gas chromatography can separate mixtures far better than distillation techniques can (see Technique 2 discussion).

Preparative GC separations, which involve perhaps 5–100 μL of material, require relatively simple instrumentation but sacrifice resolution for the ability to separate larger amounts of material.

Analytical GC separations require tiny amounts of material (often 0.1 μL of a very dilute solution), and can separate incredibly complex mixtures. The ability to work with small quantities of materials in analytical GC separations is an advantage at the microscale level. This analytical tool is used to analyze distillation fractions in Experiments [3C] and [3D].

GC Instrumentation

GC instrumentation can range from straightforward and relatively simple systems to systems with complex, highly automated, and relatively expensive components. A diagram of a common and simple GC typically used in an instructional laboratory is shown in Figure 5.1.

Injection Port. The analysis begins in a heated injection port. The sample mixture is introduced by syringe through a septum into the high-temperature chamber (injection port) through which the inert carrier gas (the mobile phase) is flowing. Helium and nitrogen are common carrier gases. The solubility of the sample in the carrier gas depends mostly on the vapor pressure of the substances in the sample. Heating the injection port

Figure 5.1   Block diagram of a dual-column gas chromatograph showing essential parts. (*Courtesy of GOW-MAC Instrument Co., 277 Brodhead Rd., Bethlehem, PA 18017.*)

helps to ensure the vaporization of less volatile samples. There are two major constraints on GC:

The sample must be stable at the temperature required to cause vaporization, and the sample must have sufficient vapor pressure to be completely soluble in the carrier gas at the column operating temperatures.

NOTE. When injecting a sample, always position your thumb or finger over the syringe plunger. This prevents a blowback of the sample by the carrier gas pressure in the injection port.

Column. The vaporized mixture is swept by the carrier gas from the injection port onto the column. Bringing the sample mixture into intimate contact with the column begins the separation process. The stationary liquid phase in which the sample will partially dissolve is physically and/or chemically bonded to inert packing material (often called the support) in the column. Gas chromatographic columns are available from manufacturers in a variety of sizes and shapes. In the diagram of the GOW-MAC instrument (Fig. 5.1), two parallel coiled columns are mounted in an oven. Considerable oven space can be saved and better temperature regulation achieved if the columns are coiled. Temperature regulation is particularly important, because column resolution degrades rapidly if the entire column is not at the same temperature. Most liquid mixtures need a column heated above ambient temperatures to achieve the vapor pressure the separation requires.

The mixture is separated as the carrier gas sweeps the sample through the column. Columns are usually made from stainless steel, glass, or fused silica. The diameter and length of the column are critical factors in separating the sample mixture.

Packed Columns. In packed columns the liquid (stationary) phase in contact with the sample contained in the mobile gas phase is maximized by coating a finely divided inert support with the nonvolatile liquid. The coated support is carefully packed into the column so as not to develop empty spaces. Packed columns are usually $\frac{1}{4}$ or $\frac{1}{8}$ inch in diameter and range from 4 to 12 ft in length. These columns are particularly useful in the microscale laboratory, since they can be used for both analytical and preparative GC. Simple mixtures of 20–80 µL of material can often be separated into their pure components and collected at the exit port of the detector. Smaller samples (0.2–2.0 µL range) will exhibit better separation.

Capillary Columns. Capillary columns have no packing; the liquid phase is simply applied directly to the walls of the column. These columns are referred to as wall-coated, open-tubular (WCOT) columns. The reduction in surface area (compared to packed columns) is compensated for by tiny column diameters (perhaps 0.1 mm) and impressive lengths (100 m is not uncommon). Capillary columns are the most powerful columns used for analytical separations. Mixtures of several hundred compounds can be completely resolved on a capillary GC column. These columns require a more sophisticated and expensive chromatography instrument. Capillary columns, because of their tiny diameters, can accommodate only very small samples, perhaps 0.1 µL or less of a dilute solution. Capillary columns cannot be used for preparative separations.

Liquid Phase. Once the sample is introduced on the column (in the carrier gas), it will undergo partition into the liquid phase. The choice of the liquid phase is particularly important because it directly affects the relative distribution coefficients.

In general, the stationary liquid phase controls the partitioning of the sample by two criteria. First, if little or no interaction occurs between the sample components and the stationary phase, the boiling point of the materials will determine the order of elution. Under these conditions, the highest boiling species will be the last to elute. Second, the functional groups of the components may interact directly with the stationary phase to establish different partition coefficients. Elution then depends on the particular binding properties of the sample components.

Some typical materials used as stationary phases are shown below.

Name	Stationary Phase	Maximum Temperature (°C)	Mechanism of Interaction
Silicone oil DC710, etc.	$R_3Si[OSiR_2]_nOSiR_3$	250	According to boiling point
Polyethylene glycol (Carbowax)	$HO[CH_2CH_2O]_nCH_2CH_2OH$	150	Relatively selective toward polar components
Diisodecyl phthalate	$o\text{-}C_6H_4[CO_2\text{-isodecyl}]_2$	175	According to boiling point

Oven Temperature. The temperature of the column will also affect the separation. In general, the elution time of a sample will decrease as the temperature is increased. That is, retention times are shorter at higher temperatures. Higher boiling components tend to undergo diffusion broadening at low column temperatures because of the increase in retention times. If the oven temperature is too high, however, equilibrium partitioning of the sample with the stationary phase will not be established. Then the components of the mixture may elute together or be incompletely separated. Programmed oven temperature increases can speed up elution of the higher boiling components, but suppress peak broadening and therefore increase resolution. Temperature-programming capabilities require more sophisticated ovens and controllers.

Flow Rate. The flow rate of the carrier gas is another important parameter. The rate must be slow enough to allow equilibration between the phases, but fast enough to ensure that diffusion will not defeat the separation of the components.

Column Length. As noted, column length is an important factor in separation performance. As in distillations, column efficiency is directly proportional to column height, which determines the number of evaporation–condensation cycles. In a similar manner, increasing the length of a GC column allows more partition cycles to occur. Difficult-to-separate mixtures, such as the xylenes (very similar boiling points: *o*-xylene, 144.4 °C; *m*-xylene, 139.1 °C; and *p*-xylene, 138.3 °C), have a better chance of being separated on longer columns. In fact, both GC and distillation resolution data are described using the same term, *theoretical plates* (see Technique 2 and Experiments [3C] and [3D]).

Detector and Exit Port. A successfully separated mixture will elute as its individual components at the instrument's exit port (also temperature controlled). To monitor the exiting vapors, a detector is placed in the gas stream just before the exit port (Fig. 5.1). After passing through the detector, the carrier gas and the separated sample components are then vented.

G One widely used detector is the nondestructive, thermal conductivity detector, sometimes called a hot-wire detector. A heated wire in the gas stream changes its electrical resistance when a substance dilutes the carrier gas and

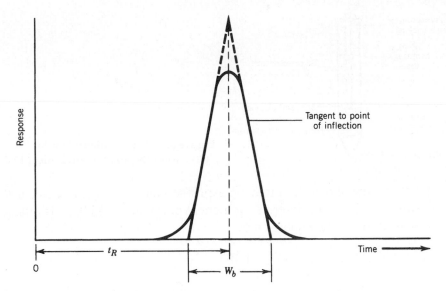

Figure 5.2   Schematic chromatogram.

thus changes its thermal conductivity. Helium has a higher thermal conductivity than most organic substances. When substances other than helium are present, the conductivity of the gas stream changes, which changes the resistance of the heated wire. The change in resistance is measured by comparing it to a reference detector mounted in a second (parallel) gas stream (Wheatstone bridge). The resulting electrical signal is plotted on a chart recorder, where the horizontal axis is time and the vertical axis is the magnitude of the resistance difference. The plot of resistance difference versus time is referred to as the gas *chromatogram*. The retention time (t_R) is defined as the time from sample injection to the time of maximum peak intensity. The baseline width (W_b) of a peak is defined as the distance between two points where tangents to the points of inflection cross the baseline (Fig. 5.2).

Capillary GC systems, and other GC systems used only for analytical separations, often use a flame-ionization detector (FID). In an FID, the gas eluting from the GC column is mixed with air (or oxygen) and hydrogen, and burned. The conductivity of the resulting flame is measured; it changes with the ionic content of the flame, which is proportional to the amount of carbon (from organic material) in the flame. The advantage of an FID is its high sensitivity; amounts of less than a microgram are easily detected. Its disadvantage is that it destroys (burns) the material it detects.

Theoretical Plates.   It is possible to estimate the number of theoretical plates (directly related to the number of distribution cycles) present in a column for a particular substance. The parameters are given in the relationship[1]

$$n = 16\left(\frac{t_R}{W_b}\right)^2$$

where the units of retention time (t_R) and baseline width (W_b) are identical (minutes, seconds, or centimeters). As in distillation columns, the larger the number of theoretical plates, *n*, the higher the resolution of the column and the better the separation.

[1]Berg, E. W. *Physical and Chemical Methods of Separation*; McGraw-Hill: New York, 1963, p. 111.

Figure 5.3 **Gas chromatographic collection tube and 0.1-mL conical vial.**

The efficiency of a system may be expressed as the *height-equivalent theoretical plate* (HETP) in centimeters (or inches) per plate. The HETP is related to the number of theoretical plates *n* by

$$\text{HETP} = \frac{L}{n}$$

where L is the length of the column, usually reported in centimeters. The smaller the HETP, the more efficient the column.

The number of theoretical plates available in fractional distillation columns is limited by column holdup (see Techniques 2, 3, and website discussion of distillation theory). Thus, distillations of less than 500 µL are generally not practical. Gas chromatographic columns, on the other hand, operate most efficiently at the microscale or submicroscale levels, where 500 µL would be an order of magnitude (even 3–8 orders of magnitude in the case of capillary columns) too large.

Fraction Collection.   Sequential collection of separated materials can be made by attaching suitable sample condensing tubes to the exit port (see Fig. 3.6 Gas chromatographic fraction collection items).

Procedure for Preparative Collection.   The collection tube (oven dried until 5 min before use) is attached to the heated exit port by the metal 5/5 ℥ joint. Sample collection is begun 30 s before detection of the expected peak on the recorder (based on previously determined retention values; refer to your local laboratory instructions) and continued until 30 s after the recorder's return to baseline. After the collection tube is detached, the sample can be analyzed directly when collected into a GC NMR collection tube[2] or transferred to the 0.1-mL conical GC collection vial. After the collection tube is joined to the vial (preweighed with cap) by the 5/5 ℥ joints, the system is centrifuged to force the sample down into the vial (Fig. 5.3). The collection tube is then removed, and the vial is capped and reweighed.

The efficiency of collection can exceed 90% with most materials, even with relatively low-boiling substances. In the latter case, the collection tube, after attachment to the instrument, is wrapped with a paper tissue. As the (oven-dried) tube is being wrapped, it is also being flushed by the carrier gas, which removes any traces of water condensation. The wrapping is then saturated with liquid nitrogen to cool the collection tube.

Preparative GC in the microscale laboratory often replaces the macroscale purification technique of fractional distillation. Distillation is impractical with less than 500 µL of liquid.

Refer to Experiment [2] for specific experimental details on preparative GC applied to the separation of a number of binary (two-component) mixtures. These are designed as practice examples to give you experience with sample collection.

---

[2]Bressette, A. R. *J. Chem. Educ.* **2001,** *78,* 366.

QUESTIONS

5-1. What is the main barrier to separating liquid mixtures of less than 500 μL by distillation?

5-2. A sample mixture of benzonitrile (bp 190 °C) and acetophenone (bp 202 °C) is injected on two GC columns. Column A has DC710 silicone oil as the stationary phase, and column B uses diisodecyl phthalate as the stationary phase. Which substance would be certain to elute first from column A and would the same material be expected to elute first from column B? Which column, A or B, would be expected to give the better separation of these two substances?

5-3. Both dodecane (bp 216 °C) and ethyl benzoate (bp 212 °C) have similar boiling points. List several GC variables, column selection (p. 62), and conditions that would make it easier to separate these substances by gas chromatography.

5-4. Capillary GC columns have better resolution than packed columns even though the enormous surface area provided by the packing material is absent in capillary columns. Why?

5-5. Preparative GC requires packed columns. Why is this technique limited to these lower resolution columns?

5-6. The method of collection involving the use of a metal joint for preparative GC has actually been granted a patent (US Patent No. 4,730,480). Locate the patent online and provide the following:

(a) Who are the inventors of the patent?

(b) On what date was the patent granted?

(c) What is the full title of the patent?

(d) On what date was the patent filed?

(e) What is the name of the attorney who drew up the patent?

(f) What pieces of the collection device are shown in Figures 3, 4, and 5?

NOTE. *Gas chromatographic purification of reaction products is suggested in the following experiments: Experiments [2], [3C], [3D], [5A], [5B], [8C], [9], [10], [13], [17], and [32].*

Simple Distillation

Distillation is the process of heating a liquid to the boiling point, condensing the heated vapor by cooling, and returning either a portion of, or none of, the condensed vapors to the distillation vessel. Distillation differs from reflux (see p. 25) only in that at least some of the condensate is removed from the boiling system. Distillations in which a fraction of the condensed vapors are returned to the boiling system are often referred to as being under "partial reflux." Two types of distillations will be described. Students are encouraged to refer to and study the more detailed discussion of distillation theory. The website also has a detailed discussion of the theory of steam distillation, which is used in Experiments [11C] and [32]. There are times when ordinary distillation may not be feasible for the separation of a liquid from dissolved impurities. The compound of interest may boil at a high temperature that is difficult to control with a simple apparatus, or it may tend to decompose or oxidize at high temperatures. If the compound is only sparingly soluble in water and any small amount of water can be removed with a drying agent, steam distillation may be the technique of choice. Compounds that are immiscible in water have very large positive deviations from Raoult's law. Therefore, the boiling temperature is generally lower than that of water and the compound.

Simple distillation involves the use of the distillation process to separate a liquid from minor components that are nonvolatile, or that have boiling points at least 30–40 °C above that of the major component. A typical setup for a macroscale distillation of this type is shown in Figure 5.4. At the microscale level,

Figure 5.4   A complete simple distillation setup. (Reproduced with permission of John Wiley and Sons, Inc., from Zubrick, J. W. *The Organic Chemistry Lab Survival Manual,* 9th ed., Copyright 2013.)

working with volumes smaller than 500 μL, GC techniques (see Technique 1) have replaced conventional microdistillation processes.[3] Semi-microscale simple distillation is an effective separation technique for volumes in the range of 0.5–2 mL. Apparatus that achieve effective separation of mixture samples in this range have been developed. One of the most significant of these designs is the classic Hickman still, shown in Figure 5.5. This still is used in several ways in the experiments described in Chapters 6 and 7 for purifying solvents, carrying out reactions, and concentrating solutions. Experiment [3] introduces the use of the Hickman still.

In a distillation where liquid is to be separated from a nonvolatile solute, the vapor pressure of the liquid is lowered by the presence of the solute, but the vapor phase consists of only one component. Thus, except for the incidental transfer of nonvolatile material by splashing, the material condensed should consist only of the volatile component (see Experiment [3A]).

We can understand what is going on in a simple distillation of two volatile components by referring to the phase diagrams shown in Figures 5.6 and 5.7. Figure 5.6 is the phase diagram for hexane and toluene. The boiling points of these liquids are separated by 42 °C. Figure 5.7 is the phase diagram for methylcyclohexane and toluene. Here the boiling points are separated by only 9.7 °C.

Imagine a simple distillation of the hexane–toluene pair in which the liquid in the pot is 50% hexane. In Figure 5.6, when the liquid reaches 80.8 °C it will be in equilibrium with vapor having a composition of 77% hexane. This result is indicated by the line *A–B*. If this vapor is condensed to a liquid of the same composition, as shown by the line *B–C*, we will have achieved a significant enrichment of the condensate with respect to hexane. This change in composition is referred to as a simple distillation. The process of evaporation

[3]Schneider, F. L. In *Monographien aus dem Gebiete der qualitativen Mikroanalyse,* Vol. II; *Qualitative Organic Microanalysis;* Benedetti-Pichler, A. A., Ed.; Springer-Verlag: Vienna, 1964; p. 31.

Figure 5.5 Hickman still (14/10 ⅀ with conical vial [3 mL]).

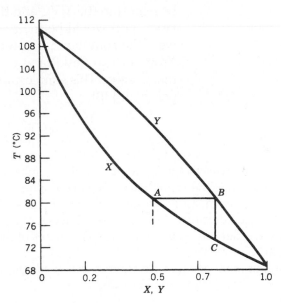

Figure 5.6 Temperature as a function of liquid composition (*X*) and vapor composition (*Y*): hexane and toluene.

and condensation is achieved by the theoretical construct known as a *theoretical plate*. When this distillation is actually done with a Hickman still, some of the mixture will go through one evaporation and condensation cycle, some will go through two of these cycles, and some may be splashed more directly into the collar. A resolution (separation) of between one and two theoretical plates is generally obtained.

Referring to Figure 5.7, if we consider the same process for a 50% mixture of methylcyclohexane and toluene, the methylcyclohexane composition will increase to 58% for a distillation with one theoretical plate. Simple distillation may thus provide adequate enrichment of the MVC (more volatile constituent) if the boiling points of the two liquids are reasonably well separated, as they are for hexane and toluene. If the boiling points are close together, as they are for methylcyclohexane and toluene, the simple distillation will not provide much enrichment.

As we continue the distillation process and remove some of the MVC by condensing it, the residue in the heated flask becomes less rich in the MVC. This means that the next few drops of condensate will be less rich in the MVC. As the distillation is continued, the condensate becomes less and less rich in the MVC.

Figure 5.7 Temperature as a function of liquid composition (*X*) and vapor composition (*Y*): methylcyclohexane and toluene.

We can improve on simple distillation by repeating the process. For example, we could collect the condensate until about one-third is obtained. Then we could collect a second one-third aliquot in a separate container. Our original mixture would then be separated into three fractions. The first third would be richest in the MVC and the final third (the fraction remaining in the distillation pot) would be the richest in the least volatile component. If the MVC were the compound of interest, we could re-distill the first fraction collected (from a clean flask!) and collect the first third of the material condensing in that process. This simplest of all fractional distillation strategies is used in Experiment [3B].

QUESTIONS

5-7. What is the major drawback of trying to distill a 500 μL mixture of liquids, all with boiling points below 200 °C?

5-8. How might you separate the mixture discussed in question 5-7 if distillation were unsuccessful? Explain your choice.

5-9. If starting with an equal mixture of hexane and toluene, approximate the composition of hexane if the vapor at 100 °C is condensed to a liquid using the data presented in Figure 5.6.

5-10. Why do simple distillations require that the components of the mixture to be separated have boiling points that are separated by 40 °C or more?

5-11. Which constituent of an equimolar mixture makes the larger contribution to the vapor pressure of the mixture, the higher or lower boiling component? Explain.

NOTE. The following experiments use Technique 2: Experiments [3A], [3B], [11C], [29], and [32].

**TECHNIQUE
3**

Fractional Distillation

Fractional distillation can occur in a distillation system containing more than one theoretical plate. This process must be used when the boiling points of the components differ by less than 30–40 °C and fairly complete separation is desired. A fractionating column is needed to accomplish this separation. As discussed previously, a liquid–vapor composition curve (Fig. 5.8) shows that the lower boiling component of a binary mixture makes a larger contribution to the vapor composition than does the higher boiling component. On condensation, the liquid formed will be richer in the lower boiling component. This condensate will not be pure, however, and, in the case of components with close boiling points, it may be only slightly enriched. If the condensate is vaporized a second time, the vapor in equilibrium with this liquid will show a further enrichment in the lower boiling component. The trick to separating liquids with similar boiling points is to repeat the vaporization–condensation cycle many times. Each cycle is one *theoretical plate*. Several column designs, which achieve varying numbers of theoretical plates, are available for use at the macro level (Fig. 5.9).

Most distillation columns are designed so that fractionation efficiency is achieved by the very large surface area in contact with the vapor phase (and very similar to the way increased resolution is obtained on a GC column (see Technique 1). This increased surface area can be accomplished by packing the fractionating column with wire gauze or glass beads. Unfortunately, a large

Figure 5.8 Liquid–vapor composition curve.

Figure 5.9 A fractional distillation setup. (Reproduced with permission of John Wiley and Sons, Inc., from Zubrick, J. W. *The Organic Chemistry Lab Survival Manual*, 9th ed., Copyright 2013.)

volume of liquid must be distributed over the column surface in equilibrium with the vapor. Furthermore, the longer the column, the more efficient it becomes (see Technique 1), but longer columns also require additional liquid phase. The amount of liquid phase required to fill the column with a liquid–vapor equilibrium is called *column holdup.* Column holdup is essentially lost from the distillation because this volume can never go past the top of the distillation column; it can only return to the distillation pot upon cooling. Column holdup can be large compared to the total volume of material available for the distillation. With mixtures of less than 2 mL, column holdup precludes the use of the most common fractionation columns. Columns with rapidly spinning bands of metal gauze or Teflon have very low column holdup and have a large number of theoretical plates relative to their height (Fig. 5.10).

Figure 5.10 Schematic of a metal-mesh, spinning-band still.

Figure 5.11 Microspinning band distillation column (3 in.).

Figure 5.12 Hickman still with thermometer adapter.

Microscale spinning-band distillation apparatus (Fig. 5.11) can achieve nearly 12 theoretical plates and are simple enough to be used in the instructional laboratory. This still has a Teflon band that fits closely inside an insulated glass tube. The Teflon band has spiral grooves which, when the band is spun (1000–1500 rpm), rapidly return condensed vapor to the distillation pot. A powerful extension of this apparatus uses a short spinning band inside a modified Hickman still head (see Fig. 3.15). These stills are called Hickman–Hinkle stills; 4-cm Hickman–Hinkle columns can have more than 10 theoretical plates. The commercially available 2.5-cm version is rated at 6 theoretical plates. Experiments [3C] and [3D] involve fractional distillation with spinning-band columns.

Experiment [3B] utilizes the Hickman still in a simple demonstration of fractional distillation. The system is arranged so that the thermometer is positioned directly down the center of the distillation column, with the bulb just at the bottom of the well. It is very important to position both the still and the thermometer as vertically as possible; the thermometer must not touch the glass walls of the column (Fig. 5.12). Experiment [3B] uses the Hickman still for fractional distillation. A two-theoretical-plate distillation is obtained with this system on a two-component mixture by carrying out two sequential fractional distillations.

For a more detailed discussion of how spinning bands work, see the discussion of distillation.

5-12. Why is it very important that the hot vapor in microscale distillations climb the column very slowly?

5-13. Why might Teflon be the material of choice for constructing microscale spinning bands?

5-14. The spinning band overcomes two major problems of microscale distillations by wiping the liquid condensate rapidly from the column walls. What are these problems?

5-15. Why are spinning bands so effective at increasing the number of theoretical plates in distillation columns?

5-16. If operating at a temperature higher than the vapor composition of a binary mixture (Fig. 5.8), what level of enrichment of either component A or B should one expect?

NOTE. *The following experiments use Technique 3: Experiments [3C] and [3D].*

Solvent Extraction

Solvent extraction is frequently used in the organic laboratory to separate or isolate a desired compound from a mixture or from impurities. Solvent extraction methods are readily adapted to microscale work because small quantities are easily manipulated in solution. Solvent extraction methods are based on the solubility characteristics of organic substances in the solvents used in a particular separation procedure. Liquid–liquid and solid–liquid extractions are the two major types of extractions used in the organic laboratory.

Intermolecular Properties: Solubility

Substances vary greatly in their solubility in various solvents, but a useful and generally true principle is that a substance tends to dissolve in a solvent that is chemically similar to itself. In other words, *like dissolves like*. The significant exceptions to this general statement are seen when solubilities are determined by acid–base properties.

Thus, to be soluble in water a compound needs to have some of the molecular characteristics of water. Alcohols, for example, have a hydroxyl group (—OH) bonded to a hydrocarbon chain or framework (R—OH). The hydroxyl group can be thought of as effectively half a water (H_2O) molecule; its polarity is similar to that of water. This polarity is due to the charge separation arising from the different electronegativities of the hydrogen and oxygen atoms. The O—H bond, therefore, is considered to have partial ionic character.

$$\overset{\delta^-}{-\ddot{\underset{\cdot\cdot}{O}}}-\overset{\delta^+}{H}$$

Partial ionic character of the hydroxyl group

This polar, or partial ionic, character leads to relatively strong hydrogen bond formation between molecules with hydroxyl groups. Strong hydrogen bonding (shown here for the ethanol–water system) occurs in molecules that have a hydrogen atom attached to an oxygen, nitrogen, or fluorine atom—all three are quite electronegative atoms.

Ethanol   Hydrogen bond formation

Table 5.1 Comparison of Boiling-Point Data			
Name	Formula	MW	bp (°C)
Ethanol	CH$_3$CH$_2$OH	46	78.3
Propane	CH$_3$CH$_2$CH$_3$	44	−42.2
Methyl acetate	CH$_3$CO$_2$CH$_3$	74	54
Diethyl ether	(CH$_3$CH$_2$)$_2$O	74	34.6
Ethene	CH$_2$=CH$_2$	28	−102
Methylamine	CH$_3$NH$_2$	31	−6

The hydroxyl end of the ethanol molecule is very similar to water. When ethanol is added to water, therefore, they are miscible in all proportions. That is, ethanol is completely soluble in water and water is completely soluble in ethanol. This degree of solubility occurs because the attractive forces between the two molecules are nearly as strong as those between two water molecules; however, the attraction in the first case is somewhat weakened by the presence of the nonpolar ethyl group, CH$_3$CH$_2$—. Hydrocarbon groups attract each other only weakly, as demonstrated by their low melting and boiling points. Three examples of the contrast in boiling points between compounds of different structure, but similar molecular weight, are summarized in Table 5.1. Molecules that attract each other weakly (lower intermolecular forces) have lower boiling points.

Ethanol is completely miscible with water, but the solubility of octanol in water is less than 1%. Why the difference in solubilities between these two alcohols? The dominant structural feature of ethanol is its polar hydroxyl group; the dominant structural feature of octanol is its nonpolar alkyl group:

CH$_3$—CH$_2$—CH$_2$—CH$_2$—CH$_2$—CH$_2$—CH$_2$—CH$_2$—$\overset{\delta-}{\ddot{O}}$:$\overset{\delta+}{\underset{H}{}}$

Octanol

CH$_3$—CH$_2$—$\ddot{O}$—CH$_2$—CH$_3$

Diethyl ether

As the size of the hydrocarbon section of the alcohol molecule increases, the intermolecular attraction between the polar hydroxyl groups of the alcohol and the water molecules is no longer strong enough to overcome the hydrophobic (lacking attraction to H$_2$O) nature of the nonpolar hydrocarbon section of the alcohol. On the other hand, octanol has a large nonpolar hydrocarbon group as its dominant structural feature. We might, therefore, expect octanol to be more soluble in less polar solvents, and, in fact, octanol is completely miscible with diethyl ether. Ethers are weakly polar solvents because a C—O bond is much less polar than an O—H bond (carbon is less electronegative than oxygen). Because both octanol and diethyl ether are rather nonpolar, each is completely soluble in the other. For compounds with both polar and nonpolar groups, in general, those compounds with five or more carbon atoms in the hydrocarbon portion of the molecule will be more soluble in nonpolar solvents, such as pentane, diethyl ether, or methylene chloride. Figure 5.13 summarizes the solubilities of a number of straight-chain alcohols, carboxylic acids, and hydrocarbons in water. As expected, most monofunctional compounds with more than five carbon atoms have solubilities similar to the hydrocarbons.

Several additional relationships between solubility and structure have been observed.

Figure 5.13   Solubility curve of acids, alcohols, and hydrocarbons. (Adapted from solubility curve from Kamm, O. *Qualitative Organic Analysis,* 2nd ed.; John Wiley and Sons, Inc. Copyright 1932.)

1. Branched-chain compounds have greater water solubility than their straight-chain counterparts, as illustrated in Table 5.2 with a series of alcohols.

2. The presence of more than one polar group in a compound will increase that compound's solubility in water and decrease its solubility in nonpolar solvents. For example, sugars, such as cellobiose, contain multiple hydroxyl and/or acetal groups and are water soluble and ether insoluble. Cholesterol, which has only a single hydroxyl group on its 27 carbon atoms, is insoluble in water and quite soluble in ether.

Cholesterol

Cellobiose

Table 5.2 Water Solubility of Alcohols		
Name	Structural Formula	Solubility (g/100 g H_2O at 20 °C)
Hexanol	$CH_3(CH_2)_4CH_2OH$	0.6
Pentanol	$CH_3(CH_2)_3CH_2OH$	2.2
2-Pentanol	$CH_3(CH_2)_2CH(OH)CH_3$	4.3
2-Methyl-2-butanol	$(CH_3)_2C(OH)CH_2CH_3$	11.0

Table 5.3 Water Solubility of Amines		
Name	Structural Formula	Solubility (g/100 g H_2O at 25 °C)
Ethylamine	$CH_3CH_2NH_2$	∞
Diethylamine	$(CH_3CH_2)_2NH$	∞
Trimethylamine	$(CH_3)_3N$	91
Triethylamine	$(CH_3CH_2)_3N$	14
Aniline	C_6H_5—NH_2	3.7
1,4-Diaminobenzene	H_2N—C_6H_4—NH_2	3.8

3. The presence of a chlorine atom, even though it lends some partial ionic character to the mostly covalent C—Cl bond, does not normally impart water solubility to a compound. In fact, compounds such as methylene chloride (CH_2Cl_2), chloroform ($CHCl_3$), and carbon tetrachloride (CCl_4) have long been used as solvents for extracting aqueous solutions. The latter two solvents are not often used nowadays, unless strict safety precautions are exercised, because they are potentially carcinogenic.

4. Most functional groups capable of forming a hydrogen bond with water increased the water solubility of a substance. For example, smaller alkyl amines have significant water solubility; the water-solubility data for a series of amines are summarized in Table 5.3.

The solubility characteristics of any given compound govern its distribution (*partition*) between the phases of two immiscible solvents (in which the material has been dissolved) when these phases are intimately mixed.

PARTITION (OR DISTRIBUTION) COEFFICIENT

A given substance X is partially soluble in each of two immiscible solvents. If X is placed in a mixture of these two solvents and shaken, an equilibrium will be established between the two phases. That is, substance X will partition (distribute) itself in a manner that is a function of its relative solubility in the two solvents:

$$X_{\text{solvent 1}} \rightleftharpoons X_{\text{solvent 2}}$$

The equilibrium constant, K_p, for this equilibrium expression is known as the *partition* or *distribution coefficient*:

$$K_p = \frac{[X_{\text{solvent 2}}]}{[X_{\text{solvent 1}}]}$$

The equilibrium constant is thus the ratio of the concentrations of the species, X, in each solvent for a given system at a given temperature. The partition coefficient can be conveniently estimated as the ratio of the solubility of X in solvent 1 vs. solvent 2:

$$K_p = \frac{\text{solubility of } X \text{ in solvent 2}}{\text{solubility of } X \text{ in solvent 1}}$$

When solvent 1 is water and solvent 2 is an organic solvent such as diethyl ether, the basic equation used to express the coefficient K_p is

$$K_p = \frac{(\text{g/100 mL})_{\text{organic layer}}}{(\text{g/100 mL})_{\text{water layer}}}$$

This expression uses grams per 100 mL for the concentration units. Note that the partition coefficient is dimensionless, so any concentration units may be used if the units are the same for both phases. For example, grams per liter (g/L), parts per million (ppm), and molarity (M) can all be used. If equal volumes of both solvents are used, the equation reduces to the ratio of the weights ($g_{organic}/g_{water}$) of the given species in the two solvents:

$$K_p = \frac{g_{organic\ layer}}{g_{water\ layer}}$$

Determination of the partition coefficient for a particular compound in various immiscible-solvent combinations often can give valuable information for isolating and purifying the compound by using extraction techniques. Thus, liquid–liquid extraction is a common separation technique used in organic as well as analytical laboratories.

Table 5.4 provides some examples of K_p values determined at room temperature for a number of compounds in the water–methylene chloride system.

Let us now look at a typical calculation for the extraction of an organic compound P from an aqueous solution using diethyl ether. We will assume that the $K_{p\ ether/water}$ value (partition coefficient of P between diethyl ether and water) is 3.5 at 20 °C. If a solution of 100 mg of P in 300 μL of water is extracted at 20 °C with 300 μL of diethyl ether, the following expression holds:

$$K_{p\ ether/water} = \frac{C_e}{C_w} = \frac{W_e/300\ \mu L}{W_w/300\ \mu L}$$

where

W_e = weight of P in the ether layer

W_w = weight of P in the water layer

C_e = concentration of P in the ether layer

C_w = concentration of P in the water layers

Since $W_w = 100 - W_e$, the preceding relationship can be written as

$$K_{p\ ether/water} = \frac{W_e/300\ \mu L}{(100 - W_e)/300\ \mu L} = 3.5$$

If we solve for the value of W_e, we obtain 77.8 mg; the value for W_w is 22.2 mg. Thus, we see that after one extraction with 300 μL of ether, 77.8 mg of P (77.8% of the total) is removed by the ether and 22.2 mg (22.2% of the total) remains in the water layer. Is it preferable to make a single extraction with the total quantity of solvent available, or to make multiple extractions with portions of the solvent? The second method is usually more efficient.

Table 5.4 Representative K_p Values in CH$_2$Cl$_2$—H$_2$O	
Compound	K_p Value
Nitrobenzene	51.5
Aniline	3.3
1,2-Dihydroxybenzene	0.2

To illustrate, consider extracting the 100 mg of P in 300 μL of water with *two 150-μL* portions of diethyl ether instead of one 300-μL portion.
For the first 150-μL extraction,

$$\frac{W_e/150\ \mu L}{W_w/300\ \mu L} = \frac{W_e/150\ \mu L}{(100 - W_e)/300\ \mu L}$$

Solving for W_e, we obtain 63.6 mg. The amount of P remaining in the water layer (W_w) is then 36.4 mg. The aqueous solution is now extracted with the second portion of ether (150 μL). We then have

$$\frac{W_e/150\ \mu L}{(36.4 - W_e)/300\ \mu L} = 3.5$$

As before, by solving for W_e, we obtain 23.2 mg for the amount of P in the ether layer; $W_w = 13.2$ mg in the water layer.

The two extractions, each with 150 μL of ether, removed a total of 63.6 mg + 23.2 mg = 86.8 mg of P (86.8% of the total). The P left in the water layer is then 100 - 86.8, or 13.2 mg (13.2% of the total).

It can be seen from these calculations that the multiple-extraction technique is more efficient. The single extraction removed 77.8% of P; the double extraction (with the same total volume of ether) increased this to 86.8%. To extend this relationship, three extractions with one-third the total quantity of ether in each portion would be even more efficient. You might wish to calculate this to prove the point. Of course, there is a practical limit to the number of extractions that can be performed.

The multiple-extraction example shown here illustrates that several extractions with small volumes is more efficient than a single extraction procedure. This is always true *provided* the partition coefficient is neither very large nor very small. If the partition coefficient K_p for a substance between two solvents is very large ($K_p > 100$) or very small ($K_p < 0.01$), multiple extractions (using the same total amount of solvent) do not significantly increase the efficiency of the extraction process.[4]

Extraction

Liquid–Liquid Extraction.   The more common type of extraction, liquid–liquid extraction, is used extensively. It is a very powerful method for separating and isolating materials at the microscale level. It is operationally not a simple process, so attention to detail is critical.

There are several important criteria to consider when choosing a solvent for the extraction and isolation of a component from a solution:

- The chosen extraction solvent must be immiscible with the solution solvent.
- The chosen extraction solvent must be favored by the distribution coefficient for the component being extracted.
- The chosen extraction solvent should be readily separated from the desired component after extraction. This usually means that it should have a low boiling point.
- The chosen organic extraction solvent must not react chemically with any component in the aqueous mixture being extracted.

---

[4]Palleros, D. R. *J. Chem. Educ.* **1995,** *72,* 319.

NOTE. The aqueous phase may be modified, as in acid–base extractions, but the organic solvent does not react with the components in the aqueous mixture. See Experiments [4B, 4C], pp. 150–154.

Microscale Extraction.   A capped conical vial or a stoppered centrifuge tube is the best container for most microscale extractions, but a small test tube may be used. Note that a conical vial and a centrifuge tube have the same inner shape. This shape has the advantage that as the lower phase (layer) is withdrawn by pipet, the interface (boundary) between the two liquid phases becomes narrower and narrower, and thus easier to see, at the bottom of a conical container. This is not the case for a test tube. The centrifuge tube has the added advantage that if a solid precipitate must be separated or an emulsion broken up, it can easily be done using a centrifuge.

A good rule of thumb is that the container to be used for the extraction should be *at least three times the volume of liquid* you wish to extract.

Regardless of the container used, in any liquid–liquid extraction, the two immiscible solvents must be completely mixed to maximize the surface area of the interface between the two and allow partitioning of the solute. This can be accomplished by shaking (carefully to avoid leakage around the cap), using a Vortex mixer, or by adding a magnetic spin vane and then stirring with a magnetic stirrer.

Another important rule in the extraction process is that you should *never discard any layer until the isolation is complete.*

Let us consider a practical example. Benzanilide can be prepared by the in situ rearrangement of benzophenone oxime in acid solution:

Benzophenone   Benzophenone oxime   Benzanilide

The benzanilide is separated from the reaction mixture by extraction with three 1.0-mL portions of methylene chloride solvent.

NOTE. Saying, for example, "extracted with three 1.0-mL portions of methylene chloride" means that three extractions are performed (one after the other, each using 1.0 mL of methylene chloride) and the three methylene chloride extracts are combined.

A microscale extraction process consists of two parts: (1) mixing the two immiscible solutions, and (2) separating the two layers after the mixing process.

1. Mixing. In the experimental procedure for the isolation of the benzanilide product, methylene chloride (1.0 mL) is added to the aqueous reaction mixture contained in a 5.0-mL conical vial (or centrifuge tube). The extraction procedure is outlined in the following steps:

Step 1. Cap the vial.
Step 2. Shake the vial gently to thoroughly mix the two phases (careful!)

Figure 5.14 Extraction devices.

Pasteur filter pipet (microscale)

Separatory funnel (semi–micro scale and macroscale)

NOTE. The mixing may be carried out using a Vortex mixer or magnetic stirrer—see previous discussion.

Step 3. Carefully vent the vial by loosening the cap to release any pressure that may have developed.

Step 4. Allow the vial to stand on a level surface to permit the two phases to separate. A sharp phase interface should appear.

NOTE. For safety reasons it is advisable to place the vial in a small beaker to prevent tipping. If a volatile solvent such as ether is used, it is advisable to place the vial or centrifuge tube in a beaker of ice water to prevent loss of solvent during the transfers.

2. Separation. At the microscale level, the two phases are separated with a Pasteur filter pipet (a simple Pasteur pipet can be used in some situations), which acts as a miniature separatory funnel. The separation of the phases is shown in Figure 5.14.

A major difference between macro and micro techniques is that when microscale volumes are used, as just discussed, the mixing and separation are done in two parts. When macroscale volumes are used in a separatory funnel, mixing and separation are both done in the funnel in one step. The separatory funnel is an effective device for extractions with larger volumes, but it is not practical for microscale extractions because of the large surface areas involved.

Benzanilide is more soluble in methylene chloride than in water. Multiple extractions are performed to ensure complete removal of the benzanilide from the aqueous phase. The methylene chloride solution is the lower layer, because it is denser than water. The following list outlines the general method for an organic solvent more dense than water.

NOTE. (a) One technique is to hold the pipet across the palm of the hand and squeeze the bulb with the thumb and index finger. (b) Remember to have an empty tared vial available in which to place the separated phase. (c) A pipet pump (Figure 5.15) may be used to replace the bulb. One advantage with using a pipet pump is the dispensing of liquids in a more controlled fashion.

The recommended procedures are shown in Figures 5.16 and 5.17.

Step 1. Squeeze the pipet bulb to force air from the pipet.

Step 2. Insert the pipet into the vial until it is close to the bottom. Be sure to hold the pipet vertically.

Step 3. Carefully allow the bulb to expand, drawing only the lower methylene chloride layer into the pipet. This should be done in a smooth, steady manner so as not to disturb the interface between the layers. With practice, you can judge the amount that the bulb must be squeezed to just separate the layers. Keep the pipet vertical. *Do not tip the pipet back and allow liquid to enter the bulb! Do not suck liquid into the bulb!*

Step 4. (Step 4 is not shown in the figure.) Holding the pipet vertical, place it over and into the neck of an empty vial (as shown in Fig. 5.16, Step 2), and gently squeeze the bulb to transfer the methylene chloride solution into the vial. A second extraction can now be performed after

Pipet pump with pipet

Figure 5.15 Pipet pump with pipet. (Reproduced with permission of John Wiley and Sons, Inc., from Szafran, Z.; Pike, R. M.; Foster, J. C. *Microscale General Chemistry Laboratory,* 2nd ed., Copyright 2003.)

Figure 5.16 Pasteur filter pipet separation of two immiscible liquid phases; the denser layer contains the product.

Figure 5.17 Pasteur filter pipet separation of two immiscible liquid phases; the less dense layer contains the product.

adding another portion of methylene chloride to the original vial. The procedure is repeated. Multiple extractions can be performed in this manner. Each methylene chloride extract is transferred to the same vial—that is, the extracts are combined. The reaction product has now been transferred from the aqueous layer (aqueous phase) to the methylene chloride layer (organic phase), and the phases have been separated.

In a diethyl ether–water extraction, the ether layer is less dense and thus is the upper layer (phase). An organic reaction product generally dissolves in the ether layer and is thus separated from water-soluble by-products and other impurities. The procedure followed to separate the water–ether phases is identical to that described above for methylene chloride-water systems, except that here the top layer (organic layer) is transferred to the new container. The following list outlines the general method for an organic solvent less dense than water (refer to Fig. 5.17).

Step 1. Squeeze the pipet bulb to force air from the pipet and insert the pipet into the vial until it is close to the bottom. Then, draw **both** phases slowly into the pipet. Keep the pipet vertical. *Do not tip the pipet back and allow liquid to enter the bulb! Do not suck liquid into the bulb!* Try not to allow air to be sucked into the pipet, as this tends to mix the phases in the pipet. If mixing does occur, allow time for the interface to re-form.

Step 2. Return the aqueous layer (bottom layer) to the **original** container by gently squeezing the pipet bulb.

Step 3. Transfer the separated ether layer (top layer) to a new tared vial.

Separatory Funnel—Semi-microscale and Macroscale Extractions. A separatory funnel (Fig. 5.14) is effective for extractions carried out at the semi-microscale and macroscale levels. The mixing and separation are done in the funnel itself in one step. Many of you may be familiar with this device from the general chemistry laboratory. The same precautions as outlined above for microscale extraction should be observed here.

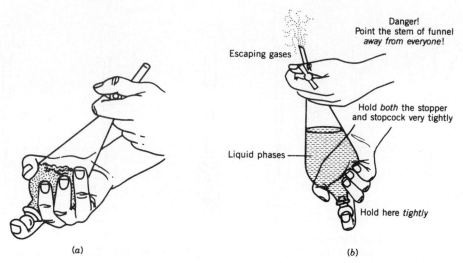

Figure 5.18 (a) Correct position for holding a separatory funnel while shaking. (b) Correct method for venting a separatory funnel.

NOTE. The funnel size should be such that the total volume of solution is less than half the total volume of the funnel. If the funnel has a ground-glass stopcock and/or stopper, the ground-glass surfaces must be lightly greased to prevent sticking, leaking, or freezing. If Teflon stoppers and stopcocks are used, grease is not necessary, because these are self-lubricating.

Step 1. Close the stopcock of the separatory funnel.

Step 2. Add the solution to be extracted, after first making sure that the *stopcock is closed*. The funnel should be supported in an iron ring attached to a ring stand or rack on the lab bench.

Step 3. Add the proper amount of extraction solvent (about one-third of the volume of the solution to be extracted is a good rule of thumb) and place the stopper on the funnel.

Step 4. Remove the funnel from the ring stand, keeping the stopper in place with the index finger of one hand, and holding the funnel in the other hand with your fingers positioned so they can operate the stopcock (Fig. 5.18*a*).

Step 5. Carefully invert the funnel (make sure its stem is pointing up, and not pointing at you or anyone else). Slowly open the stopcock to release any built-up pressure (Fig. 5.18*b*). Close the stopcock and then shake the funnel for several seconds. Position the funnel for venting (make sure the stem is pointing up, and not pointing at you or anyone else). Open the stopcock to release built-up pressure. Repeat this process 2–4 times. Then, close the stopcock and return the funnel upright to the iron ring.

Step 6. Allow the layers to separate and then remove the stopper.

Step 7. Place a suitable clean container just below the tip of the funnel. Gradually open the stopcock and drain the bottom layer into the clean container.

Step 8. Remove the upper layer by pouring it from the top of the funnel. This way it will not become contaminated with traces of the lower layer found in the stem of the funnel.

When aqueous solutions are extracted with a *less dense solvent*, such as ether, the bottom, aqueous layer can be drained *into its original container*. Once the top (organic) layer is removed from the funnel, the aqueous layer can then be returned for further extraction. Losses can be minimized by rinsing the original container with a small portion of the extraction solvent, which is then added to the funnel. When the extraction solvent is denser than the aqueous phase (e.g., methylene chloride), the aqueous phase is the top layer, and therefore is kept in the funnel for subsequent extractions.

Continuous Liquid–Liquid Extraction.   Continuous extraction of liquid–liquid systems is also possible, and is particularly valuable when the component to be separated is only slightly soluble in the extraction solvent. The advantage of using continuous extraction is that it can be carried out with a limited amount of solvent. In batchwise extractions a prohibitive number of individual extractions might have to be performed to accomplish the same overall extraction. Specialized apparatus, however, is required for continuous liquid–liquid extraction.

Two types of continuous extraction apparatus are often used to isolate various species from aqueous solutions using less dense and more dense immiscible solvents (e.g., diethyl ether and methylene chloride) (Fig. 5.19).

The extraction is carried out by allowing the condensate of the extraction solvent, as it forms on the condenser on continuous distillation, to drop through an inner tube (see Fig. 5.19*a* in the case of the less dense solvent) and to percolate up through the solution containing the material to be extracted. This inner tube usually has a sintered glass plug on its end, which generates smaller droplets of the solvent and thus increases the efficiency of the procedure. The extraction solution is then returned to the original distilling flask. Eventually, in this manner, the desired material, extracted in small increments, is collected in the boiling flask and can then be isolated by concentrating the collected solution. This method works on the premise that fresh portions of the less dense phase are continuously introduced into the system, and it is often used in those instances where the organic material to be isolated has an appreciable solubility in water. In the case of a more dense extraction solvent (see Fig. 5.19*b*) the system functions in much the same fashion, but in this case the inner tube is removed and the condensed vapors percolate directly through the lighter phase (the phase to be extracted) to form the lower layer. This layer can cycle back to the distillation flask through a small-bore tubing connection from the bottom of the receiver flask to the distillation flask. Continuous liquid–liquid extraction is useful for removing extractable components from those having partition ratios that approach zero. Note that this method requires a very long period of time.

Separation of Acids and Bases.   The separation of organic acids and bases is another important and extensive use of the extraction method. The distribution coefficients of organic acids and bases are affected by pH when one of the solvents is water. An organic acid that is insoluble in neutral water (pH 7) becomes soluble when the water is made basic with an aqueous sodium hydroxide solution. The acid and the sodium hydroxide quickly react to form a sodium carboxylate salt, RCO_2^-, Na^+. The salt is, of course, ionic and therefore readily dissolves in the water. Thus, the acid–base reaction reverses the solubility characteristics of a water-insoluble organic acid.

(a)   (b)

Figure 5.19   Early designs for single-stage extractors: (a) Kutscher-Steudel extractor; (b) Wehrli extractor.

The water phase may then be extracted with an immiscible organic solvent to remove any impurities, leaving the acid salt in the water phase. Neutralizing the water layer with hydrochloric acid (to pH ≤ 7) reprotonates the carboxylate salt to reform the carboxylic acid, and causes the purified water-insoluble organic acid to precipitate (if it's a solid). In a similar fashion, water-insoluble organic bases, such as amines (RNH_2), can be rendered water soluble by treatment with dilute hydrochloric acid to form water-soluble hydrochloride salts (e.g., Experiment [23]).

Extraction procedures can be used to separate mixtures of solids. For example, the flowchart below diagrams a sequence used to separate a mixture made up of an aromatic carboxylic acid ($ArCO_2H$), an aromatic base ($ArNH_2$), and a neutral aromatic compound (ArH). Aromatic compounds are discussed here simply because they are likely to be crystalline solids.

In this example, we assume that the organic acid and base are solids. If either or both were liquids, an additional extraction of the final acidic aqueous or alkaline solution with ether, followed by drying and concentration, would be required to isolate the acidic or basic component.

Salting Out.   Most extractions in the organic laboratory involve water and an organic solvent. Many organic compounds have partial solubility in both solvents. To extract them from water, the partition coefficient (between the organic solvent and water) can be shifted in favor of the organic layer by saturating the water layer with an inorganic salt, such as sodium chloride. Water molecules prefer to solvate the polar ions (in this case sodium and chloride ions), and thus free the neutral organic molecules to migrate into the organic phase. Another way to think of this is to realize that the ionic solution is more polar than pure water, so the less polar organic molecules are less soluble than in pure water. The process of forcing an organic material out of a water solution by adding an inorganic salt is called *salting out*.

Salting out can also be effectively used for the *preliminary drying* of the wet organic layer that results from an extraction process. (Diethyl ether, in particular, can dissolve a fair amount of water.) Washing this organic layer with a saturated salt solution removes most of the dissolved water into the aqueous phase. This makes further drying of the organic phase with solid drying agents easier and much more effective (see Drying Agents below).

Solid–Liquid Extraction

The simplest form of solid–liquid extraction involves treating a solid with a solvent and then decanting or filtering the solvent extract away from the solid. An example of this technique (see Experiment [11A]) is extracting usnic acid

Figure 5.20 A solid–liquid continuous extraction apparatus.

from its native lichen with acetone. This type of extraction is most useful when only one main component of the solid phase has appreciable solubility in the solvent. The extraction of caffeine from tea (see Experiment [11B]) is another example of this method; it is accomplished by heating the tea in an aqueous solution of sodium carbonate. This approach works well because the water swells the tea leaves and allows the caffeine to be extracted more readily.

Microscale extractions of trimyristin from nutmeg, and cholesterol from gallstones, have been described.[5] Diethyl ether was used as the solvent in both cases. A packed Pasteur pipet column was used for filtering, drying (nutmeg experiment), and decolorizing (gallstone experiment).

Herrera and Almy described a simple continuous extraction apparatus (Fig. 5.20).[6] The apparatus is constructed from a 50-mL beaker and a paper cone prepared from a 9-cm disk of filter paper (nonfluted), which rests on the lip of the beaker. A small notch is cut in the cone to allow solvent vapor to pass around it. The extraction solvent is placed in the beaker; the solid material to be extracted is placed in the cone. A watch glass containing 2–3 g of ice is placed on top of the assembly to act as the condenser and to hold the paper cone in place. As the ice melts, the water is removed and replaced with fresh ice. The beaker is heated on a hot plate in the hood (some solvent evaporates during the extraction process and may need to be replaced). The concentrated solution collected in the beaker is then cooled and the solid product is isolated by filtration or is recrystallized. This system needs to be attended at all times, but works reasonably well for brief extractions.

Various apparatus have been developed for use when longer extraction periods are required. They all use what is called a *countercurrent process*. The best-known apparatus is the Soxhlet extractor, first described in 1879 (Fig. 5.21).[7] The solid sample is placed in a porous thimble. The extraction-solvent vapor, generated by refluxing the extraction solvent contained in the distilling pot, passes up through the vertical side tube into the condenser. The liquid condensate then drips onto the solid, which is extracted. The extraction solution passes through the pores of the thimble, eventually filling the center section of the Soxhlet. The siphon tube also fills with this extraction solution, and when the liquid level reaches the top of the tube, siphoning action returns the thimbleful of extract to the distillation pot. The cycle is automatically repeated many times, concentrating the extract in the distillation pot. The advantage of this arrangement is that the process may be continued automatically and unattended for as long as necessary. The solvent is then removed from the extraction solution collected in the pot, providing the extracted compound(s). Soxhlet extractors are available from many supply houses and can be purchased in various sizes. Of particular interest to us is the microscale variety, which is effective for small amounts of material and is now commercially available.[8]

Drying Agents

Organic extracts separated from aqueous phases usually contain traces of water. Even washing with saturated salt solution (see Salting Out above)

---

[5]Vestling, M. M. *J. Chem. Educ.* **1990,** *67,* 274.
[6]Herrera, A.; Almy, J. *J. Chem. Educ.* **1998,** *75,* 83.
[7]Soxhlet, F. *Dinglers Polytech. J.* **1879,** *232, 461.*
[8]In addition to ACE Glass, Inc., a number of vendors offer microscale Soxhlet equipment.

Condenser

Vapor

Porous thimble
(to hold solid)

Siphon arm

Vertical side
tube

Distilling pot

Extraction
solvent

Figure 5.21   Soxhlet extractor.

cannot remove all the water. Organic extracts must therefore be dried to remove any residual water before the solvent is evaporated or further purification is performed. Organic extracts can be conveniently dried with an anhydrous inorganic salt, such as magnesium sulfate, sodium sulfate, or calcium sulfate. These salts readily absorb water and form insoluble hydrates, thus removing the water from the wet organic phase. The hydrated solid can then be removed from the dried solution by filtration or by decanting (pouring) the solution away from the solid. Although many drying agents are known, not every drying agent can be used in every case. The ideal drying agent should dry the solution quickly, have a high capacity for water, cost little, and not react with the material being dried.[9]

Table 5.5 summarizes the properties of some of the more common drying agents used in the laboratory.

Make sure that the solid drying agent is in its *anhydrous* form. Sodium sulfate is a good general-purpose drying agent and is usually the drying agent of choice at room temperature. Use the granular form, if at all possible.

Magnesium sulfate is supplied as a fine powder (high surface area). It has a high water capacity and is inexpensive; it dries solutions more quickly than does sodium sulfate. The disadvantage of magnesium sulfate is that the desired product (or water molecules) can become trapped on the surface of the fine particles. If it is not thoroughly washed after separation, precious product

---

[9]Quantitative studies on the efficiency of drying agents for a wide variety of solvents have been reported. See Burfield, D. R.; Smithers, R. H. *J. Org. Chem.* **1983**, *48*, 2420, and references therein. Other useful information can be found in Armarego, W. L. F. *Purification of Laboratory Chemicals*, 8th ed.; Elsevier: New York, 2017, and in Ridduck, J. A.; Bunger, W. B.; Sakano, T. K. *Organic Solvents, Physical Properties and Methods of Purification*, 4th ed.; Wiley: New York, 1986.

Table 5.5 Properties of Common Drying Agents		
Drying Agent	Formula of Hydrate	Comments
Sodium sulfate	$Na_2SO_4 \cdot 10H_2O$	Slow to absorb water and inefficient, but inexpensive and has a high capacity. Loses water above 32 °C Granular form available.
Magnesium sulfate	$MgSO_4 \cdot 7H_2O$	One of the best. Can be used with nearly all organic solvents. Usually in powder form.
Calcium chloride	$CaCl_2 \cdot 6H_2O$	Relatively fast-drying, but reacts with many oxygen- and nitrogen-containing compounds. Usually in granular form.
Calcium sulfate	$CaSO_4 \cdot \frac{1}{2} H_2O$	Very fast and efficient, but has a low dehydration capacity.
Silica gel	$(SiO_2)_m \cdot nH_2O$	High capacity and efficient. Commercially available t.h.e. SiO_2 drying agent is excellent.[a]
Molecular sieves	$[Na_{12}(Al_{12}Si_{12}O_{48})] \cdot 27H_2O$	High capacity and efficient. Use the 4 Å size.[a]

[a]Adapted from MilliporeSigma.

may be lost. Furthermore, it is usually more difficult to remove a finely powdered solid agent, which may pass through the filter paper (if used) or clog the pores of a fine porous filter. A smaller surface area translates into less adsorption of product on the surface and easier separation from the dried solution.

Molecular sieves have pores or channels in their structures. A small molecule such as water can diffuse into these channels and become trapped. The sieves are excellent drying agents, have a high capacity, and dry liquids completely. The disadvantages are that they dry slowly and are more expensive than the more common drying agents.

Calcium chloride is very inexpensive and has a high capacity. Use the granular form. Do not use it to dry solutions of alcohols, amines, or carboxylic acids, because it can react with these substances.

Calcium sulfate is often sold under the trade name of Drierite. It is a somewhat expensive drying agent. Do not use the blue Drierite (commonly used to dry gases) because the cobalt indicator (blue when dry, pink when wet) may leach into the solvent.

The amount of drying agent needed depends on the amount of water present, on the capacity of the solid desiccant to absorb water, and on its particle size (actually, its surface area). If the solution is wet, the first amount of drying agent will clump (molecular sieves and t.h.e. SiO_2 are exceptions). Add more drying agent until it appears mobile when you swirl the liquid. A solution that is no longer cloudy is a further indication that the solution is dry. Swirling the contents of the container increases the rate of drying and helps establish the equilibrium for hydration:

$$\text{Drying agent} + n H_2O \rightleftharpoons \text{Drying agent} \cdot n H_2O$$

Anhydrous solid   Solid hydrate

Most drying agents achieve approximately 80% of their drying capacity within 15 min; longer drying times are generally unnecessary. The drying agent may be added directly to the container of the organic extract, or the extract may be passed through a Pasteur filter pipet packed with the drying agent. A funnel fitted with a cotton, glass wool, or polyester plug to hold the drying agent may also be used.

As for the most common question asked with this technique—Is this "dry"?—you should be encouraged to have in your lab a series of flasks which contain a set quantity of solvent and drying agent. The difference with each flask within the series is the percentage of water which allows for a visual comparison of what is and what is not "dry."

Figure 5.22   Polyethylene solid-phase extraction column.

Solid-Phase Extraction

In the modern research laboratory, the traditional liquid–liquid extraction technique may be replaced by the solid-phase extraction method.[10] The advantages of this newer approach are that it is rapid, it uses only small volumes of solvent, it does not form emulsions, isolated solvent extracts do not require a further drying stage, and it is ideal for working at the microscale level. This technique is finding wide acceptance in the food industry and in the environmental and clinical area, and it is becoming the accepted procedure for the rapid isolation of drugs of abuse and their metabolites from urine. Solid-phase extraction is accomplished using prepackaged, disposable, extraction columns. A typical column is shown in Figure 5.22. The columns are available from several commercial sources.[11]

The polypropylene columns can be obtained packed with 100–1000 mg of 40-μm sorbent sandwiched between two 20-μm polyethylene frits. The columns are typically 5–6 cm long. Sample volumes are generally 1–6 mL.

The adsorbent (stationary phase) used in these columns is a nonpolar adsorbent chemically bonded to silica gel. In fact, they are the same nonpolar adsorbents used in the reversed-phase high-performance liquid chromatography (HPLC). More specifically, the adsorbents are derivatized silica gel where the —OH groups of the silica gel have been replaced with siloxane groups by treating silica gel with the appropriate organochlorosilanes.

Silica surface   Chemically bonded silica surface

[10]For a description of this method see Zief, M.; Kiser, R. *Am. Lab.* **1990,** *22,* 70; Zief, M. *NEACT J.* **1990,** *8,* 38; Hagen, D. F.; Markell, C. G.; Schmitt, G. A.; Blevins, D. D. *Anal. Chim. Acta* **1990,** *236,* 157; Arthur, C. L.; Pawliszyn, J. *Anal. Chem.* **1990,** *62,* 2145; Dorsey, J.; Dill, K. A. *Chem. Rev.* **1989,** *89,* 331; Zubrick, J. W. *The Organic Chem Lab Survival Manual,* 11th ed.; Wiley: New York, 2019; Simpson, N. J. K, Ed. *Solid-Phase Extraction: Principles, Techniques and Applications*; Marcel Dekker: New York, 2000; "Solid Phase Extraction." Retrieved March 5, 2022 from www.sigmaaldrich.com/analytical-chromatography/sample-preparation/spe.html.
[11]These columns are available from Analytichem International, J. T. Baker, Inc., Supelco, Inc., Aldrich Chemical, Waters Associates and Biotage (a Division of Dyax Corp).

Two of the most popular nonpolar packings are those containing R groups consisting of an octadecyl ($C_{18}H_{37}$—) or phenyl (C_6H_5—) group. These packing materials (stationary phases) can adsorb nonpolar (like attracts like) organic material from aqueous solutions. The adsorbed material is then eluted from the column using a solvent strong (nonpolar) enough to displace it, such as methanol, methylene chloride, or hexane. The analyte capacity of bonded silica gels is about 10–20 mg of analyte per gram of packing.

An example of a typical solid-phase extraction is the determination of the amount of caffeine in coffee using a 1-mL column containing 100 mg of octadecyl-bonded silica. This efficient method isolates about 95% of the available caffeine. The column is conditioned by flushing 2 mL of methanol followed by 2 mL of water through the column. One milliliter of a coffee solution (~0.75 mg of caffeine/mL) is then drawn through the column at a flow rate of 1 mL/min. The column is washed with 1 mL of water and air dried (vacuum) for 10 min. The adsorbed caffeine is then eluted with two 500-μL portions of chloroform.

QUESTIONS

5-17. You are presented a two-phase system. Of the two liquids that are immiscible, one phase is water. Please devise an experiment to **definitively** differentiate which phase is water.

5-18. Which layer (upper or lower) will each of the following organic solvents usually form when being used to extract an aqueous solution?

toluene	methylene chloride	diethyl ether
hexane	acetone	

5-19. Construct a flowchart to demonstrate how you could separate a mixture of diphenylsulfone, ethyl 4-hydroxybenzoate, and benzoic acid using an extraction procedure.

5-20. A slightly polar organic compound partitions itself between ether and water phases. The K_p (partition coefficient) value is 2.5 in favor of the ether solvent. What simple procedure could you use to increase this K_p value?

5-21. You weight out exactly 2.00 mg of benzoic acid and dissolve it in a mixture of 4.0 mL of diethyl ether and 4.0 mL of water. After mixing and allowing the layers to separate, the ether layer is removed, dried, and concentrated to yield 1.36 mg of benzoic acid. What is the K_p value (ether/water) for this system?

5-22. If asked to separate an equal mixture of 4-chlorobenzoic acid [pK_a = 3.99] and 4-methoxyphenol [pK_a = 10.2] using a liquid–liquid extraction technique, explain why an aqueous solution of $NaHCO_3$[pK_a = 6.4] would be far more effective than the stronger aqueous solution of NaOH[pK_a = 15.7].

BIBLIOGRAPHY

For overviews on extraction methods see the following general references:

Dean, J. R. *Extraction Methods for Environmental Analysis;* Wiley: New York, 1998.

Handley, A. J. *Extraction Methods for Organic Analysis;* CRC Press LLC: Boca Raton, FL, 1999.

Kirk-Othmer Encyclopedia of Chemical Technology, 4th ed.; Wiley: New York, 1993; Vol. *10,* p. 125.

Schneider, Frank L. *Qualitative Organic Microanalysis;* Vol. II of *Monographien aus dem Gebiete der qualitätiven Mikroanalysis,* A. A. Benedetti-Pichler, Ed.; Springer-Verlag: Vienna, 1964, p. 61. Shugar, G. J. *Chemical Technicians' Ready Reference Handbook,* 3rd ed.; McGraw-Hill: New York, 1900.

Zubrick, J. W. *The Organic Chem Lab Survival Manual,* 11th ed.; Wiley: New York, 2019.

NOTE. The following experiments use Technique 4: Experiments [4A], [4B], [5A], [5B], [7], [8A], [11A], [11B], [11C], [12], [13], [16], [17], [19A], [19B], [19C], [22A], [22B], [23], [27], [30], [32], [34A], [34B], [A1$_b$], [D3], [E3], [F1], [F2], [F3], and [F4].

Crystallization

This discussion introduces the basic technique of purifying solid organic substances by crystallization. The technique of crystallizing an organic compound is fundamental; it must be mastered if you are going to purify solids. *It is not an easy art to acquire.* Organic solids tend not to crystallize as easily as inorganic substances.

Legend has it that an organic chemist resisted an invitation to leave a well-worn laboratory for new quarters because he suspected that the older laboratory (in which many crystallizations had been carried out) harbored seed crystals for a large variety of substances the chemist needed. Carried by dust from the earlier work, these traces of material presumably aided the successful initiation of crystallization of reluctant materials. Further support for this legend comes from the often quoted (but never substantiated) belief that after a material was first crystallized in a particular laboratory, subsequent crystallizations of the material, regardless of its purity or origin, were always easier.

In several areas of chemistry, the success or failure of an investigation can depend on the ability of a chemist to isolate tiny quantities of crystalline substances. Often the compounds of interest must be extracted from enormous amounts of extraneous material. In one of the more spectacular examples, Reed et al. isolated 30 mg of the crystalline coenzyme lipoic acid from 10 tons of beef liver residue.[12]

Lipoic acid

General Crystallization Procedure

The following steps are the essentials of crystallization:

Step 1. Select a suitable solvent.

Step 2. Dissolve the material to be purified in the minimum amount of warm solvent. *Remember that most organic solvents are extremely flammable and that many produce very toxic vapor.*

Step 3. Once the solid mixture has fully dissolved, filter the heated solution, and then bring it to the point of saturation by evaporating a portion of the solvent.

Step 4. Cool the warm saturated solution to reduce the solubility of the solute; this usually causes the solid material to precipitate. If the material has a low melting point or is very impure it may come out of solution sometimes as an oil. If so, reheat the solution and allow it to recool slowly.

Step 5. Isolate the solid by filtration, and then remove the last traces of solvent.

The crystallization is successful if the solid is recovered in good yield and is purer than it was before the crystallization. This cycle, from solid state to solution and back to solid state, is called *recrystallization* when both the initial and final materials are crystalline.

---

[12]Reed, L. J.; Gunsalus, I. C.; Schnakenberg, G. H. F.; Soper, Q. F.; Boaz, H. E.; Kem, S. F.; Parke, T. V. *J. Am. Chem. Soc.* **1953,** *75,* 1267.

Although the technique sounds fairly simple, in reality it is demanding. Successful purification of microscale quantities of solids will require your utmost attention. Choosing a solvent system is critical to a successful crystallization. To achieve high recoveries, the compound to be crystallized should ideally be very soluble in the hot solvent, but nearly insoluble in the cold solvent. To increase the purity of the compound, the impurities should be either *very soluble* in the solvent at all temperatures or *not soluble* at any temperature. The solvent should have as low a boiling point as possible so that traces of solvent can be easily removed (evaporated) from the crystals after filtration. It is best to use a solvent that has a boiling point at least 10 °C lower than the melting point of the compound to be crystallized to prevent the solute from "oiling out" of solution. Thus, the choice of solvent is critical to a good crystallization. Table 5.6 lists common solvents used in the purification of organic solids.

When information about a suitable solvent is not available, the choice of solvent is made on the basis of solubility tests. Craig's rapid and efficient procedure for microscale solubility testing works nicely; it requires only milligrams of material and a nine-well Pyrex spot plate.[13]

Place 1–2 mg of the solid in each well and pulverize each sample with a stirring rod. Add 3–4 drops of a given solvent to the first well and observe whether the material dissolves at ambient temperature. If not, stir the mixture for 1.5–2 min and observe and record the results. Repeat this process with the chosen set of solvents, using a separate well for each solubility test. *Keep track of which well contains which solvent.* Place your test plate (containing the samples) on a hot plate (set at its lowest setting) in the **hood**; add additional solvent if necessary. Record the solubility characteristics of the sample in each hot solvent. Cool the plate and see if crystallization occurs in any of the wells. On the basis of your observations, choose an appropriate solvent or a solvent pair (see the following paragraph) to recrystallize your material.

Solubility relationships are seldom ideal for crystallization; most often a compromise is made. If there is no suitable single solvent available, it is possible to use a mixture of two miscible solvents, called a *solvent pair*. A solvent is chosen that will readily dissolve the solid. Once the solid is dissolved in the

Table 5.6 Common Solvents		
Solvent	bp (°C)	Polarity
Acetone	56	Polar
Cyclohexane	81	Nonpolar
Diethyl ether	35	Intermediate polarity
Ethanol, 95%	78	Polar
Ethyl acetate	77	Intermediate polarity
Hexane	68	Nonpolar
Ligroin	60–90	Nonpolar
Methanol	65	Polar
Methylene chloride	40	Intermediate polarity
Methyl ethyl ketone	80	Intermediate polarity
Petroleum ether	30–60	Nonpolar
Toluene	111	Nonpolar
Water	100	Polar

[13]Craig, R. E. R. *J. Chem. Educ.* **1989,** *66,* 88.

minimum amount of hot solvent, the solution is filtered. A second solvent, miscible with the first, in which the solute has much lower solubility, is then added dropwise to the hot solution to achieve saturation. In general, polar organic molecules have higher solubilities in polar solvents, and nonpolar materials are more soluble in nonpolar solvents (like dissolves like). Table 5.7 lists some common solvent pairs.

It can take a long time to work out an appropriate solvent system for a particular reaction product. In most instances, with known compounds, the optimum solvent system has been established. Most crystallizations are not very efficient because many impurities have solubilities similar to those of the compounds of interest. Recoveries of 50–70% are not uncommon.

Several microscale crystallization techniques are available.

Table 5.7 Common Miscible Solvent Pairs	
Solvent 1 (more polar)	Solvent 2 (less polar)
Acetone	Diethyl ether
Diethyl ether	Hexane
Ethanol	Acetone
Ethyl acetate	Cyclohexane
Methanol	Methylene chloride
Acetone	Water
Water	Ethanol
Toluene	Ligroin

Simple Crystallization

Simple crystallization works well with large quantities of material (100 mg and up), and it is essentially identical to that of the macroscale technique.

Step 1. Place the solid in a small Erlenmeyer flask or test tube. A beaker is not recommended because the rapid and dangerous loss of flammable vapors of hot solvent occurs much more easily from the wide mouth of a beaker than from an Erlenmeyer flask. Furthermore, solid precipitate can rapidly collect on the walls of the beaker as the solution becomes saturated because the atmosphere above the solution is less likely to be saturated with solvent vapor in a beaker than in an Erlenmeyer flask.

Step 2. Add a minimal amount of solvent and heat the mixture to the solvent's boiling point in a sand bath. Stir the mixture by twirling a spatula between the thumb and index finger. A magnetic stir bar may be used if a magnetic stirring hot plate is used.

Step 3. Continue stirring and heating while adding solvent dropwise until all of the material has dissolved.

Step 4. Add a decolorizing agent (powdered charcoal, ~2% by weight; or better, activated-carbon Norit pellets,[14] ~0.1% by weight), to remove colored minor impurities and other resinous by-products.

Step 5. Filter (by gravity) the hot solution into a second Erlenmeyer flask (preheat the funnel with hot solvent). This removes the decolorizing agent and any insoluble material initially present in the sample.

Step 6. Evaporate enough solvent to reach saturation.

Step 7. Cool to allow crystallization (crystal formation will be better if this step takes place slowly). After the system reaches room temperature, cooling it in an ice bath may improve the recovery.

Step 8. Collect the crystals by filtration on a Büchner or Hirsch funnel. Save the mother liquor (this is the term used to describe the solution that was separated from the original crystals) until the identity of the product has been established. In some cases, it is possible to recover more product by concentrating and further cooling the mother liquor. The second crop of crystals, however, is usually not as pure as the first.

Step 9. Wash (rinse) the crystals carefully.

Step 10. Dry the crystals.

---

[14]Available from MilliporeSigma.

Hirsch
funnel

Neoprene
adapter

Filter flask

(a)   (b)   (c)

Figure 5.23   Component parts for vacuum filtration.

11-mm Hirsch funnel

Clamp

To vacuum trap

Figure 5.24   Vacuum filtration apparatus.

Filtration Techniques

Use of the Hirsch Funnel. The standard filtration system for collecting products purified by recrystallization in the microscale laboratory is vacuum (suction) filtration with an 11-mm Hirsch funnel. Many reaction products that do not require recrystallization can also be collected directly by vacuum filtration. The Hirsch funnel (Fig. 5.23*a*) is composed of a ceramic cone with a circular flat bed perforated with small holes. The diameter of the bed is covered by a flat piece of filter paper of the same diameter. The funnel is sealed into a filter flask with a Neoprene adapter (Fig. 5.23*b*). Plastic and glass varieties of this funnel that have a polyethylene or glass frit are now available. It is still advisable to use the filter paper disk with these funnels to prevent the frit from clogging or becoming discolored. Regardless of the type of filter used, *always wet the filter paper disk with the solvent being used in the crystallization and then apply the vacuum.* This ensures that the filter paper disk is firmly seated on the bed of the filter.

Filter flasks have thick walls, and a side arm to attach a vacuum hose, and are designed to operate under vacuum (see Fig. 5.23*c*). The side arm is connected with *heavy-walled* rubber vacuum tubing to a water aspirator (water pump). The water pump uses a very simple aspirator based on the Venturi effect. Water is forced through a constricted throat in the pump. (See the detailed discussion of the Venturi effect and water pumps in the section on reduced pressure [vacuum] distillations.) When water flows through the aspirator, the resulting partial vacuum sucks air down the vacuum tubing from the filter flask. *Always turn the water on full force.* With the rubber adapter in place, air is pulled through the filter paper, which is held flat on the bed of the Hirsch (or Büchner) funnel by the suction. The mother liquors are rapidly forced into the filter flask, where the pressure is lower, by atmospheric pressure. The crystals retained by the filter are dried by the stream of air passing through them (Fig. 5.24).

When you use a water pump, *it is very important to have a safety trap mounted in the vacuum line leading from the filter flask.* Any drop in water pressure at the pump (easily created by other students on the same water line turning on other aspirators at the same time) can result in a backup of water into the flask. As the flow through the aspirator decreases, the pressure at that point rapidly increases and water is forced up the vacuum tubing toward the filter flask (Fig. 5.24). It is also important to vent the vacuum by opening the vent stopcock or disconnecting the rubber tubing from the filter flask, *before* the water is turned off (see Fig. 5.25).

In some cases, the precipitate collected on the Hirsch funnel is not highly crystalline. The filter cake may be too thick or pasty to dry simply by pulling air through it. A thin, flexible rubber sheet or a piece of plastic food wrap placed over the mouth of the funnel, such that the suction generated from the vacuum pulls the sheeting down onto the filter cake (collected crystals), will

From filtration
apparatus

Stopcock
vent

Vacuum
tubing

Vacuum
tubing

Aspirator

Clamp

150-mL
filter flask

Figure 5.25   Vacuum trap.

place atmospheric pressure on the solid cake. This pressure can force much of the remaining solvent from the collected material, and thus further dry it. Use a piece of sheeting large enough to cover the entire filter cake. Otherwise, a vacuum may not be created and adequate drying may not occur.

In some instances, substances may retain water or other solvents with great tenacity. To dry these materials, a *desiccator* is often used. This is generally a glass or plastic container containing a *desiccant* (a material capable of absorbing water). The substance to be dried, held in a suitable container, is then placed on a support above the desiccant. This technique is often used in quantitative analysis to dry collected precipitates. Vacuum desiccators are available (Fig. 5.26). If this method of drying is still insufficient, a *drying pistol* can be used (Fig. 5.27). The sample in an open container (vial) is placed in the apparatus, which is then evacuated. The pistol has a pocket in which a strong adsorbing agent, such as P_4O_{10} (for water), NaOH or KOH (for acidic gases), or paraffin wax (for organic solvents), is placed. The pistol is heated by refluxing vapors that surround the barrel. A simple alternative to this method is the use of a side-armed test tube (Fig. 5.28).

A Hirsch-Funnel Alternative—A Nail-Filter Funnel.

A nail-filter funnel is a low-cost substitute for a Hirsch funnel (Fig. 5.29).[15] This apparatus is easily assembled from common laboratory glassware. Obtain a soft-glass rod that fits in the stem of a small glass funnel. Cut the rod to a suitable size and heat the tip of one end over a burner flame. When the tip becomes soft, hold the rod vertically and press the hot tip against a cold metal surface to flatten it to form a flat nail-like head. *The nail head should not be perfectly round or it will block the flow of liquid.* Cut the cooled rod to a suitable length and place the "nail" inside the stem of the funnel so that the flattened head of the nail rests on the top opening of the funnel. Cut a piece of filter paper just slightly larger than the nail head, place it on the nail head, and then place the funnel in a filter flask with a neoprene adapter. Be sure to wet the filter paper before filtering.

Craig Tube Crystallizations.

The Craig tube[16] is commonly used for microscale crystallizations in the range of 10–100 mg of material (Fig. 5.30). The process consists of the following steps.

Step 1. Place the sample in a small test tube (10 × 75 mm).

Step 2. Add the solvent (0.5–2 mL), and dissolve the sample by heating in the sand bath; add drops of solvent as needed. Rapid stirring with a microspatula (roll the spatula rod between your thumb and index finger) helps dissolve the material and protects against boilover. Add several drops of solvent by Pasteur pipet after the sample has completely dissolved. It will be easy to remove this excess at a later stage, since the volumes involved are very small. The additional solvent ensures that the solute will stay in solution during the hot transfer. Norit charcoal pellets may be added at this stage, if needed to remove colored impurities.

[15]Singh, M. M.; Pike, R. M.; Szafran, Z. *Microscale & Selected Macroscale Experiments for General & Advanced General Chemistry;* Wiley: New York, 1995, pp. 47, 63; Claret, P. A. *Small Scale Organic Preparations;* Pitman: London, 1961, p. 15.
[16]Craig, L. C.; Post, O. W. *Ind. Eng. Chem. Anal. Ed.* **1944,** 16, 413.

Figure 5.26 Vacuum desiccator.

Figure 5.27 Abderhalden vacuum drying apparatus. A, refluxing heating liquid; B, vacuum drying chamber; C, desiccant.

(20 × 150 mm sidearmed test tube)

Figure 5.28 Side-arm test tube as a vacuum drying apparatus.

circular filter paper

nail inside the funnel

Figure 5.29 Nail-filter funnel.

Figure 5.30 Craig tubes.

Step 3. Transfer the heated solution to the Craig tube by Pasteur filter pipet (see Fig. 3.30) that has been preheated with hot solvent. This transfer automatically filters the solution. A second filtration is often necessary if powdered charcoal has been used to decolorize the solution.

Step 4. The hot, filtered solution is then concentrated to saturation by gentle boiling in the sand bath. Constant agitation of the solution with a microspatula during this short period can avoid the use of a boiling stone and prevent boilover. Ready crystallization on the microspatula just above the solvent surface is a good indication that saturation is close at hand.

Step 5. The upper section of the Craig tube (the "head" or stopper) is set in place and the saturated solution is allowed to cool in a safe place. As cooling commences, seed crystals, if necessary, may be added by crushing them against the side of the Craig tube with a microspatula just above the solvent line. A good routine, if time is available, is to place the assembly in a small Erlenmeyer, then place the Erlenmeyer in a beaker, and finally cover the first beaker with a second inverted beaker. This procedure will ensure slow cooling, which will enhance good crystal growth (Fig. 5.31). A Dewar flask may be used when very slow cooling and large crystal growth are required.

Step 6. After the system reaches room temperature, cooling in an ice bath may improve the recovery.

Step 7. Remove the solvent by inverting the Craig tube assembly into a centrifuge tube and spinning the mother liquors away from the crystals (Fig. 5.32). *This operation should be carried out with care.* First, fit the head with a thin copper wire (Fig. 5.32), held in place by a loop at the end of the wire that is placed around the narrow part of the neck. Some Teflon heads have a hole in the neck to anchor the wire. The copper wire should not be much longer than the centrifuge tube.

Step 8. Now insert the Craig tube into a centrifuge tube. To do this, hold the Craig tube upright (with the head portion up) and place the centrifuge tube *down over* the Craig tube. Push the Craig tube up with your finger so that the head is against the inverted bottom of the centrifuge tube, and then invert the whole assembly (Fig 5.32).

Step 9. Place the assembly into a centrifuge tube, *balance the centrifuge,* and spin the mother liquors away from the crystals (Fig. 5.32). This

Figure 5.31 Apparatus for slow crystallization.

Figure 5.32 Crystal collection with a Craig tube.

replaces the usual filtration step in simple crystallizations. It avoids another transfer of material and also avoids product contact with filter paper.

Step 10.   Remove the apparatus from the centrifuge, and then carefully remove the Craig tube from the centrifuge tube. Gently pull upward on the copper wire while at the same time applying downward pressure with your fingers to the bottom of the inverted Craig tube (this will keep the Craig tube assembly together and not let any of the crystalline product fall back into the centrifuge tube and into the mother liquors). Once the Craig tube assembly is removed from the centrifuge tube, turn it so that the neck of the tube is up, and then disassemble it. At this point scrape any crystalline product clinging to the head into the lower section. If the lower section is tared, it can be left to air dry to constant weight or placed in a warm vacuum oven (use a rubber band or thin wire to wrap a piece of filter paper over the open end to prevent dust from collecting on the product while drying).

The cardinal rule in carrying out the purification of small quantities of solids is *Keep the number of transfers to an absolute minimum!* The Craig tube is very helpful in this regard.

QUESTIONS

5-23. Is activated carbon needed in every recrystallization procedure? Explain.

5-24. List several advantages and disadvantages of using a Craig tube for recrystallization.

5-25. Why is it advisable to use a stemless or a short-stemmed funnel when carrying out a gravity filtration?

5-26. Which of the following solvent pairs could be used in a recrystallization? Why, or why not?
   (a)   Acetone and methanol
   (b)   Cyclohexane and water
   (c)   Hexane and diethyl ether

5-27. You perform a recrystallization on 75 mg of a solid material and isolate 52 mg of purified material. What is the percent recovery? Further concentration of the mother liquor provides an additional 13 mg of material. What is the total percent recovery?

5-28. Describe two techniques that can be used to induce crystallization.

5-29. When would you advise someone to use a solvent pair to carry out a recrystallization?

5-30. You are provided a solid which you suspect is not pure and the accompanying data sheet has very limited information. Two items which are recognizable are that an ethereal solvent has worked well when performing recrystallizations and the literature melting point is 61 °C. When looking at what ethereal solvents you have to choose from, you see two: *t*-butyl methyl ether and diethyl ether. Why would the latter be the far better choice knowing that aside from relative boiling points, the two ethers are very similar when considering their physical properties.

NOTE. The following experiments use Technique 5: Experiments [6], [7], [15], [16], [18], [19A], [19B], [19C], [19D], [20], [23A], [23B], [24A], [24B], [25A], [25B], [26], [28], [29A], [29B], [29C], [29D], [30], [31], [33A], [33B], [34A], [34B], [A1$_a$], [A2$_a$] [A3$_a$] [A1$_b$], [A2$_b$], [A3$_b$] [A4$_{ab}$], [B1], [C2], [D1], [D2], [E1], [E2], [F1], [F2], and [F4].

TECHNIQUE 6

Chromatography

Column, Flash, High-Performance Liquid, and Thin-Layer Chromatography

Technique 6A

The basic theory of chromatography is introduced in Technique 1 in the discussion of gas-phase separations. The word *chromatography* is derived from the Greek word for color, *chromatos*. Tswett discovered the technique in 1903 while studying ways to separate mixtures of natural plant pigments.[17] The chromatographic zones were detected simply by observing the visual absorption bands. Thus, as originally applied, the name was not an inconsistent use of terminology. Today, however, most mixtures are colorless. The separated zones in these cases are detected by other methods.

Two chromatographic techniques are discussed in this section. Both depend on adsorption and distribution between a stationary solid phase and a moving liquid phase. The first is column chromatography, which is used extensively throughout organic chemistry. It is one of the oldest of the modern chromatographic methods. The second technique, thin-layer chromatography (TLC), is particularly effective in rapid assays of sample purity. It can also be used as a preparative technique for obtaining tiny amounts of high-purity material for analysis.

Column Chromatography

Column chromatography, as its name implies, uses a column packed with a solid stationary phase. A mobile liquid phase flows by gravity (or applied pressure) through the column. Column chromatography uses polarity differences to separate materials. A sample on a chromatographic column is subjected to two opposing forces: (1) the solubility of the sample in the elution solvent system and (2) the adsorption forces binding the sample to the solid phase. These interactions comprise an equilibrium. Some sample constituents are adsorbed more tightly; other components of the sample dissolve more readily in the liquid phase and are eluted more rapidly. The more rapidly eluting materials, thus, are carried further down the column before becoming readsorbed, and thus exit the column before more tightly bound components. The longer the column, the larger the number of adsorption–dissolution cycles (much like the vaporization–condensation cycles in a distillation column), and the greater the separation of sample components as they elute down the column. A molecule that is strongly adsorbed on the stationary phase will move slowly down the column; a molecule that is weakly adsorbed will move at a faster rate. Thus, a complex mixture can be resolved into separate bands of pure materials. These bands of purified material eventually elute from the column and can be collected.

Many materials have been used as the stationary phase in column chromatography. Finely ground alumina (aluminum oxide, Al_2O_3) and silicic acid (silica gel, SiO_2) are by far the most common adsorbents (stationary phases). Many common organic solvents are used as the liquids (sometimes called *eluents*) that act as the mobile phase and elute (wash) materials through the column. Table 5.8 lists the better-known column packing and elution solvents.

---

[17]Tswett, M. *Ber. Deut. Botan. Ges.* **1906**, *24*, 235.

Table 5.8 Column Chromatography Materials			
Stationary Phase		**Moving Phase**	
Alumina Silicic acid Magnesium sulphate Cellulose paper	↑ *Increasing adsorption of polar materials*	Water Methanol Ethanol Acetone Ethyl acetate Diethyl ether Methylene chloride Cyclohexane Pentane	↑ *Increasing solvation of polar materials*

Silica gel impregnated with silver nitrate (usually 5–10%) is also a useful adsorbent for some functional groups. The silver cation selectively binds to unsaturated sites via a silver-ion π complex. Traces of alkenes are easily removed from saturated reaction products by chromatography with this system (see Experiment [12]). This adsorbent, however, must be protected from light until used, or it will quickly darken and become ineffective.

Column chromatography is usually carried out according to the procedures discussed in the following five sections.

Packing the Column. The quantity of stationary phase required is determined by the sample size. A common rule of thumb is to use a weight of packing material 30–100 times the weight of the sample to be separated. Columns are usually built with roughly a 10:1 ratio of height to diameter. In the microscale laboratory, two standard chromatographic columns are used:

1. A Pasteur pipet, modified by shortening the capillary tip, is used to separate smaller mixtures (10–100 mg). Approximately 0.5–2.0 g of packing is used in the pipet column (Fig. 5.33a).

2. A 50-mL titration buret (modified by shortening the column to 10 cm above the stopcock) is used for larger (50–200 mg) or difficult-to-separate sample mixtures. A buret column uses approximately 5–20 g of packing (Fig. 5.33b).

Figure 5.33   Chromatographic columns: (a) a Pasteur pipet column; (b) a buret column. Reproduced with permission of John Wiley and Sons, Inc., from Zubrick, J. W. *The Organic Chemistry Lab Survival Manual*, 9th ed., Copyright 2013.

Both columns are prepared by first clamping the empty column in a vertical position and then seating a small cotton or glass wool plug at the bottom. For a buret column, cover the cotton with a thin layer of sand. The Pasteur pipets are loaded by adding the dry adsorbent with gentle tapping, "dry packing." The pipet column (*dry column*) is then premoistened just prior to use.

The burets (*wet columns*) are packed by a slurry technique. In this procedure the column is filled partway with solvent; then the stopcock is opened slightly, and as the solvent slowly drains from the column a slurry of the adsorbent-solvent is poured into the top of the column. The column should be gently tapped while the slurry is added. The solvent is then drained to the top of the adsorbent level and held at that level until used. Alternatively, the wet-packed column can be loaded by sedimentation techniques rather than using a slurry. One such routine is to initially fill the column with the least-polar solvent to be used in the intended chromatographic separation. Then the solid phase is slowly added with gentle tapping, which helps to avoid subsequent channeling. As the solid phase is added, the solvent is slowly drained from the buret at the same time. After the adsorbent has been fully loaded, the solvent level is then lowered to the top of the packing as in the slurry technique.

Sample Application. Using a Pasteur pipet, apply the sample in a minimum amount of solvent (usually the least polar solvent in which the material is readily soluble) to the top of the column. *Do not disturb the sand layer!* Rinse the pipet, and add the rinses to the column just as the sample solution drains to the top of the adsorbent layer.

Elution of the Column. The critical step in resolving the sample mixture is eluting the column. Once the sample has been applied to the top of the column, the elution begins (a small layer of sand can be added to the top of the buret column after addition of the first portion of elution solvent).

NOTE. Do not let the column run dry: This can cause channels to form in the column.

In a buret column, the flow is controlled by the stopcock. The flow rate should be set to allow time for equilibrium to be established between the two phases; this will depend on the nature and amount of the sample, the solvent, and how difficult the separation will be. The Pasteur pipet column is *free flowing* (the flow rate is controlled by the size of the capillary tip and its plug); once the sample is on the column, the chromatography will require constant attention.

If necessary, it is possible to ease this restriction somewhat by modifying the pipet. Place a Tygon connector (short sleeve) at the top of the pipet column. Once the sample is on the column, insert a second pipet into this connector with its tip just below the liquid level on the top of the column. Add additional solvent through the second pipet (use a bulb, if necessary, but remove it before the elution begins), which acts as a solvent reservoir. As the solvent level in the column pipet drops below the tip of the top pipet, air is admitted, and additional solvent is automatically delivered to the chromatographic column. Thus, the solvent head on the column is maintained at a constant volume. The top pipet need be filled only at necessary intervals; larger volumes of solvent can thus be added to this reservoir. This arrangement also prevents dislodging of the absorbent as new solvent is added.

NOTE. It is exceedingly important that solvents do not come in contact with the Tygon sleeve holding the second pipet. These sleeves contain plasticizers that will readily dissolve and contaminate the sample.

The choice of solvent is dictated by a number of factors. A balance between the adsorption power of the stationary phase and the solvation power of the elution solvent governs the rate of travel of the material down the column. If the material travels rapidly down the column, then too few adsorption-elution cycles will occur and the materials will not separate. If the sample travels too slowly, diffusion broadening takes over and separation is degraded. Solvent choices and elution rates can strike a balance between these factors and maximize the separation. It can take considerable time to develop a solvent or mixture of solvents that produces a satisfactory separation of a particular mixture.

Fraction Collection. As the solvent elutes from the column, it is collected in a series of "fractions" by using small Erlenmeyer flasks or vials. Under ideal conditions, as the mixture of material travels down the column, it will separate into several individual bands (zones) of pure substances. By careful collection of the fractions, these bands can be separated as they sequentially elute from the column (similar to the collection of GC fractions in the example described in Technique 1). The bands of eluted material can be detected by a number of techniques (weighing fraction residues, colored materials, TLC, etc.).

Column chromatography is a powerful technique for the purification of organic materials. In general, it is significantly more efficient than crystallization procedures. Recrystallization is often best avoided until the last stages of purification, where it will be most efficient. Rely instead on chromatography to do most of the separation.

Column chromatography of a few milligrams of product usually takes no more than 30 min, but chromatographing 10 g of product might easily take several hours, or even all day. Large-scale column chromatography (50–100 g) of a complex mixture could take several days to complete using this type of equipment.

Flash Chromatography

Flash chromatography, first described by Still and co-workers in 1978, is a common method for separating and purifying nonvolatile mixtures of organic compounds.[18] The technique is rapid, easy to perform, and relatively inexpensive, and it gives good separations. Many laboratories routinely use flash chromatography to separate mixtures ranging from 10 mg to 10 g.

This moderate-resolution, preparative chromatography technique was originally developed using silica gel (40–63 μm). Bonded-phase silica gel of a larger particle size can be used for reversed-phase flash chromatography. Flash chromatography columns are generally packed dry to a height of approximately 6 in. Thin-layer chromatography is a quick way to choose solvents for flash chromatography. A solvent that gives differential retardation factor (DR$_f$) values (of the two substances requiring separation) ≥ 0.15 on TLC usually gives effective separation with flash chromatography. Table 5.9 lists typical experimental parameters for various sample sizes, as a guide to separations using flash chromatography. In general, a mixture of organic compounds separable by TLC can be separated preparatively using flash chromatography.

---

[18]Still, W. C.; Kahn, M.; Mitra, A. *J. Org. Chem.* **1978**, *43*, 2923.

Table 5.9   **Typical Experimental Parameters**

Column Diameter (mm)	Total Volume of Eluent (mL)[a]	Typical Sample Loading (mg)		Typical Fraction (mL)
		$DR_{f > 0.2}$	$DR_{f > 0.1}$	
10	100–150	100	40	5
20	200–250	400	160	10
30	400–450	900	360	20
40	500–550	1600	600	30
50	1000–1200	2500	1000	50

Source. Data from Majors, R. E., Enzweiler, T., *LC GC* **1998,** *6,* 1046.
[a]Required for both packing and elution.

Flash chromatography apparatus generally consists of a glass column equipped to accept a positive pressure of compressed air or nitrogen applied to the top of the column. A typical commercially available arrangement is shown in Figure 5.34.[19]

Generally, a 20–25% solution of the sample in the elution solvent is recommended, as is a flow rate of about 2 in./min. The column must be conditioned, before the sample is applied, by flushing the column with the elution solvent to drive out air trapped in the stationary phase and to equilibrate the stationary phase and the solvent.

Figure 5.34   **(a) Conventional column and (b) screw-threaded column.**

---

[19]A complete line of glass columns, reservoirs, clamps, and packing materials for flash chromatography is offered by a number of scientific supply companies (e.g., Fisher). Silica gels for use in this technique are also available from Amicon, Danvers, MA; J. T. Baker, Phillipsburg, NJ; EM Science/Merck, Gibbstown, NJ; ICN Biomedicals, Inc., Cleveland, OH.

Several modifications of the basic arrangement have been reported, especially in regard to the adaptation of the technique to the instructional laboratory. These involve inexpensive pressure control valves, use of an aquarium "vibrator" air pump, and adapting a balloon reservoir to supply the pressurized gas.

At the microscale level, a pipet bulb or pump on the pipet column can be used to supply pressure to the column. If a bulb is used, squeeze it to apply pressure, and *remove the bulb from the pipet before releasing it!* Otherwise, material may be sucked up into the bulb and most likely disturb the column. Reapply the bulb to re-create pressure. If a pump is used, do not back off the pressure once it has been applied.

An improved method, utilizing a capillary Pasteur pipet for introducing the sample onto the chromatographic column approximately doubles the effectiveness (theoretical plates) of the column.[20] Dry-column flash chromatography[21] has been adapted for use in the instructional laboratory.[22] The "column" consists of a dry bed of silica gel in a sintered glass funnel placed in a standard vacuum filtration flask; the solvent is eluted by suction. Small (16 × 150 mm) test tubes inserted into the flask below the stem of the funnel are used to collect the fractions. This technique has been used successfully to separate mixtures ranging from 150 to 1000 mg.

Thin-Layer Chromatography

Thin-layer chromatography (TLC) is closely related to column chromatography, in that the phases used in both techniques are essentially identical. Alumina and silica gel are typical stationary phases, and the usual solvents are the mobile phases. There are, however, some distinct differences between TLC and column chromatography. The mobile (liquid) phase *descends* in column chromatography; the mobile phase *ascends* in TLC. The column of stationary phase material used in column chromatography is replaced by a thin layer (100 µm) of stationary phase spread over a flat surface. A piece of window glass, a microscope slide, or a sheet of plastic can be used as the support for the thin layer of stationary phase. It is possible to prepare your own glass plates, but plastic-backed thin-layer plates are only commercially available. Plastic-backed plates are particularly attractive because they can easily be cut with scissors into strips of any size. Typical strips measure about 1 × 3 in., but even smaller strips can be satisfactory.

Thin-layer chromatography has some distinct advantages: it needs little time (2–5 min) and it needs *very* small quantities of material (2–20 µg). The chief disadvantage of this type of chromatography is that it is not very amenable to preparative scale work. Even when large surfaces and thicker layers are used, separations are most often restricted to a few milligrams of material.

NOTE. Do not touch the active surface of the plates with your fingers. Handle them only by the edge.

TLC is performed as follows:

Step 1. Draw a light pencil line parallel to the short side of the plate, 5–10 mm from the edge. Mark one or two points, evenly spaced, on the

[20]Pivnitsky, K. K. *Aldrichimica Acta* **1989,** *22,* 30.
[21]Harwood, L. M. *Aldrichimica Acta* **1985,** *18,* 25; Sharp, J. T.; Gosney I.; Rowley A. G. *Practical Organic Chemistry;* Chapman & Hall: New York, 1989.
[22]Shusterman, A. J.; McDougal, P. G.; Glasfeld, A. *J. Chem. Educ.* **1997,** *74,* 1222.

Figure 5.35   Applying a sample to a TLC plate.

line. Place the sample to be analyzed (1 mg or less) in a 100 μL conical vial and add a few drops of a solvent to dissolve the sample. Use a capillary micropipet to apply a small fraction of the solution from the vial to the plate (Fig. 5.35). (These pipets are prepared by the same technique used for constructing the capillary insert for ultramicro boiling-point determinations, see Chapter 4.) Apply the sample to the adsorbent side of the TLC plate by gently touching the tip of the filled capillary to the plate. Remove the tip from the plate before the dot of solvent grows to much more than a few millimeters in diameter. If it turns out that you need to apply more sample, let the dot of solvent evaporate and then reapply more sample to exactly the same spot.

Step 2. Place the spotted thin-layer plate in a screw-capped, wide-mouth jar, or a beaker with a watch glass cover, containing a small amount of elution solvent (Fig. 5.36). It helps if one side of the jar's (beaker's) interior is covered with a piece of filter paper that wicks the solvent up to increase the surface area of the solvent. The TLC plate must be positioned so that the spot of your sample is *above* the solvent. Cap the jar, or replace the watch glass on the beaker, to maintain an atmosphere saturated with the elution solvent. The elution solvent climbs the plate by capillary action, eluting the sample up the plate. *Do not move the developing chamber after the action has started.* Separation of mixtures into individual spots occurs by exactly the same mechanism as in column chromatography. Stop the elution by removing the plate from the jar or beaker when the solvent front nears the top of the TLC plate. Quickly (before the solvent evaporates) mark the position of the solvent front on the plate.

Figure 5.36   Developing a TLC plate. (Reproduced with permission of John Wiley and Sons, Inc., from Zubrick, J. W. *The Organic Chemistry Lab Survival Manual,* 9th ed., Copyright 2013.)

Step 3. Colorless, separated components of a mixture can often be observed in a developed TLC plate by placing the plate in an iodine-vapor chamber (a sealed jar containing solid I_2) for a minute or two. Iodine forms a reversible complex with most organic substances, and dark spots will thus appear in those areas containing sample material. Mark the spots with a pencil soon after removing the TLC plate from the iodine chamber because the spots may fade. Samples that contain a UV-active chromophore (see Chapter 8) can be observed without using iodine. TLC plates are commonly prepared with an UV-activated fluorescent indicator mixed in with the silica gel. Sample spots can be detected with a hand-held UV lamp; the sample quenches the fluorescence induced by the lamp and appears as a dark spot against the fluorescent blue-green background.

Step 4. The TLC properties of a compound are reported as R_f values (retention factors). The R_f value is the distance traveled by the substance divided by the distance traveled by the solvent front (this is why the position of the solvent front should be quickly marked on the plate when the chromatogram is terminated; see Fig. 5.37). TLC R_f values vary with the moisture content of the adsorbent. Thus, the actual R_f of a compound in a given solvent can vary from day to day and from laboratory to laboratory. The best way to determine if two samples have identical R_f values is to elute them together on the same plate.

Thin-layer chromatography is used in a number of applications. The speed of the technique makes it quite useful for monitoring large-scale column chromatography. Analysis of fractions can guide decisions on the solvent-elution sequence. TLC analysis of column-derived fractions can also determine how best to combine collected fractions. Following the progress of a reaction by periodically removing small aliquots for TLC analysis is an extremely useful application of thin-layer chromatography.

Paper Chromatography

The use of cellulose-paper as an adsorbent is referred to as *paper chromatography*. This technique has many of the characteristics of TLC in that sheets or strips

$$R_f(\text{compound 1}) = \frac{35}{70} = 0.50 \qquad R_f(\text{compound 2}) = \frac{61}{70} = 0.87$$

Figure 5.37 Determining R_f values.

of filter paper are used as the stationary phase. In this case, however, the paper is usually positioned to hang down from trays holding the paper and the elution solvent. The solvent front, therefore, descends downward rather than upward as in TLC. Paper chromatography has a distinct advantage: It is very amenable to the use of aqueous mobile phases and very small sample sizes. It is primarily used for the separation of highly polar or polyfunctional species such as sugars and amino acids. It has one major disadvantage: It is very slow. Paper chromatograms can easily take three to four hours or more to elute.

High-Performance Liquid Chromatography

Although gas chromatography is a powerful chromatographic method, it is limited to compounds that have a significant vapor pressure at temperatures up to about 200 °C. Thus, compounds of high molecular weight and/or high polarity cannot be separated by GC. High-performance liquid chromatography (HPLC) does not present this limitation.

GC and HPLC are somewhat similar, in the instrumental sense, in that the analyte is partitioned between a stationary phase and a mobile phase. Whereas the mobile phase in GC is a gas, the mobile phase in HPLC is a liquid. As shown schematically in Figure 5.38, the mobile phase (solvent) is delivered to the system by a pump capable of pressures up to about 6000 psi. The sample is introduced by the injection of a solution into an injection loop. The injection loop is brought in line between the pump and the column (stainless steel) by turning a valve; the sample then flows down the column, is partitioned, and flows out through a detector.

The solid phase in HPLC columns used for organic monomers is usually some form of silica gel. "Normal" HPLC refers to chromatography using a solid phase (usually silica gel) that is more polar than the liquid phase, or solvent, so that the less polar compounds elute more rapidly. Typical solvents include ethyl acetate, hexane, acetone, low molecular weight alcohols, chloroform, and acetonitrile. For extremely polar compounds, such as amino acids,

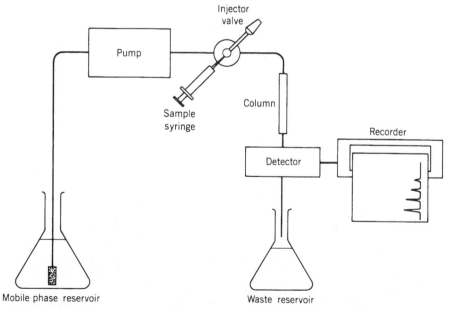

Figure 5.38   High-performance liquid chromatography system block diagram.
(Courtesy of Perkin-Elmer, Inc. All rights reserved.)

"reversed-phase" HPLC is used. Here, the liquid phase is more polar than the stationary phase, and the more polar compounds elute more rapidly. The mobile phase is usually a mixture of water and a water-miscible organic solvent such as acetonitrile, dioxane, methanol, isopropanol, or acetone. The stationary phase is usually a modified silica gel where the —OH groups of the silica gel have been replaced by —OSiR groups; R is typically a linear C_{18} alkyl chain. These so-called "bonded-phase" columns are not capable of handling as much analyte as normal silica gel columns, and are thus easily overloaded and are less useful for preparative work. (For further discussion see Solid-Phase Extraction, Technique 4, page 87.)

A wide variety of detection systems are available for HPLC. UV detection is common, inexpensive, and sensitive. The solvent flowing off the column is sent through a small cell where the UV absorbance is recorded over time. Many detectors are capable of variable wavelength operation, so the detector can be set to the wavelength most suitable to the compound or compounds being analyzed. Photodiode array detectors are available; these can obtain a full UV spectrum in a fraction of a second, so that more information can be obtained on each component of a mixture. For compounds that absorb light in the visible (vis) spectrum, many detectors can be set to visible wavelengths. The principal shortcoming of UV-vis detection is that to be detected, compounds being studied must have a UV chromophore, such as an aromatic ring or other conjugated π system (see Chapter 8).

For compounds that lack a UV-vis chromophore, refractive index (RI) detection is a common substitute. An RI detector measures the difference in refractive index between the eluant and a reference cell filled with the elution solvent. Refractive index detection is significantly less sensitive than UV-vis detection, and the detector is quite sensitive to temperature changes during the chromatographic run.

More sophisticated HPLC instruments offer the ability to mix two or three different solvents and to use solvent gradients by changing the solvent composition as the chromatographic run progresses. This allows the simultaneous analysis of compounds that differ greatly in their polarity. For example, a silica gel column might begin elution with a very nonpolar solvent, such as hexane. The solvent polarity is then continuously increased by blending in more and more ethyl acetate until the elution solvent is pure ethyl acetate. This effect is directly analogous to temperature programming in GC.

For analytical work, typical HPLC columns are about 5 mm in diameter and about 25 cm in length. The maximum amount of analyte for such columns is generally less than 1 mg, and the minimum amount is determined by the detection system. High-performance liquid chromatography can thus be used to obtain small amounts of purified compounds for infrared (IR), nuclear magnetic resonance (NMR), or mass spectrometric (MS) analysis. Larger "semi-preparative" columns that can handle up to about 20 mg without significant overloading are useful for obtaining material for ^{13}C NMR spectroscopy or further synthetic work.

HPLC has the advantage that it is rapid, it uses relatively small amounts of solvent, and it can accomplish very difficult separations.

Concentration of Solutions

The solvent can be removed from chromatographic fractions (extraction solutions, or solutions in general) by a number of different methods.

Technique 6B

Distillation

Concentration of solvent by distillation is straightforward, and the standard routine is described in Technique 2 (page 65). This approach allows for high recovery of volatile solvents and often can be done outside a hood. The Hickman still head and the 5- or 10-mL round-bottom flask are useful for this purpose. Distillation should be used primarily for concentration of the chromatographic fraction, followed by transfer of the concentrate with a Pasteur filter pipet to a vial for final isolation.

Evaporation with Nitrogen Gas

A very convenient method for removal of final solvent traces is the concentration of the last 0.5 mL of a solution by evaporation with a gentle stream of nitrogen gas while the sample is warmed in a sand bath. This process is usually done at a hood station where several Pasteur pipets can be attached to a manifold leading to a source of dry nitrogen gas. Gas flow to the individual pipets is controlled by needle valves. *Always test the gas flow with a blank vial of solvent.*

Ruekberg described an alternative way to remove solvent from solutions of compounds that are not readily oxidized.[23] The setup includes an aquarium air pump, a pressure safety valve and ballast container, a drying tube, and a manifold. Blunted hypodermic needles are used in place of Pasteur pipets.

The sample vial will cool as the solvent evaporates, and gentle warming and agitation of the vial will thus help remove the last traces of the solvent. This avoids possible moisture condensation on the sample residue, as long as the gas itself is dry. *Do not leave the heated vial in the gas flow after the solvent is removed!* This precaution is particularly important in the isolation of liquids. Tare the vial before filling it with the solution to be concentrated; constant weight over time is the best indication that all solvent has been removed.

Removal of Solvent Under Reduced Pressure

Concentration of solvent under reduced pressure is very efficient. It reduces the time for solvent removal in microscale experiments to a few minutes. In contrast, distillation or evaporation procedures require many minutes for even relatively small volumes. Several methods are available.

Filter Flask Method. This vacuum-concentration technique can be tricky and should be practiced prior to committing hard-won reaction product to this test. The procedure is most useful with fairly large chromatographic fractions (5–10 mL). The sequence of operations is as follows (see also Fig. 5.39):

11-mm Hirsch funnel
Filter paper
To vacuum trap
25-mL filter flask
Solvent

Figure 5.39   Removal of solvent under reduced pressure.

[23]Ruekberg, B. *J. Chem. Educ.* **1995,** *72,* A200.

Step 1. Transfer the chromatographic fraction to the 25-mL filter flask.

Step 2. Insert the 11-mm Hirsch funnel and rubber adapter into the flask.

Step 3. Turn on the water pump (with trap) and connect the vacuum tubing to the pressure flask side arm while holding the flask in one hand.

Step 4. Place the thumb of the hand holding the filter flask over the Hirsch funnel filter bed to shut off the air flow through the system (see Fig. 5.39). This will result in an immediate drop in pressure. The volatile solvent will rapidly come to a boil at room temperature. Thumb pressure adjusts air leakage through the Hirsch funnel and thereby controls the pressure in the system. It is also good practice to learn to manipulate the pressure so that the liquid does not foam up into the side arm of the filter flask.

The filter flask must be warmed by the sand bath during this operation; rapid evaporation of the solvent will quickly cool the solution. The air leak used to control the pressure results in a stream of moist laboratory air being rapidly drawn over the surface of the solution. If the evaporating liquid becomes cold, water will condense over the interior of the filter flask and contaminate the isolated residue. Warming the flask while evaporating the solvent will avoid this problem and help speed solvent removal. The temperature of the flask should be checked from time to time by touching it with the palm of the free hand. The flask is kept slightly above room temperature by adjusting the heating and evaporation rates. It is best to practice this operation a few times with pure solvent (blanks) to see whether you can avoid boilovers and accumulating water residue in the flask.

Rotary Evaporator Method. In most research laboratories, the most efficient way to concentrate a solution under reduced pressure is to use a **rotary evaporator**. A commercial micro-rotary evaporator is shown in Figure 5.40.

Scientific Support, Inc.

Figure 5.40 Labconco micro-rotary evaporator.

This equipment makes it possible to recover the solvent removed during the operation.

The rotary evaporator is a motor-driven device that rotates the flask containing the solution to be concentrated under reduced pressure. The rotation continuously exposes a thin film of the solution for evaporation. This process is very rapid, even well below the boiling point of the solvent being removed. Since the walls of the rotating flask are constantly rewetted by the solution, bumping and superheating are minimized. The rotating flask may be warmed in a water bath or other suitable device that controls the rate of evaporation. A suitable adapter (a "bump bulb") should be used on the rotary evaporator to guard against splashing and sudden boiling, which may lead to lost or contaminated products.

In microscale work, never pour a recovered liquid product from the rotary flask. Always use a Pasteur pipet.

Hickman Still-Rotary Evaporation Apparatus.   A simple microscale rotary evaporator for use in the instructional laboratory consists of a 10-mL round-bottom flask connected to a capped Hickman still (side-arm type), which in turn is attached to a water aspirator (with trap).[24] The procedure involves transferring the solution to be concentrated to the preweighed 10-mL flask. The flask is then attached to a Hickman still with its top joint sealed with a rubber septum and threaded compression cap. The apparatus is connected by the still side arm to the trap-vacuum source with a vacuum hose. With the aspirator on, one shakes the apparatus while warming the flask in the palm of the hand. In this manner, bumping is avoided and evaporation is expedited. The still acts as a splash guard. Heat transfer is very effective, and once the flask reaches ambient temperature, the vacuum is released by venting through the trap stopcock.

QUESTIONS

5-31. When marking the sample line on a TLC plate, consider and comment on the advantages/disadvantages when using a ball-point pen or a pencil.

5-32. A series of dyes is separated by TLC. The data are given below. Calculate the R_f value for Bismarck brown and Palisade yellow 3G.

Material	Distance moved (cm)
Solvent	6.6
Bismarck brown	1.6
Lanacyl violet BF	3.8
Palisade yellow 3G	5.6
Alizarine emerald G	0.2

5-33. Why is it important not to let the level of the elution solvent in a packed chromatographic column drop below the top of the solid-phase adsorbent?

5-34. What are some advantages of using column chromatography to purify reaction products in the microscale laboratory?

5-35. Discuss the similarities and dissimilarities of TLC, paper, and column chromatography.

[24]Maynard, D. F. *J. Chem. Educ.* **1994,** *71,* A272.

5-36. After each separation by chromatography, the solvent is evaporated and then properly discarded. Describe the good laboratory practices needed to minimize exposure to the volatiles removed.

5-37. (a) What are the main advantages of using flash chromatography?
(b) How can TLC be used in connection with flash chromatography?

5-38. Using the information presented on the right, please identify and explain which spot has an R_f value of 0.54.

NOTE. The following experiments use Technique 6: Experiments [8A], [8B], [8C], [11C], [12], [13], [16], [17], [19A], [19B], [19C], [19D], [22A], [22B], [27], [29A], [29B], [29C], [29D], [30], [33A], [33B], [35], [A2$_a$], [A1$_b$], [E1], and [E3]. [1A$_{adv}$], [1B$_{adv}$], [4$_{adv}$], and [7$_{adv}$].

Collection or Control of Gaseous Products

Water-Insoluble Gases

Numerous organic reactions lead to the formation of gaseous products. If the gas is insoluble in water, collection is easily accomplished by displacing water from a collection tube. A typical experimental setup for the collection of gases is shown in Figure 5.41.

As illustrated, the glass capillary efficiently transfers the evolved gas to the collection tube. The delivery system need not be glass; small polyethylene or polypropylene tubing may also serve this purpose. In this arrangement, a syringe needle is inserted through a septum to accommodate the plastic tubing as shown in Figure 5.42. An alternative to this connector is a shortened Pasteur pipet inserted through a thermometer adapter (Fig. 5.42). Another alternative to the syringe needle or glass pipet tip is suggested by Jacob.[25] The lower half

Figure 5.41   Microscale gas collection apparatus.

---

[25]Jacob, L. A. *J. Chem. Educ.* **1992**, *69*, A313.

Figure 5.42 Alternative arrangements for controlled gas collection.

of the tapered tip of a plastic automatic delivery pipet tip is cut off to prevent buildup of excess pressure in the reaction vessel. The pipet tip is then inserted through a previously pierced rubber septum or into a thermometer adapter. The narrow end of the tip is then inserted into the plastic tubing.

An example of a reaction leading to gaseous products that can use this collection technique is the acid-catalyzed dehydration of 2-butanol described in Experiment [9]. The products of this reaction are a mixture of alkenes: 1-butene, *trans*-2-butene, and *cis*-2-butene, which boil at -6.3, 0.9, and 3.7 °C, respectively, and *sec*-butyl ether (2,2′-oxybisbutane), which boils at 123 °C. While all four compounds are formed in the reaction mixture, the setup is uniquely designed to collect just the gases and thus driving the equilibrium in the isolation of the three volatile alkenes.

In Figure 5.41 the gas collection tube is capped with a rubber septum. This arrangement allows for convenient removal of the collected gaseous butenes using a gastight syringe, as shown in Figure 5.43. In this particular reaction, the mixture of gaseous products is conveniently analyzed at ambient temperature by GC (see Technique 1).

Trapping By-product Gases

Some organic reactions release poisonous or irritating gases as by-products. For example, hydrogen chloride, ammonia, and sulfur dioxide are typical by-products in organic reactions. In these cases, the reaction is generally run in a **hood**. A gas trap may be used to prevent the gases from being released into the laboratory atmosphere. If the evolved gas is water soluble, the trap technique works well at the microscale level. The evolved gas is directed from the reaction vessel to a container of water or other aqueous solution, wherein it dissolves (reacts). For example, a dilute solution of sodium or ammonium hydroxide is suitable for acidic gases (such as HCl); a dilute solution of sulfuric or hydrochloric acid is suitable for basic gases (such as NH_3 or low molecular weight amines). Various designs are available for gas traps. A simple, easily assembled one for a gas that is very soluble in water is shown in Figure 5.44. Note that the funnel is not immersed in the water. If the funnel is held below the surface of the water and a large quantity of gas is absorbed or dissolved, the water easily could be drawn back into the reaction assembly. If the gas to be collected is not very soluble, the funnel may be immersed just below (1–2 mm) the surface of the water.

Figure 5.43 Removal of collected gases.

Figure 5.44   Trapping a water-soluble gas.

At the microscale level, small volumes of gases are evolved that may not require the funnel. Three alternatives are available:

a. Fill the beaker (100 mL) in Figure 5.41 with moistened fine glass wool and lead the gas delivery tube directly into the wool.

b. Place moistened glass wool in a drying tube and attach the tube to the reaction apparatus . However, be careful not to let the added moisture drip into the reaction vessel; place a small section of **dry** glass wool in the tube before the moist section is added.

c. Use a water aspirator. An inverted funnel can be placed over the apparatus opening where the evolved gas is escaping (usually the top of a condenser) and connected with flexible tubing (through a water trap) to the aspirator. A second arrangement is to use a glass or plastic T-tube (open on one end) inserted in the top of the condenser, by use of a rubber stopper, in place of the funnel.[26] If the reaction must be run under anhydrous conditions, a drying tube is inserted between the condenser and T-tube. This arrangement is very efficient, easy to assemble, and inexpensive. The simplest method is to clamp a Pasteur pipet so that its tip is inserted well into the condenser, and connect it (through a water trap) to the aspirator.

QUESTIONS

5-39. In Figure 5.41 why is a septum, not just a plain cap, used on the top of the gas collection tube?

5-40. An evolved gas is directed from the reaction vessel to a container of water or other aqueous solution, wherein it dissolves (reacts). For example, a dilute solution of hydrogen peroxide is suitable for volatile thiols and sulfides. What solution would be appropriate to trap acidic gases (such as HCl)? (*Hint:* Consult a qualitative analysis text.)

5-41. One way to eliminate emissions is to place moistened glass wool in a drying tube, which is then attached to the reaction apparatus. What precautions must be taken when using this method?

5-42. In the collection of water-insoluble gases with the apparatus shown in Figure 5.41, describe how one might measure the rate at which a gas is evolved from a reaction mixture.

NOTE. The following experiments use Technique 7: Experiments [9], [10], [14], [A2ₐ], and [B2].

[26]Horodniak, J. W.; Indicator, N. *J. Chem. Educ.* **1970**, *47*, 568.

Measurement of Specific Rotation

Solutions of optically active substances, when placed in the path of a beam of polarized light, may rotate the plane of the polarized light. Enantiomers (two molecules that are nonidentical mirror images) have identical physical properties (melting points, boiling points, infrared and nuclear magnetic resonance spectra, etc.) except for their interaction with plane polarized light, their *optical activity*. Optical rotation data can provide important information concerning the absolute configuration and the enantiomeric purity of a sample.

Optical rotation is measured using a *polarimeter*. This technique is applicable to a wide range of analytical problems, from purity control to the analysis of natural and synthetic compounds. The results obtained from measuring the observed angle of rotation α are generally expressed as the *specific rotation* [α].

Figure 5.45   Oscillation of the electric field of ordinary light occurs in all possible planes perpendicular to the direction of propagation. (Reproduced with permission of John Wiley and Sons, Inc., from Solomons, T. W.; Fryhle, C. B.; Snyder, S. A. *Organic Chemistry*, 11th ed., Copyright 2014.)

Theory

Ordinary light behaves as though it were composed of electromagnetic waves in which the oscillating electric field vectors are distributed among the infinite number of possible orientations around the direction of propagation (see Fig. 5.45).

NOTE. A beam of light behaves as though it is composed of two, mutually perpendicular, oscillating fields: an electric field and a magnetic field. The oscillating magnetic field is not considered in the following discussion.

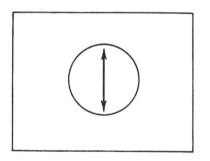

Figure 5.46   The plane of oscillation of the electric field of plane-polarized light. In this example the plane of polarization is vertical.

The planes in which the electrical fields oscillate are perpendicular to the direction of propagation of the light beam. If one separates one particular plane of oscillation from all other planes by passing the beam of light through a polarizer, the resulting radiation is plane-polarized (Fig. 5.46). In the interaction of light with matter, this plane-polarized radiation is represented as the vector sum of two circularly polarized waves. The electric vector of one of the waves moves in a clockwise direction; the other moves in a counterclockwise direction. Both waves have the same amplitude (Fig. 5.47). These two components add vectorially to produce plane-polarized light.

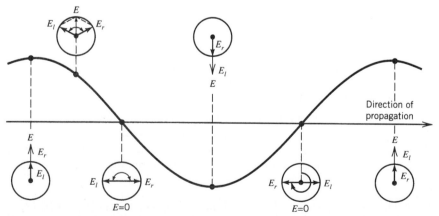

Figure 5.47   A beam of plane-polarized light viewed from the side (sine wave) and along the direction of propagation at specific times (circles) where the resultant vector E and the circularly polarized components E_l and E_r are shown. (Reproduced with permission of John Wiley and Sons, Inc., from Douglas, B.; McDaniel, D. H.; Alexander, J. J. *Concepts and Models of Inorganic Chemistry*, 3rd ed., Copyright 1994.)

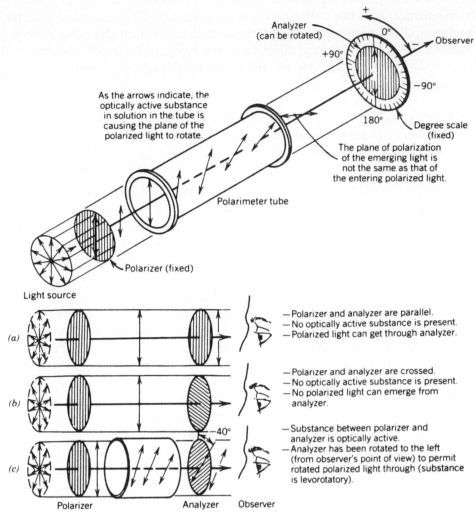

As the arrows indicate, the optically active substance in solution in the tube is causing the plane of the polarized light to rotate.

Analyzer (can be rotated)

Observer

Degree scale (fixed)

The plane of polarization of the emerging light is not the same as that of the entering polarized light.

Polarimeter tube

Polarizer (fixed)

Light source

(a) —Polarizer and analyzer are parallel.
—No optically active substance is present.
—Polarized light can get through analyzer.

(b) —Polarizer and analyzer are crossed.
—No optically active substance is present.
—No polarized light can emerge from analyzer.

(c) —Substance between polarizer and analyzer is optically active.
—Analyzer has been rotated to the left (from observer's point of view) to permit rotated polarized light through (substance is levorotatory).

Polarizer   Analyzer   Observer

Figure 5.48   Operation of a polarimeter. (Reproduced with permission of John Wiley and Sons, Inc., from Holum, J. R. *Organic Chemistry, A Brief Course.* Copyright 1975.)

If the passage of plane-polarized light through a material reduces the velocity of one of the circularly polarized components more than the other by interaction with bonding and nonbonding electrons, the transmitted beam of radiation has its plane of polarization rotated from its *original* position (Figs. 5.48 and 5.49). A polarimeter is used to measure this angle of rotation.

The Polarimeter

The polarimeter measures the amount of rotation caused by an optically active compound (in solution) placed in the beam of the plane polarized light. The principal parts of the instrument are diagrammed in Figure 5.48. Two Nicol prisms are used in the instrument. The first prism, which polarizes the original light source, is called the polarizer. The second prism, called the analyzer, is used to examine the polarized light after it passes through the solution being analyzed.

When the axes of the analyzer and polarizer prisms are parallel (0°) and no optically active substance is present, the maximum amount of light is transmitted. If the axes of the analyzer and polarizer are at right angles to

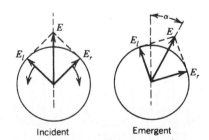

Incident   Emergent

Figure 5.49   Plane-polarized light before entering and after emerging from an optically active substance. (Reproduced with permission of John Wiley and Sons, Inc., from Douglas, B.; McDaniel, D. H.; Alexander, J. J. *Concepts and Models of Inorganic Chemistry*, 3rd ed., Copyright 1994.)

each other (90°), no transmission of light is observed. Placing an optically active solution into the path of the plane-polarized light causes one of the circularly polarized components to be slowed more than the other. The refractive indices are, therefore, different in the two circularly polarized beams. Figure 5.48 represents a case in which the left-hand component has been affected the most.

NOTE. In the simplified drawing, Figure 5.48, the effect on only one of the circularly polarized waves is diagrammed. See Figure 5.49 for a more accurate description (view from behind the figure).

This tilts the plane of polarization. The analyzer prism must be rotated to the left to maximize the transmission of light. If rotation is counterclockwise, the angle of rotation is defined as (−) and the enantiomer that caused the effect is called levorotatory (*l*). Conversely, clockwise rotation is defined as (+), and the enantiomer is dextrorotatory (*d*). Tilting the plane of polarization is called *optical activity*. Note that if a solution of equal amounts of a *d* and an *l* enantiomeric pair is placed in the beam of the polarimeter, no rotation is observed. Such a solution is *racemic*; it is an equimolar mixture of enantiomers.

The magnitude of optical rotation depends on several factors: (1) the nature of the substance, (2) the path length through which the light passes, (3) the wavelength of light used as a source, (4) the temperature, (5) the concentration of the solution used to make the measurement of optical activity, and (6) the solvent used in making the measurement.

The results obtained from the measurement of the observed angle of rotation, α_{obs}, are generally expressed in terms of *specific rotation* [α]. The sign and magnitude of [α] are dependent on the specific molecule and are determined by complex features of molecular structure and conformation; they cannot be easily explained or predicted. The specific rotation is a physical constant characteristic of a substance. The relationship of [α] to α_{obs} is as follows:

$$[\alpha]_{\lambda}^{T} = \frac{\alpha_{obs}}{lc}$$

where

> T = temperature of the sample in degrees Celsius (°C),
> l = the length of the polarimeter cell in decimeters (1 dm = 0.1 m = 10 cm),
> c = concentration of the sample in grams per milliliter (g/mL),
> λ = the wavelength of light in nanometers (nm) used in the polarimeter.

These units are traditional, though most are esoteric by contemporary standards. The specific rotation for a given compound depends on both the concentration and the solvent, and thus both the solvent and concentration used must be specified. For example, $[\alpha]_{D}^{25}$ (c = 0.4, $CHCl_3$) = 12.3° implies that the measurement was recorded in a $CHCl_3$ solution of 0.4 g/mL at 25 °C using the sodium D line (589 nm) as the light source.

For increased sensitivity, many simple polarimeters have an optical device that divides the viewed field into three adjacent parts (triple-shadow polarimeter; Fig. 5.50). A very slight rotation of the analyzer will cause one portion to become dimmer and the other lighter (Fig. 5.50*a* and 5.50*c*). The angle of rotation reading (α) is recorded when the field sections all have the same intensity. An accuracy of ± 0.1° can be obtained.

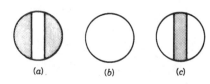

(a)   (b)   (c)

Figure 5.50   View through the eyepiece of the polarimeter. The analyzer should be set so that the intensity in all parts of the field is the same (b). When the analyzer is displaced to one side or the other, the field will appear as in (a) or (c).

Inaccurate Measurements. Several conditions may lead to inaccurate measurements, including trapped air bubbles in the cell, and solid particles suspended in the solution. Filter the solution, if necessary.

High-Performance Polarimeters and Optical Rotary Dispersion. For details of these two related topics refer to online discussion, Technique 8.

Applications to Structure Determination in Natural Products. Natural products provide interesting opportunities for measuring optical activity. An excellent example is the lichen metabolite, usnic acid, which can be easily isolated from its native source, "Old Man's Beard" lichens, as golden crystals (see Experiment [11A]).

Usnic acid

Usnic acid contains a single stereocenter (stereogenic center, or chiral center) and thus has the possibility of existing as an enantiomeric pair of stereoisomers. Generally, in a given lichen, only one of the stereoisomers (*R* or *S*) is present. Usnic acid has a very high specific rotation (~ ±460°) which makes it an ideal candidate for optical rotation measurements at the microscale level.

QUESTIONS

5-43. A solution of 250 mg of optically active 2-butanol in 13 mL of water shows an optical rotation of -0.54°. What is the specific rotation of this molecule?

5-44. Draw the structure of usnic acid and locate its stereocenter.

5-45. After drawing all stereoisomers of 3-amino-2-pentanol, identify the enantiomeric pairs.

5-46. If a solution of an equimolar mixture of an enantiomeric pair is placed in the beam path of the polarimeter, what would you observe?

5-47. The specific rotation of (+)Q is +13.20°. At identical concentration, solvent, pathlength, and light wavelength, the observed rotation of a solution containing both enantiomers of Q is 7.80°. What are the relative concentrations of each enantiomer in the solution?

NOTE. The following experiment uses Technique 8: Experiment [11A].

Sublimation

Sublimation is especially suitable for purifying solids at the microscale level. It is useful when the impurities present in the sample are nonvolatile under the conditions used. Sublimation is a relatively straightforward method; the impure solid need only be heated.

Sublimation has additional advantages: (1) It can be the technique of choice for purifying heat-sensitive materials—under high vacuum it can be effective at low temperatures; (2) solvents are not involved and, indeed, final traces of solvents are effectively removed; (3) impurities most likely to be separated are those with lower vapor pressures than the desired substance and often, therefore, lower solubilities, exactly those materials very likely to be contaminants in a recrystallization; (4) solvated materials tend to desolvate during the process; and (5) in the specific case of water of solvation, it is very effective even with substances that are deliquescent. The main disadvantage of the technique is that it can be less selective than recrystallization when the vapor pressure of the desired material being sublimed is similar to that of an impurity.

Some materials sublime at atmospheric pressure (CO_2, or dry ice, is a well-known example), but most sublime when heated below their melting points under reduced pressure. The lower the pressure, the lower the sublimation temperature. Substances that do not have strong intermolecular attractive forces are excellent candidates for purification by sublimation. Naphthalene, ferrocene, and *p*-dichlorobenzene are examples of compounds that are readily sublimed.

Sublimation Theory

Sublimation and distillation are closely related. Crystals of solid substances that sublime, when placed in an evacuated container, will gradually generate molecules in the vapor phase by the process of evaporation (i.e., the solid has a vapor pressure). Occasionally, one of the vapor molecules will strike the crystal surface or the walls of the container and be held by attractive forces. This process, condensation, is the reverse of evaporation.

Sublimation is the complete process of evaporation from the solid phase to condensation from the gas phase to directly form crystals without passing through the liquid phase.

A typical single-component phase diagram is shown in Figure 5.51, which relates the solid, liquid, and vapor phases of a substance to temperature and pressure. Where two of the areas (solid, liquid, or vapor) touch, there is a line, and along each line the two phases exist in *equilibrium*. Line *BO* is the sublimation–vapor pressure curve of the substance in question; only along line *BO* can solid and vapor exist together in equilibrium. At temperatures and pressures along the *BO* curve, the liquid state is thermodynamically unstable. Where the three lines representing pairs of phases intersect, all three phases exist together in equilibrium. This point is called the *triple point*.

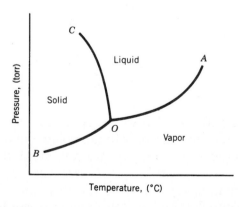

Figure 5.51   Single-component phase diagram.

Many solid substances have a sufficiently high vapor pressure near their melting point to be sublimed easily under reduced pressure in the laboratory. Sublimation occurs when the vapor pressure of the solid equals the pressure of the sample's environment.

Experimental Setup

Heating the sample with a microburner or a sand bath to just below the melting point of the solid causes sublimation to occur. Vapors condense on the cold-finger surface, whereas any less volatile residue will remain at the bottom of the flask. Apparatus for sublimation of small quantities are now commercially available (Fig. 5.52). Two examples of simple, inexpensive apparatus suitable for sublimation of small quantities of material in the microscale organic laboratory are shown in Figure 5.53.

An example of the purification of a natural product, where the sublimation technique at the microscale level is effective, is the case of the alkaloid caffeine. This substance can be isolated by extraction from tea (see Experiment [11B]).

Figure 5.52   Vacuum sublimator. (*Courtesy of Ace Glass, Inc., Vineland, NJ.*)

Precautions

Several precautions should be observed when performing a sublimation:

1. If you use the first setup in Figure 5.53, make sure you attach the hose connections to the cold finger in the proper manner. *The incoming cold water line is attached to the center tube.*

2. If you generate a vacuum using a water aspirator, make sure you place a water trap in the line. *Apply the vacuum to the system before you turn on the cooling water to the condenser.* This will keep moisture in the air in the flask from being condensed on the cold finger. Let the cold finger warm up before releasing the vacuum.

3. After the sublimation is complete, release the vacuum *slowly* so as not to disturb the sublimed material.

4. When using either of the arrangements in Figure 5.53, be careful to avoid loss of purified product as you remove the cold finger from the assembly.

5. The distance between the tip of the cold finger and the bottom of the sublimator should be less than 1 cm in most cases.

Figure 5.53   Various sublimation apparatus.

QUESTIONS

5-48. List the advantages and disadvantages of sublimation as a purification technique.

5-49. For a solid compound to evaporate at atmospheric pressure it must have an unusually high vapor pressure. What features, functional and/or molecular structural, contribute to this vapor pressure?

5-50. Why apply the vacuum to the sublimation system *before* you turn on the cooling water to the water condenser?

5-51. Why place a water trap in the vacuum line when using an aspirator to obtain the vacuum?

5-52. Why is sublimation particularly useful for purifying deliquescent compounds?

5-53. The 68% recovery after performing a sublimation translates to 53 mg of material. Calculate the amount of crude sample prior to performing the sublimation.

NOTE. The following experiments use Technique 9: Experiments [11B], [25A], and [25B].

MICROSCALE ORGANIC LABORATORY EXPERIMENTS

This chapter contains the experimental details of a collection of famous organic reactions that are at the heart of this field of chemistry. One of the great triumphs of the human intellect has been our ability to rationalize the physical transformations of organic materials. The experimental laboratory is the source of information out of which predictive organic theory has been fashioned. The microscale organic laboratory is designed to give you the opportunity to experience, first-hand, how organic reactions occur. This program will allow you to see how experimental data are directly related to the development of the structural and mechanistic theory surrounding these transformations. If successful, the microscale organic laboratory should bring the lecture portion of your course to life. What you have been studying in two dimensions in lecture now becomes alive in three dimensions in the laboratory.

The 39 experiments that make up this chapter focus largely on some of the most important of the fundamental organic reactions that have been discovered over the last two centuries. Because the application of these reactions to synthesis has been extensive, the microscale laboratory program offers a broad and practical introduction to organic chemistry. A significant number of the experiments include optional scale-up procedures to provide laboratory experience at the semi-microscale level, an inquiry-driven format allowing the student to monitor the reaction prior to workup, validation opportunities to confirm product purity by TLC analysis, and microwave procedures if alternative formats of experimental work are desired. Studying one or two of these modified formats at the beginning of the second semester can be helpful, particularly if some of the multistep syntheses covered in Chapter 7 and 7 Appendix are planned for study later in that semester. These latter sequences make extensive use of semi-microscale experimentation.

The microscale experiments are designed to enhance your ability to master the miniaturized experimental techniques used. A *Prior Reading* section highlights which techniques are to be used in a particular experiment and outlines the pages to refer to in the Techniques section (see Chapter 5). At the beginning of the semester it is important to *review these sections before proceeding* with the assigned experiment (heed the advice in Chapter 1). After examining the written material, it would be particularly helpful if you could see the video we have prepared that is accessible online. It covers the techniques in detail and can be viewed either in your spare time or in a prelab lecture period. As the basic techniques are used repeatedly, you will rapidly gain a firm grasp of the manipulations required and become comfortable working at this scale.

?	highlights an inquiry-driven format
V	highlights a validation opportunity
M	highlights a micro-wave procedure
G	highlights a green adaptation
B	highlights a process of biological significance

Chapter 6: **C₆H₆, Prismane**
Katz and Acton (1973).

You should also take advantage of the large number of graphics in the text margin that detail how the experimental apparatus are assembled. There are prompt signs (➡) in the text near the point where the equipment is to be used.

The important role of spectroscopic techniques in the modern organic laboratory is emphasized in the microscale organic laboratory. This information is set aside as a separate chapter, along with additional valuable details on the website, which are reference materials to be used while undertaking the experimental sections in Chapters 6–7 and 7 Appendix. Numerous cases of detailed analyses of the spectra are found in the *Purification and Characterization* sections of the experiments. Because infrared spectroscopy continues to play a major role in the characterization of reaction products in the introductory laboratory, and because this technique is currently given only a cursory treatment in most lecture texts, we have included a fairly detailed qualitative introduction to the theory of the effect and the instrumentation used in obtaining these observations, principally in Chapter 8. *While the IR part of the spectroscopic section (including the online material) may cover more ground than is normally found in many introductory laboratories, we feel it is important to overcome the black-box attitude that students can rapidly develop toward complex chemical instrumentation when they are turned loose on these powerful instruments with very limited knowledge.* Thus, we hope to be able to accomplish this transformation of attitude by offering students the essential details that will allow them to gain a command of the logic and the mechanics of obtaining and interpreting infrared spectral data. As mentioned earlier, *we have also included, in a number of the experiments, illustrative examples of the detailed spectral analysis* that can be used to examine starting materials, follow the progress of the reaction, and finally to assess the character and purity of the products.

A discussion section precedes each experiment. In a number of cases, especially when named reactions are involved, a brief biographical sketch of the individual so honored is included. At this point we also introduce pertinent information concerning the reaction mechanism, often in considerable detail. When appropriate, the relevance of a reaction to the life sciences and the chemical industry is explored.

Note that *Safety* and *Warning* indicators are highlighted or boxed in the experiments. We urge you to *always* adhere strictly to the safety precautions listed.

The nomenclature of organic compounds is often confusing to the beginning student and even occasionally to the experienced research chemist! To ease your introduction to the name game, the common name (sometimes referred to as the *trivial name*) of the compound to be synthesized is also listed at the beginning of each experiment. In addition, the **Chemical Abstracts** (CA) number and CA index name are also given.

Good luck! Enjoy your adventure in transforming small quantities of a large number of organic materials.

EXPERIMENT 1

Getting to Know You: Measurement of Physical Properties

Purpose.   This experiment will acquaint you with the experimental techniques used to measure certain classic physical properties of organic substances. These properties include the boiling point, density, and refractive index for liquids, and the melting point for solids. The procedures are outlined in Chapter 4.

A further objective is to study sampling techniques for obtaining the spectral characteristics of organic materials. You will observe that absorption spectra of organic substances contain, without question, the most important collection of physical constants of a material available to the investigator. Spectral information can lead to the elucidation of molecular structure and the rapid identification of unknown substances.

You will also learn how to locate the literature values for these measured properties using various chemical handbooks and online resources.

Prior Reading

> *Chapter 4:* Determination of Physical Properties
> > Ultramicro Boiling Point (pp. 48–50)
> > Density (p. 52)
> > Refractive Index (p. 57)
> > Melting Point (pp. 53–56)
> > > Capillary Method (p. 53)
> > > Evacuated Technique (pp. 54–56)
> > > Mixture Melting Points (p. 56)
>
> *Chapter 5, Technique 6:* Thin-Layer Chromatography (pp. 101–103)

DISCUSSION

Organic compounds have a number of physical properties that allow their precise characterization. These include the classical physical constants: boiling point, density, and refractive index for liquids and melting point for solids. The rapid development of modern chemical instrumentation, however, has also made easily accessible many of the spectral properties of these materials. Spectroscopy, in particular, provides information that is extremely powerful for establishing the structure of unknown molecular systems and for rapidly identifying known materials.

Not only are physical properties used to characterize a specific organic compound, but they are also often used to compare one compound to another. Examples of this approach are illustrated in Chapter 9, Qualitative Identification of Organic Compounds. The route to identification of an unknown organic species has become increasingly dependent on the measurement of the physical properties of the pure substance.

The various classical physical constants of a large number of organic substances appear in the *CRC Handbook of Chemistry and Physics, Lange's Handbook, Aldrich Catalog Handbook of Fine Chemicals*, and several online resources. The *Aldrich Catalog* also contains references to published infrared (IR) and nuclear magnetic resonance (NMR) spectral data for many of the compounds. Complementing these databases are several online resources, such as SciFinder Scholar, which allow for users to directly and efficiently access published spectral data. The *CRC Atlas of Spectral Data and Physical Constants for Organic Substances* contains IR and ultraviolet (UV) peak positions. Collections of spectra may be located in the Aldrich Libraries of both IR and NMR spectra, and in the Sadtler Library.

A detailed discussion of the spectral properties of organic compounds is given in Chapter 8.

EXPERIMENTAL PROCEDURE

NOTE. Your instructor will select which of the physical properties you are to measure. The length of your laboratory period, the size of your section, and the number of instruments available will all play a role in determining how many of these properties will be suggested.

Melting Point

1. Using the melting-point apparatus *(your instructor will provide you with the experimental details concerning the operation of the particular instrument to be used in your laboratory)*, determine the melting point of acetanilide and compare your result with the literature values (Lit. values) reported in the CRC and Aldrich references (see pp. 53–56 for the experimental details on how to proceed with this measurement).

Determined value_____

CRC: Lit. value_____; Ed_____; Page_____;
Compound No_____
Aldrich: Lit. value_____;Year_____; Page_____;
Compound No_____

2. Now determine an *evacuated* melting point (see pp. 54–56 for the experimental details on how to proceed with this measurement) of caffeine and compare it to the values in the CRC and Aldrich references. *Your instructor will provide you with the experimental details concerning the operation of the particular instrument to be used in your laboratory.*

First, determine the melting point of caffeine in an unevacuated melting-point tube. After observing the melting point, allow the temperature to drop below the melting point and observe whether the sample crystallizes again. If crystallization occurs, observe a second melting point, and then repeat this procedure a third time. Follow the same routine with an evacuated sample, and compare the results of the two sets of melting points. Do you observe any differences between these data sets?

2nd Determined value

Unevacuated	**Evacuated**
1st Determined value _____	1st Determined value _____
2nd Determined value _____	2nd Determined value _____
3rd Determined value _____	3rd Determined value _____

CRC: Lit. value _____; Ed _____; Page _____;
Compound No. _____

Aldrich: Lit. value _____;Year _____; Page _____;
Compound No. _____

3. Why is a second seal required on the evacuated melting-point tube?

4. When are evacuated melting points necessary?

5. Next consider the technique of *mixture melting points* (see p. 54 for a discussion of the details of this procedure). This technique is the classical approach for establishing a positive identification of a substance when pure reference standards are available in the laboratory. For two examples where this type of measurement is applied in the microscale organic laboratory, see Experiments [6] and [34A].

6. In the present experiment, you have obtained the melting points of acetanilide and caffeine. Using these values as reference standards observe the melting points of two mixtures: (1) caffeine 75%–acetanilide 25% and (2) caffeine 25%–acetanilide 75%. Should these observations be made in evacuated capillaries? *(Your instructor will provide you with the experimental details concerning the operation of the particular instrument to be used in your laboratory.)*

(1) caffeine 75%–acetanilide 25%   (2) caffeine 25%–acetanilide 75%

Determined value _____   Determined value _____

7. (Optional) Use the following compounds and mixtures: caffeine, acetanilide, caffeine 75%–acetanilide 25%, and caffeine 25%–acetanilide 75%; your instructor will provide no less than one duplicate set of vials. For sections that consist of an odd number of students, triplication of one set will work. While the vials will be labeled, the labels will not offer any direction as to the contents nor its match. Once the vials are distributed, your goal is to determine the melting point of your "unknown" and find your match. Caution should be exercised, since depending on how the vials are "duplicated," there may be more than one match! That is, when considering a laboratory section consisting of 18 students, the 18 vials may consist of 10 vials of caffeine and 8 vials of acetanilide. This could as well be presented as 4 vials of caffeine, 4 vials of acetanilide, 4 vials of caffeine 75%–acetanilide 25%, and 6 vials of caffeine 25%–acetanilide 75%. As you can surmise, an even more challenging exercise can be crafted upon the creative mixing of the pairings above. Success will come to those who correctly determine the melting point of their unknown, and thus multiple melting point determinations may be needed prior to seeking your match.[1]

8. Is it possible to detect the presence of impurities by melting-point measurements? Why?

9. Would it be possible to establish the composition of an unknown binary mixture of two substances from mixture melting-point data? Explain.

Ultramicro Boiling Point

1. Make several (5) glass bells (p. 48). Your instructor will demonstrate the procedure, or you can view the procedure on video, if available. Place the bells in a small glass vial and store them in your micro kit.

2. Use the technique for determining (bp) on a *melting-point apparatus hot stage* (very likely this will be the same apparatus you used in melting-point section of the experiment) as discussed in Chapter 4 (p. 49), and

---

[1]This exercise is based upon discussions led by Jerry R. Mohrig and Christina Noring Hammond during an NSF-sponsored CWCS Workshop on Teaching Guided-Inquiry Organic Chemistry Laboratories (Irvine, CA; July **2005**).

complete the following table *(your instructor will provide you with the experimental details concerning the operation of the particular instrument to be used in your laboratory).*

	Propane	Butane	Pentane	Hexane	Heptane
bp (°C)	−p (°	−p (	36.1	67	?

3. Compare your measured value for heptane to those listed in the *CRC Handbook, Aldrich Catalog,* and online.

Determined value _____; Barometric pressure _____ torr

Corrected value _____

CRC: Lit. value _____; Ed. _____; Page _____;
Compound No. _____

Aldrich: Lit. value _____; Year _____; Page _____;
Compound No. _____

4. Prepare and attach a graph of molecular weight (MW) vs. boiling point (bp) using the above data.

5. Explain the trend in boiling point as the molecular weight increases.

Density

1. Use the syringe method of measuring the density outlined in Chapter 4 (p. 51), and 0.5 mL of liquid, to complete the following table and compare your value to those listed in the CRC and Aldrich references. *(Your instructor will provide you with the experimental details concerning the operation of the particular equipment to be used in this experiment. Pay close attention to the details concerning the operation of the balances.)*

Methylene Chloride (CH_2Cl_2)   **Octane C_8H_{18}**

Density (g/mL) 1.33   ?

Determined value _____   _____

CRC: Lit. value _____; Ed. _____; Page _____;
Compound No. _____

Aldrich: Lit. value _____; Year _____; Page _____;
Compound No. _____

2. Methylene chloride is often used as a solvent in the laboratory to extract an *organic species* from an aqueous solution. Will the CH_2Cl_2 normally form the top layer, or the bottom?

3. If octane were used to extract an aqueous phase, would it form the top layer, or the bottom? How did you arrive at this answer?

Refractive Index

1. Using the lens paper disk technique outlined online, determine the refractive index of methanol and compare the value you have obtained to that found in the CRC and Aldrich data collections. *(Your instructor will provide you* *with the experimental details concerning the operation of the particular instrument used in your laboratory.)*

Determined value _____

CRC: Lit. value _____; Ed. _____; Page _____;

Compound No. _____

Aldrich: Lit. value _____; Year _____; Page _____;

Compound No. _____

2. Correct your measured value to 20 °C.

Temperature of your measurement _____.

Temperature of literature value _____.

Your measured value corrected to 20 °C _____.

Calculations:

3. Based on your observations, how does refractive index vary with temperature?

Thin-Layer Chromatography

1. Using the thin-layer chromatography technique outlined in Chapter 5 (Technique 6A, pp. 101–103), determine the R_f values for benzaldehyde, 4-chlorobenzaldehyde, and cyclohexanol using as a solvent system ethyl acetate and hexane (1:4). While a UV lamp will be sufficient for the visualization of benzaldehyde and 4-chlorobenzaldehyde when working with TLC plates with a UV-activated fluorescent indicator mixed in with the silica gel, a staining system of *p*-anisaldehyde will work when wanting to visualize cyclohexanol. *(Your instructor will provide you with the experimental details concerning the operation of staining a TLC plate using* p-anisaldehyde. *Please use for preparation of* p-anisaldehyde *solution: 135 mL EtOH, 5 mL concentrated* H_2SO_4, *1.5 mL glacial acetic acid, and 3.7 mL* p-anisaldehyde.)

2. After recording the R_f values using as a solvent system ethyl acetate and hexane (1:4), determine what solvent system is needed to obtain values between 0.3 and 0.4 for each of the following systems: benzaldehyde, 4-chlorobenzaldehyde, and cyclohexanol.

3. Using as a solvent system ethyl acetate and hexane (1:4), predict which compound from the following set would have a higher R_f value: (a) cyclohexanol and cyclohexane, (b) benzoic acid and benzaldehyde, and (c) caffeine and naphthalene.

4. (Optional) Three stock solutions, labeled solution A, solution B, and solution C, will be provided by your instructor. They will be one of the following systems: benzaldehyde, 4-chlorobenzaldehyde, or cyclohexanol. Select one (or more) and determine its identity using the thin-layer chromatography technique having already established R_f values in several solvent systems.

Infrared Spectroscopy Sampling Procedures

Sampling of Liquids

1. Use the technique for obtaining an IR spectrum of a thin-liquid film as described on page 573 *(your instructor will provide you with the experimental* *details concerning the operation of the particular instrument to be used in your laboratory and the type of windows on which your sample will be mounted).*

2. Obtain spectra of *n*-octane and 1-octanol. Compare your data with the Aldrich or Sadtler reference collections, or those obtained online using for example SciFinder Scholar, and compare the two spectra to each other.

3. What differences do you observe between the two spectra? Can you associate differences in molecular structure to the differences in the spectra (see Infrared Discussions in Chapter 8).

4. Occasionally, when an IR spectrum is obtained, some of the very strong bands will appear with flattened peaks, as if they were totally absorbing the energy at those wavenumber values. The flat bottom of the band, however, does not correspond to 0% transmission on the scale, but will indicate an energy transmission of 5–10% or even higher. Can you explain the odd shape of the band? Is the energy being totally absorbed? Explain your answers.

Sampling of Solids

1. Use the technique for obtaining an IR spectrum of a solid melting above 100 °C as described in Chapter 8 (pp. 629–630) *(your instructor will provide you with the experimental details concerning the operation of the particular instrument to be used in your laboratory and the type of KBr dye in which your sample will be pressed if an ATR FTIR is unavailable).*

2. Obtain the IR spectrum of caffeine. Compare your data with an authentic spectrum of caffeine and with the data given in Experiment [11B]. Your sample may be saved by taping it to a file card with your name, and stored by your instructor in a desiccator. If, later in the year, you isolate caffeine from its natural source (Experiment [11B]), you will be able to compare the material you have extracted and purified from the plant with your own authentic reference spectrum.

3. Occasionally, the spectral region from 4000 to 2000 cm^{-1} in solid samples is steeply sloping to lower transmission values at higher wavenumber values. Is this drop in transmission an absorption of the radiation? If so, to what process can the absorption be ascribed, and if not, what is the cause of the decreased transmission?

4. Compare the spectra of caffeine obtained in your laboratory section. Are the spectra all identical? Where do they differ? To what effect can you ascribe the differences, if there are any? (This is a good open question for the entire lab section.)

Nuclear Magnetic Resonance Spectroscopy Sampling Procedures: For NMR sampling procedures and examples, see Chapter 8 and Experiments [5B], [22A], and [28].

QUESTIONS

6-1. An unknown fatty acid as the methyl ester has a boiling point of 200 °C at 1 mmHg. Using the pressure–temperature nomograph on page 49 or searching for an interactive tool online, determine its boiling point at 40 mmHg

6-2. Discuss with those in your laboratory section the consequences of incorrect sample loading, variable sample size, and rates of heating and how these factors might lead you to obtain an incorrect value of the melting point for your solid sample.

6-3   The mass of a certain volume of an unknown liquid at 25 °C is 234 mg. The mass of an equal volume of water at the same temperature is 201 mg. Calculate the density of the unknown liquid at 25 °C.

6-4   The melting point of pure *trans*-cinnamic acid is 133 °C, and that of 2-acetoxybenzoic acid (aspirin) is 135 °C. Given pure reference standards of both acids, describe a melting-point procedure by which you could identify whether an unknown sample melting in this range could be assigned to either structure, or to neither one.

6-5   Why is filter paper a poor choice of surface on which to powder or crush a solid crystalline sample before placing it in a capillary melting-point tube?

6-6   Why is the ultramicro boiling point determined precisely at the point when the last vapor bubble has escaped and the liquid phase begins to rise in the bell cavity?

BIBLIOGRAPHY

For further information on basic laboratory techniques:

Sharp, J. T.; Gosney, I.; Rowley, A. G. *Practical Organic Chemistry: A Student Handbook of Techniques*, Chapman Hall: London, 1989.

Shriner, R. L.; Hermann, C. K. F.; Morrill, T. C.; Curtin, D.Y.; Fuson, R. C. *The Systematic Identification of Organic Compounds*, 8th ed.; Wiley: New York, 2003.

Vogel, A. I. *Vogel's Textbook of Practical Organic Chemistry*, 5th ed.; Furnis, B. S.; et al., Eds.; Wiley: New York, 1989.

Zubrick, J. W. *The Organic Chem Lab Survival Manual*, 11th ed.; Wiley: New York, 2019.

The Separation of a 25-µL Mixture of Heptanal (bp 153 °C) and Cyclohexanol (bp 160 °C) by Gas Chromatography

EXPERIMENT 2

Purpose.   This experiment illustrates the separation of a 25-µL mixture, consisting of heptanal and cyclohexanol, into the pure components. The volume of the mixture is approximately that of a single drop, and the materials boil within 7 °C of each other. This mixture would be difficult, if not impossible, to separate by the best distillation techniques available. The purity of the fractions collected from the gas chromatograph (GC) can be assessed by boiling points, refractive indexes, or infrared (IR) spectra.

Prior Reading

>   *Technique 1:*   Microscale Separation of Liquid Mixtures by Preparative Gas Chromatography (pp. 59–65)
>   *Chapter 4:*   Determination of Physical Properties
>>   Ultramicro Boiling Point (pp. 48–50)
>>   Refractive Index (Appendix)
>   *Experiment [1]:*   Measurement of Physical Properties
>>   Ultramicro Boiling Point (p. 123)
>   *Chapter 8:*   Introduction to Infrared Spectroscopy (pp. 615–637)

DISCUSSION

The efficacy of GC separations is highly dependent on the experimental conditions. For example, two sets of experimental data on the heptanal–cyclohexanol mixture are given below to demonstrate the effects of variations in oven temperature on retention times.

In Data Set A, the oven temperature was allowed to rise slowly from 160 to about 170 °C during a series of sample collections. The retention time of heptanal dropped from about 3 min to close to 2 min, whereas the retention time of cyclohexanol was reduced from about 5.5 min to nearly 4 min. The significant decrease in resolution over this series of collections is reflected in the number of theoretical plates calculated, which was over 300 for heptanal and about 500 for cyclohexanol in the first trial, but declined to below 200 for both compounds toward the last run (see Data Set A).

COLLECTION YIELD

Cyclohexanol

Density of cyclohexanol = 0.963 mg/μL.

In 25 μL of 1:1 cyclohexanol–heptanal, we have 12.5 μL of cyclohexanol.

Therefore, 125 μL × 0.963 mg/μL = 12 mg of cyclohexanol injected.

Percent recovered = (8.3 mg/12.0 mg) × 100 = 69% cyclohexanol collected.

Heptanal

Density of heptanal = 0.850 mg/μL.

Therefore, 12.5 μL × 0.85 mg/μL = 10.6 mg of heptanal injected.

Percent recovered = (8.1 mg/10.6 mg) × 100 = 76% heptanal.

In the Data Set B collections, stable oven temperatures and flow rates were maintained, and the data exhibit excellent reproducibility. Oven temperature was held at 155 °C throughout the sampling process. The retention time of heptanal was observed to be slightly longer than 3 min, with a variance of 6 s, whereas the cyclohexanol retention time was found to be slightly longer than 6 min, with a variance of 12 s. The resolution remained essentially constant

Data Set A								
	Heptanal				Cyclohexanol			
Trial No.	Retention Time (min)	Baseline Width (min)	Number of Theoretical Plates	Recovery (mg)	Retention Time (min)	Baseline Width (min)	Number of Theoretical Plates	Recovery (mg)
1	3.1	0.7	314	8.0	5.6	1.0	502	8.0
2	2.9	0.7	275	8.0	5.3	1.0	449	8.0
3	3.0	0.7	294	7.0	5.7	1.0	520	8.0
4	2.8	0.7	256	8.0	5.1	1.1	344	8.0
5	2.5	0.6	278	8.0	4.3	1.1	244	9.0
6	2.7	0.5	467	7.0	4.6	1.0	339	10.0
7	2.5	0.6	278	10.0	4.2	1.0	282	8.0
8	2.2	0.5	310	9.0	3.5	1.0	196	8.0
9	1.8	0.5	207	8.0	3.0	1.0	144	8.0
10	2.3	0.7	173	8.0	3.9	1.0	243	8.0
Av	2.5 ± 0.4	0.6 ± 0.09	285 ± 7	8.1 ± 0.8	4.5 ± 0.9	1.0 ± 0.05	326 ± 129	8.3 ± 0.7

	Heptanal				Cyclohexanol			
Trial No.	Retention Time (min)	Baseline Width (min)	Number of Theoretical Plates	Recovery (mg)	Retention Time (min)	Baseline Width (min)	Number of Theoretical Plates	Recovery (mg)
1	3.5	0.7	400	8.0	6.6	1.1	576	8.0
2	3.2	0.7	334	9.0	6.0	1.1	476	7.0
3	3.5	0.7	400	7.0	6.6	1.2	484	10.0
4	3.2	0.7	334	9.0	6.1	1.0	595	9.0
5	3.1	0.6	427	8.0	6.0	1.1	476	8.0
6	3.2	0.7	334	9.0	6.0	1.1	476	9.0
7	3.3	0.8	272	9.0	6.1	1.1	492	8.0
8	3.1	0.7	313	8.0	6.0	1.1	476	10.0
9	3.2	0.7	334	8.0	6.1	1.1	492	8.0
10	3.2	0.7	334	8.0	6.2	1.1	508	8.0
Av	3.2 ± 0.1	0.7 ± 0.05	348 ± 47	8.3 ± 0.7	6.2 ± 0.2	1.1 ± 0.05	505 ± 44	8.5 ± 1.0

Data Set B

throughout the series, and the number of theoretical plates calculated was about 350 for heptanal and about 500 for cyclohexanol (see Data Set B).

COLLECTION YIELD

Cyclohexanol

Density of cyclohexanol = 0.963 mg/µL.

In 25 µL of 1:1 cyclohexanol–heptanal, there are 12.5 µL of cyclohexanol.

Therefore, 125 µL × 0.963 mg/µL = 12 mg of cyclohexanol injected.

Percent recovered = (8.5 mg/12.0 mg) × 100 = 71% cyclohexanol collected.

Heptanal

Density of heptanal = 0.850 mg/µL.

Therefore, 12.5 µL × 0.85 mg/µL = 10.6 mg of heptanal injected.

Percent recovered = (8.3 mg/10.6 mg) × 100 = 78% heptanal.

The results just described demonstrate that the resolution of GC peaks may be very sensitive to changes in retention time resulting from instability in oven temperatures. Since the number of theoretical plates is related to resolution values, significant degradation in column plate values can occur with variations in oven temperatures. When you compare the time and effort required to obtain a two-plate fractional distillation on a 2-mL mixture (see Experiment [3B] and Technique 2) with the speed and ease used to obtain a 500 plate separation on 12.5 µL of cyclohexanol in this experiment, it is hard not to be impressed with the enormous power of this technique.

COMPONENTS

Cyclohexanol

Heptanal

EXPERIMENTAL PROCEDURE

Estimated time to complete the experiment: 2.0 h.

Physical Properties of Components					
Compound	MW	Amount	bp (°C)	Density (d)	n_D
Heptanal	114.19	12.5 µL	153	0.85	1.4113
Cyclohexanol	100.16	12.5 µL	160	0.96	1.4641

Reagents and Equipment. The procedure involves injecting a 25-µL mixture of heptanal–cyclohexanol 1:1 (v/v) into a $\frac{1}{4}$-*in.* × 8-ft stainless-steel column packed with 10% Carbowax 80/100 20M PAW-DMS. Experimental conditions (GOW-MAC series No. 350) are He flow rate, 50 mL/min; chart speed, 1 cm/min; oven temperature, 155 °C.

Procedure for Preparative Collection. The liquid effluents are collected in an uncooled, 4-mm-diameter collection tube (double reservoirs; overall tube length 40–50 mm, see Fig. 6.1).

The collection tube (oven dried until 5 min before use) is attached to the heated exit port by the 5/5 ꝸ joint. Sample collection is initiated 0.5 min prior to detection on the recorder of the expected peak (time based on previously determined retention values)[2] and continued until 0.5 min after return to baseline. After the collection tube is detached, the sample is transferred to the 0.1-mL conical GC collection vial. The transfer is facilitated by the 5/5 ꝸ joint on the conical vial. After the collection tube is joined to the vial (preweighed with stopper), the system is centrifuged (see Fig. 6.1). The collection tube is then removed and the vial is stopped and reweighed.

Characterization. Calculate the percent recovery. These amounts should range between 7 and 10 mg. Determine the boiling point of each fraction and obtain the refractive index or IR spectrum, if time permits. These latter measurements will require most, if not all, of the sample not used for boiling-point determination.

Assess the purity and efficiency of the separation from your tabulated data and the GC chromatogram.

Rubber septum

12-mL centrifuge tube

100-µL conical vial

Cotton packing

GC collection tube (heavy walled)

5/5 ꝸ and threaded

Product collects here

Figure 6.1   Gas chromatographic collection tube and 0.1-mL conical vial.

---

[2]Refer to your local laboratory instructions.

Alternative Mixture Pairs for Preparative Collection

(all mixtures are 1:1 v/v)

Mixture

a. *Separation* of a 40-μL mixture of[3]
(1S)-(–)-α-pinene (bp 156 °C, n_D = 1.4650, d = 0.855)
(1S)-(–)-β-pinene (bp 165 °C, n_D = 1.4782, d = 0.859)
Components

(1S)-(–)-α-Pinene

(1S)-(–)-β-Pinene

Chromatographic Parameters

A 40-μL injection
Flow rate: 50 mL/min
Column temperature: 120 °C
Column: 20% Carbowax
Elution time
 α-Pinene: ~ 8 min
 β-Pinene: ~ 12 min
Average recovery
 α-Pinene: 8.3 μL
 β-Pinene: 10.6 μL

Mixture

b. *Separation* of a 40-μL mixture of
2-Heptanone (bp 149–150 °C, n_D = 1.4085, d = 0.820)
Cyclohexanol (bp 160–161 °C, n_D = 1.4641, d = 0.963)
Components

H₃C—C(=O)—...—CH₃
2-Heptanone

Cyclohexanol

Chromatographic Parameters

A 40-μL injection
Flow rate: 50 mL/min
Column temperature: 145 °C
Column: 20% Carbowax
Elution time
 2-Heptanone: ~ 5.5 min
 Cyclohexanol: ~ 10.0 min

---

[3]Refractive index at D line of sodium = n_D and density = d.

Average recovery
2-Heptanone: 8.1 µL (41%)
Cyclohexanol: 11.4 µL (57%)

Mixture

c. *Separation* of a 40-µL mixture of
d-Limonene (bp 175–176 °C, $n_D = 1.4743$, $d = 0.8402$)
Cyclohexanol acetate (bp 173 °C, $n_D = 1.4401$, $d = 0.9698$)
Components

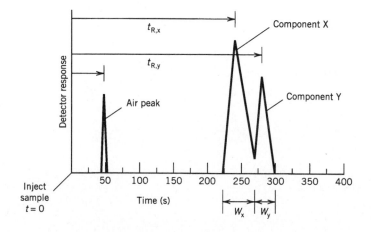

d-Limonene   Cyclohexyl acetate

Chromatographic Parameters

A 40-µL injection
Flow rate: 50 mL/min
Column temperature: 170 °C
Column: 20% Carbowax
Elution time
   d-Limonene: ~5.5 min
   Cyclohexyl acetate: ~ 7.5 min
Average recovery
   d-Limonene: 8.7 µL (44%)
   Cyclohexyl acetate: 10.0 µL (50%)

QUESTIONS

6-7. Based on the data presented in the Data Set A chromatographic separation, can you explain why there is such a steep decline in column efficiency with temperature change?

6-8. Consider the following gas chromatogram for a mixture of analytes X and Y:

    (a) Calculate the number of theoretical plates for the column in reference to the peaks of each component (X and Y).

    (b) If the column is 20 m long, calculate the height equivalent theoretical plate (HETP) (in plates per cm) for this column.

6-9. The number of theoretical plates a column has is important, but the crucial factor is the ability to separate two or more substances. That is, how well resolved are the peaks? The resolution of two peaks depends not only on how far apart they are (t_R), but also on the peak width (W). Baseline resolution (R) is defined by the following equation:

$$R = \frac{2\Delta t_R}{W_X + W_Y}$$

Because of the tailing of most species on the column, a value of 1.5 is required to give baseline resolution.

    (a) Calculate the resolution for the peaks in Question 6-8.

    (b) Do you think a quantitative separation of the mixture is possible based on your answer?

    (c) Has baseline resolution been achieved?

6-10. Discuss at least two techniques you might use to increase the resolution of the column in Question 6-9 (without changing the column).

6-11. Retention times for several organic compounds separated on a GC column are given below.

Compound	t_R (S)
Air	50
Pentane	165
Heptane	335
2-Pentene	250

    (a) Calculate the relative retention of 2-pentene with respect to pentane.

    (b) Calculate the relative retention of heptane with respect to pentane.

BIBLIOGRAPHY

Selected references on gas chromatography:

Grob, R. L.; Barry, E. F., Eds.; *Modern Practices of Gas Chromatography*; Wiley: New York, 2004.

Jennings, W.; Mittiefehidt, E.; Stremple, P., Eds.; *Analytical Gas Chromatography*, 2nd ed.; Academic Press: New York, 1997.

McNair, H. M.; Miller, J. M. *Basic Gas Chromatography*; Wiley: New York, 1997.

Sadek, P. C. *Illustrated Pocket Dictionary of Chromatography*; Wiley: New York, 2004.

EXPERIMENT
3

Distillation

In the following set of experiments, we will examine the applications of a variety of distillation techniques to the purification of liquid mixtures. In Experiments [3A] and [3B] you will conduct simple distillations. In Experiment [3A] a volatile liquid component is separated from a nonvolatile solid. Experiment [3B] illustrates the use of the Hickman still in the separation of hexane and toluene, which have boiling points 42 °C apart. The composition of the fractions is analyzed by refractive index and boiling point. Experiments [3C] and [3D] introduce the use of micro spinning-band distillation columns for the separation of cyclohexane (bp 80.7 °C) and 2-methylpentane (bp 60.3 °C). The composition of the distillate fractions are determined by gas chromatography. The number of theoretical plates is determined for the spinning-band column used. In Experiment [3D] you will be introduced to one of the simplest yet most efficient and powerful distillation techniques for the separation of liquid mixtures at the semi-microscale level, the Hickman–Hinkle still.

Experiment 3A

Simple Distillation at the Semi-microscale Level: Separation of Ethyl Acetate from *trans*-1,2-Dibenzoylethylene

Purpose. *Simple distillation* is examined using the distillation process to separate a liquid ester from minor components that are nonvolatile or that have boiling points much greater (>100 °C) than that of the major component.

Prior Reading

> *Techniques 2 and 3:* Distillation (pp. 65–71)
>
> Distillation Theory (p. 65)
>
> Simple Distillation at the Semi-microscale Level (pp. 65–68)
>
> *Chapter 4:* Determination of Physical Properties
>
> Ultramicro Boiling Point (pp. 48–50)
>
> Density (pp. 52–53)
>
> Refractive Index (p. 57)
>
> Evacuated Melting Point (pp. 54–56)

DISCUSSION

Semi-microscale simple distillation can be an effective separation technique with volumes from 0.5 to 2 mL. Apparatus have been developed that achieve effective separation of mixture samples in this range. One of the most significant of these designs is the classic Hickman still, shown in both Figures 5.5 and 6.2, because of its importance to the associated discussions. In this experiment you will effect the separation of a two-component mixture by the use of this still.

The Hickman still is used in several of the microscale experiments to purify solvents, carry out reactions, and concentrate solutions for recrystallization. An introduction to the use of the Hickman still is given in this experiment.

In a distillation where a liquid is separated from a nonvolatile solute, the vapor pressure of the liquid is lowered by the presence of the solute, but the vapor phase consists only of the pure liquid component. Thus, except for the transfer of nonvolatile material by incidental splashing, the material condensed in the collar of the Hickman still should consist only of the volatile component. In the present experiment this component is ethyl acetate. The temperature of the vial and contents being distilled will rise during the distillation process, since the concentration of the impurity is increasing as the volatile liquid is removed. This effect lowers the vapor pressure of the liquid. However, the boiling point of the liquid remains constant, since only the pure liquid component is being vaporized.

COMPONENTS

trans-1,2-Dibenzoylethylene   Ethyl acetate

EXPERIMENTAL PROCEDURE

Estimated time to complete the experiment: 2.0 h.

Physical Properties of Components						
Compound	MW	Amount	mp (°C)	bp (°C)	d	n_D
Ethyl acetate	88.12	1.0 mL		77	0.90	1.3723
trans-1,2-Dibenzoylethylene	236.27	50 mg	111			

Reagents and Equipment.   Transfer 1 mL of the yellow stock solution (*trans*-1,2-dibenzoylethylene/ethyl acetate, 50 mg/mL) to a 3-mL conical vial by automatic delivery pipet (remember to place the vial in a small beaker to prevent tipping during the transfer). Place a boiling stone (or a magnetic spin vane if desirable) into the vial and assemble the Hickman still head. The still assembly is mounted in a sand bath on a hot plate (see Fig. 6.2).

Experimental Conditions.   The temperature of the bath is raised to 90–100 °C at a rate of 5 °C/min.

CAUTION: Do not let the temperature of the still rise too rapidly.

Once boiling commences, the rate of heating should be lowered to the point where the temperature increases at 2–3 °C/min. A slow distillation rate is very important in establishing equilibrium between the vapor and liquid components in the mixture. Follow the course of the distillation by the rise of condensate on the sides of the Hickman column. When the condensate reaches the trap, adjust the bath temperature so that liquid is removed from the column slowly (~100 μL/3 min). A smooth, slow distillation will provide a

Clamp

Clamp

Clamp

Thermometer
90–100 °C

Crystallizing
dish

Sand

Thermometer

14/10 ᵀ Hickman
still head

O-ring

14/10 ᵀ and threaded
compression cap

3-mL
conical vial

Boiling stone

Heat
control

Hot plate/
Magnetic stirrer

Stir
control

trans-1,2-Dibenzoylethylene/ethyl acetate
(50 mg/mL), 1 mL

Figure 6.2   Hickman still (14/10 ᵀ with conical vial [3 mL]).

cleaner separation of the components, and will also avoid mechanical transfer of nonvolatile components via splattering to the condensate trap (if the condensate appears yellow, contamination has occurred).

Collect approximately 50–150 μL of the ester in the collar of the still (the first fraction collected is often referred to as the forerun; give the temperature range). As the distillation continues, remove the forerun with a Pasteur pipet having a slightly bent tip (bend the tip with a microburner—if you have a Hickman still with a side-arm collection port, the pipet tip will not need to be bent). Place the fraction in a clean, dry, 1-dram, screw-capped vial (use an aluminum foil liner to avoid cap contamination). Label the fraction with a marking pen. Collect a second fraction of ester (400–500 μL, which may require combining two, or even three, collections from the collar; give the temperature range), which should be clear and colorless. Remove and store as before. Discontinue the distillation. Allow the distilling flask to cool slowly by leaving it in the warm sand bath while measuring the physical properties of the distillate fractions.

Characterization.   Three physical properties of the ester will be measured to establish the identity and purity of the compound by comparison with known literature values.

Determine the refractive index of the two fractions collected. Compare the experimental values to those found in the literature for ethyl acetate. If the values are within 0.0010 unit of each other, the fractions can be considered to have the same constitution. Are the values for the two fractions the same? If not, which one deviates the most from the reference data? Attempt to explain the result.

Determine the density (see Chapter 4) of the ester, using material contained in the second fraction. This measurement is nondestructive and the material used may be recovered for use in further tests. Compare your results with those values found in the literature.

Determine the boiling point of the second fraction by the ultramicro boiling-point procedure (see Chapter 4). Compare your result with the literature value. Does this fraction appear to be pure ethyl acetate?

In the next step, disconnect and cool the 3-mL conical vial in an ice water bath for 10 min. *trans*-1,2-Dibenzoylethylene will crystallize from the concentrated solution. Remove the remaining solvent from the distillation vial with a Pasteur filter pipet and place the crystals on a porous clay plate to air-dry. The melting point of the crystalline material is obtained by the evacuated capillary method and compared with the literature value.

Reference values of the physical constants are available online and in the *CRC Handbook of Chemistry and Physics*. Submit a copy of the table prepared in your laboratory notebook to the instructor, after first tabulating the experimentally measured values of the physical properties, in addition to those reported in the literature for ethyl acetate (see *acetic acid, ethyl ester* if necessary).

Experiment 3B

Fractional Semi-microscale Distillation: Separation of Hexane and Toluene

Purpose.   This experiment effects the separation of a binary liquid mixture composed of liquids having boiling points that are relatively far apart, greater than 30 °C. It will help you develop the skills to operate a semi-microdistillation apparatus so that purifications required in later experiments can be successfully carried out.

Prior Reading

DISCUSSION

Hexane and toluene are liquid hydrocarbons that have boiling points approximately 40 °C apart. The liquid–vapor composition curve in Figure 5.6 represents this system; it is apparent that a two-plate distillation should yield nearly pure components. The procedure to be outlined consists of two parts. The first deals with the initial distillation (first plate), which separates the liquid mixture into three separate fractions. The second deals with *redistillation* of the first and third fractions (second plate). Exercising careful technique during the first distillation should provide a fraction rich in the lower boiling component, a middle fraction, and a fraction rich in the higher boiling component. Then careful *redistillation* of these fractions can be expected to complete the separation of the two components and to produce fractions of relatively pure hexane and toluene. The Hickman still used in the microscale laboratory is a simple, short-path column, and thus one would not expect complete separation of the hexane and toluene in one cycle.

COMPONENTS

$$CH_3CH_2CH_2CH_2CH_2CH_3$$

Hexane

Toluene

EXPERIMENTAL PROCEDURE

Estimated time to complete the experiment: 2.0 h.

Physical Properties of Components					
Compound	MW	Amount	bp (C°)	d	n_D
Hexane	86.18	1.0 mL	69	0.66	1.3751
Toluene	92.15	1.0 mL	111	0.87	1.4961

Reagents and Equipment.   Use an automatic delivery pipet to place 1.0 mL of hexane and 1.0 mL of toluene in a clean, dry, stoppered 5-mL conical vial.
   Place the vial in a small beaker to prevent tipping. Add a boiling stone (or a magnetic spin vane if desirable), assemble the Hickman still with the

thermometer positioned directly down the center of the column (see previous discussion), and mount the system in a sand bath (see Fig. 6.3).

Experimental Conditions. The temperature of the sand bath is raised to 80–90 °C at a maximum rate of 5 °C/min (>70 °C at 3 °C/min) using a hot plate.

CAUTION: Do not let the temperature of the still rise too rapidly.

Once gentle boiling begins, the heating rate should be lowered to a maximum of 2 °C/min. It is **absolutely crucial** that the distillation rate be kept below 100 µL/3 min to achieve the necessary fraction enrichment that will permit good separation during the second stage of the experiment. The distillate is collected in *three fractions* over the temperature ranges (1) 65–85 °C (bath temperature ~ 95–110 °C); (2) 85–105 °C (bath temperature ~ 140 °C); and (3) 105–110 °C (bath temperature ~ 170 °C) in amounts of approximately 800, 400, and 800 µL, respectively. Remove each fraction from the still with a bent-tip Pasteur pipet. Store the liquid condensate (fractions) in clean, dry, 1-dram, screw-cap vials. *Remember to number the vials in order and use an aluminum foil cap liner.*

Characterization of Crude Fractions. For each of the three fractions, record the refractive index. Fraction 1 has been enriched in one of the two components. Which one? Does the refractive index agree with that found in the literature? Fraction 3 has been enriched in the other component. Does the refractive index of that fraction support your first conclusion? If partial enrichment has been achieved, proceed to the second phase of the distillation.

Redistillation of Fraction 1   Redistill fraction 1 in a clean Hickman still with a thermometer arranged as before (Fig. 6.3), using a 3-mL conical vial and the procedure just outlined. Collect an initial fraction over the boiling range 68–71 °C (~100–200 µL). Remove it from the collar, using the Pasteur pipet, and place it in a 1-dram screw-cap vial.

Figure 6.3   Hickman still with Claisen head adapter.

Characterization of Fraction 1   Determine the ultramicro boiling point and the refractive index of this lower boiling fraction. Compare the experimental values obtained with those of pure hexane reported in the literature.

Redistillation of Fraction 3   Fraction 3 is placed in a clean Hickman still, using a thermometer and a 3-mL conical vial (Fig. 6.3), and redistilled using the procedure outlined. Collect an initial fraction over the boiling range 95–108 °C (~500 μL), and transfer this fraction by Pasteur pipet to a screw-cap vial. Collect a final fraction at 108–110 °C (~250 μL), and transfer the material to a second vial. *This second fraction is the highest boiling fraction to be collected in the three distillations and should be the richest in the high-boiling component.*

Characterization of Fraction 3   Determine the refractive index and boiling point of the second fraction and compare your results with those found in the literature for toluene. *Determine the refractive index and boiling point of pure toluene for comparison purposes.*

Fractional Semi-microscale Distillation: Separation of 2-Methylpentane and Cyclohexane Using a Spinning-Band Column

Experiment 3C

Purpose.   The purpose of this experiment is to separate two liquids with boiling points that are relatively similar: less than 20 °C apart, to learn the operation of a high-performance spinning-band distillation column, and to develop the skills for purifying small quantities of liquid mixtures.

Prior Reading

> *Technique 1:* Microscale Separation of Liquid Mixtures by GC (pp. 59–65)
> *Technique 2:* Distillation (pp. 65–68)
> > Distillation Theory (p. 65)
> > Simple Distillation at the Semi-microscale Level (pp. 65–68)
> > Fractional Semi-microscale Distillation (pp. 68–71)
> *Chapter 4:* Determination of Physical Properties
> > Ultramicro Boiling Point (pp. 48–50)
> > Refractive Index (p. 57)

DISCUSSION

In this experiment, the separation of a 2-mL mixture of 2-methylpentane and cyclohexane using a 2.5-in. spinning-band distillation column is described. The purity of the fractions is determined by gas chromatography and by measurement of the refractive index. Finally, the number of theoretical plates is estimated. You will separate a 50:50 mixture of 2-methylpentane and cyclohexane using the spinning-band distillation column shown in Figure 6.4.

Figure 6.4 Micro spinning-band distillation column (2.5 in.).

COMPONENTS

CH₃
CH₃CHCH₂CH₂CH₃

2-Methylpentane

Cyclohexane

EXPERIMENTAL PROCEDURE

Estimated time to complete the experiment: 3.0 h.

Physical Properties of Components				
Compound	MW	Amount	bp (°C)	n_D
2-Methylpentane	86.18	1.0 mL	60.3	1.3715
Cyclohexane	84.16	1.0 mL	80.7	1.4266

Reagents and Equipment.   Assemble the system as shown in Figure 6.4, making sure that the Teflon band is aligned as straight as possible in the column. In particular, the pointed section extending into the vial must be straightened to minimize vibration during spinning of the band.

Place a pipet bulb on the side arm of the collection adapter. This bulb plays an important function in the operation of the column: Attachment of the bulb creates a closed system. Suspension of the thermometer with a septum on the top of the condenser can act to release any buildup of pressure.

CAUTION: The system must be able to vent at the thermometer during operation!

Once the spinning band has been tested and rotates freely, place 1.0 mL of 2-methylpentane and 1.0 mL of cyclohexane in the vial (to be delivered with a Pasteur pipet or an automatic delivery pipet). Reassemble the system and lower the column into the sand bath or copper-tube block or aluminum metal block. The beveled edge on the air condenser should be rotated 180° from the collection arm.

NOTE. It is important to make an aluminum foil cover for the sand bath; this cover will reflect the heat and hot air away from the collection vial.

Experimental Conditions.   Gently heat (copper-tube block) the vial until boiling occurs. The magnetic stirrer is turned to a low-spin rate when heating commences. When reflux is observed at the base of the column, the magnetic stirrer is adjusted to intermediate spin rate. Once liquid begins to enter the column the spin rate is increased to the maximum (1000–1500 rpm).

NOTE. It is absolutely critical that the temperature of the vial be adjusted so that vapors in the column rise very slowly. It is possible for overheated vapors to be forced through the air condenser.

When the vapors slowly arrive in the unjacketed section of the column head, the condenser joint acts as a vapor shroud to effectively remove vapors from the receiver-cup area. During this total reflux period, maximum separation of the components is achieved. Once vapor reflux occurs in the head of the column, the system is left for 20–30 min to reach thermal equilibrium. During this period of total reflux, the head thermometer should read about 57–60 °C (at least for most of the equilibration time).

Following the equilibration period, collection of the resolved components may begin. Rotate the air condenser 180° so that the beveled edge is over the collection duct. At this point, manipulation of the pipet bulb allows drainage of the condensate from the side arm. (This procedure is repeated occasionally to continue drainage from the side arm.) Collect six drops (~0.30 mL). After removing the collection vial, transfer the contents into a covered vial using a Pasteur pipet. Label all fractions. Collect two 0.6-mL fractions (the pipet bulb may be removed during collection of these latter fractions); then turn off the heat and stirring motor, and remove the vial from the sand bath. Transfer the material remaining in the vial, using a Pasteur pipet, to a fourth covered vial.

Characterization of the Fractions.   The composition of each of the fractions may be determined by gas chromatographic analysis and measurement

Figure 6.5   Composition of the first 15% and the first 40% of the volume collected in the distillation of a 50% (v/v) mixture of 2-methylpentane and cyclohexane.

of the refractive index. A GOW-MAC gas chromatograph should be set up as follows:

Column	DC 710
Injection	10 μL
Temperature	80 °C
Flow rate (He)	55 mL/min
Chart speed	1 cm/min

If we assume that the refractive index is a linear function of the volume fraction, the following relationship gives us the volume fraction of 2-methylpentane in a mixture. The volume fraction is X and the measured refractive index is n_D:

$$X = \frac{1.4266 - n_D}{1.4266 - 1.3715}$$

The curve shown in Figure 6.5 may be used to estimate the number of theoretical plates from the composition (mole fraction) of the *first* 0.30-mL fraction. For example, if the composition of the *first* 0.30 mL is 0.89, we would infer that the system had a resolution equivalent to about four theoretical plates. Note that the number of plates cannot be determined with confidence if the composition is greater than about 0.97. If we really wanted to determine the number of theoretical plates for a system with more than five plates, we could start with a mixture containing only 10 or 20% of the most volatile component (MVC), rather than the 50% used in this example.

Experiment 3D

Fractional Semi-microscale Distillation: The Separation of 2-Methylpentane and Cyclohexane Using a Spinning Band in a Hickman–Hinkle Still

Purpose.   In this exercise you will become familiar with a powerful modification of the classic Hickman still: the Hickman–Hinkle spinning-band distillation apparatus. This small still is one of the most efficient techniques developed for the purification of small quantities of liquids. You will develop the skills for handling small quantities of liquids and their purification by distillation, and become familiar with these techniques, so that they may be used in the purification of reaction products formed in later experiments.

DISCUSSION

This distillation separates the same two compounds used in Experiment [3C]. The distillate can be analyzed to determine the number of theoretical plates. If careful attention is given to the procedure, the spinning Hickman–Hinkle is capable of more than six theoretical plates.

As in Experiment [3C] the separation of a 2-mL mixture of 2-methylpentane and cyclohexane is achieved. The purity of the fractions can be determined by gas chromatography and by measurement of the refractive index.

COMPONENTS

CH₃
|
CH₃CHCH₂CH₂CH₃

2-Methylpentane

Cyclohexane

EXPERIMENTAL PROCEDURE

Estimated time to complete the experiment: 3.0 h.

Physical Properties of Components				
Compound	MW	Amount	bp (°C)	n_D
2-Methylpentane	86.18	1.0 mL	60.3	1.3715
Cyclohexane	84.16	1.0 mL	80.7	1.4266

Reagents and Equipment.   The system is assembled as shown in Figure 3.15 on page 27. In the process, make sure that the Teflon band is aligned as straight as possible in the column. In particular, the pointed section extending into the vial must be straightened to minimize vibration during spinning of the band.

Once the spinning band has been tested and rotates freely, place 1.0 mL of 2-methylpentane and 1.0 mL of cyclohexane in the vial (to be delivered with a Pasteur pipet or an automatic delivery pipet). Reassemble the system and lower the column into the sand bath.

Cover the sand bath with aluminum foil during the distillation to prevent the collar of the still from overheating. It is easier to regulate the temperature of the bath when it is covered. However, for distillations, a more efficient heating technique is the recently developed copper-tube block or aluminum metal block.

Experimental Conditions.   Gently heat the vial until boiling occurs. When heating commences, turn on the magnetic stirrer at a low setting. When reflux commences at the base of the column, the magnetic stirrer is raised to intermediate settings. Once liquid begins to enter the column, the spin rate is increased to the maximum (1000–1500 rpm).

NOTE. It is extremely important that careful temperature control be exercised at this stage so that the condensing vapors ascend the column very slowly.

Vapor-phase enrichment by the most volatile component is limited mainly to this period, as fraction collection commences immediately on arrival of the vapor column at the annular ring. Once condensation occurs, fractions are collected by the same technique used in Experiment [3B]. Characterization of the fractions, however, follows the procedure given in Experiment [3C].

Characterization of the Fractions. The composition of each of the fractions may be determined by gas chromatography, the refractive index, or both. See Experiment [3C] for details.

An alternative approach to the procedures discussed in Experiment [3C] is to establish the fraction volume by weight. The curves shown in Figure 6.5 may again be used to estimate the number of theoretical plates. The volume of the first fraction can be estimated, or determined more accurately, by weighing the fraction in a tared screw-cap vial. The composition of this fraction then may be determined and the fraction of the total represented by this portion calculated. If, for example, the first fraction has a volume of 0.4 mL (20% of the total) and has a composition 0.89 by volume of 2-methylpentane, we would infer that the system had a resolution equivalent to about four theoretical plates.

QUESTIONS

6-12. The boiling point of a liquid is affected by several factors. What effect does each of the following conditions have on the boiling point of a given liquid?
   (a) Pressure of the atmosphere
   (b) Use of an uncalibrated thermometer
   (c) Rate of heating of the liquid in a distillation flask

6-13. Calculate the vapor pressure of a solution containing 45 mol% hexane and 55 mol% octane at 90 °C assuming that Raoult's law is obeyed.
   Given: vapor pressure of the pure compounds at 90 °C: hexane = 1390 torr, octane = 253 torr.

6-14. In any distillation for maximum efficiency of the column, the distilling flask should be approximately one-half full of liquid. Comment on this fact in terms of (a) a flask that is too full and (b) a flask that is nearly empty.

6-15. Occasionally during a distillation, a solution will foam rather than boil. A way of avoiding this problem is to add a *surfactant* to the solution.
   (a) What is a surfactant?
   (b) What is the chemical constitution of a surfactant?
   (c) How does a surfactant reduce the foaming problem?

6-16. Explain why packed and spinning-band fractional distillation columns are more efficient at separating two liquids with close boiling points than are unpacked columns.

6-17. Explain what effect each of the following mistakes would have had on the simple distillation carried out in this experiment.
   (a) You did not add a boiling stone.
   (b) You heated the distillation flask at too rapid a rate.

6-18. In the ultramicro boiling-point determination, why is the boiling point taken just as bubbles cease emerging from the bell?

6-19. Define each of the following terms, which are related to the distillation process:
   (a) Distillate
   (b) Normal boiling point
   (c) Forerun

6-20. How does the refractive index of a liquid vary with temperature? What corrective factor is often used to determine the value at a specific temperature—for example, 20 °C—if the measurement were made at 25 °C?

6-21. A patent (US Patent No. 4,770,746) has been granted for the spinning-band column shown in Figure 6.4. Locate the patent online and provide the following:
   (a) Who are the inventors of the patent?
   (b) On what date was the patent granted?
   (c) On what date was the patent filed?
   (d) What chemicals are used in the two examples given in the patent?

BIBLIOGRAPHY

General references on distillation:

Lodwig, S. N. *J. Chem. Educ.* **1989**, *66*, 77.

Stichlmair, J. G.; Fair, J. R. *Distillation: Principles and Practice*; Wiley: New York, 1998.

Technique of Organic Chemistry, Vol. IV, Distillation, 2nd ed.; Perry, E. S.; Weissberger, A., Eds.; Wiley-Interscience: New York, 1967.

Vogel, A. I. *Vogel's Textbook of Practical Organic Chemistry*, 5th ed.; Furnis, B. S., et al., Eds.; Wiley: New York, 1989.

Zubrick, J. W. *The Organic Chem Lab Survival Manual*, 11th ed.; Wiley: New York, 2019.

Solvent Extraction

Determination of a Partition Coefficient for the System Benzoic Acid, Methylene Chloride, and Water

Purpose.   This exercise illustrates the general procedures that are used to determine a partition coefficient at the microscale level. Experience in weighing milligram quantities of materials on an electronic balance, the use of automatic delivery pipets for accurately dispensing microliter quantities of liquids, the transfer of microliter volumes of solutions with the Pasteur filter pipet, and the use of a Vortex mixer, are techniques encountered in this experiment.

Prior Reading

    Chapter 3: Experimental Apparatus

                Pasteur Filter Pipet (pp. 38–39)

                Automatic Delivery Pipet (pp. 39–40)

                Weighing of Solids in Milligram Quantities (p. 41)

    Technique 4: Solvent Extraction

                Liquid–Liquid Extraction (p. 76)

                Partition Coefficient Calculations (pp. 74–76)

                Drying of the Wet Organic Layer (pp. 84–87)

                Separation of Acids and Bases (pp. 81–83)

DISCUSSION

Solubility.   Substances vary greatly in their solubility in various solvents, but based on observations, many of which were made in the very early days of chemical experimentation, a useful principle has evolved that allows the chemist to predict rather accurately the solubility of a particular substance. It is generally true that *a substance tends to dissolve in a solvent that is chemically similar to itself. In other words, like dissolves like.*

Thus, for a particular substance to exhibit solubility in water requires that species to possess some of the characteristics of water. For example, an important class of compounds, the organic alcohols, have the hydroxyl group (—OH) bonded to a hydrocarbon chain or framework (R—OH). The hydroxyl group can be viewed as being effectively one-half a water (HOH) molecule, and it has a similar polarity to that of water. This results from a charge separation arising from the difference in electronegativity between the hydrogen and oxygen and oxygen atoms. The O—H bond, therefore, is considered to have *partial ionic character*:

$$\overset{\delta-}{\underset{..}{\overset{..}{O}}}-\overset{\delta+}{H}$$

Partial ionic character of the hydroxyl group

This *polar*, or partial ionic, character leads to relatively strong hydrogen bond formation between molecules having this entity. Strong hydrogen bonding is evident in molecules that contain a hydrogen atom attached to an oxygen, nitrogen, or fluorine atom, as shown here for the ethanol–water system. This polar nature of a functional group is present when there are sufficient differences in electronegativity between the atoms making up the group:

$$CH_3-CH_2-\overset{\delta-}{\underset{..}{O}:}\underset{H}{\overset{}{\diagdown}}_{\delta+}$$

Ethanol

Hydrogen bond formation

In ethanol (CH_3CH_2OH), it is apparent that the hydroxyl end of the molecule is very similar to water (HOH). When ethanol is added to water, therefore, they are miscible in all proportions. That is, ethanol is completely soluble in water, and water is completely soluble in ethanol. This solubility results because the attractive forces set up between the two molecules (CH_3CH_2OH and H_2O) are nearly as strong as between two water molecules; however, the attraction in the first case is somewhat weakened by the presence of the nonpolar alkyl ethyl group, CH_3CH_2—. Hydrocarbon groups attract each other only weakly, as evidenced by their low melting and boiling points. Three examples of the contrast in boiling points between compounds of different structure but similar molecular weight are summarized in Table 6.1. Clearly, those molecules that attract each other weakly have lower boiling points.

When we compare the water solubility of ethanol (a two-carbon [C_2] alcohol that, as we have seen, is completely miscible with water) with that of octanol (a straight-chain eight-carbon [C_8] alcohol), we find that the solubility of octanol is less than 2% in water. Why the difference in solubilities between these two alcohols? The answer lies in the fact that the dominant structural feature of octanol has become the nonpolar nature of its alkyl group:

$$CH_3-CH_2-CH_2-CH_2-CH_2-CH_2-CH_2-CH_2-\overset{\delta-}{\underset{..}{O}:}\underset{H}{\overset{}{\diagdown}}_{\delta+}$$

Octanol

$$CH_3-CH_2-\overset{..}{\underset{..}{O}}-CH_2-CH_3$$

Diethyl ether

As the bulk of the hydrocarbon section of the alcohol molecule increases, the intramolecular attraction between the polar hydroxyl groups of the alcohol and the water molecules is no longer sufficiently strong to overcome the *hydrophobic character* (lack of attraction to water) of the nonpolar hydrocarbon section of the alcohol. On the other hand, octanol has a large nonpolar hydrocarbon group as its dominant structural feature. We might, therefore, expect octanol to exhibit enhanced solubility in less polar solvents. In fact, octanol is found to be completely miscible with diethyl ether. Ethers are solvents of weak polarity. Since the nonpolar characteristics are significant in both molecules, mutual solubility is observed. It has been empirically demonstrated that, in general, if a compound has both polar and nonpolar groups present in its structure, those

Table 6.1 Comparison of Boiling Point Data			
Name	Condensed Structural Formula	MW (g/mol)	bp (°C)
Ethanol	CH_3CH_2OH	46	78.3
Propane	$CH_3CH_2CH_3$	44	−42.2
Methyl acetate	$CH_3CO_2CH_3$	74	54
Diethyl ether	$(CH_3CH_2)_2O$	74	34.6
Ethylene	$CH_2{=}CH_2$	28	−102
Methylamine	CH_3NH_2	31	−6

Table 6.2 Water Solubility of Alcohols		
Name	Condensed Structural Formula	Solubility (g/100 g H_2O)
Pentanol	$CH_3(CH_2)_3CH_2OH$	4.0
2-Pentanol	$CH_3(CH_2)_2CH(OH)CH_3$	4.9
2-Methyl-2-butanol	$(CH_3)_2C(OH)CH_2CH_3$	12.5

Note. Data at 20 °C.

compounds having five or more carbon atoms in the hydrocarbon portion of the molecule will be more soluble in nonpolar solvents, such as pentane, diethyl ether, or methylene chloride. Figure 5.13, on page 73, summarizes the solubilities of a number of straight-chain alcohols, carboxylic acids, and hydrocarbons in water. As expected, those compounds with more than five carbon atoms are shown to possess solubilities similar to those of the hydrocarbons.

Several additional relationships between solubility and structure have been observed and are pertinent to the discussion.

1. Branched-chain compounds have greater water solubility than their straight-chain counterparts, as illustrated in Table 6.2 with a series of alcohols.

2. The presence of more than one polar group in a compound will increase that compound's solubility in water and decrease its solubility in nonpolar solvents. For example, high molecular weight sugars, such as cellobiose, which contain multiple hydroxyl and/or acetal groups, are water-soluble and etherinsoluble; cholesterol (C_{27}), which possesses only a single hydroxyl group, is water-insoluble and ether-soluble:

Cholesterol

Cellobiose

3. The presence of a chlorine atom, even though it lends some partial ionic character to the covalent C—Cl bond, does not normally impart water solubility to a compound. In fact, such compounds as methylene chloride (CH_2Cl_2), chloroform ($CHCl_3$), and carbon tetrachloride (CCl_4) have long been used as solvents for the extraction of aqueous solutions. It should be noted

Table 6.3 Water Solubility of Amines		
Name	Condensed Structural Formula	Solubility (g/100 g H$_2$O)
Ethylamine	CH$_3$CH$_2$NH$_2$	∞
Diethylamine	(CH$_3$CH$_2$)$_2$NH	∞
Trimethylamine	(CH$_3$)$_3$N	91
Triethylamine	(CH$_3$CH$_2$)$_3$N	14
Aniline	C$_6$H$_5$NH$_2$	3.7
p-Phenylenediamine	H$_2$NC$_6$H$_4$NH$_2$	3.8
Note. Data at 25 °C.		

that use of the latter two solvents is no longer recommended unless strict safety precautions are exercised, because of their carcinogenic nature.

4. Most functional groups that are capable of forming a hydrogen bond with water if it constitutes the dominant structural feature of a molecule will impart increased water solubility characteristics to a substance (the five-carbon rule obviously applies here in determining just what is a dominant feature). For example, certain alkyl amines (organic relatives of ammonia) might be expected to have significant water solubility. This finding is indeed the case, and the water-solubility data for a series of amines is summarized in Table 6.3.

The solubility characteristics of any given compound will uniquely govern that substance's distribution (*partition*) between the phases of two immiscible solvents (in which the material has been dissolved) when these phases are intimately mixed. In this experiment we determine the partition coefficient (distribution coefficient) of benzoic acid between two immiscible solvents, methylene chloride and water.

COMPONENTS

Benzoic acid   Methylene chloride   Water

EXPERIMENTAL PROCEDURE

Estimated time to complete the experiment: 1.5 h.

Physical Properties of Components						
Compound	MW	Amount	mmol	mp (°C)	bp (°C)	d
Benzoic acid	122.13	50 mg	0.41	122		
Methylene chloride		1.20 mL			40	1.33
Water		600 µL			100	1.00

Equipment Setup and Addition of Reagents. Weigh and add to a 5.0-mL conical vial fitted with a screw cap 50 mg (0.41 mmol) of benzoic acid. Now add 600 µL of methylene chloride followed by 600 µL of water.

The solvents are transferred to the vial with automatic delivery pipets (use a different pipet or pipet tip for each solvent). The methylene chloride addition should be carried out in the **hood.**

HOOD

Procedure for Establishing Equilibrium Distribution.

Cap the vial and shake (or use a Vortex mixer) until the benzoic acid dissolves and the two phases have been thoroughly mixed. Vent the vial (by releasing the Capseal) and then allow the two layers to separate.

Carefully draw the lower methylene chloride layer into a Pasteur filter pipet and transfer it to a 5-mL conical vial containing 100 mg of anhydrous, magnesium sulfate. If the amount of the methylene chloride layer is insufficient to properly transfer into the 5-mL conical vial, carefully add more so that a proper transfer can occur. Once complete, recap the vial.

NOTE. If the volume of the methylene chloride layer is so large that it cannot be transferred completely in one operation, a second transfer may be required. Be careful not to overfill the pipet to ensure that solvent does not come in contact with the rubber bulb. The technique of removing the last traces of water from the methylene chloride solution is often referred to as drying the solution. It can involve any one of a number of insoluble anhydrous salts, which convert the moisture retained in the organic phase to water of crystallization. In this case, we are using sodium sulfate.

Isolation of the Benzoic Acid.

After drying the methylene chloride solution for 10–15 min, transfer the anhydrous solution to a previously **tared** vial (the term **tare** means to preweigh the empty vial) using a Pasteur filter pipet (the use of the filter pipet is a convenient way of separating the solid hydrated sodium sulfate from the dried solution). Rinse the sodium sulfate with an additional 600 μL of methylene chloride and combine the rinse with the solution in the tared 5.0-mL conical vial. Evaporate the solvent under a gentle stream of nitrogen gas in a warm sand bath in the **hood.** (It is important to warm the solution while evaporating the solvent; otherwise the heat of vaporization will rapidly cool the solution. In this latter case, as the cold, solid acid precipitates from the saturated solution, moisture will condense from the air entrained in the evaporation process, and contaminate the surface of the recovered material.)

HOOD

NOTE. If a hot sand bath is used, a boiling stone is placed in the vial before it is tared. The boiling stone will help to avoid explosive, sudden boiling of the solvent when the vial is placed in the sand bath.

Weight Data and Calculations.

Weigh the vial and determine the weight of benzoic acid that remains following removal of the methylene chloride. Break up the hard cake of precipitated benzoic acid with a microspatula and *briefly* reheat the vial and contents in a sand bath to remove the last traces of solvent and any water that remains in the system. Cool and reweigh. Repeat this operation until a constant weight is obtained. This weight represents the benzoic acid that dissolved in the *methylene chloride layer.*

The *original weight* of benzoic acid used minus the amount of benzoic acid recovered in the methylene chloride layer equals the weight of the benzoic acid that dissolved and still remains in the *water layer.*

Since equal volumes of both solvents were used, the partition coefficient may be simply determined from the ratio of the weight of benzoic acid in the methylene chloride solvent to the weight of benzoic acid in the water layer.

Calculate the partition coefficient for benzoic acid in the solvent pair used in this exercise.

Solvent Extraction I: The System; Benzoic Acid, Methylene Chloride, and 10% Sodium Bicarbonate Solution; An Example of Acid–Base Extraction Techniques

Purpose.   This exercise illustrates an extensively used extraction technique in which a reversible reaction is employed to alter the solubility characteristics of the substance of interest.

Prior Reading

> *Chapter 3:* Experimental Apparatus
>> Pasteur Filter Pipet (pp. 38–39)
>> Automatic Delivery Pipet (pp. 39–40)
>> Weighing of Solids in Milligram Quantities (p. 41)
> *Technique 4:* Solvent Extraction
>> Liquid–Liquid Extraction (p. 76)
>> Partition Coefficient Calculations (pp. 74–76)
>> Drying of the Wet Organic Layer (pp. 84–87)
>> Separation of Acids and Bases (pp. 81–83)

REACTION

$$C_6H_5\overset{\ddot{O}}{\overset{\|}{C}}-\ddot{O}H + Na^+HCO_3^- \rightleftharpoons C_6H_5\overset{\ddot{O}}{\overset{\|}{C}}-\ddot{O}:^-Na^+ + H_2CO_3$$

Benzoic acid   Sodium benzoate   Carbonic acid

$$H_2CO_3 \rightleftharpoons CO_2 + H_2O$$

DISCUSSION

Benzoic acid reacts readily with sodium bicarbonate to form sodium benzoate, carbon dioxide, and water. The sodium salt of benzoic acid has highly ionic characteristics and thus, unlike the free acid, the salt is very soluble in water and nearly insoluble in methylene chloride. This salt is characterized by a full ionic bond between the carboxylic acid group of the acid and the sodium ion. It is thus a new substance exhibiting many of the solubility properties commonly associated with inorganic ionic salts.

EXPERIMENTAL PROCEDURE

Estimated time to complete the experiment: 1.0 h.

Physical Properties of Reactants						
Compound	MW	Amount	mmol	mp (°C)	bp (°C)	*d*
Benzoic acid	122.13	50 mg	0.41	122		
Methylene chloride		1.20 mL			40	1.33
Sodium bicarbonate (10% solution)		600 μL				

Procedure for Establishing Distribution. Repeat the procedures carried out in Experiment [4A], but replace the 600 μL of water with 600 μL of 10% sodium bicarbonate solution. A good estimate of the efficiency of the conversion of benzoic acid to the sodium salt of the acid, which because of its ionic character is found almost exclusively in the aqueous phase, can be made by recovering any unreacted acid from the methylene chloride layer and using the distribution coefficient established in Experiment [4A]. Also, be sure to obtain a melting point of any recovered residue (assumed above to be benzoic acid) from the organic phase, since contamination of free acid by the acid salt can be detected by this measurement. Sodium benzoate has a melting point above 300 °C, whereas benzoic acid melts near 122 °C.

Test for a Carboxylic Acid. As illustrated in the above reaction, when a carboxylic acid comes in contact with a solution containing bicarbonate ion, carbon dioxide is generated. Once saturation of the solution by carbon dioxide occurs, bubbles of carbon dioxide gas are observed to form in the liquid phase. This effervescence may be used as a qualitative test for the presence of the carboxylic acid functional group in an unknown substance.

Place 1–2 mL of 10% sodium or potassium bicarbonate on a small watch glass. Add the pure acid, one drop from a Pasteur pipet if the sample is a liquid (~5 mg if it is a solid), to the bicarbonate solution. Evolution of bubbles (CO_2) from the mixture indicates the presence of an acid.

Perform the above test for the presence of carboxyl groups on several organic acids, such as acetic, benzoic, propanoic, or chloroacetic acid.

Solvent Extraction II: A Three-Component Mixture; an Example of the Separation of an Acid, a Base, and a Neutral Substance

Experiment 4C

Purpose. This exercise investigates how solvent extraction techniques can be applied effectively to problems that require the separation of mixtures of organic acids, bases, and neutral compounds in the research or industrial laboratory.

Prior Reading

>   *Chapter 3:* Experimental Apparatus
>>   Pasteur Filter Pipet (pp. 38–39)
>>   Automatic Delivery Pipet (pp. 39–40)
>>   Weighing of Solids in Milligram Quantities (p. 41)
>   *Technique 4:* Solvent Extraction
>>   Liquid–Liquid Extraction (p. 76)
>>   Drying of the Wet Organic Layer (pp. 84–87)
>>   Separation of Acids and Bases (pp. 81–83)
>>   Salting Out (p. 83)

DISCUSSION

As implied in the discussions of Experiments [4A] and [4B], the solubility characteristics of organic acids in water can be shown to be highly dependent on the pH of the solution. By extending this extraction approach to include organic bases, it has been possible to develop a general procedure for the separation of mixtures of organic acids, bases, and neutral substances.

NOTE. Refer to Technique 4, p. 78–79 for a flowchart outlining the procedure.

The components of the mixture to be separated in this experiment are benzoic acid, ethyl 4-aminobenzoate (a base), and 9-fluorenone (a neutral compound, which may be prepared in Experiment [33A]).

COMPONENTS

Benzoic acid   Ethyl 4-aminobenzoate   9-Fluorenone

EXPERIMENTAL PROCEDURE

Estimated time to complete the experiment: 2.5 h.

Physical Properties of Reactants						
Compound	MW	Amount	mmol	mp (°C)	bp (°C)	d
Benzoic acid	122.1	50 mg	0.41	122		
Ethyl 4-amino-benzoate	165.19	50 mg	0.31	89		
9-Fluorenone	180.22	50 mg	0.27	84		
Diethyl ether	74.12	4 mL			35	0.7184
3 M HCl		4 mL				
3 M NaOH		4 mL				
6 M HCl						
6 M NaOH						

NOTE. In carrying out the separation, you should keep a record or flowchart of your procedure (as suggested in the prior reading assignment) in your laboratory notebook. You should also be particularly careful to label all flasks.

Reagents and Equipment.   Weigh and add to a stoppered or capped 15-mL centrifuge tube the following: 50 mg (0.41 mmol) of benzoic acid, 50 mg (0.31 mmol) of ethyl 4-aminobenzoate, and 50 mg (0.27 mmol) of 9-fluorenone.
HOOD   Now, in the **hood**, add 4 mL of diethyl ether using a 10-mL graduated cylinder for the transfer. The solids may be dissolved by either stirring with a glass rod or mixing on a Vortex mixer (capped vial).

Separation of the Basic Component.   Cool the solution in an ice bath. Now, using a calibrated Pasteur pipet, add 2 mL of 3 M HCl dropwise to the cooled solution with swirling. Cap and thoroughly mix the resulting two-phase system for several minutes (a Vortex mixer works well). Vent carefully and after the layers have separated remove the bottom (aqueous) layer using a Pasteur filter pipet and transfer this phase to a **labeled**, 10-mL Erlenmeyer flask.

Repeat this step with an additional 2 mL of the 3 M acid solution. As before, transfer the aqueous layer to the same labeled Erlenmeyer flask. Stopper or cap this flask. *Save the ether solution.* The aqueous acidic solution is to be used in the next step.

NOTE. A small amount of crystalline material may form at the interface between the layers. A second extraction generally dissolves this material.

Isolation of Ethyl 4-Aminobenzoate: The Basic Component.   To the aqueous acidic solution separated and set aside in the previous step, add 6 M NaOH dropwise until the solution is distinctly alkaline to litmus paper. Cool the flask in an ice bath for about 10–15 min. Collect the solid precipitate that forms in the basic solution by reduced-pressure filtration using a Hirsch funnel. Wash the precipitate with two 1-mL portions of distilled water. Air-dry the washed microcrystals by spreading them on a clay plate, filter paper, or in a vacuum drying oven. Weigh the material and calculate the percent recovery. Obtain the melting point of the dry ethyl 4-aminobenzoate and compare your result to the literature value. This material is used as a topical anesthetic.

Separation of the Acidic Component.   Add 2 mL of 3 M NaOH to the ether solution that was set aside earlier in the experiment. At this point, if necessary, add additional ether (~1–2 mL) so that the volume of the organic layer is at least equal to, or somewhat larger than, that of the aqueous phase. This adjustment in volume should allow an efficient distribution to take place when the two phases are mixed. Then carry out the extraction as before, allow the layers to separate, and finally transfer the bottom aqueous basic layer to a **labeled** 10-mL Erlenmeyer flask.

Repeat this routine and again remove the aqueous layer and transfer it to the same Erlenmeyer flask. Stopper this flask (containing the extracted aqueous basic phase) and set it aside for later use.

Separation of the Neutral Component.   Wash (extract) the remaining ether solution contained in the centrifuge tube with two 1-mL portions of water. Separate the lower aqueous layer in each sequence. *Save the aqueous wash layer temporarily; it will be discarded at the very end of the experiment. (It is good practice to never discard any layer until you have recovered or accounted for all of the material.)* Now add about 300 mg of anhydrous granular sodium sulfate to the wet ether (ether saturated with water) solution. Cap the tube and set it aside while working up the alkaline extraction solution. This procedure will allow sufficient time for the traces of moisture to be removed from the ether solution by hydration of the insoluble drying agent. If the sodium sulfate initially forms large clumps, you may add a further quantity of the anhydrous salt.

Isolation of Benzoic Acid: The Acidic Component.   Add 6 M HCl dropwise to the aqueous alkaline solution, which was separated and set aside earlier, until the solution becomes distinctly acidic to litmus paper. Then cool the flask in an ice bath for about 10 min. If only a small amount (10–25 mg) of precipitate is obtained on acidification, add a small amount of a saturated aqueous solution of sodium chloride (salting out effect; see Prior Reading assignment) to help promote further precipitation of the benzoic acid. Collect the precipitated benzoic acid by filtration under reduced pressure using a Hirsch funnel. Wash (rinse) the filter cake (precipitated acid) with two 1-mL portions of cold distilled water. Dry the solid product using one of the techniques described earlier for ethyl 4-aminobenzoate. Weigh the dry benzoic acid and calculate your percent recovery.

Obtain a melting point of this material and compare your result to the literature value.

Isolation of 9-Fluorenone: The Neutral Component. Use a Pasteur filter pipet to transfer the dried ether solution collected earlier to a tared 10-mL Erlenmeyer flask containing a boiling stone. Rinse the drying agent with an additional 1 mL of ether and combine the ether wash with the anhydrous organic phase.

Concentrate the ether solution on a warm sand bath using a *slow* stream of nitrogen gas in the **hood**. Obtain the weight of the residue (9-fluorenone) after removal of the solvent and calculate the percent recovery. Obtain a melting point of the material and compare your result to the literature value.

HOOD

QUESTIONS

6-22. **(a)** Explain why diethyl ether would be expected to be a satisfactory solvent for the straight-chain hydrocarbons hexane and heptane. **(b)** Explain why *t*-butyl methyl ether would not be expected to be an ideal solvent for the polyhydroxylated carbocycles glucose and fructose.

6-23. The solubility of *p*-dibromobenzene in benzene is 80 μg/100 μL at 25 °C. Would you predict the solubility of this compound to be greater, less, or approximately the same in acetone solvent at this temperature? Explain.

p-Dibromobenzene   Acetone

6-24. **(a)** Each of the solvents listed below is used in experiments in this text to extract organic compounds from aqueous solutions.

(i) Methylene chloride
(ii) Pentane
(iii) Toluene
(iv) Diethyl ether

Will the organic phase be the upper or lower layer when each of these solvents is mixed with water? Explain your answer for each case. **(b)** If you placed an ice cube in each of the solvents i–iv listed above in (a), would you expect an ice cube placed in each to float? Explain your answer for each case. **(c)** After identifying a solute and a solvent in Experiment [4B] and/or Experiment [4C], calculate, compare, and comment on the mmol used of each.

6-25. A 42-mg sample of an organic compound (MW = 84) is dissolved in 15 mL of water. This aqueous solution is extracted with 8.0 mL of hexane. Separation and analysis of the aqueous phase shows that it now contains 19 mg of the organic compound. Calculate the partition coefficient for the compound.

6-26. A qualitative method often used to determine whether an organic compound contains oxygen is to test its solubility in concentrated sulfuric acid. Almost all oxygen-containing compounds are soluble in this acid. Explain.

6-27. In the discussion of multiple extractions (p. 75), it was suggested that in the example given you might extend the relationship to the next step by using one-third of the total quantity of the ether solvent in three portions. The reason for increasing the number of extractions was to determine whether this expansion would increase the efficiency of the process even further. To determine if this next step is worth the effort, perform the calculations for the extraction of 150 mg of P in 450 μL of water with three 150-μL portions of ether. Assume the partition coefficient is 3.5 (as before).
(a) Compare the amounts of P extracted from the water layer using one, two, or three extractions.
(b) Do you think that the additional amount of P extracted from the water layer using three extractions is justified? Might it be justified if P were valuable and you were working on the industrial scale of 150 kg of P in 4500 L of water?

BIBLIOGRAPHY

Lo, T. C.; Baird, M. H. I.; Hanson, C., Eds. *Handbook of Solvent Extraction*; Krieger: Melbourne, FL, 1991.

Rydberg, J.; Cox, M.; Musikas, C.; Choppin, G. R., Eds. *Solvent Extraction Principles and Practices, Revised and Expanded*, 2nd ed.; CRC Press: Boca Raton, FL, 2020.

Shriner, R. L.; Hermann, C. K. F.; Morrill, T. C.; Curtin, D.Y.; Fuson, R. C. *The Systematic Identification of Organic Compounds*, 8th ed.; Wiley: New York, 2003.

Vogel, A. I. *Vogel's Textbook of Practical Organic Chemistry*, 5th ed.; Furnis, B. S., et al., Eds.; Wiley: New York, 1989.

Zubrick, J. W. *The Organic Chem Lab Survival Manual*, 11th ed.; Wiley: New York, 2019.

EXPERIMENT 5

Reduction of Ketones Using a Metal Hydride Reagent: Cyclohexanol and cis- and trans-4-tert-Butylcyclohexanol

Common name: cyclohexanol

CA number: [108-95-0]

CA name as indexed: cyclohexanol

Common name: 4-*tert*-butylcyclohexanol

CA number: [98-52-2]

CA name as indexed: cyclohexanol, 4-(1,1-dimethylethyl)-

Purpose.   The reduction of a ketone carbonyl to the corresponding alcohol is carried out using sodium borohydride, a commercially available metal-hydride reducing agent. The alcoholic reaction products are isolated by extraction techniques and purified by preparative gas chromatography. Cis and trans diastereomers are formed in the reduction of the 4-*tert*-butylcyclohexanone. These diastereomeric products can be separated during the preparative GC isolation. The stereochemistry of the structures can be deduced, once the mixture is separated into its pure components, using either NMR or IR spectroscopy.

Prior Reading

> *Technique 1:* Gas Chromatography (pp. 59–65)
> *Technique 4:* Solvent Extraction
>> Liquid–Liquid Extraction (p. 76)
>> Drying of a Wet Organic Layer (pp. 84–87)
>> Concentration of Solutions (pp. 105–108)
> *Technique 6:* Thin-Layer Chromatography (pp. 101–103)
> *Chapter 8:* Infrared Spectroscopy (pp. 615–630)
>> Nuclear Magnetic Resonance Spectroscopy (pp. 580–606)

REACTION (EXPERIMENT [5A])

Cyclohexanone $\xrightarrow[\substack{CH_3OH \\ CH_3O^-Na^+}]{NaBH_4}$ Cyclohexanol

DISCUSSION

An important route for the synthesis of primary and secondary alcohols is the reduction of aldehydes and ketones, respectively. Reduction involves the addition of the equivalent of molecular hydrogen H—H to the carbonyl group (C=O).

A variety of pathways have been discovered to accomplish this conversion, but the most common method used in the research laboratory involves complex metal-hydride reagents. Two reagents that enjoy wide application are lithium aluminum hydride (LiAlH$_4$) and sodium borohydride (NaBH$_4$).

Lithium aluminum hydride is a powerful reducing agent that reacts not only with aldehydes and ketones, but also with many other carbonyl-containing functional groups as well. Examples include esters, lactones, carboxylic acids, anhydrides, and amides because, in part, the high reactivity of the first of the four deliverable hydrides. It also reduces noncarbonyl systems, such as alkyl halides, alkyl azides, alkyl isocyanates, and nitriles. Note that LiAlH$_4$ can be used safely only in aprotic solvents (a solvent that does not contain an ionizable [acidic] proton), such as diethyl ether or tetrahydrofuran (THF). In protic solvents, lithium aluminum hydride reacts *violently* with the acidic hydrogen of the solvent to rapidly generate hydrogen gas:

$$LiAlH_4 + 4CH_3OH \rightarrow LiAl(OCH_3)_4 + H_2$$

CAUTION: The hydrogen gas often ignites. This particular hydride reagent should not be used unless specific instructions are made available for its proper use under anhydrous conditions.

Sodium borohydride is a much more selective reducing reagent partly due to the fact that the fourth hydride, not first as is the case with lithium aluminum hydride, is the most reactive, and when used in excess, it is a much milder reagent than LiAlH$_4$. For this reason sodium borohydride is usually used for the reduction of aldehydes and ketones. It does not react with the vast majority of the less reactive organic functional groups, such as C=C, C≡C, nitro, cyano, and even some carbonyl-containing systems, such as amides and carboxylic acids. Sodium borohydride does react at an appreciable rate with water, but only slowly with aqueous alkaline solution (no available protons), methanol, α, β-unsaturated ketones, and esters. For small-scale reactions an excess of reagent is generally used to compensate for the amount of borohydride that reacts with the protic solvent (methanol). This approach is preferred to that of using a solvent in which the sodium borohydride is less soluble (it is insoluble in ether), because the reaction is driven more rapidly to completion under the former conditions. On the other hand, sodium borohydride can react rapidly with strong acids to generate hydrogen gas. This reaction may be used to advantage as a source of in situ hydrogen for the reduction of C=C bonds (see Experiment [12]). The relatively high cost of the metal hydride reducing agents is offset by their low molecular weight (more moles per gram) and the fact that 1 mol of reducing agent reduces 4 mol of aldehyde or ketone.

The key step in the reduction of a carbonyl group by sodium borohydride is the transfer of a hydride ion (:H$^-$) from boron to the carbon atom of the polarized carbonyl group:

$$\overset{\delta+}{\underset{}{C}} \cdots \overset{\delta-}{\ddot{O}}:$$

Table 6.4 Reduction of 4-*tert*-Butylcyclohexanone		
Reagent	Trans (%)	Cis (%)
Sodium borohydride	80	20
Lithium aluminum hydride	92	8
Lithium tri-*sec*-butylborohydride	7	93

In the reaction, the electron-rich hydride ion is acting as a *nucleophile* (nucleus-seeking), which attacks the *electrophilic* (electron-seeking) carbon atom of the carbonyl group:

The hydride may make an equatorial (e) or axial (a) attack, depending on steric factors.

The overall reduction process requires two hydrogen atoms, but only one comes directly from the borohydride reagent. The other hydrogen atom is derived from the protic solvent (methanol).

In the 4-*tert*-butylcyclohexanone example, the steric environment is different on either face of the carbonyl group. In this case, the hydride reducing agent attacks more rapidly from the axial direction, and thus the equatorial alcohol (axial H) is the major product. This reaction pathway is preferred with the relatively small sodium borohydride and lithium aluminum hydride reagents. When one stereoisomeric product is preferentially formed, the reaction is called a *stereoselective* reaction.

Sterically large hydride reducing reagents, such as lithium tri-*sec*-butylborohydride, are forced to make an equatorial attack, due to steric factors (these hydride reagents run into the 1,3-diaxial hydrogen atoms), and thus the cis isomer of 4-*tert*-butylcyclohexanol is the major product. The data are summarized in Table 6.4, which relates the stereochemistry of the reduction to the metal hydride reagent used.

Cyclohexanol

The reaction is shown above.

Experiment 5A

EXPERIMENTAL PROCEDURE

Estimated time to complete the experiment: 1.5 h.
For the GC analysis, 15 min per student.

Physical Properties of Reactants						
Compound	MW	Amount	mmol	bp (°C)	d	n_D
Cyclohexanone	98.15	100 μL	0.97	156	0.95	1.4507
Methanol	32.04	250 μL	10.3	65	0.79	1.3288
Sodium borohydride reducing solution		300 μL				

HOOD

Cyclohexanone, 100 μL
+ CH₃OH, 250 μL +
NaBH₄ solution, 300 μL

Reagents and Equipment. With the aid of an automatic delivery pipet, place 100 μL (95 mg, 0.97 mmol) of cyclohexanone in an oven-dried and tared (preweighed) 5.0-mL conical vial equipped with an air condenser. Now add 250 μL of methanol and gently stir the contents in the vial using a glass stirring rod (or a magnetic spin vane if desirable) to obtain a homogeneous solution (◄). In the **hood**, add 300 μL of sodium borohydride reducing solution dropwise, with stirring, to the solution of the ketone.

NOTE. *The cyclohexanone, methanol, and sodium borohydride solutions are dispensed using automatic delivery pipets. Weigh the cyclohexanone after delivery to get an accurate weight for the yield calculations.*

The stock reducing solution should be prepared just prior to conducting the experiment.

INSTRUCTOR PREPARATION. *In a 10-mL Erlenmeyer flask place 50 mg of anhydrous sodium methoxide and 2.5 mL of methanol. To this solution add 100 mg of sodium borohydride. Stopper the flask tightly and swirl the contents gently to dissolve the solid phase (100 μL of this solution provides ~2.0 mg of NaOCH₃ and 4.0 mg of NaBH₄).*

NOTE. *Test for activity of the reducing solution: Add 1–2 drops of the freshly prepared reducing solution to about 200 μL of concentrated hydrochloric acid. Generation of hydrogen gas bubbles is a positive test.*

Reaction Conditions. After allowing the resulting solution to stand 10 min at room temperature, begin monitoring the resulting solution by TLC. Use as a solvent system ethyl acetate:hexane (1:4), the R_f of cyclohexanol is 0.4 (R_f of cyclohexanone is 0.6) when stained with a solution of *p*-anisaldehyde (135 mL ethanol, 5 mL H₂SO₄, 1.5 mL glacial acetic acid, and 3.7 mL *p*-anisaldehyde).

Anhydrous
Na₂SO₄, 500 mg

Cyclohexanol
product +
2.0 mL CH₂Cl₂

HOOD

Isolation of Product. When the reaction is complete as judged by TLC, use a calibrated Pasteur pipet to add dropwise 1.0 mL of cold dilute hydrochloric acid (0.1 M HCl). Extract the aqueous mixture with three 0.5-mL portions of methylene chloride. Upon each addition of methylene chloride, cap the vial, shake it gently, and then carefully vent it by loosening the cap (a Vortex mixer may be used if available). After the layers have separated, remove the bottom methylene chloride layer using a Pasteur filter pipet and transfer it to a Pasteur filter pipet containing about 500 mg of anhydrous sodium sulfate.

Collect the dried eluate in a tared 5.0-mL conical vial containing a boiling stone. Use an additional 0.5 mL of methylene chloride to rinse the sodium sulfate and collect the rinse in the same conical vial (◄).

An additional rinse of the sodium sulfate may be made if desired. Remove the methylene chloride solvent by careful evaporation in the **hood** by gentle warming in a sand bath (constantly agitate the surface of the solution with a microspatula to prevent superheating and subsequent boilover). *In this instance, do not use a stream of nitrogen gas to hasten the evaporation. The volatility of the product alcohol is such that a substantial loss of product will occur if this technique is used.*

Purification and Characterization. The crude cyclohexanol reaction product remaining after evaporation of the methylene chloride solvent is usually of sufficient purity for direct characterization.

Determine the weight of the liquid residue and calculate the percent crude yield. Determine the refractive index (3 μL, optional) and boiling point (4 μL) of the material. Compare your values with data given in the literature.

Obtain the infrared spectrum of the crude (dry) cyclohexanol product by the capillary film sampling technique. Compare the spectrum of the starting ketone in Figure 6.6 to that of your isolated material. Is there evidence of the unreacted starting material in your product? The spectrum of cyclohexanol crude product is shown in Figure 6.7.

NOTE. Most of the infrared spectra referred to in the experimental analysis sections are derived by Fourier transform and plotted on a slightly different scale than the other spectra presented in the text. These spectra utilize a 12.5-cm⁻¹/division format below 2000 cm⁻¹ and undergo a 2:1 compression above 2000 cm⁻¹ (25 cm⁻¹/division).

Infrared Analysis: A Comparison Reactant and Product.   The key absorption bands to examine in the spectrum of cyclohexanone occur at 3420, 3000–2850, 1715, and 1425 cm⁻¹. The lack of significant absorption between 3100–3000 and 1400–1350 cm⁻¹ also should be noted. The sharp weak band at 3420 cm⁻¹ is not a fundamental vibration (not O—H or N—H stretching), but arises from the first overtone of the very intense carbonyl

Figure 6.6   IR spectrum: cyclohexanone.

stretching mode found at 1715 cm^{-1}. The overtone does not fall exactly at double the frequency of the fundamental, but usually occurs slightly below that value because of anharmonic effects. The lack of absorption in the region near 3100–3000 cm^{-1} and the presence of a series of very strong absorption bands at 3000–2850 cm^{-1} indicate that the only C—H stretching modes present are part of sp^3 systems. Thus, the spectrum is typical of an aliphatic ketone. The occurrence of a 1425-wavenumber band suggests the presence of at least one methylene group adjacent to the carbonyl group, while the 1450-wavenumber band requires other methylene groups more remote to the C=O group. The lack of absorption in the 1400- to 1375-wavenumber region indicates the absence of any methyl groups (a good indication of a simple aliphatic ring system) and further suggests that the absorption at 1450 cm^{-1} must arise entirely from methylene scissoring modes. The value of 1715 cm^{-1} for the C=O stretch supports the presence of a six-membered ring.

Now examine the spectrum of your reaction product (a typical example is given in Fig. 6.7). The spectrum is rather different from that of the starting material. The major changes are a new very strong broad band occurring between 3500 and 3100 cm^{-1} and the large drop in the intensity of the band found at 1715 cm^{-1}.

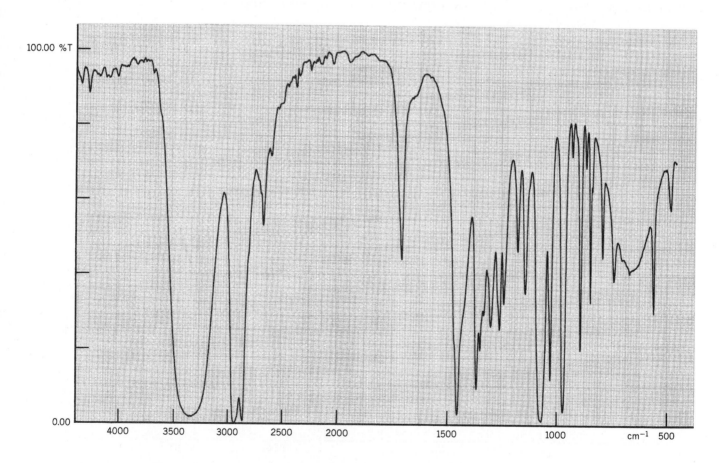

Sample	Cyclohexanol (crude product)		
%T _X_ ABS _ Background Scans _4_		Scans _16_	
Acquisition & Calculation Time _42 sec_		Resolution _4.0 cm-1_	
Sample Condition _liquid, neat_		Cell Window _KBr_	
Cell Path Length _capillary film_		Matrix Material _____	

Figure 6.7 IR spectrum: cyclohexanol (crude product).

These changes indicate the reductive formation of an alcohol group from the carbonyl system. The band centered near 3300 cm^{-1} results from the single polarized O—H stretching mode. The drop in intensity of the 1715-wavenumber band indicates the loss of the carbonyl function. The exact amount of cyclohexanone remaining can be determined by carrying out a Beer's law type analysis, but in this case we will use gas chromatographic techniques to determine this value. Other bands of interest in the spectrum of cyclohexanol occur at 1069 and 1031 cm^{-1}. These bands can be assigned, respectively, to the equatorial and axial C—O stretching of the rotational conformers of this alicyclic secondary alcohol. A broad band (width ~300 cm^{-1}) can be found near 670 cm^{-1}. This absorption arises from an O—H bending, out-of-plane mode, of the associated alcohol. This band is generally identified only in neat samples where extensive hydrogen bonding occurs. Also note that the band at 1425 cm^{-1} has vanished because there are no methylene groups alpha to carbonyl systems in the product.

Separation of Small Quantities of Cyclohexanone from Cyclohexanol   Now proceed with purification of the reaction product by preparative gas chromatography. Use the following conditions and refer to Experiment [2] for the collection technique. If time permits, or perhaps in a later laboratory period, determine the infrared spectrum of the purified product. Describe and explain the changes observed in the new spectrum compared to that of the crude product.

Example

9:1 (v/v) cyclohexanol/cyclohexanone
10% Carbowax 20 M (stationary phase)
Injection volume: 15 μL
Temperature: 130 °C
He flow rate: 50 mL/min
Column: $\frac{1}{4}$ in. × 8 ft stainless steel
Chart speed: 1 cm/min

Run	Cyclohexanol Retention Time (min)	Yield (mg)	Cyclohexanone Retention Time (min)
1	15.3	6.5	11.6
2	17.3	7.3	12.5
3	17.2	8.5	12.6
4	16.0	9.6	12.0
5	14.6	7.3	11.2
6	14.5	8.7	11.2
7	15.5	8.9	11.7
8	15.5	8.9	11.8
9	16.4	8.8	12.3
10	15.4	8.4	12.7
Av	15.8 ± 1	8.3 ± 0.9	12.0 ± 0.6

Cyclohexanol injected = 0.9(15 μL)(0.963 mg/μL) = 13.0 mg

Percent yield = 8.3/13.0 × 100 = 63.8%

NOTE. Collection efficiencies approaching 90% can be obtained by cooling the collection tube. Liquid nitrogen-soaked tissues work best, but methanol-soaked tissues or ice water can give a significant improvement.

Chemical Tests. Several chemical tests (see Chapter 9) may also be used to establish that an alcohol has been formed by the reduction of a ketone. Perform the ceric nitrate and 2,4-dinitrophenylhydrazine test on both the starting ketone and the alcohol product. Do your results demonstrate that an alcohol was obtained? You may also wish to prepare a phenyl- or α-naphthylurethane derivative of the cyclohexanol. Before the development of chemical instrumentation, the formation of solid derivatives was used extensively to identify reaction products.

Experiment 5B

cis- and *trans-*4-*tert*-Butylcyclohexanol

REACTION

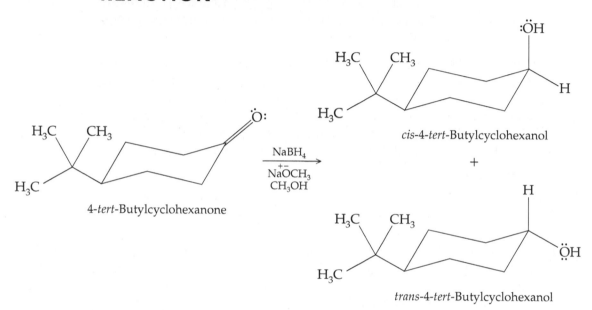

4-*tert*-Butylcyclohexanone

$\xrightarrow[\substack{\text{NaOCH}_3 \\ \text{CH}_3\text{OH}}]{\text{NaBH}_4}$

cis-4-*tert*-Butylcyclohexanol

+

trans-4-*tert*-Butylcyclohexanol

EXPERIMENTAL PROCEDURE

Estimated time to complete the experiment: 2.0 h.
For the GC analysis, 15 min per student.

Physical Properties of Reactants					
Compound	MW	Amount	mmol	mp (°C)	bp (°C)
4-*tert*-Butylcyclohexanone	154.25	50 mg	0.33	47–50	
Methanol	32.04	50 μL			65
Sodium borohydride reducing solution		100 μL			

4-*tert*-Butylcyclohexanone,
50 mg + CH₃OH, 50 μL +
NaBH₄ solution, 100 μL

Reagents and Equipment. In a tared 3.0-mL conical vial equipped with an air condenser weigh and place 50 mg (0.33 mmol) of 4-*tert*-butylcyclohexanone followed by 50 μL of methanol (◄). Gently stir using a glass stir rod (or a magnetic spin vane if desirable) the contents of the vial to obtain a homogeneous solution.

Now slowly add 100 μL of the sodium borohydride reducing solution while stirring.

NOTE. The liquid reagents are dispensed by use of automatic delivery pipets. The preparation of the reducing solution is given in Experiment [5A], Reagents and Equipment (p. 158).

Reaction Conditions.   After allowing the resulting solution to stand 10 min at room temperature, begin monitoring the resulting solution by TLC. Using as a solvent system of ethyl acetate:hexane (1:4), the R_f of 4-*tert*-butylcyclohexanol is 0.3 (R_f of 4-*tert*-butylcyclohexanone is 0.6) when stained with a solution of *p*-anisaldehyde (135 mL ethanol, 5 mL H_2SO_4, 1.5 mL glacial acetic acid, and 3.7 mL *p*-anisaldehyde).

Isolation of Product.   When the reaction is complete as judged by TLC, work up the resulting solution using the procedure described in Experiment [5A], Isolation of Product (p. 158), with the exception that 250 mg of sodium sulfate is placed in the Pasteur filter pipet (➡).

In the **hood**, remove the dried methylene chloride solvent from the final solution by directing a gentle stream of nitrogen gas onto the liquid surface while at the same time externally warming the vial in a sand bath. The use of a heating bath will help to avoid moisture condensation on the residue during solvent evaporation.

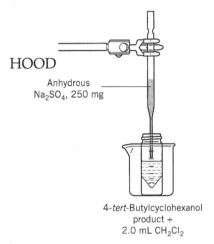

HOOD

Anhydrous
Na_2SO_4, 250 mg

4-*tert*-Butylcyclohexanol
product +
2.0 mL CH_2Cl_2

Purification and Characterization.   The product mixture remaining after removal of the methylene chloride is normally of sufficient purity for direct characterization. Weigh the solid product and calculate the percent yield. Determine the melting point of your material. 4-*tert*-Butylcyclohexanol (mixed isomers) has a melting point of 62–70 °C.

Obtain the IR and NMR spectra of the crude mixture of isomers. Infrared sampling in this instance is best accomplished by the capillary film-melt (use the heat lamp) technique (see Chapter 8).

Infrared Analysis.   Refer to the discussion in Experiment [5A] for an interpretation of the absorption bands found at 3435, 3000–2850, 1717, and 1425 cm^{-1} in the starting material (Fig. 6.8), and at 3250, 3000–2850 (1717 variable relative intensity—may be quite weak, why?), 1069, and 1031 cm^{-1} in the crude alcohol (Fig. 6.9).

In addition, the ketone has bands at 1396 (weak) and 1369 (strong), and the alcohol has bands at 1399 (weak) and 1375 (strong) cm^{-1}. These two pairs of bands establish the presence of the tertiary butyl group in these compounds.

Note that a weak band (3495 cm^{-1}) is present on the high-wavenumber side of the 3250-cm^{-1} O—H stretching mode and that even in neat samples of the tertiary butyl derivative, the 670-wavenumber band, clearly evident in cyclohexanol, is difficult to observe.

The mixture of two diastereomeric alcohols that have been synthesized provides an ideal opportunity to introduce you to nuclear magnetic resonance (NMR) spectroscopy. This technique is an extremely powerful tool for the discrimination and characterization of diastereomeric compounds. As you will see, the two diastereomers have quite different NMR spectra. An interpretation of these spectra will allow you to determine the ratio of the two isomers and to make an unambiguous assignment of their relative stereochemistry.

This experiment gives you an opportunity to obtain and interpret NMR data if you have access to NMR equipment. The two diastereomeric alcohols exhibit different splitting patterns for the proton on the carbon bearing

Sample ___4-tert-Butylcyclohexanone___

%T _X_ ABS _ Background Scans _4___   Scans _____16_____

Acquisition & Calculation Time _42 sec_   Resolution _4.0 cm⁻¹_

Sample Condition ___solid – melt___   Cell Window _____KBr_____

Cell Path Length ___capillary film___   Matrix Material _____

Figure 6.8 IR spectrum: 4-*tert*-butylcyclohexanone.

the —OH group. These signals can be integrated to determine the ratio of the diastereomeric alcohols in the sample.

Nuclear Magnetic Resonance Analysis. Refer to the expanded NMR spectrum in Figure 6.10. The signals at about 4.04 and 3.52 ppm correspond to the proton on the carbon bearing the O—H group in the two diastereomers of 4-*tert*-butylcyclohexanol shown. On closer inspection, the downfield signal (4.04 ppm) is a pentet and the upfield signal (3.52 ppm) is a triplet of triplets. The pentet implies that the proton in question is coupled with equal coupling constants (*J*) to four adjacent protons. The triplet of triplets implies that the proton in question is coupled to two adjacent protons with a large coupling constant and to two other adjacent protons with a smaller coupling constant. Specifically, the proton in the first case must be equatorial and the

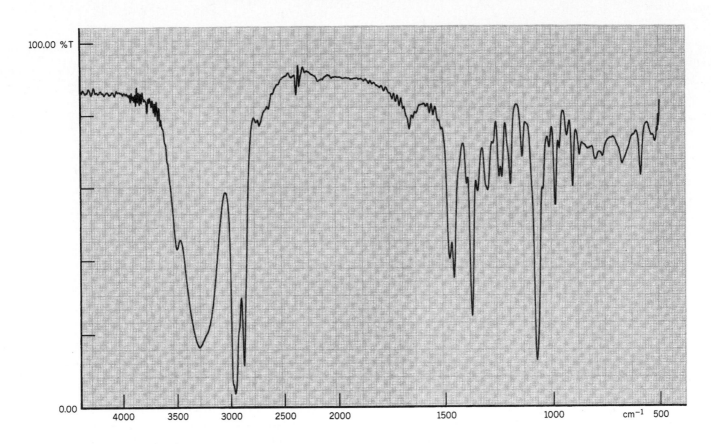

Sample	4-tert-Butylcyclohexanol (crude product and mixed isomers)		
%T _X_ ABS __ Background Scans _4_		Scans _16_	
Acquisition & Calculation Time _42 sec_		Resolution _4.0 cm⁻¹_	
Sample Condition _solid_		Cell Window __	
Cell Path Length __		Matrix Material _KBr_	

Figure 6.9 IR spectrum: 4-*tert*-butylcyclohexanol (crude product and mixed isomers).

proton in the second case must be axial, because the dihedral angle between an equatorial proton and each of the four adjacent protons is the same, about 60°, J = 2–3 Hz. When a proton is axial, the dihedral angle to the two adjacent equatorial protons is about 60° ($J \approx 3$ Hz) and the dihedral angle to the two adjacent axial protons is about 180° ($J \approx 13$ Hz), thus producing a triplet of triplets.

Cis

Trans

Gas Chromatographic Analysis. The cis and trans isomers of 4-*tert*-butylcyclohexanol may be separated by gas chromatography using a $\frac{1}{4}$-in. × 8-ft,

Figure 6.10   NMR spectrum: *cis*- and *trans-4-tert*-butylcyclohexanol.

10% FFAP column at 110 °C. Prepare a methylene chloride solution of the alcohol mixture having a concentration of 0.5 mg/μL and inject a 5.0-μL sample into the GC apparatus. At a flow rate of 50 mL/min (He), the retention time of the cis isomer is ~13 min, and the retention time of the trans isomer is ~16 min.

Determine the percentage of each isomer present in the sample by determining the area under each curve.

NOTE. If a Carbowax column (170 °C) is used, the starting ketone has a retention time similar to that of the cis alcohol. Therefore, if the reaction does not go to completion, the apparent cis/trans ratio may not be accurate. It has recently been demonstrated that 10% FFAP columns will resolve all components present in the product mixture. Thus, the starting ketone and the cis isomer concentrations may be effectively established in addition to the trans isomer.

The latter separation scheme was developed by T. J. Dwyer and S. Jones at the University of California, San Diego.

NOTE. Several techniques may be used to calculate the area under a curve (mm²), but the following method is satisfactory for your needs and gives reproducible results of ±3–4%: Multiply the peak height (mm) by the width at one-half height (mm), measured from the base line of the curve.

QUESTIONS

6-28. Suggest a chemical test that would allow you to distinguish between *tert*-butyl alcohol and 1-butanol, both of which give a positive ceric nitrate test.

6-29. Which of the isomeric pentanols ($C_5H_{12}O$) can be prepared by reduction of a ketone with sodium borohydride?

6-30. Why are there axial and equatorial hydroxyl isomers for 4-*tert*-butylcyclohexanol, but not for cyclohexanol itself?

6-31. What aldehyde or ketone would you reduce to prepare the following alcohols?
    (a) 3-Hydroxymethylfuran   **(b)** 3,3-Dimethyl-1-butanol   **(c)** 3-Methyl-1-pentanol

6-32. The diastereomers *cis*- and *trans*-4-*tert*-butylcyclohexanol prepared in Experiment [5B] each have a plane of symmetry. Draw this symmetry element for each of the diastereomers.

6-33. In the spectrum of the crude product obtained from the reduction of 4-*tert*-butylcyclohexanone, the fingerprint region appears to possess bandwidths that are slightly broader than those found in cyclohexanol. Explain.

6-34. The reduction of 4-*tert*-butylcyclohexanone is a stereoselective reaction. Which isomer predominates? If you do not have NMR data available, it is still possible to arrive at a rough estimate of the product ratio. How would you go about this measurement? Suggest a value.

6-35. In the spectrum of the crude 4-*tert*-butylcyclohexanols one observes: **(a)** a weak band (3495 cm^{-1}) located on the high side of the 3250-cm^{-1} O—H stretching mode. **(b)** Even in neat samples of this alcohol, the 670-cm^{-1} band, clearly evident in cyclohexanol, is difficult to observe. Explain these observations. Is the same effect operating in both cases?

6-36. Sketch the proton NMR spectrum you would expect to observe for the following compounds.
    (a) 2-Butanone
    (b) 1,1,2-Trichloroethane
    (c) 1-Chloropropane
    (d) 3,3-Dimethyl-2-butanone
    (e) 1-Bromo-4-methoxybenzene

BIBLIOGRAPHY

General references on metal hydride reduction:

Itsuno, S. *Org. React.* **1998**, *52*, 395.

Seyden-Penne, J. *Reductions by Alumino- and Borohydrides in Organic Synthesis*; VCH-Lavoisier: Paris, 1997.

Walker, E. R. H. *Chem. Soc. Rev.* **1976**, *5*, 23.

Sodium borohydride as a reducing agent:

Brown, H. C. *Organic Synthesis via Boranes*; Wiley: New York, 1975.

Cragg, G. M. W. *Organoboranes in Organic Synthesis*; Marcel Dekker: New York, 1973.

Ems-Wilson, J. *J. Chem. Educ.* **1996**, *73*, A171.

Fieser, L. F.; Fieser, M. *Reagents for Organic Synthesis*; Wiley: New York, 1967; Vol. I, p. 1050 and subsequent volumes.

Paquette, L., Ed. *Encyclopedia of Reagents for Organic Synthesis*; Wiley: New York, 2004.

Lithium aluminum hydride as a reducing agent:

Brown, W. G. *Org. React.* **1951**, *6*, 469.

Fieser, L. F.; Fieser, M. *Reagents for Organic Synthesis*; Wiley: New York, 1967; Vol. I, p. 581 and subsequent volumes.

trans-4-tert-Butylcyclohexanol has been prepared from the ketone using LiAlH₄ as the reducing agent:

Eliel, E. L.; Martin, R. J. L.; Nasipuri, D. *Organic Syntheses*; Wiley: New York, 1973; Collect. Vol. V, p. 175.

See *Organic Syntheses*, Collect. Vols., for the use of these reagents in specific reductions.

EXPERIMENT
6

Photochemical Isomerization of an Alkene: cis-1,2-Dibenzoylethylene

Common names: *cis*-1,2-dibenzoylethylene
cis-1,4-diphenyl-2-butene-1,4-dione
CA number: [959-27-3]
CA name as indexed: 2-butene-1,4-dione, 1,4-diphenyl-, (Z)-

Purpose. This exercise illustrates the ease of cis–trans isomerization in organic molecules and, specifically in this case, demonstrates the isomerization of a trans alkene to the corresponding cis isomer via photochemical excitation.

B BIOLOGICALLY IMPORTANT PHOTOCHEMICAL REACTIONS

A number of important biochemical reactions are promoted by the adsorption of UV–vis radiation (see Appendix, UV–Vis Discussion). Vitamin D_3, which regulates calcium deposition in bones, is biosynthesized in just such a photochemical reaction. This vitamin is formed when the provitamin, 7-dehydrocholesterol, is carried through fine blood capillaries just beneath the surface of the skin and exposed to sunlight. The amount of radiation exposure, which is critical for the regulation of the concentration of this vitamin in the blood stream, is controlled by skin pigmentation and geographic latitude. Thus, the color of human skin is an evolutionary response to control the formation of vitamin D_3 via a photochemical reaction.

7-Dehydrocholesterol   Vitamin D_3

Another set of significant photochemical reactions in human biochemistry is contained in the chemistry of vision. These reactions involve vitamin A_1 (retinol), which is a C20 compound belonging to a class of compounds known as diterpenes. These compounds are molecules formally constructed by the biopolymerization of four isoprene, CH_2=$C(CH_3)$—CH=CH_2, molecules. Retinol (an all-trans pentene) is first oxidized via liver enzymes (biological catalysts) to vitamin A aldehyde (*trans*-retinal). The *trans*-retinal, which is present in the light-sensitive cells (the retina) of the eye, undergoes further enzymatic transformation (retinal isomerase) to give *cis*-retinal (a second form of vitamin A aldehyde) in which one of the double bonds of the all-trans compound is isomerized.

Vitamin A (retinol)

1. liver enzymes (—CH_2OH → —CHO)
2. retinal isomerase (trans → cis)

cis linkage

Vitamin A aldehyde (*cis*-retinal)

The cis isomer of vitamin A aldehyde (retinal) possesses exactly the correct dimensions to become coupled to opsin, a large protein molecule (MW ~38,000) (coupling involves a reaction of the retinal aldehyde group, —C(H)=O, with an amine group [—NH$_2$] of the protein to form an imine linkage [RCH=NR]), to generate a light-sensitive substance, rhodopsin. This material is located in the rodlike structures of the retina. When the protonated form of rhodopsin (—CH=N$^+$HR), which absorbs in the blue-green region of the visible spectrum near 500 nm, is exposed to radiation of this wavelength, isomerization of the lone cis double bond of the diterpene group occurs and *trans*-rhodopsin is formed:

cis-Rhodopsin   *trans*-Rhodopsin

This photoreaction (a fast reaction, 10^{-12} s) involves a significant change in the geometry of the diterpene group, which eventually (10^{-9} s) results in both a nerve impulse and the separation of *trans*-retinal from the opsin. The trans isomer is then enzymatically reisomerized back to the cis compound, which then starts the initial step of the visual cycle over again.

There are two interesting facts about this reaction: (1) This reaction is incredibly sensitive. A single photon will cause the visual nerve to fire. (2) All known visual systems use *cis*-retinal, regardless of their evolutionary trail.

The photoreaction that we study next is very similar to the cis–trans double-bond isomerism found in the vitamin A visual pigments. The only difference is that in our case we will be photochemically converting a trans double-bond isomer to the cis isomer.

Prior Reading

    Technique 5: Crystallization

                Introduction (pp. 89–91)

                Use of the Hirsch Funnel (pp. 92–93)

                Craig Tube Crystallization (pp. 93–95)

                Recrystallization Pipet

    Technique 6A: Thin-Layer Chromatography (pp. 101–103)

REACTION

trans-1,2-Dibenzoylethylene   *cis*-1,2-Dibenzoylethylene

DISCUSSION

The π bond of an alkene (C=C) is created by overlap of the sp^2 hybridized carbons' p orbitals. Rotation about the axis of the C=C requires a good deal of energy because it destroys the p orbital overlap, and thus the π bond. Unless the material is irradiated with light of the appropriate wavelength, absorption of the radiation does not take place, and the isomers do not interconvert unless sufficient thermal energy (60–65 kcal/mol, typically >200 °C) is supplied to break the π bond. If one of the electrons in the π bond (π molecular orbital) is photochemically excited into an antibonding, π^*, molecular orbital, as occurs in this experiment, the π bond is weakened significantly. Rotation about this bond can then occur rapidly at room temperature.

It is the high-energy barrier to rotation about the C=C that gives rise to the possibility of alkene stereoisomers. The cis and trans isomers of an alkene system are called *diastereoisomers*, or *diastereomers*. Like all stereoisomers, these isomers differ only in the arrangement of the atoms in space. These isomers have all the same atoms bonded to each other. Diastereomers are not mirror images of each other and, of course, are not superimposable (identical). This particular type of diastereomer often was referred to in the older literature as a "geometric isomer." Diastereomers, therefore, would be expected to possess different physical properties, such as melting points, boiling points, dipole moments, densities, and solubilities, as well as different spectroscopic properties. Because of these differences in physical properties, diastereomers are amenable to separation by chromatography, distillation, crystallization, and other separation techniques. We use both chromatographic and crystallization methods in the present experiment.

The course of the isomerization may be followed using *thin-layer chromatography* (Experiment [6B]) or by spectroscopic techniques using NMR analysis (Experiment [6C]).

The photochemical isomerization of a diazabicyclohexene system is presented in Experiment [F4].

Experiment 6A

Purification of *trans*-1,2-Dibenzoylethylene

EXPERIMENTAL PROCEDURE

Estimated time to complete the purification: 1.5 h of laboratory time.

Physical Properties of Components					
Compound	MW	Amount	mmol	mp (°C)	bp (°C)
trans-1,2-Dibenzoylethylene	236.27	100 mg	0.42	111	
Ethanol (95%)		6.0 mL			78.5
Methylene chloride		4.0 mL			40

Purification Conditions.   Purify the starting alkene by recrystallization.

Weigh and add to a 10-mL Erlenmeyer flask 100 mg (0.42 mmol) of *trans*-1,2-dibenzoylethylene and 3.0 mL of methylene chloride.

NOTE. If the melting point of the alkene was not supplied to you, set aside a small sample (1–2 mg) of the weighed sample so that the evacuated melting point of this initial material can be obtained later.

Add decolorizing charcoal pellets (10 mg) to this solution and swirl the mixture gently for several minutes. Use a Pasteur filter pipet to transfer the methylene chloride solution to a second 10-mL Erlenmeyer flask containing a boiling stone (remember to hold the necks of the two flasks close together with the fingers of one hand during the transfer).

Trans isomer + CH₂Cl₂, 3 mL

NOTE. If powdered charcoal is used instead of pellets, filter the solution using a Pasteur pipet packed with Celite, sand, and a cotton plug (Technique 6A). Rinse the filter paper with an additional 1 mL of methylene chloride and collect this rinse in the same flask.(➡)

Concentrate the filtered solution to dryness in a warm (50 °C) sand bath under a slow stream of nitrogen gas in the **hood**.

HOOD

Now add 95% ethanol (1–3 mL) to the flask and dissolve the yellow solid residue by warming in a sand bath, adding hot ethanol dropwise, until a homogeneous solution is obtained (➡).

Allow the solution to cool slowly to room temperature over a period of 15 min and then place it in an ice bath for an additional 10 min. Collect the yellow needles by vacuum filtration using a Hirsch funnel (➡) and then air-dry them on a porous clay plate or on filter paper.

trans–1,2-Dibenzoylethylene, 150 mg +CH₂Cl₂, 3.0 mL, charcoal pellets

Weigh the product and calculate the percent recovery. Determine the evacuated melting point and compare your result with both the literature value and that obtained with the material saved prior to recrystallization. If Experiment [3A] has been completed, compare the melting point of that sample of the trans alkene, which was obtained by concentration, via distillation, of an ethyl acetate solution. In the latter experiment, a simple crystallization was performed without the aid of decolorizing charcoal.

Isomerization of an Alkene: Thin-Layer Chromatographic Analysis

Experiment 6B

EXPERIMENTAL PROCEDURE

Estimated time to complete the experiment: 2.5 h of laboratory time. The reaction requires approximately 1 h of irradiation; the actual time to completion is quite sensitive to both radiation flux and temperature. These factors are largely determined by the distance the reaction vessel is positioned from the source of radiation.

Physical Properties of Reactants					
Compound	MW	Amount	mmol	mp (°C)	bp (°C)
trans-1,2-Dibenzoylethylene	236.27	25 mg	0.11	111	
Ethanol (95%)		3.5 mL			78.5

Reagents and Equipment. To a 13 × 100-mm test tube weigh and add 25 mg (0.11 mmol) of recrystallized *trans*-1,2-dibenzoylethylene and 3.0 mL of 95% ethanol.

Reaction Conditions. Use a sand bath to warm the mixture ***gently*** until a homogeneous solution is obtained. Stopper the test tube *loosely*, or cover

GENTLY

Filter paper + rubber band

← 2 in. →

$(C_6H_5CO)HC=CH(COC_6H_5)$, 25 mg + 95% ethanol, 3 mL

it with filter paper held in place by a rubber band, and then place it approximately 2–4 in. from a 275-W sun lamp. Irradiate the solution for approximately 1 h (◄).

NOTE. If a lower wattage lamp is used, longer irradiation times or shorter distances will be necessary. In either case, solvent evaporation can be significantly reduced by directing a flow of cool air (fan) over the reaction tubes. An alternative procedure is to allow the (sealed) tube to stand in sunlight at room temperature for several days.

The progress of the isomerization may be followed by thin-layer chromatography (TLC) analysis.

HOT

*INFORMATION. The TLC analysis is carried out using Eastman Kodak silica gel–polyethylene terephthalate plates with a fluorescent indicator. **Activate the plates at an oven temperature of 100 °C for 30 min** and then place them in a desiccator to cool until needed. After spotting, elute the plates using methylene chloride as the solvent. Visualize the spots with a UV lamp. The course of the reaction is followed by removing small samples (2–3 drops) of solution from the **hot** test tube at set time intervals with a Pasteur pipet and placing them in separate $\frac{1}{2}$-dram vials. See Technique 6A for the method of TLC analysis and the determination of R_f values. Approximate R_f values: trans = 0.72; cis = 0.64.*

Sample	cis-Dibenzoylethylene		
%T X ABS — Background Scans 4		Scans 16	
Acquisition & Calculation Time 42 sec		Resolution 4.0 cm⁻¹	
Sample Condition solid		Cell Window	
Cell Path Length		Matrix Material KBr	

Figure 6.11 **IR spectrum:** *cis-dibenzoylethylene.*

Isolation of Product. Remove the **hot** test tube from the light source and   HOT
allow the solution to cool to room temperature.

Place the resulting mixture in an ice bath to complete crystallization of
the *colorless* cis-1,2-dibenzoylethylene product. Collect the crystals by vacuum
filtration using a Hirsch funnel, wash them with 0.5 mL of cold 95% ethanol,
and then air-dry them on a porous clay plate or on filter paper.

Purification and Characterization. Weigh the dried product and calculate the percent yield. Determine the melting point (evacuated) and compare
your result with the literature value. The purity of the crude isolated product
may be further determined by TLC analysis (if not used above). Finally, if necessary, further purify a portion of the isolated product by recrystallization from
95% ethanol using a Craig tube.

Obtain IR spectra of the cis and trans isomers and compare them to Figures 6.11 and 6.12. Alternatively, or in addition to the IR analysis, the UV–vis
spectra may be observed in methanol solution and the results compared to
Figures 6.13 and 6.14.

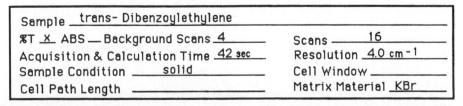

Figure 6.12   IR spectrum: *trans*-dibenzoylethylene.

Figure 6.13 UV spectrum: *trans*-dibenzoylethylene.

Ultraviolet–Visible Analysis. The bright yellow color of the *trans*-dibenzoylethylene rapidly fades as the conversion to the colorless cis compound progresses under irradiation. This visual observation may be supported by an examination of the absorption spectra of the isomers in a methanol solution from 225 to 400 nm (see Figs. 6.13 and 6.14). The λ_{max} of the trans isomer drops from 268 to 259 nm in the cis compound. This shift to shorter wavelengths is just enough to move the long-wavelength end of the absorption band in the trans isomer out of the visible region and into the near-ultraviolet (thus, the cis compound does not absorb light to which the eye is sensitive and the compound appears colorless). This observation is consistent with the theory that the spatial contraction of extended π systems moves the associated electronic transitions to higher energy gaps (higher frequencies or

Figure 6.14 UV spectrum: *cis*-dibenzoylethylene.

shorter wavelengths), which is exactly what takes place during the trans–cis isomerization.

Infrared Analysis. The infrared spectral changes are consistent with the proposed reaction product. Consider the spectrum of the trans starting material (Fig. 6.12). It contains the following:

1. The macro group frequency train (see Strategies for Interpreting Infrared Spectra, p. 542) for conjugated aromatic ketones is 3080 and 3030 (C—H, aromatic), 1652 (doubly conjugated carbonyl), 1599, and 1581 (v_{8a}, v_{8b} degenerate ring stretch, strong intensity of 1581-wavenumber peak confirms ring conjugation), 1495 and 1450 (v_{19a}, v_{19b} degenerate ring stretch, v_{19a} weak) cm^{-1}.
2. The monosubstituted phenyl ring macro group frequency train is 1980(d), 1920(d),[4] 1820, 1780 (mono combination-band pattern), 708 (C—H out-of-plane bend, C=O conjugated), 686 (ring puckering) cm^{-1}.
3. The presence of the trans double bond is indicated by the single medium intensity 970 cm^{-1} band, as the C—H stretching region is overlapped by the aromatic ring C—H stretches.

The spectrum of the cis photoproduct (Fig. 6.11) possesses the same macro group frequencies as the starting material:

1. The conjugated aromatic ketone frequency train is 3335 (overtone of C=O stretch), 3080 and 3040, 1667, 1601, 1581, 1498 (weak), and 1450 cm^{-1}.
2. The monosubstituted phenyl ring macro frequency train is assigned peaks of 1990, 1920, 1830, 1795, 710, and 695 cm^{-1}.
3. The cis double bond is clearly present and utilizes a macro frequency train of 3030 (overlapped by aromatic C—H stretch), 1650 (C=O stretch, shoulder on the low-wavenumber side of conjugated carbonyl; resolved in spectra run on thin samples), 1403 (=C—H out-of-phase, in-plane bending mode, strong band not present in trans compound), 970 (trans, in-phase, out-of-plane bend missing), 820 (cis, in-phase out-of-plane bend; band not found in spectrum of trans isomer) cm^{-1}.

Discuss the similarities and differences of the experimentally derived spectral data to the reference spectra (Figs. 6.11 and 6.12).

Optional Isolation of the Thermodynamically Most Stable Reaction Product Under the Conditions Used to Carry Out the Above Reaction. If the photo reaction exposure is continued, the cis isomer, which is formed quickly, slowly undergoes conversion to a more stable, new product, ethyl 4-phenyl-4-phenoxy-3-butenoate. This conversion of the intermediate cis isomer may be followed conveniently by thin-layer chromatography. Maximum yields are obtained over approximately 24 h under the above reaction conditions. The product has $R_f = 0.8$ in 1:1 hexane/methylene chloride. Evaporation of the solvent yields a yellow oil.

---

[4]d = double peak.

Scan Time: 12 min
Slit Program: normal
Concentration: neat
Cell Path length: capillary film
Operator: E.M.
Date 7/20/87
Sample: Ethyl 4-phenyl-4-phenoxy-3-butenoate

Figure 6.15   IR spectrum: ethyl 4-phenyl-4-phenoxy-3-butenoate.

The infrared spectrum of this new material (Fig. 6.15) indicates that significant changes have occurred in the structure of the material. The proposed structure involves a rather spectacular molecular rearrangement of the *cis*-unsaturated ketone to yield an ethyl ester containing a phenoxy-substituted double bond. Can you rationalize the data to fit this structure? Suggest possible macro group frequencies that are present in the IR spectrum of the rearranged product.

Mixture Melting-Point Measurements.   Observe a series of evacuated mixture melting points (see Chapter 4) with isomer ratios of 75:25, 50:50, and 25:75. These values will allow you to *estimate* the eutectic temperature of this system. This same technique is used in Experiment [29A] to aid in identifying the isolated product, 2,5-dichloronitrobenzene, which has a melting point only 3 °C higher than 1,4-dichlorobenzene, the starting material in this nitration reaction.

Isomerization of an Alkene: Nuclear Magnetic Resonance Analysis

EXPERIMENTAL PROCEDURE

Estimated time to complete the experiment: 0.5 h.

Physical Properties of Reactants					
Compound	MW	Amount	mmol	mp (°C)	bp (°C)
trans-1,2-Dibenzoylethyene	236.27	5 mg	0.02	111	
Chloroform-*d*	120.39	500 µL			62

Reagents and Equipment.   Prepare a sample of 4–5 mg of recrystallized *trans*-1,2-dibenzoylethylene in about 500 µL of CDCl$_3$ in an NMR tube. Using a fine capillary, spot a silica gel TLC plate (see Experiment [6B]) with a sample of this solution. Use TLC to track the results of the NMR experiment.

Obtain an NMR spectrum of this solution.

Reaction Conditions

NOTE. A 15-min exposure is normally adequate, but, as can be seen by TLC analysis, complete isomerization of this alkene may not occur in this short a time.

Clamp the NMR tube 3–4 in. (see comments above, Experiment [6A]) from a 275-W sun lamp for 15–20 min (see Experiment [6B]).

Analysis of the Results.   After irradiation, spot the original TLC plate with this irradiated solution and elute the plate with methylene chloride/hexane (1:1) solvent (see Experiment [6B]). Obtain an ^{1}H NMR spectrum of the irradiated solution.

Does the TLC analysis correlate with the NMR data? Is there evidence that the isomerization did occur?

The data in the following table were taken on a 300-MHz NMR instrument. Which signal in each isomer comes from the hydrogen atoms attached to the C=C group?

^{1}H Chemical Shift Data	
Trans Isomer (ppm)	Cis Isomer (ppm)
8.06 doublet	7.9 doublet
8.01 singlet	7.55 triplet
7.63 triplet	7.43 triplet
7.52 triplet	7.14 singlet

QUESTIONS

6-37. What properties should an ideal recrystallization solvent have?

6-38. What is meant by the term *solvent pair* when it is used in reference to the recrystallization of solids?

6-39. What is the purpose of adding powdered charcoal to the solvent system during a recrystallization sequence?

6-40. Why do we obtain larger crystals if a solution of an organic compound is allowed to cool slowly?

6-41. List several advantages of using the Craig tube to purify crystalline products in contrast to using the Hirsch funnel filtration method.

6-42. The stereochemistry of the more highly substituted alkenes is difficult to define using the cis and trans designations. Therefore, a more systematic manner of indicating stereochemistry in these systems has been developed that uses the E and Z nomenclature system. Draw the structures of the E and Z stereoisomers of 1,4-diphenyl-2-butene-1,4-dione used in this experiment. In this case, which is cis and which is trans?

6-43. The cis H—C≡C—H in-phase out-of-plane bending frequency normally is found in the range 740–680 cm^{-1} and is broad. In addition, this vibrational mode can be quite variable in intensity. On conjugation with C═O groups, the band rises to near 820 cm^{-1}.

    (a) Can you explain the underlying cause of this observation? (*Hint:* See the discussion of cis double-bond group frequencies in Chapter 8.)

    (b) Can you explain why a broad band near 690 cm^{-1} assigned to this out-of-plane mode is found in the cis product obtained in this reaction?

6-44. In *trans*-1,2-dibenzoylethylene, even though the conjugated C═O group frequency coincides directly with the expected C═C stretching frequency, we would not have expected to observe the latter stretching mode in the infrared spectrum. Why?

6-45. Explain how this photochemical isomerization allows the production of the thermodynamically *less* stable cis isomer. In other words, why is the trans isomer exclusively converted to the cis isomer during short reaction periods and not vice versa? Is it possible, under these conditions, that the trans and cis isomers are in equilibrium with one another?

BIBLIOGRAPHY

The photochemical isomerization was adapted from the following references:

Pasto, D. J.; Ducan, J. A.; Silversmith, E. F. *J. Chem. Educ.* **1974**, *51*, 277.

Silversmith, E. F.; Dunsun, F. C. *J. Chem. Educ.* **1973**, *50*, 568.

Reviews on photochemical isomerization reactions may be found in the following references:

Arai, T.; Tokumaru, K. *Chem. Rev.* **1993**, *93*, 23.

Coxton, J. M.; Halton, B. *Organic Photochemistry*, 2nd ed.; Cambridge University Press: New York, 1987.

Coyle, J. D. *Introduction to Organic Photochemistry*; Wiley: New York, 1991.

Kagan, J. *Organic Photochemistry: Principles and Applications*; Academic Press: Orlando, FL, 1993.

Kopecky, J. *Organic Photochemistry: A Visual Approach*; Wiley/VCH: New York, 1992.

Prasad, P. N. *Introduction to Biophotonics*; Wiley: New York, 2003.

EXPERIMENT 7

The Cannizzaro Reaction with 4-Chlorobenzaldehyde: 4-Chlorobenzoic Acid and 4-Chlorobenzyl Alcohol[5]

Common name: 4-chlorobenzoic acid

CA number: [74-11-3]

CA name as indexed: 4-chlorobenzoic acid

Common name: 4-chlorobenzyl alcohol

CA number: [873-76-7]

CA name as indexed: benzenemethanol, 4-chloro-

---

[5]Portions of this experiment were previously published: Mayo, D. W.; Butcher, S. S.; Pike, R. M.; Foote, C. M.; Hotham, J. R.; Page, D. S. *J. Chem. Educ.* **1985**, *62*, 149.

Stanislao Cannizzaro (1826–1910).   This famous reaction is named for Stanislao Cannizzaro. He was born in Palermo, Italy, where he became interested in chemistry at an early age and studied with Professor Piria at the University of Pisa. He arrived in Paris after participating in the 1847 Sicilian uprising. By 1851 he had moved to Egypt where he became Professor of Chemistry at the National School of Alexandria. It was here that he discovered the reaction that carries his name. In the process he also was the first person to identify benzyl alcohol, which he obtained from the reaction of potash (KOH) with benzaldehyde. It is interesting to note that benzyl alcohol was the first alcohol of the aromatic series to be isolated and characterized. Cannizzaro devoted extensive work to the study of aromatic alcohols and he was the first to propose the term *hydroxyl* for the ——OH group. Later he demonstrated the conversion of aromatic alcohols of the benzyl class into their corresponding halides and further to their phenyl acetic acid derivatives. In 1855 he moved to Genoa and in 1861 he moved back to his birthplace to become Professor of Chemistry at Palermo. Finally, 10 years later, at the young age of 45, he assumed the chemistry chair at the University of Rome, which he held for 39 years! Upon his move to Rome his old political interests surfaced and he became a member of the Italian Senate that same year. He later carried the honor of serving as vice president of that august body. Cannizzaro was also honored by the Royal Society of London in 1891 when he was awarded the Copley Medal for his investigations of atomic and molecular weights. He was quick to recognize the importance of Avogadro's hypothesis, and his studies played a key role in placing this hypothesis on a sound scientific basis, which ultimately was the spark that ushered chemistry into the modern era.[6]

Purpose.   This experiment illustrates the simultaneous oxidation and reduction of an aromatic aldehyde to form the corresponding benzoic acid and benzyl alcohol. The experiment further demonstrates the techniques for separation of a carboxylic acid from a neutral alcohol. For a detailed discussion of this extraction procedure, see Experiment [4C].

Prior Reading

> *Technique 4:* Solvent Extraction
>> Liquid–Liquid Extraction (p. 76)
>> Separation of Acids and Bases (pp. 81–83)
>
> *Technique 5:* Crystallization
>> Use of the Hirsch Funnel (pp. 92–93)
>> Craig Tube Crystallization (pp. 93–95)
>
> *Technique 6A:* Thin-Layer Chromatography (pp. 101–103)
>
> *Chapter 8:* Infrared Spectroscopy (pp. 615–630)

---

[6]See Newell, L. C. *J. Chem. Educ.* **1926,** *3*, 1361; Parravano, N. *J. Chem. Educ.,* **1927,** *4*, 836; Tilden, W. A. *J. Chem. Soc.* **1912,** 1677; *Dictionary of Scientific Biography*; Gillespie, C. C., Ed.; Scribner's: New York, 1971; Vol. III, p. 45; Surrey, A. R. *Name Reactions in Organic Chemistry*; Academic Press: New York, 1954; p. 27.

REACTION

$$2 \ Cl\text{—}\langle\bigcirc\rangle\text{—}\overset{\overset{\cdot\cdot}{O}\text{:}}{\underset{}{C}}\text{—H} \xrightarrow[\text{2. H}^+]{\text{1. KOH}} Cl\text{—}\langle\bigcirc\rangle\text{—}CH_2\text{—}\overset{\cdot\cdot}{O}H \ + \ Cl\text{—}\langle\bigcirc\rangle\text{—}\overset{\overset{\cdot\cdot}{O}\text{:}}{\underset{\overset{\cdot\cdot}{O}H}{C}}$$

| 4-Chlorobenzaldehyde | 4-Chlorobenzyl alcohol | 4-Chlorobenzoic acid |

DISCUSSION

The carbonyl group of an aldehyde represents the intermediate stage of oxidation (ox) (or reduction [red]) between an alcohol and a carboxylic acid:

$$\overset{H}{\underset{H}{\text{—C—}\overset{\cdot\cdot}{O}H}} \underset{\text{red}}{\overset{\text{ox}}{\rightleftharpoons}} \overset{\overset{\cdot\cdot}{O}\text{:}}{\underset{H}{\text{—C}}} \underset{\text{red}}{\overset{\text{ox}}{\rightleftharpoons}} \overset{\overset{\cdot\cdot}{O}\text{:}}{\underset{\overset{\cdot\cdot}{O}H}{\text{—C}}}$$

| Primary alcohol | Aldehyde | Carboxylic acid |

It is not surprising, then, to find a reaction in which a specific type of aldehyde will undergo an oxidation–reduction sequence to form the corresponding alcohol and carboxylic acid. Such a reaction is the **Cannizzaro reaction**. In the presence of hydroxide ion, aldehydes that lack (acidic) *alpha (α)-hydrogen* atoms undergo a self-oxidation–reduction reaction. Thus, under the influence of strong base, one molecule of the aldehyde reduces a second molecule of aldehyde to the primary alcohol and in process is itself oxidized to the corresponding carboxylate anion. Aldehydes with α-hydrogen atoms, however, do not undergo this reaction, because in the presence of base these α-carbon atoms are deprotonated. The resulting enolate generally leads to an aldol reaction (see Experiments [20] and [A3ₐ]).

The first step in the mechanism is the nucleophilic attack of the hydroxide anion on the carbonyl group of the aldehyde. This attack is followed by the key step in the reaction, the transfer of a hydrogen atom with its pair of electrons (a hydride ion) to the carbonyl group of a second molecule of aldehyde:

$$R\text{—}\overset{\overset{\cdot\cdot}{O}\text{:}}{\underset{H}{C}} + \text{:}\overset{\cdot\cdot}{O}H^- \rightleftharpoons R\text{—}\overset{\text{:}\overset{\cdot\cdot}{O}\text{:}^-}{\underset{\text{:}\overset{\cdot\cdot}{O}H}{C}}\text{—H}$$

$$R\text{—}\overset{\text{:}\overset{\cdot\cdot}{O}\text{:}^-}{\underset{\text{:}\overset{\cdot\cdot}{O}H}{C}}\text{—H} + R\text{—}\overset{\overset{\cdot\cdot}{O}\text{:}}{\underset{H}{C}} \xrightarrow[\text{transfer}]{\text{hydride}} R\text{—}\overset{\overset{\cdot\cdot}{O}\text{:}}{\underset{\overset{\cdot\cdot}{O}H}{C}} + R\text{—}\overset{\text{:}\overset{\cdot\cdot}{O}\text{:}^-}{\underset{H}{C}}\text{—H}$$

$$(R = Ar, \ R_3C, \ H)$$

$$\Updownarrow$$

$$R\text{—}\overset{\overset{\cdot\cdot}{O}\text{:}}{\underset{\overset{\cdot\cdot}{O}\text{:}^-}{C}} + R\text{—}\overset{\text{:}\overset{\cdot\cdot}{O}H}{\underset{H}{C}}\text{—H}$$

The strong electron-donating character of the negatively charged oxygen atom in the first intermediate anion greatly facilitates the ability of the aldehydic hydrogen to be transferred as a hydride ion—that is, *with* its pair of electrons (:H⁻) to a second molecule of aldehyde. As seen in the preceding diagram, this nucleophilic attack on the carbon of a carbonyl group leads to the formation of a carboxylic acid and an alkoxide anion. The equilibrium established in the final step lies far to the right, and involves a fast acid–base reaction that yields both an alcohol and a carboxylate anion. *Thus, even though the Cannizzaro reaction is an equilibrium reaction, it proceeds nearly to completion.* The proposed mechanism is supported by evidence obtained by carrying out the reaction in D_2O. It was found that under these conditions the product alcohol did not contain any α-deuterium substitution. This result suggests that the transferred hydride ion must come from the aldehyde group and not from the solvent.

Since few aldehydes lack α-hydrogen atoms, the Cannizzaro reaction is of limited use in modern synthetic sequences, although a variation called the *crossed* Cannizzaro is occasionally used for the reduction of aromatic aldehydes. In this reaction excess formaldehyde is mixed with the aromatic aldehyde and becomes preferentially activated by base since it is present in excess. Thus, the aldehyde present to the lesser extent is reduced to the primary alcohol in high yield. For example,

| *p*-Methyl-benzaldehyde | Formaldehyde | | *p*-Methylbenzyl alcohol (90%) | Formic acid |

In addition, it is known that α-ketoaldehydes undergo an internal Cannizzaro reaction to yield α-hydroxy carboxylic acids:

(α-Ketoaldehyde) (α-Hydroxyacid)

EXPERIMENTAL PROCEDURE

Estimated time to complete the experiment: two laboratory periods.

Physical Properties of Reactants						
Compound	MW	Amount	mmol	mp (°C)	bp (°C)	n_D
4-Chlorobenzaldehyde	140.57	150 mg	1.1	47.5		
Methanol	32.04	400 μL			65	1.3288
Potassium hydroxide (11 M)	56.11	550 μL	6.05			

65–75 °C
thermometer

Speed   Heat

4–Cl(C₆H₄)CHO, 150 mg
+ KOH/CH₃OH, 0.4 mL

NOTE. This reaction may be carried out in a centrifuge tube containing a boiling stone. The tube should be covered by a piece of filter paper held in place by a rubber band.

Reagents and Equipment. Weigh and add to a 5.0-mL conical vial containing a magnetic spin vane and equipped with a reflux condenser, 150 mg (1.1 mmol) of 4-chlorobenzaldehyde followed by 0.4 mL of methanol (◂). With gentle swirling, add 0.55 mL of an 11 M aqueous solution of potassium hydroxide.

NOTE. It is convenient to dispense the methanol and KOH solution using automatic delivery pipets. The glass equipment should NOT be rinsed or cleaned with acetone followed by air-drying. The residual acetone undergoes an aldol reaction that forms high-boiling contaminants, which interfere with the isolation of the desired products. Also, because **strong alkali** *is being used in the reaction, it is important to lightly grease the ground joint of the condenser.*

CAUTION: The concentrated methanolic KOH solution is very caustic. Do not allow it to come into contact with the skin or eyes.

Reaction Conditions. Once the reaction mixture reaches the temperature range of 65–75 °C for a period of 30 min, begin monitoring the reaction progress by TLC. Using a solvent system of ethyl acetate:hexane (1:4) and visualization by a UV lamp, the R_f values of 4-chlorobenzalcohol and 4-chlorobenzoic acid are 0.4 and 0.3 respectively (R_f of 4-chlorobenzaldehyde is 0.6).

NOTE. Initially the mixture is very thick. It will not stir well until the reaction temperature is reached.

Isolation and Purification. When the reaction is complete as judged by TLC, cool the reaction mixture to room temperature and add 2.0 mL of chilled distilled water. Extract the resulting solution with three 0.5-mL portions of methylene chloride, using a Pasteur filter pipet to transfer the extracts to a 3.0-mL conical vial. On each addition of methylene chloride, cap the vial, shake gently, and then carefully vent by loosening the cap. A Vortex mixer, if available, is convenient for this extraction step. After separation of the layers, remove the lower methylene chloride layer using a Pasteur filter pipet.

IMPORTANT. Save both the alkaline and methylene chloride phases for further workup.

Alcohol product + CH₂Cl₂, 1.5 mL

Anhydrous Na₂SO₄, 150 mg

1. 4-Chlorobenzyl alcohol. Wash the combined methylene chloride extracts with two 0.25-mL portions of saturated sodium bicarbonate solution followed by one 0.5-mL portion of distilled water. Remove the aqueous upper phase (Pasteur filter pipet) used in each washing step and save this combined material in a separate 10-mL Erlenmeyer flask. (*This material will be discarded at the end of the experiment [see Rule 10 for the Microscale Laboratory, Chapter 1, p. 4.]*) Dry the methylene chloride layer over 150 mg of granular anhydrous sodium sulfate (◂).

After drying the solution, transfer it by use of a Pasteur filter pipet to a tared 3.0-mL conical vial. Be particularly careful not to let any granules of the hydrated sodium sulfate adhere to the surface of the pipet and become transferred with the dried organic phase. Rinse the sodium sulfate drying agent with 0.3 mL of fresh methylene chloride, and combine the rinse with the dried organic phase. Evaporate the methylene chloride using a stream of dry

nitrogen gas in a warm sand bath in the **hood**. If formation of the alcohol took   HOOD
place during the reaction, crude 4-chlorobenzyl alcohol will remain in the vial
following removal of the solvent.

To purify the crude alcohol, recrystallize the residue using a solution of
4% acetone in hexane (0.25 mL). Collect the product under reduced pressure
using a Hirsch funnel, and wash the filter cake (*product crystals packed on the
funnel are often referred to as filter cake*) with 0.2 mL of ice cold hexane to give
the desired 4-chlorobenzyl alcohol. Air-dry the product on a porous clay plate
or on filter paper.

Weigh the 4-chlorobenzyl alcohol and calculate the percent yield. Determine the melting point and compare your value with that found in the literature. Obtain the IR spectrum and compare it with that in Figure 6.16.

*This point in the procedure is a logical place to divide the experiment into two
laboratory sessions, if this seems appropriate. A second melting point may be obtained at the beginning of the next laboratory period after the sample of aromatic*

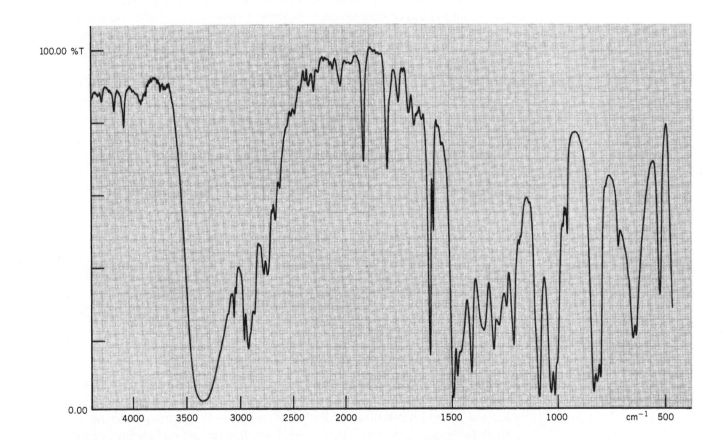

Figure 6.16   IR spectrum: 4-chlorobenzyl alcohol.

alcohol has had additional time to air-dry. A convenient way to let the sample dry for several days is to transfer the crystals from the porous plate and spread them over the bottom of a small Erlenmeyer flask (either 5 or 10 mL). Then, cover the mouth of the flask with filter paper held in place by a rubber band. This procedure will allow the last traces of moisture and solvent to evaporate, and at the same time protect the sample from dirt and dust particles. This is a good technique for drying most crystalline samples. If moisture is a particularly difficult problem, the sample stored in this fashion can be placed in a desiccator in the presence of a hygroscopic material, such as calcium chloride.

Acid product collected here

H₂0, 6 mL + HCl, 0.4 mL

2. 4-Chlorobenzoic acid. Dilute the alkaline phase, obtained and saved during the original extraction procedure, by adding 2.0 mL of water. The dilute aqueous phase is then acidified by the addition of 0.4 mL of concentrated hydrochloric acid. Collect the voluminous white precipitate of the product that forms on addition of the acid, under reduced pressure by use of a Hirsch funnel. Rinse the filter cake with 2.0 mL of distilled water (◄).

NOTE. It is advisable to check the pH of the solution to insure that it is acidic. Blue litmus or pH paper may be used. Remove a small drop of solution using a glass stirring rod and touch the rod to the paper. **Never** *place the paper into the reaction solution.*

Air-dry the solid product on a porous clay plate or on filter paper to obtain the crude 4-chlorobenzoic acid product. If a white precipitate is not obtained upon acidification, add a small amount of a saturated sodium chloride solution (salting out technique [see Technique 4, p. 79]) to aid the process.

To purify the crude acid, recrystallize with methanol in a Craig tube. Weigh the dried material and calculate the percent yield. Determine the melting point and compare your value to that found in the literature. Obtain the IR spectrum in a potassium bromide disk and compare your spectrum with that in Figure 6.17.

Infrared Analysis. The molecule is 4-chlorobenzaldehyde. The infrared spectrum of this aromatic aldehyde (Fig. 6.18, page 186) is rich and interesting. The aromatic aldehyde macro group frequency train (see Strategies for Interpreting Infrared Spectra) consists of peaks near 3090, 3070, 2830 and 2750, 1706, 1602, 1589, and 1396 cm^{-1}.

a. 3070 cm^{-1}: A C—H stretch on sp^2 carbon.

b. 2840 and 2740 cm^{-1}: This pair of bands is a famous example of powerful Fermi coupling (see Chapter 8 and Infrared Discussions). The unperturbed C—H stretching of the aldehyde C—H group would be expected to occur near 2790 cm^{-1} (the C—H stretch on an sp^2 carbon would have been expected to be found at higher values, but the oxygen system in this case significantly lowers the observed values). The in-plane bending mode of this C—H group occurs at 1390 cm^{-1}. Thus, the first harmonic should fall near 2780 cm^{-1}, very close (~10 cm^{-1}) to the stretching frequency of this oscillator. The essential conditions are, therefore, satisfied and strong Fermi coupling occurs and gives rise to the split peaks at 2840 and 2740 cm^{-1}. The latter band is moved well away from the normal sp^3 C—H symmetric stretching modes. Thus, the lower wavenumber component of the Fermi-coupled

Sample	4-Chlorobenzoic acid		
%T X ABS ___ Background Scans 4		Scans 16	
Acquisition & Calculation Time 42 sec		Resolution 4.0 cm⁻¹	
Sample Condition solid		Cell Window	
Cell Path Length		Matrix Material KBr	

Figure 6.17   IR spectrum: 4-chlorobenzoic acid.

aldehyde C—H mode leads to easy identification of an aldehyde, even when it represents a very small fraction of an aliphatic system. In the present case where no aliphatic C—H oscillators are present, both components are observed.

c. 1704 cm⁻¹: The carbonyl stretch of the aldehyde group. The frequency observed in aliphatic aldehydes falls in the range 1735–1720 cm⁻¹, but when conjugated, the value drops 15–25 cm⁻¹ (see Chapter 8) and is found in the range 1720–1700 cm⁻¹.

d. 1601 and 1578 cm⁻¹: This pair of bands is related to the degenerate ring stretching vibrations v_{8a} and v_{8b} of benzene (see also Infrared Discussions in Experiment [1B$_{adv}$]).

e. 1390 cm⁻¹: The aldehyde C—H in-plane bending vibration. The first harmonic of this vibration is Fermi coupled to the aldehyde C—H stretching mode, as discussed earlier.

Figure 6.18 IR spectrum: 4-chlorobenzaldehyde.

This compound also possesses a second powerful macro group frequency train that assesses the substitution pattern of the aromatic ring system (see Chapter 8, Infrared Discussions). The para-disubstituted benzene ring macro group frequency train requires peaks in the following regions: 1950, 1880, 1800, 1730, 750, and 690 cm^{-1}.

a. 1905, 1795 cm^{-1}: This pair of weak bands, with the higher wavenumber band more intense than the lower member, arises from combination bands (see Chapter 8, Infrared Discussions) which involve the out-of-plane bending frequencies of the ring C—H bonds (see below). The exact wavenumber positions are not very important, but the overall shape of the pattern can be used to determine the ring substitution pattern (see Chapter 8).

b. 832 cm^{-1}: This strong band is very characteristic of para-disubstituted benzene rings. The 832-cm^{-1} peak arises from the in-phase out-of-plane bending vibration of the two pairs of C—H groups on opposite sides of the six-membered ring (see Chapter 8).

The C—Cl stretching mode when substituted on an aromatic ring often becomes heavily coupled with ring vibrations and, as in this case, does not give rise to an identifiable group frequency. The presence of this group must be determined by other methods, such as a Beilstein or sodium fusion test (see Chapter 9, pp. 657–658). The presence of a hidden group (very likely halogen), however, is strongly indicated. Because the system must be para substituted (see above paragraphs), one of the substituents on the ring must exhibit very little identifiable absorption from 4000 to 500 cm^{-1}.

The reaction involves the formation of two products, one neutral and one acidic, both of which incorporate large portions of the substrate molecule. These materials can be characterized by their IR spectra. The neutral product is proposed to be 4-chlorobenzyl alcohol. An examination of the spectrum (Fig. 6.16) supports the presence of two macro group frequency trains. One (a) is the same one found in the starting aldehyde, the para-disubstituted benzene ring frequency train. The other (b) is a primary aliphatic alcohol macro frequency train. Macro (a) has been expanded to include all aromatic ring-specific group frequencies, as the alcoholic side chain is decoupled from the ring and the carbon–chlorine frequencies fall outside the range of the instrumentation normally used in these measurements.

The macro group frequencies for 4-chlorobenzyl alcohol are as follows:

a. 3055 (aromatic C—H stretch), 1906 and 1793 (para combination band pattern), 1596 and 1583 (ring stretch degenerate pair, ν_{8a} and ν_{8b}; 1583 intensity indicates weak conjugation with the ring), 1495 and 1475 (ring stretch degenerate pair, ν_{19a} and ν_{19b}), 834 (C—H, out-of-plane bend) cm^{-1}.

b. 3340 (broad, O—H stretch), 2965–2925 (C—H, aliphatic sp^3), 1450–1300 (broad, O—H bend, associated), 1015 (C—O, stretch, primary alcohol), 630 (broad, weak O—H bend, associated) cm^{-1}.

The acidic product generated in the reaction is assumed to be 4-chlorobenzoic acid. This material also is a para-substituted aromatic compound; thus, the spectrum of this compound (Fig. 6.17) possesses a macro frequency train (a), similar to those of the aldehyde and alcohol. In addition, this benzoic acid derivative exhibits an extended aromatic acid macro group frequency train (b). The macro frequencies are as follows:

a. 1935 and 1795 (para combination band pattern), 852 (ring C—H, in-phase, out-of-plane bend) cm^{-1}.

b. 340–2200 (very broad, very strong, O—H stretch, associated acid), 1683 (C—O, stretch, out-of-phase, associated acid dimer), 1596 and 1578 (degenerate ring stretch), 1500 and 1432 (degenerate ring stretch), 1320–1280 (C—O, stretch), 885 (O—H, out-of-plane, ring dimer bend) cm^{-1}.

The carbon–chlorine stretch is not observed.

Examine the spectra of the reaction products you have obtained in a potassium bromide matrix. Discuss the similarities and differences of the experimentally derived spectral data to the reference spectra (Figs. 6.16–6.18).

It is a useful exercise to compare the results of organic qualitative analysis reaction tests, which were used historically to classify the preceding compounds, with the data now available from modern spectroscopic instrumentation.

Chemical Tests.   Perform each of the following tests (see Chapter 9). Do the results confirm that you have isolated an aromatic carboxylic acid and an aromatic alcohol?

1. The ignition test.
2. The Beilstein or the sodium fusion test for the presence of halogen.
3. The ceric nitrate and/or the Jones oxidation test for the alcohol.
4. The solubility of the acid in sodium bicarbonate and sodium hydroxide solutions. Is carbon dioxide evolved in the bicarbonate test?

If you were to prepare a derivative for the alcohol and acid products, which one would you choose? See Chapter 9, Preparation of Derivatives.

Experiment 7-1*

M

4-Chlorobenzoic Acid and 4-Chlorobenzyl Alcohol: Preparation Using a Monomode Microwave Apparatus

EXPERIMENTAL PROCEDURE

Estimated time to complete the experiment: 2.0 h.

Physical Properties of Reactants						
Compound	MW	Amount	mmol	mp (°C)	bp (°C)	n_D
4-Chlorobenzaldehyde	140.57	150 mg	1.1	47		
Methanol	32.04	0.4 mL			65	1.3288
Potassium hydroxide (11 M)	56.11	0.4 mL				

Reagents and Equipment.   This experiment is designed for use in the CEM Discover and Biotage Initiator microwave units.

In a 10.0-mL glass microwave reaction vessel containing a magnetic stir bar, place 150 mg (1.1 mmol) of 4-chlorobenzaldehyde followed by 0.4 mL of methanol. With gentle swirling, add 0.4 mL of an 11 M aqueous solution of potassium hydroxide.

HOOD

CAUTION: Dispense liquid reagents in the **hood** using automatic delivery pipets. Concentrated methanolic potassium hydroxide is *highly caustic.* Since the reaction requires heating reagents to above their boiling point in sealed vessels, *adherence to the microwave manufacturer's guidelines is essential.*

Reaction Conditions.   Place the reaction vessel in the microwave cavity and, depending on the equipment used, position the pressure device on top. Program the microwave unit to heat the reaction mixture to 120 °C using no more than 50 W of microwave power, and hold at this temperature for 1 min. After heating, allow the reaction mixture to cool to 50 °C or below before removing the tube from the microwave unit.

*This section has been written by Dr. Nicholas E. Leadbeater from the Department of Chemistry at the University of Connecticut and Dr. Cynthia B. McGowan from the Department of Chemistry at Merrimack College, MA.

Isolation of Product. Add 2.0 mL of cold distilled water to the reaction vessel. Swirl or stir on a stirplate to dissolve the solid material. Transfer the reaction mixture with a Pasteur pipet into a 5.0-mL conical vial. Rinse the microwave reaction vessel with 0.5 mL of dichloromethane and add the washings to the conical vial. Cap the vial, shake gently, and then carefully vent by loosening the cap. After allowing separation of the layers, transfer the bottom (organic) layer to a 3.0-mL conical vial with a Pasteur filter pipet. Extract the aqueous solution in the 5-mL conical vial with two additional 0.5 mL portions of dichloromethane. After each addition carefully cap the vial, shake gently, and vent the solution by loosening the cap. On each occasion, transfer the bottom (organic) layer to the 3.0-mL conical vial with a Pasteur filter pipet. At the end of the extractions, label the 5.0-mL conical vial containing the aqueous alkaline layer "alkaline phase." This phase will be used later in the experiment.

IMPORTANT. Save both the aqueous and dichloromethane layers for further workup.

The remainder of the workup for the two products is the same as for Experiment 7 (pp. 178–188).

4-Chlorobenzoic Acid and 4-Chlorobenzyl Alcohol: Preparation Using a Multimode Microwave Apparatus

Experiment 7-2*

EXPERIMENTAL PROCEDURE

M

Estimated time to complete the experiment: 2.0 h.

Physical Properties of Reactants						
Compound	MW	Amount	mmol	mp (°C)	bp (°C)	n_D
4-Chlorobenzaldehyde	140.57	465 mg	3.3	47		
Methanol	32.04	2.5 mL			65	1.3288
Potassium hydroxide (11 M)	56.11	2.5 mL				

Reagents and Equipment. This experiment is designed for use in the CEM MARS, Milestone START, and Anton Paar Synthos 3000 microwave units. When using the Anton Paar Synthos 3000 unit with the 24-position silicon carbide plate rotor containing glass vials, the reagent and solvent quantities cited in the monomode procedure should be used in conjunction with the reaction conditions here in the multimode procedure.

In a microwave reaction vessel containing a magnetic stir bar, place 465 mg (3.3 mmol) of *p*-chlorobenzaldehyde and 2.5 mL of methanol. With stirring, slowly add 2.5 mL of an 11 M aqueous solution of potassium hydroxide. Cap the vessel with the microwave pressure cap and adjust the tightness to the manufacturer-specified level. Place the sealed vessel into its outer protective jacket.

*This section has been written by Dr. Nicholas E. Leadbeater from the Department of Chemistry at the University of Connecticut and Dr. Cynthia B. McGowan from the Department of Chemistry at Merrimack College, MA.

HOOD

CAUTION: Dispense liquid reagents in the **hood** using automatic delivery pipets. Concentrated methanolic potassium hydroxide is *highly caustic*. Since the reaction requires heating reagents to above their boiling point in sealed vessels, *adherence to the microwave manufacturer's guidelines is essential.*

Reaction Conditions. Insert the loaded vessels into the reaction carousel ensuring they are evenly spaced and then place the carousel into the microwave cavity. If provided by the manufacturer, connect a temperature probe to the control vessel. Program the microwave unit to heat the reaction vessels to 120 °C and hold at this temperature for 1 min. After heating, allow the reaction mixture to cool to 50 °C or below before removing the carousel from the microwave unit.

Isolation or Product. Add 5 mL of distilled water to the reaction vessel to dissolve any solids, transfer the solution to a 60-mL separatory funnel, and clamp the funnel to a ring stand. Rinse the microwave reaction vessel with an additional 5 mL of distilled water and add it to the separatory funnel. Extract the alkaline solution with three 3.0-mL portions of dichloromethane. After each addition of dichloromethane carefully cap and invert the funnel. Immediately vent the funnel by opening the stopcock. Close the stopcock, mix the two layers several times by inverting the funnel, repeatedly vent, and then place the funnel back on the ring stand and remove the stopper. Drain the lower (organic) layer into a 25-mL Erlenmeyer flask after each extraction. At the end of the extractions, pipet the alkaline layer into a separate 25-mL Erlenmeyer flask and label it "alkaline phase." (*This phase will be used in part 2.*)

IMPORTANT. Save both the aqueous and organic layers for further workup.

1. 4-Chlorobenzyl alcohol. Pipet the organic layer back into the separatory funnel. Wash the combined organic layers with two 5-mL portions of saturated sodium bicarbonate solution. Remove the lower (organic) layer by draining it through the stopcock into a 25-mL Erlenmeyer flask and pipet the remaining aqueous phase into a 25-mL Erlenmeyer flask after each washing. Save the aqueous waste until the experiment is complete and then discard as directed. Return the organic layer to the separatory funnel by pipet. Wash the organic layer once with 5 mL of distilled water. Dry the lower (organic) layer by draining it into a clean 25-mL Erlenmeyer flask containing 200 mg of anhydrous sodium sulfate. Transfer the anhydrous dichloromethane solution, using a Pasteur filter pipet, to a clean tared 10-mL pear-shaped flask. Remove the dichloromethane on a rotary evaporator or by evaporation in the **hood** using a gentle stream of nitrogen gas with warming in a sand bath. A white precipitate will remain in the flask. Reweigh the flask plus product and calculate the crude yield.

HOOD

To further purify the crude alcohol, recrystallize the residue using a solution of 4% acetone in hexane (0.75 mL). Collect the product under reduced pressure using a Hirsch funnel, and wash the filter cake with 0.5 mL of ice cold hexanes. Air-dry the product on a porous clay plate.

Weigh the 4-chlorobenzyl alcohol and calculate the % yield. Determine the melting point of the product and compare it to the literature value. Obtain the IR spectrum and compare it with that in Figure 6.16.

2. 4-Chlorobenzoic acid. Cool the "alkaline phase" in an ice bath and add dropwise 4 mL of 6M aqueous hydrochloric acid. A white precipitate of the product will form. Allow the solution to chill for a further 10 min and then collect the product under reduced pressure using a Hirsch funnel. Rinse the filter cake with 2.0 mL of cold distilled water.

Air-dry the solid on a porous clay plate or on filter paper to obtain the crude 4-chlorobenzoic acid. If the product is very pasty it may be left to dry overnight. The product can be further purified by recrystallization with methanol. Weigh the dried material and calculate the % yield. Determine the melting point of your product and compare your value to the literature value. Obtain the IR spectrum and compare it to the spectrum given in Figure 6.17.

QUESTIONS

6-46. The discussion mentions that the crossed Cannizzaro reaction can be realized when one of the components is formaldehyde. Predict the product(s) of the reaction below and give a suitable name to the reactants and products.

6-47. One group of investigators has suggested that a dianion might be the source of hydride in the Cannizzaro reaction. Explain why this species would be a better source of hydride than the species in the mechanism presented in the discussion.

6-48. The Cannizzaro reaction is an oxidation–reduction sequence. What type of reagent is formaldehyde acting as in Question 6-46?

6-49. Propose a mechanism for the internal Cannizzaro reaction depicted below.

BIBLIOGRAPHY

Cannizzaro, S. *Annalen* **1853**, *88*, 129.

Reviews on the Cannizzaro reaction:

Bergens, S. H.; Fairlie, D. P.; Bosnich, B. *Organometallics* **1990**, *9*, 566.
Geissman, T. A. *Org. React.* **1944**, *2*, 94.
McDonald, R. S.; Sibley, C. E. *Can. J. Chem.* **1981**, *59*, 1061.
Swain, C. G.; Powell, A. L.; Sheppard, L. A.; Morgan, C. R. *J. Am. Chem. Soc.* **1979**, *101*, 3576.

Examples of the Cannizzaro reaction:

Davidson, D.; Weiss, M. *Organic Syntheses*; Wiley: New York, 1943; Collect. Vol. II, p. 590.
Esteb, J. J.; O'Reilly, S.; Richter, J. M. *J. Chem. Educ.* **2004**, *81*, 1794.
Wilson, W. C. *Organic Syntheses*; Wiley: New York, 1941; Collect. Vol. I, p. 256.

EXPERIMENT
8

B

The Esterification Reaction: Ethyl Laurate, Isopentyl Acetate, and the Use of Acidic Resins

Common names: ethyl laurate, ethyl dodecanoate
CA number: [106-33-2]
CA name as indexed: dodecanoic acid, ethyl ester

Common names: isopentyl acetate, isoamyl acetate
CA number: [123-92-2]
CA name as indexed: 1-butanol, 3-methyl-, acetate

Common name: butyl acetate
CA number: [123-86-4]
CA name as indexed: acetic acid, butyl ester

Common names: pentyl acetate, amyl acetate
CA number: [628-63-7]
CA name as indexed: acetic acid, pentyl ester

Common name: hexyl acetate
CA number: [142-92-7]
CA name as indexed: acetic acid, hexyl ester

Common name: octyl acetate
CA number: [112-14-1]
CA name as indexed: acetic acid, octyl ester

Prior Reading

Technique 4: Solvent Extraction
    Liquid–Liquid Extraction (p. 76)
    Drying of the Wet Organic Layer (pp. 84–87)
Technique 6A: Chromatography
    Packing the Column (p. 97)
    Elution of the Column (p. 98)

Purpose. This exercise explores the classic reactions of carboxylic acids (RCO_2H) with alcohols ($R'OH$), in the presence of acid catalyst, to yield esters (RCO_2R') plus water (H_2O, a small stable molecule). The physical properties of these esterification products are examined and the techniques of distillation and column chromatography are applied to the purification of these materials.

REACTION

$$R-\overset{\overset{\displaystyle \ddot{O}:}{\|}}{C}{\diagdown}_{\ddot{O}H} \; + \; R'-OH \; \underset{}{\overset{H^+}{\rightleftharpoons}} \; R-\overset{\overset{\displaystyle \ddot{O}:}{\|}}{C}{\diagdown}_{\ddot{O}R'} \; + \; H_2O$$

Carboxylic   Alcohol   Ester
acid

DISCUSSION

Esters are among the most important of the carboxylic acid (and alcohol) derivatives. Substances possessing this functional group are widely distributed in nature in the form of waxes, essential oils, fatty acid esters, and aromas. The ester functionality plays a significant role in biochemistry, both in primary metabolism and in a variety of substances exhibiting remarkable physiological activity in humans (hormones and neurotransmitters). Esters find extensive use in commercial products from fingernail polish remover and artificial sweeteners to polymeric fibers, plasticizers, and surfactants.

B

Biosynthesis of Esters.   Fatty acids are naturally occurring, long, straight-chain, C_{12}—C_{40} carboxylic acids; most contain an even number of carbon atoms. Their biosynthesis provides an important and interesting example of a primary metabolic pathway in which a special type of ester is the essential link between the enzyme and the substrate (acetic acid). The enzyme-bound substrate grows by repeated addition of two-carbon (C_2) units and, when eventually released from the enzyme, has undergone an extension of the fatty acid hydrocarbon chain.

   The first step in fatty acid biosynthesis involves the formation of a thiol ester, acetyl coenzyme A (acetyl CoA), from acetic acid (present in the primary metabolic pool) and the thiol group (mercapto, or —SH group) of the coenzyme (HSCoA). A thiol ester is an ester in which the single-bonded oxygen (from the alcohol component) is replaced by a sulfur atom,

$$-\overset{\overset{\displaystyle \cdot\cdot}{\overset{\displaystyle O}{\|}}}{C}-\overset{\cdot\cdot}{\underset{\cdot\cdot}{S}}-C-$$

from the coenzyme. Coenzymes are loosely bound, nonprotein factors attached to the enzyme that play an important role in the catalytic function of the enzyme. These coenzymes are distinguished, in an ill-defined manner, from prosthetic groups, which are intimately attached to active sites of enzymes. Part of the role of the CoA is to facilitate the transfer of the substrate (the C_2 unit) to a new thiol group of the enzyme (protein), where the next stage of the biosynthesis takes place:

Thiol group   Coenzyme A (HSCoA)

Acetyl thiolester

Acetyl coenzyme A ($CH_3COSCoA$)

This reactive thiol ester is capable of undergoing aldol-type (see Experiment [20]) condensations under physiological conditions. Acetyl-CoA is first carboxylated with the help of the enzyme, acetyl CoA-carboxylase, to yield a thiolmalonyl derivative. The resulting intermediate possesses an activated methylene group,

$$-\overset{\overset{\ddot{O}}{\|}}{C}-\underline{CH_2}-\overset{\overset{\ddot{O}}{\|}}{C}-\ddot{S}-$$

which with further enzymatic support undergoes a Claisen condensation (see Experiment [3A$_{adv}$]) with an acetyl group that has been also transferred via acetyl CoA to an appropriate acyl-carrier protein, fatty acid synthetase. Reduction, dehydration, further reduction, and finally hydrolysis of the thiol ester, yields the fatty acid extended by a C_2 hydrocarbon group,

$$CH_3-\overset{\overset{\ddot{O}}{\|}}{C}-\ddot{O}H + HSCoA \longrightarrow CH_3-\overset{\overset{\ddot{O}}{\|}}{C}-SCoA + H\ddot{O}H$$

CoA   Acetyl CoA

$$CH_3-\overset{\overset{\ddot{O}}{\|}}{C}-\ddot{S}CoA + HCO_3^- \xrightarrow{\text{aldol}} H\ddot{O}-\overset{\overset{\ddot{O}}{\|}}{C}-CH_2-\overset{\overset{\ddot{O}}{\|}}{C}-SCoA + H\ddot{O}H$$

Malonyl CoA

$$CH_3-\overset{\overset{\ddot{O}}{\|}}{C}-\ddot{S}CoA + H\ddot{O}-\overset{\overset{\ddot{O}}{\|}}{C}-CH_2-\overset{\overset{\ddot{O}}{\|}}{C}-\ddot{S}CoA \longrightarrow \text{both HS–Protein bound}$$

reduction
dehydration
reduction
$\xleftarrow{\hspace{1.5cm}}$
2nd reduction
may be skipped
yields unsaturated
fatty acid product

Claisen

$$CH_3CH_2CH_2-\overset{\overset{\ddot{O}}{\|}}{C}-\ddot{S}-\text{Protein} \qquad CH_3-\overset{\overset{\ddot{O}}{\|}}{C}-CH_2-\overset{\overset{\ddot{O}}{\|}}{C}-\ddot{S}-\text{Protein}$$

hydrolysis

$$CH_3CH_2CH_2-\overset{\overset{\ddot{O}}{\|}}{C}-\ddot{O}H \longrightarrow \text{repeats to yield long-chain fatty acids}$$

LIPIDS

B   ***Oils, Fats, Waxes, and Aromas.*** Fatty acids derived from primary metabolism play a key role in the formation of naturally occurring oils, fats, and waxes. Fats and oils are esters of these acids with a triol, glycerol ($HOCH_2CHOHCH_2OH$). The common fats and oils are formed from mixtures of C_4—C_{26} saturated fatty acids with the vast majority derived from C_{12}—C_{18} acids. The oils are more likely to include significant contributions from mixtures of unsaturated fatty acids.

Since fats and oils are triesters of glycerol, they are generally called *triglycerides*. In plants and animals, triglycerides function as energy reserves that can be used in primary metabolism when food (energy) is not available to the organism.

Although the fats have high molecular weights, they are generally found to be very low-melting solids, particularly if they contain unsaturated fatty acids. The bent chains, which result from incorporating cis-alkene (C=C) groups into the chain, prevent close packing in a solid and, as a result, such molecules exhibit lower melting points. For example, compare oleic acid ($C_{18}H_{34}O_2$, mp 4 °C) to its saturated analog, stearic acid($C_{18}H_{36}O_2$, mp 69–70 °C). The former melts more than 60 °C lower because of the cis carbon–carbon double bond.

Oleic acid is the simplest of the unsaturated C_{18} fatty acids, because it has a single C=C group located in the middle of the chain. This unsaturated fatty acid is the most widely distributed of all fatty acids. It is the dominant component (76–86%) of the triglycerides in olive oil. Highly saturated fats, on the other hand, are generally solids at room temperature, because the straight-chain fatty acids pack together well (see Fig. 6.19).

Oleic acid
[(Z)-9-Octadecenoic acid]

Stearic acid
[Octadecanoic acid]

Many vegetable and fish oils are liquid triglycerides. Because these organisms operate at ambient temperatures, evolution dictated that low-melting fats were required to avoid solidification. In warm-blooded animals, higher melting fats can be tolerated and are used.

The cheap and plentiful unsaturated oils can be converted to solid fats by hydrogenation of the alkene groups, which gives straight-chain alkyl groups. As consumers historically have desired to cook with solid, white, and creamy fats (such as lard) derived from animal triglycerides (low in unsaturation), hydrogenation of vegetable oils, such as peanut, soybean, and cottonseed oils, has been carried out on a large scale (this process is referred to as **hardening** the fat).

Unfortunately, the relationship between saturated fats in the human diet and the formation of cholesterol (a simple lipid, see below), plaque, and coronary heart disease has been established. The dietary switch to less saturated fats is currently under way.

Cholesterol

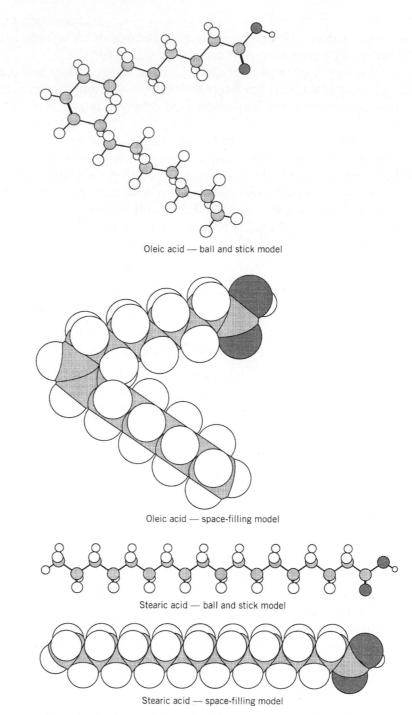

Oleic acid — ball and stick model

Oleic acid — space-filling model

Stearic acid — ball and stick model

Stearic acid — space-filling model

Figure 6.19   Molecular models of fatty acids.

The triglycerides obtained from animal fats have been used for a very long time as a source of soap. When fats are boiled with lye (sodium hydroxide), the ester linkages are cleaved by a process known as *saponification* (the term originates from the Latin word for soap, *sapon*, as does the modern French word for soap, *savon*) to yield the sodium salt of the fatty acid and the esterifying alcohol (glycerol).

$$\text{CH}_3(\text{CH}_2)_{16}\overset{\overset{\cdot\cdot}{\text{O}}}{\underset{}{\text{C}}}\overset{\cdot\cdot}{\underset{\cdot\cdot}{\text{O}}}-\text{CH}_2$$
$$\text{CH}_3(\text{CH}_2)_{16}\overset{\overset{\cdot\cdot}{\text{O}}}{\underset{}{\text{C}}}\overset{\cdot\cdot}{\underset{\cdot\cdot}{\text{O}}}-\text{CH} \xrightarrow{\text{3 NaOH}} \text{H}\overset{\cdot\cdot}{\underset{\cdot\cdot}{\text{O}}}-\text{CH} + 3\ \text{CH}_3(\text{CH}_2)_{16}\overset{\overset{\cdot\cdot}{\text{O}}}{\underset{}{\text{C}}}\text{O}^-\ \text{Na}^+$$
$$\text{CH}_3(\text{CH}_2)_{16}\overset{\overset{\cdot\cdot}{\text{O}}}{\underset{}{\text{C}}}\overset{\cdot\cdot}{\underset{\cdot\cdot}{\text{O}}}-\text{CH}_2$$

H$\overset{\cdot\cdot}{\underset{\cdot\cdot}{\text{O}}}$—CH$_2$

H$\overset{\cdot\cdot}{\underset{\cdot\cdot}{\text{O}}}$—CH$_2$

Soap

Tristearin   Glycerol

Saponification of a triglyceride found in animal fat

Salts of fatty acids function effectively as soaps because one end of the straight-chain system has the highly polar carboxylate ion and is readily solvated in water. The rest of the fatty acid molecule has all the characteristics of a nonpolar hydrocarbon and readily dissolves in hydrocarbons, such as greases and oils. We refer to the polar end (head) as being hydrophilic (attracted to water) and the hydrocarbon end (tail) as being lipophilic (attracted to oils). When dispersed in an aqueous solution, fatty acids tend to form micelles (spherical clusters of molecules). The lipophilic ends of the fatty acids occupy the interior of the cluster, while the polar ends, which are heavily solvated by water molecules, form the outer surface of the spherical micelle. Micelles absorb the hydrocarbon chains of the triglycerides, and thus soaps break up and help to dissolve the fats and oils that tend to coat skin, clothes, and the surfaces of eating and cooking utensils (see Figs. 6.20 and 6.21).

Waxes are naturally occurring esters of fatty acids (in waxes, chain lengths can reach as high as C$_{36}$) and a variety of other alcohols that often possess relatively complicated structures (steroid alcohols) and/or long chains. For example, n-octacosanol, CH$_3$(CH$_2$)$_{26}$CH$_2$OH, has been isolated from the esters in wheat

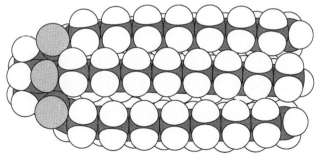

A space-filling model of a saturated triglyceride

A saturated triglyceride

Figure 6.20   A space-filling model of a saturated triglyceride.

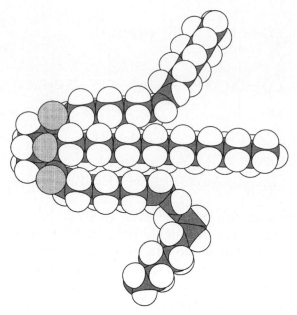

A space-filling model of an unsaturated triglyceride.

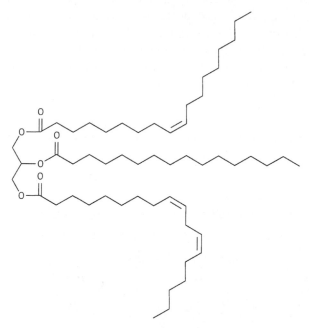

An unsaturated triglyceride

Figure 6.21 A space-filling model of an unsaturated triglyceride.

waxes, and a component of carnauba wax (traditionally an automobile wax) has 62 carbons, $CH_3(CH_2)_{33}CO_2(CH_2)_{26}CH_3$. The biological role of carnauba wax is as a leaf coating involved in the conservation of plant moisture. Animal waxes include cetyl palmitate (spermaceti) found in sperm whales and beeswax (one constituent of which has been identified as $CH_3(CH_2)_{29}CO_2(CH_2)_{29}CH_3$, which is used in the construction of the honeycomb).

Lower molecular weight, naturally occurring esters make major contributions to the pleasant aromas of fruits and flowers. These odors have been shown to be composed generally of complex mixtures of materials that have been separated only since the development of modern chemical instrumentation. Single components, however, may play a dominant role in an individual plant or animal. Propyl acetate (pears), ethyl butyrate (pineapples), and 3-methylbutyl acetate (bananas) are examples of simple esters responsible for a particular plant odor. Odors derived from esters are not limited just to esters of straight-chain carboxylic acids, as is demonstrated by oil of wintergreen, methyl salicylate:

Oil of wintergreen

Phospholipids. *Lipid* is a term applied to those natural substances that are more soluble in nonpolar solvents than in water. In its most general sense, it is a broad definition that includes fats, waxes, hydrocarbons, and so on. In biochemistry, lipids are more narrowly defined as substances that yield fatty acids upon hydrolysis.

Another class of glycerides are those substances in which one of the fatty acid groups has been replaced by a phosphoric acid residue: the phospholipids, or more accurately, the phosphoglycerides. The phosphate group is almost always further esterified, usually with a biological amino alcohol, such as choline (the lecithins) or ethanolamine (the cephalins):

Choline

Ethanolamine

Phosphoglyceride, a cephalin

Phosphoglyceride, a lecithin

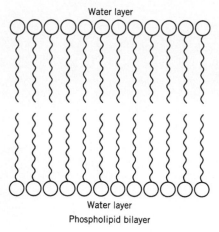

Water layer

Water layer
Phospholipid bilayer

Figure 6.22 Phospholipid bilayer.

These latter groups significantly increase the polarity of the glycerol section of the molecule so that phosphoglycerides undergo strong self-association. In aqueous solutions, this intermolecular attraction can lead to lipid bilayer formation (Fig. 6.22). In a lipid bilayer, the molecules organize themselves to form sheets that contain a double layer of the molecules formed by tail-to-tail association within the interior of the sheet; the outer surface of the lipid bilayer contains the polar heads, which are heavily solvated by water molecules. This association of phosphoglycerides is the key feature in the construction of cell membranes. Thus, esters must have played a vital role at the very earliest stages as cell structures evolved in the development of living systems.

Preparation of Esters. Esters are generally synthesized by one of four fundamental routes:

1. Esterification of a carboxylic acid with an alcohol in the presence of an acid catalyst
2. Alcoholysis of acid chlorides, anhydrides, or nitriles
3. Reaction of a carboxylate salt with an alkyl halide or sulfate
4. Transesterification reactions

The first of these pathways, known as Fischer esterification, is the method used for the preparation of ethyl laurate in Experiment [8A] and of isopentyl acetate in Experiment [8B]. A modern variation of the Fischer esterification is used in Experiment [8C]. The development of this esterification reaction represents just one of a number of major discoveries in organic chemistry by Emil Fischer.

Emil Fischer (1852–1919).[7] In 1874 Fischer obtained his Ph.D. from the University of Strasbourg, studying with Adolf von Baeyer. He later had appointments as Professor of Chemistry at Erlangen, Würzburg, and Berlin universities.

In 1875, at the age of 23, and one year after completing his graduate studies, he synthesized phenyl hydrazine (C_6H_5—$NHNH_2$) for the first time. This highly reactive reagent later played a key role in Fischer's work on elucidating structures of a large majority of the sugars (carbohydrates), an entire class of important and complex organic molecules. Sugars, or carbohydrates, represent the prime pathway for the storage of radiant energy from the sun, through photosynthesis, as chemical energy. In the short period from 1891 to 1894, Fischer established not only the basic structures, but also the configurations of all the known sugars. In addition, he predicted all the theoretically possible isomers and, in the process, developed a method of representing the three-dimensional molecular structures in two-dimensional drawings that became known as Fisher projection formulas. These representations are still in use today, and have been widely applied beyond sugar chemistry. This work by Fischer led directly to proving the existence of the asymmetric carbon atom, a concept proposed by Vant Hoff and LeBel in 1874.

Fischer was also active in the area of protein chemistry. He demonstrated that amino acids are the basic subunits from which proteins are constructed.

[7]See *Chem. Ind.* **1919**, *42*, 269; Darmstaedter, L.; Oester, R. E. *J. Chem. Educ.* **1928**, *5*, 37; Ratman, C.V. *ibid.* **1942**, *38*, 93; Kauffman, G. B.; Priebe, P. M. *ibid.* **1990**, *67*, 93; *Chem. Eng. News* **1992**, June p. 25. Recommended reading,"The Emil Fischer–William Ramsay Friendship: The Tragedy of Scientists in War."*J. Chem. Educ.* **1990**, *67*, 451.

Fischer devised methods for the synthesis of many of the known amino acids. Perhaps his most ingenious contribution was the "lock and key" hypothesis of how proteins bind with substrates of complementary shapes. This work ultimately led to our understanding of how enzymes, the catalysts of biochemical reactions, function.

Fischer carried out extensive work on the chemistry of purine and on those compounds containing its nucleus. Purine is one of the two nitrogen base ring systems present in DNA. Fischer synthesized approximately 150 members of this class of heterocyclic compounds (including the first synthesis of the alkaloid caffeine (see Experiment [11B]), uric acid, and the xanthines (also see Experiment [11B]). He developed a general synthesis of another nitrogen heterocycle, indole, which was so effective that it has become one of the classic synthetic methods of organic chemistry and is known today as the "Fischer indole synthesis":

Indole

Fischer's work essentially laid the foundation of modern biochemistry. Regarded as the greatest organic chemist of his time, Fischer became the second chemist to receive the Nobel Prize (1902). Depressed by the loss of his young wife at the age of 33, by the loss of two of his three sons (one by suicide, the other in World War I), suffering from the advanced stages of intestinal cancer, and saddened by the socioeconomic conditions of postwar Germany, Fischer committed suicide.

Mechanism of the Fischer Esterification Reaction. The Fischer esterification proceeds by nucleophilic attack of the alcohol on the protonated carbonyl group of the carboxylic acid to form a tetrahedral intermediate. Collapse of the tetrahedral intermediate regenerates the carbonyl group and produces the ester and water. The overall sequence is outlined here:

In the Fischer esterification with primary alcohols, the products are only slightly favored by the equilibrium, and thus to obtain substantial yields of the ester, the equilibrium must be shifted toward the products. This result can be accomplished in a number of ways. For example, an excess of the starting alcohol can be used to shift the position of equilibrium toward the products. This technique is used in the preparation of ethyl laurate (Experiment [8A]). An analogous alternative is to use an excess of the carboxylic acid. A third option to drive the reaction is the removal of one or both of the products (the ester or water) as they are formed during the reaction. The preparation of isopentyl acetate, synthesized in Experiment [8B], depends on two of these strategies: (1) the reaction is run in an excess of the carboxylic acid (it doubles as the solvent) and (2) the water generated as one of the products is removed by a drying agent. The acid catalyst used in Fischer esterifications is generally dry hydrogen chloride, concentrated sulfuric acid, or a strong organic acid, such as p-toluenesulfonic acid.

When the carboxyl group and hydroxyl group are present in the same molecule, an intramolecular esterification may occur and a cyclic ester (called a lactone) may be formed. Lactonization requires an acceptable conformation: the two groups must be close and spatially positioned to react. Ring closure is especially favorable if lactone formation yields five- or six-membered (stable and rapidly formed) ring systems.

γ-Hydroxybutyric acid   γ-Butyrolactone

As noted earlier, Fischer esterification is an equilibrium reaction and is thus reversible. Thus, heating an ester in aqueous solution, in the presence of an acid catalyst, regenerates the corresponding carboxylic acid and alcohol. This latter reaction is called *acid hydrolysis* of an ester. The rate-determining step in both the forward esterification reaction and the reverse reaction, acid hydrolysis, is the formation of the tetrahedral intermediate. It is, therefore, evident that the rate of the reaction will be determined by the ease with which the nucleophile (alcohol on esterification and water on hydrolysis) approaches the carbonyl group. Steric and electronic factors have been shown to have large effects on the rate of esterification. An increase in the number of bulky substituents substituted on the α and β positions of the carbonyl-containing compound decreases the rate (steric effects). Electron-withdrawing groups near the carbonyl group, on the other hand, tend to increase the rate by increasing the electrophilicity (partial positive charge) of the carbonyl carbon (electronic effects). Conversely, electron-donating groups act to retard the rate of esterification (electronic effects).

You are to synthesize the ethyl ester of lauric acid in Experiment [8A]. Lauric acid, $CH_3(CH_2)_{10}CO_2H$ (dodecanoic acid), is one of the four most common fatty acids found in naturally occurring triglycerides. It is named for the laurel botanical family from which it was first isolated in 1842. It is the most abundant of the fatty acids isolated from the vegetable oils of palm kernel oil (52%), the seed fat of *Elaeis guineensis*; of coconut oil (48%), *Cocos nucifera*; and of babassu oil (46%), *Attalea funifera*.

In the preparation given for ethyl laurate (Experiment [8A]), acetyl chloride is used to generate the HCl catalyst in situ. Notice that the other product of this step is a molecule of the desired ester:

In Experiment [8B] you will be synthesizing the isopentyl alcohol (3-methylbutanol) ester of acetic acid (the basic building block of the fatty acids), isopentyl acetate (isoamyl acetate). This low molecular weight ester has a distinct banana- or pear-like odor, and the liquid product is often referred to as banana oil or pear oil (see above). Isopentyl acetate has a wide variety of uses: as a flavoring agent in mineral waters and syrups; a solvent for oil paints, tannins, nitrocellulose, lacquers, and a number of other commercial products; a perfume ingredient in shoe polish; and in the manufacture of artificial silk, leather, and pearls. You are very likely to find this experiment to be a pleasant olfactory experience!

In Experiment [8C] you will use a polymer-bound acid reagent to catalyze the esterification reaction. Polymer-bound reagents are becoming increasingly useful in organic synthesis, both in the research laboratory and in industrial-scale reactions.

Ethyl Laurate

Experiment 8A

REACTION

EXPERIMENTAL PROCEDURE

Estimated time to complete the experiment: 3.0 h.

Physical Properties of Reactants							
Compound	MW	Amount	mmol	mp (°C)	bp (°C)	d	n_D
Lauric acid	200.3	70 mg	0.35	44			
Ethanol	46.07	1.0 mL	17		78.5	0.71	1.3611
Acetyl chloride	78.50	30 μL	0.42		50.9	1.11	1.3898

HOOD

110–120 °C,
thermometer

HOOD

Lauric acid, 70 mg
+ CH₃CH₂OH, 1.0 mL
+ CH₃COCl, 30 μL

Reagents and Equipment. Weigh and add 70 mg (0.35 mmol) of lauric acid to a 3.0-mL conical vial containing a magnetic spin vane. Then equip the reaction vial with a reflux condenser that is protected by a calcium chloride drying tube (◄). Use a graduated 1.0-mL pipet to add 1.0 mL of absolute ethanol to the vial. In the **hood**, add 30 μL (0.43 mmol) of acetyl chloride, using an automatic delivery pipet. (Remember not to turn on the cooling water in the condenser until the system is completely assembled. Otherwise, the cold inner surface will condense moisture from the laboratory atmosphere. It takes only a very small amount of water on the condenser walls to completely deactivate your acid chloride catalyst.)

CAUTION: Acetyl chloride is an irritant. Dispense this reagent in the **hood** using an automatic delivery pipet. (Be sure to quickly reassemble the reaction vial and condenser following addition of the acid chloride, because this reagent will rapidly react with moist laboratory air and lose its activity.)

Reaction Conditions. Heat the reaction mixture for 1 h at gentle reflux with stirring, using a sand bath temperature of 110–120 °C. Cool the resulting mixture to room temperature.

Isolation of Product. After removing the reflux condenser and drying tube, remove the spin vane with forceps. Add a boiling stone to the cooled vial and then concentrate the reaction solution to a volume of about 0.25 mL by warming in a sand bath in the **hood**. Remove the vial from the sand bath and allow the vial to cool. Use a 1.0-mL graduated pipet to add 0.5 mL of diethyl ether and 0.25 mL of 5% sodium bicarbonate solution to the concentrated product mixture. Cap the conical vial and shake gently (or mix on a Vortex mixer). Loosen the cap carefully to vent the two-phase mixture. Remove the bottom aqueous layer using a Pasteur filter pipet and set it aside in a labeled Erlenmeyer flask. Extract the ether phase with three additional 0.25-mL portions of 5% sodium bicarbonate solution. Save the aqueous wash after each extraction (you may combine them with the initial aqueous basic phase in the Erlenmeyer) and do not discard them until you have successfully purified and characterized the ethyl laurate.

Anhydrous
Na₂SO₄,
500 mg

SiO₂, 500 mg

Sand, 50 mg

Ethyl laurate
+ CH₂Cl₂,
3.0 mL

HOOD

Purification and Characterization. Dry and purify the wet, crude ether solution of ethyl laurate by column chromatography. In a Pasteur filter pipet, place 500 mg of activated silica gel followed by 500 mg of anhydrous sodium sulfate (◄). Wet the column first with 0.5 mL of methylene chloride and then transfer the crude ether solution of ethyl laurate to the column using a Pasteur filter pipet. Use a **tared** 5-mL conical vial containing a boiling stone as a collection flask for the column eluant. Rinse the reaction vial with methylene chloride (0.5 mL). Transfer the rinse to the column. Repeat both rinse and transfer with a second aliquot of methylene chloride (0.5 mL). Add an additional 1.0 mL of methylene chloride directly to the column to ensure complete elution of the ester.

HOOD

Remove the ether–methylene chloride solvent by evaporation in the **hood** by using a stream of nitrogen gas with gentle warming in a sand bath. Make sure the vial remains warm to your fingers during the evaporation process—a warm vial ensures that condensation of moisture on the liquid product is avoided during solvent concentration.

The recovered ethyl laurate is a clear, viscous, pleasant-smelling ester, and is usually fully characterized without further purification. Weigh the product and calculate the percent yield. Determine the refractive index (optional) and boiling point and compare your data to the literature values. Obtain an IR spectrum of the ester using the capillary film technique and compare it to the spectrum of an authentic sample or to one found in the literature (*Aldrich Library of IR spectra* and/or SciFinder Scholar).

Chemical Tests. Does this ester give a positive hydroxamate test (see Chapter 9)? Check the solubility of ethyl laurate in water. Would you have predicted the result? Is the ester soluble in 85% phosphoric acid or in concentrated sulfuric acid?

Isopentyl Acetate: Semi-microscale Preparation

Experiment 8B

Isopentyl acetate is prepared by the following procedure.

REACTION

Isopentyl alcohol   Acetic acid   Isopentyl acetate

EXPERIMENTAL PROCEDURE

Estimated time to complete the experiment: 3.0 h.

Physical Properties of Reactants						
Compound	MW	Amount	mmol	bp (°C)	d	n_D
Isopentyl alcohol	88.15	800 μL	7.4	132	0.81	1.4053
Acetic acid	60.1	1.5 mL	26.2	116	1.05	1.3720
Sulfuric acid, concd		4 drops				

160–180 °C thermometer

Al foil

Reagents and Equipment. In a 5.0-mL conical vial containing a magnetic spin vane and equipped with a reflux condenser protected by a calcium chloride drying tube, place 800 μL (7.4 mmol) of isopentyl alcohol, 1.5 mL (26.2 mmol) of glacial acetic acid, 4 drops (Pasteur pipet) of concentrated sulfuric acid, and approximately 100 mg of silica gel beads (➡).

Speed   Heat

Isopentyl alcohol, 800 μL
+ CH₃CO₂H, 1.5 mL
+ 4 drops concd H₂SO₄
+ silica gel beads, 100 mg

CAUTION: Cap the vial immediately after addition of each reagent. Dispense the reagents in the **hood** using automatic delivery pipets. Concentrated acetic and sulfuric acids are *corrosive.*

HOOD

NOTE. *The silica gel beads (the desiccant) are used to absorb the water as it is generated in the reaction.*

Reaction Conditions.   Heat and stir the reaction mixture using a sand bath temperature of 160–180 °C for 1 h. Cool the resulting mixture to room temperature and remove the spin vane with forceps. Add 0.5 mL of diethyl ether to increase the volume of the organic phase.

Isolation of Product.   Extract the crude organic product with three 2-mL portions of 5% sodium bicarbonate solution, followed by 1 mL of water. During each extraction, cap the vial, shake gently, vent carefully, and then allow it to stand so the layers may separate. A Vortex mixer may be used to good advantage in this sequence. Remove the aqueous layer and be sure not to discard it until the final product is purified and characterized.

Following the aqueous extraction, dry the organic phase by adding anhydrous sodium sulfate to the vial. Transfer the anhydrous solution, using a Pasteur filter pipet, to a clean, dry 3-mL conical vial containing a boiling stone. Remove the added ether by evaporation in the **hood** by using a gentle stream of nitrogen gas with gentle warming in a sand bath. Next, attach the vial containing the crude ester residue to a Hickman still equipped with an air condenser and arranged in a sand bath for distillation (◀). Cover the sand bath with aluminum foil during the procedure.

HOOD

160–180 °C thermometer

Al foil

Speed   Heat

Distillation of reaction products

Purification and Characterization.   Distill the isopentyl acetate product at a sand bath temperature of 160–180 °C. As the material collects in the collar of the still, transfer it by Pasteur pipet (9 in.) to a **tared** $\frac{1}{2}$-*dram* vial.

Weigh the clear, viscous, pleasant-smelling isopentyl acetate, and calculate the percent yield. Determine the refractive index (optional) and boiling point, and compare your values with those found in the literature.

Obtain an IR spectrum of the ester using the capillary film technique and compare it with that shown in Figure 6.23.

Infrared Analysis.   The conversion of a branched-chain aliphatic primary alcohol to an acetate ester results in significant changes in the infrared spectrum of the molecule. These changes are similar to those observed in straight-chain systems. The two macro group frequency trains in the present example are

1. Isopentyl alcohol (Fig. 6.24): 3350 (broad), 3000–2850, 1460–1300, 1060, and 660 cm^{-1}

2. Isopentyl acetate (Fig. 6.23): 3490 (weak), 3000–2850, 1746 (strong), 1367 (broad), 1250, and 1040 cm^{-1}

If the macro group frequencies are further constrained to include only those systems in which the chain branching involves the presence of an isopropyl group, two additional bands near 1385 and 1365 cm^{-1} are required. These peaks, which are present in both the alcohol and the ester, arise from the spatially coupled and split symmetric methyl bending vibrations of the aliphatic backbone. In both isopentyl alcohol and isopentyl acetate, this pair of bands are found at identical locations, 1386 and 1367 cm^{-1}.

Chemical Tests.   Does the product give a positive hydroxamate test for an ester (Chapter 9)? Check the solubility of this ester in water, ether, and 85% H_3PO_4. In which solubility group (see Chapter 9, pp. 658–659) does isopentyl acetate fall?

Figure 6.23   IR spectrum: isopentyl acetate.

Isopentyl Acetate: Preparation Using a Monomode Microwave Apparatus

EXPERIMENTAL PROCEDURE

Estimated time to complete the experiment: 2.25 h.

Physical Properties of Reactants						
Compound	MW	Amount	mmol	bp (°C)	d	n_D
Isopentyl alcohol	88.15	0.8 mL	7.4	132	0.81	1.4053
Acetic acid	60.1	1.5 mL	26.2	116	1.05	1.3720
Sulfuric acid, concd		4 drops				

*This section has been written by Dr. Nicholas E. Leadbeater from the Department of Chemistry at the University of Connecticut and Dr. Cynthia B. McGowan from the Department of Chemistry at Merrimack College, MA.

Sample	Isopentyl alcohol		
%T _X_ ABS — Background Scans _4_		Scans _16_	
Acquisition & Calculation Time _42 sec_		Resolution _4.0 cm⁻¹_	
Sample Condition _liquid, neat_		Cell Window _KBr_	
Cell Path Length _capillary film_		Matrix Material _____	

Figure 6.24   IR spectrum: isopentyl alcohol.

Reagents and Equipment.   This experiment is designed for use in the CEM Discover and Biotage Initiator microwave units.

In a 10.0-mL glass microwave reaction vessel containing a magnetic stir bar, place 0.8 mL (7.4 mmol) of isopentyl alcohol, 1.5 mL (26.2 mmol) of glacial acetic acid, 4 drops (Pasteur pipet) of concentrated sulfuric acid, and approximately 100 mg of silica beads. Immediately cap the vessel with the microwave pressure cap.

HOOD

CAUTION: Dispense reagents in the **hood** using automatic delivery pipets. Concentrated acetic and sulfuric acids are *corrosive*. Since the reaction requires heating reagents to above their boiling point in sealed vessels, *adherence to the microwave manufacturer's guidelines is essential.*

Reaction Conditions.   Place the reaction vessel in the microwave cavity and, depending on the equipment used, position the pressure device on top. Program the microwave unit to heat the reaction mixture to 130 °C and hold at this temperature for 15 min. After heating, allow the reaction mixture to cool to 50 °C or below before removing the tube from the microwave unit. Transfer the reaction mixture with a Pasteur pipet into a 5.0-mL conical vial. Rinse the microwave reaction vessel with 0.5 mL of diethyl ether and add the washings to the conical vial.

Isolation of Product. The remainder of the procedure is identical to experiment 8B (p. 205).

Isopentyl Acetate: Preparation Using a Multimode Microwave Apparatus

EXPERIMENTAL PROCEDURE

Estimated time to complete the experiment: 2.25 h.

Physical Properties of Reactants						
Compound	MW	Amount	mmol	bp (°C)	d	n_D
Isopentyl alcohol	88.15	1.6 mL	14.7	132	0.81	1.4053
Acetic acid	60.1	3 mL	52.4	116	1.05	1.3720
Sulfuric acid, concd		8 drops				

Reagents and Equipment. This experiment is designed for use in the CEM MARS, Milestone START, and Anton Paar Synthos 3000 microwave units. When using the Anton Paar Synthos 3000 unit with the 24-position silicon carbide plate rotor containing glass vials, the reagent and solvent quantities cited in the monomode procedure should be used in conjunction with the reaction conditions here in the multimode procedure.

In a microwave reaction vessel containing a magnetic stir bar, place 1.6 mL (14.7 mmol) of isopentyl alcohol, 3.0 mL (52.4 mmol) of glacial acetic acid, 8 drops (Pasteur pipet) of concentrated sulfuric acid, and approximately 200 mg of silica beads. Immediately cap the vessel with the microwave pressure cap and adjust the tightness to the manufacturer-specified level. Place the sealed vessel into its outer protective jacket.

CAUTION: Dispense reagents in the **hood** using automatic delivery pipets. Concentrated acetic and sulfuric acids are *corrosive*. Since the reaction requires heating reagents to above their boiling point in sealed vessels, *adherence to the microwave manufacturer's guidelines is essential.*

HOOD

Reaction Conditions. Insert the loaded vessels into the reaction carousel ensuring they are evenly spaced and then place the carousel into the microwave cavity. If provided by the manufacturer, connect a temperature probe to the control vessel. Program the microwave unit to heat the reaction vessels to 120 °C and hold at this temperature for 5 min. After heating, allow the reaction mixture to cool to 50 °C or below before removing the carousel from the microwave unit.

Isolation of Product. Place 10 mL of 10% sodium bicarbonate solution in a 30-mL separatory funnel. Clamp the funnel to a ring stand. Pipette the reaction mixture from the microwave reaction vessel to the separatory funnel. An

HOOD

*This section has been written by Dr. Nicholas E. Leadbeater from the Department of Chemistry at the University of Connecticut and Dr. Cynthia B. McGowan from the Department of Chemistry at Merrimack College, MA.

effervescent reaction will occur as the excess acetic acid is neutralized. After the effervescence has subsided, add water to dissolve any solids. Add 5 mL of diethyl ether and carefully cap and invert the funnel. Immediately vent the funnel by opening the stopcock. Close the stopcock, place the funnel back on the ring stand, and remove the stopper. Drain the lower (aqueous) layer into a 50-mL Erlenmeyer flask. Extract the crude organic layer with a further two 5-mL portions of 10% sodium bicarbonate solution, followed by 5 mL of water. During each extraction, cap and invert the funnel several times, releasing the pressure by opening the stopcock, and then allow the funnel to stand so the layers will separate. Remove the lower (aqueous) layer after each extraction into the 50-mL Erlenmeyer flask. Save the aqueous waste until the experiment is complete and then discard as directed. Dry the organic layer by pipetting it into a clean 25-mL Erlenmeyer flask containing 200 mg of anhydrous sodium sulfate. Transfer the anhydrous solution, using a Pasteur filter pipet, to a clean tared 10-mL pear-shaped flask containing a boiling stone. Remove the ether on a rotary evaporator or by evaporation in the **hood** using a gentle stream of nitrogen gas with warming in a sand bath. Next, attach the flask containing the crude ester residue to a Hickman still equipped with an air condenser and arrange in a sand bath for distillation. Cover the sand bath with aluminum foil during the procedure.

Purification and Characterization.   Distill the isopentyl acetate product at a sand bath temperature of 160–180 °C. As the material collects in the collar of the still, transfer it by Pasteur pipet (9 in.) to a **tared** $\frac{1}{2}$-*dram* vial.

Weigh the clear, viscous, pleasant-smelling isopentyl acetate, and calculate the percent yield. Determine the refractive index (optional) and boiling point, and compare your values with those found in the literature.

Obtain an IR spectrum of the ester using the capillary film technique and compare it with that shown in Figure 6.23.

Esterification by Use of Acidic Resins: Semi-microscale Preparations

Experiment 8C

G

Modern synthetic reactions are making increased use of reagents that are heterogeneous in character. The use of resins as the support material for reactive compounds has become very popular because of their ease of removal and reliability. In this experiment you will utilize one of the recent additions to the arsenal of resin catalysts, Nafion 417. This catalyst is a powerfully acidic resin. It can approach the acidities of 100% sulfuric acid and of trifluoromethanesulfonic acid in trifluoroacetic anhydride solution. The "superacidity" of the sulfonic acid group in Nafion 417 is attributable to the electron-withdrawing ability of the perfluorocarbon backbone of the resin to which it is attached.

$$-[CF_2CF_2]_n[CFCF_2]_x-$$
$$|$$
$$(OCF_2CF)_mOCF_2CF_2SO_3H$$
$$|$$
$$CF_3$$

Nafion 417 resin with an equivalent weight of approximately 1200 contains tetrafluoroethylene (the monomer used to make Teflon) and perfluorovinyl ether units in a ratio of 7:1. These resins have been used successfully by Olah, as in this experiment, for the esterification of carboxylic acids.[8]

---

[8]Olah, G. A.; Keumi, T.; Meidar, D. *Synthesis* **1978**, 929.

REACTION

R—CH₂—ÖH + CH₃—C(=Ö:)(ÖH) $\xrightleftharpoons{\text{Nafion resin 417}}$ CH₃—C(=Ö:)(Ö—CH₂—R) + H₂O

Alcohol   Acetic acid   Ester

EXPERIMENTAL PROCEDURE

Estimated time to complete the experiment: 1.5 h.

Physical Properties of Reactants						
Compound	MW	Amount	mmol	bp (°C)	d	n_D
1-Butanol	74	365 μL	4.0	118	0.810	1.3990
Isopentyl alcohol	88	500 μL	4.6	132	0.81	1.4053
1-Pentanol	88	500 μL	4.6	138	0.811	1.4090
1-Hexanol	102	570 μL	4.55	156	0.814	1.4180
1-Octanol	130	700 μL	4.45	196	0.827	1.4290
Acetic acid	60	275 μL	4.8	116	1.049	1.3720
Nafion 417 resin	Equivalent weight 1100			1.5 × 0.5-cm strip		

Reagents and Equipment. In a 5.0-mL conical vial containing a magnetic spin vane and equipped with a reflux condenser protected by a calcium chloride drying tube, place the alcohol (*see table for correct molar quantities for your particular alcohol*), 275 μL (4.8 mmol) of glacial acetic acid, and a strip of Nafion 417 resin (1.5 × 0.5 cm). (If the strip of resin is placed so as to encircle the spin vane, mixing of the reaction solution is particularly effective.)

NOTE. The Nafion 417 resin has a significant number of hydrophilic centers that function in exactly the same way as the silica gel does in Experiment [8B]. Thus, Nafion 417, in addition to functioning as an acid catalyst, also acts as a desiccant, which helps to drive the esterification to completion by absorbing water as it is generated in the reaction.

Reaction Conditions. The reaction mixture is heated with stirring at a sand bath temperature of 160–180 °C for 30 min. The mixture is then cooled to room temperature.

Isolation of Product. Remove the Nafion strip and the spin vane (with forceps) from the cool reaction mixture and carefully (to avoid foaming) rinse them with 1 mL of 5% NaHCO₃. The rinse is added to the vial (to form a two-phase system) and swirled with the reaction mixture. The bicarbonate wash is then separated and the aqueous phase is transferred to a 10-mL Erlenmeyer flask for temporary storage. The reaction mixture is washed with a further 1-mL portion of 5% NaHCO₃, and then by two 1-mL distilled water washes (2 × 1 mL). All the aqueous washes are transferred to the storage Erlenmeyer. Dry the crude and wet reaction product by passing it down a Pasteur pipet containing a cotton plug and 1 g of anhydrous granular sodium sulfate (Na₂SO₄). Collect the product residue from the column in a tared 3-mL conical vial and weigh it to determine the crude yield. Obtain an IR spectrum and measure the refractive index (optional) of the crude material. Does the IR spectrum indicate the presence of any unreacted starting material in the crude product? What would be the most likely contaminant?

Purification and Characterization. A 25-μL sample of the crude ester is purified by prep-GC (see GC conditions for each product mixture and Experiment [2] for details on the collection procedure). Following collection of the purified ester, obtain an IR spectrum and boiling point, and compare these with the data obtained on the crude residue. The ester is the first compound to elute from the column. It may be followed by a small amount of a second component. You can calculate the purity of the crude product by measuring the area under the two elution bands. (Only in the case of the butyl acetate synthesis does the workup procedure distort the apparent crude yields: Can you explain why this occurs?) If you want to identify the contaminant (second band) directly, it may be possible to also collect a second fraction from the GC if the contaminant comprises a reasonable fraction of the product mixture (5–10%). It may, however, require a second injection of 30–40 μL.

Gas Chromatographic Conditions	
General	
Stainless steel columns in 8ft × $\frac{1}{4}$ in. packed with 20% Carbowax	
Flow rates: 50 mL/min (He); sample size: 25 μL	
Specific	
Ester 1	**Ester 2**
Butyl acetate	*Pentyl acetate*
Column temperature: 120 °C	Column temperature: 130 °C
Retention time	Retention time
Ester: 8.5 min	Ester: 8.5 min
Impurity: 11 min	Impurity: 11 min
Ester 3	**Ester 4**
Isopentyl acetate	*Hexyl acetate*
Column temperature: 140 °C	Column temperature: 160 °C
Retention time	Retention time
Ester: 6 min	Ester: 10 min
Impurity: 8 min	Impurity: 12 min
Ester 5	
Octyl acetate	
Column temperature: 185 °C	
Retention time	
Ester: 10 min	
Impurity: 13 min	

QUESTIONS

6-50. In the preparation of the esters given in this experiment, the reaction product was extracted with 5% sodium bicarbonate solution (NaHCO₃) in the isolation step. Why? What gas was evolved during this washing step? Write a balanced equation for the reaction that produced it.

6-51. **(a)** Why is a large excess of acetic acid used in the preparation of isopentyl acetate?
    (b) Write a mechanism for the preparation of isopentyl acetate using isopentyl alcohol and acetic acid.

6-52. Concentrated sulfuric acid is used as a catalyst for the esterification of acetic acid in the preparation of isopentyl acetate. Why is the sulfuric acid needed if another acid, acetic acid, is already present?

6-53. Fatty acids are long-chain carboxylic acids, usually of 12 or more carbon atoms, isolated from saponification of fats and oils (esters of glycerol). Draw the structure of each of the fatty acids named below and also determine its common name:

(a) Hexadecanoic acid (Z)-9-octadecenoic acid

(b) Octadecanoic acid (Z,Z)-9,12-octadecadienoic acid

6-54. Write a mechanism for the acid-catalyzed transesterification reaction of propyl acetate with 1-pentanol, which gives pentyl propionate.

6-55. In the infrared spectra of acetates, two intense bands are usually observed in the 1270- to 1000-wavenumber region. These peaks are related to the stretching vibrations of the two C—O bonds of the ester group. Should we expect to be able to assign the C—O bonds to individual peaks, and if so, which mode is associated with which band? Why? (*Hint:* The peak that occurs at higher wavenumbers is consistently close to 1250 cm^{-1}.)

BIBLIOGRAPHY

These references are selected from the large number of examples of esterification given in Organic Syntheses:

Bailey, D. M.; Johnson, R. E.; Albertson, N. F. *Organic Syntheses*; Wiley: New York, 1988; Collect. Vol. VI, p. 618.

Bowden, E. *Organic Syntheses*; Wiley: New York, 1943; Collect. Vol. II, p. 414.

Eliel, E. L.; Fisk, M. T.; Posser, T. *Organic Syntheses*; Wiley: New York, 1963; Collect. Vol. IV, p. 169.

Emerson, W. S.; Longley, R. I., Jr. *Organic Syntheses*; Wiley: New York, 1963; Collect. Vol. IV, p. 302.

Fuson, R. C.; Wojcik, B. H. *Organic Syntheses*; Wiley: New York, 1943; Collect. Vol. II, p. 260.

McCutcheon, J. W. *Organic Syntheses*; Wiley: New York, 1955; Collect. Vol. III, p. 526.

Mic'ovic', V. M. *Organic Syntheses*; Wiley: New York, 1943; Collect. Vol. II, p. 264.

Peterson, P. E.; Dunham, M. *Organic Syntheses*; Wiley: New York, 1988; Collect. Vol. VI, p. 273.

Stevenson, H. B.; Cripps, H. N.; Williams, J. K. *Organic Syntheses*; Wiley: New York, 1973; Collect. Vol. V, p. 459.

For an overall review of esterification, see:

Euranto, E. K. *in The Chemistry of Carboxylic Acids and Esters; Patai,* S. Ed.; Interscience: New York, 1969, p. 505.

The E1 Elimination Reaction: Dehydration of 2-Butanol, 2-Methylcylohexanol, and 4-Methylcyclohexanol

EXPERIMENT 9

Common name: 1-butene
CA number: [06-98-9]
CA name as indexed: 1-butene

Common name: *trans*-2-butene
CA number: [624-64-6]
CA name as indexed: 2-butene, (*E*)-

Common name: *cis*-2-butene
CA number: [590-18-1]
CA name as indexed: 2-butene, (*Z*)-

Common name: 2-methylcyclohexanol
CA number: [583-59-5]
CA name as indexed: cyclohexanol, 2-methyl-

Common name: 4-methylcyclohexanol
CA number: [589-91-3]
CA name as indexed: cyclohexanol, 4-methyl-

Common name: 4-methylcyclohexene
CA number: [591-47-9]
CA name as indexed: cyclohexene, 4-methyl-

Common name: 3-methylcyclohexene
CA number: [591-48-0]
CA name as indexed: cyclohexene, 3-methyl-

Common name: 1-methylcyclohexene
CA number: [591-49-1]
CA name as indexed: cyclohexene, 1-methyl-

Purpose.   This experiment illustrates the variety of pathways that are available to acid-catalyzed elimination reactions of secondary (2°) alcohols via carbocation intermediates. With each system, the dehydration forms a mixture of alkene products. The alkenes formed in these reactions can be separated and/or identified by using some of the most powerful instrumental techniques available to the modern research chemist for the separation of complex mixtures: gas chromatography (GC) and nuclear magnetic resonance (NMR).

THE DEVELOPMENT OF CARBOCATION THEORY

The dehydration reaction that you are about to study is representative of the large collection of reactions that are classified as E1 elimination reactions. These reactions all form an intermediate in which one of the carbon atoms bears, if not a *full* positive charge, at least a significant fractional positive charge. It is this fleeting, high-energy intermediate, an aliphatic carbocation, that makes these reactions so interesting.

The development of bonding theory in organic chemistry during the late nineteenth and early twentieth centuries did not accept the existence of carbocations, except in a few esoteric instances. This position was reasonable because aliphatic compounds show little ionic character. The first proposal that these nonpolar substances might actually form cations came from Julius Stieglitz (1867–1937), of the University of Chicago, in a paper published in 1899. Eight years later, James F. Norris (1871–1940) at Massachusetts Institute of Technology produced compelling evidence that these substances might be intermediates in reactions of certain *tert*-butyl halides.

These two papers were the origin of the concept that organic carbon cations, which in those days became known as *carbonium ions*, were far more widespread in organic reactions than previously anticipated. Indeed, these early investigations caused more controversy and more experimental work, to unambiguously prove the existence of what are now called *carbocations*, than any other single problem in American chemistry.

It was, however, the English chemists Arthur Lapworth, Sir Robert Robinson, C. K. Ingold (University of London), and E. D. Hughes who, from 1920 to 1940, undertook a massive effort to develop the experimental and theoretical data to place these early postulates on a solid scientific foundation. Between 1920 and 1922, Hans Meerwein in Bonn, Germany, demonstrated that carbon rearrangements in the camphene series could be best explained by postulating the presence of carbocation intermediates. Perhaps the most important contribution to the entire subject was published in 1932 by Frank C. Whitmore of Pennsylvania State College, where he was dean of the School of Chemistry and Physics. His paper "The Common Basis of Intramolecular Rearrangements" brought together a vast array of data in a beautifully consistent interpretation that essentially cemented the carbocation into contemporary organic chemical theory.

A reaction not too distant (in fact, rather close if you overheat your own!) from the one that is to be carried out below, but one that also included a rearrangement along with the dehydration, was studied by Dorothy Bateman and C. S. "Speed" Marvel at the University of Illinois as early as 1927.

Within 2 years of the publication of the Whitmore paper, Robinson proposed the formation of the steroids (including cholesterol) from squalene (a C_{30} polyunsaturated polyisoprene molecule) via an incredible series of intermediates and rearrangements. Later, following the elucidation of the structure of lanosterol, R. B. Woodward and K. Bloch made a brilliant proposal that at once rationalized the biosynthetic origin of both lanosterol and cholesterol and implicated lanosterol as an intermediate in cholesterol biosynthesis. Their mechanism involved the concerted (bonds made and broken simultaneously) cyclization of four rings, as well as four rearrangements following the generation of the initial carbocation intermediate, to ultimately yield lanosterol.[9]

The elucidation of these biochemical pathways is further evidence of the major impact that our understanding of carbocation chemistry has had on related fields, such as biochemistry.

Squalene

Lanosterol

Cholesterol

[9]Stieglitz, J. *Am. Chem. J.* **1899**, *21*, 101. Norris, J. F. *Am. Chem. J.* **1907**, *38*, 627. Meerwein, H.; van Emster, K. *Berichte* **1922**, *55*, 2500. Whitmore, F. C. *J. Am. Chem. Soc.* **1932**, *54*, 3274. Bateman, D. E.; Marvel, C. S. *J. Am. Chem. Soc.* **1927**, *49*, 2914. Robinson, R. *Chem. Ind.* **1934**, *53*, 1062. Woodward, R. B.; Bloch, K. *J. Am. Chem. Soc.* **1953**, *75*, 2023. See also: Tarbell, D. S.; Tarbell, T. *The History of Organic Chemistry in the United States, 1875–1955*; Folio: Nashville, TN, 1986.

Prior Reading
  Technique 1: Gas Chromatography (pp. 59–65)
  Technique 3: Fractional Distillation (pp. 68–71)
  Technique 4: Solvent Extraction (pp. 71–88)
  Technique 7: Collection or Control of Gaseous Products (pp. 109–111)

Experiment 9A

Dehydration of 2-Butanol

REACTION

DISCUSSION

The formation of an alkene (or alkyne) frequently involves loss of a proton and a leaving group from adjacent carbon atoms. The generalized reaction scheme is shown below.

Such a reaction is called an *elimination reaction,* because a small molecule (HL) is *eliminated* from the organic molecule. If the molecule eliminated is water, the reaction may also be referred to as a *dehydration.* One of the common synthetic routes to alkenes is the dehydration of an alcohol.

Dehydration of alcohols is an acid-catalyzed elimination reaction. Experimental evidence shows that alcohols react in the order tertiary (3°) > secondary (2°) > primary (1°); this reactivity relates directly to the stability of the carbocation intermediate formed in the reaction. Generally, sulfuric or phosphoric acid is used as the catalyst in the research laboratory. A Lewis acid, such as aluminum oxide or silica gel, is usually the catalyst of choice at the fairly high temperatures used in industrial scale reactions.

The mechanism for this reaction is classified as E1 (elimination, unimolecular). The elimination, or dehydration reaction, proceeds in several steps:

The first step (1) involves the very rapid, though reversible, protonation of the oxygen atom of the alcohol to form an *oxonium ion* (an oxygen cation with a full octet of electrons). This protonation step is important because it produces a good leaving group, water. Without acid, the only available leaving group is hydroxide ion (HO⁻), which, as a strong base, is a poor leaving group. This finding is the reason why acid plays such an important role in the mechanism of this dehydration. The second step (2) of the reaction is the dissociation of the oxonium ion to form an intermediate *carbocation* and water. This step is the rate-determining (and therefore the slowest) step of the reaction. In the third step (3), the carbocation is deprotonated by a ubiquitous water molecule (or other base, such as bisulfate ion, present in the system) in another rapid equilibration. The carbocation gains stability (lower energy) by releasing a proton (H⁺) from a carbon atom adjacent (α) to the carbocation (route (a) or (b) shown) to the attacking base. Thus, in step 4 the catalyst is regenerated as a protonated molecule of water (H₃O⁺), and the electron pair, previously comprising the C—H bond adjacent to the carbocation, flows toward the positive charge, generating a stable and neutral alkene. A variety of isomeric products may be formed, since different protons adjacent to the carbocation may be removed, and different conformations of the intermediate carbocation are also possible (see routes (a) and (b) above and further discussion below).

E1 elimination reactions, as we have just seen, involve equilibrium conditions and, thus, to maximize the yield (drive the reaction to completion), the alkene is usually removed from the reaction while it is in progress. A convenient technique for accomplishing this task is distillation, which is often used because alkenes *always* have a lower boiling point than the corresponding alcohol. In the present reaction, the alkenes are gases and, therefore, are easily removed and collected as described in the experimental section.

Many 1° (primary) alcohols also undergo dehydration, but usually by a different route, the E2 (elimination, bimolecular) mechanism. This step is governed mainly by the fact that the 1° carbocation that would be required in an E1 process is a relatively unstable (very high energy) intermediate. In this case, attack by the base occurs directly on the oxonium ion:

$$CH_3CH_2CH_2\ddot{O}H \underset{}{\overset{H_2SO_4}{\rightleftharpoons}} CH_3CH_2CH_2\overset{+}{\ddot{O}}H_2 + HSO_4^-$$

Oxonium ion

$$HSO_4^- + CH_3 - \overset{\overset{\displaystyle H}{|}}{\underset{\underset{\displaystyle H}{|}}{C}} - CH_2 - \overset{+}{\ddot{O}}H_2 \rightleftharpoons CH_3CH{=}CH_2 + H_2SO_4 + H_2O$$

Oxonium ion

E1 elimination is usually accompanied by a competing S_N1 substitution reaction that involves the *same* carbocation intermediate. Since both reaction mechanisms are reversible in this case, and since the alkene product gases are easily removed from the reaction, the competing substitution reaction (which predominantly regenerates the starting alcohol by attack of water, as a nucleophile, at the carbocation) is not troublesome, and the equilibrium eventually leads to gas evolution.

As discussed earlier, many alcohols can dehydrate to yield more than one isomeric alkene. In the present reaction involving the dehydration of 2-butanol, at least three alkenes are usually formed—1-butene, *trans*-2-butene, and *cis*-2-butene:

1-Butene   *trans*-2-Butene   *cis*-2-Butene

We would expect the alkene generated in the largest amount to be the one possessing the highest degree of substitution, since it is the most stable product (lowest energy). This is exactly what is observed: more than 90% of the products are isomers of 2-butene. Because trans alkenes are thermodynamically more stable than their cis counterparts, and since the reaction is reversible, one might expect the dominant isomer in the 2-butene mixture to be trans. Indeed, nearly twice as much trans as cis isomer is formed. An empirical rule, originally formulated by the Russian chemist Alexander Zaitsev (or Saytzeff), for base-catalyzed E2 eliminations states that the alkene with the largest number of alkyl substituents on the double bond will be the major product. This rule can also correctly predict the relative ratio of substituted alkenes to be expected from a given E1 elimination reaction, and it obviously applies in the case of 2-butanol.

Rearrangements of alkyl groups (such as methide, $^-$:CH_3 and ethide, $^-$:CH_2CH_3) and hydrogen (as hydride, :H^-) are often observed during the dehydration of alcohols, especially in the presence of very strong acid where carbocations can exist for longer periods of time. For example, when 3,3-dimethyl-2-butanol is treated with sulfuric acid, the elimination reaction yields the mixture of alkenes shown below. Can you predict which alkene is the principal product?

3° Carbocation   2° Carbocation

2,3-Dimethyl-2-butene   3,3-Dimethyl-1-butene

Reactions carried out under these conditions are very susceptible to alkide or hydride shifts if a more stable carbocation intermediate can be formed. In the example above, a 2° carbocation rearranges into the more stable 3° carbocation. This intramolecular rearrangement involves the transfer of an entire alkyl group (in this case a methyl substituent), together with its bonding pair of electrons, to an adjacent carbon atom. This migration, or shift, of the methyl group to an adjacent carbocation is called a *1,2-methide shift*. It commonly occurs in aliphatic systems involving carbocation intermediates that have alkyl substituents adjacent to the cation. Hydride shifts are also frequently observed and actually appear in a wider range of molecules.

EXPERIMENTAL PROCEDURE

Estimated time to complete the experiment: 2.0 h.

NOTE. Staggering the starting times of the reaction will allow more students easier access to the GC instrument when the product gases are analyzed.

Physical Properties of Reactants					
Compound	MW	Amount	mmol	bp (°C)	d
2-Butanol	74.12	100 μL	1.1	99.5	0.81
Concd sulfuric acid	98.08	50 μL	0.9		1.84

100 °C thermometer

See below

Speed   Heat

2-Butanol, 100 µL, + concd H₂SO₄, 50 µL

3 mL
4 mL

Reagents and Equipment. Assemble the gas collection apparatus shown in the figure *before* the reactants are mixed (◀).

The capacity of the gas collection reservoir is determined by the following procedure. Seal the collection tube with a septum cap and invert it. Then add 3.0 mL of water and mark the 3.0-mL level. Finally, add an additional 1.0 mL, and also mark the 4.0-mL level.

NOTE. Use of a Teflon-lined septum cap on the gas collection reservoir is necessary to prevent loss of the collected butene gases by permeation.

To position the gas collection reservoir, carry out the following steps: (1) fill the reservoir with water; (2) place your finger (index finger is usually used) over the open end of the reservoir; (3) invert it; (4) place it, with the open end down, into a beaker (250 mL) filled with water. When your finger is removed, the column of water should remain in the reservoir.

Place 100 µL (81 mg, 1.1 mmol) of 2-butanol and 50 µL of concentrated sulfuric acid in a **clean**, dry, 1.0-mL conical vial containing a magnetic spin vane.

CAUTION: Sulfuric acid is a strong, corrosive material. Contact with the skin or eyes can cause severe burns. It is best to dispense the reactants using automatic delivery pipets.

Cap-seal the 1-mL vial to the gas delivery tube and position the tube under water into the open end of the gas collection reservoir as shown in the figure. Clamp the reservoir in place.

Reaction Conditions. Heat the reaction mixture, with gentle stirring, using a sand bath until the evolution of gas takes place (sand bath temperature ~110–120 °C). The mixture should be warmed slowly through the upper temperature range to prevent foaming.

Isolation of Product. Collect about 3–4 mL of gas in the collection reservoir. Remove the gas delivery tube from the gas collection reservoir, and then from the water bath, before removing the reaction vial from the heat. Use this sequence of steps in shutting down the reaction to prevent water from being sucked back into the hot reaction flask while it is cooling down.

CAUTION: If water is drawn back into hot concentrated sulfuric acid, a very dangerous situation can occur, particularly if larger quantities of the reagents than recommended above are used in the experiment.

NOTE. Do not remove the gas collection reservoir from the water bath. The beaker containing the reservoir is carried to the gas chromatograph for analysis of the collected gas.

With the aid of a gas-tight syringe, withdraw a 0.5-cm³ sample through the rubber septum for GC analysis (➡).

Purification and Characterization. Analyze the collected gas by GC without further purification.

Gas Chromatographic Conditions
Column: 0.25 in. × 8 ft packed with 20% silicone DC 710
Room temperature
Flow rate: 20 mL/min (He gas)
Sample size: 0.5 cm³ of collected gas

Record the literature values of the physical properties of the products.

The butenes have been determined to elute from the DC 710 column in the following order: 1-butene, *trans*-2-butene, and *cis*-2-butene. *If the reaction mixture is heated above the recommended temperatures, a rearrangement can occur to yield isobutene, which will be detected as an additional isomeric product.*

If we make the assumption that the amount of each substance in the gaseous mixture is proportional to the area under its corresponding GC peak, determining the relative amounts of the three components of the gas sample becomes a straightforward calculation. The accuracy is, of course, dependent on how well the three peaks are resolved in the gas chromatogram.

NOTE. In calculating relative quantities of alkenes formed in the reaction, several techniques may be used to quantitatively determine the composition of the gas mixture if an integrating recorder is not available. Two methods are described here:

1. *Determination of the areas under the peaks gives reproducible results of ±3–4% when these areas are assumed equal to the peak heights (mm) × the peak widths at half-height (mm), measured from the baseline of the curve.*

2. *An alternative way to determine the areas under the peaks is to cut out the peaks from the chromatogram and weigh them on an analytical balance (sensitivity to 0.1 mg). The weights of the peaks are directly proportional to the relative amount of each compound in the gas sample.*[10]

Dehydration of 2-Methylcyclohexanol

Experiment 9B

REACTION

2-Methylcyclohexanol + H₃PO₄ ⇌ H₂O + 1-Methylcyclohexene + 3-Methylcyclohexene

EXPERIMENTAL PROCEDURE

Estimated time to complete the experiment: 2.0 h.

[10]See Pecsok, R. L. *Principles and Practices of Gas Chromatography*; Wiley: New York, 1961, p. 145.

NOTE. Staggering the starting times of the reaction will allow more students easier access to the GC instrument when the isomeric alkenes are analyzed.

Physical Properties of Reactants					
Compound	MW	Amount	mmol	bp (°C)	d
2-Methylcyclohexanol	114.19	2.0 mL	16.2	163–166	0.93
Concd phosphoric acid	97.99	0.5 mL	8.6		1.68

Reagents and Equipment. The system is assembled as shown in Figure 3.15 with the following modifications: (1) replace the magnetic-driven spinning band with a magnetic spin vane and (2) position a Claisen head protected by a calcium chloride drying tube placed on top of a water-jacketed condenser between the thermometer adapter equipped with a thermometer and the Hickman–Hinkle still head. The thermometer will need to be adjusted so that the bulb is positioned at the bottom of the well as shown in Figure 3.15.

Prior to reactant transfer, make sure that the Teflon magnetic spin vane is centered on the hot plate/magnetic stirrer. Once tested and free rotation is observed, place 2.0 mL of 2-methylcyclohexanol and 0.5 mL concentrated phosphoric acid in the vial (to be delivered with a Pasteur pipet or an automatic delivery pipet). Reassemble the system and lower the setup into the sand bath or alternative setups involving a copper-tube block or aluminum metal block. If electing to work with a sand bath, cover the sand bath with aluminum foil during the distillation as it is easier to regulate the temperature of the bath when it is covered.

Gently heat the vial after turning on the magnetic stirrer at a low setting. When reflux commences at the base of the Hickman–Hinkle still head, the magnetic stirrer is raised to intermediate settings. Upon heating the still and collecting the distillate, the distillate will appear as a cloudy white liquid and occur around 100 °C in the still while the sand bath is around 170 °C.

Fraction collection commences immediately on arrival of the vapor column at the still head. The distillate is transferred to a screw-cap 5.0-mL conical vial using a bent-tip Pasteur pipet. Continue the distillation until the volume remaining in the reaction conical vial is ~0.5 mL or the boiling in the reaction conical vial has ceased. Neither overheat nor distill to dryness.

Once the setup has cooled to room temperature, carefully rinse the inside wall of the Hickman–Hinkle still head with 0.5-mL saturated aqueous sodium chloride with a Pasteur pipet. Transfer this material to the 5.0-mL conical vial. Allow the vial to stand on a level surface to permit the two phases to separate. A sharp phase interface should appear.

As illustrated in Figure 5.17, draw both phases into the pipet, return the aqueous phase (bottom layer) to the original 5.0-mL conical vial by gently squeezing the pipet bulb, and transfer the top layer to a new screw-cap vial.

As there is no solvent rinse, add a *minimum* amount of anhydrous sodium sulfate using a microspatula, allow the mixture to stand for 5 min, and transfer the dried mixture of isomeric methylcyclohexenes to a tared screw-cap vial. Weigh the screw-cap vial and calculate the crude yield.

Characterization. Analyze the isomeric methylcyclohexenes by GC without further purification. The procedure involves the injection of 10 μL of the liquid material onto a 0.25 in. × 8-ft steel column packed with 20% silicone

DC 710. Experimental conditions are He flow rate, 20 mL/min; temperature, 190 °C.

The methylcyclohexenes have been determined to elute from the DC 710 column in the following order: 3-methylcyclohexene (5.1 min) and 1-methylcyclohexene (5.7 min). The peak ratios favoring the formation of the more substituted alkene 1-methylcyclohexene are approximately 3:1.

If we make the assumption that the amount of each substance in the mixture is proportional to the area under its corresponding GC peak, determining the relative amounts of the two components of the sample becomes a straightforward calculation. The accuracy is, of course, dependent on how well the two peaks are resolved in the gas chromatogram.

NOTE. In calculating relative quantities of methylcyclohexenes formed in the reaction, several techniques may be used to quantitatively determine the composition of the mixture if an integrating recorder is not available. Two methods are described here:

1. *Determination of the areas under the peaks gives reproducible results of ±3–4% when these areas are assumed equal to the peak heights (mm) × the peak widths at half-height (mm), measured from the baseline of the curve.*

2. *An alternative way to determine the areas under the peaks is to cut out the peaks from the chromatogram and weigh them on an analytical balance (sensitivity to 0.1 mg). The weights of the peaks are directly proportional to the relative amount of each compound in the gas sample.*

Dehydration of 4-Methylcyclohexanol

Experiment 9C

REACTION

EXPERIMENTAL PROCEDURE

Estimated time to complete the experiment: 2.0 h.

Physical Properties of Reactants					
Compound	MW	Amount	mmol	bp (°C)	d
4-Methylcyclohexanol	114.19	1.5 g	13.1	172	
Concd phosphoric acid	97.99	0.5 mL	8.6		1.68
Concd sulfuric acid	98.08	4 drops			1.84

Reagents and Equipment. The system is assembled as shown in Figure 3.15 with the following modifications: (1) replace the magnetic-driven spinning band with a magnetic spin vane and (2) position a Claisen head protected by a calcium chloride drying tube placed on top of a water-jacketed condenser between the thermometer adapter equipped with a thermometer and the Hickman–Hinkle still head. The thermometer will need to be adjusted so that the bulb is positioned at the bottom of the well as shown in Figure 3.15.

Prior to reactant transfer, make sure that the Teflon magnetic spin vane is centered on the hot plate/magnetic stirrer. Once tested and free rotation is observed, place 1.5 g of 4-methylcyclohexanol, 0.5 mL concentrated phosphoric acid, and 4 drops of concentrated sulfuric acid in the vial (to be delivered with a Pasteur pipet or an automatic delivery pipet). Reassemble the system and lower the setup into the sand bath or alternative setups involving a copper-tube block or aluminum metal block. If electing to work with a sand bath, cover the sand bath with aluminum foil during the distillation as it is easier to regulate the temperature of the bath when it is covered.

Gently heat the vial after turning on the magnetic stirrer at a low setting. When reflux commences at the base of the Hickman–Hinkle still head, the magnetic stirrer is raised to intermediate settings. Upon heating the still and collecting the distillate, the distillate will appear as a cloudy white liquid and occur around 100 °C in the still while the sand bath is around 170 °C.

Fraction collection commences immediately on arrival of the vapor column at the still head. The distillate is transferred to a screw-cap 5.0-mL conical vial using a bent-tip Pasteur pipet. Continue the distillation until the volume remaining in the reaction conical vial is ~0.5 mL or the boiling in the reaction conical vial has ceased. Neither overheat nor distill to dryness.

Once the setup has cooled to room temperature, carefully rinse the inside wall of the Hickman–Hinkle still head with 0.5-mL saturated aqueous sodium chloride with a Pasteur pipet. Transfer this material to the 5.0-mL conical vial. Allow the vial to stand on a level surface to permit the two phases to separate. A sharp phase interface should appear.

As illustrated in Figure 5.17, draw both phases into the pipet, return the aqueous phase (bottom layer) to the original 5.0-mL conical vial by gently squeezing the pipet bulb, and transfer the top layer to a new screw-cap vial.

As there is no solvent rinse, add a *minimum* amount of anhydrous sodium sulfate using a microspatula, allow the mixture to stand for 5 min, and transfer the dried mixture of isomeric methylcyclohexenes to a screw-cap tared vial. Weigh the screw-cap vial and calculate the crude yield.

Characterization. Analyze the crude reaction mixture by IR and NMR without purification. While the IR analysis of the isomeric mixture will offer clear evidence of alkene formation as noted by the peak at 3022 cm^{-1}, an NMR analysis using band deconvolution offers one the opportunity to delve deeper and both qualify and quantify the isomeric composition of all methylcyclohexenes generated.[11] Standard operating procedures for performing band deconvolutions are both known and readily available for Agilent/Varian, Bruker, and JEOL (see Bibliography).

---

[11]Lopansri, L. S.; Letson, J. N.; O'Brien, R. A.; Battiste, D. R.; Forbes, D. C. *World Journal of Chemical Education*, **2022**, *10*, 51.

NOTE. For Experiment 9B, the two isomers 1-methylcyclohexene and 3-methylcy-clohexene generated are baseline resolved using GC analysis. If electing to subject the crude reaction mixture of Experiment 9B to NMR analysis, the diagnostic peaks found in the vinylic region in both the ^{1}H and ^{13}C NMR are as well baseline resolved. When introducing a third methylcyclohexene isomer, 4-methylcyclohexene, as is the case in Experiment 9C where not 2-methylcyclohexanol but 4-methylcyclohexanol is dehydrated using acid catalysis, band deconvolution, even with its limitations, is a viable option to both qualify and quantify the isomers generated.

Applying NMR band deconvolution to this system results in an isomeric composition (4-methylcyclohexene:3-methylcyclohexene:1-methylcyclohexene) of 68:14:18 using the ^{1}H NMR data and a composition of 79:13:8 when using ^{13}C NMR data. Solutions using CH_2Cl_2 as solvent for the NMR analyses were used.

Isomer	Peak, vinylic (ppm)	
	Filled	Unfilled
1-Methylcyclohexene	120.3	133.0
3-Methylcyclohexene	125.4	132.6
4-Methylcyclohexene	125.8	125.7

QUESTIONS

6-56. Gas chromatographic analysis of a mixture of organic compounds gave the following peak areas (cm²): pentane = 2.6, octane = 1.7, heptanol = 2.1, *o*-xylene = 0.8.
  (a) Calculate the mole percent composition of the mixture. Assume that the response of the detector (area per mole) is the same for each component.
  (b) Calculate the weight percent composition of the mixture, using the same assumptions as in (a).

6-57. (a) The major isomer in the dehydration of 2-methylcyclohexanol is 1-methylcyclohexene. Suggest a mechanism to account for the formation of this compound.
  (b) The major isomer when performing the same transformation starting with 4-methylcyclohexanol is 4-methylcyclohexene. Suggest a mechanism to account for the formation of this compound.
  (c) Comment on why the major isomer of (a) happens to be the minor of (b).

6-58. When *tert*-pentyl bromide is treated with 80% ethanol, the following amounts of alkene products are detected on analysis:

Offer an explanation of why compound **I** is formed in far greater amount than the terminal alkene.

6-59. The —⁺SR₂ group is easily removed in elimination reactions, but the —SR group is not. Explain.

6-60. Why is sulfuric acid, rather than hydrochloric acid, used to catalyze the dehydration of alcohols?

BIBLIOGRAPHY

Several dehydration reactions of secondary alcohols using sulfuric acid as the catalyst are given in *Organic Syntheses*:

Adkins, H.; Zartman, W. *Organic Syntheses*; Wiley: New York, 1943; Collect. Vol. II, p. 606.

Bruce, W. F. *Organic Syntheses*; Wiley: New York, 1943; Collect. Vol. II, p. 12.

Coleman, G. H.; Johnstone, H. F. *Organic Syntheses*; Wiley: New York, 1941; Collect. Vol. I, p. 183.

Grummitt, O.; Becker, E. I. *Organic Syntheses*; Wiley: New York, 1963; Collect. Vol. IV, p. 771.

Norris J. F. *Organic Syntheses*; Wiley: New York, 1941; Collect. Vol. I, p. 430.

Wiley, R. H.; Waddey, W. E. *Organic Syntheses*; Wiley: New York, 1955; Collect. Vol. III, p. 560.

For an overview of elimination reactions:

March, J. *Advanced Organic Chemistry*, 4th ed.; Wiley: New York, 1992, p. 982.

Smith, M. B. *March's Advanced Organic Chemistry*, 8th ed.; Wiley: New York, 2019, Chap. 17.

For details on carbonium ion formation:

Olah, G. A.; Schleyer, P. von R., Eds. *Carbonium Ions*; Wiley: New York, Vol. I, 1968; Vol. II, 1970; Vol. III, 1972.

Olah, G. A.; Prakash, G. K. S. *Carbocation Chemistry*; Wiley: New York, 2004.

For additional examples illustrating NMR band deconvolution:

Lopansri, L. S., Letson, J. N., O'Brien, R. A., Battiste, D. R., Forbes, D. C. *World Journal of Chemical Education*, **2022**, *10*, 51.

For detailed instructions in the area of NMR band deconvolution, see:

Agilent (Varian), NMR Spectroscopy, User Guide. https://www.agilent.com/cs/library/usermanuals/public/0199934300c.pdf (accessed 10 June 2022).

Bruker, New In Bruker User Library: DREAMTIME NMR. https://www.bruker.com/en/resources/library/application-notes-mr/dreamtime_nmr.html (accessed 10 June 2022).

JEOL, Delta™ NMR Data Processing Software. https://www.jeolusa.com/PRODUCTS/Nuclear-Magnetic-Resonance/Delta-NMR-Software (accessed 10 June 2022).

Experiment [9A] is adapted from the method given by Helmkamp, G. K.; Johnson, H. W. Jr. *Selected Experiments in Organic Chemistry*, 3rd ed.; W. H. Freeman: New York, 1983; p. 99.

EXPERIMENT 10

The E2 Elimination Reaction: Dehydrohalogenation of 2-Bromobutane to Yield 1-Butene, trans-2-Butene, and cis-2-Butene

Common name: 1-butene
CA number: [06-98-9]
CA name as indexed: 1-butene

Common name: *trans*-2-butene
CA number: [624-64-6]
CA name as indexed: 2-butene, (*E*)-

Common name: *cis*-2-butene
CA number: [590-18-1]
CA name as indexed: 2-butene, (*Z*)-

Purpose. This reaction illustrates the base-induced dehydrohalogenation of alkyl halides with strong base and is used extensively for the preparation of alkenes. The stereo- and regiochemical effects of the size of the base are investigated, and the product mixture is analyzed by the use of gas chromatography.

Prior Reading

   Technique 1: Gas Chromatography (pp. 59–65)

   Technique 7: Collection or Control of Gaseous Products (pp. 109–111)

REACTION

1-Butene

cis-2-Butene

trans-2-Butene

DISCUSSION

Base-induced elimination (dehydrohalogenation) of alkyl halides is a general reaction and is an excellent method for preparing alkenes. This process is often referred to as β *elimination*, since a hydrogen atom is always removed β to the halide (leaving group):

A high concentration of a strong base in a relatively nonpolar solvent is used to carry out the dehydrohalogenation reaction. Such combinations as sodium methoxide in methanol, sodium ethoxide in ethanol, potassium isopropoxide in isopropanol, and potassium *tert*-butoxide in *tert*-butanol or dimethyl sulfoxide (DMSO) are often used.

Elimination reactions almost always yield an isomeric mixture of alkenes, where this is possible. Under the reaction conditions, the elimination is *regioselective* and follows the Zaitsev rule when more than one route is available for the elimination of HX from an unsymmetrical alkyl halide. That is, the reaction proceeds in the direction that yields the most highly substituted alkene. For example,

$$CH_3-\overset{\overset{\displaystyle CH_3}{|}}{\underset{\underset{\displaystyle Br}{|}}{C}}-\overset{\overset{\displaystyle H}{|}}{\underset{\underset{\displaystyle CH_3}{|}}{C}}-CH_3 \xrightarrow[HOC_2H_5]{K^+,\,^-OC_2H_5}$$

2-Bromo-
2,3-dimethylbutane

2,3-Dimethyl-
2-butene
(major product)

$+ \quad CH_2=\overset{\overset{\displaystyle CH_3}{|}}{C}-\overset{\overset{\displaystyle H}{|}}{\underset{\underset{\displaystyle CH_3}{|}}{C}}-CH_3$

2,3-Dimethyl-
1-butene
(minor product)

In cases where cis or trans alkenes can be formed, the reaction exhibits stereoselectivity, and the more stable trans isomer is the major product.

$$CH_3CH_2CH_2\overset{\overset{\textstyle }{}}{\underset{\underset{\displaystyle Br}{|}}{C}}HCH_2CH_2CH_3 \xrightarrow[HOC_2H_5]{K^+,\,^-OC_2H_5}$$

4-Bromoheptane

cis-3-Heptene (22%)

+

trans-3-Heptene (78%)

Experimental evidence indicates that the five atoms involved in the E2 elimination reaction must lie in the same plane; the anti-periplanar conformation is preferred. This conformation is necessary for the orbital overlap that must occur for the π bond to be generated in the alkene. The sp³-hybridized atomic orbitals on carbon that comprise the C—H and C—X σ bonds broken in the reaction develop into the p orbitals comprising the π bond of the alkene formed:

Anti-periplanar conformation

There is a smooth transition between reactant and product. Analogous to the S$_N$2 reaction, no intermediate has been isolated or detected. Furthermore, no rearrangements occur under E2 conditions. This situation is in marked contrast to E1 elimination reactions, where carbocation intermediates are generated and rearrangements are frequently observed (see Experiment [9]).

The alkyl halide adopts the anti-periplanar conformation in the transition state, and experimental evidence demonstrates that if the size of the base is increased, then it must be difficult for the large base to abstract an internal β-hydrogen atom. In such cases, the base removes a less hindered β-hydrogen atom, leading to a predominance of the thermodynamically **less stable (terminal) alkene** in the product mixture. This type of result is often referred to as **anti-Zaitsev** or **Hofmann elimination**. Thus, in the reaction of 2-bromo-2,3-dimethylbutane given above, the 2,3-dimethyl-1-butene would be the major product (anti-Zaitsev) if the conditions used a bulkier base. The anti-periplanar arrangements are illustrated in the Newman projections below.

Zaitsev product
internal alkene

Anti-Zaitsev product
terminal alkene

Two anti-periplanar conformations
of 2-bromo-2,3-dimethylbutane

Dehydrohalogenation of alkyl halides in the presence of strong base (E2) is often accompanied by the formation of substitution (S_N2) products. The extent of the competitive substitution reaction depends on the structure of the alkyl halide. Primary alkyl halides give predominantly substitution products (the corresponding ether), secondary alkyl halides give predominantly elimination products, and tertiary alkyl halides give exclusively elimination products. For example, the reaction of 2-bromopropane with sodium ethoxide proceeds as follows:

2-Bromopropane   Sodium
ethoxide

Ethyl isopropyl   Propene
ether (21%)   (79%)

In general, for the reaction of alkyl halides with strong base, the following trendlines exist and are contingent upon the degree of alkyl substitution (primary, secondary, and tertiary as a function of eliminative or substitutive outcome):

1°   2°   3°

ease of S_N2 reaction

ease of E2 reaction

EXPERIMENTAL PROCEDURE

Estimated time to complete the experiment: 2.5 h, if two reactions are run by each student.

Physical Properties of Reactants

Compound	MW	Amount	mmol	bp (°C)	d	n_D
2-Bromobutane	137.03	100 µL	0.92	91.2	1.26	1.4366
Methanol	32.04	3.5 mL		64.9	0.791	1.3288
2-Propanol	60.09	3.5 mL		82.4	0.785	1.3776
2-Methyl-2-propanol (*tert*-butanol)	74.12	3.5 mL		82–83	0.786	1.3838
3-Ethyl-3-pentanol	116.20	3.5 mL		140–142	0.839	1.4266
Sodium	22.98	60 mg	2.6	883	0.97	
Potassium	39.10	60 mg	1.5	760	0.86	

Table 6.5 Reagent Combinations

Alcohol Solvent	Metal	Alkoxide Base Produced
Methanol	Sodium	Sodium methoxide
2-Propanol	Potassium	Potassium 2-propoxide
2-Methyl-2-propanol (*tert*-butanol)	Potassium	Potassium 2-methyl-2-propoxide (potassium *tert*-butoxide)
3-Ethyl-3-pentanol	Potassium	Potassium 3-ethyl-3-pentoxide

Reagents and Equipment. The combinations of reagents in Table 6.5 may be used to prepare the alkoxide base. Students should compare results to observe a total picture of the effect.

Preparation of the Alkoxide Base. Measure and add to a 5.0-mL conical vial containing a magnetic spin vane 3–3.5 mL of the anhydrous alcohol to be used (see Table 6.5). Add a 60-mg piece of potassium (or sodium) metal and immediately attach the vial to a reflux condenser protected by a calcium chloride drying tube. Place the arrangement in a sand bath and with stirring heat the mixture gently (~50 °C) (⬇).

Assembly for preparation
of alkoxide base

NOTE. If the sodium/methanol combination is used, it is not necessary to heat the mixture. A fairly vigorous reaction occurs at room temperature. It is recommended that the instructor cut the Na/K metal before commencing the laboratory.

CAUTION. Handle sodium and potassium with care. These metals react vigorously with moisture and are kept under paraffin oil or xylene. Remove the small piece of the metal from the oil using a pair of forceps or tongs—*never use your fingers!* Dry the metal quickly by pressing it with filter paper (to soak up the oil), and immediately add it to the alcohol in the reaction vial. Any residual pieces of sodium/potassium should be stored in a bottle marked "sodium/potassium residues." **Never** throw small pieces of these metals in the sink or in water. To destroy, in the **hood**, add small amounts of the metal to absolute ethanol.

HOOD

When all the metal has reacted, remove the assembly from the sand bath and cool to near room temperature (do not remove the drying tube).

Table 6.6 Temperature Conditions	
Base	Temperature (°C)
$NaOCH_3$	100–110
$KOCH(CH_3)_2$	130–140
$KOC(CH)_3$	140–150
$KOC(CH_2CH_3)_3$	175–180

Reaction Conditions.   Remove the drying tube from the condenser and use a calibrated Pasteur pipet to introduce 100 μL of 2-bromobutane down through the condenser into the vial. Replace the drying tube and place the assembly in the **preheated** sand bath (see Table 6.6). Remove the drying tube and attach the gas delivery tube to the top of the condenser so that the open end of the tube is beneath the water level of the reservoir (♦). If the connection to the top of the condenser is not made with an O-ring Capseal connection, lightly grease the ground-glass joint to insure a gas-tight seal. After about 10–15 air bubbles emerge, place the water-filled gas collection tube over the open end of the gas delivery tube.

Assembly for gas delivery

Arrangement for gas collection

Isolation of Product. Collect about 6–7 mL of gas in the collection reservoir and then use a hypodermic syringe to withdraw a 0.7- to 0.8-mL sample through the rubber septum for GC analysis.

NOTE. Remove the gas delivery tube from the collecting reservoir and then from the water before discontinuing the heat on the reaction vial. This order of events prevents water from being sucked back into the reaction flask.

Purification and Characterization. The collected gas is analyzed by gas chromatography without further purification.

> *Gas Chromatographic Conditions*
> Column: $\frac{1}{4}$ in. × 8 ft packed with 20% DC 710
> Room temperature
> Flow rate (He gas): 30 mL/min
> Sample size: 0.7–0.8 mL of collected gas
> Chart speed: 1 cm/min

Assuming that the amount of each substance in the gas is proportional to the areas of its corresponding peak, determine the ratio of the three components in the gas sample.

Area Under a Curve. *Several techniques may be used. The following method gives reproducible results of ± 3–4%:* Area = peak height (mm) × width at half-height (mm), measured from the baseline of the curve.

The order of elution of the butenes is 1-butene, *trans*-2-butene, and *cis*-2-butene. Record the literature values of their physical properties.

QUESTIONS

6-61. Outline a complete mechanistic sequence for the reaction of 2-bromobutane with potassium 2-methyl-2-propoxide in 2-methyl-2-propanol solvent to form the three alkenes generated in the reaction (1-butene, *trans*-2-butene, and *cis*-2-butene). Include a clear drawing of the anti-periplanar transition state for the formation of each alkene.

6-62. Does the mixture of gases collected in this experiment consist only of alkenes? If not, what other gases might be present?

6-63. Predict the predominant alkene product that would form when 2-bromo-2-methylpentane is treated with sodium methoxide in methanol. If the base were changed to $KOC(CH_2CH_3)_3$ would the same alkene predominate? If not, why? What would be the structure of this alternate product, if it formed?

6-64. Predict the more stable alkene of each of the following pairs:
 (a) 2-Methyl-1-butene or 2-methyl-2-butene
 (b) *trans*-3-Hexene or *cis*-3-hexene
 (c) 2-Methyl-2-hexene or 2,3-dimethyl-2-pentene

6-65. Starting with the appropriate alkyl halide and base–solvent combination, outline a synthesis that would yield each of the following alkenes as the major or only product:
 (a) 1-Pentene
 (b) 3-Methyl-1-butene
 (c) 2,3-Dimethyl-1-butene
 (d) 3-Methylcyclohexene

6-66. (a) When *cis*-1-bromo-4-*tert*-butylcyclohexane reacts with sodium ethoxide in ethanol, it reacts rapidly to yield 4-*tert*-butylcyclohexene. Under similar conditions, *trans*-1-bromo-4-*tert*-butylcyclohexane reacts very slowly. Using conformational structures, explain the difference in reactivity of these cis–trans isomers. (b) Under E2 reaction conditions, formation of 3-methylcyclohexene is exclusively formed when working with one of the isomers of 1-bromo-2-methylcyclohexane. Consider both cis–trans isomers and rationalize how one isomer cannot and can result in formation of 3-methylcyclohexene.

BIBLIOGRAPHY

Several dehydrohalogenation reactions of alkyl halides using alkoxide bases are given in Organic Syntheses:

Allen, C. F.; Kalm, M. J. *Organic Syntheses*; Wiley: New York, 1963; Collect. Vol. IV, p. 398.

McElvain, S. M.; Kundiger, D. *Organic Syntheses*; Wiley: New York, 1955; Collect. Vol. III, p. 506.

Paquette, L. A.; Barrett, J. H. *Organic Syntheses*; Wiley: New York, 1973; Collect. Vol. V, p. 467.

Schaefer, J. P.; Endres, L. *Organic Syntheses*; Wiley: New York, 1973; Collect. Vol. V, p. 285.

For an overview of elimination reactions:

March, J. *Advanced Organic Chemistry*, 4th ed.; Wiley: New York, 1992, p. 982.

Smith, M. B. *March's Advanced Organic Chemistry*, 8th ed.; Wiley: New York, 2019, Chap. 17.

Experiment [10] is adapted from the method given by:

Leone, S. A.; Davis, J. D. *J. Chem. Educ.* **1992**, *69*, A175.

The Isolation of Natural Products

EXPERIMENT 11

These experiments are designed to acquaint you with the procedures used to isolate naturally occurring and often biologically active organic compounds. These substances are known as *natural products* because they are produced by living systems. The particular natural products you are going to study come from the plant kingdom. At the end of the nineteenth century more than 80% of all medicines in the Western world were natural substances found in roots, barks, and leaves. There was a widespread belief at that time that in plants there existed cures for all diseases. As Kipling wrote, "Anything green that grew out of the mold/Was an excellent herb to our fathers of old." Even as the power of synthetic organic chemistry has grown during this century, natural materials still constitute a significant fraction of the drugs used in modern medicine. For example, in the mid-1960s when approximately 300 million new prescriptions were written each year, nearly half were for substances of natural origin. These materials have played a major role in successfully combating the worst of human illnesses, from malaria to high blood pressure, diseases that affect hundreds of millions of people.

B

Unfortunately, during the latter half of this century a number of very powerful natural products that subtly alter the chemistry of the brain have been used in vast quantities by our society. The ultimate impact on civilization is of grave concern. Evidence clearly demonstrates that these natural substances disrupt the exceedingly complex and delicate balance of biochemical reactions that lead to normal human consciousness. How well the brain is able to repair the damage from repetitive exposure is unknown. We are currently conducting experiments to answer that question.

The natural products that you may isolate in the following experiments include a bright-yellow crystalline antibiotic (Experiment [11A]), a white crystalline alkaloid that acts as a stimulant in humans (Experiment [11B]), and an oily material with a pleasant odor and taste (Experiment [11C]).

Isolation and Characterization of an Optically Active Natural Product: Usnic Acid

Experiment 11A

Common name: usnic acid

CA number: [7562-61-0]

CA name as indexed: 1,3(2*H*,9b*H*)-dibenzofurandione, 2,6-diacetyl-7,9-dihydroxy-8,9b-dimethyl-

Purpose.   In this exercise you will extract the active principle, usnic acid, from one of the lichens that produce it. Usnic acid is a metabolite found in a variety of lichens. For this experiment we utilize a local (in Maine) species of lichen, *Usnea hirta* (often referred to as *old-man's beard*), which is a fruticose lichen (a lichen that possesses erect, hanging, or branched structures). The extraction technique illustrated here is often used to isolate natural products from their native sources (see also Experiment [11B], on page 238, for another extraction strategy). Because usnic acid possesses a single chiral center (stereocenter) and only one of the enantiomers is produced in old-man's beard, this experiment also functions as an introduction to the methods used to measure the specific rotation of optically active substances.

B   LICHENS AND NATURAL PRODUCTS

Lichens, of which there are estimated to be more than 15,000 species, are an association between an algae and a fungus that live together in an intimate relationship. This association is often called symbiosis. Symbiosis requires that two different organisms live together in both close structural proximity and interdependent physiological combination. The term ordinarily is applied to situations where the relationship is advantageous, or even required, for one or both, but not harmful to either. In the case of lichens, the algae can be grown independently of the fungi that obtain nutrients from the algae cells. The fungi are, therefore, considered to be parasitic and their contribution to the union has been viewed historically only as an aid in the absorption and retention of water and perhaps to provide a protective structure for the algae. It appears, however, that the fungi may play a far more important role in the life of the lichen than earlier appreciated. The fungi appear to generate a metabolite, usnic acid, which is the most common substance found in these primitive systems. This acid can comprise up to 20% of the dry weight of some lichens! Even more intriguing is the original belief that usnic acid appears to have *no biological function* in these plants. Why would a living system channel huge amounts of its precious energy into making an apparently useless substance? Recently, with our increased understanding of the role of chemical communication substances in ecology, it has been recognized that usnic acid very likely makes a major symbiotic contribution as a *chemical defense agent*. Indeed, in 1945 Burkholder demonstrated that several New England lichens possess antibiotic properties, and usnic acid was subsequently shown to be the active agent against several kinds of bacteria, including staphylococcus. The Finnish company, Lääke Oy Pharmaceutical, has prepared from reindeer lichen a broad-spectrum usnic acid antibiotic for treating tuberculosis and serious skin infections. There is, in fact, evidence that lichens were used in medicine by the ancient Egyptians, and from 1600 to 1800 C.E. these plants were considered an outstanding cure for tuberculosis. Usnic acid has been investigated for use as an antibiotic by the U.S. Public Health Service. It proved to be effective in dilutions between 1 part in 100,000 and 1 part in 1,000,000 against several Gram-positive organisms.

This widespread lichen metabolite is the material isolated in this experiment. Usnic acid was first isolated and identified in 1843 by Rochleder, but a molecule of this complexity was beyond the structural knowledge of organic chemistry in those days. The structure was finally determined in 1941 by Schöpf, and in 1956 it was synthesized in the laboratory by Sir D. H. R. Barton (Nobel Laureate). Barton's route involved a spectacular one-step dimerization of a simple precursor, a synthesis that very closely mimicked the actual biogenetic pathway (see chemistry). The key step was the one-electron ($1\ e^-$) oxidation of methylphloracetophenone, which leads directly to the dimerization. The mechanism of this reaction, both in the plant and in the laboratory synthesis, is essentially identical to the oxidative coupling of 2-naphthol to give 1,1′-bi-2-naphthol, which is explored in detail in Experiment [5$_{adv}$]:

Methylphloracetophenone

dimerization

Usnic acid

The chiral center (stereocenter) (*) is bonded to a highly conjugated aromatic ring system (see structure), which gives rise to a very large specific rotation. This enhanced interaction with polarized radiation makes this compound a particularly interesting molecule to examine for optical activity. The production of a single enantiomer in the natural product, which, as discussed above, is formed by an oxidative coupling process, implies that there must be an intimate association between the substrate and an enzyme (a biological catalyst that itself is optically active) during the crucial coupling process.[12]

Prior Reading

Technique 6A: Thin-Layer Chromatography (pp. 101–103)

Technique 8: Measurement of Specific Rotation

Optical Rotation Theory (pp. 112–115)

[12]Dean, F. M.; Halewood, P.; Mongkolsuk, S.; Robertson, A.; Whally, W. B. *J. Chem. Soc.* **1953,** 1250. Kreig, M. B. *Green Medicine*; Rand McNally: New York, 1964. Lewis, W. H.; Elvin-Lewis, M. P. F. *Medical Botany*; Wiley: New York, 1977. Hendrickson, J. B. *The Molecules of Nature*; W. A Benjamin,. New York: 1965. Richards, J. H.; Hendrickson, J. B. *The Biosynthesis of Steroids, Terpenes, and Acetogenins*; W. A. Benjamin: New York, 1964. Schöpf, C.; Ross, F. *Annalen* **1941,** *546,* 1 (see further references cited in Experiment [5$_{adv}$]).

DISCUSSION

Nonracemic solutions of chiral substances, when placed in the path of a beam of polarized light, may rotate the plane of the polarized light clockwise or counterclockwise and are thus referred to as *optically active*. This angle of optical rotation is measured using a *polarimeter*. This technique is applicable to a wide range of analytical problems varying from purity control to the analysis of natural and synthetic products in the medicinal and biological fields. The results obtained from the measurement of the observed angle of rotation (α_{obs}) are generally expressed in terms of *specific rotation* [α]. The sign and magnitude of [α] are dependent on the specific molecule and are determined by complex features of molecular structure and conformation, and thus cannot be easily explained or predicted. The relationship of [α] to α_{obs} is $[\alpha]_{\lambda}^{T} = \dfrac{\alpha_{obs}}{l \cdot c}$ where T is the temperature of the sample **in degrees Celsius** (°C), l is the length of the polarimeter cell **in decimeters** (1 dm = 0.1 m = 10 cm), c is the concentration of the sample **in grams per milliliter** (g/mL), and λ is the wavelength of the light used in the polarimeter **in nanometers** (nm). These units are traditional, though most are esoteric by contemporary standards. Thus, the specific rotation for a given compound is normally reported in terms of temperature, wavelength, concentration, and the nature of the solvent. For example, $[\alpha]_{D}^{25} = +12.3°$ ($c = 0.4$, CHCl$_3$) implies that the measurement was recorded in a CHCl$_3$ solution of 0.4 g/mL at 25 °C using the sodium D line (589 nm) as the light source. Unless indicated, the path length is assumed to be 1 dm in these observations.

Usnic acid contains a single stereocenter (see structure), and therefore it can exist as a pair of enantiomers. In nature, however, only one of the enantiomers (*R* or *S*) would be expected to be present. Usnic acid has a very high specific rotation, $[\alpha]_{D}^{25} = +488°$ ($c = 0.4$, CHCl$_3$), which will give a large α_{obs} even at low concentrations, and for this reason it is an ideal candidate to measure rotation in a microscale experiment.

Racemic (equimolar amounts of each enantiomer) usnic acid has been resolved (separated into the individual enantiomers) through preparation and separation of the diastereomeric (–) brucine salts. This procedure was the route followed to obtain an authentic synthetic sample for comparison with the natural material. The separation was required because the dimerization step in the synthesis, which was carried out in the absence of enzymatic, or other chiral, influence, gave a racemic product.

A common method of extracting chemical constituents from natural sources is presented in this experiment. In this case, only one chemical compound, the usnic acid, is significantly soluble in the extraction solvent, acetone. For this reason, the isolation sequence is straightforward.

EXPERIMENTAL PROCEDURE

Isolation of Usnic Acid.   Estimated time to complete the experiment: 2.5 h.

Physical Properties of Components			
Compound	MW	Amount	bp (°C)
Lichen		1.0 g	
Acetone	58.08	15.0 mL	56.2

Reagents and Equipment. Weigh and place about 1.0 g of oven-dried (40 °C) crushed or cut-up lichens and 15.0 mL of acetone in a 50-mL Erlenmeyer flask containing a magnetic stirrer. Loosely cap the flask with a cork stopper (➡). The lichens used in this experiment are *Usnea hirta*.

Reaction Conditions. Stir or occasionally swirl the mixture for no less than 30 min at room temperature. If necessary, periodically push the lichens below the surface of the acetone solvent using a glass rod.

Isolation of Product. Prior to filtering the resulting mixture, remove a 0.5-mL aliquot for final analysis. Filter by gravity and collect the filtrate in a 25-mL Erlenmeyer flask (➡). A Pasteur filter pipet may be used to make this transfer, if desired. In the **hood**, remove the acetone solvent under a slow stream of air or nitrogen on a warm sand bath nearly to dryness. Allow the remainder of the acetone to evaporate at room temperature to obtain the crude bright yellow or orange usnic acid crystals.

Purification and Characterization. Recrystallize all but 5 mg of the crude extract from acetone/95% ethanol (10:1). Dissolve the crystals in the minimum amount of hot acetone, keeping the recrystallization vessel hot, and add the appropriate volume of 95% ethanol. Allow the mixture to cool to room temperature and then place the flask in an ice bath to complete the recrystallization. Collect the golden-yellow crystals by vacuum filtration (➡) and wash them with *cold* acetone. Dry the crystals on a porous clay plate or on a sheet of filter paper. As an alternative and more efficient procedure, the crude material may be recrystallized using a Craig tube, avoiding the filtration step with the Hirsch funnel.

Weigh the yellow needles of usnic acid and calculate the percentage of the acid extracted from the dry lichen. Determine the melting point (use the evacuated melting-point technique) and compare your value to that found in the literature. Using a solvent system of ethyl acetate:hexane (1:4) and a UV lamp for visualization, compare the crude extract with the purified usnic acid by TLC (R_f value for usnic acid is 0.32). Obtain an IR spectrum and compare it with that of an authentic sample or one from the literature (*The Aldrich Library of IR Spectra* and/or *SciFinder Scholar*).

Chemical Tests. Chemical tests can assist in establishing the nature of the functional groups in usnic acid. Perform the 2,4-dinitrophenylhydrazine test and the ferric chloride test (see Chapter 9). Are the results significant?

Determination of the Specific Rotation. Though usnic acid is an optically active compound with a very high specific rotation, a low-volume, long-pathlength cell must be used to successfully determine its specific rotation with microscale quantities.

Dissolve usnic acid (80 mg) in 4.0 mL of tetrahydrofuran (THF) solvent and transfer the solution to the polarimeter cell using a Pasteur pipet.

NOTE. To obtain this quantity (80 mg) of usnic acid will very likely require pooling the recrystallized product of eight or nine students. Spectral grade THF should be used as the solvent. Many of the early specific rotation values on these substances were recorded with chloroform as the solvent, but because it possesses some toxicity, it is now avoided if possible.

Place the cell in the polarimeter and measure the angle of rotation. Calculate the specific rotation using the equation given in the discussion section.

QUESTIONS

6-67 (a) Determine the correct R or S designation for each of the following molecules. (b) Clearly highlight which group is considered third priority:

6-68 The structure originally proposed for cordycepic acid, which has $[\alpha] = +40.3°$, was

Why is this not a plausible structure?

6-69 A sample of 210 mg of an organic compound is dissolved in 8.2 mL of water. The solution is placed in a 20-cm polarimeter tube and the rotation is measured in a polarimeter. The rotation observed is +2.676°. Distilled water, in the same tube, gave a reading of +0.016°. Calculate the specific rotation for the compound.

6-70 Compound A is optically active and has the molecular formula $C_5H_{10}O$. On catalytic hydrogenation (addition of hydrogen) of A, compound B is obtained. Compound B has the molecular formula $C_5H_{12}O$ and is optically inactive. Give the structure for compounds A and B.

6-71 Which of the following compounds have a meso form?
(a) 2,3-Dibromopentane
(b) 2,4-Dibromopentane
(c) 2,3-Dibromobutane

BIBLIOGRAPHY

This experiment is adapted from that given by:

Todd, D. *Experimental Organic Chemistry*; Prentice-Hall: Englewood Cliffs, NJ, **1979**, p. 57.

Synthesis of usnic acid:

Barton, D. H. R.; DeFlorin, A. M.; Edwards, O. E. *J. Chem. Soc.* **1956**, 530.

Penttila, A., Fales, H. M. *Chem. Commun.* **1966**, 656.

A large-scale method of isolation of usnic acid has been reported:

Stark, J. B.; Walter, E. D.; Owens, H. S. *J. Am. Chem. Soc.* **1950**, 72, 1819.

Optical, crystallographic and X-ray diffraction data have been reported for usnic acid:

Jones, F. T.; Palmer, K. J. *J. Am. Chem. Soc.* **1950**, 72, 1820.

Experiment 11B	**Isolation and Characterization of a Natural Product: Caffeine and Caffeine 5-Nitrosalicylate**

Product

Common names: caffeine, 1,3,7-trimethyl-2,6-dioxopurine

CA number: [58-08-2]

CA name as indexed: 1*H*-purine-2,6-dione, 3,7-dihydro-1,3,7-trimethyl-

Purpose.   To extract the active principle, an alkaloid, caffeine, from a native source, tea leaves. Caffeine is a metabolite (a product of the living system's biochemistry) found in a variety of plants. We will use ordinary tea bags as our source of raw material. This experiment illustrates an extraction technique often used to isolate water-soluble, weakly basic natural products from their biological source (see also Experiment [11A] for another extraction strategy). The isolation of caffeine will also give you the opportunity to use sublimation as a purification technique, since caffeine is a crystalline alkaloid that possesses sufficient vapor pressure to make it a good candidate for this procedure. In addition, the preparation of a derivative of caffeine, its 5-nitrosalicylate salt, will be carried out. This latter conversion takes advantage of the weakly basic character of this natural product.

ALKALOIDS

B

Caffeine belongs to a rather amorphous class of natural products called alkaloids. This collection of substances is unmatched in its variety of structures, biological response on nonhost organisms, and the biogenetic pathways to their formation.

The history of these fascinating organic substances begins at least 4000 years ago. They were incorporated into poultices, potions, poisons, and medicines, but no attempt was made to isolate and identify the substances responsible for the physiological response until the very early 1800s.

The first alkaloid to be obtained in the pure crystalline state was morphine. Friedrich Wilhelm Sertürner (1783–1841) isolated morphine in 1805. He recognized that the material possessed basic character and he thus classified it as a vegetable alkali (that is, a base with its origin in the plant kingdom). Thus, compounds with similar properties ultimately became known as alkaloids. The term "alkaloid" was introduced for the first time by an apothecary, Meissner, in Halle in 1819.

Sertürner, also a pharmacist, lived in Hamelin, another city in Prussia. He isolated morphine from opium, the dried sap of the poppy. Since the analgesic and narcotic effects of the crude resin had been known for centuries, it is not surprising that with the emerging understanding of chemistry, the interest of Sertürner became focused on this drug, which is still medicine's major therapy for intolerable pain. He published his studies in detail in 1816, and very quickly two French professors, Pierre Joseph Pelletier (1788–1842) and Joseph Caventou (1795–1877) at the Ecole de Pharmacie in Paris, recognized the enormous importance of Sertürner's work.

In the period from 1817 to 1820, these two men and their students isolated many of the alkaloids, which continue to be of major importance. Included in that avalanche of purified natural products was caffeine, which they obtained from the coffee bean. This substance is the target compound that you will be isolating directly from the raw plant in this experiment. A little more than 75 years later, caffeine was first synthesized by Fischer in 1895 from dimethylurea and malonic acid.

THE CLASSIFICATION OF ALKALOIDS

These compounds are separated into three general classes of materials.

1. **True alkaloids:** These compounds contain nitrogen in a heterocyclic ring; are almost always basic (the lone pair of the nitrogen is responsible for

this basic character); are derived from amino acids in the biogenesis of the alkaloid; invariably are toxic and possess a broad spectrum of pharmacological activity; are found in a rather limited number of plants (of the 10,000 known genera only 8.7% possess at least one alkaloid); and normally occur in a complex with an organic acid (this helps to make them rather soluble in aqueous media). As we will see, there are numerous exceptions to these rules. For example, there are several very well-known quaternary alkaloids. These are compounds in which the nitrogen has become tetravalent and positively charged (as in the ammonium ion). Thus, they are not actually basic.

2. **Protoalkaloids:** These compounds are simple amines, derived from amino acids, in which the basic nitrogen atom is not incorporated into a ring system; they are often referred to as *biological amines*. An example of a protoalkaloid is mescaline.

3. **Pseudoalkaloids:** These compounds contain nitrogen atoms usually *not* derived from amino acids. There are two main classes into which pseudoalkaloids are divided, the steroidal alkaloids and the *purines*. Caffeine has been assigned to this latter class of alkaloids.

Morphine

Mescaline

Caffeine

Prior Reading

    Technique 4: Solvent Extraction

        Solid–Liquid Extraction (p. 83)

        Liquid–Liquid Extraction (p. 76)

    Technique 9: Sublimation

        Sublimation Theory (pp. 116–117)

DISCUSSION

Caffeine (1,3,7-trimethylxanthine) and its close relative theobromine (3,7-dimethyl-xanthine) both possess the oxidized purine skeleton (xanthine). These compounds are often classified as pseudoalkaloids, since only the nitrogen atom at the 7 position can be traced to an amino group originally derived from an amino acid (in this case glycine). This classification emphasizes the rather murky problem of deciding just what naturally occurring nitrogen bases are *true* alkaloids. We will simply treat caffeine as an alkaloid.

Xanthine

Purine

Pyrimidine

Although the pyrimidine ring (present in caffeine's purine system) is a significant building block of nucleic acids, it is rare elsewhere in nature.

These two methylated xanthines are found in quite a number of plants and have been extracted and widely used for centuries. Indeed, they very likely have been, and remain today, the predominant stimulant consumed by humans. Every time you make a cup of tea or coffee, you perform an aqueous extraction of plant material (tea leaves, *Camellia sinensis,* 1–4%, or coffee beans, *Coffea* spp., 1–2%) to obtain a dose of 25–100 mg of caffeine. Caffeine is also the active substance (~2%) in maté (used in Paraguay as a tea) made from the leaves of *Ilex paraguariensis.* In coffee and tea, caffeine is the dominant member of the pair, whereas in *Theobroma cacao,* from which we obtain cocoa, theobromine (1–3%) is the primary source of the biological response. Caffeine acts to stimulate the central nervous system with its main impact on the cerebral cortex, and as it makes one more alert, it is no surprise that it is the chief constituent in No-Doz® pills.

Caffeine is readily soluble in hot water, because the alkaloid is often bound in thermally labile, partially ionic complexes with naturally occurring organic acids, such as with 3-caffeoylquinic acid in the coffee bean. For this reason it is relatively easy to separate caffeine from black tea leaves by aqueous extraction.

3-Caffeoylquinic acid

Other substances, mainly tannic acids, are also present in the tea leaves and are also water-soluble. The addition of sodium carbonate, a base, during the aqueous extraction helps to increase the water solubility of these acidic substances by forming ionic sodium salts and liberating the free base.

Subsequent extraction of the aqueous phase with methylene chloride, in which free caffeine has a moderate solubility, allows the transfer of the caffeine from the aqueous extract to the organic phase. At the same time, methylene chloride extraction leaves the water-soluble sodium salts of the organic acids behind in the aqueous phase.

Extraction of the tea leaves directly with nonpolar solvents (methylene chloride) to remove the caffeine gives very poor results—since, as we have seen, the caffeine is bound in the plant in a partially ionic complex that will not be very soluble in nonpolar solvents. Thus, water is the superior extraction solvent for this alkaloid. The water also swells the tea leaves and allows for easier transport across the solid–liquid interface.

Following extraction and removal of the solvent, sublimation techniques are applied to the crude solid residues to purify the caffeine. This technique is especially suitable for the purification of solid substances at the microscale level, if they possess sufficient vapor pressure. Sublimation techniques are particularly advantageous when the impurities present in the sample are nonvolatile under the conditions used.

Sublimation occurs when a substance goes directly from the solid phase to the gas phase upon heating, bypassing the liquid phase. Sublimation is technically a straightforward method for purification in that the materials need only be heated and therefore mechanical losses can be kept to a minimum (the target substance must, of course, be thermally stable at the required temperatures). Materials sublime only when heated *below* their melting points, and reduced pressure is usually required to achieve acceptable sublimation rates. Obviously, substances that lend themselves best to purification by sublimation are those that do not possess strong intermolecular attractive forces. Caffeine and ferrocene (used as a reactant in Experiment [27]) meet these criteria because they present large flat surfaces occupied predominantly with repulsive π electrons. For other isolation methods, see the discussion of *solid-phase* extraction in Technique 4, where caffeine is extracted from coffee beans (pp. 87–88).

EXPERIMENTAL PROCEDURE

Estimated time to complete the experiment: 2.5 h.

Physical Properties of Constituents				
Compound	MW	Amount	mmol	mp (°C)
Tea		1.0 g		
Water		10 mL		
Sodium carbonate	105.99	1.1 g	10	851

Reagents and Equipment.   Carefully open a commercial tea bag (2.0–2.5 g of tea leaves) and empty the contents. Weigh out 1.0 g of tea leaves and place them back in the empty tea bag. Close and secure the bag with staples.

Weigh, and add to a 50-mL Erlenmeyer flask, 1.1 g (0.01 mol) of anhydrous sodium carbonate followed by 10 mL of water (◄). Heat the mixture with occasional swirling on a hot plate to dissolve the solid. Now add the 1.0 g of tea leaves (in the tea bag) to the solution. Place the bag in the flask so that it lies flat across the bottom. Anhydrous Na_2CO_3, 1.1 g + H_2O, 10 mL + tea bag with tea leaves, 1.0 g

Anhydrous Na_2CO_3, 1.1 g + H_2O, 10 mL + tea bag with tea leaves, 1.0 g

GENTLY

Reaction Conditions.   Place a small watch glass over the mouth of the Erlenmeyer flask and then heat the aqueous suspension to a **gentle** boil for 30 min on the hot plate.

Isolation of Product.   Cool the flask and contents to room temperature. Transfer the aqueous extract from the Erlenmeyer flask to a 12- or 15-mL centrifuge tube using a Pasteur filter pipet. In addition, gently squeeze the tea bag by pressing it against the side of the Erlenmeyer flask to recover as much of the basic extract as possible. Set aside the tea bag and its contents.

Extract the aqueous solution with 2.0 mL of methylene chloride.

NOTE. The tea solution contains some constituents that may cause an emulsion. If you obtain an emulsion during the mixing of the aqueous and organic solvent layers (by shaking or using a Vortex mixer), it can be broken readily by centrifugation.

Separate the lower (methylene chloride) layer (check to make sure that the lower layer is, indeed, the organic layer by testing the solubility of a few drops of it in a test tube with distilled water) using a 9-in. Pasteur pipet. Drain the wet extracts through a filter funnel containing a small plug of cotton that is covered with about 2.0 g of anhydrous sodium sulfate, previously "moistened" with a small amount of methylene chloride (➡). (The organic phase will be saturated with water following the extraction; therefore it is referred to as "wet." It also may contain a few droplets of the aqueous phase, which become entrained during the phase separation; this can be particularly troublesome if an emulsion forms during the mixing.)

Anhydrous
Na$_2$SO$_4$,
2 g

Caffeine +
CH$_2$Cl$_2$,
12 mL

Collect the dried filtrate in a 25-mL filter flask. Extract the remaining aqueous phase with four additional 2.0-mL portions of methylene chloride (4 × 2 mL). Each extraction (an extraction is often referred to as a washing) is separated, dried as above, and transferred to the same filter flask. Finally, rinse the sodium sulfate with an additional 2.0 mL of methylene chloride and combine this wash with the earlier organic extracts.

Add a boiling stone to the flask and concentrate the solution to dryness in the **hood** by warming the flask in a sand bath. The crude caffeine should be obtained as an off-white crystalline solid.

HOOD

Purification and Characterization. Purify the crude solid caffeine by sublimation.

Assemble a sublimation apparatus as shown in Figure 5.53, on page 117; either arrangement is satisfactory. Using an aspirator, apply a vacuum to the system through the filter flask (remember to install a water-trap bottle between the sidearm flask and the aspirator). After the system is evacuated, run cold water gently through the cold finger or add ice to the centrifuge tube. By cooling the surface of the cold finger *after* the system has been evacuated, you will minimize the condensation of moisture on the area where the sublimed sample will collect.

NOTE. Less caffeine will be lost if the bottom of the cold finger is positioned less than 5 mm from the bottom of the filter flask.

Once the apparatus is evacuated and cooled, begin the sublimation by gently heating the flask with a microburner or sand bath. If you use a gas burner, always keep moving the flame back and forth around the bottom and sides of the flask.

NOTE. Be careful. Do not melt the caffeine. If the sample does begin to melt, remove the flame for a few seconds before heating is resumed. Overheating the crude sample will lead to decomposition and the deposition of impurities on the cold finger. High temperatures are not necessary since the sublimation temperature of caffeine (and of all solids that sublime) is below the melting point. It is generally worthwhile to carry out sublimations as slowly as possible, as the purity of the material collected will be enhanced.

When no more caffeine will sublime onto the cold finger, remove the heat, shut off the aspirator and the cooling water to the cold finger, and allow the apparatus to cool to room temperature under reduced pressure. Once cooled, carefully vent the vacuum and return the system to atmospheric pressure. *Carefully* remove the cold finger from the apparatus.

NOTE. If the removal of the cold finger is done carelessly, the sublimed crystals may be dislodged from the sides and bottom of the tube and drop back onto the residue left in the filter flask.

Scrape the caffeine from the cold finger onto weighing paper using a microspatula and a sample brush. Weigh the purified caffeine and calculate its percent by weight in the original tea leaves. Determine the melting point and compare your value to that in the literature.

If your melting-point apparatus uses capillary tubes to determine the melting point, an evacuated sealed tube is necessary, since caffeine sublimes; the melting point is above the sublimation temperature (see Chapter 4). The melting point may be obtained using the Fisher–Johns apparatus without this precaution.

Obtain an IR spectrum and compare it with that of an authentic sample.

Chemical Test.   Does the soda lime or the sodium fusion test (see Chapter 9) confirm the presence of nitrogen in your caffeine product?

DERIVATIVE: CAFFEINE 5-NITROSALICYLATE

It is not completely surprising to find that caffeine in the coffee bean is bound in a thermally labile complex with acid, since this alkaloid is a weakly basic substance possessing a base strength somewhat greater than that of an aryl amide. Because the purine ring system of caffeine has little reactive functionality, the formation of simple chemical derivatives is limited. The ease of association with high-melting carboxylic acids, however, offers a route to a variety of materials with well-defined melting points.

One such acid is 5-nitrosalicylic acid, which is prepared by nitration of salicylic acid in Experiment [29]. The purified caffeine obtained in this experiment may be further characterized by preparation of the 5-nitrosalicylate complex. This association is similar to the natural one formed with 3-caffeoylquinic acid:

Caffeine 5-nitrosalicylate

EXPERIMENTAL PROCEDURE

Estimated time to complete the experiment: 0.5 h.

Physical Properties of Reactants and Product					
Compound	MW	Amount	mmol	mp (° C)	bp (° C)
Caffeine	194.20	11 mg	0.06	238	
5-Nitrosalicylic acid	183.12	10 mg	0.06	229–230	
Petroleum ether (60–80 °C)		0.5 mL			60–80
Ethyl acetate	88.12	0.7 mL			77
Caffeine 5-nitrosalicylate	377.32			180	

Reagents and Equipment.   Weigh and add to a 3.0-mL conical vial, containing a magnetic spin vane and equipped with an air condenser, 11 mg (0.06 mmol) of caffeine, 10.0 mg (0.06 mmol) of 5-nitrosalicylic acid, and 0.7 mL of ethyl acetate.

Reaction Conditions.   *Gently* warm the mixture on a hot plate, stirring to dissolve the solids. Add 0.5 mL of petroleum ether (bp 60–80 °C) to the warm ethyl acetate solution, mix, and warm for several seconds. Remove the spin vane using forceps.

Isolation of Product.   Cool the mixture to room temperature and then place it in an ice bath for 10–15 min. Collect the crystals under reduced pressure using a Hirsch funnel, and then wash the filter cake with 0.5 mL of cold ethyl acetate. Dry the product on a porous clay plate or filter paper.

Purification and Characterization.   The product normally is sufficiently pure for direct characterization. Weigh the caffeine 5-nitrosalicylate and calculate the percent yield. Determine the melting point and compare your value with that reported above. Obtain an IR spectrum and compare it with those shown in Figures 6.25 and 6.26. The infrared spectrum reveals some of the details of the derivative formation (see below).

Infrared Analysis.   The infrared spectrum of 5-nitrosalicylic acid (Fig. 6.26 on page 247) is characteristic of aromatic carboxylic acids (see discussion in Experiment [7]). Note that (1) the substitution of the ring is revealed by the presence of the 1,2,4-combination band pattern with peaks at 1940, 1860, and 1815 cm^{-1}; (2) the strongest band in the spectrum below 1750 cm^{-1} is assigned to the symmetric stretch of the —NO$_2$ group found at 1339 cm^{-1}, and (3) the conjugated carboxyl C═O stretch is located at 1675 cm^{-1}.

In the spectrum of the complex (Fig. 6.25) we do not find evidence for ionized carboxylate. This group, if present, would give rise to two very strong *broad* bands at 1600–1550 and 1400–1330 cm^{-1}. What is observed is the carboxylate C═O stretch at 1665 cm^{-1} overlapped with a caffeine band. Evidence for very strong hydrogen bonding, however, is indicated by the series of very broad bands extending from 3550 to 2000 cm^{-1}. The complex association, therefore, very likely does not involve a complete proton transfer as is indicated in the above simplified chemical structures of the complex. The unambiguous formation of the 5-nitrosalicylic acid complex is best ascertained by the identification of the presence of the —NO$_2$ symmetric stretching vibration from a strong band located at 1345 cm^{-1}. Caffeine does not possess a band in this region.

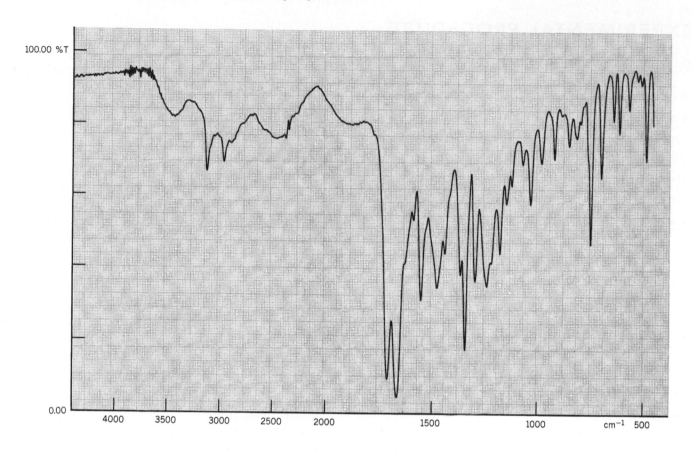

Sample	Caffeine 5-nitrosalicylate		
%T X ABS — Background Scans 4		Scans 16	
Acquisition & Calculation Time 42 sec		Resolution 4.0 cm⁻¹	
Sample Condition solid		Cell Window	
Cell Path Length		Matrix Material KBr	

Figure 6.25 **IR spectrum: caffeine 5-nitrosalicylate.**

QUESTIONS

6-72 Compounds such as naphthalene and 1,4-dichlorobenzene find use as mothballs since they sublime at a slow rate at atmospheric pressure. Explain this behavior in terms of the structure of the molecules.

6-73 **(a)** How many peaks would you expect to find in the 13C NMR spectrum of caffeine? **(b)** What characteristic absorption bands would you expect to find in the infrared spectrum of caffeine? **(c)** Are the NMR spectra (proton and carbon) of caffeine when compared to caffeine 5-nitrosalicylate as characteristic to the comparisons found using IR?

6-74 The vapor pressures of 1,2-diphenylethane, p-dichlorobenzene, and 1,3,5-trichlorobenzene are 0.06, 11.2, and 1.4 torr, respectively, at their melting point (52–54 °C). Which compound is likely to be sublimed most rapidly at a reduced pressure of 15 torr and a temperature of 40 °C?

6-75 To color spots on TLC plates for easier visualization after elution with solvent, the plates can be "developed" in a sealed chamber containing solid iodine. Explain how the solid–vapor equilibrium operates in this instance.

6-76 Caffeine is soluble in ethyl acetate. Do you think that the purity of your product could be checked by TLC using ethyl acetate as an elution solvent? Explain.

6-77 The infrared spectrum of 5-nitrosalicylic acid (Fig. 6.26) possesses the typical broad medium band found in *acid dimers* (908 cm⁻¹). In the caffeine 5-nitrosalicylate complex, however, this band is missing. Suggest a reason why the 908-wavenumber peak vanishes.

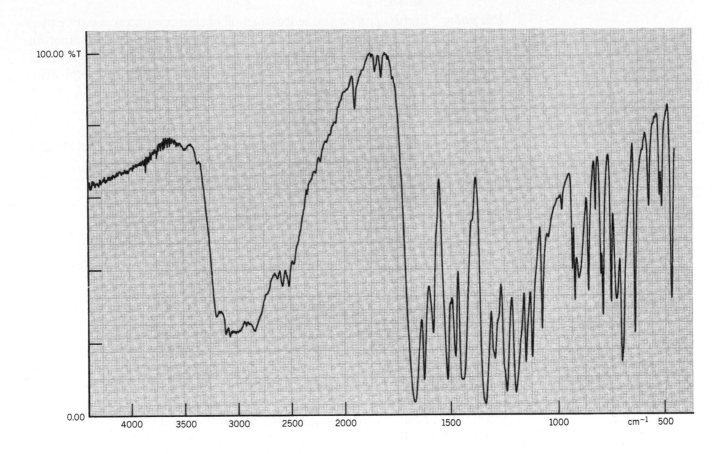

Sample	5-Nitrosalicylic acid		
%T _X_ ABS _Background Scans_ 4		Scans _16_	
Acquisition & Calculation Time _42 sec_		Resolution _4.0 cm⁻¹_	
Sample Condition _solid_		Cell Window _____	
Cell Path Length _____		Matrix Material _KBr_	

Figure 6.26   IR spectrum: 5-nitrosalicylic acid.

BIBLIOGRAPHY

Several extraction procedures for isolating caffeine from tea, some designed for the introductory organic laboratory, have been reported:

Ault, A.; Kraig, R. *J. Chem. Educ.* **1969,** *46,* 767.
Mitchell, R. H.; Scott, W. A.; West, P. R. *J. Chem. Educ.* **1974,** *51,* 69.
Moyé, A. L. *J. Chem. Educ.* **1972,** *49,* 194.
Murray, S. D.; Hansen, P. J. *J. Chem. Educ.* **1995,** *72,* 851.

Onami, T.; Kanazawa, H. *J. Chem. Educ.* **1996,** *73,* 556.

For additional information on the Fisher-Johns melting-point apparatus, see:

Morhrig, J. R.; Neckers, D. C. *Laboratory Experiments in Organic Chemistry,* 3rd ed.; Van Nostrand: New York, **1979,** p. 92.
Zubrick, J. W. *The Organic Chem Lab Survival Manual,* 11th ed.; Wiley: New York, 2019.

Isolation of a Natural Product by Steam Distillation: Cinnamaldehyde from Cinnamon

Experiment 11C

Common name: cinnamaldehyde

CA number: [14371-10-9]

CA name as indexed: 2-propenal, 3-phenyl-, (*E*)-

Purpose. In this exercise you will extract oil of cinnamon from a native plant source, such as *Cinnamomum zeylanicum,* and then purify the principal flavor and odor component of the oil, cinnamaldehyde. The experiment demonstrates the importance of steam distillation techniques (at the semi-micro level) to the collection of essential oils.

Prior Reading

> *Technique 2:* Simple Distillation at the Semi-microscale Level (pp. 65–68)
> *Technique 3:* Steam Distillation
> *Technique 4:* Solvent Extraction
> > Liquid–Liquid Extraction (p. 76)
> > Drying of the Wet Organic Layer (pp. 84–87)
> *Technique 6:* Chromatography
> > Concentration of Solutions with Nitrogen Gas (p. 106)

B ESSENTIAL OILS

Let us begin by defining what we mean by the term "metabolite." The metabolism of an organism is composed of the biochemical reactions and pathways in that living system. The products (most of them organic molecules) derived from this array of molecular transformations are the *metabolites.* This vast collection of substances that are generally referred to as *natural products* are, in fact, the metabolites of the natural living world.

Natural products are divided into two large families of compounds. Those metabolites that are *common* to the large majority of all organisms are known as the *primary metabolites.* In general, they have well-defined roles in the biochemistry of the system. For example, the amino acids are the building blocks for protein synthesis in all organisms. The second great category of natural products is known as the *secondary metabolites.* Individual secondary metabolites are far less widely distributed in nature and may be unique to single species (or even limited to a variety of a particular species). While the biochemical role of some of these compounds was established early and easily, the majority of these materials were believed to be of little importance to the functioning of the living system, and their presence was unexplained until very recently. With the development of chemical communication theory over the last few decades, however, the vitally important roles of many of the secondary metabolites in the life cycles of their particular host organisms have been revealed.

You may have had the opportunity to become acquainted with secondary metabolites in Experiments [11A] and [11B]. In Experiment [11A] an acetogenin (this term refers to the biochemical origin of this material from eight acetic acid residues), usnic acid, was isolated from a lichen where it can occur in dramatically high concentrations. Only recently has the role of usnic acid as a defense mechanism come to be fully appreciated.

In Experiment [11B], the alkaloid caffeine was obtained from tea. This compound is a very unusual example of a purine ring system in a secondary metabolite. The ecological significance of the presence of caffeine in both tea and coffee seeds has been established and it has been shown that caffeine acts against both predators and competitors.

We will now examine, in Experiment [11C], a third class of secondary metabolites: the *essential oils*. The majority of these materials are high-boiling liquids that can be extracted from plant material via steam distillation techniques. The value of codistilling high-boiling substances was learned early in the days of alchemy. Because these oils often gave pleasant odors and flavors, they were considered to be the "essence" of the original plant material. Eventually, they became known as essential oils.

These materials were used as flavorings, perfumes, and medicines, and as both insect repellents and attractants. By the early 1800s, as it became possible to establish the carbon/hydrogen ratio in organic substances, many of the oils possessing pine-type odors (the oil of turpentine) were shown to have identical C/H ratios. These materials ultimately became known as terpenes. The terpenes all have their origin in mevalonic acid, from which they utilize, as their building block, a branched five-carbon unit as in isoprene. The terpenes of the essential oils occur as C_{10} (monoterpenes), C_{15} (sesquiterpenes), C_{20} (diterpenes), C_{30} (triterpenes), and C_{40} (tetraterpenes) compounds. Today, this collection of substances represents a large fraction of the known secondary metabolites, including the steroids. Terpenes may be polymerized, extending to much higher molecular weights, with between 1000 and 5000 repeating isoprene units (MW = 60,000–350,000) to yield polymers known as the natural rubbers.

Mevalonic acid   Isoprene

As we have seen, many of the compounds found in the essential oils possess pleasing properties of taste and odor, and we now know that many of these systems contain either ketone or aldehyde functional groups. Our senses of taste and smell, however, possess a wide range of responses to the shape and dimensions of the carbon skeleton supporting the main functionality that triggers the odor signal. Thus, our sense of odor may involve simultaneous multiple stimulations by many different molecular species or, as in a number of cases, the principal response may be to a single component. Since the shape of the odor- or taste-inducing molecule plays a significant role in the effect, it is not surprising that chirality can have a dramatic impact on our perception of a particular odor.

One of the classic examples of this type of response is the case of the cyclic ketone carvone, which contains a single stereocenter (*). The *S* enantiomer is the principal odor and flavor component in caraway seed, whereas the *R* enantiomer gives rise to the odor and flavor of spearmint!

Carvone

What we find pleasant may be offensive to others. A constituent of the oil of lemon grass is citronellal, a C_{10} unsaturated aldehyde. While we find this compound to have a pleasant fragrance, it is a potent alarm signal in ants that is shunned by many other insects. Thus, this terpene aldehyde has been used effectively by both ants and humans as an insect repellent.

Citronellal

In this experiment we will isolate the principal component of the oil of cinnamon, another naturally occurring aldehyde, cinnamaldehyde. The oil is first extracted from the dried parts of the *Cinnamomum* plant by steam distillation. Although this aromatic aldehyde is a component of an essential oil, it is not formed from mevalonic acid and is not a terpene. Cinnamaldehyde is also not an acetogenin, nor is it related to usnic acid. The origin of this fragrant material is shikimic acid, which is part of the plant's *primary metabolism*.

Shikimic acid

Cinnamaldehyde's formation from shikimic acid utilizes one of only two biogenetic routes in nature that lead to the aromatized benzene ring (the other pathway is found in the acetogenins and produces secondary metabolites like usnic acid; see Experiment [11A]). The shikimic acid route contributes to a class of metabolites called the phenylpropanes (Ph—C_3), of which cinnamaldehyde is one of a limited number of simple end products. Another close relative is, for example, eugenol from oil of cloves.

Eugenol

The principal metabolic fate of the phenylpropanes is the formation of lignin polymers that are the fundamental basis of the structural tissue in all plants. Thus, cinnamaldehyde itself is a relatively rare example of a primary metabolite which has been expressed in an *essential oil*.

DISCUSSION

Steam Distillation. The process of steam distillation can be a valuable technique in the laboratory for the separation of thermally labile, high-boiling substances from relatively nonvolatile materials. For steam distillation to be successful, however, the material to be isolated must be nearly immiscible with water. (For details of the theory of steam distillation refer to Technique 3 discussions.)

Steam distillation is in essence the codistillation (or simultaneous distillation) of two immiscible liquid phases. By definition, one of these liquid phases is water and the other phase is usually a mixture of organic substances that have a low solubility in water. Though steam distillation is widely used as a separation technique for natural products, and occasionally for the isolation and/or purification of synthetic products that decompose at their normal boiling points, it has several limitations. For example, it is not the method of choice when a dry product is required or if the compound to be isolated reacts with water. Obviously, steam distillation is not feasible if the compound to be isolated decomposes upon contact with steam at 100 °C.

Although cinnamaldehyde decomposes at its normal boiling point, it may be extracted from the plant without degradation by boiling water. Steam distillation, therefore, is the method of choice for the isolation of this pleasant smelling and tasting aldehyde.

COMPONENT

Cinnamaldehyde

EXPERIMENTAL PROCEDURE

Estimated time to complete the experiment: 2.5 h.

Physical Properties of Components		
Compound	MW	Amount
Cinnamon		1 g
Water	18.02	4.0 mL

Components and Equipment. Place 1 g of chopped stick cinnamon (or powder) and 4 mL of water in a 10-mL round-bottom flask containing a boiling stone. Attach the flask to a Hickman still head equipped with an air condenser.

Distillation Conditions. Place the apparatus on a **sand bath maintained at 150–160 °C.** Use an aluminum foil shield or baffle (not shown in drawing) to cover the sand bath. This procedure will prevent the collection collar of the Hickman still from overheating.

*NOTE. The mixture tends to foam during the distillation, so care must be taken to prevent contamination of the distillate by cinnamon particles, especially if powdered cinnamon is used. If stick cinnamon is used, first evacuating the flask containing the water and cinnamon for a few minutes and then returning the system to atmospheric pressure will allow water to fill the pores of the bark and greatly aids in reducing the foaming problem (see Taber, D. F.; Weiss, A. J. J. Chem. Educ. **1998**, 75, 633).*

Position a thermometer in the throat of the Hickman still to record the distillation temperature, which should be very close to 100 °C. In early Hickman stills (without sidearm access) this thermometer makes pipetting of the condensate difficult. It is suggested that in the latter setup, the thermometer be removed after the initial distillation temperature is established and recorded.(♦)

Cinnamon stick + H₂O, 4 mL

Isolation of Cinnamaldehyde.

Remove the milky cinnamaldehyde–water (two-phase) distillate that collects in the collar of the still using a 9-in. Pasteur pipet (or 6-in. in the case of sidearm stills). Transfer this material to a 12- or 15-mL centrifuge tube. Continue the distillation for approximately 1 h or until about 5–6 mL of distillate is collected in the centrifuge tube.

NOTE. Add additional water during the course of the distillation to maintain the original volume in the flask. Add this water using a 9-in. Pasteur pipet inserted down the neck of the Hickman still, after first removing the thermometer if it is still in place (the thermometer need not be replaced in the still following the addition).

Extract the combined distillate fractions, which you collected in the centrifuge tube, with three successive 2-mL portions of methylene chloride (3 × 2 mL). Use the first portion of methylene chloride to rinse the collection collar of the Hickman still. After each extraction, transfer the lower methylene chloride layer (Pasteur filter pipet) to a 25-mL Erlenmeyer flask.

Dry the combined extracts over anhydrous sodium sulfate.

Purification and Characterization.

Transfer the dried methylene chloride solution **in at least two portions**, using a Pasteur filter pipet, to a **tared** 5-mL conical vial. Evaporate the solvent in a warm sand bath using a slow stream of nitrogen gas. After all the solution has been transferred and the solvent evaporated, rinse the sodium sulfate with an additional two 0.5-mL portions of methylene chloride. Transfer the rinses to the same vial and concentrate as before.

Weigh the flask and calculate the percentage of crude cinnamaldehyde extracted from the original sample of cinnamon.

Record the infrared spectrum and compare it to that reported in the literature (*Aldrich Library of Infrared Spectra* and/or *SciFinder Scholar*).

To further characterize the aldehyde, prepare its semicarbazone derivative (see Chapter 9, Preparation of Derivatives: Aldehydes and Ketones). The semicarbazone derivative has a melting point of 215 °C.

QUESTIONS

6-78. List several advantages and disadvantages of steam distillation as a method of purification.

6-79. Explain why the distillate collected from the steam distillation of cinnamon is cloudy.

6-80. Calculate the weight of water required to steam distill 500 mg of bromobenzene at 95 °C. The vapor pressure of water at this temperature is 640 torr; that of bromobenzene is 120 torr.

6-81. Steam distillation may be used to separate a mixture of *p*-nitrophenol and *o*-nitrophenol. The ortho isomer distills at 93 °C; the para isomer does not. Explain.

6-82. A mixture of nitrobenzene and water steam distills at 99 °C. The vapor pressure of water at this temperature is 733.2 torr. What weight of water is required to steam distill 300 mg of nitrobenzene?

BIBLIOGRAPHY

Several steam distillation procedures have been reported for isolation of essential oils from native plant sources:

Bell, C. E., Jr.; Taber, D. I.; Clark, K. A. *Organic Chemistry Laboratory: Standard and Microscale Experiments,* 3rd ed.; Harcourt College Pubs.: Orlando, FL, 2001.

Taber, D. F.; Weiss, A. J. *J. Chem. Educ.* **1998,** *75,* 633.

Vogel, A. I. *Vogel's Textbook of Practical Organic Chemistry,* 5th ed.; Furnis, B. S., et al., Eds.; Wiley: New York, 1989.

Zubrick, J. W. *The Organic Chem Lab Survival Manual,* 11th ed.; Wiley: New York, 2019.

Reductive Catalytic Hydrogenation of an Alkene: Octane

EXPERIMENT 12

Common name: octane

CA number: [1111-65-9]

CA name as indexed: octane

Purpose.   This experiment shows you how to reduce the carbon–carbon double bond of an alkene by addition of molecular hydrogen (H_2). You will gain an understanding of the important role that metal catalysts play in the stereospecific reductions of alkenes (and alkynes), to form the corresponding alkanes, by the activation of molecular hydrogen. You can observe the powerful influence on column chromatography of heavy metal ions, such as silver (Ag^+), which lead to effective separation of mixtures of alkenes from alkanes. Finally, you will appreciate the enormous importance and breadth of application of these reduction reactions in both industrial synthesis and basic biochemistry (for important examples, see Experiment [8])

REACTION

$$CH_3-(CH_2)_5-CH=CH_2 \xrightleftharpoons[\substack{NaBH_4 \\ C_2H_5OH \\ HCl(6\ M)}]{H_2PtCl_2} CH_3-(CH_2)_5-CH_2-CH_3$$

1-Octene   Octane

Prior Reading

    Technique 4: Solvent Extraction

             Liquid–Liquid Extraction (p. 76)

    Technique 6: Chromatography

             Column Chromatography (pp. 96–99)

             Concentration of Solutions (pp. 105–108)

DISCUSSION

The addition of hydrogen to an alkene (or to put it another way, the saturation of the double bond of an alkene with hydrogen) to produce an alkane is an important reaction in organic chemistry. Alkanes are also called saturated hydrocarbons, because the carbon skeletons of alkanes contain the greatest possible number of hydrogen atoms permitted by tetravalent carbon atoms; alkenes are thus unsaturated hydrocarbons. Hydrogenation reactions have widespread use in industry. For example, we all consume vegetable fats hardened by partial hydrogenation (see Experiment [8]) of the polyunsaturated oils that contained several carbon–carbon double bonds per molecule as isolated from their original plant sources. Partially hydrogenated fats represented a consumer market of over 2.9 billion pounds in 1992. Since that time major restaurants, fast food chains, and food producers have greatly lowered the amount of trans fats in their products. Furthermore, batch processes involving hydrogenation reactions are nondiscriminating when considering the geometric isomers cis and trans.

    The hydrogenation reaction is exothermic; the energy released is approximately 125 kJ/mol for most alkenes, but on the kinetic side of the ledger, this reductive pathway requires a significant activation energy to reach the transition state. Thus, alkenes can be heated in the presence of hydrogen gas at high temperatures for long periods without any measurable evidence of alkane formation. However, when the reducing reagent (H_2) and the substrate (the alkene) are in intimate contact with each other in the presence of a finely divided metal catalyst, rapid reaction does occur at room temperature. Under these conditions, successful reduction is generally observed at pressures of 1–4 atm. For this reason, this reaction is often referred to as low-pressure catalytic hydrogenation. These reactions are called heterogeneous reactions, since they occur at the boundary between two phases—in this case a solid and a liquid.

    The main barrier to the forward progress of the reaction is the very strong H—H bond that must be broken. Molecular hydrogen, however, is adsorbed by a number of metals in substantial quantities; indeed, in some instances the amount of hydrogen contained in the metal lattice can be greater than that in an equivalent volume of pure liquid hydrogen! In this adsorption process the H—H bond is broken or severely weakened. (This adsorption process, which necessarily involves a large exchange of chemical energy between the metal lattice and the adsorbed hydrogen, may in some, as yet unexplained way, be related to the "cold fusion" problem in which palladium saturated with a heavy isotope of hydrogen [deuterium] allegedly exhibits apparent excess thermal energies on electrolysis.)

The π system of the alkene is also susceptible to adsorption onto the metal surface, and when this occurs the barrier to reaction between the alkene and the activated hydrogen drops dramatically.

Catalytic hydrogenations are also a representative example of a class of organic reactions known as *addition reactions,* which are reactions in which two new substituents are *added* to a molecule (the alkene substrate in this case) across a π system. Usually addition is 1,2, but in extended π systems such as 1,3 dienes, the addition may occur 1,4. In catalytic hydrogenations, formally, one hydrogen atom of a hydrogen molecule adds to each carbon of the alkene linkage, C═C. It is not at all clear that both hydrogen atoms must come from the same original hydrogen molecule even though they are added stereospecifically in syn (cis) fashion while both systems are coordinated with the metal surface. A representation of the stereochemistry of this addition is given below.

The metals most often used as catalysts in low-pressure (1–4 atm) hydrogenations in the laboratory are nickel, platinum, rhodium, ruthenium, and palladium. In industry, high-pressure, large-scale processes are more likely to be found. For example, Germany had little or no access to naturally formed petroleum deposits during World War II, but did possess large coal mines. The Germans mixed powdered coal with heavy tar (from previous production runs) and 5% iron oxide and heated this in the presence of H_2 at a pressure of 3000 lb/in.^{2} at about 500 °C for 2 h to yield synthetic crude oil. Thirteen German plants operating in 1940 produced 24 million barrels that year, with an average of 1.5–2 tons of coal ultimately converted to about 1 ton of gasoline.

In the present experiment the metal catalyst, platinum, is generated in situ by the reaction of chloroplatinic acid with sodium borohydride. The reduced platinum metal is formed in a colloidal suspension, which provides an enormous surface area, and therefore excellent conditions, for heterogeneous catalysis. The molecular hydrogen necessary for the reduction can also be conveniently generated in situ by the reaction of sodium borohydride with hydrochloric acid:

$$4\ NaBH_4 + 2\ HCl + 7\ H_2O \rightarrow Na_2B_4O_7 + 2\ NaCl + 16\ H_2$$

This reduction technique does not require equipment capable of safely withstanding high pressures. The use of chloroplatinic acid, therefore, is particularly attractive for saturating easily reducible groups, such as unhindered alkenes or alkynes, in the laboratory. The potential limitation to the use of this reagent is that other reducible functional groups, such as aldehydes and ketones, normally inert to catalytic hydrogenations of alkenes and alkynes, may be reduced by the sodium borohydride. Thus, with chloroplatinic acid and sodium borohydride, we accept, as a compromise, a more limited set of potential reactants (substrates) for the convenience inherent in the reagent.

The platinum catalyst generated in the reaction medium adsorbs both the internally generated molecular hydrogen and the target alkene on its surface. The addition of the hydrogen molecule (H_2) (evidence strongly suggests that it is actually atomic hydrogen that attacks the alkene π system) to the alkene system while they are both adsorbed on the metal surface results in the reduction of the substrate and the formation of an alkane. The addition is, as mentioned earlier, syn (or cis), since both hydrogen atoms add to the same face of the alkene plane. The mechanistic sequence is outlined below:

This experiment also provides an opportunity for you to study a powerful aspect of column chromatography in which heavy metal ions have a particularly important role in the purification of the product. Unreacted alkene has the potential to be a problem contaminant during the isolation and purification of the relatively low-boiling saturated reaction product. The successful removal of the remaining 1-octene from the desired *n*-octane in the product mixture is achieved by using column chromatography with silver nitrate/silica gel as the stationary phase. Complex formation between the silver ion (on the silica gel surface) and the π system of the unreacted alkene acts to retard the rate of elution of the alkene relative to that of the alkane.

The ability of alkenes to form coordination complexes with certain metal ions having nearly filled d orbitals was established some time ago. In the case of the silver ion complex with alkenes, the orbital nature of the bonding is believed to involve a σ bond formed by overlap of the filled π orbital of the alkene with the free s orbital of the silver ion plus a π bond formed by overlap of the vacant antibonding π* orbitals of the alkene together with the filled d orbitals of the metal ion.

EXPERIMENTAL PROCEDURE

Estimated time to complete the experiment: 2.5 h.

Physical Properties of Reactants						
Compound	MW	Amount	mmol	bp (°C)	d	n_{D}
1-Octene	112.22	120 µL	0.76	121.3	0.72	1.4087
Ethanol (absolute)	46.07	1.0 µL		78.5		
Chloroplatinic acid (0.2 M)	517.92	50 µL				
Sodium borohydride (1 M)	37.83	125 µL				
Dilute HCl (6 M)		100 µL				

Reagents and Equipment. Equip a 5.0-mL conical vial containing a magnetic spin vane with a Claisen head fitted with a rubber balloon and Teflon-lined rubber septum cap (good *GC septa* work best—this is an important point, because several injections through the septum are required and the seal must remain gas-tight). Mount the assembly in a sand bath on a magnetic stirring hot plate (➡).

NOTE. 1. No residual acetone (perhaps from cleaning the equipment) can be present since it reacts with the NaBH₄. 2. The balloon must make a gas-tight seal to the Claisen head, so be sure to secure it with copper wire or a rubber band.

Remove the 5.0-mL vial from the Claisen head and add the following reagents. *Recap the vial after each addition.*

a. Add 50 µL of a 0.2 M solution of chloroplatinic acid (H_2PtCl_6) (automatic delivery pipet).

b. Add 1.0 mL of absolute ethanol (calibrated Pasteur pipet).

c. Add 125 µL of the sodium borohydride reagent (automatic delivery pipet).

*NOTE. Reattach the vial **immediately** to the Claisen head after the NaBH₄ solution is added.*

Stir the mixture vigorously. The solution should turn black immediately as the finely divided platinum catalyst is formed.

INSTRUCTOR PREPARATIONS. 1. The 0.2 M H_2PtCl_6 solution is prepared by adding 41 mg (0.1 mmol) of the acid to 0.5 mL of deionized water. 2. The sodium borohydride reagent is prepared by adding 0.38 g (0.01 mol) of NaBH₄ to a solution of 0.5 mL of 2.0 M aqueous NaOH in 9.5 mL of absolute ethanol.

After 1 min, use a syringe to add 100 µL of 6 M HCl solution through the septum cap. In a like manner using a fresh syringe, add *immediately* to the acid solution, a solution of 120 µL (86 mg, 0.76 mmol) of 1-octene dissolved in 250 µL of absolute ethanol. (This solution is conveniently prepared in a 1-mL conical vial; the reagents are best dispensed using automatic delivery pipets.)

Now add dropwise (clean syringe) 1.0 mL of the NaBH₄ reagent solution over a 2-min interval.

NOTE. At this point the balloon should inflate and remain inflated for at least 30 min. If it does not, the procedure must be repeated.

Reaction Conditions. Stir the reaction mixture vigorously at a sand bath temperature of 50 °C for 45 min.

Isolation of Product. Cool the reaction to ambient temperature and dropwise add 1 mL of water. Extract the resulting mixture in the reaction vial with three 1.0-mL portions of pentane. Transfer each pentane extract to a stoppered 25-mL Erlenmeyer flask containing 0.5 g of anhydrous sodium sulfate.

The reaction mixture is extracted as follows. Upon addition of each portion of pentane, cap the vial, shake, vent carefully, and then allow the layers to separate. A Vortex mixer may be used if available. The transfers must be made using a Pasteur filter pipet, because the pentane solvent is particularly volatile.

Syringe — Balloon

50–55 °C thermometer

Speed   Heat

Step I:
0.2 *M* H₂PtCl₆ solution,
50 µL +
CH₃CH₂OH, 1.0 mL
+ NaBH₄ solution, 125 µL
Step II:
6 *M* HCl, 100 µL
+ 1-octene, 120 µL
+ NaBH₄ solution, 1.0 mL

Using a Pasteur filter pipet, transfer the dried solution to a second 25-mL Erlenmeyer flask. Rinse the drying agent with an additional 1.0 mL of pentane (calibrated Pasteur pipet) and add the rinse to this second flask. Add a boiling stone and concentrate the solution to a volume of about 1.0–1.5 mL by warming it gently in a sand bath in the **hood**.

HOOD

Anhydrous
Na$_2$SO$_4$, 50 mg

10% AgNO$_3$/SiO$_2$,
500 mg

Sand, 50 mg

Cotton plug

Octane +
pentane,
~2.5 mL

Purification and Characterization. The saturated product, *n*-octane, is purified by column chromatography. In a Pasteur filter pipet place about 50 mg of sand, 500 mg of 10% silver nitrate on activated silica gel (200 mesh), and then 50 mg of anhydrous sodium sulfate (◄).

CAUTION. Silver nitrate stains the skin. Protective gloves should be worn during this operation. The ~10% silver nitrate/silica gel used in this separation is commercially available.

Wet the column with 0.5 mL of pentane (calibrated Pasteur pipet) and then transfer the concentrated crude product, as obtained above, to the column by Pasteur filter pipet. Elute the octane from the column using 1.5 mL of pentane and collect the eluate in a tared 5.0-mL conical vial containing a boiling stone.

HOOD

Fit the vial with an air condenser and then place the assembly in the **hood** in a sand bath maintained at a temperature of 90–100 °C to evaporate the pentane solvent.

OPTIONAL. The evaporation is continued until a constant weight of product is obtained. This procedure is the best approach, but has to be done very carefully or a considerable amount of product can be lost.

Record the weight of product and calculate the percent yield. Determine the boiling point and refractive index (optional) of your material and compare your results with those reported in the literature for octane. Obtain an IR spectrum. Compare your results with those reported in the literature (*Aldrich Library of IR Spectra* and/or *SciFinder Scholar*). Also, compare your IR spectrum to that of the 1-octene starting material. Can you establish from the above data if your sample is contaminated by traces of the pentane extraction solvent? If not, how would you go about determining the presence of this potential impurity?

QUESTIONS

6-83. Squalene, first isolated from shark oil and a biological precursor of cholesterol, is a long-chain aliphatic alkene (C$_{30}$H$_{50}$). The compound undergoes catalytic hydrogenation to yield an alkane of molecular formula C$_{30}$H$_{62}$. How many double bonds does a molecule of squalene have?

6-84. A chiral carboxylic acid A (C$_5$H$_6$O$_2$) reacts with 1 mol of hydrogen gas on catalytic hydrogenation. The product is an achiral carboxylic acid B (C$_5$H$_8$O$_2$). What are the structures of compounds A and B?

6-85. Two hydrocarbons, A and B, each contain six carbon atoms and one C=C. Compound A can exist as both *E* and *Z* isomers but compound B cannot. However, both A and B on catalytic hydrogenation give only 3-methylpentane. Draw the structures and give a suitable name for compounds A and B.

6-86. What chemical test would you use to distinguish between the 1-octene starting material and the octane product?

6-87. Give the structure and names of three alkenes having the molecular formula C$_5$H$_{10}$ that produce simple and branched alkanes on catalytic hydrogenation.

BIBLIOGRAPHY

Selected references on catalytic hydrogenation:

Birch, A. J.; Williamson, D. H. *Org. React.* **1976,** *24,* 1.

Carruthers, W. Some *Modern Methods of Organic Synthesis,* 3rd ed.; Cambridge Univ. Press: New York, 1986; p. 411.

De, S; Gambhir, G.; Krishnamurty, H. G. *J. Chem. Educ.* **1994,** *71,* 922.

Kabalka, G. W.; Wadgaonkar, P. P.; Chatla, N. *J. Chem. Educ.* **1990,** *67,* 975.

Parker, D. in *The Chemistry of the Metal-Carbon Bond,* Hartley, F. R., Ed.; Wiley: New York, 1987, Vol. IV, Chap. 11, p. 979.

Pelter, A.; Smith, K.; Brown, H. C. *Borane Reagents (Best Synthetic Methods);* Academic Press: Orlando, FL, 1988.

Smith, M. B. *March's Advanced Organic Chemistry,* 8th ed.; Wiley: New York, 2019, Chap. 15.

Selected examples of catalytic hydrogenation of alkenes in Organic Syntheses:

Adams, R.; Kern, J. W.; Shriner, R. L. *Organic Syntheses;* Wiley: New York, 1941; Collect. Vol. I, p. 101.

Bruce, W. F.; Ralls, J. O. *Organic Syntheses;* Wiley: New York, 1943; Collect. Vol. II, p.191.

Cope, A. C.; Herrick, E. C. *Organic Syntheses;* Wiley: New York, 1963; Collect. Vol. IV, p. 304.

Herbst, R. M.; Shemin, D. *Organic Syntheses;* Wiley: New York, 1943; Collect. Vol. II, p. 491.

Ireland, R. E.; Bey, P. *Organic Syntheses;* Wiley: New York, 1988; Collect. Vol. VI, p. 459.

McMurry, E. *Organic Syntheses;* Wiley: New York, 1988; Collect. Vol. VI, p. 781.

Meyers, A. I.; Beverung, W. N.; Gault, R. *Organic Syntheses;* Wiley: New York, 1988; Collect. Vol. VI, p. 371.

Hydroboration–Oxidation of an Alkene: Octanol

EXPERIMENT
13

Common name: octanol

CA number: [111-87-5]

CA name as indexed: 1-octanol

Purpose.   The oxidation of an alkene to an alcohol is investigated via the in situ formation of the corresponding trialkylborane, followed by the oxidation of the carbon–boron bond with hydrogen peroxide. The conditions required for hydroboration (a reduction) of unsaturated hydrocarbons are explored. Alkylboranes are particularly useful synthetic intermediates for the preparation of alcohols. The example used in this experiment is the conversion of 1-octene to 1-octanol in which an anti-Markovnikov addition to the double bond is required to yield the intermediate, trioctylborane. Since it is this alkylborane that subsequently undergoes oxidation to the alcohol, hydroboration offers a synthetic pathway for introducing substituents at centers of unsaturation that are not normally available to the anti-Markovnikov addition reactions that are based on radical intermediates.

Prior Reading

> *Technique 4:* Solvent Extraction
>> Liquid–Liquid Extraction (p. 76)
> *Technique 6:* Chromatography
>> Column Chromatography (pp. 96–99)
>> Concentration of Solutions (pp. 105–108)

REACTION

$$3 \text{ CH}_3\text{—(CH}_2)_5\text{—CH}=\text{CH}_2 \xrightarrow{\text{THF·BH}_3} [\text{CH}_3(\text{CH}_2)_7]_3\text{B} \xrightarrow[\text{OH}^-]{\text{H}_2\text{O}_2} 3 \text{ CH}_3\text{—(CH}_2)_7\text{—OH}$$

1-Octene   Trioctylborane   1-Octanol

DISCUSSION

The course of this reaction depends (1) on the *stereospecific* reductive addition of diborane (B_2H_6, introduced as the borane • tetrahydrofuran complex ($BH_3 \cdot$ THF)) to an alkene to form an intermediate trialkylborane and (2) on oxidation of the borane with alkaline hydrogen peroxide to yield the corresponding alcohol.

The first step in the reaction sequence is generally called a *hydroboration*.[13] The addition of diborane is a rapid, quantitative, and general reaction for all alkenes (as well as alkynes) when carried out in a solvent that can act as a Lewis base. The ether solvation of the diborane, for example, is the key to the success of this reaction. In the absence of a Lewis base, borane (BH_3) exists as a dimer (B_2H_6), which is much less reactive than the monomer (BH_3). Borane, however, does exist in coordination with ether type solvents. It is the monomer (BH_3) that functions as the active reagent in the reductive addition.

As depicted in the following mechanism, the boron hydride rapidly adds successively to three molecules of the alkene to form a trialkylborane.

$$\text{CH}_3\text{—(CH}_2)_5\text{—HC}=\text{CH}_2 \longrightarrow \text{CH}_3\text{—(CH}_2)_5\text{—CH}_2\text{—CH}_2\text{—BH}_2$$

$$\underset{\delta^-}{\text{H}}\text{—}\underset{\delta^+}{\text{B}}\overset{\text{H}}{\underset{\text{H}}{<}}$$

repeat $\Big|$ $2 \text{ CH}_3\text{—(CH}_2)_5\text{—CH}=\text{CH}_2$

$$[\text{CH}_3(\text{CH}_2)_5\text{—CH}_2\text{—CH}_2]_3\text{B}$$

Note that the transition state of this addition reaction is generally considered to be a *four-center* one, and that the 1-octene substrate is oriented such that the boron becomes bonded to the least-substituted carbon atom of the double bond. Thus, the reaction can be classified as *regioselective,* and it will be sensitive to substitution on the carbon–carbon double bond.

In the developing transition state, the alkene π electrons (the least tightly held, and most nucleophilic) flow to the electron-deficient boron atom (the vacant p orbital is the electrophile). The formation of the *transition state* is controlled in large part by the polarization of the alkene π system during the early stages of formation of the transition state. At this point a partial positive charge begins to form on the more highly substituted carbon (the more stable carbocation), and a partial negative charge on the least substituted carbon. The orientation of the polarization, therefore, is to a large extent controlled by the electron-releasing effects of the alkyl substituents on the alkene, which enables the more highly substituted of the sp^2 carbon atoms to better accommodate the positive charge. As the reaction proceeds, the boron acquires a partial negative charge in response to the incoming electron density.

---

[13]For references relating to the use of diborane as a hydroboration reagent, see Experiment [1$_{\text{adv}}$].

The ease of hydride (:H$^-$) transfer from the boron to the more highly substituted carbon atom of the alkene, therefore, increases. Thus, hydroboration involves simultaneous hydride release and boron-carbon bond formation, and is a concerted reaction. The reaction can be conveniently considered as passing through a four-centered transition state, wherein the atoms involved undergo simultaneous changes in bonding (i.e., electron redistribution [see below]).

Hydroboration, as we have seen, can be classified as a **concerted addition reaction** in which no intermediate is formed. The mechanism is characteristic of a group of reactions called **pericyclic** (from the Greek, meaning *around the circle*) **reactions**, which involve a cyclic shift of electrons in and around the *transition state*. The mechanism proposed is further supported by the fact that rearrangements are not normally observed in hydroboration reactions, which implies that there are no carbocationic intermediates.

When alkenes with varying degrees of substitution undergo hydroboration, the boron ends up on the least substituted sp^2 carbon atom. While it might appear from the products that the regioselectivity is controlled by steric factors, this assumption is probably too simplistic. Steric and electronic factors both favor, and are likely responsible for, the observed regioselectivity in hydroboration reactions.

Accumulated evidence demonstrates that the reaction occurs by **syn** addition, which is a consequence of the four-centered transition state. Therefore, the new C—B and C—H bonds are necessarily formed on the same face of the C=C bond, as shown in the following example:

Organoboranes are important in organic synthesis as reactive intermediates. Reactions have been developed by which the boron atom may be replaced by a wide variety of functional groups, such as —H, —OH, —NH$_2$, —Br, —I, and —COOH. The present experiment demonstrates the conversion of an organoborane to an alcohol by oxidation with alkaline hydrogen peroxide. It is not necessary to isolate the organoborane prior to its oxidation. This simplification is particularly fortuitous in this case, since most alkylboranes, when not in solution, are pyrophoric (spontaneously flammable in air).

With regard to the second stage of the hydroxylation process, there is now conclusive evidence that oxidation of the C—B bond proceeds with retention of configuration at the carbon atom bearing the boron. That is, the hydroxyl group that replaces the boron atom has the identical orientation in the molecule as the boron:

Thus, in unsymmetrical alkenes the hydroboration–oxidation sequence of reactions leads to the addition of the elements of H—OH to the original C=C in an anti-Markovnikov manner.

In the oxidation step a hydroperoxide anion (HOO⁻) is generated in the alkaline medium. This species makes a nucleophilic attack on the boron atom to form a boron hydroperoxide. A 1,2 migration of an alkyl group from boron to oxygen occurs to yield a boron monoester (a borate). Hydrolysis of the boron triester, generated by successive rearrangement of all three alkyl groups, produces the desired alcohol. The mechanism of the oxidation sequence is given below.

$$B(OR)_3 + OH^- \xrightarrow{H_2O} 3\ ROH + B(OH)_4^-$$

In the final step, alkaline hydrolysis of the trialkyl borate ester yields 3 moles of the alcohol.

The effective use of B_2H_6 in the hydroboration reaction was discovered in 1955 by H. C. Brown, and is just one of the many important hydride reagents developed by Professor Brown and his coworkers at Purdue University.

Herbert Charles Brown (1912). Brown obtained his B.S. in chemistry from the University of Chicago (1936) and his Ph.D. from the same institution in 1938. He later became a Professor of Chemistry at Wayne State and Purdue Universities.

Working with H. I. Schlesinger at the University of Chicago, Brown developed practical routes for the synthesis of diborane (B_2H_6). He discovered that diborane reacted rapidly with LiH to produce lithium borohydride ($LiBH_4$), discovering and opening a synthetic route to the metal borohydrides. These compounds proved to be powerful reducing agents. Later, he developed an effective route to $NaBH_4$, which led to the commercial production of this material. Metal borohydrides, particularly $LiAlH_4$ (developed by Schlesinger and Albert Finholt), have revolutionized how organic functional groups are reduced in both the research laboratory and the industrial plant.

In 1955, Brown discovered that alkenes can be converted to organoboranes by reaction with diborane (actually the monomer in ether solution) and with organoboranes containing a B—H bond (the hydroboration reaction). The organoboranes are valuable intermediates in organic synthesis because the boron substituent can be quickly and quantitatively replaced by groups such as —OH, —H, —NH₂, or —X (halogen). Thus, organoboranes have become an attractive pathway for the preparation of alcohols, alkanes, amines, and organohalides.

Brown's investigation of the addition compounds of trimethyl borane, diborane, and boron trifluoride with amines has provided a quantitative estimation for steric strain effects in chemical reactions. He also investigated the role of steric effects in solvolytic, displacement, and in elimination reactions. His results demonstrate that steric effects can assist, as well as hinder, the rate of a chemical reaction.

Brown has published over 700 scientific papers and is the author of several texts. For his extensive work on organoboranes, Brown (with G. Wittig [organophosphorus compounds]) received the Nobel Prize in Chemistry in 1979.[14]

EXPERIMENTAL PROCEDURE

Estimated time to complete the experiment: 4.0 h.

Physical Properties of Reactants						
Compound	MW	Amount	mmol	bp (°C)	d	n_D
1-Octene	112.22	210 μL	1.34	121	0.72	1.4087
Borane • THF (1M)		500 μL	0.50			
Sodium hydroxide (3 M)	40.00	300 μL				
Hydrogen peroxide (30%)	34.01	300 μL				

Reagents and Equipment.   Equip a 5.0-mL conical vial, containing a spin vane, with a Claisen head fitted with a rubber septum and calcium chloride drying tube (➡). Through the rubber septum add 210 μL (150 mg, 1.34 mmol) of 1-octene (in one portion) with a 1.0-cm³ syringe.

NOTE. Dry the glassware and syringe in a 100 °C oven for at least 30 min before use. An alternate method is to "flame-out" the glassware with a microburner and a flow of dry nitrogen, and to then add the drying tube and caps.

Cool the reaction vessel in an ice bath and, using the same syringe, add 500 μL (0.5 mmol) of the 1 M borane · THF solution through the septum over a 5-min period.

1-Octene, 210 μL
+ 1*M* BH₃•THF, 500 μL

CAUTION. The $BH_3 \cdot THF$ reagent reacts *violently* with water.

[14]See *McGraw-Hill Modern Scientists and Engineers*; S. P. Parker, Ed.; McGraw-Hill: New York, 1980, Vol. 1, p. 150.

Reaction Conditions. Allow the reactants to warm to room temperature and then stir for 45 min. Using a Pasteur pipet, carefully add two drops of water to hydrolyze any unreacted borane complex.

NOTE. At this stage of the procedure the vial may be removed from the Claisen head, capped, and allowed to stand until the next laboratory period.

If the experiment is not interrupted, proceed by first removing the reaction vial from the Claisen head. Then use a graduated 1.0-mL pipet to add 300 μL of 3 M NaOH solution, followed by the dropwise addition of 300 μL of 30.0% hydrogen peroxide solution over a 10-min period using another graduated 1.0-mL pipet. Stir the reaction vial gently after each addition.

CAUTION. Hydrogen peroxide blisters the skin. Concentrated solutions of hydrogen peroxide can explode!

Attach the vial to a reflux condenser and warm the reaction mixture with stirring, for 1 h in a sand bath at 40–50 °C (◄).

Cool the resulting two-phase mixture to room temperature and use forceps to remove the spin vane. Add 0.5 mL of diethyl ether to establish a reasonable volume for extraction of the organic phase.

Isolation of Product. Using a Pasteur filter pipet, separate the bottom aqueous layer and transfer it to a 3.0-mL reaction vial. *Save* the organic phase in the 5.0-mL conical vial.

Extract the aqueous phase placed in the 3.0-mL conical vial with two 1.0-mL portions of diethyl ether. Upon the addition of each portion of ether, cap, shake, and carefully vent the vial and allow the layers to separate. The top ether layer is then separated using a Pasteur filter pipet and the ether extracts are combined with the previously saved organic phase in the 5.0-mL conical vial. *If a solid forms during the extraction, add a few drops of 0.1 M HCl.*

Extract the combined organic phases with 750 μL of 0.1 M HCl solution, followed by extraction with several 0.5-mL portions of distilled water or until the aqueous extract is neutral to pH paper. Transfer the neutral organic phase to a 10-mL Erlenmeyer flask. Rinse the conical vial with a further 0.5 mL of ether and combine the rinse with the ether solution in the Erlenmeyer flask. Add granular anhydrous sodium sulfate (Na_2SO_4) (200 mg) to the combined organic phases and let it stand with occasional swirling for 20 min. If large clumps of drying agent form, add an additional 100 mg of Na_2SO_4. The solution should be clear at the end of the drying period. Then transfer the solution to a tared 3.0-mL conical vial in 1-mL aliquots, add a boiling stone to the vial, and concentrate by warming in a sand bath (60–65 °C) in the **hood** to yield the crude product residue.

HOOD

Weigh the vial and calculate the crude yield.

Gas Chromatographic Analysis. This crude product is easily analyzed by gas chromatography. The procedure involves the injection of 10 μL of the liquid material onto a $\frac{1}{4}$-in. × 8-ft steel column packed with 10% Carbowax 80/100 20M PAW-DMS. Experimental conditions are He flow rate, 50 mL/min; chart speed, 1 cm/min; temperature, 190 °C.

The liquid components elute in the order: unreacted 1-octene, a small amount of 2-octanol, and the major product, 1-octanol. Approximate retention times (conditions above) are 1.3, 3.1, and 4.2 min, respectively.

100–110 °C
thermometer

Speed   Heat

3 *M* NaOH, 300 μL +
30% H_2O_2, 300 μL +
[$CH_3(CH_2)_7$]$_3$B intermediate +
0.5 mL THF

Collect the eluted 1-octanol in an uncooled, 4-mm-diameter collection tube. Transfer the collected material to a 0.1-mL conical vial (see Technique 1). This material may then be analyzed by IR spectroscopy and/or used in the procedure described in the following section on Purification and Characterization.

Anhydrous
Na$_2$SO$_4$, 100 mg

10% AgNO$_3$–SiO$_2$,
400 mg

Sand, 50 mg

Cotton plug

1-Octanol in
2–3 mL 1:4
ether–pentane
solution

Purification and Characterization. Pack a Pasteur filter pipet with 400 mg of 10% silver nitrate-treated activated silica gel followed by 100 mg of anhydrous sodium sulfate. Dissolve the organic residue isolated above in 500 µL of pentane (spectral or HPLC grade) and then transfer this solution by Pasteur filter pipet to the column (➡). Elute the material from the column with 3 mL of a 1:4 diethyl ether–pentane solution. Collect the eluate in a tared 5.0-mL conical vial containing a boiling stone.

Concentrate the collected eluate to a constant weight by warming on a sand bath (60–65 °C.) in the **hood**. Weight the octanol product and calculate the percent yield.   HOOD

This product may again be analyzed by gas chromatography. Follow the procedure and experimental conditions outlined above. The 1-octene impurity should have been removed during the column chromatography step. The small percentage of 2-octanol byproduct, however, more than likely will still be detected. The actual percent composition of the mixture can be calculated by determination of the areas under the chromatographic peaks (see Technique 1).

Obtain an IR spectrum of the alcohol and compare your result to that recorded in the literature (*Aldrich Library of IR Spectra* and/or SciFinder Scholar).

Chemical Tests. A positive ceric nitrate test (Chapter 9) should confirm the presence of the alcohol grouping. The ignition test may be used to establish that the material is an aliphatic species. The phenyl or α-naphthylurethane derivative may also be prepared to further characterize the alcohol (Chapter 9).

It might also be of interest to determine the solubility characteristics of this C$_8$ alcohol in water, ether, concentrated sulfuric acid, and 85% phosphoric acid (Chapter 9). Do your results agree with what you would predict for this alcohol?

What chemical tests would you perform to determine the difference between the starting alkene and the alcohol product?

QUESTIONS

6-88. Using the hydroboration reaction, outline a reaction sequence for each of the following conversions:
   (a) 1-Pentene to 1-pentanol
   (b) 1-Methylcyclopentene to *trans*-2-methylcyclopentanol
   (c) 2-methylpropene to 2-methyl-1-propanol

6-89. When diborane (B$_2$H$_6$) dissociates in ether solvents, such as tetrahydrofuran (THF), a complex between borane (BH$_3$) and the ether is formed. For example,

$$B_2H_6 \ + \ 2 \ \ddot{O} \bigcirc \ \longrightarrow \ 2 \ \bigcirc \overset{+}{\ddot{O}} - \bar{B}H_3$$

   (a) In the Lewis sense, what is the function of BH$_3$ as it forms the complex? Explain.
   (b) Write the Lewis structure for BH$_3$. Diagram its expected structure indicating the bond angles in the molecule.

6-90. In reference to Question 6-89a:
   (a) Explain why borane (BH$_3$) reacts readily with the π-electron system of an alkene.
   (b) Explain why diborane (B$_2$H$_6$) reacts only very slowly with C=C groups.

6-91. In an unsymmetrical alkene, the boron atom adds predominantly to the least substituted carbon atom. For example, 2-methyl-2-butene gives the products indicated below:

$$CH_3-CH=CH-CH_3 \xrightarrow[\text{diglyme}]{BH_3} \underset{\substack{| \\ BH_2 \\ 2\%}}{CH_3-\overset{\overset{\displaystyle CH_3}{|}}{C}-CH_2-CH_3} + \underset{\substack{| \\ BH_2 \\ 98\%}}{CH_3-\overset{\overset{\displaystyle CH_3}{|}}{CH}-\overset{\overset{\displaystyle H}{|}}{C}-CH_3}$$

(Diglyme: $CH_3\ddot{O}-CH_2CH_2-\ddot{O}-CH_2CH_2-\ddot{O}CH_3$,
diethyleneglycol dimethyl ether)

Offer a reasonable explanation to account for the ratio obtained.

NOTE. The above solvent (diglyme) has been shown to cause a significant increase in the number of miscarriages by workers who come in contact with it.

6-92. An advantage of the hydroboration reaction is that rearrangement of the carbon skeleton does not occur. This lack of migration contrasts with results obtained upon the addition of hydrogen chloride to the double bond. For example,

$$CH_3-\overset{\overset{\displaystyle CH_3}{|}}{CH}-CH=CH_2 \xrightarrow{HCl} CH_3-\overset{\overset{\displaystyle CH_3}{|}}{CH}-\underset{\substack{| \\ Cl}}{CH}-CH_3 + CH_3-\overset{\overset{\displaystyle CH_3}{|}}{\underset{\substack{| \\ Cl}}{C}}-CH_2CH_3$$

$$CH_3-\overset{\overset{\displaystyle CH_3}{|}}{CH}-CH=CH_2 \xrightarrow{BH_3} CH_3-\overset{\overset{\displaystyle CH_3}{|}}{CH}-CH_2CH_2BH_2$$

Offer an explanation for the difference in these results.

Offer an explanation for the difference in these results.

BIBLIOGRAPHY

Brown, H. C.; Subba, B. C. *J. Am. Chem. Soc.* **1956,** *78,* 5694.

Matteson, D. S. in *The Chemistry of the Metal-Carbon Bond,* Hartley, F. R., Ed.; Wiley: New York, 1987, Vol. 4, Chap. 3, p. 307.

Pelter, A.; Smith, K.; Brown, H. C. *Boron Reagents: Best Synthetic Methods*; Academic Press: New York, 1988.

Smith, K. *Organoboron Chemistry in Organometallics in Synthesis—A Manual*; Schlosser, M., Ed.; Wiley: New York, 1994.

Also see:

Kabalka, G. W.; Maddox, J. T.; Shoup, T.; Bowers, K. R. *Org. Synth.* **1996,** *73,* 116.

Experiment [13] was based on the work reported by:

Kabalka, G. W. J. *Chem. Educ.* **1975,** *52,* 745; also see Kabalka, G. W; Wadgaonkar, P. P.; Chatla, N. J. *Chem. Educ.* **1990,** *67,* 975.

Smith, M. B. *March's Advanced Organic Chemistry*, 8th ed.; Wiley: New York, 2019, Chap. 15.

EXPERIMENT 14

Diels–Alder Reaction: 4-Cyclohexene-cis-1, 2-dicarboxylic Acid Anhydride

Common name: 4-cyclohexene-*cis*-1,2-dicarboxylic acid anhydride
CA number: [85-43-8]
CA name as indexed: 1,3-isobenzofurandione, 3a,4,7,7a-tetrahydro-

Purpose. This experiment demonstrates the use of the Diels–Alder reaction in the preparation of six-membered carbocyclic rings. The cyclic products are obtained by reaction of a conjugated diene with an alkene. The overall process is considered a cycloaddition and represents one of three major classes of pericyclic reactions. The illustration given here involves the treatment of 1,3-butadiene (generated in situ) with maleic anhydride to form the corresponding Diels–Alder product. These *add*ition prod*ucts* are often called *adducts*.

Prior Reading

> *Technique 7:* Collection or Control of Gaseous Products (pp. 109–111)
> *Chapter 8:* Infrared Spectroscopy (pp. 615–637)
> Nuclear Magnetic Resonance Spectroscopy (pp. 580–612)

REACTION

3-Sulfolene   *s-cis*-1,3-Butadiene

s-cis-1,3-Butadiene   Maleic anhydride   Diels–Alder product: 4-cyclohexene-*cis*-1,2-dicarboxylic anhydride

Otto Paul Hermann Diels (1876–1954). Diels obtained his Ph.D. in 1899 while studying with Emil Fischer at the University of Berlin. He later became associate professor of chemistry at the University of Berlin, and in 1916 he moved to the University of Kiel. In 1906 Diels discovered carbon suboxide gas (C_3O_2), obtained from the dehydration of malonic acid. He did extensive studies on saturated fats and fatty acids. Diels also developed the use of selenium as a mild dehydrogenation agent. This latter work led to the commercial production of polyunsaturated oils.

In the same year that he identified carbon suboxide, Diels began to investigate cholesterol with E. Abderhalden. The structure of this lipid had not yet been determined. He was the first to study the products of the selenium dehydrogenation of cholesterol and isolated a hydrocarbon ($C_{18}H_{16}$), which became known as "Diels' hydrocarbon." This substance proved to possess the basic steroidal ring structure; its subsequent synthesis by Diels in 1935 (almost 30 years after he started this work) led to the rapid elucidation of the structures of a vast array of steroidal sex hormones, saponins, cardiac glycosides, bile pigments, and adrenal cortical hormones, such as cortisone.

Diels, however, is best known for his discovery (with his student Kurt Alder) of the reaction that now bears his name (it is now known as the *Diels–Alder cycloaddition reaction*), which was first published in 1928 (see Diels, O.; Alder, K. *Liebigs Ann. Chem.* **1928,** *460,* 98). This reaction involves the 1,4 addition of dienophile reagents to diene substrates to produce six-membered cycloalkenes. The reaction has found extensive application in the synthesis of terpenes and other natural products, since six-membered rings abound in the metabolites of living systems, and because for some time it was one of the few methods available for the synthesis of these cyclic structures. Because of the impact of their work in the field of organic synthesis, Diels shared the 1950 Nobel Prize (in chemistry) with Alder. He also was the author of a popular textbook (*Einfuhrung in die organische Chemie*), first published in 1907, which went through 19 editions by 1962.[15]

DISCUSSION

The Diels–Alder reaction is one of the most useful synthetic reactions in organic chemistry because, in a *single* step, it produces two new carbon–carbon bonds and up to four stereocenters. It is an example of a [4 + 2] cycloaddition reaction (4 π electrons + 2 π electrons) between a *conjugated* 1,3-diene and an alkene (dienophile; to have an affinity for dienes, from the Greek *philos*, meaning loving), which leads to the formation of cyclohexenes. Because the conversion efficiency of this process which is defined as the percentage of mass throughput when comparing reactants with that of product is exceptionally high, 100% to be exact, the Diels–Alder reaction is often referenced as the benchmark standard of atom economy. Experiment 5 (Reduction of Ketones Using a Metal Hydride Reagent), Experiment 7 (The Cannizzaro Reaction), and Experiment 19 (Alkene Preparation by the Wittig Reaction) are very poor examples of atom economy as most of the total molecular mass of the reactants used is not part of the desired product. Thus when using 1,3-butadiene and ethylene as reactants in the Diels–Alder reaction, this cycloaddition reaction results in the formation of cyclohexene—a transformation of 100% atom economy. The same would be true with Experiment 15, the total molecular mass of anthracene and maleic anhydride is part of the cycloadduct. Alkynes may also be used as dienophiles, in which case the reaction produces 1,4-cyclohexadienes. The reaction proceeds faster if the dienophile bears electron-withdrawing groups and if the diene bears electron-donating groups. Thus, α,β-unsaturated esters, ketones, nitriles, and so on, make excellent dienophiles, which are often used in the Diels–Alder reaction. By varying the nature of the diene and dienophile, a very large number of compounds can be prepared. Unsubstituted alkenes, such as ethylene, are poor dienophiles and react with 1,3-butadiene only at elevated temperatures and pressures. These high activation energies (slow reactions) pose a particular problem for the Diels–Alder reaction. The Diels–Alder reaction is a reversible, equilibrium reaction that is not very exothermic. Since the equilibrium constant (K_{eq}) is temperature dependent [$K_{eq} = e^{-\Delta G/RT}$], K_{eq} decreases with increasing temperature and eventually can become quite small at the high temperature needed for the reaction of an unactivated dienophile to proceed at a reasonable rate. Elevating the temperature will increase the rate of the reaction, but this will also reduce the amount of product formed, and therefore the lowest possible temperature must often be used.

The reaction is a thermal *cycloaddition* (a ring is formed), which occurs in one step and is thus a *concerted* reaction. Both new C—C single bonds and the new C=C π bond are formed simultaneously, as the three π bonds in the reactants break. The electron flow for the reaction is shown below. The reaction is thus classified as a *pericyclic* reaction (from the Greek meaning "around the circle").

[15]See Newett, L. C. *J. Chem. Educ.* **1931,** *8,* 1493; *Dictionary of Scientific Biography,* C. C. Gillespie, Ed.; Scribner's: New York, 1971, Vol. IV, p. 90; *McGraw-Hill Modern Scientists and Engineers,* S. P. Parker, Ed.; McGraw-Hill: New York, 1980, Vol. 1, p. 289.

The diene component must be in the s-cis (the "s" refers to the conformation about a single bond) conformation to yield the cyclic product with the cis C=C required by the six-membered ring. For this reason, cyclic dienes usually react more readily than acyclic species. For example, 1,3-cyclopentadiene, which is locked in the s-cis configuration, reacts with maleic anhydride about 1000 times faster than 1,3-butadiene, which prefers an s-trans conformation.

s-trans-1,3-Butadiene   s-cis-1,3-Butadiene

The reaction is highly stereospecific and the orientation of the groups on the dienophile are *retained* in the product; thus, the addition must be suprafacial–suprafacial. That is, by having the stereochemical information preserved, both new bonds are formed on the same face of the diene and on the same face of the dienophile. Thus, two groups that are cis on the dienophile will be cis in the product (and trans will give a trans product).

Carbomethoxy groups cis   Carbomethoxy groups cis

Carbomethoxy groups trans   Carbomethoxy groups trans

The reaction of cyclopentadiene with maleic anhydride demonstrates the further stereochemical consequence of the relative orientation of the reactants in Diels–Alder reactions. In this situation, there are two possible ways in which the reactants may bond. This reaction leads to the formation of two products: the endo and exo stereoisomers:

Exo adduct   Endo adduct

Generally, the endo form of the product predominates (endo-rule which is a result of secondary orbital overlap between the diene and dienophile), but endo/exo ratios may vary, depending on several steric and electronic factors and with reaction conditions.

The diene used in this experiment, 1,3-butadiene, is a gas at room temperature (bp –5 °C), which makes it a difficult reagent to measure and handle in the laboratory. Fortunately, 1,3-butadiene can be generated in situ from 3-sulfolene, a solid reagent that is easily handled. In an example of a retro-cycloaddition reaction, 3-sulfolene decomposes at a moderate temperature to yield sulfur dioxide and 1,3-butadiene.

$$\text{3-Sulfolene} \xrightarrow{\text{heat}} \text{s-cis-1,3-Butadiene} + SO_2$$

3-Sulfolene   s-cis-1,3-Butadiene

EXPERIMENTAL PROCEDURE

Estimated time to complete the experiment: 1.5 h.

200 °C
thermometer

3-Sulfolene, 170 mg,
+ maleic anhydride, 90 mg
+ xylene, 80µL

Physical Properties of Reactants					
Compound	MW	Amount	mmol	mp (°C)	bp (°C)
3-Sulfolene	118.15	170 mg	1.42	66	
Maleic anhydride	98.06	90 mg	0.92	60	
Xylene		80 µL			137–140

Reagents and Equipment. Weigh and place 80 µL of xylene, 90 mg (0.92 mmol) of maleic anhydride, and 170 mg (1.42 mmol) of 3-sulfolene in a 3.0-mL conical vial equipped with an air condenser protected with a calcium chloride drying tube and containing a boiling stone (◄).

HOOD

*NOTE. The maleic anhydride should be finely ground and protected from moisture to prevent hydrolysis to the corresponding acid. A mixture of xylenes in the boiling-point range 137–140 °C will suffice. Use freshly distilled solvent or dry it over molecular sieves before use. Dispense the xylene in the **hood** using an automatic delivery pipet. In large laboratory sections it is recommended that the evolved SO₂ be trapped (see Prior Reading).*

Reaction Conditions. Heat the reaction mixture to reflux using a sand bath for 20 min. Avoid overheating.

CAUTION. The reaction is exothermic. Avoid overheating. Sulfur dioxide is evolved in the process and adequate ventilation should be provided.

HOT   Carefully remove the **hot** conical vial from the sand bath and allow the contents to cool to room temperature.

Isolation of Product. Add 0.5 mL of toluene to the cooled solution, and then add petroleum ether (60–80 °C) dropwise until a slight cloudiness persists. Roughly 0.25–0.35 mL of petroleum ether will be needed.

Reheat the solution until it becomes clear and then cool it in an ice bath.
During the recrystallization step, the sides of the vial may have to be scratched with a glass rod to induce crystallization.

Collect the crystalline product by vacuum filtration and wash the filter cake on the Hirsch funnel with 0.5 mL of *cold* petroleum ether (60–80 °C) (➡).

NOTE. Do not wash with an excess of the petroleum ether, or a loss of product will result.

Anhydride product collected here

Xylene, 80 μL
+ toluene, 0.5 mL
+ petroleum ether
(60–80 °C),
~0.8 mL

Purification and Characterization. The product is of sufficient purity for direct characterization. Weigh the material and calculate the percent yield. Determine the melting point and compare your result with the value found in the literature. Obtain an IR spectrum of the material.

The reaction involves two reactants (butadiene and maleic anhydride), which both contribute functional groups to the product. The infrared spectrum of the isolated material reflects this observation. Compare the infrared spectrum of your product with that of the reference spectrum.

Infrared Analysis. The spectrum of one of the starting materials, 3-sulfolene (Fig. 6.27), is representative of an alkene sulfone. The macro group frequency train for an unconjugated five-membered ring alkene fits the data reasonably

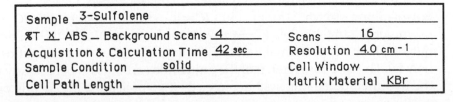

Sample __3-Sulfolene__

%T _X_ ABS _ Background Scans _4___   Scans ___16___

Acquisition & Calculation Time _42 sec_   Resolution _4.0 cm⁻¹_

Sample Condition _____solid____   Cell Window _____

Cell Path Length _____   Matrix Material _KBr_

Figure 6.27 IR spectrum: 3-Sulfolene.

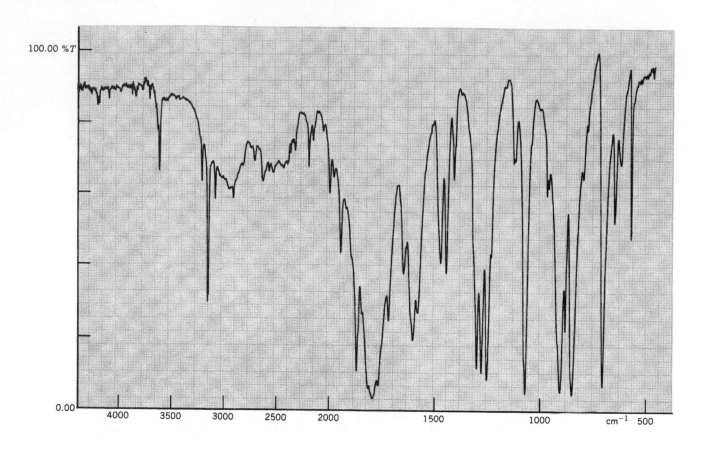

100.00 %T

0.00

4000   3500   3000   2500   2000   1500   1000   cm⁻¹ 500

Sample ___Maleic anhydride___

%T _X_ ABS __ Background Scans _4_   Scans _____16_____
Acquisition & Calculation Time _42 sec_   Resolution _4.0 cm⁻¹_
Sample Condition _____solid_____   Cell Window _____
Cell Path Length _____   Matrix Material _KBr_

Figure 6.28   IR spectrum: maleic anhydride.

well: 3090 ($=$C—H, stretch), 1635 C$=$C, stretch), and 657 (H—C$=$C—H, cis, out-of-plane bend) cm⁻¹. The presence of the sulfone group is convincingly identified by the very strong bands at 1287 and 1127 cm⁻¹, which are assigned to the coupled in-phase and out-of-phase S—O stretching vibrations. The antisymmetric and symmetric C—H stretching vibrations of the methylene groups are observed at 2975 and 2910 cm⁻¹. Note that the influence of the heterocyclic five-membered ring system has raised all of the C—H stretching modes by 40–50 cm⁻¹.

The spectrum of the other reactant, maleic anhydride (Fig. 6.28), possesses all the peaks representative of a conjugated five-membered ring anhydride: 3110 ($=$C—H, stretch), 1858 (C$=$O, in-phase stretch, weak), 1777 (C$=$O, out-of-phase stretch, strong), 1595 (C$=$C, stretch, weak), 1060 (C—O, antisymmetric stretch), 899 (C—O, symmetric stretch), 835 (H—C$=$C—H, out-of-plane bend, normally near 700 cm⁻¹, but raised by conjugation to carbonyls). The in-phase stretch of the anhydride carbonyls is particularly weak as the five-membered ring system forces the two oscillators into nearly opposing positions.

Sample	cis-4-Cyclohexene-1,2-dicarboxylic acid anhydride	
%T _X_ ABS __ Background Scans _4_		Scans ____16____
Acquisition & Calculation Time _42 sec_		Resolution _4.0 cm⁻¹_
Sample Condition ____solid____		Cell Window _____
Cell Path Length _____		Matrix Material _KBr_

Figure 6.29   IR spectrum: 4-cyclohexene-*cis*-1,2-dicarboxylic acid anhydride.

The Diels–Alder product exhibits all of the spectral properties of a non-conjugated, alkenyl five-membered anhydride (Fig. 6.29). The macro group frequency train for the anhydride portion of the molecule involves peaks at 1844, 1772, 998, and 935 cm^{-1}:

a. 1844 cm^{-1}: This weak band is similar to the 1858-wavenumber band found in maleic anhydride, and involves the in-phase stretch of the two coupled carbonyl groups. The band is somewhat more intense in the case of this saturated five-membered ring example, because the ring system is more easily distorted from planarity. This band often exhibits some secondary splitting on the high-wavenumber side (see discussion of the 935-wavenumber band below).

b. 1772 cm^{-1}: This band arises from the out-of-phase stretch of the anhydride carbonyls and is the most intense band in the spectrum. The separation, or splitting, between the two carbonyl stretching modes of anhydrides is relatively constant and falls in the range of 60–90 cm^{-1}.

c. **998 and 935 cm^{-1}:** These two bands are identified with the antisymmetric and symmetric C—O—C stretching modes. It is the overtone of the symmetric stretch that is Fermi coupled (see Infrared Discussions) to the symmetric carbonyl stretch ($935 \times 2 = 1870$ cm^{-1}). The frequency match is variable, with the harmonic generally falling on the high-wavenumber side of the fundamental.

The alkenyl section of the product possesses the following macro group frequency bands: 3065 (=C—H stretch), 1635 (C=C stretch, unsubstituted C=C, 5,6-membered fused ring system), and 685 (H—C=C—H, cis, out-of-plane bend) cm^{-1}.

The saturated C—H section of the cyclohexene ring possesses two identical methylene groups that contain the short macro frequency train of 2972, 2929 and 2860, and 1447 cm^{-1}:

a. **2972 cm^{-1}:** Antisymmetric C—H stretch of the —CH$_2$— group.

b. **2929 and 2860 cm^{-1}:** Split symmetric C—H stretch of the —CH$_2$— group. This mode is split by Fermi coupling (see Chapter 8) with the overtone of the symmetric methylene scissoring vibration found at 1447 cm^{-1} ($1447 \times 2 = 2894$). The uncoupled fundamental mode would be expected to occur at or near 2890 cm^{-1}, and thus falls very close to the predicted location of the overtone.

c. **1447 cm^{-1}:** Symmetric deformation vibration (scissoring motion) of the methylene group present in the cyclohexene system.

Examine the spectrum of your reaction product. Discuss the similarities and differences of the experimentally derived spectral data to the reference spectra (Figs. 6.27–6.29).

Nuclear Magnetic Resonance Analysis. The 4-cyclohexene-*cis*-1, 2-dicarboxylic acid anhydride provides an ideal example of the utility of ^{13}C NMR when only limited information is available from the ^{1}H NMR spectrum. The 300-MHz ^{1}H spectrum is shown in Figure 6.30, page 275. Due in part to the presence of two stereocenters, as well as long-range coupling through the π system of the alkene, the entire ^{1}H spectrum is second-order and no information is available from the coupling constants because the spectrum is too complex. Limited assignments to peaks could be made on the basis of chemical shift, but it would be difficult to make any statements regarding the purity of your sample based on the ^{1}H NMR spectrum, since an impurity could well be hidden beneath any of the complex signals.

On the other hand, the fully ^{1}H-decoupled ^{13}C spectrum (Fig. 6.31, page 276) of 4-cyclohexene-*cis*-1,2-dicarboxylic acid anhydride is much less complex. Because of the mirror plane of symmetry in the compound, there are only four unique carbon atoms and thus only four peaks are seen in the ^{13}C NMR spectrum. The 1:1:1 triplet centered at 77 ppm is due to the solvent, CDCl$_3$. Although no ^{1}H–^{13}C coupling is seen as a result of the ^{1}H decoupling, ^{2}H–^{13}C coupling is observed because ^{1}H and ^{2}H (D) resonate at different frequencies.

Sample preparation for ^{13}C NMR is essentially the same as for ^{1}H NMR spectroscopy except that significantly more material is required to obtain a ^{13}C NMR spectrum in a reasonable amount of time. In this case, acceptable signal-to-noise levels can be obtained on 40 mg of material in about 10 min. Although tetramethylsilane (TMS) can be added as a reference, it is often more

300 MHz NMR OF 4-CYCLOHEXENE-CIS-1,2-DICARBOXYLIC ACID ANHYDRIDE

Figure 6.30 ¹H-NMR spectrum: 4-cyclohexene-*cis*-1,2-dicarboxylic acid anhydride.

convenient to reference the spectrum relative to the known chemical shift of the solvent signal. Do not hesitate to use all of your material for the ^{13}C spectrum; it can be easily recovered later by emptying the NMR tube into a small vial, rinsing once with solvent, and evaporating the solvent in a hood under a gentle stream of dry nitrogen.

Chemical Tests.   Selected chemical classification tests can also be used to aid in characterization of this compound. Is the compound soluble in water? If so, does the aqueous solution turn blue litmus paper red? Is the compound soluble in 5% NaOH and 5% NaHCO$_3$? Is there evidence of CO$_2$ evolution with the bicarbonate solution? If so, what does this test indicate? Give the structure of the product formed when the material is added to the sodium hydroxide solution.

Perform the Baeyer test for unsaturation (Chapter 9). Is there evidence for the presence of a C=C bond?

OPTIONAL SEMI-MICROSCALE PREPARATION

The microscale Diels–Alder addition may be scaled up by a factor of 5. The procedure is similar to that outlined above, with the exceptions noted below.

1. Use a 10-mL round-bottom flask containing a magnetic stirrer, equipped with a reflux condenser protected by a calcium chloride drying tube. Place an insulating aluminum foil guard between the bottom of the reflux condenser and the round-bottom flask (➡).

Reflux conditions
10-mL RB flask

Figure 6.31 ^{13}C-NMR spectrum: 4-cyclohexene-*cis*-1,2-dicarboxylic acid anhydride.

HOOD

CAUTION. At this scale, due to the generation of a considerably larger quantity of sulfur dioxide than at the microscale level, the reaction must be run in the **hood** or provisions should be made to trap the evolved gas.

---

2. Increase the reagent and solvent amounts about 5-fold:

Physical Properties of Reactants					
Compound	MW	Amount	Mmol	mp (°C)	bp (°C)
3-Sulfolene	118.15	850 mg	7.2	66	
Maleic anhydride	98.06	450 mg	4.6	60	
Xylene		400 µL			137–140

3. Once in solution, heat the reaction mixture to reflux for 20 min. *Avoid overheating.* Then cool the reaction to room temperature.

4. After cooling, add 5 mL of toluene and transfer the resulting clear solution to a 25-mL Erlenmeyer flask. The toluene solution may require heating and stirring to dissolve any crystals that may have formed. Now add petroleum ether (60–80 °C) (2.5–3.5 mL).

5. After isolation by vacuum filtration, wash the product with 5 mL of cold petroleum ether (60–80 °C).

6. Characterize the product as outlined in the microscale procedure (above).

QUESTIONS

6-93. Predict the product in each of the following Diels–Alder reactions.

6-94. Cyclopentadiene reacts as a diene in the Diels–Alder reaction a great deal faster than does 1,3-butadiene. Explain.

6-95. Predict the diene and dienophile that would lead to each of the following products.

6-96. The decomposition of 3-sulfolene to form 1,3-butadiene generates 1 mol of sulfur dioxide gas per mol of 1,3-butadiene. Substantial quantities of SO_2 would be generated if this decomposition were carried out on a large scale. Suggest a method for trapping the gas to prevent its escape into the environment.

6-97. Two structural isomers are formed when 2-methyl-1,3-butadiene reacts with ethyl acrylate (ethyl 2-propenoate). Draw structures for these isomers.

6-98. In general, on going from the unsaturated fully conjugated anhydrides to the saturated systems, the two C—O stretching modes begin to coalesce, and in many cases only a single band is observed. In maleic anhydride, these modes occur at 1060 and 899 cm^{-1}, while in cis-4-cyclohexene-1,2-dicarboxylic acid anhydride they are found at 998 and 935 cm^{-1}. Explain this observation.

6-99. Acyclic anhydrides exhibit a reversal of the intensity relationship of the carbonyl stretching vibrations found in the cyclic anhydrides. That is, the lower wavenumber band is now the weaker member of the pair. Explain why this intensity exchange occurs.

6-100. If in Experiment [14] you had prepared the other possible diastereomer of the product, 4-cyclohexene-trans-1,2-dicarboxylic acid anhydride, how many lines would you expect to see in the ^{13}C NMR spectrum? Note that this isomer does not have a mirror plane of symmetry.

6-101. Both 3-methylenecyclohexene and toluene satisfy the criteria of conjugation when asked about the Diels–Alder reaction yet neither will react. Explain.

6-102. In ^1H spectra, the relative intensities of lines within a triplet are 1:2:1. The signal for CDCl$_3$ is a triplet with relative intensities of 1:1:1. Why?

BIBLIOGRAPHY

The preparation of 4-cyclohexene-cis-1,2-dicarboxylic acid anhydride by the reaction of 1,3-butadiene with maleic anhydride is recorded:

Cope, A. C.; Herrick, E. C. *Organic Syntheses*; Wiley: New York, 1963; Collect. Vol. IV, p. 890.

Other Diels–Alder reactions reported in *Organic Syntheses* include:

Carlson, R. M.; Hill, R. K. *Organic Syntheses*; Wiley: New York, 1988; Collect. Vol. VI, p. 196.

Greico, P.; Larsen, S. D. *Organic Syntheses*; Wiley: New York, 1933; Collect. Vol. VIII, p. 31.

Jung, M. E.; McCombs, C. A. *Organic Syntheses*; Wiley: New York, 1988; Collect. Vol. VI, p. 445.

Kozmin, S. A.; He, S.; Rawal, V. H. *Organic Syntheses*; Wiley: New York, 2004; Collect. Vol. X, p. 442.

For selected reviews of the reaction, see:

Butz, L. W.; Rytina, A. W. *Org. React.* **1949**, *5*, 136.
Ciganek, E. *Org. React.* **1984**, *32*, 1.

Pindur, U.; Lutz, G.; Otto, C. *Chem. Rev.* **1993**, *93*, 741.
Rappoport, Z., Ed. *The Chemistry of Dienes and Polyenes*; Wiley: New York, 1997, Vol. I.
Weinreb, S. M. *Comp. Org. Syn.* **1991**, *5*, 513.

For selected reviews on atom economy, see:

Abhyankar, S. B. Introduction to Teaching Green Organic Chemistry, in *Green Organic Chemistry in Lecture and Laboratory*; Dicks, A. P., Ed.; CRC Press: Boca Raton, FL, 2012, pp. 1–28
Sheldon, R. A. *Pure and Applied Chemistry* **2000**, *72*, 1233.
Trost, B. M. *Angew. Chem. Int. Ed.* **1995**, *34*, 259.
Trost, B. M. *Science* **1991**, *254*, 1471.

The conditions of this reaction were adapted from those reported by:

Sample, T. E., Jr.; Hatch, L. F. *J. Chem. Educ.* **1968**, *45*, 55.

EXPERIMENT 15

Diels–Alder Reaction: 9,10-Dihydroanthracene-9,10-α,β-succinic Acid Anhydride

Common name: 9,10-dihydroanthracene-9,10-**α**,β-succinic acid anhydride
CA number: [85-43-8]
CA name as indexed: 1,3-isobenzofurandione, 3a,4,7,7a-tetrahydro-

Purpose. The Diels–Alder reaction is investigated. You will explore the role of an aromatic ring system as the diene substrate in this addition reaction. The reaction studied in this experiment is an example of a 100% atom economic 1,4 addition by an activated alkene dienophile across the 9,10 positions of anthracene.

Prior Reading

  Technique 5: Crystallization
        Use of the Hirsch funnel (pp. 92–93)

REACTION

Anthracene   Maleic anhydride   9,10-Dihydroanthracene-9,10-α,β -succinic acid anhydride

This famous class of reactions are named for **Otto Paul Hermann Diels** and **Kurt Alder**, who were primarily responsible for its development. Diels and Alder received the Nobel Prize in 1950 for this work. See Experiment [14] for a biography of Diels. A short biography of his student Kurt Alder follows.[16]

Kurt Alder (1902–1958). Alder obtained his Ph.D. in 1926 while studying with Otto Diels at the University of Kiel. His dissertation was titled *Causes of the Azoester Reaction*. He became professor of chemistry at the University of Kiel in 1934 and later became head of the Chemical Institute at the University of Cologne (1940). For a few years (1936–1940) he was research director of the Baeyer dye works.

Together with Diels, Alder was responsible for the development of what came to be known as the Diels–Alder reaction. This reaction typically involves the reaction of a 1,3-conjugated diene with an activated alkene (dienophile) to form a six-membered cycloalkene. While reactions of this type had been reported as early as 1893, Diels and Alder were the first to recognize their great versatility. Alder continued to focus his academic research in this area following his graduate work with Diels. Over a number of years, Alder carried out a systematic study of the reactivity of a large number of dienes and dienophiles and established the structure and stereochemistry of many new adducts. He also expanded his doctoral research, studying the condensation of azoesters with dienes to yield the corresponding heterocyclic adducts.

Alder demonstrated that successful addition required that the diene double bonds possess an s-cis conformation (s refers to the single bond connecting the two double bonds). Furthermore, he realized that the bridged ring adducts formed by using cyclic dienes were closely related to natural products, such as camphor, and that this reaction offered a powerful route for the synthesis of a wide variety of naturally occurring compounds, particularly the terpenes. The Diels–Alder reaction also has been invaluable in the industrial synthesis of thousands of new organic materials from insecticides and dyes to lubricating oils and pharmaceuticals.

Alder investigated autooxidation and polymerization processes particularly during his industrial years. For example, he was involved in an extensive study of polymerizations related to the formation of Buna-type synthetic rubbers.[17]

DISCUSSION

This experiment is a further example of the Diels–Alder reaction. *For a discussion of the basic aspects of this reaction see Experiment [14].* In the present case, the central ring of anthracene is shown to possess the characteristic properties of a diene system. Thus, this aromatic compound reacts to form stable Diels–Alder adducts with many dienophiles at the 9 and 10 positions (the two positions on the central ring where new bonds can be made without destroying the aromaticity of the other two rings). Maleic anhydride, a very reactive dienophile, is used here in the reaction with anthracene. Note that as this reaction is reversible, it is usually best carried out at the lowest possible temperatures consistent with an acceptable reaction rate (see Experiment [14]).

---

[16]For references to the Diels–Alder reaction, see Experiment [14].
[17]See Allen, C. F. H. *J. Chem. Educ.* **1933,** *10,* 494; *Dictionary of Scientific Biography*, C. C. Gillespie, Ed.; Scribner's: New York, 1970, Vol. I, p. 105; *McGraw-Hill Modern Scientists and Engineers*, S. P. Parker, Ed., McGraw-Hill: New York, 1980, Vol. I, p. 8.

Higher molecular weight polynuclear aromatic hydrocarbons (PAHs) containing the anthracene nucleus have also been found to react with maleic anhydride. These ring systems, however, can differ widely in reaction rates. Typical examples of those systems that undergo the Diels–Alder reaction are 1,2,5,6-dibenzanthracene (a), 2,3,6,7-dibenzanthracene (pentacene) (b), and 9,10-diphenylanthracene (c).

(a)   (b)   (c)

EXPERIMENTAL PROCEDURE

Estimated time to complete the experiment: 1.5 h.

Physical Properties of Reactants and Product					
Compound	MW	Amount	mmol	mp (°C)	bp (°C)
Anthracene	178.24	80 mg	0.44	216	
Maleic anhydride	98.06	40 mg	0.40	60	
Xylenes		1.0 mL			137–140
9,10-Dihydroanthracene-9,10-α,β-succinic acid anhydride	276			261–262	

200 °C
thermometer

Anthracene, 80 mg
+ maleic anhydride, 40 mg
+ xylene, 1.0 mL

HOOD

Reagents and Equipment.   Weigh and place 80 mg (0.44 mmol) of anthracene and 40 mg (0.40 mmol) of maleic anhydride in a 3.0-mL conical vial containing a boiling stone and equipped with an air condenser protected by a calcium chloride drying tube. Now, while in the **hood**, add 1.0 mL of xylene to the solid mixture using an automatic delivery pipet (◂).

NOTE. High-purity grades of anthracene and maleic anhydride are strongly recommended. Anthracene may be recrystallized from 95% ethanol. A mixture of xylenes with a boiling-point range of 137–140 °C is sufficient, but the solvent (xylenes) should be dried over molecular sieves before use.

Reaction Conditions.   Heat the reaction mixture at reflux for 30 min in a sand bath at about 200 °C. During this time the initial yellow color of the reaction mixture gradually disappears. (Why?) Allow the resulting bleached solution to cool to room temperature and then place it in an ice bath for 10 min to complete the crystallization of the product.

Diels–Adler
product
collected
here

Xylene,
0.75 mL

Isolation of Product.   Collect the crystals by vacuum filtration using a Hirsch funnel and wash the filter cake with two 300-µL portions of cold ethyl acetate (◂). Partially dry the filter cake under suction using plastic food wrap (see Prior Reading). Transfer the filtered, washed, and partially dried product to a porous clay plate or filter paper to complete the drying.

Purification and Characterization.   The product is often of sufficient purity for direct characterization at this point. The adduct, however, may be further purified by recrystallization from ethyl acetate using a Craig tube. Weigh the dried Diels–Alder product and calculate the percent yield. Determine the melting point and compare your result with the value listed above.

NOTE. It is suggested that a hot-stage melting-point apparatus, such as the Fisher–Johns, be used due to the high melting point of the product.

Chemical Tests.   The *ignition test* is often used as a preliminary method to categorize hydrocarbon materials (see Chapter 9). Aromatic compounds give a yellow, sooty flame. Perform this test on anthracene. Based on your results can anthracene be classified as aromatic?

Carry out the ignition test on the following compounds and determine whether they should be classified as aromatic materials.

a. Toluene

b. Octane

c. Isopropyl alcohol

d. Nitrobenzene

e. *trans*-Stilbene

OPTIONAL SEMI-MICROSCALE PREPARATIONS

This Diels–Alder reaction may be scaled up by a 5- or 50-fold increase in reactants. The scaled-up procedures are nearly identical in either case to that given for the microscale preparation; changes are noted below.

1. Fivefold Scaleup

a. In place of the conical vial as the reaction vessel, use a 10-mL round-bottom flask containing a magnetic stirrer connected to the water-jacketed reflux condenser that is protected by a calcium chloride drying tube (➡).

b. The reagent and solvent quantities are given in the following table:

Compound	MW	Amount	mmol	mp (°C)	bp (°C)
Anthracene	178.24	400 g	2.2	216	
Maleic anhydride	98.06	200 g	2.0	60	
Xylene	106.16	5.0 mL			137–140
Ethyl acetate	88.11	3.0 mL			77

c. The reaction flask is **heated at 185–190 °C** bath temperature for 30 min to obtain optimized yields.

2. Fiftyfold Scaleup

a. In place of the conical vial as the reaction vessel, use a 100-mL round-bottom flask connected to a water-cooled reflux condenser that is fitted with a calcium chloride drying tube.

185–190 °C thermometer

Al foil

Speed   Heat

Reflux conditions
10- or 100-mL RB flask

b. The required reagent and solvent quantities are given in the following table:

Compound	MW	Amount	mol	mp (°C)	bp (°C)
Anthracene	178.24	4.0 g	0.2	216	
Maleic anhydride	98.06	2.0 g	0.2	60	
Xylene	106.16	50 mL			137–140
Ethyl acetate	88.11	30 mL			77

c. Heat the reaction mixture at **vigorous** reflux for **2 h.**

d. Recrystallize the crude product from ethyl acetate.

Experiment 15-1* 9,10-Dihydroanthracene-9,10-α,β-succinic Acid Anhydride: Preparation Using a Monomode Microwave Apparatus

EXPERIMENTAL PROCEDURE

Estimated time to complete the experiment: 1.0 h.

Compound	MW	Amount	mmol	mp (°C)	bp (°C)
Anthracene	178.24	178 mg	1.00	216	
Maleic anhydride	98.06	98 mg	1.00	60	
Toluene		2 mL			111
9,10-Dihydroanthracene-9,10-α,β-succinic acid anhydride	276			261–262	

Reagents and Equipment. This experiment is designed for use in the CEM Discover and Biotage Initiator microwave units.

In a 10.0-mL glass microwave reaction vessel containing a magnetic stir bar, place 178 mg (1.00 mmol) of anthracene, 98 mg (1.00 mmol) of maleic anhydride, and 2 mL of toluene. Immediately cap the vessel with a microwave pressure cap.

CAUTION: Since the reaction requires heating toluene (solvent for the reaction) to above its boiling point in sealed vessels, *adherence to the microwave manufacturer's guidelines is essential.*

Reaction Conditions. Place the reaction vessel in the microwave cavity and, depending on the equipment used, position the pressure device on top. Program the microwave unit to heat the reaction mixture at maximum power to 180 °C and hold at this temperature for 10 min. After heating, allow the reaction mixture to cool to 50 °C or below before removing the tube from the microwave unit. Allow the resulting solution to cool to room temperature and then place it in an ice bath for 10 min to complete the crystallization of the product.

*This section has been written by Dr. Nicholas E. Leadbeater from the Department of Chemistry at the University of Connecticut and Dr. Cynthia B. McGowan from the Department of Chemistry at Merrimack College, MA.

Isolation of Product. The remainder of the procedure is identical to Experiment 15 (pages 278–279).

9,10-Dihydroanthracene-9,10-α,β-succinic Acid Anhydride: Preparation Using a Multimode Microwave Apparatus

Experiment 15-2*

M

EXPERIMENTAL PROCEDURE

Estimated time to complete the experiment: 1.0 h.

Compound	MW	Amount	mmol	mp (°C)	bp (°C)
Anthracene	178.24	214 mg	1.20	216	
Maleic anhydride	98.06	118 mg	1.20	60	
Toluene		5 mL			111
9,10-Dihydroanthracene-9,10-α,β-succinic acid anhydride	276			261–262	

Reagents and Equipment. This experiment is designed for use in the CEM MARS, Milestone START, and Anton Paar Synthos 3000 microwave units. When using the Anton Paar Synthos 3000 unit with the 24-position silicon carbide plate rotor containing glass vials, the reagent and solvent quantities cited in the monomode procedure should be used in conjunction with the reaction conditions here in the multimode procedure.

In a microwave reaction vessel containing a magnetic stir bar, add 214 mg (1.20 mmol) of anthracene, 118 mg (1.20 mmol) of maleic anhydride, and 5 mL of toluene. Immediately cap the vessel with the microwave pressure cap and adjust the tightness to the manufacturer-specified level. Place the sealed vessel into its outer protective jacket.

CAUTION: Since the reaction requires heating toluene (solvent for the reaction) to above its boiling point in sealed vessels, *adherence to the microwave manufacturer's guidelines is essential.*

Reaction Conditions. Insert the loaded vessels into the reaction carousel ensuring they are evenly spaced and then place the carousel into the microwave cavity. If provided by the manufacturer, connect a temperature probe to the control vessel. Program the microwave unit to heat the reaction vessels to 180 °C using maximum power and hold at this temperature for 10 min. After heating, allow the reaction mixture to cool to 50 °C or below before removing the carousel from the microwave unit. Allow the resulting solution to cool to room temperature and then place it in an ice bath for 10 min to complete the crystallization of the product.

Isolation of Product. The remainder of the procedure is identical to Experiment 15 (pages 278–279).

*This section has been written by Dr. Nicholas E. Leadbeater from the Department of Chemistry at the University of Connecticut and Dr. Cynthia B. McGowan from the Department of Chemistry at Merrimack College, MA.

QUESTIONS

6-103. Given the data tabulated below for the rate of reaction of maleic anhydride with a series of substituted 1,3-butadienes, offer a reasonable explanation to account for the trend in the rates.

R	k (rel) at 25 °C
—H	1
—CH$_3$	4.2
—C(CH$_3$)$_3$	<0.05

6-104. Predict the structure of the product formed in the following reactions.

(a)

(b)

6-105. Offer an explanation of why anthracene preferentially forms a Diels–Alder adduct at the 9,10 positions.

6-106.. Experiment [31] demonstrates the monobromination of anthracene to yield one specific monosubstituted product. Other conditions may be used to form other monobrominated anthracenes. Draw the structures of, and name, the possible monobromo-substituted anthracenes that could be prepared in the laboratory.

6-107. There are four reasonable resonance structures for phenanthrene. Draw them.

6-108. A large number of polycyclic benzenoid aromatic hydrocarbons are known. One of these, benz[a]pyrene, is a powerful carcinogen found in tobacco smoke. From the literature, locate and then draw the structure of this hydrocarbon. Can you suggest other sources where this material might be expected to be present?

6-109. Anthracene undergoes a *suprafacial* [$\pi 4_s + \pi 4_s$] cycloaddition reaction at the 9,10 positions when subjected to irradiation to form a cyclic dimer. *Suprafacial* is a term used to describe in detail that the addition of the dienophile to the 1,3-π-system has occurred all from the same side (or face) of this planar structure. It comes from the Latin meaning above, or the dorsal side, and it is symbolized by the subscript "s" in the terminology: [$\pi 4_s + \pi 4_s$]. This is a theoretically forbidden thermal pericyclic reaction, but it can occur under photochemical conditions. Draw the structure of this product.

BIBLIOGRAPHY

For a review of diastereoselectivity in the reaction, see:

Coxon, J. M., et al., "Diastereofacial Selectivity in the Diels-Alder Reaction" in *Advances in Detailed Reaction Mechanisms* **1994,** 3, 131.

Waldmann, H. *Synthesis* **1994,** 535.

EXPERIMENT 16

Grignard Reaction with a Ketone: Triphenylmethanol

Common name: triphenylmethanol

CA number: [76-84-6]

CA name as indexed: benzenemethanol, α,α -diphenyl-

Purpose. The techniques required to prepare Grignard reagents are developed. The reaction of these reagents with ketones to form *tertiary* alcohols is investigated. You will gain experience, first-hand, working with these highly air- and moisture-sensitive materials. And you will observe the formation of this famous reagent, in which magnesium metal is transformed (heterogeneous conditions) into organometallic salts of enormous value to the synthetic chemist.

This exercise is the first of a number of experiments in this chapter in which the Grignard reaction is studied at the microscale level. Even if you do not actually do more than one of them, it is well worth the time to read through the other Experiments [17], [21], and [4$_{adv}$] to get a sense of the breadth of the applications of the reaction.

Prior Reading

Technique 4: Solvent Extraction
Liquid–Liquid Extraction (p. 76)
Drying of the Wet Organic Layer (pp. 84–87)
Technique 5: Crystallization
Craig Tube Crystallizations (pp. 93–95)
Technique 6A: Chromatography
Thin-Layer Chromatography (pp. 101–103)

REACTION

Phenylmagnesium bromide   Benzophenone   Triphenylmethanol

François Auguste Victor Grignard (1871–1935). Born in Cherbourg (France), Grignard was professor of Chemistry at the Universities of Lyons and Nancy. After studying for one year with Bouveault, Grignard became a graduate student of Philippe Antoine Barbier, a professor at the University of Lyons. Barbier, who was working in the area of terpene chemistry, had found that magnesium could be used in place of zinc in the reaction of methyl iodide with an unsaturated ketone (methylheptenone) to yield the corresponding tertiary alcohol. This route was much preferred since the zinc reagents were difficult to work with because they were pyrophoric (spontaneously flammable in air). The use of magnesium in the formation of tertiary alcohols was reported in 1899. Barbier suggested to Grignard that it might be interesting to further investigate the reaction of magnesium with alkyl halides. This study was to form the basis of Grignard's doctoral dissertation. Grignard discovered that treatment of alkyl iodides with magnesium in *diethyl ether* produced an alkylmagnesium iodide by a spontaneous reaction at ambient temperatures. His initial results, reported in 1900, were followed by seven papers the following year. His doctoral thesis on organomagnesium compounds and their application to synthetic organic chemistry was presented in 1901, when Grignard was 30 years old.

Grignard continued this work, having recognized the enormous potential of the alkylmagnesium halides in organic synthesis. These species are now known as *Grignard reagents* and when the reagent is used in synthesis, the reaction is called a *Grignard reaction. These* reagents have found great utility in the preparation of many kinds of organic compounds, including alcohols, ketones, esters, and carboxylic acids. As mentioned above, these reagents contain a carbon–metal bond, and therefore they are classed in a large group of substances called *organometallic* compounds.

For this work, Grignard received the Nobel Prize in 1912. In 1919 Grignard returned to Lyons where he succeeded Barbier as chairman of the Department. By the end of his life, the scientific literature contained over 6000 papers dealing with Grignard reagents and their application.

Grignard also did extensive work in the areas of the terpenes, quantitative ozonolysis of alkenes, aldol reactions, catalytic hydrogenation, and dehydrogenation and cracking of hydrocarbons.[18]

DISCUSSION

Grignard reagents possess significant nucleophilic character because of the highly polarized carbon–metal bond that results in considerable carbon-ionic character at carbon. Grignard discovered that these reactive materials readily attack the electrophilic carbon of a carbonyl group. It is this direct attack on carbon by a carbon nucleophile, resulting in carbon–carbon bond formation, that makes these such important reactions. Furthermore, as the carbonyl is the most ubiquitous functionality in all of organic chemistry, Grignard reagents have found great utility and widespread use in organic synthesis.

The formation of the organomagnesium halide (Grignard reagent) involves a heterogeneous reaction between magnesium metal and an alkyl, alkenyl, or aryl halide in ether solution. The solvent may be any one of a number of ethers, but diethyl ether and tetrahydrofuran are by far the most popular.

$$R(Ar) - X + Mg \xrightarrow{\text{ether}} R(Ar) - Mg - X$$
$$R = \text{alkyl, alkenyl}$$
$$\text{and } Ar = \text{aryl}$$

The reaction between an alkyl, alkenyl, or aryl halide and magnesium takes place on the surface of the metal and is an example of a *heterogeneous* (across two phases) reaction. The reactivity of the alkyl halides is in the order Cl < Br < I; fluorides do not generally react. Substituted alkyl halides react in the order 1° > 2° > 3° alkenyl and aromatic halides also form Grignard reagents to varying degrees.

It is important to understand the role of the ether solvent in the formation of the Grignard reagents. The reaction at the surface of the metal is essentially an oxidation–reduction reaction. The metal is partially oxidized to the greater than 1+ state, and the organohalide is reduced to a halide ion and a highly

---

[18]See Gordon, N. E. *J. Chem. Educ.* **1930,** *7,* 1487; Rheineoldt, H. *J. Chem. Educ.* **1950,** *27,* 476; Kauffman, G. B. *J. Chem. Educ.* **1990,** *67,* 569; Gilman H. *J. Am. Chem. Soc. (Proc.)* **1937,** *59,* 17; Gibson, C. S.; Pope, W. J. *J. Chem. Soc.* **1937,** 171; *Dictionary of Scientific Biography,* C. C. Gillespie, Ed., Scribner's: New York, 1972, Vol. V, p. 540.

polarized carbon–metal bond with the magnesium. The overall reaction can be viewed as forming the species $R\overset{\delta-}{\underline{\quad}}\overset{\delta+}{}Mg^+X^-$ as the Grignard reagent. This highly polarized material is insoluble in most nonpolar organic solvents. The reaction will proceed at the surface of the metal until a layer of the insoluble organometallic reagent has formed. At this point, the surface reaction with the magnesium will immediately cease. If protic solvents are used, they would instantly react with the highly basic Grignard reagent, R—MgX, to form the corresponding hydrocarbon, R—H. Thus, the use of either nonpolar or protic solvents does not lead to successful Grignard reagent formation. Why, then, is ether, a relatively nonpolar solvent, essential to the preparation of Grignard reagents?

The magnesium is essentially divalent, and electron deficient, when it reacts with the halide to form the RMgX species. A full octet around the metal atom requires two additional pairs of electrons. It is the energy gained by filling this octet that drives the coordination of the magnesium with two molecules of the ether solvent. This association in turn dramatically increases the solubility of the Grignard reagent in the relatively nonpolar ether solvent, and thus promotes further Grignard reagent formation.

$$\overset{\delta+}{CH_3CH_2\overset{..}{\underset{..}{O}}CH_2CH_3}$$
$$\downarrow$$
$$R{-}X + Mg \xrightarrow{\text{ether}} (R{-}Mg{-}X)^{\delta-}$$
$$\uparrow$$
$$\underset{\delta+}{CH_3CH_2\overset{..}{\underset{..}{O}}CH_2CH_3}$$

This interaction of RMgX with ether solvent may also be described as a Lewis acid–base interaction in which the coordinating solvent molecules are usually not written. When a Grignard reagent is described, it is important to remember that this vital solvation is always taking place.

The reactions of Grignard reagents with different types of carbonyl groups yield a number of important functional groups. For example, reaction with formaldehyde yields 1° alcohols; with higher aldehydes, 2° alcohols; with ketones, 3° alcohols; with esters, 3° alcohols; with acyl halides, ketones; with *N,N*-dialkylformamides, aldehydes; and with carbon dioxide, carboxylic acids.

In this experiment you will study the addition of the aryl Grignard reagent (phenylmagnesium bromide) to a diaryl ketone (benzophenone) to yield the corresponding tertiary (3°) alcohol. Because it is possible to vary both the structure of the Grignard reagent and the ketone, a wide variety of 3° alcohols may be obtained by this synthetic route.

The mechanism, as discussed above, can be thought of as involving rapid nucleophilic attack by the Grignard reagent at the carbon of the carbonyl group. Hydrolysis of the resulting alkoxide ion intermediate with dilute acid yields the desired alcohol. The reaction sequence is outlined here:

By using Grignard reagents, it is theoretically possible to synthesize a very large number of alcohols. Indeed, there is often more than one synthetic pathway open to a desired product. The choice of route is generally dictated by the availability of starting materials and the associated costs of these compounds.

EXPERIMENTAL PROCEDURE

Estimated time to complete the experiment: Two laboratory periods when starting with bromobenzene. If using commercial phenylmagnesium bromide, one laboratory period (experiment starting on p. 287). (**The Benzophenone Reagent**)

CAUTION: Ether is a flammable liquid. All flames must be extinguished during the time of this experiment.

Step I:
BrC$_6$H$_5$, 76 µL
+ Mg, 18 mg
+I$_2$, 1 crystal
+ (CH$_3$CH$_2$)$_2$O, 700 µL
Step II:
(C$_6$H$_5$)$_2$CO, 105 mg
+ (CH$_3$CH$_2$)$_2$O, 600 µL

Physical Properties of Reactants							
Compound	MW	Amount	mmol	mp (°C)	bp (°C)	d	n_D
Bromobenzene	157.02	76 µL	0.72		156	1.50	1.5597
Diethyl ether	74.12	1.3 mL			34.5	0.73	
Magnesium	24.3	18 mg	0.74				
Iodine	253.8	1 crystal					
Benzophenone	182.21	105 mg	0.58	48			

Reagents and Equipment
Preparation of Phenylmagnesium Bromide

NOTE. *All the glassware used in the preparation of the Grignard reagent should be cleaned and dried in an oven at 110 °C for at least 30 min. After removal from the drying oven, the hot glassware should be placed in a desiccator and cooled before being assembled. Flame drying of the apparatus with a microburner annealing flame (high gas mixture) is an alternative, if ovens are not available. This latter method preferably should be carried out prior to assembly and, as with the oven-dried equipment, the hot glassware should be placed in a desiccator to cool. Care must be taken when flaming the reaction vial, because thermal shock can easily crack this heavy-walled vessel. If the flame-drying procedure is performed on the assembled apparatus, care also must be taken not to overheat the O-rings and plastic Capseals.*

HOOD   In a 3.0-mL conical vial containing a magnetic spin vane and equipped with a Claisen head fitted with a calcium chloride drying tube and a rubber septum, weigh and place 18 mg (0.74 mmol) of polished magnesium ribbon, a small crystal of iodine, and 100 µL of anhydrous ether (using an automatic delivery pipet in the **hood**)(◄).

NOTE. *Scrape a 2- to 3-in. piece of magnesium ribbon clean of surface oxide (MgO) coating and then cut it into 1-mm long sections.*

Alternate Procedure: Place the (polished) magnesium metal and an iodine crystal in the reaction vial and quickly assemble the apparatus. Warm the
GENTLY   mixture (using a microburner or hot plate) **gently** until evidence of purple iodine vapor is observed.

NOTE. *If a microburner is used in this step, all flames must be extinguished before proceeding.*

Now add 100 µL of anhydrous ether using a 1.0-mL syringe inserted through the septum.

In a dry, screw-capped vial, prepare a solution of 76 µL (113 mg, 0.72 mmol) of bromobenzene in 400 µL of anhydrous diethyl ether. *Automatic delivery pi-* HOOD *pets are used to transfer these reagents in the* **hood**. Recap the vial until just before injecting its contents into the reaction vial.

Draw the bromobenzene–ether solution into a 1.0-mL syringe and insert the syringe through the rubber septum on the Claisen head. Place an additional 300 µL of anhydrous diethyl ether rinse in the empty vial (which contains traces of bromobenzene), cap it, and set it aside for later use. With slow stirring, add 6–8 drops of the bromobenzene solution to induce the initial formation of Grignard reagent. The evolution of tiny bubbles from the surface of the magnesium (the heat of reaction is vaporizing the low-boiling ether solvent) is evidence of successful reaction initiation.

Once the reaction gives evidence of initiation, add the remainder of the bromobenzene dropwise *slowly* over a 3- to 5-min period. (*In macroscale re-actions it is extremely important to make sure that initiation of the reaction has occurred prior to adding large quantities of the organohalide, as Grignard reactions often go through an induction period before starting up. If significant quantities of the halide are present when the reaction commences, the sudden and rapid rate of reaction can produce a very rapid evolution of heat, and the reaction may well erupt out of control.) Warm the reactants gently to maintain gentle reflux.*

After adding the bromobenzene, draw the ether rinse from the capped vial into the syringe and also add it to the reaction vial through the septum in a single portion.

Heat the resulting heterogeneous reaction mixture gently with stirring for 15 min.

CAUTION: *Do not overheat!* This overheating will cause loss of ether solvent and promote formation of byproducts. Small fragments of magnesium may remain at the end of the reaction. Maintain no more than a gentle reflux at all times. If solvent volume decreases rapidly, check for leaks around the Capseals and add additional anhydrous ether through the septum to make up for the lost volume of ether.

Cool the gray-brown mixture of the Grignard reagent, phenylmagnesium bromide, to room temperature.

The Benzophenone Reagent. Prepare a solution of 105 mg (0.58 mmol) of benzophenone in 300 µL of anhydrous diethyl ether in a dry vial with a cap. *The ether is measured using a graduated 1-mL syringe and is dispensed in the* **hood**. HOOD

Draw the solution immediately into a 1.0-mL syringe, and then insert the syringe needle through the rubber septum on the Claisen head. Place an additional 300 µL of the anhydrous diethyl ether in the empty vial, cap it, and set it aside for later use.

Reaction Conditions. *Carefully,* with stirring, add the benzophenone solution to the Grignard reagent (1.2 equiv) over a period of approximately 30 s or at a rate that maintains the temperature of the ether solvent at a no more than gentle reflux.

Upon completion of this addition, add the rinse from the capped vial, in like manner, in a single portion.

Stir the reaction mixture for 2–3 min and then allow it to cool to room temperature. Remove the reaction vial from the Claisen head and cap it.

During this cooling period the reaction mixture generally solidifies. *Once the reaction vial is detached from the Claisen head, it is recommended that the vial be placed in a 10-mL beaker to prevent loss of product by accidental tipping.*

NOTE. *If the laboratory is done in two periods, you should stop either at this point, or following the hydrolysis sequence described in the next step.*

Isolation of Product.   Hydrolyze the magnesium alkoxide salt by the *careful,* dropwise addition of 3 M HCl from a Pasteur pipet, while at the same time using a small stirring rod to break up the solid residue. Continue the addition until the aqueous phase tests acidic with litmus paper. A two-layer reaction mixture forms (ether–water) as the solid gradually dissolves.

CAUTION: The addition of the acid may be accompanied by the evolution of heat and some frothing of the reaction mixture. An ice bath should be handy to cool the solution, if necessary. Additional ether may be added, if required, to maintain the volume of the organic phase. Check the acidity of the mixture periodically. The total reaction mixture must be acidic; both insufficient and excess amounts of hydrochloric acid will result in a decreased yield of product during the subsequent workup.

Now remove the magnetic spin vane with forceps and set it aside to be rinsed with an ether wash. Cap the vial tightly, shake, carefully vent, and allow the layers to separate.

Using a Pasteur filter pipet, transfer the lower aqueous layer to a clean 5.0-mL conical vial.

NOTE. *Save the ether layer—it should contain your product.*

Wash the acidic aqueous layer with three 0.5-mL portions of diethyl ether (calibrated Pasteur pipet). Rinse the spin vane with the first portion of ether as it is added to the vial. Cap the vial, shake (or use a Vortex mixer, if available), vent carefully, and allow the layers to separate. After each extraction, combine the organic phase with the ether solution saved above. The bottom (aqueous) layer is set aside in a 10-mL Erlenmeyer flask until the experiment is completed.

Now extract the combined ether layers with 0.5 mL of cold water to remove any acidic residue. Combine the aqueous rinse with the previously extracted and stored aqueous layers in a 10-mL Erlenmeyer flask. Dry the ether solution (capped) over 250–300 mg of anhydrous granular sodium sulfate for approximately 10 min. Stir the drying agent intermittently with a glass rod or swirl the flask. If large clumps of sodium sulfate begin to develop and the solution remains cloudy, it may be necessary to transfer the ether extracts to another vial for a second treatment with the drying agent, or you may be able to simply add more Na_2SO_4 to the original vial. The ether solution should be *clear* following treatment with the anhydrous sodium sulfate. Make all these transfers with Pasteur filter pipets.

HOOD   Transfer the dried ether solution to a *previously tared* Craig tube containing a boiling stone. Carry out the transfer in 0.5-mL portions, concentrating each ether aliquot by warming the vial in a sand bath in the **hood** between the transfers. Rinse the vial and drying agent with an additional 0.5 mL of ether, add the rinse to the Craig tube, and, finally, concentrate the solution to dryness to yield the product residue.

Determine the weight of the crude triphenylmethanol product.

Purification and Characterization.   The major impurity usually present in the triphenylmethanol is biphenyl, which is formed by a coupling reaction,

$$2RX + Mg \xrightarrow{\text{ether}} R{-}R + MgX_2$$

The purity of your crude product may be determined using *thin-layer chromatography*.

TLC CONDITIONS. Use Eastman Kodak fluorescent silica gel sheets (1 × 4 cm). Develop the plates with methylene chloride and visualize the spots by UV light. Reference R_f values for triphenylmethanol and biphenyl are about 0.6 and 0.9, respectively.

The coupled byproduct can be separated from the desired alcohol by taking advantage of the difference in solubilities of the hydrocarbon and the alcohol in ligroin. Ligroin, a nonpolar alkyl solvent, readily dissolves the nonpolar biphenyl, whereas the polar tertiary alcohol is much less soluble.

Add 0.5 mL of cold ligroin to the crude product contained in a Craig tube and scrape and agitate the solid material into a suspension with a small stirring rod. Swirl and stir the solid product with the solvent for several minutes.

Recover the solid triphenylmethanol using the Craig tube in the usual manner. *Save* the ligroin solution that will contain any biphenyl by transferring it, using a Pasteur filter pipet, to a tared 10-mL Erlenmeyer flask.

Repeat the above extraction with a second 0.5-mL portion of ligroin, again stirring the solid suspension and combining the recovered ligroin solution with that saved above. Place the Erlenmeyer flask (cover the mouth with filter paper held by a rubber band) in the **hood** overnight, or warm it in a sand bath to allow the ligroin to evaporate. Estimate the amount of biphenyl (and any other impurities) produced in the reaction.   HOOD

Heat the Craig tube containing the solid triphenylmethanol in a 100 °C oven for 5 min, and then place the crystals on a clay plate to complete the drying process. A further check of the product purity should be carried out by TLC, as described above. Comparisons of the two, pure product and impure product, will validate that upon trituration followed by recrystallization, TLC analysis offers the experimentalist sufficient evidence of product purity.

Weigh the purified triphenylmethanol product and calculate the percent yield. Determine the melting point of the material and compare your result to that recorded in the literature. If desired, the product may be purified further by recrystallization from isopropanol using the Craig tube.

Characterization of the triphenylmethanol is best done by obtaining the IR and NMR spectra and comparing the spectral data to that of an authentic sample or with that recorded in the literature (*The Aldrich Library of IR Spectra* and/or SciFinder Scholar).

Triphenylmethanol (0.1 g/L in methanol) and biphenyl (0.05 g/L in methanol) also have markedly different UV spectra. This electronic absorption data can further help to establish the identity of the products formed in this Grignard reaction.

UV Spectral Data

Triphenylmethanol λ_{max} 240 nm (log ε_{max} 3.16, dioxane)   (Chapter 8)

λ_{max} 253 nm (log ε_{max} 3.26, dioxane)

λ_{max} 260 nm (log ε_{max} 3.28, dioxane)

Biphenyl λ_{max} 247 nm (log ε_{max} 4.24, ethanol)

QUESTIONS

6-110. Predict the product formed in each of the following reactions and give each reactant and product a suitable name:

(a) $\underline{\quad}\diagup$MgBr + CO_2 $\xrightarrow[\text{2) H}^+]{\text{1) ether}}$

(c) [structure with MgBr] + CH_2O $\xrightarrow[\text{2) H}^+]{\text{1) ether}}$

(b) [mesityl-MgBr structure] + D_2O $\xrightarrow[\text{2) H}^+]{\text{1) ether}}$

(d) PhMgBr + [aldehyde structure] $\xrightarrow[\text{2) H}^+]{\text{1) ether}}$

6-111. Provide the necessary Grignard reagent to conduct each of the following reactions.

(a) [cyclopentanone structure] + **?** $\xrightarrow[\text{2) H}^+]{\text{1) ether}}$ [HO-substituted cyclopentane with phenyl]

(b) [epoxide structure] + **?** $\xrightarrow[\text{2) H}^+]{\text{1) ether}}$ [OH structure]

6-112. Outline a synthetic reaction scheme for the preparation of triphenylmethanol from
  (a) Benzoyl chloride
  (b) Diethyl carbonate

6-113. In the experiment, ligroin may be used as a solvent for the separation of the product from biphenyl.
  (a) What is ligroin?
  (b) Can you suggest an alternative solvent that might be used in this step?

6-114. Give the reaction scheme, showing the products formed (before hydrolysis), when one equivalent of ethylmagnesium bromide is treated with one equivalent of 5-hydroxy-2-pentanone. Does addition of two equivalents of the Grignard reagent to this ketone yield a different product(s)? If so, give the structure(s).

BIBLIOGRAPHY

The alternative method of preparing the Grignard reagent is adapted from:

Eckert, T. S. *J. Chem. Educ.* **1987,** *64,* 179.

General references on Grignard reagents:

Coates, G. E.; Green, M. L. H.; Wade, K. *Organometallic Compounds,* 3rd ed.; Methuen: London; Vol. II, 1968.
Grignard, V. *Compt. Rend.* **1900,** *130,* 1322.
Huryn, D. M. *Comp. Org. Syn.* **1991,** *1,* 49.
Lai, Y. H. *Synthesis* **1981,** 585.
Raston, C. L.; Salem, G. in *The Chemistry of the Metal-Carbon Bond;* Hartley, F. R., Ed.; Wiley: New York, 1987, Vol. IV, p. 159.

Shirley, D. A. *Org. React.* **1954,** *8,* 28.

A synthesis of triphenylmethanol (triphenylcarbinol) is reported in *Organic Syntheses*:

Bachmann, W. E.; Hetzner, H. P. *Organic Syntheses;* Wiley: New York, 1955; Collect. Vol. III, p. 839.

The preparation of a series of tertiary alcohols by the addition of Grignard reagents to diethyl carbonate is reported in *Organic Syntheses*:

Moyer, W. W.; Marvel, C. S. *Organic Syntheses;* Wiley: New York, 1943; Collect. Vol. II, pp. 602–603. Also see Braun, M.; Graf, S.; Herzog, S. *Org. Synth.* **1995,** *72,* 32.

Grignard Reaction with an Aldehyde: 4-Methyl-3-heptanol

Common name: 4-methyl-3-heptanol
CA number: [14979-39-6]
CA name as indexed: 3-heptanol, 4-methyl-

Purpose. You will carry out a classic method for the synthesis of *secondary* alcohols: the addition of a Grignard reagent to an aldehyde (other than formaldehyde).[19]

Prior Reading

Technique 4: Solvent Extraction
Liquid–Liquid Extraction (p. 76)
Drying of the Wet Organic Layer (pp. 84–87)
Technique 6: Chromatography
Column Chromatography (pp. 96–99)
Technique 1: Gas Chromatography (pp. 59–65)

REACTION

$$CH_3CH_2CH_2-\underset{\underset{H}{|}}{\overset{\overset{CH_3}{|}}{C}}-Br \xrightarrow{Mg/Et_2O} CH_3CH_2CH_2-\underset{\underset{H}{|}}{\overset{\overset{CH_3}{|}}{C}}-MgBr$$

$$CH_3CH_2CH_2-\underset{\underset{H}{|}}{\overset{\overset{CH_3}{|}}{C}}-MgBr + CH_3CH_2-\overset{\overset{\overset{..}{O}..}{||}}{C}-H \xrightarrow[\text{2. H}^+, H_2O]{\text{1. ether}} CH_3CH_2CH_2-\underset{\underset{H}{|}}{\overset{\overset{H_3C}{|}}{C}}-\underset{\underset{:OH}{|}}{\overset{\overset{H}{|}}{C}}-CH_2CH_3$$

1-Methylbutylmagnesium   Propanal   4-Methyl-3-heptanol
bromide

NOTE. See Experiment [16] for a biography of François Auguste Victor Grignard, Nobel Laureate, who discovered and developed the Grignard reagents. This experiment also contains further details about the mechanism and use of these reagents that have had such a powerful influence on synthetic organic chemistry.

DISCUSSION

In this experiment, the addition of a nucleophilic Grignard reagent (1-methylbutylmagnesium bromide), to the electrophilic carbonyl carbon of an aldehyde (propanal) is described. The product obtained is a 2° alcohol, 4-methyl-3-heptanol. Because it is possible to vary the structure of both the

---

[19]For references relating to the preparation of Grignard reagents, see Experiment [16].

Grignard reagent and the aldehyde, a wide variety of 2° alcohols can be prepared by this route. Primary alcohols result when formaldehyde is used as the aldehyde. Secondary alcohols may also be obtained with these reagents when ethyl formate, an ester, acts as the electrophile. This latter reaction, however, requires two molar equivalents of the Grignard reagent. The mechanism for the reaction of an aldehyde with a Grignard reagent follows.

EXPERIMENTAL PROCEDURE

Estimated time to complete the experiment: 3.0 h. The chromatographic separation requires approximately an additional 15 min per student.

CAUTION: Ether is a flammable liquid. All flames must be extinguished during this experiment.

Physical Properties of Reactants						
Compound	MW	Amount	mmol	bp (°C)	d	n_D
2-Bromopentane	151.05	125 μL	1.0	117	1.2	1.4413
Diethyl ether	74.12	700 μL		34.5		
Magnesium	24.31	36 mg	1.48			
Iodine	253.81	1 crystal				
Propanal	58.08	50 μL	0.69	49	0.81	1.3636

Reagents and Equipment

Preparation of 1-Methylbutylmagnesium Bromide

NOTE. All the glassware used in the preparation of the Grignard reagent should be cleaned and dried in an oven at 110 °C for at least 30 min. After removal from the drying oven, the hot glassware should be placed in a desiccator and cooled before being assembled. Flame drying of the apparatus with a microburner annealing flame (high gas mixture) is an alternative, if ovens are not available and if ether is not in use. This latter method preferably should be carried out prior to assembly, and as with the oven-dried equipment, the hot glassware should be placed in a desiccator to cool. Care must be taken when flaming the reaction vial, as thermal shock can easily crack this heavy-walled vessel. If the flame-drying procedure is performed on the assembled apparatus, care also must be taken not to overheat the O-rings and plastic Capseals.

NOTE. Scrape a 2- to 3-in. piece of magnesium ribbon clean of oxide coating and cut it into sections 1 mm in length. This freshly cut material should be handled only with forceps.

Prepare a 3.0-mL conical vial containing a magnetic spin vane and equipped with a Claisen head fitted with a calcium chloride drying tube and a rubber septum. Weigh and place 36 mg (1.5 mmol) of magnesium in the vial, and then add a small crystal of iodine, followed by 100 μL of anhydrous ether (➡).

Alternative Procedure: Place the magnesium metal and the iodine crystal in the vial and assemble the apparatus. *Gently* warm the mixture (microburner or hot plate) until evidence of purple vapor from the iodine is seen.

NOTE. If a microburner is used in this step, all flames must be extinguished before proceeding.

Now add the 100 μL of anhydrous ether, using a 1.0-mL syringe inserted through the septum.

Prepare a solution of 125 μL (153 mg, 1.0 mmol) of 2-bromopentane in 300 μL of anhydrous diethyl ether in a dry, screw-capped vial. *Use an automatic delivery pipet to deliver these reagents.*

After the assembly has cooled to room temperature, draw the 2-bromopentane solution into a 1.0-mL syringe and then insert the syringe needle through the rubber septum on the Claisen head. Place an additional 100 μL of diethyl ether in the empty vial, cap it, and set it aside for later use.

While stirring the heterogeneous mixture, add 6–8 drops of the 2-bromopentane–ether solution to initiate the formation of the Grignard reagent. The evolution of tiny bubbles from the surface of the magnesium is evidence of reaction.

When the reaction has started, **slowly** add the remainder of the 2-bromopentane–ether solution dropwise over a 3- to 5-min period. Warm the reactants slightly. Upon completion of this addition, draw the rinse in the capped vial into the syringe and add it through the septum in a single portion to the reaction vial. Gently warm the resulting solution for 15 min.

Step I:
Mg, 36 mg + I$_2$, 1 crystal
+(CH$_3$CH$_2$)$_2$O, 500 μL
+ CH$_3$CHBr(CH$_2$)$_2$CH$_3$, 125 μL

Step II:
CH$_3$CH$_2$CHO, 50 μL
+ (CH$_3$CH$_2$)$_2$O, 200 μL

SLOWLY

CAUTION: *Do not overheat.* Overheating will cause loss of ether solvent. Small fragments of magnesium may remain at the end of the addition of the alkyl halide.

Cool the gray-colored solution of Grignard reagent to room temperature.

The Propanal Reagent. Prepare a solution of the aldehyde by weighing 50 μL (40 mg, 0.7 mmol) of propanal into a tared, oven-dried, capped vial followed by the addition of 100 μL of anhydrous diethyl ether. The propanal is the limiting reagent, so an accurate weight should be recorded for the yield calculations. *Dispense the aldehyde and diethyl ether by automatic delivery pipets in the* **hood**.

HOOD

Immediately draw the aldehyde solution into a 1.0-mL syringe and insert the syringe needle through the rubber septum on the Claisen head.

Place an additional 100 μL of the anhydrous diethyl ether in the empty vial, cap it, and set it aside for later use.

Reaction Conditions.   Now add the propanal solution *carefully*, with stirring, to the Grignard reagent over a period of about 30 s at such a rate as to keep the ether solvent at a steady reflux.

Following this addition, add the rinse in the capped vial in one portion in a similar manner.

Stir the reaction mixture for 5 min and then allow it to cool to room temperature. Remove the conical vial and cap. It is recommended that the vial be placed in a beaker to prevent tipping and loss of product.

NOTE. If the laboratory is done in two periods, you should stop at this point (recap the vial for storage) or after the hydrolysis sequence in the next step.

Isolation of Product.   Hydrolyze the magnesium alkoxide salt by the *careful*, dropwise addition of 2–3 drops of water from a Pasteur pipet. Stir the resulting mixture for 5 min. A two-phase (ether–water) reaction mixture develops as the magnesium salt is hydrolyzed.

CAUTION: The addition of water causes the evolution of heat. An ice bath should be handy to cool the solution if it begins rapid reflux.

Now add 2–3 drops of 3 M HCl. Remove the vial, cap it, and allow it to stand at room temperature for 5 min. Test the aqueous layer with litmus paper. The solution should be slightly acidic. *Too much or too little aqueous HCl will cause problems in the subsequent workup.*

Remove the magnetic spin vane with forceps and set it aside to be rinsed with an ether wash. Cap the vial tightly, shake (or use a Vortex mixer), vent *carefully*, and allow the layers to separate.

Using a Pasteur filter pipet, transfer the aqueous (lower) layer to a clean 5.0-mL conical vial. *Save* the ether layer since it contains the crude reaction product.

Now wash the aqueous layer, previously transferred to the 5.0-mL vial, with three 0.5-mL portions of diethyl ether. Rinse the magnetic spin vane with the first portion as it is added to the vial. Upon addition of each portion of ether (using a calibrated Pasteur pipet), cap the vial, shake (or use a Vortex mixer), vent carefully, and allow the layers to separate. With the aid of a Pasteur filter pipet, remove each ether layer and combine it with the ether solution retained above. After the final extraction, save the aqueous (lower) layer until you have isolated and characterized the final product. Extract the combined ether fractions with 0.5 mL of cold water to remove any acidic material. Save the aqueous rinse until you have isolated and characterized the final product.

Dry the ether solution by transferring it, using a Pasteur filter pipet, to a shortened Pasteur filter pipet containing 500 mg of anhydrous sodium sulfate. Collect the eluate in a tared 10 × 75-mm test tube. In the **hood**, remove the ether solvent from the eluate by warming in a sand bath to concentrate the solution to a weight less than 90 mg.

Purification and Characterization.   The product, 4-methyl-3-heptanol, is isolated and purified using gas chromatography.

Use a 100-μL syringe to inject the entire sample of crude material obtained above onto the GC column. Collect the components of interest as they elute from the column, using the chromatography technique described in the Prior Reading section (◄).

HOOD

Rubber septum

12-mL
Centrifuge tube

GC collect
(heavy wall)

Product vial, 0.1 mL

5/5 and threaded

100-μL conical vial

Product collects here

Cotton packing

Gas Chromatographic Conditions

GOW-MAC Series-150, Thermal Conductivity Detector
20% Carbowax 20M column, $\frac{1}{4}$ in. × 8 ft
Temperature, 145 °C
Flow rate of 50 mL/min (He gas)

The retention time for 4-methyl-3-heptanol under these conditions is 7–8 min (2 min after any other peak).

Determine the weight of the 4-methyl-3-heptanol collected and calculate the percent yield. Determine the boiling point and refractive index (optional) of the alcohol, and compare these with values in the literature.

Obtain an IR spectrum of the product as a thin film and compare it with that recorded in the literature (*The Aldrich Library of IR Spectra* and/or SciFinder Scholar). If enough of the hydrocarbon byproduct is collected, also obtain an IR spectrum to aid in its identification.

Chemical Tests. The ignition test should indicate that this compound is an aliphatic species. Does your result confirm this fact? Perform the ceric nitrate test to demonstrate the presence of the —OH group and the Lucas test to demonstrate that a secondary alcohol has been prepared. If you were required to prepare a solid derivative of this alcohol, which one would you select? It may be of interest to determine the solubility of this product in water, ether, concentrated sulfuric, and 85% phosphoric acids. Do your results agree with what you would predict? What test(s) would you perform to establish that one of the starting reagents was an aldehyde?

OPTIONAL ANALYSIS USING NMR BAND DECONVOLUTION

REACTION

Reagents and Equipment. Duplicate the procedure above in its entirety with the following modification: Replace propanal with benzaldehyde. The solution of the aldehyde will consist of 70 μL (73 mg, 0.7 mmol) benzaldehyde in a tared, oven-dried, capped vial dissolved 100 μL of anhydrous diethyl ether.

As was the case with propanal, benzaldehyde is the limiting reagent, so an accurate weight should be recorded.

NOTE. The aldehyde must be free from the corresponding acid or lower yields of product will result. Dispense the ethereal solution of aldehyde by automatic delivery pipette in the **hood**.

HOOD

Reaction Conditions and Isolation of Product. The reaction conditions, workup, and isolation steps are identical to what is detailed above. The product, 2-methyl-1-phenylpentan-1-ol, is not purified and subjected to

NMR analysis directly using the band deconvolution technique. Solutions using $CDCl_3$ as solvent for the NMR analyses were used. Standard operating procedures for performing band deconvolutions are both known and readily available for Agilent/Varian, Bruker, and JEOL (see **Bibliography**).

Applying NMR band deconvolution to this system results in an isomeric composition of 53:47 using the integrals from the diagnostic peaks of 4.53 ppm and 4.43 ppm of the respective erythro/threo isomers and the ratios ranging from 53:47 to 51:49 using the integrals obtained from the deconvoluted peaks specified in the table below.

entry	peak assignment	splitting pattern	peak (ppm)	
			major	minor
1	CH at C_1	dd	4.43	4.53
2	CH_3 at C_2	d	0.75	0.91
3	terminal CH_3 (C_5)	t	0.90	0.85

DISCUSSION

While NMR is one of the most powerful analytical tools the organic chemist has at their disposal, rarely is it the case where all the peaks generated in the ^1H and ^{13}C NMR for both the starting material and product are baseline resolved. Often, as seen in Experiment 9C, only the diagnostic regions offer clarity in structural confirmation.

This is especially true when working with diastereomeric isomers where the element of complexity can increase exponentially. Because of the equivalent yet inequivalent environments represented by each proton and carbon atom of the diastereomeric mixture, unresolved peaks create challenges if wanting to quantify let alone qualify product distribution.

For this particular sequence, reaction of 1-methylbutylmagnesium bromide with benzaldehyde to afford 2-methyl-1-phenylpentan-1-ol results in unresolved peaks when working with both the GC and NMR. Outside of the proton at C-1 of the product, an opportunity is missed when working off of the NMR spectra. NMR band deconvolution, as illustrated in Experiment 9C as well, allows the experimentalist to collect accurate and reliable quantitative data.[20]

[20]Lopansri, L. S.; Letson, J. N.; O'Brien, R. A.; Battiste, D. R.; Forbes, D. C. *World Journal of Chemical Education*, **2022,** *10*, 51.

QUESTIONS

6-115. Show how one could carry out each of the following transformations using the Grignard reaction. Any necessary organic or inorganic reagents may be used.

(a)

(b)

(c)

6-116. Explain why Grignard reagents cannot be prepared from an organic halide that also contains a hydroxyl (—OH), a carboxyl (—CO_2H), a thiol (—SH), or an amino (—NH_2) group.

6-117. What would be the final product of the reaction between methyl 4-chlorobenzoate and two equivalents of methylmagnesium bromide?

6-118. Consider the same reaction as in Question 6-117 except that in this case it is carried out with ethyl 4-chlorobenzoate. What product would be expected in this case?

6-119. Grignard reagents may be used to prepare other organometallic reagents—for example, ethylmagnesium bromide reacts with cadmium chloride to yield diethylcadmium:

$$2\ CH_3CH_2MgCl + CdCl_2 \rightarrow (CH_3CH_2)_2Cd + 2\ MgCl_2$$

Indicate the products from each of the following reactions and name each organometallic product:

$$4\ CH_3MgCl + SiCl_4 \rightarrow$$
$$2\ C_6H_5MgCl + HgCl_2 \rightarrow$$

BIBLIOGRAPHY

List of secondary alcohol preparations presented in *Organic Syntheses:*

Boeckman, R. K., Jr.; Blum, D. M.; Ganer, B.; Halvey, N. *Organic Syntheses*; Wiley: New York, 1988; Collect. Vol. VI, p. 1033.

Coburn, E. R. *Organic Syntheses*; Wiley: New York, 1955; Collect. Vol. III, p. 696.

Coleman, G. H.; Craig, D. *Organic Syntheses*; Wiley: New York, 1943; Collect. Vol. II, p. 179.

Drake, N. L. *Organic Syntheses*; Wiley: New York, 1943; Collect. Vol. II, p. 406.

Overberger, C. G.; Saunders, J. H.; Allen, R. E.; Gander, R. *Organic Syntheses*; Wiley: New York, 1955; Collect. Vol. III, p. 200.

Skattebol, L.; Jones, E. R. H.; Whiting, M. C. *Organic Syntheses*; Wiley: New York, 1963; Collect. Vol. IV, p. 792.

Trust, R. I.; Ireland, R. E. *Organic Syntheses*; Wiley: New York, 1988; Collect. Vol. VI, p. 606.

For additional examples illustrating NMR band deconvolution:

Lopansri, L. S.; Letson, J. N.; O'Brien, R. A.; Battiste, D. R.; Forbes, D. C. *World J. Chem. Educ.*, **2022**, *10*, 51.

For detailed instructions in the area of NMR band deconvolution, see:

Agilent (Varian), NMR Spectroscopy, User Guide. https://www.agilent.com/cs/library/usermanuals/public/0199934300c.pdf (accessed 10 June 2022).

Bruker, New In Bruker User Library: DREAMTIME NMR. https://www.bruker.com/en/resources/library/application-notes-mr/dreamtime_nmr.html (accessed 10 June 2022).

JEOL, Delta™ NMR Data Processing Software. https://www.jeolusa.com/PRODUCTS/Nuclear-Magnetic-Resonance/Delta-NMR-Software (accessed 10 June 2022).

EXPERIMENT 18

The Perkin Reaction: Condensation of Rhodanine with an Aromatic Aldehyde to Yield o-Chlorobenzylidene Rhodanine

Common name: *o*-chlorobenzylidene rhodanine
CA number: [6318-36-1]
CA name as indexed: 4-thiazolidinone, 5-[(2-chlorophenyl)methylene]-2-thioxo-

Purpose. This experiment explores the use of the interesting heterocyclic compound, rhodanine, as the source of an active (acidic) —CH$_2$— group. (The methylene group contained in the thiazolidinone ring system possesses the capacity to participate in base-catalyzed condensation reactions similar to those of the **aldol reaction.**) You will carry out a base-catalyzed condensation reaction with an aromatic aldehyde. You will examine the properties of this condensation product, which has the capacity to function as an intermediate in a number of synthetic pathways. Indeed, one of these routes yields the important class of aromatic amino acids, the phenylalanines.

Prior Reading

Technique 5: Crystallization
Use of the Hirsch Funnel (pp. 92–93)
Craig Tube Crystallization (pp. 93–95)

REACTION

o-Chlorobenzaldehyde   Rhodanine   o-Chlorobenzylidene rhodanine

Sir William Henry Perkin (1838–1907). Perkin came under the influence of the German chemist A. W. von Hofmann at the Royal College of Chemistry (London), where Perkin was a student. In his second year at the school, he gained the title of Hofmann's honorary assistant, and following a publication the next year, when he was 17, he was advanced to the rank of assistant. At home, he set up his own laboratory where he worked evenings and on vacations. This laboratory is where he undertook, at Hofmann's suggestion, the synthesis of quinine.

Working with his friend Arthur Church in his home laboratory, he and Church prepared one of the first azo dyes derived from naphthalene (nitrosonaphthlene). This effort resulted in his first patent (with Church). In 1856 at the age of 18, Perkin discovered the first commercially significant synthetic coal-tar dye, *mauve* or *aniline purple*. He also obtained a patent on the synthetic method for preparing this material. The method involved the treatment of an aniline salt with bichromate of potash ($K_2Cr_2O_7$) (this was an outgrowth of his quinine studies). Against Hofmann's wishes, Perkin withdrew from college, designed a factory, and started commercial production of this dye, which was marketed as Perkin's Tyrian Purple. The name *Tyrian Purple* was originally used for a prized purple dye from the Phoenician city of Tyre, which today is part of Lebanon. Tyrian Purple was produced in small quantities from a material isolated from a snail. The production of this dye became a lost art during the Dark Ages, but a species of mollusk containing this very rare pigment was rediscovered in Ireland in 1684. The natural dye is obtained by the air oxidation of a colorless fluid expressed from the glands of the snail. It required the contents of 10,000 snails to obtain a single gram of the fluid.

Perkin's dye synthesis was the beginning of the coal-tar dyestuffs industry. He later developed and manufactured magenta (violet dye) and alizarin (a red dye). The rapid acceptance of these dyes by fabric manufacturers was demonstrated by the fact that annual alizarin production reached 220 tons per year by 1871. At age 37 and a wealthy man, Perkin sold his commercial holdings and devoted the rest of his life to pure chemical research.

Perkin's later years were as productive as his earlier ventures. He developed methods for the preparation of aminoacetic acid. He established the structural relationships between tartaric, fumaric, and maleic acids, and synthesized cinnamic acid. This last endeavor led to the development of what is now known as the *Perkin reaction*. The reaction is widely used to prepare unsaturated acids from aromatic aldehydes. These studies led to his synthesis of coumarin (actually it was the first condensation product he obtained with the classic reaction; see also Experiment [3A$_{adv}$]). Other areas investigated by Perkin dealt with the relationship of physical properties and chemical structure.

Perkin was knighted in 1906 in recognition of his contributions to chemistry and to the practical application of many of his discoveries. In the United States, the Society of Chemical Industry awards a Perkin Medal each year to an outstanding industrial chemist.[21]

DISCUSSION

The classic Perkin reaction is the base-catalyzed condensation of an aromatic aldehyde with a carboxylic acid anhydride to yield and an α,β-unsaturated carboxylic acid. The initial stages of the condensation can be viewed as an aldol-type reaction (see Experiments [20], [3A$_{adv}$], [3B$_{adv}$], and [A3a]).

A variation of the Perkin reaction is the condensation of aromatic aldehydes with rhodanine, which plays a similar role to that of the anhydride in the original reaction. Rhodanine, a derivative of the thiazolidinone ring system

[21]See Edelstein, S. M. *American Dyestuff Reporter* **1956,** *45,* 598; Mendola, R. *J. Chem. Soc.* **1908,** *93,* 2214; Levinstein, H. *Chem. Ind.* **1938,** 1137; Rose, R. E. *Ind. Eng. Chem.* **1938,** *16,* 608; *Dictionary of Scientific Biography,* C. C. Gillespie, Ed., Scribner's: New York, 1974, Vol. X, p. 515.

was first synthesized in 1935 by Percy Julian (a future president of Howard University) and Bernard Sturgis (an undergraduate at DePauw University at the time). This heterocyclic molecule has an active (acidic) methylene group that can be deprotonated with a relatively mild base (in this case acetate, CH_3COO^- or AcO^-, ion) to generate the nucleophile that attacks the carbonyl group of the aldehyde.

Under the conditions used in the current experiment, dehydration-elimination rapidly follows the initial nucleophilic addition with formation of the benzylidene intermediate. The mechanism is shown here:

B   While the resulting rhodanine derivatives have exhibited antibacterial, antitubercular, antimalarial, antifungal, and antiparasitic activity, the principal focus of attention on these interesting compounds has been as reactive synthetic intermediates. For example, these particular compounds can be converted, in high yield, in three steps, to nitriles in which the side chain of the original aldehyde has been extended by an additional carbon atom:

EXPERIMENTAL PROCEDURE

Estimated time to complete the experiment: 1.5 h.

Physical Properties of Reactants and Product							
Compound	MW	Amount	mmol	mp (°C)	bp (°C)	d	n_D
Rhodanine	133.19	30 mg	0.23	170			
Sodium acetate	82.03	52 mg	0.63	324			
Acetic acid, glacial	60.05	1.0 mL			118		
o-Chlorobenzaldehyde	140.57	58 mg	0.41		212	1.25	1.5662
o-Chlorobenzylidene rhodanine	259.76			191			

Reagents and Equipment. In a 3.0-mL conical vial containing a boiling stone and equipped with an air condenser, weigh out and place 30 mg (0.23 mmol) of rhodanine and 52 mg (0.63 mmol) of anhydrous sodium acetate. Now, in the **hood**, add 1.0 mL of glacial acetic acid dispensed from a graduated pipet. To this mixture, measure and add 58 mg (0.41 mmol) of o-chlorobenzaldehyde (➡).

NOTE. Dry the sodium acetate in the oven for 1 h before use. The aldehyde must be free from the corresponding acid or lower yields of product will result. It is recommended that the purity of the aldehyde be checked by IR analysis. The reaction vial may be weighed before and after the addition of aldehyde to obtain an accurate weight.

Reaction Conditions. Heat the reaction mixture in a sand bath at 140–150 °C for 30 min.

NOTE. Immerse the vial in the sand up to the level of the top of the reaction mixture.

As the reaction progresses, the mixture becomes homogeneous and turns yellow. At the end of the reaction period, the resulting solution is cooled to room temperature. When the conical vial has reached ambient temperature, place it in an ice bath to complete crystallization of the condensation product.

Isolation of Product. Collect the yellow crystals by vacuum filtration using a Hirsch funnel. Rinse the reaction vial with two 1.0-mL portions of cold glacial acetic acid transferred by a calibrated Pasteur pipet, and then use each rinse to wash the filter cake on the Hirsch funnel (➡). Complete the removal of the acid solvent by air-drying the crystals on a porous clay plate.

Purification and Characterization. Weigh the dried product and calculate the percent yield of the crude material. Recrystallization (Craig tube) of a 5-mg portion of the benzylidene product from 0.5 mL of glacial acetic acid yields fine, bright yellow needles that can be used to complete the characterization of this interesting substance.

HOOD

140–150 °C
thermometer

Speed   Heat

Rhodanine, 30 mg
+ NaOAc, 52 mg
+ CH₃CO₂H, 1.0 mL
+ o–ClC₆H₄CHO, 58 mg
Note *It is important that the vial be immersed in sand to the level of the reaction mixture.*
Ac = acetyl group

Perkin
condensation
product
collected
here

CH₃CO₂H, 3 mL

Determine the melting point and compare your results with the value given in the table at the beginning of the experiment.

Chemical Tests. This compound is an interesting reaction product because it contains three different types of heteroatoms: chlorine, nitrogen, and sulfur. The sodium fusion test (see Chapter 9) may be used to substantiate the presence of these elements.

It would also be of interest to establish if a positive Beilstein test for chlorine is found when in the presence of sulfur and nitrogen, or whether the soda–lime test for nitrogen is observed with the elements of chlorine and sulfur also located within this molecule (see Chapter 9).

Does ignition of the material (see Chapter 9) indicate that the compound contains an aromatic ring?

150 °C
thermometer

Speed   Heat

10-mL RB flask

OPTIONAL SEMI-MICROSCALE PREPARATION

This experiment may be scaled up and carried out at a level 5 or 10 times greater than the amounts used in the above microscale experiment. The data given below are for the 10-fold procedure. The procedure is identical to that given above in the microscale section, but with the following modifications:

1. A 10-mL round-bottom flask containing a spin bar and fitted with an air-cooled condenser is used in place of the conical reaction vial (◄).

2. The reagent and solvent amounts are given in the following table.

Physical Properties of Reactants and Products							
Compound	MW	Amount	mmol	mp (°C)	bp (°C)	d	n_D
Rhodanine	133.19	300 mg	2.3	170			
Sodium acetate	82.03	520 mg	6.3	324			
Acetic acid, glacial	60.05	5.0 mL		118			
o-Chlorobenzaldehyde	140.57	580 mg	4.1		212	1.25	1.5662
o–Chlorobenzylidene rhodanine	259.76		191				

3. After the product is air-dried on a clay plate until no acetic acid (vinegar) odor remains, it is placed in a 100 °C oven to dry overnight. It may also be dried in a vacuum drying oven until no acetic acid (vinegar) odor remains.

QUESTIONS

6-120. In which of the following two species, (a) or (b), is the underlined hydrogen atom more acidic? Explain.

(a)   (b)

6-121 A number of compounds similar to rhodanine and possessing active methylene groups have been used in the Perkin condensation. Two are shown below:

(An azlactone)   Hydantoin

Draw the structure of the product that would be formed if each underwent the Perkin condensation with *p*-chlorobenzaldehyde.

6-122 para-Nitrobenzaldehyde reacts at a faster rate than benzaldehyde in the Perkin reaction, while *p-N,N*--dimethylaminobenzaldehyde is much less reactive toward the same nucleophile. Explain.

6-123 Explain the fact that the C=O group in $-\overset{\overset{\ddot{O}}{\|}}{C}-CH-$ is effective in increasing the acidity of the **α**-hydrogen atom.

6-124 A large number of condensations that are closely related to the Perkin reaction. Among these are the aldol (Experiments [20], [3B_{adv}], and [A3a]), Knoevenagel, Claisen (Experiment [3A_{adv}]), and Dieckmann condensations. What general class of compounds can be prepared using each of these well-known reactions?

BIBLIOGRAPHY

For a review on the general Perkin reaction, see:

Johnson, J. R. *Org. React.* **1942,** *1,* 210.
Koepp, E.; Vögtle, F. *Synthesis* **1987,** 177.
Rosen, T. *Comp. Org. Syn.* **1991,** *2,* 395.

For examples of the general Perkin reaction, see:

Corson, B. B. *Organic Syntheses*; Wiley: New York, 1943; Collect. Vol. II, p. 229.
Herbst, R. M.; Shemin, D. *Organic Syntheses*; Wiley: New York, 1943; Collect. Vol. II, p. 1.
Johnson, J. R. *Organic Syntheses*; Wiley: New York, 1955; Collect. Vol. III, p. 426.

Thayer, F. K. *Organic Syntheses*; Wiley: New York, 1941; Collect. Vol. I, p. 398.
Weiss, R. *Organic Syntheses*; Wiley: New York, 1943; Collect. Vol. II, p. 61.

For references using rhodanine in the Perkin reaction, see:

Andreasch, R. *Monatsh. Chem.* **1928,** *49,* 122.
Brown, F. C. *Chem. Rev.* **1961,** *61,* 463.
Campbell, N.; McKail, J. E. *J. Chem. Soc.* **1948,** 1215.
Foye, W. O.; Tovivich, P. *J. Pharm. Sci.* **1977,** *66,* 1607.
Julian, P. L.; Sturgis, B. M. *J. Am. Chem. Soc.* **1935,** *57,* 1126.

EXPERIMENT 19

Alkene Preparation by the Wittig Reaction: (E)-Stilbene; Methylene-4-tert-butylcyclohexane; and trans-9-(2-Phenylethenyl)anthracene

Common names: (*E*)-stilbene; *trans*-1,2-diphenylethene
CA number: [103-30-0]
CA name as indexed: benzene, 1,1'-(1,2-ethenediyl)bis-, (*E*)-

Common name: 1-methylene-4-*tert*-butylcyclohexane
CA number: [13294-73-0]
CA name as indexed: cyclohexane, 1-(1,1-dimethylethyl)-4-methylene-

Common name: *trans*-9-(2-phenylethenyl)anthracene
CA number: [42196-97-4]
CA name as indexed: anthracene, 9-(2-phenylethenyl)-, (*E*)-

Purpose. The conditions under which the Wittig reaction is carried out are investigated. The Wittig reaction involves the reaction of a *phosphorus ylide* (Experiments [19A] and [19C]) with an aldehyde or ketone, and is used extensively in organic synthesis to synthesize alkenes. The use of the Horner–Wadsworth–Emmons modified Wittig reaction between an aldehyde and a phosphonate ester is investigated (Experiments [19B] and [19D]) using phase-transfer catalysis.

Prior Reading

 Technique 4: Solvent Extraction
    Liquid–Liquid Extraction (p. 76)
    Drying of a Wet Organic Layer (pp. 84–87)
    Concentration of Solutions (pp. 105–108)
 Technique 5: Crystallization
    Use of the Hirsch Funnel (pp. 92–93)
    Craig Tube Crystallization (pp. 93–95)
 Technique 6: Chromatography
    Packing the Column (p. 97)
    Elution of the Column (p. 98)
    Thin-Layer Chromatography (pp. 101–103)

Georg Friedrich Karl Wittig (1897–1987). Wittig obtained his Ph.D. in 1923 at the University of Marburg under von Auwers. He later became Professor of Chemistry at the Universities of Braunschweig, Freiburg, Tübingen, and Heidelberg. His initial research work was concerned with the concept of ring strain, diradical formation, and valence tautomerism. However, he soon became involved in carbanion chemistry. Wittig discovered halogen–metal exchange reactions and then moved extensively into research involving the chemistry of the ylides. In 1953, he discovered the reactive phosphonium ylides and subsequently their reaction with aldehydes and ketones to give alkenes under very mild conditions; this reaction is now known as the *Wittig reaction*. This discovery allowed the introduction of the C=C linkage at a specific location in the product. In recognition for this work, Wittig received the Nobel Prize in 1979 (with H. C. Brown [organoboron compounds]).

In other work, Wittig postulated the dehydrobenzene intermediate and proved its existence through trapping reactions of the Diels–Alder type. He discovered sodium tetraphenylborate, which is now used in the analytical determination of potassium and ammonium ions. His later work involved the chemistry of metalated Schiff bases. This work subsequently led to the development of the concept of directed aldol condensations.

Wittig coined a large number of technical terms—valence tautomerism, ylide, onium complexes, halogen–metal exchange, dehydrobenzene, and umpolung (polarity reversal). Wittig died in 1987 at the age of 90.[22]

REACTION

$$C_6H_5CHO + [(C_6H_5)_3\overset{+}{P}CH_2C_6H_5, Br^- + NaNH_2] \xrightarrow{\text{THF}}$$

Benzaldehyde   "Instant ylide"

benzyltriphenylphosphonium
bromide–sodium amide

E-Stilbene   Triphenylphosphine oxide

DISCUSSION

The Wittig reaction constitutes a valuable method for the preparation of alkenes. The major advantages of this approach are that (1) there is no ambiguity in the location of the C=C generated by the reaction, as there can be in many elimination reactions, and (2) there is no potential for rearrangements, as there can be in E1 elimination reactions.

While the advantages clearly make this the method of choice when needing to assemble alkene functionality, this process is very poor when calculating atom economy.

An *ylide* is a neutral species whose Lewis structure contains opposite charges on adjacent atoms. The atoms involved are carbon and an element from either group 15 (VA) or 16 (VIA) of the periodic table, such as N, P, or S. The Wittig reaction uses phosphorus ylides, which are obtained by deprotonation of a phosphonium salt with a strong base. Phosphorus ylides are relatively stable, but reactive species, for which the following resonance structures may be written; the phosphorus atom can exceed an octet by accommodating electron donation into its 3d orbitals.

$$\left[R_3 \overset{+}{P} - \overset{-}{C}H_2 : \leftrightarrow R_3 P = CH_2 \right]$$

The phosphonium salts are available through a nucleophilic displacement reaction of triphenylphosphine with various alkyl halides. Triphenylphosphine is a good nucleophile, and thus most phosphonium salts are easily prepared.

$$(C_6H_5)_3P : + RCH_2X \rightarrow (C_6H_5)_3 \overset{+}{P} CH_2R, X^-$$
$$(X = I, Br, Cl)$$

The hydrogen atoms on the carbon attached to the resulting phosphorus cation are somewhat acidic, because they are adjacent to a positive charge, a significant electron-withdrawing group. Thus, treatment of the phosphonium salt with a strong base, such as butyllithium in THF or sodium hydride in DMSO, removes one of these protons and produces the ylide.

$$(C_6H_5)_3 \overset{+}{P} CH_2R, X^- + C_4H_9Li \xrightarrow{\text{THF}} (C_6H_5)_3 \overset{+}{P} - \overset{\cdot\cdot}{C}H - R + C_4H_{10} + LiX$$

An ylide

---

[22]See Parker, S. P. Ed., *McGraw-Hill Modern Scientists and Engineers*; McGraw-Hill: NY, 1980, Vol. III, p. 341; *The Annual Obituary*; St. James Press: Chicago, 1987, p. 460; Oesper, R. E. *J. Chem. Educ.* **1954**, *31*, 357.

The ylides are generally not isolated, but rather are generated in situ and reacted directly with the carbonyl compound.

The "instant ylides" are solid-phase mixtures of a phosphonium salt with sodium amide ($NaNH_2$, a strong base and the conjugate base of ammonia) and are now commercially available. In the solid phase, no reaction between the strong base and the phosphonium salt occurs. Thus, to generate the desired ylide, the "instant ylide" mixture need only be placed in a suitable solvent. This process is a marked advantage over the usual methods used to obtain these species. Ylides, because of the significant negative charge density on carbon, are nucleophilic enough to attack electrophiles as reactive as the carbon of a carbonyl group. When the ylide is reacted with an aldehyde or a ketone, an intermediate *oxaphosphetane* (a four-membered ring containing an oxygen and a phosphorus atom) is formed, which decomposes to give the alkene product. The mechanistic sequence is outlined here.

Generation of the Ylide

$(C_6H_5)_3P: + H_3C—Br \longrightarrow (C_6H_5)_3P^+—CH_3, Br^-$

Triphenylphosphine A phosphonium salt

A phosphonium salt

An ylide

$+ NH_3 + Na^+, Br^-$

An ylide

An oxaphosphetane

An alkene

+

Triphenylphosphine oxide

The Wittig reaction is very general. The significant limitation is that the carbonyl compound cannot contain acidic hydrogen atoms, such as in an alcohol or carboxyl group, or electrophiles more reactive than the aldehyde or ketone itself. The Wittig reaction can be quite stereoselective for the formation of either Z or E alkenes, but the factors involved in such stereoselectivity are sometimes difficult to predict and often difficult to explain.

An important modification (often called the Horner–Wadsworth–Emmons reaction) of the Wittig reaction makes use of phosphonate esters, $RPO(OR')_2$. It is highly stereoselective for the formation of E-alkenes. The reaction and mechanism are depicted below for the preparation of (E)-stilbene. Instead of using a phosphorus cation to stabilize the negative charge, as in the phosphonium ylide above, a phosphonate ester group is used to stabilize an adjacent carbanion.

E-Stilbene

Diethyl benzylphosphonate

E-Stilbene

The use of the phosphonate ester (Horner–Wadsworth–Emmons reaction) allows much easier separation of the product alkene, since the sodium phosphate byproduct is water-soluble; the byproduct of the Wittig reaction, triphenylphosphine oxide, is not water-soluble. In the Horner–Wadsworth–Emmons modification, a conjugated, or electron-withdrawing, substituent (such as a phenyl or carbonyl group) on the nucleophilic carbon is used to assist in the stabilization of the carbanion. This modification (Experiment [19B]) may be used as an alternative to Experiment [19A] for the preparation of (E)-stilbene. The "instant-ylide" Wittig reaction yields predominantly the E isomer of stilbene (70%). The Horner–Wadsworth–Emmons reaction yields exclusively the *E* isomer. Both procedures are given below. The synthesis described in Experiment [19D] also uses the Horner–Wadsworth–Emmons modification.

Horner–Wadsworth–Emmons reactions lend themselves to the use of *phase-transfer catalysis* in Experiments [19B] and [19D]. Phase-transfer catalysis allows the use of an aqueous base (NaOH in H_2O) with the organic compounds dissolved in an organic solvent (hexane in Experiment [19B], methylcyclohexane in Experiment [19D]) immiscible with water. The reaction system, as the name implies, involves two phases—an aqueous phase and an organic phase. The phase-transfer catalyst plays a very important role: without it, no reaction would occur, since the initial reactants (hydroxide ion and diethyl benzylphosphonate) are dissolved in different, immiscible phases, and NaOH is insoluble in hexane (or methylcyclohexane), and diethyl benzylphosphonate is insoluble in water.

The key features of the phase-transfer catalyst, Aliquat 336, are that it is a quaternary ammonium salt, with long-chain alkyl groups attached to the nitrogen.

Aliquat 336

The phase-transfer catalyst is soluble in both water and the organic solvent. It is water-soluble because it is an ion, and it is hexane-soluble because of the three long-chain alkyl groups. Thus, the phase-transfer catalyst distributes itself in both phases, and freely shuttles back and forth through the phase boundary between solvent layers. In aqueous NaOH, the chloride anion exchanges with hydroxide anion, as the counterion to the ammonium cation. When it does this, the catalyst carries the hydroxide ion from the aqueous phase, as an ion-pair, across the phase boundary into the organic phase, where the base then reacts with the diethyl benzylphosphonate. The Horner–Wadsworth–Emmons reaction then occurs, producing the alkene and diethyl phosphate anion. This anion becomes associated with the ammonium cation of the phase-transfer catalyst and is transported to the aqueous layer, where the catalyst picks up another hydroxide ion and repeats the entire process.

Experiment 19A

(*E*)-Stilbene by the "Instant Ylide" Method

CAUTION: Tetrahydrofuran is a flammable liquid. All flames must be extinguished in the laboratory when this solvent is used.

REACTION

C₆H₅CHO + [(C₆H₅)₃⁺PCH₂C₆H₅, Br⁻ + NaNH₂] →THF→

Benzaldehyde   "Instant ylide"
benzyltriphenylphosphonium
bromide–sodium amide

E-Stilbene   Triphenylphosphine oxide

EXPERIMENTAL PROCEDURE

Estimated time to complete the reaction: Two 3-h laboratory periods.

Physical Properties of Reactants						
Compound	MW	Amount	mmol	bp (°C)	d	n_D
Benzaldehyde	106.13	100 μL	0.95	178	1.04	1.5463
"Instant ylide" benzyltriphenyl phosphonium bromide–sodium amide		600 mg	~1.2			
Tetrahydrofuran	72.12	1.0 mL		67		

Step I:
$(C_6H_5)_3\overset{+}{-}P-CH_2-C_6H_5, \bar{B}r + NaNH_2$
600 mg + THF, 1.0 mL

Step II:
$C_6H_5 CHO$, 100, μL

Reagents and Equipment.   Weigh and place 600 mg (~1.2 mmol) of benzyl-triphenylphosphonium bromide-sodium amide (*instant ylide*) mixture in a dry 5.0-mL conical vial containing a magnetic spin vane. Add *freshly distilled* dry tetrahydrofuran (1.0 mL) using a calibrated Pasteur pipet, and immediately attach the vial to an air condenser protected by a calcium chloride drying tube (➡). Stir the mixture for 15 min at room temperature. During this time, the mixture turns orange. Following generation of the ylide, remove the air condenser for as short a time as possible and quickly add 100 μL (0.95 mmol) of benzaldehyde (*freshly distilled*) to the reaction flask, using an automatic delivery pipet, and immediately reattach the air condenser.

Reaction Conditions.   Stir the resulting heterogeneous mixture at room temperature for an additional 15 min. The system develops a light brown color during this time.

Isolation of Product.   Work up the reaction by adding 1.0 mL of a 25% aqueous NaOH solution (calibrated Pasteur pipet), and transfer the resulting mixture to a 15-mL centrifuge tube using a Pasteur pipet. Rinse the reaction vial with three 2.0-mL portions of diethyl ether, each of which is also transferred to the centrifuge tube (Pasteur filter pipet). Partially neutralize the resulting two-phase mixture by careful addition of 3.0 mL of 0.1 N HCl. Transfer the ether layer (top) by Pasteur filter pipet to a short microcolumn prepared from a Pasteur filter pipet containing 1.5 g of anhydrous sodium sulfate (➡).

Collect the dried eluate in a 25-mL Erlenmeyer flask containing a boiling stone. Extract the remaining aqueous layer with two additional 2.0-mL portions of ether and transfer these ether extracts, as before, to the microcolumn containing anhydrous sodium sulfate. Concentrate the eluate solution (~10 mL) to dryness using a gentle stream of nitrogen, or by warming it in a sand bath in the **hood** to give a white solid.

HOOD

Anhydrous Na₂SO₄, 1.5 g

Wittig product in $(CH_3CH_2)_2O$, 10 mL + THF, ~0.5 mL

Purification and Characterization.   Purify the crude product by chromatography, using a silica gel column.

The triphenylphosphine oxide is relatively insoluble in the hexane solvent. First, separate the triphenylphosphine oxide byproduct from the mixture of stilbenes by extracting the solid obtained above with hexane, using the following procedure.

To the 25-mL Erlenmeyer flask containing the crude product, add 2.0 mL of hexane and agitate the solution with swirling. Some breakup of the material with a microspatula may be necessary. Transfer the hexane solution by Pasteur filter pipet to a 10-mL Erlenmeyer flask containing a boiling stone. Extract the remaining crude solid with two additional 2.0-mL portions of hexane, and transfer these extracts to the 10-mL Erlenmeyer flask as before. Concentrate the hexane solution (~6 mL) to about 1.5 mL using a gentle stream of nitrogen gas, or by warming on a sand bath in the **hood**.

HOOD

Pack a short (1 × 10 cm) chromatography column with 2.0 g of activated silica gel, and premoisten the column with ligroin (60–80 °C) (◀). Add the above hexane solution directly to the column. Now elute the column with 15 mL of a 1:1 ligroin (60–80 °C)/methylene chloride solution. Collect the eluate, in a single fraction, in a tared 25-mL **filter flask** containing a boiling stone. Concentrate the solution to dryness under reduced pressure in a warm sand bath to yield the pure product mixture (◀).

- Sand, 50 mg
- SiO$_2$, 2.0 g
- Sand, 50 mg

Wittig product in 15 mL, 1:1 ligroin/CH$_2$Cl$_2$

NOTE. A 25-mL sidearm filter flask, equipped with a Hirsch funnel and filter paper disks to control the pressure, is a convenient system for the removal of a small volume (5–20 mL) of solvent. A rotary evaporator, if available, is a nice alternative.

Weigh the crude product residue and calculate the percent yield.

The isolated product mixture may be analyzed by gas chromatography and/or thin-layer chromatography. *This constitutes the second week of laboratory for this experiment.*

Gas Chromatographic Analysis. Dissolve the crude product mixture isolated above in the minimum amount of 1:1 ligroin (60–80 °C)/methylene chloride solution (~ 0.5–0.75 mL). Inject a 10-μL sample into a gas chromatograph, set up according to the following conditions:

Column: $\frac{1}{4}$-in. × 8-ft, 20% Carbowax 20M on Chromosorb P (80/100 mesh)

Temperature: 220 °C

Flow rate: 30 mL/min (He gas)

Chart speed: 1 cm/min

Thumb controls pressure; continuous shaking

Ligroin (60–80 °C)/CH$_2$Cl$_2$, 15 mL + Wittig product

The compounds elute in the following order: (Z)-stilbene, followed by (E)-stilbene. Measure the ratio of peak heights and calculate the isomeric composition of your mixture.

Thin-Layer Chromatographic Analysis. Spot a TLC plate with a sample from the product solution used for GC analysis, and also with a standard solution.

Use hexane as the elution solvent, silica gel (containing a fluorescent indicator) as the stationary phase, and UV light for visualization.

Typical R_f values are (E)-stilbene 0.21; (Z)-stilbene, 0.27.

Concentrate the ligroin–methylene chloride product mixture as before. Separate, and purify, the (E)-stilbene by recrystallization from a minimum amount of 95% ethanol, using the Craig tube.

Weigh the dried and purified (E)-stilbene product and calculate the percent yield. The purity of this material may be determined by TLC, using the conditions outlined above.

Obtain a melting point and IR spectrum of the material and compare your results with those reported in the literature.

(E)- and (Z)-Stilbene also exhibit different absorptions in the ultraviolet region. The data are summarized below:

(Z)-Stilbene: (1-mm cell)

$$\lambda_{max} 223 \text{ nm } (\varepsilon_{max} = 20{,}600 \text{ methanol}, 0.05 \text{ g/L})$$

$$\lambda_{max} 276 \text{ nm } (\varepsilon_{max} = 10{,}900, \text{methanol}, 0.1 \text{ g/L})$$

(E)-Stilbene: (0.05 g/L, 1 mm cell)

$$\lambda_{max} = 229 \text{ nm } (\varepsilon_{max} = 21{,}000, \text{ methanol})$$
$$\lambda_{max} = 294 \text{ nm } (\varepsilon_{max} = 33{,}200, \text{ methanol})$$
$$\lambda_{max} = 307 \text{ nm } (\varepsilon_{max} = 32{,}100, \text{ methanol})$$

These ultraviolet absorption data illustrate an interesting example of steric effects on the absorption pattern exhibited by geometrical isomers of an alkene. There is significant steric hindrance between the two phenyl groups in the Z isomer, which causes the phenyl groups to twist out of coplanarity with the alkene. Thus, conjugation is diminished. This result is reflected in the lower intensity of the 276-nm band as compared to the 294-nm band in the E isomer.

Chemical Tests.   Further characterization may be accomplished by performing the Br_2/CH_2Cl_2 test for unsaturation. Note that the dibromo compound is prepared in Experiment [A2_b]. It may be used here as a reference sample in the characterization of (E)-stilbene. The ignition test (Chapter 9) may be used to confirm the presence of the aromatic portion of the molecule.

(E)-Stilbene by the Horner–Wadsworth–Emmons Reaction

Experiment 19B

REACTION

$$C_6H_5CHO + (C_2H_5O)_2\overset{\cdot\ddot{O}\cdot}{\underset{\|}{P}}{-}CH_2C_6H_5 \xrightarrow[\substack{\text{hexane} \\ \text{Aliquat 336}}]{\substack{40\% \\ \text{NaOH}}} C_6H_5CH{=}CHC_6H_5 + (C_2H_5O)_2\overset{\cdot\ddot{O}\cdot}{\underset{\|}{P}}{-}O^-, Na^+$$

Benzaldehyde   Diethylbenzyl   E-Stilbene   Sodium diethyl
                              phosphonate   phosphate

EXPERIMENTAL PROCEDURE

Estimated time to complete the experiment: 2.5 h.

Physical Properties of Reactants						
Compound	MW	Amount	mmol	bp (°C)	d	n_D
Benzaldehyde	106.12	100 μL	0.98	178	1.04	1.5463
Diethyl benzylphosphonate	228.23	200 μL	0.96	106–108	1.095	
Aliquat 336 (tricaprylmethylammonium chloride)	404.17	88 mg	0.22			
Hexane	86.18	2.0 mL				
40% Sodium hydroxide		2.0 mL				

Reagents and Equipment.   Weigh and place 88 mg (100 μL) of tricaprylmethylammonium chloride (Aliquat 336) in a 10-mL round-bottom flask containing a magnetic stirrer. Add 100 μL (0.98 mmol) of benzaldehyde, 200 μL (0.96 mmol) of diethyl benzylphosphonate, 2.0 mL of hexane, and 2 mL of 40% sodium hydroxide solution. Attach the flask to a reflux condenser.

*NOTE. The benzaldehyde, diethyl benzylphosphonate, hexane, and NaOH solution are dispensed in the **hood** using automatic delivery pipets. Aliquat 336 is very viscous and is best measured by weighing. A medicine dropper is used to dispense this material. It is advisable to lightly grease the bottom joint of the condenser since strong base is being used.*

HOOD

? **Reaction Conditions.** Warm the two-phase mixture to reflux on a sand bath (temperature ~ 90–100 °C) for 1 h. Stir the reaction mixture vigorously during this period. Allow the resulting solution to cool to nearly room temperature. Crystals of product may appear as cooling occurs. Confirm by TLC (plates containing a fluorescent indicator) that the reaction is complete using as a solvent system ethyl acetate: hexane (0.1:9.9) and a UV lamp for visualization. The R_f value of benzaldehyde (0.4) is lower than both the R_f values of *E*- and *Z*-stilbene (◀).

Isolation of Product. Once all the evidence shows that benzaldehyde is no longer present, add methylene chloride (700 µL), which will dissolve any crystalline material that may have formed. Then, using a Pasteur pipet, transfer the contents of the round-bottom flask to a 15-mL centrifuge tube. Rinse the flask with an additional 300 µL of methylene chloride and transfer the rinse to the same centrifuge tube. Remove the aqueous layer carefully, using a Pasteur filter pipet, and save it in a vial until you have successfully isolated and characterized the product.

NOTE. Use care when determining which layer is the "aqueous layer." At room temperature, the density of a 40% sodium hydroxide solution is 1.40 g/mL, whereas the density of methylene chloride is 1.33 g/mL.

Wash the organic layer with two 1-mL portions of water. Stir the mixture with a small glass rod after each addition (or a Vortex mixer may be used) and then remove the water layer and add it to the one collected before. Dry the methylene chloride solution by adding a small amount of anhydrous sodium sulfate. Use a Pasteur filter pipet to transfer the dried solution to a 10-mL Erlenmeyer flask. Wash the sodium sulfate remaining in the centrifuge tube with two 1-mL portions of methylene chloride. Remove these washings, using the Pasteur filter pipet, and transfer them to the same Erlenmeyer flask.

Concentrate the solution, which contains the desired product, to dryness on a warm sand bath under a stream of nitrogen.

Purification and Characterization. The (*E*)-stilbene obtained is, in most cases, sufficiently pure for characterization. However, this should be confirmed by thin-layer chromatography as outlined earlier (see Purification and Characterization, Experiment [19A]). If only a trace of the *Z* isomer is detected, recrystallize the product directly, using absolute ethanol. Collect the recrystallized material by vacuum filtration using a Hirsch funnel. Maintain the vacuum for an additional 10 min to partially dry the crystalline product. Then place it on a clay plate, or on filter paper, and allow it to dry thoroughly. As an alternative, the product may be dried in a vacuum drying oven (or pistol) for 10–15 min at 30 °C (1–2 mm Hg).

Weigh the dried (*E*)-stilbene product and calculate the percent yield. The (*E*)-stilbene may be characterized as outlined in Experiment [19A].

90–100 °C
thermometer

Speed   Heat

C₆H₅CHO, 100 µL
diethyl benzylphosphonate, 200 µL,
+ Aliquat 336, 88 mg
40% NaOH, 2 mL, hexane, 2 mL
10–mL RB flask

Experiment 19C

Methylene-4-*tert*-butylcyclohexane

REACTION

4-*tert*-Butylcyclohexanone $\quad+\quad$ [(C₆H₅)₃P̟CH₃, Br⁻ + NaNH₂] $\xrightarrow{\text{THF}}$ Methylene-4-*tert*-butylcyclohexane $\quad+\quad$ (C₆H₅)₃PO

4-*tert*-Butylcyclohexanone

"Instant ylide"
methyltriphenylphosphonium
bromide–sodium amide

Methylene-
4-*tert*-butylcyclohexane

Triphenylphosphine
oxide

EXPERIMENTAL PROCEDURE

Estimated time to complete the experiment: 4.0 h.

Physical Properties of Reactants and Product						
Compound	MW	Amount	mmol	mp (°C)	bp (°C)	n_D
4-*tert*-Butylcyclohexanone	154.26	100 mg	0.64	50		
"Instant ylide" (methyltriphenylphosphonium bromide–sodium amide)		320 mg	~0.72			
Tetrahydrofuran		1.0 mL			67	
Methylene-4-*tert*-butylcyclohexane	151.27				185	1.4630

Reagents and Equipment.   In a dry 5.0-mL conical vial containing a magnetic spin vane, and equipped with an air condenser protected by a calcium chloride drying tube, weigh and place 320 mg (~0.72 mmol) of methyltriphenylphosphonium bromide–sodium amide (instant ylide) mixture (➡). Now add *freshly distilled* tetrahydrofuran (1.0 mL), using a calibrated Pasteur pipet, and stir the mixture for 15 min at room temperature. During this period it turns a bright yellow color.

Following generation of the ylide, weigh and add 100 mg (0.64 mmol) of 4-*tert*-butylcyclohexanone to the reaction flask.

Reaction Conditions.   Stir the resulting heterogeneous mixture, at room temperature, for an additional 90 min. The mixture develops a light-tan color over this period of time.

Isolation of Product.   Work up the reaction by adding 1.0 mL of a 25% aqueous NaOH solution (calibrated Pasteur pipet), and then transfer the resulting mixture to a 12-mL centrifuge tube using a Pasteur filter pipet. Rinse the reaction flask with three 2.0-mL portions of diethyl ether. Transfer each rinse to the same centrifuge tube. Partially neutralize the resulting two-phase system by the careful addition of 2.0 mL of 0.1 N HCl and mix briefly.

Separate the ether layer by Pasteur filter pipet and place it in a 25-mL Erlenmeyer flask. Extract the remaining aqueous layer with three additional 5-mL portions of diethyl ether. Separate these extracts as before and combine them with the original ether layer.

Transfer the combined ether fractions, by Pasteur filter pipet, to a short microcolumn prepared from a Pasteur filter pipet containing 1.5 g of anhydrous sodium sulfate (➡). Collect the dried eluate in a 25-mL Erlenmeyer flask containing a boiling stone. Concentrate this solution to dryness, using a gentle stream of nitrogen, or by warming on a sand bath in the **hood**, to yield a colorless liquid residue.

Step I:
$(C_6H_5)_3\overset{+}{P}$-CH_3, $\bar{B}r$ + $NaNH_2$,
320 mg + THF, 1.0 mL
Step II:
4-*tert*-Butycyclohexanone, 100 mg

HOOD

Anhydrous
Na_2SO_4, 1.5 g

Wittig product in
$(CH_3CH_2)_2O$, ~ 21 mL
+ THF, ~ 0.5 mL

HOOD

Purification and Characterization.   Purify the crude product isolated above by chromatography on an alumina column.

Pack a short (1 × 10 cm) buret column with 4.0 g of activated basic alumina and premoisten the alumina with ligroin. Dissolve the crude material in 0.5 mL of 9:1 ligroin (60–80 °C)/methylene chloride solvent and transfer the solution to the column using a Pasteur filter pipet. Elute the product in a single fraction using 8.0 mL of 9:1 ligroin (60–80 °C)/methylene chloride. Collect the eluate in a tared 25-mL filter flask containing a boiling stone. Evaporate the solvent in the **hood** under vacuum, with swirling, in a warm sand bath, leaving a liquid residue. Product loss may occur if the concentrated residue is heated excessively.

Al$_2$O$_3$, 4.0 g

Sand, 50 mg

Wittig product in 8 mL,
9:1 ligroin/CH$_2$Cl$_2$

The product is of sufficient purity for characterization. Weigh the methylene-4-*tert*-butylcyclohexane and calculate the percent yield. Determine the boiling point and compare your result with that listed in the Reactants and Product table.

Obtain an IR spectrum of the alkene product, using the capillary film technique. Compare your result with Figure 6.32.

Infrared Analysis. The spectrum (Fig. 6.8 on page 186) of the ketone was discussed in Experiment [5B].The major change observed in the spectrum of the product (Fig. 6.32), when compared with the starting material, is the loss of the carbonyl absorption band (1717 cm^{-1}), and its replacement by a less-intense new band at 1654 cm^{-1}. This latter absorption is associated with the stretching motion of the C═C system exocyclic to the six-membered ring. The key bands associated with group frequencies present in the reaction product are identified at 3075, 3000–2850, 1783, 1654, 1394, 1368, and 883 cm^{-1}.

The presence of these bands confirms the projected structure of the Wittig reaction product. The data may be interpreted as follows: the bands at 3075 (sharp spike), 1783 (weak), 1654 (sharp-medium), and 883 (strong) cm^{-1} form a frequency train that is defined as a "terminal alkene macrofrequency." The overall interpretation requires the presence of all four bands in the spectrum if the specific assignment of any one of them is to be correct. Thus, these four

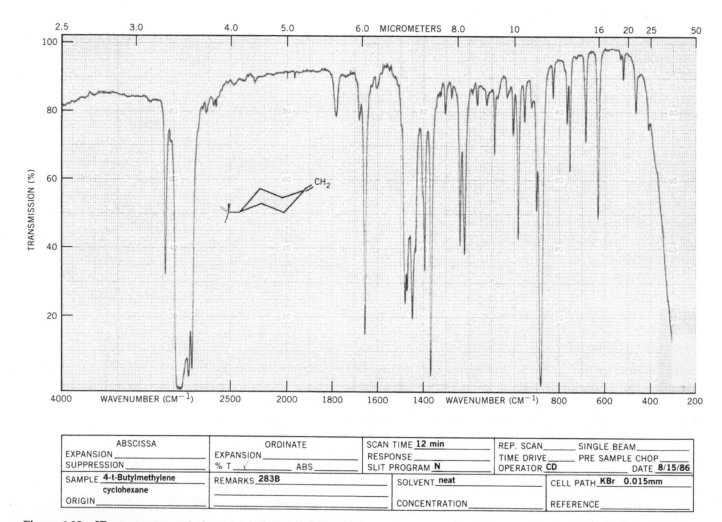

Figure 6.32   IR spectrum: methylene-4-*tert*-butylcyclohexane.

data points lead to a structural interpretation with very high confidence limits. The 3075-wavenumber band arises from the coupled antisymmetric stretch of the two C—H oscillators on the terminal alkenyl methylene group. The weak 1783-wavenumber peak is an overtone of the strong band observed near 883 cm^{-1} (883 × 2 = 1766 cm^{-1}) and is unusually intense for an isolated harmonic. The fundamental can be assigned as the =CH$_2$ out-of-plane wag (C—H deformation). In this fairly rare example, the observed harmonic occurs at a higher wavenumber value than twice the fundamental frequency. A situation of this type is termed *negative anharmonicity* (see Infrared Discussions). Finally, the C=C stretching mode is assigned to the 1654-wavenumber band, which is consistent with the requirements of this frequency train as it is found below 1660 cm^{-1} (see Infrared Discussions of cis, terminal, or vinyl carbon–carbon double bonds). The bands identified at 1396 and 1368 cm^{-1} indicate that the tertiary butyl group, as expected, has been preserved during the conversion of the ketone to a terminal alkene (see also Discussion, Experiment [5B]).

Thumb controls pressure; continuous shaking

9:1 Ligroin/CH$_2$Cl$_2$, 8 mL
+ THF ~ 0.5 mL
+ Wittig product

trans-9-(2-Phenylethenyl)anthracene

Experiment 19D

REACTION

9-Anthraldehyde + (C$_2$H$_5$O)$_2$P—CH$_2$C$_6$H$_5$ (Diethylbenzyl phosphonate) $\xrightarrow[\text{methylcyclohexane}]{\text{40\% NaOH} \atop \text{Aliquat 336}}$ (C$_2$H$_5$O)$_2$P—Ö:$^-$, Na$^+$ + (Sodium diethyl phosphate) *trans*-9-(2-Phenylethenyl)-anthracene

EXPERIMENTAL PROCEDURE

Estimated time to complete the reaction: 2.5 h.

Physical Properties of Reactants and Product							
Compound	MW	Amount	mmol	mp (°C)	bp (°C)	d	n_D
9-Anthraldehyde	206.24	106 mg	0.51	104–105			
Diethyl benzylphosphonate	228.23	120 μL	0.58		106–108 @ 1mm	1.095	1.4970
Aliquat 336 (tricaprylmethyl ammonium chloride)	404.17	2 drops					
Methylcyclohexane	98.2	2.0 mL			101	0.77	1.4215
40% Aqueous sodium hydroxide *trans*-(2-phenylethenyl) anthracene	280.4	2.0 mL		130–132			

Reagents and Equipment. Weigh and place 106 mg (0.51 mmol) of 9-anthraldehyde in a 10-mL round-bottom flask containing a magnetic stirrer. Add 2 drops of Aliquat 336 (tricaprylmethylammonium chloride), 120 μL

(0.58 mmol) of diethyl benzylphosphonate, 2.0 mL of methylcyclohexane, and 2 mL of 40% sodium hydroxide solution. Attach the vial to a reflux condenser (◄).

HOOD

*NOTE. The diethyl benzylphosphonate, methylcyclohexane, and NaOH solution are dispensed in the **hood** using automatic delivery pipets. The Aliquat 336 is very viscous. A medicine dropper is used to dispense this material. It is advisable to lightly grease the bottom joint of the condenser, since strong base is being used.*

Reaction Conditions. Heat the two-phase mixture, on a sand bath at a temperature of about 125–130 °C, for 45 min. Stir the reaction mixture vigorously during this period; the upper (organic) layer turns deep red. Allow the mixture to cool to room temperature.

Isolation of Product. Transfer the two-phase solution to a 15-mL glass centrifuge tube using a Pasteur filter pipet. Rinse the flask with 1 mL of methylene chloride and transfer this rinse to the centrifuge tube. Carefully remove the aqueous layer, using a Pasteur filter pipet, and save it in a vial until you have successfully isolated and characterized the product. Wash the organic layer with two 2-mL portions of water. Stir the mixture with a small glass rod after each addition (or a Vortex mixer may be used) and then remove the water layer and save it in the same vial as before. Dry the methylene chloride solution by adding sodium sulfate. Using a Pasteur filter pipet, transfer the dried solution to a tared 10-mL Erlenmeyer flask containing a boiling stone. Wash the sodium sulfate remaining in the centrifuge tube with two 1-mL portions of methylene chloride. Also transfer these washings to the Erlenmeyer flask.

Concentrate the solution, which contains the desired product, almost to dryness on a sand bath under a slow stream of nitrogen gas. Now add approximately 1–2 mL of 2-propanol (isopropanol) to the flask and allow the resulting solution to stand at room temperature for 10–15 min, and then place it in an ice bath to complete the crystallization of the product.

Purification and Characterization. Collect the yellow crystals by vacuum filtration, using a Hirsch funnel, and wash the filter cake with two 1-mL portions of *cold* methanol. Maintain the vacuum for an additional 10 min to partially dry the crystalline product. Then place the material on a clay plate, or on filter paper, and allow it to dry thoroughly. As an alternative, the product may be dried in a vacuum drying oven (or pistol) for 10–15 min at 30 °C (1–2 mm).

The purity of the product may be checked using *thin-layer chromatography* (see Technique 6A). Dissolve a small amount of the starting aldehyde and the product in ethanol and apply a sample to the TLC plate. Use toluene as the elution solvent, silica gel (containing a fluorescent indicator) as the stationary phase, and UV light for visualization. Typical R_f values are 0.92 for *trans*-9-(2-phenylethenyl)anthracene and 0.50 for 9-anthraldehyde.

Weigh the product and calculate the percent yield. A portion of the material may be further purified by recrystallization from 2-propanol using the Craig tube.

Determine the melting point and compare it with the value given in the Reactants and Product table. Obtain the IR spectrum of the alkene product. The material should have a strong absorption band at 962 cm^{-1}, which confirms the presence of a trans double bond.

125–130°C
thermometer

Speed   Heat

9-Anthraldehyde, 106 mg,
+ diethylbenzyl phosphonate, 120 µL,
+ Aliquat 336, 2 drops, 40% NaOH, 2 mL
+ methylcyclohexane, 2 mL

Figure 6.33   UV–visible spectrum: *trans*-9-(2-phenylethenyl)anthracene.

This material has a characteristic UV spectrum in methanol. The following data were obtained at a concentration of 3.1×10^{-6} M (see Fig. 6.33):

$$\lambda_{max} \, 255 \, \text{nm} \; (\varepsilon_{max} = 35,207, \; \text{methanol})$$

$$\lambda_{max} \, 383 \, \text{nm} \; (\varepsilon_{max} = 3618, \; \text{methanol})$$

QUESTIONS

6-125. Complete each of the following reactions by giving a suitable structure for the species represented by the letters.

(a)

$(Ph)_3\overset{\oplus}{P}$—CH$\overset{\ominus}{Cl}$(Ph) $\xrightarrow[\text{EtOH}]{\text{NaOEt}}$ **A**

A + [benzaldehyde] $\longrightarrow$ **B** + $(Ph)_3PO$

(b)

$(Ph)_3\overset{\oplus}{P}$—CH$_3$ $\overset{\ominus}{Cl}$ $\xrightarrow[\text{DMSO}]{\text{NaH}}$ **A**

A + H_3C—C(=O)—CH$_3$ $\longrightarrow$ **B** + $(Ph)_3PO$

6-126. Why is it important that any aldehyde used in the Wittig reaction be free of carboxylic acid impurities?

6-127. Reaction of triphenylphosphine with benzyl bromide produces the corresponding phosphonium salt. Suggest a suitable mechanism for this reaction.

6-128. Heteroatoms other than P are also capable of stabilizing the negative charge on C to yield ylides. For example, nitrogen is capable of forming such a system:

$$\left[\begin{array}{c} (C_6H_5)_3\overset{+}{P}=\overset{..}{C}H_2 \\ \updownarrow \\ (C_6H_5)_3P=CH_2 \end{array} \right] \qquad (CH_3)_3\overset{+}{N}=\overset{..}{C}H_2$$

A phosphorus ylide   A nitrogen ylide

Why are resonance structures not drawn for the nitrogen ylide as in the phosphorus system?

6-129. Would you expect that the sulfonium salt, $(C_6H_5)_2SCH_3{}^+,Br^-$, is capable of forming an ylide when reacted with a strong base? If so, would its structure be best represented in a manner resembling the P or N ylide? Explain.

6-130. Explain why the C==C stretching mode gives rise to a rather weak IR band in 1-methylcyclohexene, while in its isomer, methylenecyclohexane, the band is of medium to strong intensity.

6-131. Calculate and then compare the atom economy percentages for Experiments 19A–D. A challenge exercise would ask for a comparison of the atom economy percentages for all the scheduled experiments performed in your organic laboratory experience.

6-132. Predict the C=C stretching frequencies of the alkenes formed when cyclopentanone, cyclobutanone, and cyclopropanone undergo the Wittig reaction with methyltriphenylphosphonium bromide–sodium amide reagent.

6-133. What concentration of a given compound (mol/L) in methanol, having a molecular weight of 165, would have been used if the solution gave an absorbance of 0.68 with a calculated $\varepsilon_{max} = 14{,}800$?

BIBLIOGRAPHY

Of the many articles on the Wittig reaction, four are listed:

Maryanoff, B. E.; Reitz, A. B. *Chem. Rev.* **1989,** *89*, 863.

Silversmith, E. F. *J. Chem. Educ.* **1986,** *63*, 645.

Vedejs, E.; Marth, C. F. *J. Am. Chem. Soc.* **1990,** *112*, 3905.

Wadsworth, W. S., Jr. *Org. React.* **1977,** *25*, 73.

Selected references pertaining to the Horner–Wadsworth–Emmons modification of the Wittig reaction:

Boutagy, J.; Thomas, R. *Chem. Rev.* **1974,** *74*, 87.

Denmark, S. C.; Chen, C.-T. *J. Am. Chem. Soc.* **1992,** *114*, 10674.

Horner, L.; Hoffmann, H.; Wippel, H. G.; Klahre, G. *Chem. Ber.* **1959,** *92*, 2499.

Wadsworth, W. S., Jr.; Emmons, W. D. *J. Am. Chem. Soc.* **1961,** *83*, 1733.

Below are selected examples of the Wittig reaction in *Organic Syntheses*:

Campbell, T. W.; McDonald, R. N. *Organic Syntheses*; Wiley: New York, 1973; Collect. Vol. V, p. 985.

Jorgenson, M. J.; Thacher, A. F. *Organic Syntheses*; Wiley: New York, 1973; Collect. Vol. V, p. 509.

Lang, R. W.; Hansen, H. *Organic Syntheses*; Wiley: New York, 1990; Collect. Vol. VII, p. 232.

McDonald, R. N.; Campbell, T. W. *Organic Syntheses*; Wiley: New York, 1973; Collect. Vol. V, p. 499.

Nagata, W.; Wakabayashi, T.; Hayase, Y. *Organic Syntheses*; Wiley: New York, 1988; Collect. Vol. VI, p. 358.

Wadsworth, W. S., Jr.; Emmons, W. D. *Organic Syntheses*; Wiley: New York, 1973; Collect. Vol. V, p. 547.

Wittig, G.; Schöllkopf, U. *Organic Syntheses*; Wiley: New York, 1973; Collect. Vol. V, p. 751.

For a historical perspective of the Wittig reaction, see:

Hofmann, R. W. "Wittig and His Accomplishments: Still Relevant Beyond His 100th Birthday." *Angew. Chem. Int. Ed.* **2001,** *40*, 3915.

EXPERIMENT 20

The Aldol Reaction: trans-4-Nitrochalcone and Dibenzalacetone

Common name: 4-nitrochalcone
CA number: [2960-55-6]
CA name as indexed: 2-propen-1-one, 3-(4-nitrophenyl)-1-phenyl-, (E)-

Common names: dibenzalacetone, dibenzylideneacetone
CA number: [35225-79-7]
CA name as indexed: 1,4-pentadien-3-one, 1,5-diphenyl-, (*E*,*E*)-

Prior Reading

Technique 4: Solvent Extraction
          Liquid–Liquid Extraction (p. 76)
Technique 5: Crystallization
          Introduction (pp. 89–91)
          Use of the Hirsch Funnel (pp. 92–93)
          Craig Tube Crystallization (pp. 93–95)

Purpose. The synthetically useful **aldol reaction** is investigated as a method of forming carbon–carbon bonds. It is a general reaction of aldehydes that may also be extended to ketones. The specific case outlined in this experiment is known as the Claisen–Schmidt reaction as the enolate generated is reacting with an aryl carbonyl compound lacking any α-hydrogen atoms.

DISCUSSION

The aldol reaction (aldol condensation) is one of the fundamental reactions of organic chemistry, since it leads to the formation of a new carbon–carbon bond and is broadly applicable. A condensation reaction is one in which two molecules are joined with the concomitant expulsion of a small stable molecule, usually water or an alcohol. The aldol reaction may be used to condense various combinations of aldehydes and ketones. The mixed aldol condensation of an aldehyde having no α-hydrogen atom with a ketone is specifically known as the Claisen–Schmidt reaction. This variation of the aldol condensation is illustrated here in the synthesis of dibenzalacetone.

The reaction conditions of this aldol condensation favor the formation of the product, dibenzalacetone. This product is insoluble in the aqueous ethanol solvent and precipitates from the reaction as it is formed, whereas the starting materials and the intermediate, benzalacetone, are all soluble in aqueous ethanol. These experimental conditions assist in driving the equilibrium reaction to completion.

The aldol condensation involves generation of an *enolate* by removal of an acidic proton from a carbon to the carbonyl group of an aldehyde or ketone, and subsequent nucleophilic addition of this enolate to the carbonyl carbon of an aldehyde or ketone. The reaction is usually base-catalyzed and involves several mechanistic steps.

$$R-\overset{\overset{\ddot{O}\!:}{\|}}{C}-CH_3 + :\ddot{O}H^- \rightleftharpoons \left[R-\overset{\overset{\ddot{O}\!:}{\|}}{C}-\bar{C}H_2: \longleftrightarrow R-\overset{\overset{:\ddot{O}:^-}{|}}{C}=CH_2 \right] + H_2O$$

(an enolate)

$$R'-\overset{\overset{\ddot{O}\!:}{\|}}{C}-H + :\bar{C}H_2-\overset{\overset{\ddot{O}\!:}{\|}}{C}-R \rightleftharpoons R'-\overset{\overset{:\ddot{O}:^-}{|}}{\underset{H}{C}}-CH_2-\overset{\overset{\ddot{O}\!:}{\|}}{C}-R \xrightarrow{HOH} R'-\overset{\overset{:\ddot{O}H}{|}}{\underset{H}{C}}-CH_2-\overset{\overset{\ddot{O}\!:}{\|}}{C}-R$$

(nucleophilic attack)   (protonation)

The reaction involves several steps: (1) base-catalyzed generation of an enolate, (2) nucleophilic attack of this anion on a carbonyl carbon, and (3) protonation of the resulting anion to yield the initial aldol product, a β-hydroxy carbonyl compound. Note that each step in the sequence is in equilibrium, and the entire reaction is thus reversible. Treatment of the β-hydroxy carbonyl compound with base causes the reverse aldol (**retro-aldol**) reaction to occur.

$$:\ddot{O}H^- + R'-\overset{\overset{:O:}{|}}{\underset{R'}{C}}-CH_2-\overset{\overset{:O:}{\|}}{C}-R \longleftrightarrow R'-\overset{\overset{\ddot{O}\!:}{\|}}{\underset{R'}{C}} + CH_2=\overset{\overset{:\ddot{O}:^-}{|}}{C}-R + HOH$$

$$R'-\overset{\overset{\ddot{O}\!:}{\|}}{\underset{R'}{C}} + CH_3-\overset{\overset{\ddot{O}\!:}{\|}}{C}-R + HO^-$$

The β-hydroxy carbonyl product may be isolated in most cases, if desired, as the subsequent dehydration is generally much slower than the addition reaction that precedes it. The final stage, as in the present reaction, is a hydroxide-catalyzed dehydration of this initial product by way of its enolate. Though hydroxide ion (HO⁻) is generally not a good leaving group, the hydrogen α to the ketone is quite acidic, the elimination produces a rather stable and conjugated α, β-unsaturated ketone, and under strongly basic conditions the hydroxide ion is an adequate leaving group.

Enolate anion

NOTE. This reaction may be carried out in a 10 × 75-mm test tube. However, the reagents must be stirred efficiently with a glass rod at frequent intervals. If a larger test tube is used, a small magnetic stirring bar or vane is more efficient as an agitator.

In the present reaction, a double aldol condensation occurs, which yields the dibenzalacetone product. An additional example of the aldol reaction is shown in Experiment [A3ₐ], where tetraphenylcyclopentadienone is prepared.

Experiment 20A

trans-4-Nitrochalcone

REACTION

4-Nitrobenzaldehyde   Acetophenone   *trans*-4-Nitrochalcone

EXPERIMENTAL PROCEDURE

Estimated time to complete the experiment: 3.0 h.

Physical Properties of Reactants						
Compound	MW	Amount	mmol	mp (°C)	bp (°C)	*d*
Acetophenone	120.16	488 μL	4.16	20.5	202.6	
Ethanol (95%)		20 mL			78.5	0.789
4-Nitrobenzaldehyde	151.12	500 mg	3.31	106		
Sodium hydroxide (aq., 10%)		2 mL				

NOTE. Upon isolation, purification, and characterization, this material can be part of a sequence yielding a photochromic substance. A detailed experimental procedure on this sequence (Sequence F) can be found online.

Reagents and Equipment.   Weigh and place 500 mg (3.31 mmol) of 4-nitrobenzaldehyde in a 50-mL Erlenmeyer flask containing a magnetic stir bar. Now add 20 mL of 95% ethanol and 488 μL of acetophenone.

Reaction Conditions.   Warm the reaction mixture, while stirring in a sand bath at 65–70 °C, until the aldehyde dissolves to yield a clear light yellow solution.

At this point, cool the Erlenmeyer flask in an ice bath for at least 5 min and then, while continuing to cool the system, *carefully* add, dropwise with stirring, 2 mL of 10% (aq.) sodium hydroxide over a second 5-min period. During this time, the reaction mixture often turns a dark orange color and considerable precipitation may occur. Cool the Erlenmeyer flask for an additional 15 min following the last addition of base.

Isolation of Product.   Collect the solid tan precipitate, which formed during the reaction, on the filter bed of a Hirsch funnel under reduced pressure (➡).

If the solid continues to form in the filtrate, refilter the reaction solution and combine the second collection of crystals with the first batch. You want to maximize your total yield of the aldol product, because this is the first step of a four-step synthesis. Thus, you will need efficient recovery of product at each intermediate stage of the synthesis to successfully obtain a reasonable quantity of the photochromic target molecule. Rinse the Erlenmeyer flask once or twice with ice-cold water to effect as closely as possible a quantitative transfer of the chalcone to the Hirsch funnel.

Purification and Characterization.   Wash the tan filter cake containing the reaction product, dropwise with an ice-cold 80:20 ethanol/water solution until the product appears as pale yellow crystals. The wash dissolves and removes a red-brown amorphous material that contaminates the crude product in many instances. Transfer the purified and partially dried chalcone to a watch glass, and then place it in a desiccator for final drying. Characterize the anhydrous intermediate accurately by an evacuated melting point and infrared spectrum; the latter can be compared to that recorded in the literature (*The Aldrich Library of IR Spectra* and/or SciFinder Scholar).

Nitrochalcone product

20 mL ethanol }+
−2 mL 10% NaOH

Dibenzalacetone

Experiment 20B

REACTION

$$2 \text{ } C_6H_5{-}\overset{\overset{\ddot{O}}{\|}}{C}{-}H + CH_3{-}\overset{\overset{\ddot{O}}{\|}}{C}{-}CH_3 \xrightarrow[\text{C}_2\text{H}_5\text{OH}]{\text{NaOH, H}_2\text{O}} C_6H_5{-}CH{=}CH{-}\overset{\overset{\ddot{O}}{\|}}{C}{-}CH{=}CH{-}C_6H_5$$

Benzaldehyde   Acetone   Dibenzalacetone

EXPERIMENTAL PROCEDURE

Estimated time to complete the experiment: 2.0 h.

Physical Properties of Reactants						
Compound	MW	Amount	mmol	bp (°C)	d	n_D
Benzaldehyde	106.13	80 μL	0.79	178	1.04	1.5463
Acetone	58.08	29 μL	0.40	56	0.79	1.3588
NaOH catalyst solution		1 mL				

NOTE. It is recommended that the purity of the benzaldehyde be checked by IR. The presence of benzoic acid in the benzaldehyde can substantially lower the yield of product. The benzaldehyde may be purified by distillation under reduced pressure (bp 178–179 °C; 57–59 °C @ 8 torr).

The benzaldehyde may be added to the vial by weight, or by volume (using an automatic delivery pipet). To prevent loss of acetone by evaporation, fit the reaction vial with a septum cap and cool it in an ice bath. Add the acetone (by volume) through the septum using a GC syringe.

Stoichiometric quantities of the reagents are used. An excess of benzaldehyde results in a more intractable product; excess acetone favors the formation of benzalacetone.

Reagents and Equipment.
In a 3.0-mL conical vial containing a magnetic spin vane and equipped with an air condenser place 80 μL (84 mg, 0.79 mmol) of benzaldehyde and 29.0 μL (23 mg, 0.40 mmol) of acetone (◀).

Now add to this reaction mixture 1.0 mL of the aqueous, ethanolic sodium hydroxide catalyst solution, delivered from a calibrated Pasteur pipet.

CAUSTIC

C₆H₅CHO, 80 μL, + acetone, 29 μL + ethanolic NaOH, 1.0 mL

*INSTRUCTOR PREPARATION. The catalyst solution is prepared by dissolving 0.4 g of sodium hydroxide (which is **caustic**) in 4.0 mL of water. To this solution add 3.0 mL of 95% ethanol.*

Reaction Conditions.
Stir the reaction mixture at room temperature for 30 min. During this time the solid yellow product precipitates from solution.

Isolation of Product.
Collect the crude yellow dibenzalacetone by vacuum filtration using a Hirsch funnel (◀). Remove the magnetic spin vane from the reaction vial with forceps. *Some of the product adheres to the magnetic spin vane. This material should be removed by carefully scraping the vane with a microspatula. The material is added to the product collected by filtration.*

Wash the filter cake with three 1.0-mL portions of water. The filtrate should be nearly neutral, as indicated by pH test paper. If not, repeat the washing until the test indicates that the filtrate is neutral.

Aldol product collected here

H₂O/CH₃CH₂OH, ~ 5 mL

NOTE. It is essential to remove the NaOH completely. If it is not removed, the recrystallization step will be difficult.

Air-dry the product by maintaining the suction on the Hirsch funnel for approximately 10 min. During this operation a piece of plastic food wrap may be placed over the mouth of the Hirsch funnel to aid in the drying process (see Prior Reading).

NOTE. An alternative procedure may be used to isolate and purify the dibenzalacetone. Transfer the reaction product directly to a large Craig tube. The washings and the ethanol recrystallization step (see Purification and Characterization section) may then be carried out with no transfer of the material. In this way, loss of product is minimized.

Purification and Characterization.
The crude dibenzalacetone may be purified by recrystallization from 95% ethanol, using a Craig tube.

Weigh the dried product and calculate the percent yield. Determine the melting point and compare your result with those in the literature.

Obtain an infrared spectrum of the material and compare your spectrum with that shown in Figure 6.34. A comparison of the infrared spectra of the starting reagents with that of the product is given below.

Figure 6.34   IR spectrum: dibenzalacetone.

Infrared Analysis

Acetone: This simple aliphatic ketone possesses the short macro group frequency train: 3415 (overtone of C=O stretch, 2 × 1712 = 3424 cm^{-1}), 3000–2850 (sp^3 C—H stretch), 1712 (C=O stretch), and 1360 cm^{-1} (symmetric methyl bend α to a carbonyl. See discussion of methyl bends in acetates in Experiment [8B]).

Benzaldehyde: The infrared spectrum of this aromatic aldehyde (Fig. 6.35) is rich and interesting. The aromatic aldehyde macro group frequency consists of the following peaks: 3070, 2830 and 2750, 1706, 1602, 1589, and 1396 cm^{-1}.

a. **3070 cm^{-1}:** C—H stretch on the sp^2 carbon.
b. **2830 and 2750 cm^{-1}:** This pair of bands is another example of Fermi coupling (see discussion in Experiment [7] and Infrared Discussions).
c. **1706 cm^{-1}:** The carbonyl stretch of the aldehyde group. The frequency observed in aliphatic aldehydes falls in the range 1735–1720 cm^{-1}, but when conjugated, the value drops (see Infrared Discussions) and is found in the range 1720–1700 cm^{-1}.

d. **1602 and 1589 cm^{-1}:** This pair of bands is related to the degenerate ring stretching vibrations ν_{8a} and ν_{8b} of benzene (see Infrared Discussions in Chapter 10, Experiment [1B$_{adv}$] and Infrared Discussions).

e. **1396 cm^{-1}:** The aldehyde C—H in-plane bending vibration. The first harmonic of this vibration is Fermi coupled to the aldehyde C—H stretching mode.

Benzaldehyde also possesses a second powerful macro group frequency train that reflects the substitution pattern of the aromatic ring system (see Infrared Discussions). The monosubstituted benzene ring macro group frequency train contains peaks in the following regions: 1950, 1880, 1800, 1730, 750, and 690 cm^{-1}.

a. **1990, 1920, 1830, and 1770 cm^{-1}:** This series of four weak bands with generally decreasing intensity from high-to-low wavenumber values (the third peak near 1800 cm^{-1} may be intensified if the ring is conjugated as in this case) arise from combination bands (see Infrared Discussions), which involve the out-of-plane bending frequencies of the ring C—H bonds (see below). The exact wavenumber positions are not very important, but the overall shape of the pattern can be used to determine the ring substitution pattern (see Infrared Discussions).

b. **750 and 690 cm^{-1}:** This pair of bands is very characteristic of monosubstituted phenyl groups. The 750-wavenumber peak arises from the all-in-phase out-of-plane bending vibration of the five C—H groups adjacent to each other on the ring (see Infrared Discussions). The 690 cm^{-1} companion band involves an out-of-plane displacement of the ring carbon atoms (see Infrared Discussions).

Dibenzalacetone: The infrared spectrum of the product (Fig. 6.35) contains many features of the starting materials, plus new and shifted bands unique to the newly formed structure.

a. The monosubstituted aromatic macro group frequency remains: 1960, 1890, 1815, 1770, 760, and 690 cm^{-1}. In addition, the two pairs of degenerate ring stretching bands are present: 1598, 1580 and 1500, 1450 cm^{-1}. The 1500-wavenumber band is rather weak in benzaldehyde, but intensifies in the product.

b. The ketone carbonyl remains the most intense band in the spectrum but is shifted to 1653 cm^{-1} because of the carbonyl's conjugation to two aromatic rings. The aldehyde's carbonyl stretching mode at 1706 cm^{-1} has vanished.

c. The methyl C—H stretching bands between 3000 and 2850 cm^{-1}, and the coupled aldehyde C—H stretching peaks at 2830 and 2750 cm^{-1} have disappeared.

d. New bands appear at 3058 and 3035 cm^{-1} (alkene C—H, stretch) which are overlapped with the aromatic ring C—H stretching) peaks (3075 cm^{-1}). In addition, bands appear at 1627 (conjugated C=C) and 985 cm^{-1} (*trans*-substituted C=C, C—H in-phase out-of-plane bend). The latter band occurs slightly above its usual location near 965 cm^{-1}. This rise in frequency is the result of conjugation of the double bond to the carbonyl group. For an example of a much more dramatic rise in frequency, refer to the discussion of the cis C—H bending modes in the spectrum of maleic anhydride (Experiment [14]).

Sample _Benzaldehyde_

%T _X_ ABS _ Background Scans _4_   Scans _16_
Acquisition & Calculation Time _42 sec_   Resolution _4.0 cm^{-1}_
Sample Condition _liquid, neat_   Cell Window _KBr_
Cell Path Length _capillary film_   Matrix Material _____

Figure 6.35   IR spectrum: benzaldehyde.

If performed in a potassium bromide matrix, examine the spectrum of your product. Discuss the similarities and differences of the experimentally derived spectral data to the reference spectra (Figs. 6.34 and 6.35).

Dibenzalacetone is a compound that can be characterized by classical chemical tests.

Chemical Tests.   An ignition test (See Table 9.1, p. 633) should indicate that dibenzalacetone contains an aromatic ring system. Perform the test to confirm this.

Several classification tests might also be of assistance in classifying this compound. Does the 2,4-dinitrophenylhydrazine test for an aldehyde or ketone give a positive result? Isolate the 2,4-dinitrophenylhydrazone derivative and determine its melting point. Does it correspond to the literature value of 180 °C? What further test could be run to determine whether the carbonyl is present as an aldehyde or ketone?

A test for unsaturation should be enlightening. Would you perform the bromine–methylene chloride or the Baeyer test (Chapter 9, Classification Tests). Did the correct test give a positive result?

OPTIONAL SEMI-MICROSCALE PREPARATION

This experiment may be scaled up to be carried out at 5 or 10 times the amounts used in the above micropreparation. The data summarized below are for the 10-fold procedure.

The procedure is identical to the above with the following exceptions.

1. Use a 10-mL round-bottom flask fitted with an air condenser (◄).
2. Increase the reagent and solvent amounts.

Room temp.

Speed   Heat

10-mL RB flask

Properties of Reactants						
Compound	MW	Amount	mmol	bp (°C)	d	n_D
Benzaldehyde	106.13	800 µL	7.9	178	1.04	1.5463
Acetone	58.08	300 µL	4.0	56	0.79	1.3588
NaOH catalyst solution		5 mL				

3. The collected filter cake is transferred to a 10-mL beaker, stirred with 5.0 mL of water, and then recollected by vacuum filtration. Repeat this process, usually about three times, until the filtrate is neutral to litmus paper.

QUESTIONS

6-134. In a number of cases, it is possible to successfully isolate the β-hydroxyketone intermediate prior to the dehydration that forms the α,β-unsaturated ketone.
   (a) For this case, does the isolation of the β-hydroxyketone suggest which step in the aldol condensation is the rate-determining step?
   (b) Suggest what reaction conditions may have a significant impact on determining which step becomes rate determining.
   (c) What structural changes might lead to a change in the rate-determining step?

6-135. Ketones also undergo the aldol condensation, although a successful reaction often requires "enhanced" conditions, since the addition involves an unfavorable equilibrium constant. This is the situation in the reaction in which 4-nitrochalcone is synthesized. With the odds against it, why is the reaction successful in this case?

6-136. If you had obtained both the *cis*- and *trans*-chalcone products, and had purified them by recrystallization, how could you instantly know which one was cis and which one was trans without any further characterization?

Trans   Cis

6-137. The mixed aldol reaction between propionaldehyde (propanal) and acetone gives an 85% yield of 4-hydroxy-2-hexanone when run in THF at −78 °C with lithium diisopropylamide (LDA, a powerful base). The reaction is carried out by first adding the ketone to the base, cooling the solution, and then adding the aldehyde.

    (a) Why does this mixed system give essentially a single product?

    (b) Why is there no self-condensation of the acetone?

    (c) Why does the system not rapidly go on to dehydrate?

    (d) Why is the ketone added to the base rather than vice versa?

6-138. Aldol condensations are extremely practical if one of the aldehydes (or ketones) has no α-hydrogen atoms. This is the case for Experiments 20A and 20B, which involved reactive carbonyls bearing no α-hydrogen atoms and arguably the primary reason for why you observed a high yielding process. When working with systems with multiple α-hydrogen atoms and similar reactivity, both "crossed" and "self" condensations are possible. Provide all possible aldol products from the following reaction:

$$CH_3CH_2CH_2CH_2CH_2CHO + CH_3CH_2CH_2CH_2CHO \xrightarrow{\text{base}}$$

6-139. Both of the systems below are high yielding and have multiple α-hydrogen atoms. Comment on why with so many α-hydrogen atoms only one system is generated. One approach to consider is to outline a suitable mechanism to account for all cyclization routes and compare the resulting annulations.

(a)

(b)

BIBLIOGRAPHY

Review articles:

Cowden, C. J.; Paterson, I. *Org. React.* **1997,** *51,* 1.

Mahrwald, R. *Modern Aldol Reactions,* Vols. I, II; Wiley-VCH: New York, 2004.

Mestres, R. "A Green Look at the Aldol Reaction." *Green Chemistry;* **2004,** *12,* 586.

Mukaiyama, T. *Org. React.* **1982,** *28,* 203.

Mukaiyama, T.; Kobazachi, S. *Org. React.* **1994,** *46,* 1.

Patai, S.; Rappoport, Z., Eds. *The Chemistry of the Enones*; Wiley: New York, 1989, Part I, Part II.

Smith, M. B. *March's Advanced Organic Chemistry*, 8th ed.; Wiley: New York, 2019.

Examples of the aldol and Claisen–Schmidt reactions:

Auerbach, R. A.; Crumrine, G. S.; Ellison, D. L.; House, H. O. *Organic Syntheses*; Wiley: New York, 1988; Collect. Vol. VI, p. 692.

Conrad, C. R.; Dolliver, M. A. *Organic Syntheses*; Wiley: New York, 1943; Collect. Vol. II, p. 167.

Hill, G. A.; Bramann, G. M. *Organic Syntheses*; Wiley: New York, 1941; Collect. Vol. I, p. 81.

Kohler, E. P.; Chadwell, H. M. *Organic Syntheses*; Wiley: New York, 1941; Collect. Vol. I, p. 78.

Leuck, G. J.; Cejka, L. *Organic Syntheses*; Wiley: New York, 1941; Collect. Vol. I, p. 283.

Russell, A.; Kenyon, R. L. *Organic Syntheses*; Wiley: New York, 1955; Collect. Vol. III, p. 747.

Wittig, G.; Hesse, A. *Organic Syntheses*; Wiley: New York, 1988; Collect. Vol. VI, p. 901.

G

EXPERIMENT 21

Quantitative Analysis of Grignard Reagents: 1-Methylbutylmagnesium Bromide and Phenylmagnesium Bromide

Common names: 1-methylbutylmagnesium bromide,
   2-pentylmagnesium bromide
CA number: [57325-22-1]
CA name as indexed: magnesium, bromo(1-methylbutyl)-
Common name: phenylmagnesium bromide
CA number: [100-58-3]
CA name as indexed: magnesium, bromophenyl-

Purpose.   In this experiment you will generate a Grignard reagent, a common and synthetically useful source of a nucleophilic carbanion, and use an aqueous titration method to determine the amount prepared.

Prior Reading

   Moisture-Protected Reaction Apparatus (pp. 27–28)

REACTION

CH$_3$CH$_2$CH$_2$—C(CH$_3$)(H)—Br + Mg $\xrightarrow{\text{ether}}$ CH$_3$CH$_2$CH$_2$—C(CH$_3$)(H)—MgBr

2-Bromopentane   Magnesium   1-Methylbutylmagnesium bromide

or

—Br + Mg $\xrightarrow{\text{ether}}$ —MgBr

Bromobenzene   Phenylmagnesium bromide

then

CH$_3$CH$_2$CH$_2$—C(CH$_3$)(H)$\overset{\delta^-\ \delta^+}{:}$MgBr + HÖH $\longrightarrow$ CH$_3$CH$_2$CH$_2$CH$_2$CH$_3$ + Mg(OH)Br

1-Methylbutylmagnesium bromide   Pentane

or

—$\overset{\delta^-\ \delta^+}{}$MgBr + H—Ö—H $\longrightarrow$ —H + Mg(OH)Br

Benzene

DISCUSSION

This experiment demonstrates the formation of a Grignard reagent (1-methyl-butylmagnesium bromide or phenylmagnesium bromide) and a titration method by which the amount of the reagent prepared can be analyzed.

The discovery by Victor Grignard in 1900 that organic halides react with magnesium metal to give organomagnesium compounds was a landmark in organic chemistry. Grignard reagents are among the most useful and versatile reagents in organic synthesis.

The reaction of the Grignard reagent with water is the basis of the analytical method used in this experiment.

$$RMgX + HOH \longrightarrow RH + Mg(OH)X$$

In the Grignard reagent, the carbon atom bound to the electropositive magnesium atom has a high negative charge density, which is responsible for the strong nucleophilic and basic character exhibited by this organometallic reagent. The carbon, acting as a base, can abstract even a weakly acidic proton from protic reagents, such as water, carboxylic acids, alcohols, and so on. In this process, the corresponding hydrocarbon (the conjugate acid of the R group carbanion) and the basic magnesium halide species are produced. This reaction sequence can be used in the laboratory as a synthetic method to convert organohalides to hydrocarbons. In the examples given in this experiment, 1-methylbutylmagnesium bromide would yield pentane, while phenylmagnesium bromide gives benzene upon protonation.

Titration of the Mg(OH)X species with standardized acid solution makes it possible to determine the amount of Grignard reagent originally formed in the solution.

$$2\,Mg(OH)X + H_2SO_4 \longrightarrow + 2\,HOH + MgSO_4 + MgX_2$$

An excess of the sulfuric acid is generally added to ensure that the Mg(OH)X is completely reacted. The excess acid is then neutralized with standard sodium hydroxide solution. The difference between the total amount of sulfuric acid used and the amount of sodium hydroxide required corresponds to the number of equivalents of the acid actually used to neutralize the Mg(OH)X species. This value is then directly related to the equivalents of Grignard reagent by the reactions given above.

Preparation of the Grignard Reagent[23]

CAUTION: Ether is a flammable liquid. Extinguish all flames during this experiment.

NOTE. All the glassware used in the experiment should be cleaned, dried in an oven at 110 °C for at least 30 min, and then cooled in a desiccator before use.

EXPERIMENTAL PROCEDURE

Estimated time to complete the experiment: 1.5 h.

---

[23]For references relating to the preparation of Grignard reagents, see Experiment [16].

Part 1 1-Methylbutylmagnesium Bromide

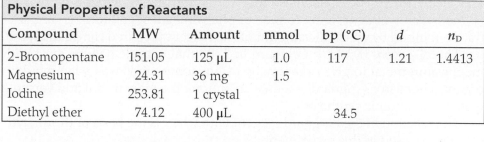

Physical Properties of Reactants						
Compound	MW	Amount	mmol	bp (°C)	d	n_D
2-Bromopentane	151.05	125 μL	1.0	117	1.21	1.4413
Magnesium	24.31	36 mg	1.5			
Iodine	253.81	1 crystal				
Diethyl ether	74.12	400 μL		34.5		

Mg, 36 mg + I$_2$, 1 crystal
+ (CH$_3$CH$_2$)$_2$O, 400 μL +
CH$_3$CHBr(CH$_2$)$_2$CH$_3$, 125 μL

Prepare the Grignard reagent exactly as described in Experiment [16]. The reagents, amounts of reagents, order of addition, workup manipulations, and precautions are the same. The equipment is also identical (◄).

Cool the gray-colored Grignard reagent mixture to room temperature, and then assay it by the titration method outlined in Part 2.

Part 2 Phenylmagnesium Bromide[24]

Physical Properties of Reactants						
Compound	MW	Amount	mmol	bp (°C)	d	n_D
Bromobenzene	157.02	76 μL	0.72	156	1.50	1.5597
Diethyl ether	74.12	1.3 mL		34.5	0.73	
Magnesium	24.3	17.5 mg	0.73			
Iodine		1 crystal				

NOTE. If not used for analysis, the solution of Grignard reagent may be treated with various reagents to prepare a wide variety of compounds. For example, see Experiments [16], [17], and [4$_{adv}$].

Analysis of the Grignard Reagent: Place 10 mL of freshly boiled distilled water and one drop of phenolphthalein indicator in a 50-mL Erlenmeyer flask. Using a syringe, transfer the cool Grignard reagent solution to the Erlenmeyer flask. Rinse the reaction vial with 0.5 mL of diethyl ether and add the rinse to the Erlenmeyer flask.

NOTE. The addition of water to the Grignard reagent results in the hydrolysis of this reagent to form the corresponding hydrocarbon and a basic magnesium halide, Mg(OH)X. The water is initially boiled to remove any dissolved carbon dioxide that might interfere with the titration.

Make sure that in the transfer of the Grignard reagent solution all small pieces of unreacted magnesium are excluded.

Analysis by Titration: Add, from a 10-mL buret, 5 or 6 mL of standard 0.2 N H$_2$SO$_4$ solution to the ethereal Grignard solution. *The resulting solution should be acidic and colorless. If not, add an additional portion of the acid.*

---

[24]This Grignard reagent is prepared exactly as described in Experiment [16]. The reagents, amount of reagents, order of addition, workup manipulations, and precautions are the same. The equipment is also identical.

Add a boiling stone to the flask and heat the mixture at a sand bath temperature of 90–95 °C in the **hood** for 5 min.

HOOD

While the solution is still warm, add a drop of phenolphthalein indicator solution and neutralize the excess acid by back titration with 0.1 M NaOH solution. Back titration produces a very light-colored, pink end point. *It may be necessary to add an additional drop of acid, and then more base, to get the best possible end point.*

Data and Calculations: The difference between the initial and final buret readings is the volume of standard acid and base used in the titration of the Grignard reagent.

From the data, calculate the equivalents of Grignard reagent formed. Also, as a percentage, determine the amount of Grignard reagent analyzed compared to its theoretical yield of formation.

QUESTIONS

6-140. Technical grade ether often contains ethanol. Would you recommend this material as a suitable solvent for the preparation of Grignard reagents? If not, why not?

6-141. It is likely that the amount of Grignard reagent your analysis indicates was formed is greater than the amount of Grignard reagent actually present just before you added water. Explain.

6-142. What hydrocarbon would you expect to obtain by the action of water on each of the Grignard reagents listed below?
   (a) Pentylmagnesium bromide
   (b) Phenylmagnesium bromide
   (c) Isopropylmagnesium bromide
   (d) *tert*-Butylmagnesium bromide

6-143. What product would each of the Grignard reagents in Question 6-142 yield when treated with D_2O?

6-144. (a) What product would each of the Grignard reagents listed in Question 6-142 yield when treated with ethanol?
   (b) With isopropyl alcohol?
   (c) Explain why 4-hydroxycyclohexanone is not a viable candidate when considering Grignard reagents?

6-145. The solubility of Grignard reagents in ether plays a crucial role in their formation. The reagents are soluble because the magnesium is coordinated to the ether oxygen in a Lewis acid–base interaction. Each ether molecule donates an electron pair to the magnesium to complete an octet.

$$(CH_3CH_2)_2\ddot{O}: \rightarrow \overset{\overset{\displaystyle R}{|}}{\underset{\underset{\displaystyle Br}{|}}{Mg}} \leftarrow :\ddot{O}(CH_2CH_3)_2$$

Grignard reagents are normally insoluble in hydrocarbon solvents. However, they can be rendered soluble by the addition of a tertiary amine to the hydrocarbon–Grignard reagent mixture. Explain.

BIBLIOGRAPHY

For the many references related to the preparation and use of the Grignard reagent cited in *Organic Syntheses*, see the Reaction Indexes in Collected Volumes I–X under Grignard Reactions

EXPERIMENT
22

Williamson Synthesis of Ethers

Common names: propyl *p*-tolyl ether, 4-propoxytoluene
CA number: [5349-18-8]
CA name as indexed: benzene, 1-methyl-4-propoxy-

Common names: methyl *p*-ethylphenyl ether, *p*-ethylanisole
CA number: [1515-95-3]
CA name as indexed: benzene, 1-ethyl-4-methoxy-

Common names: butyl *p*-nitrophenyl ether
CA number: [7244-78-2]
CA name as indexed: benzene, 1-butoxy-4-nitro-

Purpose. The conditions under which ethers are prepared are explored by the well-known Williamson ether synthesis. You will prepare alkyl aryl ethers by S_N2 reactions of alkyl halides with substituted phenoxide anions. The use of phase-transfer catalysis is demonstrated.

Prior Reading

> *Technique 4:* Solvent Extraction
> Liquid–Liquid Extraction (p. 76)
> *Technique 6:* Chromatography
> Column Chromatography (pp. 96–99)
> *For Optional Scaleup:* Separatory Funnel Extraction (pp. 79–81)

REACTION

DISCUSSION

The compounds whose preparations are described in Experiments [22A], [22B], [22C], and [22D] are alkyl aryl ethers. The general method of preparation is the Williamson synthesis, an S_N2 reaction specifically between a phenoxide ion (ArO⁻) nucleophile and an alkyl halide. This reaction is often used for the synthesis of symmetrical and unsymmetrical ethers where at least one of the ether carbon atoms is primary or methyl, and thus amenable to an S_N2 reaction. Elimination (E2) is generally observed if secondary or tertiary halides are used, since phenoxide ions are also bases.

The conditions under which these reactions are conducted lend themselves to the use of phase-transfer catalysis. The reaction system involves two phases: the aqueous phase and the organic phase. In the present case, the alkyl halide reactant acts as the organic solvent, as does the product formed. The phase-transfer catalyst plays a very important role. In effect, it carries the phenoxide ion, as an ion-pair, from the aqueous phase, across the phase boundary into the organic phase, where the S_N2 reaction then occurs.

The ether product and the corresponding halide salt of the catalyst are produced in this reaction. The halide salt then migrates back into the aqueous phase, where the halide ion is exchanged for another phenoxide ion, and the process repeats itself. The catalyst can play this role, since the large organic groups (the four butyl groups) allow the solubility of the ion-pair in the organic phase, while the charged ionic center of the salt renders it soluble in the aqueous phase. For further discussions of phase-transfer catalysis, see Experiments [19B] and [19D].

In the reactions described below, the mechanism is a classic S_N2 process, and involves a backside nucleophilic attack of the phenoxide anion on the alkyl halide.

It is of interest to contrast the acidity of phenols with that of simple alcohols. A phenol is more acidic than an alcohol. In a typical aliphatic alcohol (e.g., ethanol) loss of the proton forms a strong anionic base, alkoxide ion (ethoxide ion).

$$R-CH_2-OH \rightleftharpoons H^+ + R-CH_2-O^-$$
Alkoxide ion

The strongly basic characteristics of the alkoxide species are due to the fact that the negative charge is localized on the oxygen atom. Ethanol has a $pK_a = 16$. In contrast, the conjugate base of a phenol can delocalize its negative charge.

Phenoxide ion

The phenoxide ion is stabilized by this resonance delocalization; therefore, it is a weaker base than the alkoxide ion. Conversely, the phenol is a stronger acid than a typical aliphatic alcohol. Phenol has a $pK_a = 10$ and is thus 1 million times more acidic than ethanol.

Propyl p-Tolyl Ether

Experiment 22A

The reaction for Experiment [22A] is shown above.

EXPERIMENTAL PROCEDURE

Estimated time to complete the experiment: 2.5 h.

Physical Properties of Reactants

Compound	MW	Amount	mmol	mp (°C)	bp (°C)	d	n_D
p-Cresol	108.15	160 µL	1.56	32–34	202	1.02	1.5312
25% NaOH solution		260 µL					
Tetrabutylammonium bromide	322.38	18 mg	0.056	103–104			
Propyl iodide	169.99	150 µL	1.54		102	1.75	1.5058

Reagents and Equipment.   Weigh and place 160 µL (168 mg, 1.56 mmol) of *p*-cresol in a 5.0-mL conical vial containing a magnetic spin vane. Now add 260 µL of 25% aqueous sodium hydroxide and thoroughly mix the resulting solution (◄). To this solution weigh and add the tetrabutylammonium bromide (Bu₄N⁺Br⁻) catalyst (18 mg), followed by 150 µL (262 mg, 1.54 mmol) of propyl iodide. Immediately attach the vial to a reflux condenser.

HOOD

NOTE. Warm the cresol in a hot water bath to melt it. Dispense this reagent and the propyl iodide in the **hood** *using an automatic delivery pipet.*

CAUTION: Propyl iodide is a suspected carcinogen.

100–115 °C
thermometer

Speed   Heat

p-CH₃C₆H₄OH, 160 µL +
25% NaOH, 260 µL +
Bu₄N⁺ Br⁻, 18 mg
+ PrI, 150 µL

Reaction Conditions.   Place the reaction vessel in a sand bath and stir vigorously at 110–115 °C for 45–60 min.

Isolation of Product.   Cool the resulting two-phase mixture to room temperature, and remove the spin vane with forceps. Rinse the spin vane with 1.0 mL of diethyl ether, adding the rinse to the two-phase mixture. Cap the vial, agitate, vent, and transfer the bottom aqueous layer, using a Pasteur filter pipet, to a 3.0-mL conical vial. A Vortex mixer, if available, can be used to good advantage in this extraction step. Wash this aqueous fraction with 1.0 mL of diethyl ether. Save this and all subsequent aqueous fractions together in a small Erlenmeyer flask until your final product has been isolated and characterized. Now transfer this diethyl ether wash to the 5-mL conical vial containing the ether solution of the product. Extract the resulting ether solution with a 400-µL portion of 5% aqueous sodium hydroxide solution. Cap the vial, agitate, vent, and remove and save the bottom aqueous layer, using a Pasteur filter pipet. Wash the product-ether solution with 200 µL of water. Remove, and save, the aqueous phase to obtain the crude, wet ether solution of the product. Add a boiling stone to the vial and concentrate the solution in a warm sand bath under a gentle stream of nitrogen to isolate the crude product.

Anhydrous
Na₂SO₄, 50 mg

SiO₂, 500 mg

Sand, 50 mg

Ether product
in CH₂Cl₂,
~ 2.0 mL

Purification and Characterization.   The crude product is purified by chromatography on silica gel. Prepare a microchromatographic column by placing 500 mg of activated silica gel in a Pasteur filter pipet, followed by 50 mg of anhydrous sodium sulfate (◄). Dissolve the crude product in 250 µL of methylene chloride and transfer the resulting solution to the dry column by use of a Pasteur pipet. Elute the material with 2.0 mL of methylene chloride and collect the eluate in a tared 3.0-mL conical vial containing a boiling stone. Evaporate the solvent by placing the vial in a sand bath maintained at a temperature of 60–65 °C.

Weigh the pure propyl *p*-tolyl ether and calculate the percent yield. Determine the boiling point and density (optional) and compare the experimental values with those in the literature.

Obtain an IR spectrum of the compound and compare it to that shown in Figure 6.36 for 4-propoxytoluene (propyl *p*-tolyl ether).

Nuclear Magnetic Resonance Analysis. For this and every experiment conducted, it is without question to the benefit of the student to obtain "more" experimental data than "less" for any material generated within the time frame permitted. The reason is simple, any postlab will benefit by having a deeper evidence narrative supporting the claims of a successful experience, which in organic chemistry is the acquisition of data.

For this sequence, obtaining both ^{1}H and ^{13}C NMR spectra of your propyl *p*-tolyl ether in CDCl$_3$ and comparing your spectra with those in Figures 6.37 and 6.38 will be crucial when confirming product formation.

As discussed in your lecture section, when working with deuterated solvents, you need to factor the extraneous peaks. There are two possibly three extraneous peaks in the ^{1}H spectrum: the small singlet at 7.26 ppm is due to residual CHCl$_3$ in the CDCl$_3$, and the small singlet at 1.55 ppm is probably due to a trace amount of H$_2$O in either the sample or the NMR solvent. The third peak depends upon if an internal standard is part of the NMR solvent. The most common standard is tetramethylsilane (0.00 ppm). There are possibly two extraneous peaks in the 13C spectrum: the 1:1:1 triplet at 77 ppm in the ^{13}C spectrum is from the CDCl$_3$ solvent and the second is dependent upon the internal standard, if present, you will see a peak at 0 ppm.

Since the ^{1}H spectrum is entirely first-order, it can be readily interpreted. You should be able to use the splitting patterns to assign peaks to each of the different groups of protons in the molecule. The integration can assist you.

Chemical Tests. Qualitative chemical tests can also be used to assist in characterizing this compound as an ether. Perform the ignition test (Table 9.1) to determine whether the material contains an aromatic ring.

A key factor to investigate is the solubility characteristics of this material (see Chapter 9). Determine its solubility in water, 5% sodium hydroxide, 5% hydrochloric acid, concentrated sulfuric acid, and 85% phosphoric acid. Do the results place this compound in the solubility class of an ether containing less than 8, or more than 8, carbon atoms? Does the ferrox test (Chapter 9) confirm the presence of oxygen in the compound?

OPTIONAL MACROSCALE PREPARATION

This ether may be prepared on a larger scale ($\sim$ 100-fold increase) using a procedure similar to that just outlined, with the following modifications.

1. Use a 100-mL round-bottom flask containing a magnetic stirrer and equipped with a reflux condenser ($\rightarrow$).

2. The reagent and solvent amounts are summarized in the following table. Note that propyl bromide is used in place of propyl iodide at this scale.

reflux
thermometer

Speed   Heat

Reflux conditions
100-mL RB flask

Physical Properties of Reactants							
Compound	MW	Amount	mmol	mp (°C)	bp (°C)	*d*	n_D
p-Cresol	108.15	16.3 g	0.15	32–34	202	1.02	1.5312
NaOH	40.0	6.05 g	0.15				
Tetrabutylammonium bromide	322.38	0.41 g	0.0012	103–104			
Propyl bromide	122.99	12.71 g	0.1		71	1.35	1.4341
Water		25 mL					

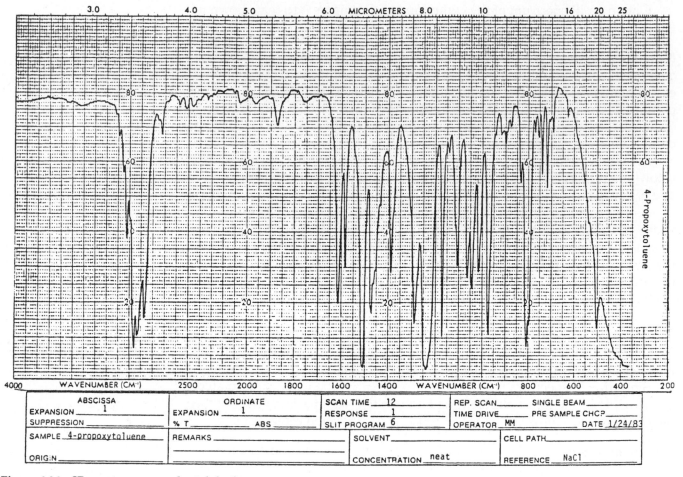

ABSCISSA	ORDINATE	SCAN TIME ___12___	REP. SCAN_____ SINGLE BEAM _____
EXPANSION ___1___	EXPANSION ___1___	RESPONSE ___1___	TIME DRIVE_____ PRE SAMPLE CHOP_____
SUPPRESSION _____	% T _____ ABS _____	SLIT PROGRAM _6_	OPERATOR __MM_____ DATE _1/24/83_
SAMPLE _4-propoxytoluene_	REMARKS _____	SOLVENT_____	CELL PATH_____
ORIGIN _____		CONCENTRATION _neat_	REFERENCE __NaCl__

Figure 6.36   IR spectrum: propyl p-tolyl ether.

300 MHz NMR SPECTRUM OF PROPYL p-TOLYL ETHER IN CDCl₃

Figure 6.37   ¹H-NMR spectrum: propyl-p-tolyl ether.

Figure 6.38 ^{13}C-NMR spectrum: propyl *p*-tolyl ether.

3. Stir the reaction mixture at reflux temperature for 90 min.

Isolation of Product. Allow the solution to cool to room temperature, transfer it to a 125-mL separatory funnel, and remove the aqueous layer. Store the aqueous layer in a 125-mL Erlenmeyer flask. Wash the organic layer successively with 5% NaOH solution (20 mL) and distilled H_2O (20 mL). After each washing, remove the aqueous phase and add it to the 125-mL Erlenmeyer flask, which should be kept until the final product has been purified and characterized. Collect the remaining deep-red organic layer in a 125-mL Erlenmeyer flask, and dry it over anhydrous sodium sulfate.

Remove the drying agent by filtration through a glass wool plug and collect the product ether in a tared container. Weigh, calculate the percent yield, and then purify and characterize a small amount of the material as described in the microscale procedure.

| Experiment 22B |

Methyl *p*-Ethylphenyl Ether

REACTION

$$CH_3CH_2-\!\!\!\!\langle\ \rangle\!\!\!\!-\ddot{O}H + CH_3-I \xrightarrow[(C_4H_9)_4N^+, Br^-]{NaOH} CH_3CH_2-\!\!\!\!\langle\ \rangle\!\!\!\!-\ddot{O}-CH_3 + Na^+, I^-$$

4-Ethylphenol   Methyl iodide   Methyl *p*-ethylphenyl ether

EXPERIMENTAL PROCEDURE

Estimated time to complete the experiment: 2.5 h.

Physical Properties of Reactants							
Compound	MW	Amount	mmol	mp (°C)	bp (°C)	d	n_D
4-Ethylphenol	122.17	150 mg	1.2	42–45			
25% NaOH solution		250 µL					
Tetrabutylammonium bromide	322.38	15 mg	0.047	103–104			
Methyl iodide	141.94	90 µL	1.45		41–43	2.28	1.5304

60–65 °C
thermometer

HOOD

p-CH₃CH₂C₆H₄OH, 150 mg
+ 25% NaOH, 250 µL +
Bu₄N⁺ Br⁻, 15 mg
CH₃I, 205 mg

Speed   Heat

Reagents and Equipment. Use the same apparatus as in Experiment [22A] for this synthesis. Weigh and add 150 mg (1.2 mmol) of 4-ethylphenol to the reaction vial, followed by 250 µL of 25% aqueous sodium hydroxide solution (◄). Stir the mixture at room temperature until dissolution occurs. The phase-transfer catalyst (tetrabutylammonium bromide (Bu₄N⁺Br⁻), 15 mg, 0.05 mmol) is now added, followed by 90 µL (205 mg, 1.45 mmol) of methyl iodide.

NOTE. Methyl iodide is toxic and must be dispensed in the **hood***. Dispense both the alkaline solution and the methyl iodide using an automatic delivery pipet. Because of its volatility, methyl iodide is used in slight excess.*

Reaction Conditions. Place the reaction assembly in a sand bath maintained at 60–65 °C and stir the mixture for 1 h.

Isolation of Product. Work up the resulting product mixture as described in Experiment [22A].

Purification and Characterization. Purify the crude product by chromatography on silica gel as described in Experiment [22A], Purification and Characterization (◄).

Weigh the pure methyl *p*-ethylphenyl ether and calculate the percent yield. Determine the boiling point and density (optional). Compare your results with the values reported in the literature.

Obtain the IR spectrum of the compound and compare it with that in Figure 6.39.

Nuclear Magnetic Resonance Analysis. Obtain both ¹H and ¹³C NMR spectra of your methyl *p*-ethylphenyl ether in CDCl₃ and compare your spectra with those in Figures 6.40 and 6.41. For additional information on analyses conducted using NMR, please review the identical section from Experiment [22A] above.

Chemical Tests.   Qualitative chemical tests can also be used to assist in characterizing this compound as an ether. Perform the ignition test (Table 9.1) to determine whether the material contains an aromatic ring.

A key factor to investigate is the solubility characteristics of this material (see Chapter 9). Determine its solubility in water, 5% sodium hydroxide, 5% hydrochloric acid, concentrated sulfuric acid, and 85% phosphoric acid. Do the results place this compound in the solubility class of an ether containing fewer than 8, or more than 8, carbon atoms?

Does the ferrox test (Chapter 9) confirm the presence of oxygen in the compound?

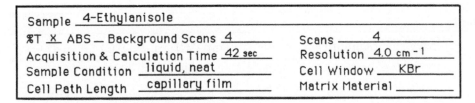

Figure 6.39   IR spectrum: methyl *p*-ethylphenyl ether.

300 MHz ¹H NMR SPECTRUM OF METHYL *p*-ETHYLPHENYL ETHER IN CDCl₃

Figure 6.40 ¹H-NMR spectrum: methyl *p*-ethylphenyl ether.

OPTIONAL SEMI-MICROSCALE AND MACROSCALE PREPARATIONS

If desired, this experiment can be scaled up by a factor of 10 or more.

Tenfold Scaleup. The reagent and solvent amounts are given in the following table.

Physical Properties of Reactants							
Compound	MW	Amount	mmol	mp (°C)	bp (°C)	d	n_D
4-Ethylphenol	122.17	1.5 g	12.3	42–45			
NaOH	40.0	625 mg	15.6				
Tetrabutylammonium bromide	322.38	150 mg	0.47	103–104			
Methyl iodide	141.94	2.05 g	14.5		41–43	2.28	1.5304
Water		2.5 mL					

Figure 6.41   ¹³C-NMR spectrum: methyl *p*-ethylphenyl ether.

Reagents and Equipment.   In a 25-mL round-bottom flask equipped with a stirring bar and a reflux condenser, place 2.5 mL of water and 1.5 g (12 mmol) of *p*-ethylphenol (the phenol will not be water-soluble). Cool the mixture by immersing the flask in a beaker of cold water, and then, with stirring, **cautiously** add 625 mg of sodium hydroxide.

After dissolution of the sodium hydroxide, weigh and add 150 mg of the phase-transfer catalyst, tetrabutylammonium bromide, followed by 2.05 g (900 µL) of methyl iodide (in the **hood**) using an automatic delivery pipet (➡).

Reaction Conditions.   Place the reaction vessel in a sand bath, and heat the reaction mixture, with stirring, at 60–65 °C for 1 h.

Isolation of Products.   Cool the resulting two-phase mixture to room temperature and then transfer it by Pasteur pipet to a 15-mL centrifuge tube. Wash the reaction flask with three 1-mL portions of diethyl ether and transfer the washings to the centrifuge tube. Cap the tube, shake, and vent (this operation may be done using a Vortex mixer if available), and allow the layers to separate. Remove the lower aqueous layer using a Pasteur filter pipet, and save it in a

HOOD

60–65 °C
thermometer

Speed   Heat

25-mL RB flask

10-mL Erlenmeyer flask until the final product has been purified and characterized.

NOTE. Do not remove any precipitated material that settles between the two layers.

Extract the ether layer with one 2-mL portion of 5% sodium hydroxide solution and then with 1 mL of water. Add these washings to the 10-mL Erlenmeyer flask.

Purify the crude product by chromatography on silica gel. Prepare a column by placing 5 g of activated silica gel in a 25-mL buret, followed by 0.5 g of anhydrous sodium sulfate. Dissolve the wet ether extract obtained above in 2.5 mL of methylene chloride and transfer the resulting solution by Pasteur pipet to the dry column. Elute the sample with an additional 20 mL of methylene chloride. Collect the eluate in a tared 50-mL Erlenmeyer flask containing a boiling stone. Evaporate the solvent by placing the flask in a sand bath maintained at 60–65 °C in the **hood**. Use a gentle stream of nitrogen or dry air to hasten the process.

HOOD

Weigh the resulting product and calculate the percent yield. Characterize the product as outlined above in the microscale procedure.

Eightyfold Scaleup. The reagent and solvent amounts are given in the following table.

Physical Properties of Reactants							
Compound	MW	Amount	mol	mp (°C)	bp (°C)	d	n_D
4-Ethylphenol	122.17	12.5 g	0.102	42–45			
NaOH	40.0	4.0 g	0.10				
Tetrabutylammonium bromide	322.38	0.40 g	0.0012	103–104			
Methyl iodide	141.94	14.5 g	0.102		41–43	2.28	1.5304
Water		25 mL					

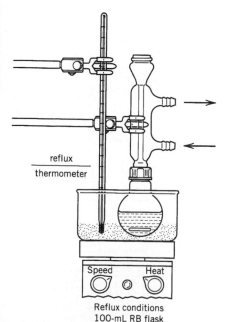

reflux
thermometer

Speed   Heat

Reflux conditions
100-mL RB flask

Reagents and Equipment. Use a 100-mL round-bottom flask containing a magnetic stirring bar and equipped with a reflux condenser (◄).

Reaction Conditions. Heat the reaction mixture at reflux, with stirring, for 90 min.

Isolation of Product. Allow the two-phase mixture to cool to room temperature. Transfer this mixture to a 125-mL separatory funnel and remove the aqueous layer. Store this in a 125-mL Erlenmeyer flask until the final product has been purified and characterized. Wash the organic layer successively with 5% sodium hydroxide solution (20 mL) and distilled water (20 mL). After each washing remove the aqueous layer and add it to the 125-mL Erlenmeyer flask. Finally, collect the reddish-brown organic layer in another 125-mL Erlenmeyer flask and dry it over anhydrous sodium sulfate.

Remove the drying agent by filtration through a glass wool plug and collect the product ether in a tared container. Weigh, calculate the percent yield, and then characterize the material as described in the microscale procedure.

Butyl *p*-Nitrophenyl Ether: Preparation Using a Monomode Microwave Apparatus

REACTION

$$O_2N-\langle\ \rangle-OH + CH_3CH_2CH_2CH_2-I \xrightarrow[\text{(C}_4\text{H}_9\text{)}_4\text{N}^+\text{Br}]{\text{NaOH}} O_2N-\langle\ \rangle-OCH_2CH_2CH_2CH_3$$

EXPERIMENTAL PROCEDURE

Estimated time to complete the experiment: 2.0 h.

Compound	MW	Amount	mmol	mp (°C)	bp (°C)	d	n_D
p-Nitrophenol	139	215 mg	1.55	110–115			
1-Iodobutane	184	0.17 mL	1.49		130–131	1.617	1.498
Tetrabutylammonium bromide	322	20 mg	0.06	103–104			
15% NaOH solution		2 mL					

Reagents and Equipment. This experiment is designed for use in the CEM Discover and Biotage Initiator microwave units.

In a 10.0-mL glass microwave reaction vessel containing a magnetic stir bar, place 215 mg (1.55 mmol) of *p*-nitrophenol, 20 mg (0.06 mmol) of tetrabutylammonium bromide, 2 mL of 15% sodium hydroxide solution, and 0.17 mL (1.49 mmol) of 1-iodobutane. Immediately cap the vessel with the microwave pressure cap.

CAUTION: 1-Iodobutane is toxic and must be dispensed in the **hood**. Dispense both the 1-iodobutane and alkaline solution using an automatic delivery pipet. Since the reaction requires heating the reaction mixture to above the boiling point of some components in sealed vessels, *adherence to the microwave manufacturer's guidelines is essential.*

HOOD

Reaction Conditions. Place the reaction vessel in the microwave cavity and, depending on the equipment used, position the pressure device on top. Program the microwave unit to heat the reaction mixture to 150 °C using no more that 50 W of microwave power, and hold at this temperature for 5 min. After heating, allow the reaction mixture to cool to 50 °C or below before removing the tube from the microwave unit.

Isolation of Product. Transfer the reaction mixture with a Pasteur pipet into a 30-mL separatory funnel. Clamp the funnel to a ring stand. Rinse the microwave reaction vessel with 2.0 mL of diethyl ether and add the washings to the separatory funnel. Carefully cap and invert the funnel. Immediately vent the funnel by opening the stopcock. Close the stopcock, place the funnel back on the ring stand and remove the stopper. Drain the lower (aqueous) layer into a 50-mL Erlenmeyer flask. Extract the crude organic layer with an additional two 5-mL portions of 5% sodium hydroxide solution, followed by 5 mL of water. During each extraction, cap and invert the funnel several times and each time release the pressure by opening the stopcock and then allow the funnel to stand so the layers will separate. Remove the lower (aqueous) layer

*This section has been written by Dr. Nicholas E. Leadbeater from the Department of Chemistry at the University of Connecticut and Dr. Cynthia B. McGowan from the Department of Chemistry at Merrimack College, MA.

after each extraction into the 50-mL Erlenmeyer flask. Save the aqueous waste until the experiment is complete and then discard as directed. Dry the organic layer by pipetting it into a clean 25-mL Erlenmeyer flask containing 200 mg of anhydrous sodium sulfate. Transfer the anhydrous solution, using a Pasteur filter pipet, to a clean tared 10-mL pear-shaped flask. Remove the ether on a rotary evaporator or by evaporation in the **hood** using a gentle stream of nitrogen gas with warming in a sand bath to isolate the crude product. Reweigh the flask and calculate the crude yield.

HOOD

Purification and Characterization.   The crude product can be further purified by recrystallization from 95% ethanol using a Craig tube.

Weigh the pure butyl *p*-nitrophenyl ether and calculate the percent yield. Determine the melting point and compare the experimental values with those in the literature.

Obtain an IR spectrum of the compound and compare it to that shown in Figure 6.42 for 4-butoxy nitrobenzene (butyl *p*-nitrophenyl ether).

Nuclear Magnetic Resonance Analysis.   If facilities permit, you can obtain both ^{1}H and ^{13}C NMR spectra of your butyl *p*-nitrophenyl ether in CDCl$_3$, and compare your spectra with those in Figures 6.43 and 6.44.

^{13}C NMR (CDCl$_3$): δ 164.3, 141.4, 125.8, 114.5, 68.6, 31.1, 19.2, 13.8
^{1}H NMR (CDCl$_3$): δ 8.2 (d, J = 9.1, 2H), 7.93 (d, J = 9.2, 2H), 4.05
(t, J = 6.5, 2H), 1.8 (m, J = 8.0, J = 6.5, 2H), 1.6 (m, J = 8.0, J = 7.4, 2H), 1.0
(t, J = 7.4, 3H)
IR (neat, HATR) 3114, 2955, 2873, 1594, 1509, 1261, 1175, 1109, 846 cm^{-1}

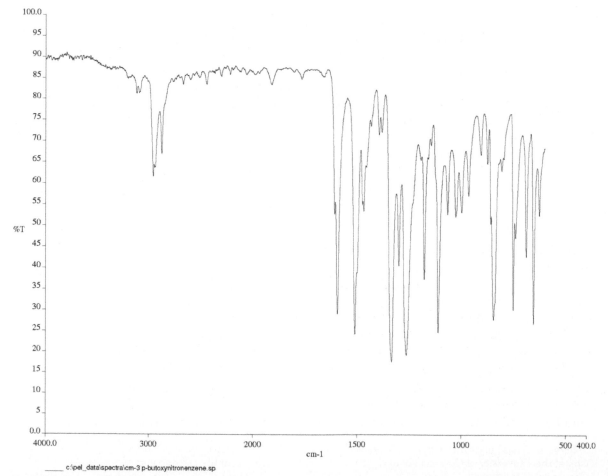

c:\pel_data\spectra\cm-3 p-butoxynitronenzene.sp

Figure 6.42   HATR-IR spectrum: butyl *p*-nitrophenyl ether.

Figure 6.43 ¹H-NMR spectrum: butyl *p*-nitrophenyl ether in CDCl₃.

Butyl *p*-Nitrophenyl Ether: Preparation Using a Multimode Microwave Apparatus

Experiment 22D*

M

REACTION

$$O_2N-\langle\bigcirc\rangle-OH + CH_3CH_2CH_2CH_2\text{-}I \xrightarrow[\text{(C}_4\text{H}_9)_4\text{N}^+\text{Br}^-]{\text{NaOH}} O_2N-\langle\bigcirc\rangle-OCH_2CH_2CH_2CH_3$$

EXPERIMENTAL PROCEDURE

Estimated time to complete the experiment: 2.0 h.

Compound	MW	Amount	mmol	mp (°C)	bp (°C)	*d*	n_D
p-Nitrophenol	139	215 mg	1.55	110–115			
1-Iodobutane	184	0.17 mL	1.49		130–131	1.617	1.498
Tetrabutylammonium bromide	322	20 mg	0.06	103–104			
15% NaOH solution		5 mL					

*This section has been written by Dr. Nicholas E. Leadbeater from the Department of Chemistry at the University of Connecticut and Dr. Cynthia B. McGowan from the Department of Chemistry at Merrimack College, MA.

Figure 6.44 ^{13}C-NMR spectrum: butyl *p*-nitrophenyl ether in CDCl$_3$.

Reagents and Equipment.   This experiment is designed for use in the CEM MARS, Milestone START, and Anton Paar Synthos 3000 microwave units. When using the Anton Paar Synthos 3000 unit with the 24-position silicon carbide plate rotor containing glass vials, the reagent and solvent quantities cited in the monomode procedure should be used in conjunction with the reaction conditions here in the multimode procedure.

In a microwave reaction vessel containing a magnetic stir bar, place 215 mg (1.55 mmol) of *p*-nitrophenol, 20 mg (0.06 mmol) of tetrabutylammonium bromide, 5 mL of a 15% sodium hydroxide solution, and 0.17 mL (1.49 mmol) of 1-iodobutane. Immediately cap the vessel with the microwave pressure cap and adjust the tightness to the manufacturer-specified level. Place the sealed vessel into its outer protective jacket.

HOOD

CAUTION: 1-iodobutane is toxic and must be dispensed in the **hood**. Dispense both the 1-iodobutane and alkaline solution using an automatic delivery pipet. Since the reaction requires heating the reaction mixture to above the boiling point of some components in sealed vessels, *adherence to the microwave manufacturer's guidelines is essential.*

Reaction Conditions.   Insert the loaded vessels into the reaction carousel, ensuring they are evenly spaced, and then place the carousel into the microwave cavity. If provided by the manufacturer, connect a temperature probe to the control vessel. Program the microwave unit to heat the reaction vessels to 150 °C and hold at this temperature for 5 min. After heating, allow the reaction mixture to cool to 50 °C or below before removing the carousel from the microwave unit.

Isolation of Product. Transfer the reaction mixture with a Pasteur pipet into a 30-mL separatory funnel. Clamp the funnel to a ring stand. Rinse the microwave reaction vessel with 2.0 mL of diethyl ether and add the washings to the separatory funnel. Carefully cap and invert the funnel. Immediately vent the funnel by opening the stopcock. Close the stopcock, place the funnel back on the ring stand and remove the stopper. Drain the lower (aqueous) layer into a 50-mL Erlenmeyer flask. Extract the crude organic layer with an additional two 5-mL portions of 5% sodium hydroxide solution, followed by 5 mL of water. During each extraction, cap and invert the funnel several times and each time release the pressure by opening the stopcock and then allow the funnel to stand so the layers will separate. Remove the lower (aqueous) layer after each extraction into the 50-mL Erlenmeyer flask. Save the aqueous waste until the experiment is complete and then discard as directed. Dry the organic layer by pipetting it into a clean 25-mL Erlenmeyer flask containing 200 mg of anhydrous sodium sulfate. Transfer the anhydrous solution, using a Pasteur filter pipet, to a clean tared 10-mL pear-shaped flask. Remove the ether on a rotary evaporator or by evaporation in the **hood** using a gentle stream of nitrogen gas with warming in a sand bath to isolate the crude product. Reweigh the flask and calculate the crude yield.

HOOD

Purification and Characterization. The crude product can be further purified by recrystallization from 95% ethanol using a Craig tube.

Weigh the pure butyl *p*-nitrophenyl ether and calculate the percent yield. Determine the melting point and compare the experimental values with those in the literature.

Obtain an IR spectrum of the compound and compare it to that shown in Figure 6.42 for 4-butoxy nitrobenzene (butyl *p*-nitrophenyl ether).

Nuclear Magnetic Resonance Analysis. If facilities permit, you can obtain both ^{1}H and ^{13}C NMR spectra of your butyl *p*-nitrophenyl ether in CDCl$_3$ and compare your spectra with those in Figures 6.43 and 6.44.

^{13}C NMR (CDCl$_3$): δ 164.3, 141.4, 125.8, 114.5, 68.6, 31.1, 19.2, 13.8

^{1}H NMR (CDCl$_3$): δ 8.2 (d, J = 9.1, 2H), 7.93 (d, J = 9.2, 2H), 4.05 (t, J = 6.5, 2H), 1.8 (m, J = 8.0, J = 6.5, 2H), 1.6 (m, J = 8.0, J = 7.4, 2H), 1.0 (t, J = 7.4, 3H)

IR (neat, HATR) 3114, 2955, 2873, 1594, 1509, 1261, 1175, 1109, 846 cm^{-1}

QUESTIONS

6-146. Sulfides are often prepared using an S$_N$2 reaction. For example,

$CH_3\ddot{\underset{..}{O}}$—⟨ ⟩—$\ddot{\underset{..}{S}}$:$^-$, Na$^+$ + isopropyl bromide $\xrightarrow{C_2H_5OH}$ $CH_3\ddot{\underset{..}{O}}$—⟨ ⟩—$\ddot{\underset{..}{S}}$CH(CH$_3$)$_2$ + NaBr

$CH_3\ddot{\underset{..}{O}}$—⟨ ⟩—$\ddot{\underset{..}{S}}$:$^-$, Na$^+$ + 2-bromo-1-nitropropane $\xrightarrow{C_2H_5OH}$ $CH_3\ddot{\underset{..}{O}}$—⟨ ⟩—$\ddot{\underset{..}{S}}$CH(CH$_3$)CH$_2NO_2$ + NaBr

The reaction with isopropyl bromide is 16 times faster than the reaction with 2-bromo-1-nitropropane. Explain.

6-147. If 4-bromo-1-butanol is treated with NaH, a compound of molecular formula C_4H_8O is formed. Suggest a structure for this product.

6-148. Arrange the substituted phenols given below in order of increasing reactivity toward ethyl iodide in the Williamson reaction. After arranging and explaining your order, does this order match that of rates of deprotonation?
   (a) 4-(Trifluoromethyl)phenol
   (b) *p*-Ethoxyphenol
   (c) 4-Methylphenol

6-149. *trans*-2-Chlorocyclohexanol reacts readily with NaOH to form cyclohexene oxide, but the cis isomer will not undergo this reaction. Explain.

cis-2-Chlorocyclohexanol

trans-2-Chlorocyclohexanol   Cyclohexene oxide

6-150. *tert*-Butyl ethyl ether might be prepared two ways using different starting materials.

Which route would you choose to prepare the above ether, and why?

6-151. What product(s) would you expect to form when tetrahydrofuran is treated with excess hydroiodic acid (HI)?

6-152. Write a suitable mechanism for the cleavage of butyl propyl ether with excess HBr at 100 °C to form exclusively 1-bromopropane and 1-bromobutane. Explain why neither butyl alcohol nor propyl alcohol is formed in the reaction.

6-153. There are only four lines in the aromatic region of the fully ^1H decoupled ^{13}C NMR spectrum of propyl *p*-tolyl ether (110–160 ppm), yet there are six aromatic carbon atoms. Explain.

BIBLIOGRAPHY

Selected review articles on the Williamson synthesis:

Dermer, O. C. *Chem. Rev.* **1934,** *14,* 409.

Feuer, H.; Hooz, J. in *The Chemistry of the Ether Linkage;* Patai, S., Ed.; Wiley: New York, 1967, p. 446.

Smith, M. B.; March J. *Advanced Organic Chemistry,* 7th ed.; Wiley-Interscience: New York, 2013, Chap. 10, p. 459.

Examples of the Williamson reaction in *Organic Syntheses* include:

Allen, C. F. H.; Gates, J. W., Jr. *Organic Syntheses;* Wiley: New York, 1955; Collect. Vol. III, p. 140. *Ibid.,* p. 418.

Boehme, W. R. *Organic Syntheses;* Wiley: New York, 1963; Collect. Vol. IV, p. 590.

Fuson, R. C.; Wojcik, B. H. *Organic Syntheses*; Wiley: New York, 1943; Collect.Vol. II, p. 260.

Gassman, P. G.; Marshall, J. L. *Organic Syntheses*; Wiley: New York, 1973; Collect.Vol.V, p. 424.

Gokel, G. W.; Cram, D. J.; Liotta, C. L.; Harris, H. P.; Cook, F. L. *Organic Syntheses*; Wiley: New York, 1988; Collect.Vol. VI, p. 301.

Kuryla, W. C.; Hyve, J. E. *Organic Syntheses*; Wiley: New York, 1973; Collect.Vol.V, p. 684.

Mirrington, R. N.; Feutrill, G. I. *Organic Syntheses*; Willey: New York, 1988; Collect.Vol.VI, p. 859.

Pedersen, C. J. *Organic Syntheses*; Wiley: New York, 1988; Collect.Vol.VI, p. 395.

Vyas, G. N.; Shah, N. M. *Organic Syntheses*; Wiley: New York, 1963; Collect.Vol. IV, p. 836.

Review articles on phase transfer catalysis:

Dehmlow, E.V.; Dehmlow, S. S. *Phase Transfer Catalysis*, 3rd ed.; VCH: New York, 1993.

Gokel, G. W.; Weber, W. P. *J. Chem. Educ.* **1978**, *55*, 350; *ibid.*, 429.

Smith, M. B. *March's Advanced Organic Chemistry*, 8th ed.; Wiley: New York, 2019, Chap. 10.

The procedures used in these experiments for the preparation of the ethers were adapted from the work of:

McKillop, A.; Fiaud, J. C.; Hug, R. P. *Tetrahedron* **1974**, *30*, 1379.

Rowe, J. E. *J. Chem. Educ.* **1980**, *57*, 162.

Amide Synthesis: Acetanilide and *N,N'*-Diacetyl-1,4-phenylenediamine

EXPERIMENT
23

Common name: acetanilide

CA number: [103-84-4]

CA name as indexed: acetamide, *N*-phenyl-

Common name: *N,N'*-diacetyl-1,4-phenylenediamine

CA number: [140-50-1]

CA name as indexed: acetamide, *N,N'*-1,4-phenylenebis-

Purpose.   You will carry out one of the major synthetic routes used in the preparation of amides; the method involves the reaction of ammonia, or a primary or secondary amine, with an active acylating reagent. You will also explore the use of acetic anhydride as an acylating agent. The acetanilide product (Experiment [23A]) may be used in Experiment [28].

Prior Reading

Technique 5: Crystallization

Use of the Hirsch Funnel (pp. 92–93)

Craig Tube Crystallizations (pp. 93–95)

REACTION

Aniline + Acetic anhydride ⇌ Acetanilide + Acetic acid

DISCUSSION

A number of important chemical and biochemical synthetic sequences are initiated by the addition of a nitrogen nucleophile to a carbonyl carbon atom to yield **carboxylic amides**. Amides are classified as **primary (1°), secondary (2°)**, or **tertiary (3°)** based on the number of groups attached to the nitrogen.

1° Amide	2° Amide	3° Amide
Methanamide	N-Methylethanamide	N,N-Dimethylbenzamide
(formamide)	(N-methylacetamide)	

Cyclic amides are called **lactams** and are classified by ring size. Imides contain a nitrogen bonded to two carbonyl groups and are nitrogen analogues of anhydrides.

An α-lactam	A β-lactam	A γ-lactam	A δ-lactam	An imide

B Amides appear in such diverse compounds as penicillin V (a β-lactam and an amide) and polypeptides (α-amino acids linked by amide bonds); and an imide, 1,2-benzenecarboxylic imide, is used in the Gabriel synthesis of amines. An imide is prepared in Experiment [24] and anhydrides are prepared in Experiments [25A] and [25B]. The polyamide polymer, nylon, is prepared in Chapter 7, Sequence B.

Penicillin V   A polypeptide   1,2-Benzenecarboxylic imide

The experiments outlined here illustrate the preparation of 2° amides. The process involves the attack of a primary amine on the acetyl group of acetic anhydride. Ammonia or secondary amines also react readily with this reagent to yield 1° and 3° amides, respectively. The mechanism shown here is an example of the attack of a nucleophilic reagent on a carbonyl carbon of the anhydride:

In the preparations given below, the amine reagents (aniline and *p*-phenylenediamine) are purified as their hydrochloride salts. Arylamines are relatively weak bases (K_b values in the order of 10^{-10}) but when treated with a strong acid, such as HCl, they are completely protonated, yielding the corresponding water-soluble hydrochloride salt:

$$C_6H_5{-}NH_2 + HCl \rightarrow C_6H_5{-}NH_3^+, Cl^-$$

Aniline   Anilinium
                                      hydrochloride salt
                                        (water-soluble)

As directed in the experiment, decolorizing charcoal is added to the aqueous solution of the arylamine salt. The charcoal absorbs impurities and subsequent filtration of the mixture, which removes the charcoal, yields an aqueous solution of the purified arylamine salt.

The second stage of the reaction sequence requires that a solution of sodium acetate be added to the reaction mixture after initial addition of acetic anhydride to the purified anilinium hydrochloride salt solution:

$$C_6H_5{-}NH_3^+, Cl^- + CH_3COO^-, Na^+ \rightleftharpoons C_6H_5{-}NH_2 + CH_3COOH + Na^+, Cl^-$$

Addition of the sodium acetate solution serves to liberate the arylamine so that the desired nucleophilic substitution reaction may occur; ammonium cations are not nucleophilic, since they are positively charged and do not even possess a lone pair of electrons.

Sodium acetate is the conjugate base of acetic acid, which is a weak acid. Furthermore, the anilinium ion ($pK_a = 4.6$) is a slightly stronger acid than acetic acid ($pK_a = 4.8$). Thus, the equilibrium reaction is shifted to the right, producing the arylamine.

In Experiment [23B], the *p*-phenylenediamine forms the corresponding dihydrochloride salt, $Cl^- \cdot H_3N^+{-}C_6H_4{-}^+NH_3 \cdot Cl^-$ As in Experiment [23A], the aqueous solution of the salt is purified with charcoal, and upon addition of the sodium acetate solution, the free amine is regenerated.

This overall process illustrates an important transformation for most amines. These amines can be converted to water-soluble ionic salts by reaction with acids and can be recovered from these acid salts by treatment with base. This technique was used in Experiment [4C] to extract ethyl 4-aminobenzoate, as its water-soluble salt, from a mixture.

Experiment 23A

Acetanilide

The reaction is shown above.

EXPERIMENTAL PROCEDURE

Estimated time to complete the experiment: 1.5 h.

Physical Properties of Reactants							
Compound	MW	Amount	mmol	mp (°C)	bp (°C)	d	n_D
Aniline	93.13	100 μL	1.09		184	1.02	1.5863
Concd HCl		3 drops					
Sodium acetate trihydrate	136.08	150 mg	1.10	58			
Acetic anhydride	102.09	150 μL	1.59		140	1.08	1.3901

HOOD ***Reagents and Equipment.*** In the **hood**, place 100 μL of aniline in a tared 10 × 75-mm test tube (standing in a small beaker or Erlenmeyer flask). Fit the tube with a cork stopper.

CAUTION: Aniline is a toxic material and a suspected carcinogen.

NOTE. Dispense the aniline using an automatic delivery pipet. Again weigh the test tube and container to determine the exact amount of aniline delivered.

HOOD Now, using a 1.0-mL graduated pipet, add, with swirling, 0.5 mL of water; then, in the **hood**, add 3 drops of concentrated hydrochloric acid using a Pasteur pipet. Add 10 mg of powdered decolorizing charcoal, or the pelletized form (Norit), to the resulting solution.

Using a Pasteur pipet, transfer the well-mixed suspension to a 25-mm funnel fitted with fast-grade filter paper to remove the charcoal by gravity filtration. *Wet the filter paper in advance with distilled water and blot the excess water from the stem of the funnel.*

Collect the filtrate in a 3.0-mL conical vial. Use an additional 0.5 mL of water to rinse the test tube and the collected charcoal. Combine the rinse with the original filtrate. Place a magnetic spin vane in the vial and attach it to an air condenser (◂).

C₆H₅NH₂, 100 μL
+ H₂O, 1.0 mL
+ concd HCl, 3 drops
+ NaOAc·3H₂O, 150 mg
+ (CH₃CO)₂O, 150 μL

NOTE. If the pelletized form of charcoal is used, transfer through a Pasteur filter pipet directly to the 3.0-mL conical vial should be sufficient.

NOTE. Tap all the filtrate from the funnel stem into the collecting vial. As a result of this purification step, a clear, colorless solution of aniline hydrochloride should be obtained.

Dissolve 150 mg (1.10 mmol) of sodium acetate trihydrate in 0.5 mL of distilled water in a 10 × 75-mm test tube. Cap the tube and set the solution aside for use in the next step.

Remove the air condenser, and then use an automatic delivery pipet in the **hood** to add, with stirring, 150 µL of acetic anhydride to the solution of aniline hydrochloride, followed quickly by addition (Pasteur pipet) of the previously prepared solution of sodium acetate. Reattach the air condenser.

HOOD

Reaction Conditions.   The reaction is very rapid, and the product begins to precipitate immediately upon mixing of the reagents. Stir to thoroughly mix the reagents. Allow the reaction mixture to stand at room temperature for approximately 5 min and then place it in an ice bath for an additional 5–10 min to complete the crystallization process.

Isolation of Product.   Collect the acetanilide product by filtration under reduced pressure using a Hirsch funnel (➡). Rinse the conical vial with two 0.5-mL portions of water (using calibrated Pasteur pipet) and use the rinse to wash the collected filter cake. Place a piece of plastic food wrap over the mouth of the funnel and continue the suction for 5–8 min (see Prior Reading). The snow-white crystals are further dried on a porous clay plate or on filter paper in a desiccator.

Acetanilide collected here

H$_2$O, ~ 2 mL + water-soluble reaction products

Purification and Characterization.   Further purification of the product is generally not required. However, the acetanilide may be recrystallized from hot water or from ethanol–water using the Craig tube.

NOTE. Acetanilide (150 mg) can be recrystallized from approximately 3 mL of water, or from 2 mL of ethanol–water (1:10 v/v), with better than 80% recovery.

Weigh the dried crystals and calculate the percent yield. Determine the melting point of the material and compare your result to that reported in the literature.

Obtain an IR and/or NMR spectrum of the product and compare it with that of an authentic sample or to that recorded in the literature (*The Aldrich Library of IR Spectra, The Aldrich Library of NMR Spectra*, and/or SciFinder Scholar).

Chemical Tests.   Characterization of the product may be enhanced by performing several chemical tests given in Chapter 9.

Check the solubility of acetanilide in water. Is the aqueous solution acidic, basic, or does it remain neutral as indicated by pH paper? Does the ignition test indicate that an aromatic group is present? Does the soda lime or sodium fusion test indicate the presence of nitrogen? Does the hydroxamate test for amides give a positive result?

OPTIONAL SEMI-MICROSCALE PREPARATION

This experiment can be scaled up to be run at five times the amounts used in the above microscale preparation.

Room temp
thermometer

Speed   Heat

A 10-mL RB flask

The procedure is identical to the above with the following exceptions.

1. Use two 15 × 100-mm test tubes.

2. Carry out the reaction in a 10-mL round-bottom flask containing a magnetic spin bar and fitted with an air condenser (←).

3. Increase the amounts of all reagents and the solvent by a factor of 5.

Physical Properties of Reactants							
Compound	MW	Amount	mmol	mp (°C)	bp (°C)	d	n_D
Aniline	93.13	500 μL	5.45		184	1.02	1.5863
Concd HCl		0.75 mL					
Sodium acetate trihydrate	136.08	750 mg	5.50	58			
Acetic anhydride	102.09	750 μL	7.93		140	1.08	1.3901

Experiment 23B

N,N'-Diacetyl-1,4-phenylenediamine

REACTION

$$H_2\ddot{N}-\text{⟨benzene ring⟩}-\ddot{N}H_2 + 2 \quad \begin{array}{c} CH_3-C \\ | \\ CH_3-C \end{array} \ddot{O}: \rightleftharpoons CH_3-\overset{\ddot{O}}{C}-\underset{H}{\ddot{N}}-\text{⟨benzene ring⟩}-\underset{H}{\ddot{N}}-\overset{\ddot{O}}{C}-CH_3 + 2\,CH_3-\overset{\ddot{O}:}{C}\underset{\ddot{O}H}{}$$

1,4-Phenylenediamine   Acetic anhydride   *N,N'*-Diacetyl-1,4-phenylenediamine   Acetic acid

EXPERIMENTAL PROCEDURE

Estimated time to complete the experiment: 1.5 h.

Physical Properties of Reactants							
Compound	MW	Amount	mmol	mp (°C)	bp (°C)	d	n_D
1,4-Phenylenediamine	108.14	117 mg	1.08	138			
Concd HCl		6 drops					
Sodium acetate trihydrate	136.08	300 mg	2.20	58			
Acetic anhydride	102.09	350 μL	3.71		140	1.08	1.3901

Reagents and Equipment. Weigh and place 117 mg (1.08 mmol) of 1,4-phenylenediamine in a 10 × 75-mm test tube (standing in a small beaker or Erlenmeyer flask).

CAUTION: This reagent is toxic and a suspected carcinogen.

With gentle swirling, add 1.0 mL of distilled water and, using a Pasteur pipet, 6 drops of concentrated hydrochloric acid. After dissolution, add 30 mg of either powdered decolorizing charcoal (Norit) or the pelletized form. Transfer the well-mixed suspension, by use of a Pasteur pipet, to a 25-mm funnel fitted with fast-grade filter paper previously wet with water. The charcoal is removed by gravity filtration. Collect the filtrate, which is clear to slightly yellow, in a 5-mL conical vial containing a magnetic spin vane. Use three 0.5-mL portions of water (calibrated Pasteur pipet) to rinse the test tube, and in turn use the rinse to wash the collected charcoal. The rinse is combined with the original filtrate. *Blot the excess water from the stem of the funnel.* Attach the vial to an air condenser (➡).

p-NH$_2$C$_6$H$_4$NH$_2$, 117 mg
+ H$_2$O, 2.5 mL
+ concd HCl, 6 drops
+ NaOAc, 300 mg
+ (CH$_3$CO)$_2$O, 350 μL

NOTE. If the pelletized form of charcoal is used, transfer through a Pasteur filter pipet directly to the 5.0-mL conical vial should be sufficient.

Dissolve 300 mg (2.20 mmol) of sodium acetate trihydrate in 0.5 mL of distilled water in a 10 × 75-mm test tube. Stir the mixture with a spatula to aid the dissolution process. Cap the tube and set it aside for use in the next step.

Reaction Conditions.   Remove the air condenser, and use an automatic delivery pipet to add 350 μL of acetic anhydride to the solution of 1,4-phenylenediamine dihydrochloride, and stir the mixture briefly using a magnetic stirrer.

NOTE. Dispense the acetic anhydride in the **hood** *using an automatic delivery*   HOOD
pipet. A slight amount of white precipitate may be observed at this stage.

Now add the previously prepared sodium acetate solution by Pasteur pipet to the reaction mixture with stirring. Reattach the air condenser.

Isolation of Product.   The reaction is very rapid and the desired product begins to precipitate almost immediately. After stirring briefly, allow the mixture to stand at room temperature for a few minutes and then place it in an ice bath for an additional 5–10 min.

Purification and Characterization.   Collect the crude *N,N'*-diacetyl-1,4-phenylenediamine by vacuum filtration using a Hirsch funnel (➡). Rinse the vial with two 0.5-mL portions of water and use the rinse to wash the filter cake. Place a piece of plastic food wrap over the mouth of the funnel and continue the suction for an additional 5–8 min. Place the collected material on a porous clay plate or filter paper to dry further.

Diamide product collected here

H$_2$O, ~ 3.5 mL + water-soluble reaction products

NOTE. If this material is to be used in Experiment [29B], recrystallization from methanol is suggested.

Weigh the dried crystals and calculate the percent yield. Determine the melting point and compare it with the value in the literature. Obtain an IR and/or the NMR spectrum of the product. The infrared spectrum is shown in Figure 6.45.

Chemical characterization tests might also be run as outlined in Experiment [23A].

ABSCISSA	ORDINATE	SCAN TIME 12 min	REP. SCAN_____ SINGLE BEAM_____
EXPANSION_____	EXPANSION_____	RESPONSE_____	TIME DRIVE_____ PRE SAMPLE CHOP_____
SUPPRESSION_____	% T_____ ABS_____	SLIT PROGRAM normal	OPERATOR_____ DATE_____
SAMPLE N,N-Diacetyl-1,4-	REMARKS Perkin-Elmer 283 B	SOLVENT KBr pellet	CELL PATH_____
ORIGIN phenylenediamine		CONCENTRATION_____	REFERENCE_____

Figure 6.45   IR spectrum: *N,N'*-diacetyl-1,4-diphenylenediamine.

QUESTIONS

6-154. What is the function of the sodium acetate in the reactions outlined in this experiment?

6-155. Which is the stronger base: aniline or cyclohexylamine? Explain.

6-156. Arrange the following substituted anilines in increasing order of reactivity toward acetic anhydride:

6-157. Suggest a mechanism for the preparation of acetic anhydride from acetic acid and acetyl chloride in the presence of pyridine (an amine base).

6-158. Anhydrides generally react more slowly with an amine than acid chlorides, though both reactions produce amides. Explain this observation.

6-159.

$$CH_3CH_2-\overset{\overset{\ddots\overset{\displaystyle \cdot\cdot}{O}}{\parallel}}{\underset{\underset{\ddot N H_2}{|}}{C}}$$ is less basic than $CH_3CH_2-\ddot N H_2$

Explain.

BIBLIOGRAPHY

Review articles:

Beckwith, A. L. J. In *The Chemistry of the Amides*; J. Zabicky, Ed.; Wiley: New York, 1970; p. 73.

Satchell, D. P. N. *Q. Rev.* **1963**, *17*, 160.

Smith, M. B. *March's Advanced Organic Chemistry*, 8th ed.; Wiley: New York, 2019.

Selected acylation reactions between anhydrides and amines in Organic Syntheses:

Cava, M. P.; Deana, A. A.; Muth, K.; Mitchell, M. J. *Organic Syntheses*; Wiley: New York, 1973; Collect. Vol. V, p. 944.

Fanta, P. E.; Tarbell, D. S. *Organic Syntheses*; Wiley: New York, 1955; Collect. Vol. III, p. 661.

Herbst, R. M.; Shemin, D. *Organic Syntheses*; Wiley: New York, 1943; Collect. Vol. II, p. 11.

Jacobs, T. L.; Winstein, S.; Linden, G. B.; Robson, J. H.; Levy, E. F.; Seymour, D. *Organic Syntheses*; Wiley: New York, 1955; Collect. Vol. III, p. 456.

Noyes, W. A.; Porter, P. K. *Organic Syntheses*; Wiley: New York, 1941; Collect. Vol. I, p. 457.

Wiley, R. H.; Borum, O. H. *Organic Syntheses*; Wiley: New York, 1963; Collect. Vol. IV, p. 5.

Imide Synthesis: *N*-Phenylmaleimide

Common name: *N*-phenylmaleimide
CA number: [941-69-5]
CA name as indexed: 1*H*-pyrrole-2,5-dione, 1-phenyl-

Purpose.   To extend the amide synthesis (Experiment [23]) to the preparation of imides. In this experiment, the condensation of a cyclic anhydride with aniline to form an imide is described. The initial reaction to give the carboxylic amide is followed by an intramolecular condensation to produce the desired imide derivative.

Prior Reading

Technique 5: Crystallization
                   Use of the Hirsch Funnel (pp. 92–93)
                   Craig Tube Crystallizations (pp. 93–95)
Technique 6A: Thin-Layer Chromatography (pp. 101–103)

REACTION

Maleic anhydride   Aniline   *N*-Phenylmaleimide

DISCUSSION

Imides are diacyl derivatives of ammonia or primary amines. The reaction is similar in its scope and mechanism to the acetylation of aniline or 1,4-phenylene-diamine presented in Experiments [23A] and [23B]. As illustrated in the present experiment, cyclic anhydrides produce cyclic imides. Cyclic anhydrides are prepared in Experiments [25A] and [25B]. Derivatives of imides have been suggested for use in the treatment of arthritis, tuberculosis, and epilepsy. Several also have been found to be growth stimulants. Imide-based polymers are used in many applications, including fire-resistant woven fabrics. The N-phenylmaleimide prepared in this experiment is also a good dienophile in the Diels–Alder reaction (see Experiments [14] and [15]), and in fact has been used as a reagent to characterize 1,3 dienes.

B

The first step in the reaction between the primary amine and the cyclic anhydride is an addition–elimination reaction, which involves a nucleophilic attack by the amine on a carbonyl carbon of the maleic anhydride. This results in the formation of an amide and a carboxylic acid, which are linked together to constitute maleanilic acid. Since anhydrides are considerably more reactive toward nucleophiles, to promote ring closure to the imide, the carboxylic acid is then converted to another (mixed) anhydride by reaction with acetic anhydride. This anhydride then undergoes an intramolecular nucleophilic addition–elimination by the amide nitrogen, which gives the desired imide, N-phenylmaleimide. The second acylation of a nitrogen nucleophile is much slower than the first. That is, attack of an amide nitrogen on the carbonyl carbon of the anhydride is slower than the attack of the amine nucleophile on the anhydride carbonyl carbon. The mechanistic sequence is given below (R = phenyl):

Maleic anhydride

Maleanilic acid

Maleanilic acid

Acetic anhydride

(see next page)

where R = Phenyl

N-Phenylmaleimide

Maleanilic Acid

EXPERIMENTAL PROCEDURE

Estimated time to complete the experiment: 1.5 h.

REACTION

Maleic anhydride   Aniline   Maleanilic acid

NOTE. It is recommended that the purity of the maleic anhydride be checked by mp. The presence of maleic acid can substantially lower the yield of product.

Physical Properties of Reactants							
Compound	MW	Amount	mmol	mp (°C)	bp (°C)	d	n_D
Maleic anhydride	98.06	60 mg	0.61	60			
Acetonitrile	41.05	1.2 mL			81		
Aniline	93.13	56 µL	0.62		184	1.02	1.5863

Maleic anhydride, 60 mg +
$C_6H_5NH_2$, 56 μL +
$(CH_3CO)_2O$, 1.2 mL

Reagents and Equipment. In a 3.0-mL conical vial containing a magnetic spin vane, and equipped with an air condenser protected with a drying tube, place 60 mg (0.61 mmol) of maleic anhydride and 1.0 mL of anhydrous acetonitrile (➤). Stir the mixture at room temperature until all the maleic anhydride has dissolved.

HOOD

CAUTION: Dispense this reagent in the **hood** using a calibrated Pasteur pipet.

In a separate, dry $\frac{1}{2}$-*dram* vial prepare a solution of 56 μL (57 mg, 0.62 mmol) of aniline in 100 μL of anhydrous acetonitrile.

HOOD

CAUTION: Dispense these reagents in the **hood** using a calibrated Pasteur pipet. Aniline is highly toxic and is a suspected carcinogen.

Using a Pasteur pipet, add the aniline–acetonitrile solution in one portion to the stirred maleic anhydride–acetonitrile solution. Rinse the $\frac{1}{2}$-*dram* vial with 100 μL of anhydrous acetonitrile and also transfer this rinse to the reaction solution.

Reaction Conditions. After stirring the reaction mixture at room temperature for 15 min, begin monitoring the system by TLC (plates with fluorescent indicator). Using as a solvent system ethyl acetate:hexane (3:2) and a UV lamp for visualization, the R_f value for maleanilic acid is 0.2. The R_f values for aniline and maleic anhydride are 0.73 and 0.67, respectively. Once complete as judged by TLC, cool the reaction mixture in an ice bath for 5–10 min.

Isolation of Product. Collect the deposit of fine, cream-colored powder by vacuum filtration using a Hirsch funnel (➤). Wash the maleanilic acid crystals with 0.5 mL of cold diethyl ether (calibrated Pasteur pipet), and air-dry them in the funnel for 5 min while maintaining the suction.

Maleanilic acid product collected here

$(CH_3CH_2)_2O$, 1.7 mL

Purification and Characterization. Weigh the maleanilic acid and calculate the percent yield. Determine the melting point and compare your value to that given in Cava et al. (**Bibliography**). Obtain an IR spectrum using the KBr pellet technique and compare it with an authentic sample. The air-dried product is suitable for use in the next step without further purification.

N-Phenylmaleimide

Experiment 24B

REACTION

Maleanilic acid $\xrightarrow[\text{(CH}_3\text{CO)}_2\text{O}]{\text{CH}_3\text{COONa}}$ N-Phenylmaleimide $+ H_2O$

EXPERIMENTAL PROCEDURE

Estimated time to complete the experiment: 1.5 h.

Physical Properties of Reactants							
Compound	MW	Amount	mmol	mp (°C)	bp (°C)	d	n_D
Maleanilic acid	191.18	100 mg	0.52	201–202			
Sodium acetate	82.03	25 mg	0.30	324			
Acetic anhydride	102.09	200 µL	2.12		140	1.08	1.3901

80–90 °C
thermometer

Speed   Heat

CH_3CO_2Na, 25 mg
+ $(CH_3CO)_2O$, 200 µL +
maleanilic acid, 100 mg

Reagents and Equipment.   In a 3.0-mL conical vial containing a magnetic spin vane, and equipped with an air condenser protected by a drying tube, place 25 mg (0.30 mmol) of anhydrous sodium acetate and 200 µL (216 mg, 2.12 mmol) of acetic anhydride (➡).

CAUTION: Acetic anhydride is corrosive and a lachrymator. It should be dispensed in the **hood** by use of an automatic delivery pipet.

Now add 100 mg (0.52 mmol) of maleanilic acid (prepared in Experiment [24A]) to the reaction vial.

Reaction Conditions.   Heat the reaction mixture, with stirring, at a sand bath temperature of 80–90 °C for 30 min. Then cool the resulting mixture to room temperature, add 1.0 mL of cold water (calibrated Pasteur pipet), stir for a few minutes, and then place the vial in an ice bath for 5–10 min to complete crystallization.

Isolation of Product.   Collect the solid product by vacuum filtration using a Hirsch funnel and then wash the filter cake with three 0.5-mL portions of cold water (calibrated Pasteur pipet) (➡). Cover the mouth of the funnel with plastic food wrap and continue the suction for an additional 5–10 min.

Purification and Characterization.   Recrystallize the crude *N*-phenyl-maleimide from cyclohexane using the Craig tube, to yield canary-yellow needles. After drying the product on filter paper, or on a porous clay plate, weigh the crystals and calculate the percent yield. Determine the melting point and compare your result with the value given by Cava et al. (**Bibliography**). Obtain an IR spectrum and compare it with that of an authentic sample or with that shown in the literature (*The Aldrich Library of IR Spectra* and/or SciFinder Scholar).

Imide product
collected here

H_2O, ~ 2.5 mL +
water–soluble reaction
products

QUESTIONS

6-160. As stated in the discussion section, the second step in the reaction to form the imide is much slower than that of the first stage (formation of the acid amide). Explain.

6-161. Phthalimide has a $K_a = 5 \times 10^{-9}$. Write an equation for the reaction of phthalimide with potassium amide (a strong base) in N,N-dimethylformamide (DMF) solvent. Name the product.

6-162. Predict which of the following species is the most acidic. Explain.

Phthalimide   Benzamide

6-163. The phthalimide anion is a strong nucleophile. It can react easily with primary alkyl halides to form substituted phthalimides. One advantage when working with substituted phthalimides is that when treated with hydrazine, primary amines are furnished through this alkylation protocol (Gabriel synthesis). For both synthetic pathways (a) and (b), suggest a suitable mechanism.

(a) phthalimide anion $:N:^-$, K^+ + CH_3CH_2Br ⟶ $N-CH_2CH_3$ + KBr

(b) $N-CH_2CH_3$ $\xrightarrow{H_2NNH_2}$ phthalhydrazide $\overset{NH}{\underset{NH}{}}$ + $H_2NCH_2CH_3$

6-164. N-Phenylmaleimide, the product prepared in Experiment [24B], can act as a dienophile in the Diels–Alder reaction (see Experiments [14] and [15]). Draw the structure of the product that would be formed by the treatment of N-phenylmaleimide with (a) 3-sulfolene under the conditions given in Experiment [14], (b) furan, and (c) cyclopentadiene.

BIBLIOGRAPHY

Review articles on cyclic imides:

Benjamin, E.; Hijji, Y. *Molecules* **2008**, *13*, 157.

Hargreaves, M. K.; Pritchard, J. G.; Dave, H. R. *Chem. Rev.* **1970**, *70*, 439.

Naik, S.; Bhattacharjya, G.; Talukdar, R.; Patel, B. K. *J. Org. Chem.* **2004**, *69*, 1254.

Smith, M. B. *March's Advanced Organic Chemistry*, 8th ed.; Wiley: New York, 2019

Wheeler, O. H.; Rosado, O. In *The Chemistry of the Amides*; Zabicky, J., Ed.; Wiley: New York, 1970, p. 335.

Selected imide preparations in *Organic Syntheses* include:

Cava, M. P.; Deana, A. A.; Muth, K.; Mitchell, M. J. *Organic Syntheses*; Wiley: New York, 1973; Collect. Vol. V, p. 944.

Noyes, W. A.; Porter, P. K. *Organic Syntheses*; Wiley: New York, 1941; Collect. Vol. I, p. 457.

Smith, L. I.; Emerson, O. H. *Organic Syntheses*; Wiley: New York, 1955; Collect. Vol. III, p. 151.

Soine, T. O.; Buchdahl, M. R. *Organic Syntheses*; Wiley: New York, 1963; Collect. Vol. IV, p. 106.

For the preparation of maleanilic acid, also see:

Ram, R. N.; Varsha, K. *J. Chem. Educ.* **1990**, *67*, 985.

Synthesis of Cyclic Carboxylic Acid Anhydrides: Succinic Anhydride and Phthalic Anhydride

Common name: succinic anhydride
CA number: [108-30-5]
CA name as indexed: 2,5-furandione, dihydro-

Common names: phthalic anhydride, benzene-1,2-dicarboxylic anhydride
CA number: [85-44-9]
CA name as indexed: 1,3-isobenzofurandione

Purpose.   One of the important methods for preparing cyclic carboxylic acid anhydrides is carried out. The reaction demonstrates the use of acetic anhydride, an important industrial and research chemical, as a dehydrating agent.

Prior Reading

Technique 5: Crystallization
             Use of the Hirsch Funnel (pp. 92–93)
             Craig Tube Crystallizations (pp. 93–95)
Technique 9: Sublimation (pp. 115–118)

REACTION

Succinic acid   Succinic anhydride

DISCUSSION

Five- and six-membered cyclic anhydrides can be easily formed when the corresponding dicarboxylic acid is heated in the presence of a dehydrating agent. One of the most commonly used dehydrating agents is acetic anhydride. The formation of an anhydride from its corresponding acid by reaction with another anhydride is referred to as *anhydride exchange.*

   The similarity of this reaction for the preparation of anhydrides to that for the synthesis of imides (Experiment [24]) should be noted. The ring closure to form the imide is mechanistically related to that of anhydride formation: in one case an amide nitrogen makes a nucleophilic attack on a carbonyl, while in the other, an acid oxygen acts as the nucleophile.

The mechanistic sequence for anhydride exchange is

Phthalic anhydride

It is possible to prepare five- and six-membered cyclic anhydrides in the absence of acetic anhydride by direct dehydration at elevated temperatures. Maleic anhydride, for example, is easily obtained by this method in greater than 85% yield. Heating of succinic acid at 300 °C yields succinic anhydride in 95% yield.

This equilibrium is driven toward the products since formation of three molecules (two molecules of acetic acid and one molecule of anhydride) is entropically favored over the two molecules of reactants. The equilibrium could be further driven toward the products by distilling off the more volatile acetic acid as it is formed.

Acetic anhydride is an important industrial reagent. Over one-half its annual production of approximately 750,000 tons is used for the manufacture of cellulose acetate. Cellulose acetate is a widely used textile fiber and is the chief component of cigarette filters. Acetic anhydride is the acetylation reagent used for the production of aspirin (acetylsalicylic acid). Succinic anhydride finds use in the succinylation of gelatin used as a blood plasma substitute, as a food preservative in chicken against *Salmonella*, and as a dog food preservative.

B

Phthalic anhydride finds extensive use in plasticizer formulations for many resins, and in the manufacture of dyes.

Succinic Anhydride

The reaction is shown on p. 354.

Experiment 25A

EXPERIMENTAL PROCEDURE

Estimated time to complete the experiment: 1.5 h.

Physical Properties of Reactants							
Compound	MW	Amount	mmol	mp (°C)	bp (°C)	d	n_D
Succinic acid	118.09	150 mg	1.27	188			
Acetic anhydride	102.09	200 μL	2.12		140	1.08	1.3901

Reagents and Equipment

NOTE. All equipment must be dried in an oven (110 °C) for 30 min before use.

Weigh and place 150 mg (1.3 mmol) of succinic acid in a 1.0-mL conical vial containing a magnetic spin vane. Add 200 μL (2.12 mmol) of acetic anhydride and then attach the vial to a reflux condenser protected by a calcium chloride drying tube (➡).

180 °C
thermometer

Speed   Heat

Succinic acid, 150 mg +
(CH₃CO)₂O, 200 μL

CAUTION: Acetic anhydride is moisture-sensitive and an irritant. Dispense it in the **hood** using an automatic delivery pipet.

HOOD

Reaction Conditions.   Heat the reaction mixture, with stirring, in a sand bath at a temperature of 180 °C for 45 min, timing the reaction from the point at which the succinic acid is completely dissolved.

Isolation of Product.   Cool the mixture to room temperature. A voluminous precipitate of succinic anhydride deposits. Further cool the vial in an ice bath for 5 min, and collect the solid material by vacuum filtration using a Hirsch funnel (➡). Wash the white needles with three 0.5-mL portions of diethyl ether (calibrated Pasteur filter pipet) and then place them on a porous clay plate or filter paper to dry.

Purification and Characterization.   The succinic anhydride crystals should be sufficiently pure for characterization. These crystals may, however, be recrystallized from absolute ethanol using a Craig tube.

Weigh the product and calculate the percent yield. Determine the melting point and obtain an infrared spectrum. Compare your results to those listed in the literature (*The Aldrich Library of IR Spectra* and/or SciFinder Scholar). What characteristic absorptions do you observe for the anhydride group in the carbonyl region of the spectrum?

Succinic anhydride
collected here

(CH₃CH₂)₂O, 1.5 mL +
CH₃CO₂H, ~ 100 μL

Experiment 25B

Phthalic Anhydride

REACTION

Phthalic acid   Phthalic anhydride

EXPERIMENTAL PROCEDURE

Estimated time to complete the experiment: 1.5 h.

150–160 °C
thermometer

Physical Properties of Reactants							
Compound	MW	Amount	mmol	mp (°C)	bp (°C)	d	n_D
Phthalic acid	166.14	100 mg	0.60	210			
Acetic anhydride	102.09	200 μL	2.12		140	1.08	1.3901

Reagents and Equipment.   Weigh and add 100 mg (0.60 mmol) of phthalic acid to a 3.0-mL conical vial containing a magnetic spin vane. Add 200 μL (2.1 mmol) of acetic anhydride and then attach the vial to a reflux condenser protected by a calcium drying tube (◄).

Speed   Heat

Phthalic acid, 100 mg +
(CH₃CO)₂O, 200 μL

HOOD

CAUTION: Acetic anhydride is moisture-sensitive and an irritant. Dispense it in the **hood** using an automatic delivery pipet.

Reaction Conditions.   Heat the reaction solution, with stirring, at a sand bath temperature of 150–160 °C for 30 min.

NOTE. Position the vial firmly on the bottom of the sand bath vessel to maintain this reaction temperature.

Phthalic anhydride
collected here

Hexane, 0.5 mL +
CH₃CO₂H, ~ 100 μL

Isolation of Product.   Cool the mixture to room temperature, whereupon the product crystallizes from solution. Cool the vial and contents in an ice bath for 10 min and collect the solid by vacuum filtration using a Hirsch funnel (◄). Rinse the filter cake carefully by dropwise addition of 0.5 mL of cold hexane (Pasteur pipet) and continue the suction for several minutes. Complete the drying of the solid product by placing the crystals on a porous clay plate or on filter paper.

Purification and Characterization.   The phthalic anhydride should be sufficiently pure for characterization. It may be purified further by sublimation (Technique 9, p. 111) or by recrystallization from absolute ethanol using the Craig tube.

Weigh the product and calculate the percent yield. Determine the melting point and obtain an infrared spectrum. Compare your results with those reported in the literature (*The Aldrich Library of IR Spectra* and/or SciFinder Scholar). What characteristic absorptions do you observe for the anhydride group in the carbonyl region of the spectrum?

QUESTIONS

6-165. As stated in the discussion, direct dehydration can be used as a method for the preparation of five- and six-membered cyclic anhydrides. Propose a suitable mechanism for the reaction below:

$$\text{maleic acid} \xrightarrow{\Delta} \text{maleic anhydride} + H_2O$$

6-166. Propose a suitable mechanism for the formation of the mixed anhydride obtained in the following reaction:

6-167. There are two stereoisomeric 1,3-cyclobutane dicarboxylic acids. One can form a cyclic anhydride, the other cannot. Draw the structures of these compounds and indicate which one can be converted to a cyclic anhydride. Explain.

6-168. When maleic acid is heated to about 100 °C it forms maleic anhydride. However, fumaric acid requires a much higher temperature (250–300°C) before it dehydrates. In addition, it forms only maleic anhydride. Explain.

Maleic acid   Fumaric acid   Maleic anhydride

6-169. What product would you expect to obtain from reaction of one equivalent of methanol with phthalic anhydride?

BIBLIOGRAPHY

Selected references from *Organic Syntheses* in which anhydrides are prepared, using acetic anhydride as the dehydrating agent, include:

Cason, J. *Organic Syntheses*; Wiley: New York, 1963; Collect. Vol. IV, p. 630.

Clarke, H. T.; Rahrs, E. J. *Organic Syntheses*; Wiley: New York, 1944; Collect. Vol. I, p. 91.

Grummitt, O.; Egan, R.; Buck, A. *Organic Syntheses*; Wiley: New York, 1955; Collect. Vol. III, p. 449.

Horning, E. C.; Finelli, A. F. *Organic Syntheses*; Wiley: New York, 1963; Collect. Vol. IV, p. 790.

Nicolet, B. H.; Bender, J. A. *Organic Syntheses*; Wiley: New York, 1944; Collect. Vol. I, p. 410.

Shriner, R. L.; Furrow, C. L. Jr. *Organic Syntheses*; Wiley: New York, 1963; Collect. Vol. IV, p. 242.

The synthesis of succinic anhydride is described in:

Fieser, L. F.; Martin, E. L. *Organic Syntheses*; Wiley: New York, 1943; Collect. Vol. II, p. 560.

Diazonium Coupling Reaction: Methyl Red

EXPERIMENT 26

Common names: methyl red

CA number: [493-52-7]

CA name as indexed: benzoic acid, 2-[[4-(dimethylamino)phenyl]azo]-

Purpose. In this experiment you will learn the process of generating arenediazonium salts in solution. The arenediazonium salt generated will be used in an electrophilic aromatic substitution reaction (diazo coupling) to prepare an azobenzene derivative. Many azobenzene derivatives, including the one prepared here, have extensively conjugated π-electron systems. Because these are highly colored compounds, they are generally referred to as azo dyes.

Prior Reading
    Technique 5: Crystallization
                      Use of the Hirsch Funnel (pp. 92–93)
                      Craig Tube Crystallization (pp. 93–95)

REACTION

Anthranilic acid   Anthranyl diazonium chloride

Anthranyl diazonium chloride   *N,N*-Dimethylaniline   Methyl red

DISCUSSION

The coupling of a diazonium salt to a suitable aromatic substrate is an example of an aromatic electrophilic substitution reaction. When primary aromatic (and also aliphatic) amines ($ArNH_2$) are treated with nitrous acid ($NaNO_2$ + $HCl \rightarrow HONO$), they are converted into diazonium cations, ArN_2^+. In solution, nitrous acid (HONO) is in equilibrium with its anhydride, dinitrogen trioxide (N_2O_3), which is the actual diazotizing agent. The primary amine reacts with the dinitrogen trioxide to form a nitrosamine:

$$Ar—NH_2 + N_2O_3 \longrightarrow Ar—NH—N=\ddot{O} + HONO$$
(a nitrosamine)

The nitrosamine is in equilibrium with its tautomer, a diazoic acid. The diazoic acid then undergoes dehydration to form the diazonium salt. Diazonium salts are explosive when dry, and therefore are generally not isolated.

$$Ar—\ddot{N}H—\ddot{N}=\ddot{O}: + H_3O:^+ \rightleftharpoons Ar—\ddot{N}=\ddot{N}—\ddot{O}H + H_3O:^+$$
(a nitrosamine)   (a diazoic acid)

$$Ar—\ddot{N}=\ddot{N}—\ddot{O}H + H_3O:^+ \longrightarrow Ar—\overset{+}{N}\equiv N: + 2 H_2\ddot{O}:$$
Diazonium ion

Reaction of the diazonium salt with various aromatic compounds leads to the formation of azo derivatives by what is generally called a "coupling reaction," but is mechanistically simply an ordinary electrophilic aromatic substitution reaction. The mechanism of the reaction is given here:

Azo dyes find use as acid–base indicators. For example, the Methyl Red prepared in this experiment, Methyl Orange, and Congo Red are well-known acid–base indicators. Azo dyes are commonly used in the textile, food, and cosmetic industries; FD&C Yellow No. 6, a yellow azo dye is used to color candy, ice cream, beverages, and so on. Several azo dyes (including Butter Yellow and FD&C Red No. 2) have been banned by the FDA from use in foods, drugs, and cosmetics in the United States because of suspected carcinogenic properties.

Methyl Orange

Congo Red

FD&C Yellow No. 6

FD&C Red No. 2

Butter Yellow

EXPERIMENTAL PROCEDURE

Estimated time to complete the experiment: 3.0 h.
The reaction is shown on p. 357.

Physical Properties of Reactants							
Compound	MW	Amount	mmol	mp (°C)	bp (°C)	d	n_D
Anthranilic acid	137.14	65 mg	0.47	146–147			
Concd HCl		150 µL					
Water		800 µL					
Sodium nitrite	69.0	36 mg	0.52	271			
N,N-Dimethylaniline	121.18	89 µL	0.71		194	0.96	1.5582
Sodium acetate	82.03	68 mg	0.83	324			
10% aq. NaOH		100 µL					

CAUTION: When dry, benzenediazonium 2-carboxylate detonates violently upon being scraped or heated. It must, therefore, be kept in solution at all times.

< 5 °C
thermometer

Speed   Heat

Addition 1:
Anthranilic acid, 65 mg
+ concd HCl, 150 µL +
H₂O, 600 µL +
NaNO₂, 36 mg
Addition 2:
C₆H₅N(CH₃)₂, 85 mg
Addition 3:
CH₃CO₂Na, 68 mg +
H₂O, 200 µL

Reagents and Equipment.   Equip a 3.0-mL conical vial with a magnetic spin vane and an air condenser (◄). Weigh and add 65 mg (0.48 mmol) of anthranilic acid to the vial. Now add a solution of 150 µL of concentrated hydrochloric acid dissolved in 400 µL of water to the vial, using a Pasteur pipet.

CAUTION: When preparing the acid solution, the acid must be added *to* the water. Dispense these reagents using automatic delivery pipets.

If necessary, warm the mixture, with stirring, on a hot plate magnetic stirrer to obtain a homogeneous solution. Cool the solution in an ice bath, with stirring, for 10 min.

In a 10 × 75-mm test tube, or a small vial, prepare a solution of 36 mg (0.52 mmol) of sodium nitrite dissolved in 200 µL of water. Cool this solution in an ice bath.

Reaction Conditions.   When both solutions in the ice bath are cooled to a temperature below 5 °C slowly add (dropwise) the nitrite solution to the stirred anthranilic acid solution, while maintaining the temperature below 5 °C. This transfer is accomplished using a Pasteur pipet. *The solution must be kept cool so that the diazonium salt will not hydrolyze to the corresponding phenol.*

After a period of 4–5 min, check the clear solution of anthranyldiazonium chloride for the presence of excess nitrous acid by placing a drop of the solution on a piece of potassium iodide–starch test paper. If an excess is present, the test paper gives an immediate blue color. If no color is obtained, prepare additional nitrite solution and add as before until a positive test is observed.

Remove the air condenser from the reaction vial containing the solution of anthranyldiazonium chloride. Fairly rapidly, add 89 µL (85 mg, 0.71 mmol) of N,N-dimethylaniline (automatic delivery pipet). Reattach the air condenser.

HOOD

CAUTION: This aniline derivative is toxic and should be dispensed in the **hood**.

Stir the solution for an additional 15 min, keeping the temperature below 5 °C.

Prepare a solution of 68 mg (0.83 mmol) of sodium acetate dissolved in 200 μL of water in a 10 × 75-mm test tube. Transfer this solution (Pasteur pipet) to the reaction mixture. Make this addition *without* removing the air condenser. Maintain the resulting solution at 5 °C with stirring, for an additional 20 min.

Remove the reaction vial from the ice bath and allow it to stand for 15 min in order to warm to ambient temperature.

Now add 100 μL of 10% aqueous NaOH solution (automatic delivery pipet) to the solution. Allow the reaction mixture to stand at room temperature for about 30 min. *The formation of the azo compound is a very slow reaction, but the rate of formation is increased by raising the pH of the solution.*

Isolation of Product.   Collect the precipitate of crude Methyl Red dye by vacuum filtration using a Hirsch funnel (➡). Rinse the reaction flask with 0.5 mL of water and use this rinse to wash the crystals. Then wash the crystals with 0.5 mL of 3 M acetic acid, to remove unreacted N,N-dimethylaniline from the product, followed by another wash with 0.5 mL of water. *This last wash is usually pale pink in color.*

NOTE. *Dispense the small amounts of water and acetic acid using a calibrated Pasteur pipet.*

Purification and Characterization.   Dissolve the crude product in 500 μL of methanol. If necessary, warm the mixture in a beaker of hot water to aid in the dissolution. Cool the solution in an ice bath and collect the resulting crystals of Methyl Red by vacuum filtration using a Hirsch funnel. Dry the material on filter paper or under vacuum at room temperature.

Weigh the product and calculate the percent yield. Determine the melting point and compare it to the value given in the literature. If further purification is desired, recrystallize the material from toluene using a Craig tube.

Dye collected here

Aqueous acetic acid, ~ 2.5 mL + reaction byproducts

QUESTIONS

6-170. In the experiment, a point is made that the formation of the azo compound is a slow reaction, but that the rate is increased by raising the pH of the solution. Why is this necessary? In other words, how does the pH of the solution affect the reactivity of the N,N-dimethylaniline reagent?

6-171. In relation to Question 6-170, diazonium salts couple with phenols in slightly alkaline solution. What effect does the pH of the solution have on the reactivity of the phenol?

6-172. Starting with the appropriate aromatic amine and using any other organic or inorganic reagent, outline a synthetic sequence for the preparation of the following azo dyes:

(a) Chrysoidine

(b) Para Red

(c) Congo Red

6-173. What is the main structural feature of the azo dyes that causes them to be colored compounds?

6-174. Methyl Orange is an acid–base indicator. In dilute solution at pH > 4.4, it is yellow.

$$\overset{+}{N}a, \bar{O}_3S\text{---}\underset{}{\text{(ring)}}\text{---}\ddot{N}=\ddot{N}\text{---}\underset{}{\text{(ring)}}\text{---}\ddot{N}(CH_3)_2$$

At pH = 3.2 the solution appears red. Draw a structure of the species that is formed at the lower pH if the acid proton adds to the azo nitrogen atom adjacent to the aromatic ring containing —SO_3^- group. Why does the proton add to this particular nitrogen when two other nitrogen atoms are available in the molecule?

BIBLIOGRAPHY

For reviews on diazo compounds and azo dyes, see:

Gordon, P. F.; Gregory, P. *Organic Chemistry of Colour*; Springer: New York, 1983, p. 95.

Patai, S., Ed. *The Chemistry of Diazonium and Diazo Groups*; Wiley: New York, 1978, Chap. 8, 11, 14.

Saunders, H.; Allen, R. L. M. *Aromatic Diazo Compounds*, 3rd ed.; Edward Arnold: London, 1985.

Smith, M. B.; March. J. *Advanced Organic Chemistry*, 7th ed.; Wiley-Interscience: New York, 2013, Chap. 11, p. 597.

Zollinger, H. *Color Chemistry*; VCH: New York, 1987, p. 85.

Zollinger, H. *Diazo Chemistry I*; VCH: New York, 1994.

Selected coupling reactions with diazonium salts from *Organic Syntheses*:

Cleland, G. H. *Organic Syntheses*; Wiley: New York, 1988; Collect. Vol. VI, p. 21.

Conant, J. B.; Lutz, R. E.; Corson, B. B. *Organic Syntheses*; 1941; Collect. Vol. I, p. 49.

Fieser, L. F. *Organic Syntheses*; Wiley: New York, 1943; Collect. Vol. II, p. 35. *ibid.*, p. 39.

Hartwell, J. L.; Fieser, L. F. *Organic Syntheses*; Wiley: New York, 1943; Collect. Vol. II, p. 145.

Santurri, P.; Robbins, F.; Stubbins, R. *Organic Syntheses*; Wiley: New York, 1973; Collect. Vol. V, p. 341.

The synthesis of Methyl Red is also given in *Organic Syntheses*:

Clarke, H. T.; Kirner, W. R. *Organic Syntheses*; Wiley: New York, 1941; Collect. Vol. I, p. 374.

The present experiment is an adaptation of that given in:

Vogel, A. I. *A Textbook of Practical Organic Chemistry*, 5th ed.; Furnis, B. S., et al., Eds.; Wiley: New York, 1989.

EXPERIMENT
27

Friedel–Crafts Acylation: Acetylferrocene and Diacetylferrocene

Common name: acetylferrocene

CA number: [1271-55-2]

CA name as indexed: ferrocene, acetyl-

Common names: diacetylferrocene, 1,1′-diacetylferrocene

CA number: [1273-94-5]

CA name as indexed: ferrocene, 1,1′-diacetyl-

Purpose. To investigate the conditions under which the synthetically important Friedel–Crafts acylation (alkanoylation) reaction is carried out. The reaction described here illustrates electrophilic aromatic substitution on an aromatic ring contained in an organometallic compound. The highly colored products are easily separated by both thin-layer and dry-column chromatography.

Prior Reading

   Technique 4: Solvent Extraction
               Liquid-Liquid Extraction (p. 76)
               Drying of the Wet Organic Layer (pp. 84–87)
   Technique 6: Chromatography
               Column Chromatography (pp. 96–99)
               Thin-Layer Chromatography (pp. 101–103)
               Concentration of Solutions (pp. 105–107)

REACTION

Ferrocene   1-Acetylferrocene   1,1′-Diacetylferrocene

DISCUSSION

The generation of the appropriate electrophile (carbocation, carbocation complex, or acylium ion) in the presence of an aromatic ring system (nucleophile) can lead to alkylation or acylation of the aromatic ring. This set of reactions, discovered by Charles Friedel and James Crafts in 1877, originally used aluminum chloride as the catalyst. The reaction is now known to be catalyzed by a wide range of Lewis acids, including ferric chloride, zinc chloride, boron trifluoride, and strong acids, such as sulfuric, phosphoric, and hydrofluoric acids.

   Alkylation is accomplished by use of haloalkanes, alcohols, or alkenes—any species that can function as a carbocation precursor. The alkylation reaction is accompanied by two significant and limiting side reactions: polyalkylation, due to ring activation by the added alkyl groups, and rearrangement of the intermediate carbocation. These lead to diminished yields and mixtures of products that can be difficult to separate, as shown here:

Acylation reactions generally do not suffer from these limitations, and can be conducted using acid chlorides or anhydrides as the electrophilic reagents. Since the introduction of a carbonyl group onto the aromatic ring in an acylation reaction deactivates the ring, the problem of multiple substitution is avoided. The acylium cation, since it is resonance-stabilized, is unlikely to rearrange.

The mechanism involves three steps: (1) formation of a cationic electrophile, (2) nucleophilic attack on this electrophile by an aromatic ring, and (3) loss of a proton from the resulting cation to regenerate the aromatic ring system. The mechanism shown here represents the $AlCl_3$ catalyzed generation of the acylium ion electrophile from acetyl chloride (ethanoyl chloride), followed by subsequent nucleophilic attack by the ferrocene ring system:

Acylium ion

The present experiment also demonstrates the practical value of monitoring reaction progress by TLC analysis.

Charles Friedel (1832–1899). Friedel was Professor of Chemistry at the Sorbonne. He did extensive work on ketones, lactic acid, and glycerol and he discovered isopropyl alcohol. He is best known for his studies of the use of aluminum chloride in the synthesis of aromatic products (**Friedel–Crafts reaction**, 1877). Friedel prepared a series of esters of silicic acid and demonstrated the analogy between the compounds of carbon and silicon, meanwhile confirming the atomic weight of silicon. He determined the vapor densities and molecular weights of the chlorides of aluminum, iron, and gallium.[25]

James Mason Crafts (1839–1917). Crafts was Professor of Chemistry at Cornell University and later at Massachusetts Institute of Technology, where he eventually became President. Crafts studied with Bunsen (Germany) and Wurtz (France) and also worked on the organic compounds of silicon. Crafts was, of course, the codiscoverer of the **Friedel–Crafts reaction**. He also carried out investigations in the area of thermochemistry, catalytic effects in concentrated solutions, and determination of the densities of the halogens at high temperatures.[26]

EXPERIMENTAL PROCEDURE

Estimated time to complete the experiment: two 3.0-h laboratory periods.

Physical Properties of Reactants							
Compound	MW	Amount	mmol	mp (°C)	bp (°C)	d	n_D
Aluminum chloride	133.34	150 mg	1.12	190			
Acetyl chloride	78.50	80 μL	1.12		51	1.11	1.3898
Ferrocene	186.04	100 mg	0.54	173			
Methylene chloride		4.0 mL			40		

Reagents and Equipment

NOTE. Dry the glassware in an oven at 110 °C for 30 min and allow it to cool in a desiccator before starting the experiment.

Equip a tared 5.0-mL conical vial containing a magnetic spin vane with a Claisen head protected by a calcium chloride drying tube and a septum cap. Weigh and add 150 mg (1.12 mmol) of fresh, anhydrous aluminum chloride (➡).

Using a calibrated Pasteur pipet, add 2.5 mL of methylene chloride to the reaction vial. In the **hood**, with swirling, add 80 μL (1.12 mmol) of acetyl chloride from an automatic delivery pipet. Use a syringe to add a solution of 100 mg (0.54 mmol) of ferrocene dissolved in 1.5 mL of methylene chloride to the resulting mixture.

NOTE. Use a capped vial and recap it between the addition of each reagent. After addition of the acetyl chloride, attach the vial to the Claisen head and add the ferrocene solution through the septum as shown in the figure. Do this in one or two portions, depending on whether a 1- or 2-mL syringe is used.

HOOD

AlCl₃, 150 mg +
CH₂Cl₂, 5.5 mL +
CH₃COCl, 80 μL +
ferrocene, 100 mg

[25]See *Berichte* **1899**, *32*, 372; Crafts, J. M. *J. Chem. Soc.* **1900**, *77*, 993; *Bull. Soc. Chim. Fr.* **1900**, *23*, 1; Béhal, A. *ibid.*, **1932**, *51*, 1423; Willemart, A. *J. Chem. Educ.* **1949**, *26*, 3.
[26]Ashdown, A. A. *J. Chem. Educ.* **1928**, *5*, 911; Talbot, H. P. *J. Am. Chem. Soc.* **1917**, *39*, 171; Richards, T. W. *Proc. Am. Acad.* **1917–1919**, *53*, 801; Cross, C. R. *J. Natl. Acad. Sci.* **1914**, *9*, 159.

At this stage, the reaction mixture turns a deep violet.

HOOD

CAUTION: It is important to minimize the exposure to moist air during these transfers. Both the aluminum chloride and the acetyl chloride are highly moisture sensitive so that rapid, yet accurate, manipulations are necessary to minimize deactivation of these reagents, which leads to poor results. In addition, both chemicals are irritants. Avoid breathing the vapors or allowing the reagents to come in contact with skin. These reagents must be dispensed in the **hood**.

HOOD

TLC Sample Instructions.   Obtain an aliquot for TLC analysis by removing a small amount of the reaction mixture by touching the open end of a Pasteur pipet to the surface of the solution. First, remove the cap from the straight neck of the Claisen head, and then insert the pipet down the neck so as to touch the surface of the solution. Dissolve this aliquot in about 10 drops of cold methylene chloride in a small capped vial. Mark the vial and *save* it for TLC analysis.

Reaction Conditions.   Following the addition of the ferrocene solution, note the time, and begin stirring. Allow the reaction to proceed at room temperature for 15 min.

Isolation of Product.   Quench the reaction by transferring the mixture by Pasteur pipet to a 15-mL capped centrifuge tube (or a 15-mL screw-capped vial) containing 5.0 mL of ice water. Cool the tube in an ice bath and neutralize the resulting solution by dropwise addition (calibrated Pasteur pipet) of about 0.5 mL of 25% aqueous sodium hydroxide.

NOTE. *Avoid an excess of base. Use litmus or pH paper to confirm the neutralization.*

Now extract the mixture with three 3-mL portions of methylene chloride. Cap the tube, shake, vent, and allow the layers to separate (a Vortex mixer may be used in this step). Remove the lower (methylene chloride) layer using a Pasteur filter pipet. Combine the methylene chloride extracts in a 25-mL Erlenmeyer flask, and dry the wet solution over about 200 mg of granular anhydrous sodium sulfate for 20 min. Transfer the dried solution to a tared 10-mL Erlenmeyer flask, using a Pasteur filter pipet, in aliquots of 4 mL each. After each

HOOD transfer, concentrate the solution, in the **hood**, under a stream of dry nitrogen gas in a warm sand bath to a volume of about 0.5 mL. Rinse the drying agent with an additional 2.0 mL of methylene chloride and combine this rinse with the concentrate. Remove several drops of this solution by Pasteur pipet and place them in a capped vial containing 10 drops of *cold* methylene chloride. Mark the vial and *save* it for TLC analysis. Remove the remaining solvent by warming in a sand

HOOD bath in the **hood** to yield the crude, solid product (~130 mg). Weigh the residue.

OPTION. *As an option, the combined extracts and washes may be left to evaporate*

HOOD *in the* **hood** *in a 25-mL Erlenmeyer flask, with the mouth covered by filter paper, until the following week.*
    If the reaction is performed over a 2-week period, this is a convenient point at which to stop. However, if time permits, perform the TLC analysis now.

Thin-Layer Chromatographic Analysis.   Use TLC to analyze the two samples saved above. Also analyze a standard mixture of the substituted ferrocenes (supplied by the instructor) at the same time. Use the developed TLC plates as a guide to determine the product mixture obtained in the reaction and as an aid in determining the appropriate elution solvent required for separation of the mixture by dry-column chromatography.

INFORMATION. Good results have been achieved by conducting the TLC analysis with Eastman Kodak silica gel–polyethylene terephthalate plates (#13179). Activate the plates at an oven temperature of 100 °C for 30 min. Place them in a desiccator for cooling and storing until used. Elute the plates using pure methylene chloride as the elution solvent. Visualization of unreacted ferrocene can be enhanced with iodine vapor. See Prior Reading for methods of TLC analysis and determination of R_f values.

Purification and Characterization.   Now purify the reaction products formed in the reaction by dry-column chromatography. The term *dry-column chromatography* refers to the fact that the column is packed with dry alumina, rather than with a slurry (see Prior Reading). Dissolve the solid product residue isolated above in 0.5 mL (calibrated Pasteur pipet) of methylene chloride in a small vial. Mix this solution with 300 mg of alumina (activity III, see Glossary) in a tared vial, and evaporate the solvent under a stream of dry nitrogen in the **hood** to give a product–alumina mixture. Assemble a chromatographic buret column in the following order (bottom to top): prewashed cotton plug, 5 mm of sand, 60–80 mm of alumina (~5.0 g, activity III), *one-half* of the product–alumina mixture, and 10 mm of alumina (➡).

HOOD

NOTE. This procedure prevents overloading of the chromatographic column during the separation of the reaction products. If the yield of crude reaction products exceeds 75 mg (the usual case), introduce one-half of the alumina–product mixture to the column. If the crude products, however, are obtained in quantities of less than 75 mg, add the entire alumina–product mixture to the top of the column. If only one-half of the alumina–crude ferrocene acylation product mixture is placed on the column, it is important to reweigh the tared vial to establish a reasonably accurate estimate of the overall yields obtained in the reaction.

Al$_2$O$_3$, 150–300 mg
+ $\frac{1}{2}$ acylation products

Al$_2$O$_3$(III), ~ 5.0g

Sand

Collect materials by
tracking colored bands

Given the polar nature of the alumina, the products will elute in order of increasing polarity: ferrocene followed by acetylferrocene, followed by diacetylferrocene. Begin elution of the column with pure hexane if TLC analysis indicates that unreacted ferrocene is present in the product mixture. Be sure to add the initial solvent down the side of the column so as not to disturb the alumina bed. During elution, the ferrocenes will separate into two or three bands of different colors on the column. The volume of each eluted fraction should be in the range of 2–5 mL if the band is carefully tracked down the column. Once the ferrocene band has been collected, continue the elution with a 1:1 mixture of CH$_2$Cl$_2$/hexane to obtain the monosubstituted product. Further elution with a 9:1 mixture of CH$_2$Cl$_2$/CH$_3$OH will elute the disubstituted material. Collect and save each chromatographic band separately. Store the solvent that elutes without color in an Erlenmeyer flask or beaker until you have isolated all your product. In the **hood**, remove the solvent under a stream of dry nitrogen gas, using a warm sand bath. During concentration of the solvent, spot each fraction on a TLC plate to verify the separation and purity of its contents.

HOOD

Determine the melting point of each of the isolated products and compare your results to those reported in the literature.

Obtain an IR spectrum of each material and compare the results with an authentic sample or spectra found in the literature (*The Aldrich Library of IR Spectra* and/or SciFinder Scholar). Interpretation of the spectra allows an unambiguous determination of substitution based on the presence or absence of absorption in the 1100- to 900-wavenumber region of the spectrum.

Characterization of the Fractions (Total Sample) Isolated from the Reaction Workup

Acetylferrocene: mp _____ °C; _____ mg, _____ mmol

Diacetylferrocene: mp _____ °C; _____ mg, _____ mmol

Total: _____ mmol, _____ % yield

QUESTIONS

6-175. In the formation of diacetylferrocene, the product is always the one in which each ring is monoacetylated. Why is no diacetylferrocene produced in which both acetyl groups are on the same aromatic ring?

6-176. Ferrocene cannot be nitrated using the conventional HNO_3—H_2SO_4 mixed acid conditions, even though nitration is an electrophilic aromatic substitution reaction. Explain.

6-177. In contrast to nitration (Question 6-176), ferrocene undergoes the acetylation and sulfonation reaction. Explain.

6-178. The bonding in ferrocene involves sharing of the 6 electrons from each cyclopentadienyl ring with the iron atom. Based on the electronic configuration of the iron species in the compound, show that a favorable 18-electron inert gas configuration is established at the iron atom.

6-179. In a manner similar to that in Question 6-178, predict whether ruthenocene and osmocene (the ferrocene analogues of ruthenium and osmium) would be stable compounds. Explain.

6-180. Would you predict that bis(benzene)chromium (0) would be a stable compound? Explain.

6-181. Not all of the Friedel–Crafts reactions detailed below occur. For those that do, predict the major product and highlight with an explanation those that do not.

BIBLIOGRAPHY

The acetylation of ferrocene has been monitored using chromatographic techniques. Several references are listed:

Amenta, D. S.; DeVore, T. C.; Gallaher, T. N.; Zook, C. M. *J. Chem. Educ.* **1996**, *73*, 575.

Bohen, J. M.; Joullié, M. M.; Kaplan, F. A. *J. Chem. Educ.* **1973**, *50*, 367.

Nerwith, T. L.; Srouji, N. *J. Chem. Educ.* **1995**, *72*; 455.

Several reviews on the Friedel–Crafts reaction, selected from many cited in the literature:

Gore, P. H. *Chem. Rev.* **1955**, *55*, 229.

Harada, T; Ohno, T.; Kobayashi, S.; Mukaiyama, T. *Synthesis* **1991**, 1216.

Heaney, H. *Comp. Org. Syn.* **1991**, *2*, 733.

Olah, G. A., Ed., *Friedel–Crafts and Related Reactions;* Interscience: New York, 1963–1965, Vols. I–IV.

Smith, M. B. *March's Advanced Organic Chemistry*, 8th ed.; Wiley: New York, 2019, Chap. 11.

Several of the large number of Friedel–Crafts acetylation reactions given in Organic Syntheses:

Adams, R.; Noller, C. R. *Organic Syntheses*; Wiley: New York, 1941; Collect. Vol. I, p. 109.

Arsenijivic, L; Arsenijivic, V.; Horeau, A.; Jacques, J. *Organic Syntheses*; Wiley: New York, 1988; Collect. Vol. VI, p. 34.

Fieser, L. F. *Organic Syntheses*; Wiley: New York, 1941; Collect. Vol. I, p. 517.

Fieser, L. F. *Organic Syntheses*; Wiley: New York, 1955; Collect. Vol. III, p. 6.

Marvel, C. S.; Sperry, W. M. *Organic Syntheses*; Wiley: New York, 1941; Collect. Vol. I, p. 95.

Olson, C. E.; Bader, A. F. *Organic Syntheses*; Wiley: New York, 1963; Collect. Vol. IV, p. 898.

Sims, J. J.; Selman, L. H.; Cadogan, M. *Organic Syntheses*; Wiley: New York, 1988; Collect. Vol. VI, p. 744.

Halogenation: Electrophilic Aromatic Substitution to Yield 4-Bromoacetanilide

EXPERIMENT 28

Common names: 4-bromoacetanilide, *p*-bromoacetanilide

CA number: [103-88-8]

CA name as indexed: acetamide, *N*-(4-bromophenyl)-

Purpose. This experiment extends our understanding of the experimental conditions under which electrophilic aromatic substitution reactions are carried out (also see Experiments [27] and [29A]–[29D]). It deals with electrophilic aromatic halogenation. The directive influence of the acetamido, —NHCOCH$_3$, group on the bromination of acetanilide is explored.

Prior Reading

> *Technique 5:* Crystallization
>> Use of the Hirsch funnel (pp. 92–93)
>> Craig Tube Crystallization (pp. 93–95)
> *Technique 6A:* Thin-Layer Chromatography (pp. 101–103)

REACTION

Acetanilide   Bromine   4-Bromoacetanilide

DISCUSSION

Aromatic compounds may be brominated by treatment with bromine in the presence of a Lewis acid catalyst, such as ferric chloride. For very electron-rich aromatic rings, such as arylamines, the reaction may proceed in the absence of a catalyst. With amines or phenols, in many cases, it is difficult to stop the bromination at monosubstitution, and all open ortho and para positions are brominated. For this reason, primary aromatic amines are often converted to a corresponding amide derivative, if a monobrominated product is desired. This strategy is demonstrated in the present experiment. The —NHCOCH$_3$ group is a less powerful o,p-directing group than —NH$_2$ due to the presence of the electron-withdrawing carbonyl group, which renders the ring less nucleophilic. Electrophilic substitution by bromine is still, however, effectively directed electronically to the ortho and para positions on the ring. The acetamido group, —NHCOCH$_3$, effectively blocks the ortho positions by steric hindrance. For these reasons, only para substitution is observed. The acetanilide used in this experiment may be prepared using the procedure described in Experiment [23A].

The mechanism of the bromination reaction is a classic illustration of an electrophilic substitution on an aromatic ring. The mechanism shown below is presented as proceeding without the aid of a catalyst:

EXPERIMENTAL PROCEDURE

Estimated time to complete the experiment: 1.5 h.

INSTRUCTOR PREPARATION. For the bromine/AcOH solution. Prepare the bromine–acetic acid reagent by mixing 2.5 mL of bromine with 5.0 mL of glacial acetic acid in the **hood**.

HOOD

Physical Properties of Reactants					
Compound	MW	Amount	mmol	mp (°C)	bp (°C)
Acetanilide	135.17	25 mg	0.19	114	
Glacial acetic acid	60.05	4 drops			118
Bromine–acetic acid solution		3 drops			

Reagents and Equipment. Weigh and place 25 mg (0.19 mmol) of acetanilide in a 3.0-mL conical vial fitted with a cap. Add between 8–10 drops of glacial acetic acid using a medicine dropper. Stir with a glass rod to help dissolve the acetanilide. Once in solution, add a magnetic spin. Now, in the **hood**, add to the clear solution three drops of the bromine–acetic acid solution. Cap the vial immediately.

HOOD

CAUTION: Bromine is a severe irritant. Bromine burns can be severe and require a long time to heal. Always wear plastic gloves and dispense the bromine solution in the **hood**.

HOOD

Reaction Conditions. Allow the reddish-brown solution to stand at room temperature for 10 min. During this period, yellow-orange colored crystals precipitate from the solution.

Isolation of Product. Using a calibrated Pasteur pipet add 0.5 mL of water to the reaction mixture with swirling, followed by 5 drops of aqueous sodium bisulfite solution (33%). This treatment destroys the unreacted bromine (and its residual color) and results in white crystals. Cool the reaction mixture in an ice bath for 10 min to maximize the yield of product.

Bromination product collected here

Aqueous acetic acid, ~ 2.0 mL and other reaction byproducts

Collect the crude reaction mixture by vacuum filtration using a Hirsch funnel (➡). Once collected, add 0.25 mL of cold water (using a calibrated Pasteur pipet) to the 3.0-mL conical vial. Transfer this rinse to the filter cake and repeat two more times (rinse followed by washing of the filter cake with 0.25-mL portions of cold water). Once complete, partially dry the filter cake by drawing air through the crystals under reduced pressure for approximately 5 min. A sheet of plastic food wrap over the funnel mouth aids this process (see Prior Reading).

Purification and Characterization. Purify all but 10 mg of the crude 4-bromoacetanilide by recrystallization from 95% ethanol using the Craig tube. Weigh the dried product and calculate the percent yield. Determine the melting point, compare both the crude and recrystallized materials by TLC analysis (plates with fluorescent indicator, R_f value for 4-bromoacetanilide is 0.32 using as solvent system ethyl acetate:hexane (2:3), and a UV lamp for visualization purposes), and compare your observed melting point to the value given in the literature.

Obtain an IR spectrum of the material and compare it with one found in the literature (*The Aldrich Library of IR Spectra* and/or *SciFinder Scholar*). If possible, obtain ^{1}H and/or ^{13}C NMR spectra of your material in DMSO-d_6.

Nuclear Magnetic Resonance Analysis. Figures 6.46 and 6.47 are, respectively, the ^{1}H and ^{13}C NMR spectra of *p*-bromoacetanilide in DMSO-d_6. These can be used to compare with the NMR spectra you may obtain of your product.

In the ^{13}C NMR spectrum, the DMSO-d_6 appears as a septet at 39.7 ppm. The resonance from the methyl group of the *p*-bromoacetanilide occurs at 24 ppm and the amide carbonyl carbon resonates at 169 ppm. The carbon atoms of the benzene ring are observed between 110 and 140 ppm.

In the ^{1}H NMR spectrum, the peak representing trace amounts of nondeuterated DMSO in DMSO-d_6 is seen at about 2.6 ppm. The peak at 3.4 ppm is water and at 0.0 ppm, the internal standard of TMS. Note the two small peaks located equidistant to the tall singlet near 2.0 ppm. The small "satellite" peaks are the result of the 1.1% of the methyl groups that have ^{13}C instead of ^{12}C, and thus here coupling between the carbon and the protons is observed. The two doublets for the aromatic protons are observed near 7.5 ppm. The amide NH proton, which is probably hydrogen bonded to the basic (Lewis) sulfoxide

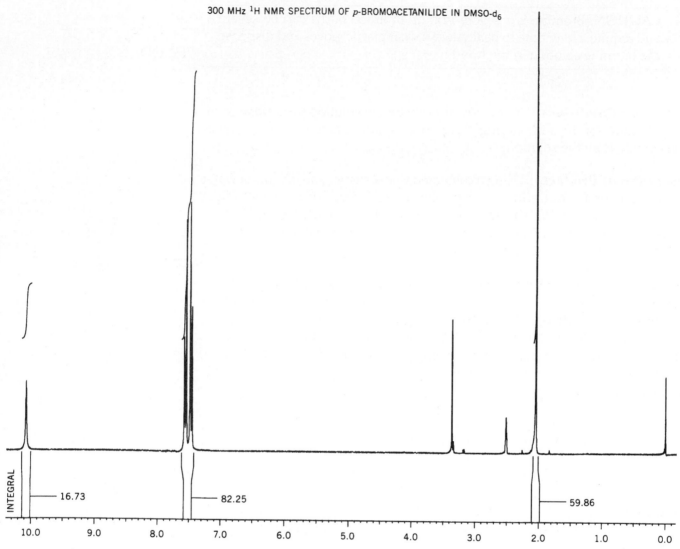

Figure 6.46 ¹H-NMR spectrum: *p*-bromoacetanilide.

functional group in DMSO-d_6, occurs rather downfield, near 10.1 ppm. This chemical shift may vary in your sample due to subtle differences in concentration, temperature, and moisture content of your DMSO-d_6.

Chemical Tests. Chemical classification tests may also be performed on the amide product. The ignition and the Beilstein test (Chapter 9) are used to confirm the presence of the aromatic ring and the halogen group, respectively. Does the hydroxamate test for amides (Chapter 9) give a positive result?

QUESTIONS

6-182. Use resonance structures to show why the group shown is a less powerful ortho–para directing group than the —NH₂ group:

Figure 6.47   ¹³C-NMR spectrum: *p*-bromoacetanilide.

6-183. Benzene is brominated in the presence of FeBr₃ catalyst:

Suggest an appropriate mechanism for this reaction.

6-184. Draw the structure of the major monobrominated product(s) formed when each of the following compounds is reacted with Br₂ in the presence of FeBr₃:

6-185. Arrange the following compounds in order of increasing reactivity toward electrophilic aromatic substitution. Explain the reason(s) for your decisions.

(a) [benzene ring with CF₃ substituent]

(b) [benzene ring with O–C(=O)–CH₃ substituent]

(c) [benzene ring with CH(CH₃)CH₃ substituent]

(d) [benzene ring with C(=O)–CF₃ substituent]

6-186. In the experiment, sodium bisulfite solution is added at the end to destroy the unreacted bromine. What reaction is occurring here? Is HSO_3^- acting as an oxidizing or reducing agent? Write a balanced equation as part of your answer.

6-187. Both the ^1H and ^{13}C NMR spectra (Figs. 6.46 and 6.47) provide unambiguous evidence that the bromination of acetanilide (1) occurred and (2) gave exclusively para substitution. Explain.

BIBLIOGRAPHY

For a general review of bromination, see:

Smith, M. B. *March's Advanced Organic Chemistry*, 8th ed.; Wiley: New York, 2019, Chap. 11.

For use of an alternate reagent for the bromination of acetanilide (in situ generation of bromine using potassium bromate and hydrobromic acid), see:

Schatz, P. F. *J. Chem. Educ.* **1996**, *73*, 267.

The following references are selected from a large number of examples given in Organic Syntheses that illustrate electrophilic aromatic substitution using bromine:

Adams, R.; Marvel, C. S. *Organic Syntheses*; Wiley: New York, 1941; Collect. Vol. I, p. 128.

Coleman, C. H.; Talbot, W. F. *Organic Syntheses*; Wiley: New York, 1943; Collect. Vol. II, p. 592.

Hartman, W. W.; Dickey, J. B. *Organic Syntheses*; Wiley: New York, 1943; Collect. Vol. II, p. 173.

Johnson, J. R.; Sandborn, L. T. *Organic Syntheses*; Wiley: New York, 1941; Collect. Vol. I, p. 111.

Robison, M. M.; Robison, B. L. *Organic Syntheses*; Wiley: New York, 1963; Collect. Vol. IV, p. 947.

Sandin, R. B.; McKee, R. A. *Organic Syntheses*; Wiley: New York, 1943; Collect. Vol. II, p. 100.

Smith, L. I. *Organic Syntheses*; Wiley: New York, 1943; Collect. Vol. II, p. 95.

EXPERIMENT 29

Nitration: 2,5-Dichloronitrobenzene; N,N′-Diacetyl-2,3-dinitro-1,4-phenylenediamine; 5-Nitrosalicylic Acid; and 2- and 4-Nitrophenol

Common name: 2,5-dichloronitrobenzene
CA number: [89-61-2]
CA name as indexed: benzene, 1,4-dichloro-2-nitro-

Common name: N,N′-diacetyl-2,3-dinitro-1,4-phenylenediamine
CA number: [7756-00-5]
CA name as indexed: acetamide, N,N′-(2,3-dinitro-1,4-phenylene)bis-

Common names: 5-nitrosalicylic acid, anilotic acid
CA number: [96-97-9]
CA name as indexed: benzoic acid, 2-hydroxy-5-nitro-

Common names: 2-nitrophenol, *o*-nitrophenol
CA number: [88-75-5]
CA name as indexed: phenol, 2-nitro-

Common names: 4-nitrophenol, *p*-nitrophenol
CA number: [100-02-7]
CA name as indexed: phenol, 4-nitro-

Purpose. Aromatic nitration is an important synthetic reaction. This experiment explores two methods used for placing a nitro group on an aromatic ring system via an electrophilic aromatic substitution reaction. In Experiments [29A], [29B], and [29C] anhydrous nitric acid is used as the nitrating agent. In Experiment [29D], nitration is accomplished using a $SiO_2 \cdot HNO_3$ reagent.

Prior Reading

Technique 2: Simple Distillation at the Semi-microscale Level (pp. 65–68)
Technique 5: Crystallization
              Use of the Hirsch Funnel (pp. 92–93)
              Craig Tube Crystallization (pp. 93–95)
Technique 6: Chromatography
              Column Chromatography (pp. 96–99)
              Thin-Layer Chromatography (pp. 101–103)
              Concentration of Solutions (pp. 105–108)
Chapter 4: Mixture Melting Points (p. 56)

GENERAL REACTION

DISCUSSION

The nitration reactions described in this experiment all demonstrate one of the classic electrophilic aromatic substitution reactions. Nitration has been used extensively in organic synthesis since a nitro group on an aromatic ring may be readily reduced to an amino group.

Once introduced onto the aromatic ring, the electron-withdrawing nitro group deactivates the ring toward further reactions with electrophiles. For example, bromination of nitrobenzene leads only to *m*-nitrobromobenzene; no dibromonitrobenzene is readily formed. However, when activating groups (π-electron donors) are present on the ring, it is possible to nitrate the ring twice. This phenomenon can be illustrated by comparing the results of the nitration of 1,4-dichlorobenzene (Experiment [29A]) with that of *N,N'*-diacetyl-1,4-phenylenediamine (Experiment [29B]). Because of the presence of the activating acetamido (CH_3CONH—) groups, the dinitro derivative forms readily. In

Experiment [29C] (the preparation of 5-nitrosalicylic acid), the directing influences of the 1-CO_2H and 2-OH substituents on the entering —NO_2 group are illustrated. In this example, these two groups complement each other since they both direct the entering nitro group to the 5 position. The 5 position and the 3 position are both electronically favored since the —CO_2H group is meta directing; the —OH group is ortho–para directing. The nitro group ends up at the 5 position, and not at the 3 position, due to steric effects.

The use of a silica gel–based reagent to accomplish nitration under fairly mild conditions is illustrated in Experiment [29D]. The nitrating reagent, $SiO_2 \cdot HNO_3$ is prepared by treatment of silica gel with nitric acid. In the experiment, phenol is nitrated to produce a mixture of products. *Thin-layer chromatography* is used to analyze the mixture, and the ortho and para nitrated phenols are separated by *column chromatography* using a silica gel column. If unreacted phenol is detected in the TLC analysis, an extraction technique is used to separate it from the para isomer. This separation technique is based on the fact that a nitrated phenol is more acidic than phenol itself.

It is generally accepted that the nitronium ion (NO_2^+) is the electrophile that adds to the aromatic ring. The overall mechanism for nitration follows:

$$HONO_2 + HONO_2 \rightleftharpoons H_2\overset{+}{O}-NO_2 + NO_3^-$$

$$H_2\overset{+}{O}-NO_2 + HONO_2 \rightleftharpoons H_3O^+ + NO_2^+ + NO_3^-$$

This mechanism illustrates two HNO_3 molecules reacting to generate the nitronium ion as when using the anhydrous nitric acid reagent. Sulfuric acid is often used to enhance the production of NO_2^+ as shown here:

$$HONO_2 + HOSO_3H \rightleftharpoons H_2\overset{+}{O}-NO_2 + HSO_4^-$$

$$H_2\overset{+}{O}-NO_2 \rightleftharpoons H_2O + \overset{+}{N}O_2$$

Thus a commonly used nitrating reagent is a mixture of concentrated sulfuric and nitric acids.

SEMI-MICROSCALE PREPARATION OF ANHYDROUS NITRIC ACID

Anhydrous nitric acid (HNO_3) is prepared by the following procedure.

HOOD

CAUTION: The reagents and the product of this preparation are highly corrosive. The distillation must be conducted in a **hood**. Appropriate gloves are strongly suggested. Prevent contact with eyes, skin, and clothing. Any spill should be neutralized using solid sodium carbonate or bicarbonate.

EXPERIMENTAL PROCEDURE

Estimated time to prepare: 0.5 h.

NOTE. Use this anhydrous nitric acid immediately for the nitration experiments given below. The amount obtained at the scale used here is sufficient for the preparation of two of the nitro compounds described in this experiment.

Concd HNO₃, 0.7 mL
+ concd H₂SO₄, 1.0 mL

Physical Properties of Reactants and Product				
Compound	MW	Amount	bp (°C)	d
Concd nitric acid (68%)		0.7 mL	120.5	1.41
Concd sulfuric acid (96–98%)		1.0 mL	338	1.84
Anhydrous nitric acid	63.01		83	1.40

Reagents and Equipment. Using two clean, dry 1.0-mL graduated pipets, add 0.7 mL of concentrated nitric acid, followed by 1.0 mL of concentrated sulfuric acid, to a 10-mL round-bottom flask containing a boiling stone. Swirl the flask gently to mix the reagents. Attach the flask to a Hickman still fitted with an air condenser (➡).

NOTE. It is useful to invert a 10-mL beaker over the air condenser to help contain the acid vapors.

CAUTION: Sulfuric acid can cause severe burns. Nitric acid is a strong oxidizing agent. Prevent contact with eyes, skin, and clothing. A spill can be neutralized using sodium carbonate or bicarbonate.

Reaction Conditions. Heat the acid solution very gently with a microburner, keeping the microburner in constant motion, until approximately 0.2 mL of anhydrous nitric acid has been collected as distillate in the collar of the still.

Purification and Characterization. Use the anhydrous nitric acid as collected. No further purification is required.

NOTE. Anhydrous nitric acid (white fuming nitric acid) is a colorless liquid, bp 83°C. It is estimated that the nitric acid obtained in this preparation is at least 99.5–100% pure. If it is necessary to store the distillate, remove the acid from the collar of the still (Pasteur pipet) and place it in a 1.0-mL conical vial fitted with a glass stopper. It may be necessary to slightly bend the end of the pipet in a flame so that it can reach the collar of a still that does not have a side port. The anhydrous nitric acid is colorless or faintly yellow.

2,5-Dichloronitrobenzene

Experiment 29A

REACTION

1,4-Dichlorobenzene   2,5-Dichloronitrobenzene

EXPERIMENTAL PROCEDURE

Estimated time to complete the experiment: 0.5 h.

Physical Properties of Reactants						
Compound	MW	Amount	mmol	mp (°C)	bp (°C)	d
1,4-Dichlorobenzene	147.01	38 mg	0.26	53		
Anhydrous nitric acid	63.01	100 μL	2.4		83	1.50

HOOD

1,4-Cl$_2$C$_6$H$_4$, 38 mg + anhydrous HNO$_3$, 100 μL

Reagents and Equipment. Equip a 3.0-mL conical vial with an air condenser (◄). Weigh and add 38 mg (0.26 mmol) of 1,4-dichlorobenzene, followed by 100 μL of anhydrous nitric acid delivered from a calibrated Pasteur pipet (9 in.).

> CAUTION: The nitric acid reagent is highly corrosive. Prevent contact with eyes, skin, and clothing. A spill is neutralized using solid sodium carbonate or bicarbonate.

Reaction Conditions. Allow the resulting solution to stand at room temperature for a period of 15 min. Next add 1.0 mL of water (calibrated Pasteur pipet) dropwise, while stirring with a thin glass rod, and then place the vial in an ice bath to cool.

Nitration product collected here

Aqueous HNO$_3$, ~ 5 mL

Isolation of Product. Collect the crystalline precipitate by vacuum filtration using a Hirsch funnel (◄). Wash the filter cake with four 1.0-mL portions of water (calibrated Pasteur pipet) and then place it on a porous clay plate or on filter paper to dry.

Purification and Characterization. The product, consisting of fine, white needles, is sufficiently pure for characterization. It may be recrystallized from ethanol–water, using a Craig tube, if desired.

Weigh the 2,5-dichloronitrobenzene and calculate the percent yield. Determine the melting point and compare your result to that reported in the literature. Notice that the starting material and the nitrated product have very close melting points. It is recommended that a mixed melting point be carried out to establish that the desired product has been isolated (see Prior Reading).

A further or alternative check on the purity of the material may be made using thin-layer chromatography (TLC).

INFORMATION. Carry out the TLC analysis with Eastman Kodak silica gel–polyethylene terephthalate plates (#13179). Activate the plates at an oven temperature of 100 °C for 30 min. Place them in a desiccator for cooling and storing until used. Elute the plates using hexane solvent. Visualization is accomplished with UV light. See Prior Reading for the methods of TLC analysis and determination of R_f values.

Chemical Tests. Additional chemical tests (Chapter 9) may also be performed to further characterize the product. Does the ignition test confirm the presence of the aromatic ring? Does the Beilstein test detect the presence of chlorine? Can the sodium fusion test detect the presence of nitrogen? Can the specific presence of the nitro group be detected by reaction with ferrous hydroxide solution?

N,N'-Diacetyl-2,3-dinitro-1,4-phenylenediamine

REACTION

$$CH_3-\overset{\overset{\cdot\cdot\cdot}{O}\cdot}{C}-\underset{H}{N}- \text{(ring)} -\underset{H}{N}-\overset{\overset{\cdot\cdot\cdot}{O}\cdot}{C}-CH_3 + HNO_3 \longrightarrow CH_3-\overset{\overset{\cdot\cdot\cdot}{O}\cdot}{C}-\underset{H}{N}- \text{(ring with } NO_2, NO_2) -\underset{H}{N}-\overset{\overset{\cdot\cdot\cdot}{O}\cdot}{C}-CH_3$$

N,N'-Diacetyl-1,4-phenylenediamine   *N,N'*-Diacetyl-2,3-dinitro-1,4-phenylenediamine

EXPERIMENTAL PROCEDURE

Estimated time to complete the experiment: 0.5 h.

Physical Properties of Reactants and Product						
Compound	MW	Amount	mmol	mp (°C)	bp (°C)	*d*
N,N'-Diacetyl-1,4-phenylenediamine	192	48 mg	0.25	312–315		
Anhydrous nitric acid	63.01	100 µL	2.4		83	1.50
N,N'-Diacetyl-2,3-dinitro-1,4-phenylenediamine	282			257		

Reagents and Equipment.   To a 3.0-mL conical vial equipped with an air condenser, weigh and add 48 mg (0.25 mmol) of *N,N'*-diacetyl-1,4-phenylenediamine. Now, using **caution,** add dropwise 100 µL of anhydrous nitric acid delivered from a calibrated Pasteur pipet (9 in.) (➡). *The N,N'-diacetyl-1,4-phenylenediamine is prepared by the procedure outlined in Experiment [23B].*

CAUTION

HOOD

EXOTHERMIC 1,4-(CH₃CONH)₂C₆H₄, 48 mg + anhydrous HNO₃, 100 µL

CAUTION: The reaction is highly **exothermic.** A vigorous reaction occurs if the acid is added too rapidly. The nitric acid reagent is highly corrosive; prevent contact with eyes, skin, and clothing. A spill is neutralized using sodium carbonate or bicarbonate.

Reaction Conditions.   Allow the resulting solution to stand at room temperature for a period of 10 min. Add 1.0 mL of water from a calibrated Pasteur pipet dropwise and then place the vial in an ice bath to cool.

Isolation of Product   Collect the resulting yellow precipitate by vacuum filtration using a Hirsch funnel (➡). Wash the filter cake with four 1.0-mL portions of water from a calibrated Pasteur pipet. Maintain suction to aid in drying of the product. A piece of plastic food wrap over the mouth of the funnel can aid this process (see Prior Reading). Dry the material further on a porous clay plate, on filter paper, or under vacuum at room temperature.

Nitration product collected here

Aqueous HNO₃, ~5 mL

Purification and Characterization.   The product needs no further purification. Weigh the dried material and calculate the percent yield. Determine

ABSCISSA		ORDINATE		SCAN TIME 12 min		REP. SCAN_____ SINGLE BEAM_____
EXPANSION_____		EXPANSION_____		RESPONSE _____		TIME DRIVE_____ PRE SAMPLE CHOP_____
SUPPRESSION_____		% T_____ ABS_____		SLIT PROGRAM normal		OPERATOR_____ DATE_____
SAMPLE N,N'-diacetyl-2,3-dinitro-1,		REMARKS Perkin-Elmer 283 B		SOLVENT KBr pellet		CELL PATH_____
ORIGIN 4-phenylenediamine				CONCENTRATION_____		REFERENCE_____

Figure 6.48   IR spectrum: *N,N'*-diacetyl-2,3-dinitro-1,4-phenylenediamine.

the melting point and compare your result with the melting point listed in the above Reactant Product table. Obtain an IR spectrum of your product and compare it with that shown in Figure 6.48.

Experiment 29C

5-Nitrosalicylic Acid

This material may be used to prepare the caffeine 5-nitrosalicylate derivative (see Experiment [11B]).

REACTION

Salicylic acid $+ HNO_3 \longrightarrow$ 5-Nitrosalicylic acid

EXPERIMENTAL PROCEDURE

Estimated time to complete the experiment: 0.5 h.

Physical Properties of Reactants						
Compound	MW	Amount	mmol	mp (°C)	bp (°C)	d
Salicylic acid	138.12	50 mg	0.36	159		
Anhydrous nitric acid	63.01	100 µL	2.4		83	1.50

CAUTION: This reaction should be conducted in a **hood**.

HOOD

CAUTION

Reagents and Equipment.   Weigh and add 50 mg (0.36 mmol) of salicylic acid to a 3.0-mL conical vial equipped with an air condenser. Place the vial in an ice bath to cool. Also place a stoppered conical vial containing 100 µL of freshly prepared **ANHYDROUS** nitric acid (use **caution**) from a Pasteur pipet in the ice bath (➡).

Salicylic acid, 50 mg +
anhydrous HNO₃, 100 µL

CAUTION: The nitric acid reagent is highly corrosive; prevent contact with eyes, skin, and clothing. A spill is neutralized using solid sodium carbonate or bicarbonate.

Reaction Conditions.   **Carefully** add the cold nitric acid dropwise from a calibrated Pasteur pipet (9 in.) to the salicylic acid. Keep the vial containing the salicylic acid in the ice bath during the addition. The evolution of a red-brown gas (NO_2, which is **toxic**) is observed during the addition, if the acid is not pure.

TOXIC

CAUTION: The reaction is highly **exothermic**. A very vigorous reaction occurs if the acid is added too rapidly.

EXOTHERMIC

Allow the vial to stand in the ice bath for an additional 20 min and then add 1.0 mL of distilled water dropwise from a calibrated Pasteur pipet to the reaction mixture.

Isolation of Product.   Collect the orange-pink solid by vacuum filtration using a Hirsch funnel (➡). Wash the filter cake with four 1.0-mL portions of cold water from a calibrated Pasteur pipet. Maintain suction to aid in drying of the product. A piece of plastic food wrap over the mouth of the funnel can aid this process (see Prior Reading). Dry the material further on a porous clay plate, on filter paper, or under vacuum at room temperature.

Nitration product
collected here

Aqueous HNO₃,~ 5 mL

Purification and Characterization.   Recrystallize the product using a Craig tube by dissolving the material in the *minimum* amount of absolute ethanol, followed by the dropwise addition of water until precipitation occurs. Cool the mixture in an ice bath and separate the light-yellow crystals. Dry them on a porous clay plate or under vacuum as noted above.

Weigh the crystals and calculate the percent yield. Determine the melting point and compare your value to that listed in the literature (*The Aldrich*

Library of IR Spectra and/or SciFinder Scholar). Obtain an IR spectrum of the material and compare it with that recorded in the literature.

2- and 4-Nitrophenol

Experiment 29D

INSTRUCTOR PREPARATION. In a 250-mL Erlenmeyer flask containing a magnetic stirring bar, weigh and place 20.0 g of silica gel (70–230 mesh; the removal of fines is not necessary). Now add 50 mL of 7.5 M nitric acid and stir the mixture for 3 h at room temperature. Remove the nitrated silica gel by gravity filtration (do not rinse), place it on a clay plate, and allow it to air-dry in a hood overnight. Store the product in an airtight container.

HOOD

G

Determine the nitric acid content of the silica gel by titration of a water suspension of the gel with a 0.1 M NaOH solution. The acid content of the gel should be in the range of 16–20% by weight. When compared to standard electrophilic aromatic nitration reaction conditions, the avoidance of sulfuric acid and the opportunity to recycle the volatile organic compound methylene chloride and silica gel provide the necessary elements of a green process.

Nitration of Phenol

REACTION

$$\text{OH} \xrightarrow{SiO_2 \cdot HNO_3} \text{OH-NO}_2 + \text{OH-NO}_2 \left[+ \text{OH} \right]$$

EXPERIMENTAL PROCEDURE

Estimated time to complete the experiment: 2.0 h.

Physical Properties of Reactants					
Compound	MW	Amount	mmol	mp (°C)	bp (°C)
Phenol	94.11	240 mg	2.55	40.5–41.5	182
$SiO_2 \cdot HNO_3$		1.0 g			
Methylene chloride		5 mL			40

Reagents and Equipment.   Weigh and add 240 mg (2.55 mmol) of phenol to a 10-mL round-bottom flask containing a stir bar. Now add 5 mL of methylene chloride. To this solution, weigh and add 1.0 g of nitrated silica gel (~16% HNO_3). Attach the flask to an air condenser (◄).

Room temp thermometer

Speed   Heat

A 10-mL RB flask

HOOD

CAUTION: Phenol is highly toxic and corrosive. Prevent contact with eyes and skin. It is best dispensed by warming the container of phenol in a warm water bath and then using an automatic delivery pipet. This should be done in the **hood**.

Reaction Conditions.   Stir the resulting mixture at room temperature for 5 min.

Isolation of Product.   Separate the silica gel from the reaction mixture by gravity filtration through a filter funnel containing a small plug of glass wool. Collect the filtrate in a 10-mL Erlenmeyer flask containing a boiling stone. Wash the collected silica gel with two 0.5-mL portions of methylene chloride, and collect these washings in the same Erlenmeyer flask.

Concentrate the filtrate to a volume of about 1.0 mL using a warm sand bath under a slow stream of nitrogen in the **hood**.   HOOD

Use thin-layer chromatography to obtain an analysis of the product mixture (see Prior Reading). Use methylene chloride as the elution solvent, silica gel (with a fluorescent indicator) as the stationary phase, and UV light for visualization. Typical R_f values are 0.04 for 4-nitrophenol, 0.15 for phenol, and 0.58 for 2-nitrophenol.

NOTE. 2,4-Dinitrophenol has a typical R_f value of 0.33 under the chromatography conditions above; it is not usually formed under these reaction conditions.

Characterization and Purification.   Column chromatography is now used to separate the mixture of products. Pack (dry) a 1.0-cm-diameter buret column with 7.5 g of activated silica gel. Place the above product solution on the column using a Pasteur pipet and then elute the column with 25 mL of 60:40 methylene chloride/pentane solvent. Collect the first 20 mL of eluate in a tared 25-mL Erlenmeyer flask. Concentrate this fraction to dryness in a warm sand bath under a slow stream of nitrogen to isolate the 2-nitrophenol. Weigh the product.

Now elute the column with about 30 mL of 1:1 ethyl acetate/methylene chloride solvent. Collect the first 20 mL of eluate in a tared 25-mL Erlenmeyer flask containing a boiling stone. Concentrate this fraction as above to a volume of about 2 mL. Use thin-layer chromatography to determine the purity of the product (see conditions outlined above). The main constituent of this fraction is 4-nitrophenol. However, the presence of unreacted phenol or possibly quinone is often detected. Now concentrate the product solution to dryness and weigh the 4-nitrophenol isolated.

If the TLC analysis indicates impurity, 4-nitrophenol may be purified as follows. Dissolve the solid residue in 7 mL of saturated sodium bicarbonate solution and transfer the resulting solution to a 15-mL centrifuge tube. Extract this solution twice with 2-mL portions of methylene chloride (a Vortex mixer may be used to good advantage here). Remove the methylene chloride layers using a Pasteur pipet and save them (together) in a small Erlenmeyer flask until you have isolated and characterized the product.

Cool the resulting aqueous solution in an ice bath and add 6 M HCl dropwise, with mixing (glass rod or Vortex mixer), until it becomes neutral or slightly acidic toward litmus or pH paper. *Do this step carefully. Too vigorous a reaction may result in loss of product.* Now extract the resulting solution with three 2-mL portions of methylene chloride. Following each extraction, remove the methylene chloride using a Pasteur filter pipet and transfer it to a 10-mL Erlenmeyer flask. Dry the wet solution over anhydrous sodium sulfate and then transfer it by Pasteur filter pipet to a tared 10-mL Erlenmeyer flask containing a boiling stone. Concentrate this solution using a warm sand bath under a slow stream of nitrogen gas to yield the 4-nitrophenol product. Weigh the product. The purity of the material may be checked again, using TLC as outlined above. It may also be recrystallized from water using the Craig tube, if necessary.

To characterize the 2- and 4-nitrophenols, determine their melting points and obtain their infrared spectra. Compare your results to those reported in the literature (*The Aldrich Library of IR Spectra* and/or SciFinder Scholar).

Chemical Tests.   Chemical classification tests (see Chapter 9) are also of value to establish the identity of these compounds. Perform the ferric chloride test for phenols. Is a positive result obtained for each compound? Does the sodium fusion test detect the presence of nitrogen? The test for nitro groups might also be performed. The phenyl- or α-naphthylurethane derivatives of the phenols may be prepared to further establish their identity (see Chapter 9, Preparation of Derivatives).

QUESTIONS

6-188. Predict the most likely mononitration product from each of the following compounds. Explain the reasons for your choice.

6-189. Write equations if not a mechanism to show how nitronium ions are formed using a mixture of nitric and sulfuric acids and how nitronium ion formation is contingent upon the use of anhydrous (concentrated) sulfuric acid.

6-190. Which ring would you expect to undergo nitration more readily? Explain.

6-191. Arrange the following compounds in order of increasing reactivity toward nitration. Explain.
   (a)   Acetanilide
   (b)   Acetophenone
   (c)   Bromobenzene
   (d)   Toluene
6-192. Offer a reasonable explanation of why nitration of 1,4-dichlorobenzene yields the mononitro derivative while N,N'-diacetyl-1,4-phenylenediamine forms the dinitro compound.
6-193. Explain why *p*-nitrophenol is a stronger acid than phenol itself. Would *p*-methoxyphenol be a stronger or weaker acid than phenol? Explain.

BIBLIOGRAPHY

Related to anhydrous nitric acid:

Stern, S. A.; Mullhaupt, J. T.; Kay, W. B. *Chem. Rev.* **1960**, *60*, 185.

The preparation used in this experiment is an adaptation of that reported by:

Cheronis, N. D.; Entrikin, J. B. *Semimicro Qualitative Organic Analysis*; Crowell: New York, 1947, p. 258.

The preparation of the nitrated silica gel was adapted from that reported by:

Tapia, R.; Torres, G.; Valderrama, J. A. *Synth. Commun.* **1986**, *16*, 681.

For a general review of nitration of aromatic species, see:

Olah, G. A.; Malhotra, R.; Narang, S. C. *Nitration: Methods and Mechanism*; VCH: New York, 1989.

Schefield, K. *Aromatic Nitration*; Cambridge Univ. Press: New York, 1980.

Smith, M. B. *March's Advanced Organic Chemistry*, 8th ed.; Wiley: New York, 2019, Chap. 11.

These references are selected from a large number of examples of nitration given in *Organic Syntheses*. None of these examples uses anhydrous nitric acid or the nitrated silica gel as the nitrating agent:

Braum, C. E.; Cook, C. D. *Organic Syntheses*; Wiley: New York, 1963; Collect. Vol. IV, p. 711.

Fanta, P. E.; Tarbell, D. S. *Organic Syntheses*; Wiley: New York, 1955; Collect. Vol. III, p. 661.

Fetscher, C. A. *Organic Syntheses*; Wiley: New York, 1963; Collect. Vol. IV, p. 735.

Fitch, H. M. *Organic Syntheses*; Wiley: New York, 1955; Collect. Vol. III, p. 658.

Howard, J. C. *Organic Syntheses*; Wiley: New York, 1963; Collect. Vol. IV, p. 42.

Huntress, E. H.; Shriner, R. L. *Organic Syntheses*; Wiley: New York, 1943; Collect. Vol. II, p. 459.

Mendenhall, G. D.; Smith, P. A. S. *Organic Syntheses*; Wiley: New York, 1973; Collect. Vol. V, p. 829.

Newman, M. S.; Boden, H. *Organic Syntheses*; Wiley: New York, 1973; Collect. Vol. V, p. 1029.

Nucleophilic Aromatic Substitution: 2,4-Dinitrophenylthiocyanate

EXPERIMENT 30

Common name: 2,4-dinitrophenylthiocyanate
CA number: [1594-56-5]
CA name as indexed: thiocyanic acid, 2,4-dinitrophenyl ester

Purpose. The conditions under which a specific type of **nucleophilic aromatic substitution reaction** can occur are investigated. You will carry out the conversion in a two-phase solvent system through the use of a phase-transfer catalyst. The experiment is an example of nucleophilic aromatic substitution on an aromatic ring having two strongly electron-withdrawing substituents. In this experiment, a bromine substituent (leaving group) is replaced by a thiocyanate, —S—C≡N, group (nucleophile).

Prior Reading

    Technique 4: Solvent Extraction
                Liquid–Liquid Extraction (p. 76)
                Drying of the Wet Organic Layer (pp. 84–87)
    Technique 5: Crystallization
                Craig Tube Crystallization (pp. 93–95)
    Technique 6: Chromatography
                Column Chromatography (pp. 96–99)
                Concentration of Solutions (pp. 105–108)

REACTION

Br

NO$_2$

NO$_2$

2,4-Dinitrobromobenzene

$+$ KSCN
Potassium
thiocyanate

$\xrightarrow[\text{(C}_4\text{H}_9)_4\text{N}^+,\text{Br}^-]{\text{toluene}}$

SCN

NO$_2$

NO$_2$

2,4-Dinitrophenyl
thiocyanate

DISCUSSION

Nucleophilic aromatic substitution reactions generally take place only if activating groups are present at the positions ortho and/or para to a good leaving group. Activating groups, in this case, are groups that are electron-withdrawing and render the aromatic ring more electron-deficient, and thus more susceptible to nucleophilic attack. On the other hand, π-electron donating groups hinder the reaction. Groups such as —NO$_2$, —SO$_2$CH$_3$, —$^+$N(CH$_3$)$_3$, —CF$_3$, —CN, —SO$_3^-$, —Br, —Cl, —I, —CO$_2^-$, especially if one —NO$_2$ group is already present, are strongly activating. These groups not only make the ring more susceptible to nucleophilic attack, they also assist in the resonance stabilization of the intermediate cyclohexadienyl anion. In the reaction illustrated in this experiment, two nitro groups are present on the ring, both ortho and para to the departing bromine substituent. While functional groups containing a carbonyl group do, in principle, activate the aromatic ring toward nucleophilic attack, the carbonyl groups themselves are usually more likely to react with a nucleophile than is the aromatic ring.

The conditions under which this reaction is conducted lend themselves to the use of phase-transfer catalysis. The system involves two phases: the aqueous phase and the organic phase (toluene). The phase-transfer catalyst plays a very important role. As an ammonium cation, it carries the SCN$^-$ ion from the aqueous phase into the organic phase where the substitution reaction then occurs. The product and the corresponding bromide salt of the phase-transfer catalyst are produced in this conversion. The bromide salt of the phase-transfer catalyst then migrates back into the aqueous phase, and the process repeats itself. The catalyst can play this role since the large organic groups (the four butyl groups) increase the solubility of the phase-transfer catalyst in the organic phase, while the charged ionic center of the catalyst renders it water-soluble. Phase-transfer catalysts can also be used in many other reactions, including the preparation of ethers by the Williamson method (see Experiments [22A] and [22B]), and the Horner–Wadsworth–Emmons modification of the Wittig reaction (see Experiments [19B] and [19D]).

The reaction in this experiment proceeds by an *addition–elimination two-step* sequence, of which the first step is generally rate determining. A tetrahedral intermediate (cyclohexadienyl anion) is formed by the attack of the nucleophile on the carbon atom to which the leaving group is attached. In the subsequent step, the leaving group then departs with its bonding electrons, regenerating the aromatic system. It is important to note that, in contrast to nucleophilic substitution reactions on alkyl carbon atoms, the reaction does not proceed by an S$_N$2 mechanism, as in this case an intermediate is formed. The mechanism is outlined below.

EXPERIMENTAL PROCEDURE

Estimated time to complete the reaction: 2.0 h.

Physical Properties of Reactants and Product					
Compound	MW	Amount	mmol	mp (°C)	bp (°C)
2,4-Dinitrobromobenzene	247.01	50 mg	0.20	75	
Tetrabutylammonium bromide	322.38	5 mg	0.02	103–104	
Aqueous potassium thiocyanate (50%)		150 μL			
Toluene	92.15	350 μL			111
2,4-Dinitrophenyl-thiocyanate	244			138–139	

110 °C
thermometer

Speed   Heat

2,4-(NO$_2$)$_2$C$_6$H$_3$Br, 50 mg
+ C$_6$H$_5$CH$_3$, 350 μL +
Bu$_4$N$^+$Br$^-$, 5 mg +
50% aqueous KSCN, 150 μL

Reagents and Equipment. Assemble a 3.0-mL conical vial containing a magnetic spin vane and equipped with a reflux condenser. Weigh and place in the vial 50 mg (0.20 mmol) of 2,4-dinitrobromobenzene followed by 350 μL of toluene (➡). To this solution add 5.0 mg (0.02 mmol) of tetrabutylammonium bromide and 150 μL of a 50% aqueous potassium thiocyanate (wt/wt) solution.

CAUTION: Tetrabutylammonium bromide is an irritant. ***Handle with care!*** It is also hygroscopic and should be protected from moisture prior to its use. Dispense the toluene and the thiocyanate solution in the **hood** using automatic delivery pipets.

HOOD

Reaction Conditions. Heat the resulting mixture at a sand bath temperature of 100 °C with stirring for 1 h. Allow the solution to cool to room temperature.

Isolation of Product. Use a Pasteur filter pipe to separate the toluene layer from the aqueous layer (**saving** both layers) and place the toluene layer in a 3.0-mL capped vial. Extract the toluene layer with two 1.0-mL portions of water and combine the water extracts with the original water layer you saved. Now extract the combined aqueous layers with two 0.5-mL portions of toluene, and combine these toluene extracts with the original toluene layer.

Anhydrous
Na₂SO₄, 700 mg

Toluene solution
with substitution product,
~ 1.0 mL

HOOD

HOOD

NOTE The volumes of liquid used in the extraction are measured using calibrated Pasteur pipets. For each extraction, shake the vial, vent it carefully, and allow the layers to separate. The transfers are made using Pasteur filter pipets.

Transfer the toluene solution by Pasteur pipet to a Pasteur filter pipet microcolumn containing 700 mg of anhydrous sodium sulfate previously wetted with toluene (). Allow the solution to elute through the column, and collect the dried toluene eluate in a 10-mL Erlenmeyer flask containing a boiling stone. Concentrate this solution by warming in a sand bath under a gentle stream of nitrogen gas in the **hood**. A yellow, solid product is obtained.

Purification and Characterization. The product is generally nearly pure. Recrystallize the material from chloroform, using a Craig tube, if desired.

CAUTION: Chloroform is toxic! Dispense and use only in the **hood**. Do not breathe the vapors.

Weigh the dried product and calculate the percent yield. Determine the melting point and compare it with that listed in the Reagent and Product table. Obtain an IR spectrum and compare your result to that of an authentic sample, or to one in the literature (*The Aldrich Library of IR Spectra* and/or SciFinder Scholar)

Experiment 30-1*

M

2,4-Dinitrophenylthiocyanate: Preparation Using a Monomode Microwave Apparatus

EXPERIMENTAL PROCEDURE

Estimated time to complete the experiment: 1.0 h.

Physical Properties of Reactants					
Compound	MW	Amount	mmol	mp (°C)	bp (°C)
1-Bromo-2,4-dinitrobenzene	247.01	67 mg	0.27	71–73	
Aqueous potassium thiocyanate (50%)		0.2 mL			
Ethanol	276	0.8 mL			78
2,4-Dinitrophenylthiocyanate	244			138–139	

Reagents and Equipment. This experiment is designed for use in the CEM Discover and Biotage Initiator microwave units.

In a 10.0-mL glass microwave reaction vessel containing a magnetic stir bar, add 67 mg (0.27 mmol) of 2,4-dinitrobromobenzene, 0.2 mL of 50% aqueous potassium thiocyanate (wt/wt) solution, and 0.8 mL of ethanol. Immediately cap the vessel with the microwave pressure cap.

---

*This section has been written by Dr. Nicholas E. Leadbeater from the Department of Chemistry at the University of Connecticut and Dr. Cynthia B. McGowan from the Department of Chemistry at Merrimack College, MA.

CAUTION: Dispense the potassium thiocyanate solution in the **hood** using an automatic delivery pipet. Since the reaction requires heating the reaction mixture to above the boiling point of some components in sealed vessels, *adherence to the microwave manufacturer's guidelines is essential.*

HOOD

Reaction Conditions.   Place the reaction vessel in the microwave cavity and, depending on the equipment used, position the pressure device on top. Program the microwave unit to heat the reaction mixture to 125 °C and hold at this temperature for 5 min. After heating, allow the reaction mixture to cool to 50 °C or below before removing the tube from the microwave unit. Allow the resulting solution to cool to room temperature and then place it in an ice bath for a 10-min period to complete the crystallization of the product.

Isolation of Product.   Collect the yellow crystals by vacuum filtration using a Hirsch funnel and wash the filter cake with cold 95% ethanol (1–2 mL). Partially dry the filter cake under suction. Transfer the filtered, washed, and partially dried product to a porous clay plate or filter paper to complete drying.

Characterization.   Weigh the dried product and calculate the percent yield. Determine the melting point and compare it with the list in the Reagent and Product table. Obtain an IR spectrum of your product, and compare it to that of an authentic sample, or to the spectrum shown in the literature (*The Aldrich Library of IR Spectra* and/or SciFinder Scholar).

2,4-Dinitrophenylthiocyanate: Preparation Using a Multimode Microwave Apparatus

Experiment 30-2*

EXPERIMENTAL PROCEDURE

Estimated time to complete the experiment: 1.0 h.

Physical Properties of Reactants					
Compound	MW	Amount	mmol	mp (°C)	bp (°C)
1-Bromo-2,4-dinitrobenzene	247.01	296 mg	1.20	71–73	
Aqueous potassium thiocyanate (50%)		1.0 mL			
Ethanol	276	4 mL			78
2,4-Dinitrophenylthiocyanate	244			138–139	

Reagents and Equipment.   This experiment is designed for use in the CEM MARS, Milestone START, and Anton Paar Synthos 3000 microwave units. When using the Anton Paar Synthos 3000 unit with the 24-position silicon carbide plate rotor containing glass vials, the reagent and solvent quantities cited in the monomode procedure should be used in conjunction with the reaction conditions here in the multimode procedure.

In a microwave reaction vessel containing a magnetic stir bar, add 296 mg (1.2 mmol) of 2,4-dinitrobromobenzene, 1.0 mL of 50% aqueous potassium

*This section has been written by Dr. Nicholas E. Leadbeater from the Department of Chemistry at the University of Connecticut and Dr. Cynthia B. McGowan from the Department of Chemistry at Merrimack College, MA.

thiocyanate (wt/wt) solution, and 4.0 mL of ethanol. Immediately cap the vessel with the microwave pressure cap and adjust the tightness to the manufacturer-specified level. Place the sealed vessel into its outer protective jacket.

HOOD

CAUTION: Dispense the potassium thiocyanate solution in the **hood** using an automatic delivery pipet. Since the reaction requires heating the reaction mixture to above the boiling point of some components in sealed vessels, *adherence to the microwave manufacturer's guidelines is essential.*

Reaction Conditions.   Insert the loaded vessels into the reaction carousel ensuring they are evenly spaced and then place the carousel into the microwave cavity. If provided by the manufacturer, connect a temperature probe to the control vessel. Program the microwave unit to heat the reaction vessels to 125 °C and hold at this temperature for 5 min. After heating, allow the reaction mixture to cool to 50 °C or below before removing the carousel from the microwave unit. Allow the resulting solution to cool to room temperature and then place it in an ice bath for 10 min to complete the crystallization of the product.

Isolation of Product.   Collect the yellow crystals by vacuum filtration using a Hirsch funnel and wash the filter cake with cold 95% ethanol (3–5 mL). Partially dry the filter cake under suction. Transfer the filtered, washed and partially dried product to a porous clay plate or filter paper to complete drying.

Characterization.   Weigh the dried product and calculate the percent yield. Determine the melting point and compare it with the list in the Reagent and Product table. Obtain an IR spectrum of your product, and compare it to that of an authentic sample, or to the spectrum shown in the literature (*The Aldrich Library of IR Spectra* and/or SciFinder Scholar).

QUESTIONS

6-194. Explain why the first reaction below requires substantially more vigorous reaction conditions than the second reaction:

6-195. What criteria of aromaticity are missing when considering the intermediate cyclohexadienyl anion in the nucleophilic aromatic substitution reaction (see Discussion section)?.

6-196. Complete each of the following reactions:

6-197. Compare the potential energy diagram of an S_N2 substitution reaction on an aliphatic halide to that of the nucleophilic substitution reaction carried out in this experiment. Discuss the main differences in the diagrams in terms of the mechanisms.

6-198. Would you expect the rate of the reaction in this experiment to depend on the concentrations of *both* the thiocyanate ion *and* 2,4-dinitrobromobenzene? Explain.

6-199. Show the intermediates, including all the resonance hybrid structures for the cyclohexadienyl anion, that are formed in the following aromatic nucleophilic substitution reactions:

6-200. On workup of each of the reactions given in Question 6-199, what product(s) would you expect to form? If more than one, indicate their relative amounts.

BIBLIOGRAPHY

Selected examples of nucleophilic aromatic substitution reactions in *Organic Syntheses* include:

Brewster, R. Q.; Groening, T. *Organic Syntheses*; Wiley: New York, 1943; Collect. Vol. II, p. 445.

Bunnett, J. F.; Conner, R. M. *Organic Syntheses*; Wiley: New York, 1973; Collect. Vol. V, p. 478.

Hartman, W. W. *Organic Syntheses*; Wiley: New York, 1941; Collect. Vol. I, p. 175.

Hartman, W. W.; Byers, J. R.; Dickey, J. B. *Organic Syntheses*; Wiley: New York, 1943; Collect. Vol. II, p. 451.

Kharasch, N.; Langford, R. B. *Organic Syntheses*; Wiley: New York, 1973; Collect. Vol. V, p. 474.

Reverdin, F. *Organic Syntheses*; Wiley: New York, 1941; Collect. Vol. I, p. 219.

Sahyun, M. R.V.; Cram, D. J. *Organic Syntheses*; Wiley: New York, 1973; Collect. Vol. V, p. 926.

Skorcz, J. A.; Kuminski, F. E. *Organic Syntheses*; Wiley: New York, 1973; Collect. Vol. V, p. 263.

Smith M. B. *March's Advanced Organic Chemistry*, 8th ed.; Wiley: New York, 2019, Chap. 13.

Halogenation Using N-Bromosuccinimide: 9-Bromoanthracene

EXPERIMENT
31

Common name: 9-bromoanthracene

CA number: [1564-64-3]

CA name as indexed: anthracene, 9-bromo-

Purpose. A free radical halogenation reaction is run using *N*-bromosuccinimide (NBS), a highly specific brominating agent. This material has the advantage that it is a solid and is therefore easier to handle than bromine, which is a toxic liquid.

Prior Reading
    Technique 5: Crystallization
                Use of the Hirsch Funnel (pp. 92–93)
                Craig Tube Crystallization (pp. 93–95)

REACTION

Anthracene   N-Bromosuccinimide   9-Bromoanthracene   Succinimide

DISCUSSION

N-Bromosuccinimide (NBS) is a highly specific brominating agent. When this reagent is used, anthracene is brominated in the 9-position. *N*-Bromosuccinimide may also be used to brominate positions α to (1) a carbonyl group, (2) a triple bond, (3) an alkene (allylic position), and (4) aromatic rings (benzylic position). Other polynuclear hydrocarbons that have been brominated using NBS include naphthalene, phenanthrene, and acenaphthene.

In the preparation of 9-bromoanthracene, the reaction progress is easily followed, since NBS (a reactant) and succinimide (a product) are both nearly insoluble in carbon tetrachloride. The NBS is more dense than the carbon tetrachloride solvent, and as the reaction proceeds this solid disappears from the bottom of the reaction flask and the less dense succinimide forms and floats on the surface of the reaction solution.

Free radicals have been identified in the mechanism of bromination using *N*-bromosuccinimide. In fact, the reaction proceeds only under conditions likely to produce radicals: photochemical conditions, by heating, or in the presence of a free radical initiator. The NBS reagent provides a steady source of small amounts of Br_2 via the rapid reaction of NBS with traces of hydrogen bromide.

The initiation step in bromination with NBS is the formation of a bromine radical by the homolytic dissociation of the Br_2 molecule.

Initiation Step

$$Br_2 \xrightarrow[\text{free radical initiators}]{h\nu \text{ or}} 2\,Br\cdot$$

The bromine radical then abstracts a hydrogen atom from the 9-position of the anthracene.

Propagation Step

The HBr so formed then reacts with NBS to produce a bromine molecule and succinimide:

The Br$_2$ molecule then reacts with the anthracenyl radical formed above to yield the product and a bromine radical:

This bromine radical starts the sequence over again. That is, a chain reaction is initiated. In the present reaction, a trace amount of an iodine–carbon tetrachloride solution is added to the reaction mixture. The iodine acts as a moderator or a retarder in the reaction (see Dauben et al. in the **Bibliography** section), and thus only the monobrominated product is formed; 9,10-dibromoanthracene is not generated under these conditions.

EXPERIMENTAL PROCEDURE

Estimated time to complete the experiment: 2.0 h.

Physical Properties of Reactants					
Compound	MW	Amount	mmol	mp (°C)	bp (°C)
Anthracene	178.24	50 mg	0.28	216	
N-Bromosuccinimide	177.99	50 mg	0.28	180–183	
Carbon tetrachloride	153.82	0.4 mL			77
Iodine–CCl$_4$ solution		1 drop			

Reagents and Equipment. Weigh and add 50.0 mg (0.28 mmol) of anthracene and 50 mg (0.28 mmol) of *N*-bromosuccinimide to a 3.0-mL conical vial containing a magnetic spin vane. To this mixture add 0.4 mL of carbon tetrachloride followed by one drop of I$_2$–CCl$_4$ solution delivered from a Pasteur pipet.

85–95 °C
thermometer

CAUTION: Carbon tetrachloride is a suspected carcinogen. Dispense it in the **hood** using an automatic delivery pipet.

Attach the vial to a reflux condenser fitted with a calcium chloride drying tube (➡).

Anthracene, 50 mg
+ NBS, 50 mg
+ CCl$_4$, 0.4 mL +
I$_2$· CCl$_4$, 1 drop

*INSTRUCTOR PREPARATION I$_2$–CCL$_4$ SOLUTION. Iodine (0.2 g, 0.01 mol) is dissolved in 10 mL of carbon tetrachloride. Place the solution in a **hood** for student use.* HOOD

Reaction Conditions. Heat the reaction mixture, with stirring, to reflux in a sand bath at 85–95 °C for 1 h. During this time the solution turns brown and crystals of succinimide appear at the surface of the reaction solution.

Succinimide product collected here

Isolation of Product. Collect the succinimide product by vacuum filtration of the warm solution using a Hirsch funnel (➡). Wash the filter cake of succinimide with three or four 0.5-mL portions of cyclohexane from a calibrated Pasteur pipet. Combine the washings with the original filtrate.

Concentrate the filtrate to dryness in a **hood** under reduced pressure to obtain yellow-green crystals of 9-bromoanthracene. *Accelerate evaporation of the solvent by immersing the flask in warm water.*

Cyclohexane, 2.0 mL +
CCl$_4$, 0.4 mL +
HOOD 9-bromoanthracene product

Purification and Characterization. Weigh the air-dried succinimide and calculate the percent yield. Determine the melting point and compare your value to that in the literature. Obtain an IR spectrum and compare your spectrum with that shown in the literature (*The Aldrich Library of IR Spectra* and/or SciFinder Scholar).

The crude 9-bromoanthracene is purified by recrystallization from 95% ethanol using the Craig tube. Weigh the dried product and calculate the percent yield. Determine the melting point and compare your result with that listed in the literature.

Chemical Tests. The ignition test should establish the presence of the aromatic nature of the substituted anthracene compound. Does the Beilstein test indicate the presence of a halogen?

QUESTIONS

6-201. Predict and give a suitable name for the product(s) formed in the following reactions:

(a) [cyclohexene structure] + NBS $\xrightarrow{hv}$

(b) 1-Propene + NIS $\xrightarrow{hv}$

(note: NIS is *N*-iodosuccinimide)

(c) [3-methylthiophene structure] + NBS $\xrightarrow{hv}$

6-202. When 1-octene is treated with NBS, three monobromo straight-chain alkenes with the molecular formula $C_8H_{15}Br$ are isolated from the reaction mixture. Identify these compounds and give each a suitable name.

6-203. Benzyl bromide ($C_6H_5CH_2Br$) can be prepared by treating toluene with NBS in the presence of a peroxide initiator. Suggest a suitable mechanism to account for this reaction.

6-204. The benzyl radical has unusual stability. Account for this fact by drawing appropriate resonance structures.

[benzyl radical structure, CH₂•]

6-205. Suggest a suitable mechanism for the following reaction:

[reaction scheme: cyclohexene + NBS → three bromocyclohexene products (50%, 25%, 25%) + succinimide]

* = ^{13}C label   50%   25%   25%

BIBLIOGRAPHY

Dauben, H. J. Jr.; McCoy, L. L. *J. Am. Chem. Soc.* **1959**, *81*, 4863.

Djerassi, C. *Chem. Rev.* **1948**, *43*, 271.

Horner, L.; Winkelmann, E. H. *Angew. Chem.* **1959**, *71*, 349.

Horner, L.; Winkelmann, E. H. *Newer Methods Prep. Org. Chem.* **1964**, *3*, 151.

For an overview of the scope of the reaction, see:

Smith, M. B. *March's Advanced Organic Chemistry*, 8th ed.; Wiley: New York, 2019, Chap. 11.

Also see Mauthen, H. A. *J. Org. Chem.* **1992**, *57*, 2740 for alternative brominating agents.

Brominations in Organic Syntheses using NBS:

Amat, M.; Hadida, S.; Sathyanarayana, S.; Bosch, J. *Org. Synth.* **1997**, *74*, 248.

Campaigne, E.; Tullar, B. F. *Organic Syntheses*; Wiley: New York, 1963; Collect. Vol. IV, p. 921.

Corbin, T. F.; Hahn, R. C.; Schechter, H. *Organic Syntheses*; Wiley: New York, 1973; Collect. Vol. V, p. 328.

Greenwood, F. L.; Kellert, M. D.; Sedlak, J. *Organic Syntheses*; Wiley: New York, 1963; Collect. Vol. IV, p. 108.

Kalir, A. *Organic Syntheses*; Wiley: New York, 1973; Collect. Vol. V, p. 825.

Nakagawa, N.; Saegusa, J.; Tomozuka, M.; Ohi, M.; Kiuchi, M.; Hino, T.; Ban, Y. *Organic Syntheses*; Wiley: New York, 1988; Collect. Vol. VI, p. 462.

Hypochlorite Oxidation of an Alcohol: Cyclohexanone

Common name: cyclohexanone

CA number: [108-94-1]

CA name as indexed: cyclohexanone

Purpose. The oxidation of a secondary alcohol to a ketone is explored using the hypochlorite reagent. Steam distillation is used to separate the product from the reaction mixture, and chromatographic techniques are used to purify and isolate the cyclohexanone product.

Prior Reading

Technique 1: Gas Chromatography (pp. 59–65)

Technique 2: Simple Distillation (pp. 65–68)

Steam Distillation

Technique 4: Solvent Extraction

Liquid–Liquid Extraction (p. 76)

Drying of the Wet Organic Layer (pp. 84–87)

Salting Out (p. 83)

Technique 6A: Thin-Layer Chromatography (pp. 101–103)

REACTION

$$\text{Cyclohexanol} \xrightarrow[\text{NaOCl}]{\text{CH}_3\text{CO}_2\text{H}} \text{Cyclohexanone}$$

DISCUSSION

Sodium hypochlorite solutions (1.8–2.0 M) sold as liquid bleach are often described as having 11.5–12.5% available chlorine. The term "available chlorine" compares the oxidizing capacity of the solution relative to that of the same weight of chlorine, Cl_2.

Sodium hypochlorite solutions are used extensively in swimming pool sanitation and as a bleach in the pulp and textile industries. A less concentrated product (5% available chlorine) is used in laundries and as household bleach. The reaction described in this experiment illustrates the use of liquid bleach (11.5–12.5% available chlorine) as an oxidizing agent in the organic laboratory.

Sodium hypochlorite is prepared commercially by passing chlorine gas through a solution of aqueous sodium hydroxide:

$$Cl_2 + NaOH \rightleftharpoons NaOCl + NaCl$$

The actual oxidizing agent in the present experiment is the chloronium ion (Cl^+), which is reduced in the reaction to chloride ion (Cl^-). The cyclohexanol acts as a reducing agent, and thus becomes oxidized to cyclohexanone:

Chromium-based oxidants are commonly used to accomplish these transformations in the research laboratory (see Experiments [33A] and [E1]); the hypochlorite oxidation presents an opportunity to perform an oxidation using the much less costly and much less toxic hypochlorite solution as the oxidant.

In this experiment, a Hickman still is used to isolate the crude cyclohexanone product from the reaction mixture in an example of the steam distillation technique (see Prior Reading). The crude mixture collected in the collar of the still consists of cyclohexanone, water, and acetic acid. If any unreacted cyclohexanol is present in this mixture, it is removed in the subsequent chromatographic purification sequence using alumina. Gas chromatographic analysis may be used to determine the purity of the cyclohexanone product.

EXPERIMENTAL PROCEDURE

Estimated time to complete the experiment: 3.0 h.

Physical Properties of Reactants						
Compound	MW	Amount	mmol	bp (°C)	d	n_D
Cyclohexanol	100.16	100 mg	1.0	161	0.96	1.4641
Glacial acetic acid	60.05	250 µL		118		
Sodium hypochlorite soln (~12.5% available chlorine)		2 mL				

Reagents and Equipment. Weigh and place 100 mg (1.0 mmol) of cyclohexanol in a 5.0-mL conical vial containing a magnetic spin vane. Add 250 μL of glacial acetic acid and then attach the vial to a Hickman still (➡). *Dispense the glacial acetic acid in the* **hood** *by use of an automatic delivery pipet.*

Cool the resulting solution in an ice bath and add dropwise, with stirring, 2.0 mL of aqueous sodium hypochlorite (NaOCl) solution (~12.5% available chlorine, 1.8–2.0 M) by use of a graduated pipet. Remove the ice bath following the addition. *Add the NaOCl solution by inserting the pipet down the neck of the still just into the throat of the vial.*

NOTE. *Solid calcium hypochlorite (65% available chlorine) may be used as a source of oxidant in place of the sodium hypochlorite solution.*

Reaction Condition. After 30 min, begin monitoring the reaction mixture by TLC. Using as a solvent system ethyl acetate:hexane (1:4), the R_f value of cyclohexanone is 0.6 (R_f value of cyclohexanol is 0.4) when stained with a solution of *p*-anisaldehyde (135 mL ethanol:5 mL H$_2$SO$_4$:1.5 mL glacial acetic acid:3.7 mL *p*-anisaldehyde). *If incomplete as judged by TLC, maintain an excess of hypochlorite oxidizing agent for an additional 30 min and check for the disappearance of starting material by TLC. Overoxidation is not likely. Monitor the aqueous layer periodically using KI–starch paper If a positive test is not obtained (the paper should turn blue if an oxidant is present), add additional sodium hypochlorite solution (1–3 drops) to ensure that an excess of the oxidizing agent is present. Use a Pasteur pipet inserted down the neck of the still to add the reagent and a clean pipet to remove a drop of solution for testing with KI–starch paper.*

Isolation of Product. Using a Pasteur pipet, add a saturated, aqueous sodium bisulfite solution dropwise to the reaction mixture until the solution gives a negative KI–starch test.

Separate the crude product from the reaction mixture by steam distillation. With stirring, heat the mixture in a sand bath at a temperature of 120–130 °C (➡). Collect the first 0.5–1.0 mL of distillate in the ring collar of the condenser. Using a Pasteur pipet, transfer this material to a 3.0-mL conical vial containing a spin vane, and remove the heat from the still pot. After the distillation apparatus has cooled, rinse the condenser collar with 0.5 mL of diethyl ether and also transfer this rinse to the conical vial.

NOTE. *The distillate collected in the condenser collar consists of cyclohexanone, water, and acetic acid.*

To neutralize the acetic acid present in the separated product mixture, add anhydrous sodium carbonate (~100 mg) in small portions, with stirring, to the solution until evolution of CO$_2$ gas ceases. Now add 50 mg of NaCl to the mixture. The resulting two-phase system is stirred until all the solid material dissolves.

Use a Pasteur filter pipet to separate the ether layer containing the cyclohexanone product from the aqueous phase and transfer it to a microscale drying and chromatography column. Assemble the column using a Pasteur filter pipet packed first with about 50 mg of sand, then with 300 mg of alumina (Activity I, see Glossary), followed by 200 mg of anhydrous magnesium sulfate (➡). Wet the column with diethyl ether before the transfer.

HOOD

Cyclohexanol, 100 mg +
CH$_3$CO$_2$H, 250 μL +
11.5–12.5% NaOCl, 2 mL

120–130 °C
thermometer

Cyclohexanol, 100 mg +
CH$_3$COOH, 250 μL +
11.5–12.5% NaOCl, 2 ml +
NaHSO$_3$(sat soln),
~8–10 drops

Anhydrous
MgSO$_4$, 200 mg

Al$_2$O$_3$(I), 300 mg

Sand, 50 mg

Oxidation product
in (CH$_3$CH$_2$)$_2$O,
~ 1.0 mL

Collect the eluate in a tared 3.0-mL conical vial containing a boiling stone. Extract the aqueous layer remaining in the conical vial with three 0.5-mL portions of ether, and also pass each extract through the column and combine these fractions with the original eluate.

Fit the 3.0-mL conical vial with an air condenser and remove the ether by HOOD   gentle warming on a sand bath in the **hood**.

Purification and Characterization. The liquid residue of cyclohexanone isolated upon evaporation of the ether solvent is sufficiently pure for characterization. It may be further purified by prep-GC (see below). Weigh the product and calculate the percent yield.

The purity of the cyclohexanone product may be determined using *gas chromatographic analysis (see Prior Reading and Experiment [5A] to review this technique and the experimental conditions for its use).* Determine the boiling point and density of your product; compare your results to those given in the literature.

Obtain the IR spectrum of your product and compare it with one in the literature (*The Aldrich Library of IR Spectra* and/or *SciFinder Scholar*).

Chemical Tests. Chemical classification tests may also be run to assist in the characterization of this material. The 2,4-dinitrophenylhydrazine test (Chapter 9) should give a positive result. Isolation of this derivative and the determination of its melting point would further establish the identity of the product as cyclohexanone.

QUESTIONS

6-206. In the experiment, why is a solution of sodium bisulfite added to the reaction product mixture (see Isolation of Product)? Write a reaction to account for what is happening. Is the bisulfite ion acting as an oxidizing or reducing agent?

6-207. When thinking about COVID-19 and the global demand for disinfectants, speculate then research/comment on the 2022 market statistics using the following baseline values: Consumption statistics for 1982 indicate that 210×103 tons of sodium hypochlorite were consumed in the United States alone. World consumption of sodium hypochlorite for household use is estimated to be 426×103 metric tons annually in 2005.

6-208. Predict the product(s) for each of the following oxidation reactions. Give a suitable name for each product.

(a)

$$H_3C \quad OH \qquad \text{1) NaOBr}$$
$$\qquad CH_3 \qquad \text{2) H}^{\oplus}$$

(b)

$$H_3C \qquad OH \qquad \text{1) NaOI}$$
$$\qquad \qquad \text{2) H}^{\oplus}$$

6-209. 2,3-Dimethyl-2,3-butanediol (*pinacol*), upon heating in aqueous acid, rearranges to form 3,3-dimethyl-2-butanone (*pinacolone*). Suggest a mechanism for this reaction.

6-210. What analytical/spectroscopic tools might be used to distinguish between pentanal and 2-pentanone? Between benzyl alcohol and diphenylmethanol?

BIBLIOGRAPHY

As a general reference for the use of sodium hypochlorite as an oxidant, see:

Fieser, L. F; Fieser, M. *Reagents for Organic Synthesis*; Wiley: New York, 1967; Vol. I, p. 1084.

Nohrig, J. R.; Nienhuis, D. M.; Linck, C. F.; Van Zoeren, C.; Fox, B. G. *J. Chem. Educ.* **1985**, *62*, 519.

Experiment [32] is based on work reported by:

Stevens, R. V.; Chapman, K. T.; Weller, H. N. *J. Org. Chem.* **1980**, *45*, 2030.

See also Zuczek, N. M.; Furth, P. S. *J. Chem. Educ.* **1981**, *58*, 824.

EXPERIMENT
33

Chromium Trioxide–Resin or Hypochlorite Oxidation of an Alcohol: 9-Fluorenone

Common name: 9-fluorenone

CA number: [486-25-9]

CA name as indexed: 9*H*-fluoren-9-one

Purpose. The experiment investigates the use of a polymer-bound chromium trioxide oxidizing agent in the oxidation of a secondary (2°) alcohol to a ketone. An alternative oxidizing agent, sodium hypochlorite, may be used instead. The progress of the reaction is monitored by thin-layer chromatography (TLC). The product ketone can be characterized by formation of its 2,4-dinitrophenylhydrazone (2,4-DNP) derivative.

Prior Reading

    Technique 5: Crystallization

                Use of the Hirsch Funnel (pp. 92–93)

                Craig Tube Crystallizations (pp. 93–95)

    Technique 6: Chromatography

                Thin-Layer Chromatography (pp. 101–103)

                High-Performance Liquid Chromatography (pp. 104–105)

                Concentration of Solutions (pp. 105–108)

9-Fluorenone: CrO$_3$ Oxidation of 9-Fluorenol

Experiment 33A

REACTION

9-Fluorenol $\xrightarrow[\text{Amberlyst resin}]{\text{CrO}_3}$ 9-Fluorenone

DISCUSSION

This experiment illustrates the oxidation of a 2° alcohol to a ketone. The oxidizing agents commonly used for this purpose are sodium dichromate or chromic oxide in sulfuric acid. In the present case, a convenient and advantageous polymer-bound chromium trioxide reagent is used. It is not only easy to prepare, but is also easy to separate from the product mixture and can be recycled. Today, the use of polymer-bound reagents in organic synthesis is developing at a rapid pace. The mechanism of the oxidation is outlined here:

G

(Chromate ester)

(Chromate ester)

(Ketone)

As seen, the oxidation proceeds through the formation of a chromate ester. Note that the oxidation state of chromium, Cr(VI), at this stage remains the same as in the starting reagent. The second stage is equivalent to an E2 elimination reaction, with the water molecule acting as a base. The donation of an electron pair to the metal atom changes its oxidation state to Cr(IV).

Note that a solution of chromic oxide in aqueous sulfuric acid (the Jones reagent) is used as a test reagent for 1° and 2° alcohols. A positive test is observed when the clear orange test reagent gives a greenish opaque solution upon addition of the alcohol (see Chapter 9).

EXPERIMENTAL PROCEDURE

Estimated time to complete the experiment: 3–4 h.

INSTRUCTOR PREPARATION. Prepare the CrO₃ resin by adding 35 g of Amberlyst A-26 resin to a solution of 15 g of CrO₃ in 100 mL of water. Stir the mixture for 30 min at room temperature. Then collect the resin by vacuum filtration using a Büchner funnel and successively rinse it with water and acetone. Partially dry the material on the Büchner funnel by drawing air through the resin under vacuum for 1 h. Allow it to air-dry overnight.

Physical Properties of Reactants					
Compound	MW	Amount	mmol	mp (°C)	bp (°C)
9-Fluorenol	182.23	100 mg	0.55	154	
Chromic oxide–resin		500 mg			
Toluene	92.15	3.5 mL			111

NOTE. CrO₃ has been listed as a known carcinogen (see The Merck Index, 12th ed.; 1996; p. 375, no. 2293).

Reagents and Equipment. Weigh and add 100 mg (0.55 mmol) of 9-fluorenol to a 10-mL round-bottom flask containing a magnetic stirring bar. Add 3.5 mL of toluene, and sample the resulting solution for TLC analysis (see Prior Reading). Now add 500 mg of the oxidizing resin and attach the flask to a reflux condenser (➡). If the longest Pasteur pipet available to you will not reach down the condenser and allow you to sample the solution, consider placing a Claisen head between the condenser and the flask (using the neck above the bend to connect the condenser. If you do this, be sure to cap the other neck of the Claisen head, and be sure to open it for the minimum possible amount of time when sampling—otherwise toluene vapors will escape into the air you breathe.

~130 °C
thermometer

9-Fluorenol, 100 mg +
CrO$_3$ resin, 500 mg
CH$_3$C$_6$H$_5$, 3.5 mL,
10-mL RB flask

Reaction Conditions.   Heat the contents of the flask to reflux, with stirring, using a sand bath temperature of approximately 130 °C.

Sample the solution for TLC analysis after a period of 5 min and every 15–20 min thereafter. In this manner, the progress of the reaction may be monitored until the conversion is complete (~35–40 min). This point is reached when the TLC analysis shows that the 9-fluorenol has been completely consumed.

SUGGESTED TLC CONDITIONS. Use Eastman Kodak fluorescent silica gel sheets (1 × 6 cm). Elute them with methylene chloride solvent and visualize the spots by UV light. Determine reference R$_f$ values for 9-fluorenol and 9-fluorenone using known reference samples under the same conditions.

Isolation of Product.   Cool the reaction mixture and remove the resin by gravity filtration through a cotton plug placed in a small funnel. Transfer the solution to the filter funnel using a Pasteur pipet, and collect the filtrate in a tared 10-mL Erlenmeyer flask containing a boiling stone. Rinse the reaction flask and resin with two 1.0-mL portions of methylene chloride using a calibrated Pasteur pipet. Combine the rinse with the original filtrate.

In the **hood** remove the solvent from the filtrate under a stream of nitrogen by warming in a sand bath to yield a yellow colored residue of crude 9-fluorenone product.

HOOD

Purification and Characterization.   Obtain the weight of the crude product and calculate the percent yield.

Recrystallize a 30- to 50-mg portion of the 9-fluorenone from hexane (~1.0 mL of hexane/50 mg of ketone) using a Craig tube. Determine the melting point of this purified material and compare your data to those reported in the literature. Calculate the percent recovery on recrystallization.

NOTE. High-performance liquid chromatography (see Prior Reading) is effective in separating the pure ketone. Use a C$_{18}$ reversed-phase column and methanol–water as the elution solvent.

Chemical Tests.   The 2,4-dinitrophenylhydrazine test for aldehydes and ketones (see Chapter 9) may be done to confirm the presence of the carbonyl group. The isolation of this derivative and the determination of its melting point (Lit. value = 283 °C) would aid in establishing the identity of the product.

Does the ignition test indicate that this compound is aromatic?

9-Fluorenone has a characteristic UV spectrum. The following data were obtained at a concentration of 8.66×10^{-6} M (see Appendix, Ultraviolet–Visible Discussions):

$$\lambda_{max}248 \text{ nm}(\varepsilon_{max} = 5.27 \times 10^4, \text{methanol})$$
$$\lambda_{max}255 \text{ nm}(\varepsilon_{max} = 7.83 \times 10^4, \text{methanol})$$
$$\lambda_{max}290 \text{ nm}(\varepsilon_{max} = 3.93 \times 10^3, \text{methanol})$$

Experiment 33B

9-Fluorenone: NaOCl Oxidation of 9-Fluorenol

REACTION

9-Fluorenol → 9-Fluorenone

DISCUSSION

This experiment is a further illustration of the oxidation of a 2° alcohol to a ketone. The reagent used is sodium hypochlorite solution. This reagent is also used in oxidation reactions described in Experiments [32] and [34A, B]. The chemistry of the reagent, sodium hypochlorite, is discussed in Experiment [32]. This alternative reagent for the oxidation of 9-fluorenone offers a material that is cheap (Clorox, a household bleach), easy to handle, and environmentally safe.

As in experiment [33A], the progress of the reaction is followed using TLC.

EXPERIMENTAL PROCEDURE

Estimated time to complete the experiment: 3.0 h.

Physical Properties of Reactants					
Compound	MW	Amount	mmol	mp (°C)	bp (°C)
9-Fluorenol	182.23	50 mg	0.27	154	
Acetone	58.08	3 mL			56.5
Acetic acid, glacial	60.05	120 µL		16.7	118
Sodium hypochlorite (5.25%)	74.44	~1.3 mL			

Room temp
thermometer

Speed   Heat

A 10-mL RB flask

HOOD

Reagents and Equipment. Weigh and add 50 mg (0.27 mmol) of 9-fluorenol to a 10-mL round-bottom flask containing a magnetic stir bar. Add 3 mL of acetone and *sample the resulting solution (2–4 µL) for TLC analysis* (see Prior Reading). Add 120 µL of glacial acetic acid and then attach the flask to an air condenser (◄).

CAUTION: Dispense the glacial acetic acid in the **hood** by use of an automatic delivery pipet. This acid burns skin!

Use a graduated 1-mL pipet to slowly add, with gentle stirring, 0.4 mL of aqueous sodium hypochlorite solution (5.25%—commercial bleach). Add the NaOCl solution by inserting the pipet down the neck of the condenser just to the throat of the flask.

Reaction Conditions.   Gently stir the reaction mixture at room temperature. After a period of 5 min, sample the solution (2–5 µL) for TLC analysis. *Sample the reaction solution by inserting a 9-in. Pasteur pipet down the air condenser to obtain a drop of solution by capillary action.* If starting material is present as shown by TLC analysis, add an additional 0.4 mL of the hypochlorite solution as before. Stir for 5 min and sample again. Continue this process until TLC analysis shows that the 9-fluorenol has been completely consumed. Approximately 1.2–1.4 mL of hypochlorite solution should be sufficient, if the reagent is fresh.

SUGGESTED TLC CONDITIONS. Use Eastman Kodak Fluorescent silica gel sheets (1.0 cm × 6 cm). Elute the plates with 30% acetone/70% hexane solvent, and visualize the spots by UV light. R_f values are 0.56 for 9-fluorenol and 0.80 for 9-fluorenone.

Isolation of Product.   Using a Pasteur pipet, transfer the reaction solution to a stoppered 15-mL centrifuge tube. Rinse the flask with 300 µL of acetone and transfer the rinse to the same centrifuge tube. Extract the solution with two 2-mL portions of hexane. Transfer the hexane extracts (upper layer) to a second centrifuge tube. Wash the combined hexane extracts with one 1-mL portion of 5% sodium bicarbonate solution followed by one 2-mL portion of water. Dry the hexane solution by addition of a small amount of anhydrous sodium sulfate. Use a Pasteur filter pipet to transfer approximately 3.0 mL of the dried solution to a tared 5-mL conical vial. Concentrate the solution, which contains a portion of the desired product, to dryness on a warm sand bath under a stream of nitrogen or dry air in the **hood**. Transfer the remaining hexane   HOOD
solution to the same vial. Rinse the sodium sulfate with an additional 1 mL of hexane and transfer this rinse to the same vial. Concentrate the solution as before.

Purification and Characterization.   See this section in Experiment [33A].

QUESTIONS

6-211. Suggest a suitable mechanism for the reaction of 9-fluorenone with 2,4-dinitrophenylhydrazine to form the corresponding 2,4-dinitrophenylhydrazone.

6-212. It is also possible to characterize 9-fluorenone by preparation of an oxime or semicarbazone. Formulate equations showing clearly the formation of these two derivatives and name each reagent used in the preparation.

6-213. As indicated in the Discussion, a solution of chromic oxide in aqueous sulfuric acid is used as a test reagent for 1° and 2° alcohols.
(a)   What is this test (consult Chapter 9)?
(b)   After identifying each system as a 1°, 2°, or 3° alcohol, predict which of the following alcohols will give a positive test with the chromic oxide reagent: 1-pentanol, 2-methyl-2-propanol, cyclohexanol, 2-methyl-2,4-pentanediol..

6-214 There are actually **two** isomeric 2,4-dinitrophenylhydrazones of 2-pentanone. Draw the structures of these isomers.

6-215   2-Pentanone, in reference to Question 6-214, also forms a derivative on treatment with semicarbazide:

$$CH_3CH_2CH_2\overset{\overset{\displaystyle \cdot\cdot O \cdot\cdot}{\|}}{C}CH_3 + H_2\ddot{N}-\ddot{N}H-\overset{\overset{\displaystyle \cdot\cdot O \cdot\cdot}{\|}}{C}-\ddot{N}H_2 \longrightarrow CH_3CH_2CH_2\overset{\overset{\displaystyle \ddot{N}-\ddot{N}H-\overset{\overset{\displaystyle \cdot\cdot O \cdot\cdot}{\|}}{C}-\ddot{N}H_2}{}}{C}CH_3 \qquad + H_2O$$

2-Pentanone   Semicarbazide   The semicarbazone

Note that semicarbazide has **two** —NH$_2$ groups that might react with the carbonyl of the ketone to form the semicarbazone. Explain why it reacts as depicted above.

BIBLIOGRAPHY

Experiment [33] is an adaptation of that reported by:

Wade, L. G., Jr.; Stell, L. M. J. Chem. Educ. **1980,** 57, 438; Experiment [33B] from the work of Jones, C. S.; Albizati, K. J. Chem. Educ. **1994,** 71, A271.

Introduction to the use of polymer-bound reagents:

Hodge, P. Chem. in Britain **1978,** 14, 237.
Leznoff, C. C. Acc. Chem. Res. **1978,** 1, 327.
McKillopp, A.; Young, D. W. Synthesis **1979,** 401.
Smith, K., Ed. Solid Supports and Catalysts in Organic Synthesis; Ellis Harwood: Chichester, UK, 1992.

Selected chromate oxidations reported in Organic Syntheses include:

Boeckman, R. K., Jr.; Blum, D. M.; Ganem, B.; Halvey, N. Organic Syntheses; Wiley: New York, 1988; Collect. Vol. VI, p. 1033.
Conant, J. B.; Quayle, O. R. Organic Syntheses; Wiley: New York, 1941; Collect. Vol. I, p. 211.

Eisenbraun, E. J. Organic Syntheses; Wiley: New York, 1973; Collect. Vol. V, p. 310.
Fieser, L. F. Organic Syntheses; Wiley: New York, 1963; Collect. Vol. IV, p. 189.
Krumpolc, M.; Rocek, J. Organic Syntheses; Wiley: New York, 1990; Collect. Vol. VII, p. 114.
Zibuck, R.; Streiber, J. Organic Syntheses **1993,** 71, 236.

For reviews of chromium catalyzed oxidations in organic synthesis, see:

Cainelli, G.; Cardillo, G. Chromium Oxidations in Organic Chemistry; Springer-Verlag: New York, 1984.
Muzart, J. Chem. Rev. **1992,** 92, 113.

For an overview of the oxidation of alcohols using sodium hypochlorite solutions, see:

Mohrig, J. R.; Nienhuis, D. M.; Linck, C. F.; Van Zoeren, C.; Fox, B. G. J. Chem. Educ. **1985,** 62, 519.

EXPERIMENT 34

Hypochlorite Oxidation of Methyl Ketones by the Haloform Reaction: Benzoic Acid and p-Methoxybenzoic Acid

Common name: benzoic acid

CA number: [65-86-0]

CA name as indexed: benzoic acid

Common names: *p*-methoxybenzoic acid, 4-methoxybenzoic acid, *p*-anisic acid

CA number: [100-09-4]

CA name as indexed: benzoic acid, 4-methoxy-

Purpose.   The well-known haloform reaction is explored as a synthetic route to the preparation of organic acids. You will investigate the use of a basic aqueous solution of hypochlorite ion as a source of molecular chlorine (Cl$_2$).

Prior Reading

Technique 4: Solvent Extraction
        Liquid–Liquid Extraction (p. 76)
        Separatory Funnel Extraction (Scaleup) (pp. 79–81)
Technique 5: Crystallization
        Use of the Hirsch Funnel (pp. 92–93)
        Craig Tube Crystallizations (pp. 93–95)

REACTION

Acetophenone   Benzoic acid   Chloroform

DISCUSSION

The reaction of methyl ketones with a halogen in an alkaline medium is known as the haloform reaction. In this experiment, halogen and base are present because of the equilibrium reaction shown here:

$$H_2O + NaOCl + NaCl \rightleftharpoons 2\,NaOH + Cl_2$$

In the haloform reaction, two products are formed: (1) a haloform ($CHCl_3$, $CHBr_3$, or CHI_3, depending on the halogen used) and (2) the carboxylic acid having one less carbon atom than the starting ketone. It is the formation of the carboxylic acid that gives the reaction its synthetic utility. An advantage of this oxidation is that it does not affect carbon–carbon double bonds, as illustrated here:

If I_2 in a basic solution is used, iodoform (CHI_3) is generated in the reaction. This compound is a yellow solid that precipitates from the reaction medium. This observation has been used extensively as a chemical test **(iodoform test)** for methyl ketones and the $RCH(OH)CH_3$ structural group (see Chapter 9).

In the present experiment, a solution of sodium hypochlorite (NaOCl) is used as the source of the chlorine. Sodium hypochlorite solutions are marketed with the hypochlorite concentration described in terms of the "available chlorine" content, which is a term comparing the oxidizing potential of the solution with that of the equivalent mass of chlorine. For the purposes of this experiment, a solution of NaOCl that has 5% available chlorine is needed, and common household bleach will suffice. For another example of an oxidation with an aqueous hypochlorite solution, see Experiment [32].

The reaction takes place in two stages. In the first stage, the methyl group is trihalogenated in a stepwise fashion. In the second stage, hydroxide ion attacks the carbonyl carbon of the trihaloketone to generate the haloform along with the alkali metal salt of the acid; acidification yields the carboxylic acid. The mechanistic sequence follows:

$$R-\overset{\overset{\ddot{O}}{\|}}{C}-\overset{\overset{H}{|}}{\underset{H}{C}}-H + {}^-\!\ddot{O}H, Na^+ \rightleftharpoons R-\overset{\overset{\ddot{O}}{\|}}{C}-\overset{\overset{H}{|}}{C}{:}^- \overset{\curvearrowright}{:\!\ddot{C}l-\ddot{C}l:} R-\overset{\overset{\ddot{O}}{\|}}{C}-\overset{\overset{H}{|}}{\underset{H}{C}}-Cl + {:}\ddot{C}l{:}^-, Na^+$$

$$R-\overset{\overset{\ddot{O}}{\|}}{C}-\overset{\overset{H}{|}}{\underset{H}{C}}-Cl \xrightarrow[2.\ Cl_2]{1.\ {}^-OH} R-\overset{\overset{\ddot{O}}{\|}}{C}-\overset{\overset{Cl}{|}}{\underset{H}{C}}-Cl \xrightarrow[2.\ Cl_2]{1.\ {}^-OH} R-\overset{\overset{\ddot{O}}{\|}}{C}-\overset{\overset{Cl}{|}}{\underset{Cl}{C}}-Cl$$

$$R\overset{\overset{\ddot{O}}{\|}}{C}-\overset{\overset{Cl}{|}}{\underset{Cl}{C}}-Cl + {}^-\!\ddot{O}H, Na^+ \rightleftharpoons R\overset{\overset{\ddot{O}}{\|}}{C}{-}\overset{\overset{Cl}{|}}{\underset{\underset{H}{\ddot{O}:}}{C}}-Cl \longrightarrow R\overset{\overset{\ddot{O}}{\|}}{C}\diagdown_{\overset{\ddot{O}}{\text{OH}}} + {}^-\!\text{CCl}_3 \rightleftharpoons R\overset{\overset{\ddot{O}}{\|}}{C}\diagdown_{\overset{\ddot{O}:{}^-, Na^+}{}} + HCCl_3$$

Benzoic Acid

The reaction is shown above.

EXPERIMENTAL PROCEDURE

Estimated time to complete the experiment: 1.5 h.

Physical Properties of Reactants							
Compound	MW	Amount	mmol	mp (°C)	bp (°C)	d	n_D
Acetophenone	120.16	60 μL	51	20.7	202.6	1.03	1.5372
Aqueous NaOCl (household bleach, 5% available chlorine)		2.1 mL					
Sodium sulfite	120.6	15 mg					

HOOD

$C_6H_5COCH_3$, 20 μL + 5% NaOCl, 700 μL

Reagents and Equipment. To a 5-mL conical vial containing a magnetic spin vane and equipped with an air condenser, add 6 μL (63 mg, 51 mmol) of acetophenone and 2.1 mL of household bleach (NaOCl, 5% available chlorine) (◄).

CAUTION: Both reagents are irritants and should be dispensed in the **hood** using automatic delivery pipets.

Reaction Conditions. Stir the mixture at room temperature for 30 min.

Isolation of Product. Add approximately 15 mg of sodium sulfite to destroy any unreacted bleach, and stir the reaction mixture briefly. Extract the resulting mixture with two 0.5-mL portions of ether (calibrated Pasteur filter pipet). Separate each portion of ether extract using a Pasteur filter pipet.

NOTE. *The ether layer is the top layer; the lower, alkaline, layer contains the acid product. The ether extraction removes the chloroform generated in the reaction and any unreacted acetophenone. Save the ether extracts in a small vial until you have successfully isolated and characterized the final product.*

Acidify the aqueous layer (check with pH paper) by adding dropwise 3 M HCl from a Pasteur pipet. A thick, white precipitate of benzoic acid then appears. Collect the solid by vacuum filtration using a Hirsch funnel, and wash the filter cake with three 0.5-mL portions of water (calibrated Pasteur pipet) (➡). Maintain the vacuum for approximately 5 min by covering the funnel with plastic food wrap (see Prior Reading) to partially dry the product. Transfer the material to a porous clay plate or filter paper to complete the drying process.

Oxidation product collected here

Aqueous HCl, ~ 2 mL + reaction byproducts

Purification and Characterization. The product is of sufficient purity for characterization. Weigh the material and calculate the percent yield.

Determine the melting point and compare your result with the value listed in the literature. If desired, obtain an IR spectrum and compare it with an authentic sample of benzoic acid, or with that given in the literature (*The Aldrich Library of IR Spectra* and/or SciFinder Scholar).

Chemical Tests. Chemical classification tests may also be used to assist in product analysis. You might wish to perform the following:

1. The ignition test should indicate the presence of the aromatic ring (Chapter 9).

2. The solubility of the compound in water, 5% NaOH, and 5% $NaHCO_3$ should be checked (see Chapter 9). Does the water solution turn blue litmus paper red? Is there evidence of CO_2 evolution when the benzoic acid is added to the bicarbonate solution? If positive results are obtained in these tests, how do they confirm that a carboxylic acid is present?

3. The preparation of an amide derivative (see Chapter 9) would also aid in establishing the identity of the acid product.

p-Methoxybenzoic Acid

Experiment 34B

REACTION

$$CH_3\ddot{O}-\langle\rangle-\overset{\overset{\cdot\cdot}{O}\cdot}{\underset{}{C}}-CH_3 \xrightarrow[\text{2. H}^+]{\text{1. NaOCl}} CH_3\ddot{O}-\langle\rangle-\overset{\overset{\cdot\cdot}{O}\cdot}{\underset{}{C}}-\ddot{O}H + HCCl_3$$

p-Methoxyacetophenone   *p*-Methoxybenzoic acid   Chloroform

EXPERIMENTAL PROCEDURE

Estimated time to complete the experiment: 1.5 h.

Physical Properties of Reactants					
Compound	MW	Amount	mmol	mp (°C)	bp (°C)
p-Methoxyacetophenone	150.8	29 mg	0.19	38–39	258
Aqueous NaOCl (household bleach, 5% available chlorine)		700 μL			
Sodium sulfite	120.6	5 mg			

Reagents and Equipment. Weigh and add 29 mg (0.19 mmol) of *p*-methoxyacetophenone to a 3.0-mL conical vial containing a magnetic spin vane. Now add 700 μL of household bleach (5% available chlorine), and then attach the vial to an air condenser (◀).

CAUTION: The bleach is an irritant to skin and eyes. It should be dispensed using an automatic delivery pipet in the **hood**.

Reaction Conditions. Stir the mixture for 30 min with very gentle heating. *Use the lowest possible setting on a hot plate magnetic stirrer.*

Isolation of Product. Weigh and add approximately 5 mg of sodium sulfite. Stir the reaction medium briefly and then cool it to room temperature. Extract the resulting mixture with two 0.5-mL portions of ether (calibrated Pasteur pipet). Separate each portion of ether extract using a Pasteur filter pipet.

NOTE. The ether layer is the top layer; the bottom aqueous layer contains the product. The ether extraction removes the chloroform generated in the reaction and any unreacted p-methoxyacetophenone. Save the ether extracts in a small vial until you have successfully isolated and characterized the final product.

Acidify the aqueous layer by the dropwise addition (check with pH paper) of 3 M HCl from a Pasteur pipet until a thick, white precipitate of *p*-methoxybenzoic acid is formed. Cool the mixture in an ice bath for 5 min and collect the solid by vacuum filtration using a Hirsch funnel (◀). Wash the filter cake with three 0.5-mL portions of cold water (calibrated Pasteur pipet). Maintain the vacuum for approximately 5 min and cover the funnel with plastic food wrap (see Prior Reading) to partially dry the product. Transfer the material to a porous clay plate or filter paper to complete the drying process.

Purification and Characterization. The product is reasonably pure but may be recrystallized from ethanol–water, if desired, using a Craig tube. Weigh the product and calculate the percent yield. Determine the melting point and obtain an IR spectrum. Compare your results with those reported in the literature (*The Aldrich Library of IR Spectra* and/or *SciFinder Scholar*). Chemical characterization tests can be used to assist in the classification of this product, as outlined in Experiment [34A].

OPTIONAL SEMI-MICROSCALE PREPARATION

This preparation may be scaled up by a 15-fold increase in reagent amounts. The procedure is similar to that given above for the micropreparation with the exceptions noted below. *A major difference is that a separatory funnel is used in the extraction step.*

1. Use a 25-mL round-bottom (RB) flask containing a magnetic stirring bar and equipped with an air condenser. The reaction conditions are as given in the microscale experiment (◀).

2. The reagent and solvent amounts are increased 15-fold.

HOOD

p-CH₃OC₆H₄COCH₃,29 mg
+ 5% NaOCl, 700 μL

Oxidation product
collected here

Aqueous HCl, ~2 mL +
reaction byproducts

Room temp
thermometer

25-mL RB flask

Physical Properties of Reactants					
Compound	MW	Amount	mmol	mp (°C)	bp (°C)
p-Methoxyacetophenone	150.8	435 mg	2.88	38–39	258
Aqueous NaOCl (household bleach, 5% available chlorine)			10.5 mL		
Sodium sulfite	120.6	75 mg			

3. Add the sodium sulfite, stir, cool, and then transfer the product mixture to a 125-mL separatory funnel. Extract this mixture with two 7.5-mL portions of diethyl ether. Save the ether extracts in a small Erlenmeyer flask until you have successfully isolated and characterized the final product. The product is in the aqueous layer.

NOTE. *Carry out the extraction by first returning the lower aqueous layer to the reaction flask and then removing the ether layer. Then return the aqueous layer to the separatory funnel and rinse the reaction flask with the second portion of ether. This ether rinse is then added to the separatory funnel to complete the extraction.*

4. Collect the product by vacuum filtration, then wash the filter cake with three 7-mL portions of cold water.

5. The pale yellow product is characterized as described in the microscale procedure.

QUESTIONS

6-216. In the haloform reaction, once the first **α**-hydrogen atom is replaced by a halogen atom, each successive hydrogen is more easily substituted until the trihalo species is obtained. Explain.

6-217. The haloform reaction using I_2 and NaOH is referred to as the "iodoform" test for methyl ketones (see Chapter 9). The test also gives positive results for compounds containing the —$CH(OH)CH_3$ group. This results from the oxidation of the alcohol to the methyl ketone in the first stage. Write a balanced equation for the conversion of $C_6H_5CHOHCH_3$ to the methyl ketone in the presence of I_2 and NaOH. Identify which species is being oxidized and which is being reduced.

6-218. If you were carrying out an industrial scale synthesis in which one step involved a haloform reaction to convert a methyl ketone into the corresponding acid having one less carbon atom, would you use NaOH and Cl_2, NaOH and Br_2, or NaOH and I_2 as the reagent? Give reasons for your choice.

6-219. Can you explain the fact that even though dibenzoylmethane ($C_6H_5COCH_2COC_6H_5$) is not a methyl ketone, it gives a positive iodoform test when treated with the NaOH and I_2.

6-220. Do you think that acetaldehyde would give a positive iodoform test? Explain your reasoning.

6-221. A water-soluble phenol is also acidic toward litmus paper as is a water-soluble carboxylic acid. How would you distinguish the difference between an aromatic acid and a phenol using a chemical test?

BIBLIOGRAPHY

Selected examples of the haloform reaction from Organic Syntheses:

Newman, M. S.; Holmes, H. L. *Organic Syntheses*; Wiley: New York, 1943; Collect. Vol. II, p. 428.

Sanborn, L. T.; Bousquet, E. W. *Organic Syntheses*; Wiley: New York, 1941; Collect. Vol. I, p. 526.

Smith, L. I.; Prichard, W. W.; Spillane, L. J. *Organic Syntheses*; Wiley: New York, 1955; Collect. Vol. III, p. 302.

Smith, W. T.; McLeod, G. L. *Organic Syntheses*; Wiley: New York, 1963; Collect. Vol. IV, p. 345.

Staunton, J.; Eisenbraun, E. J. *Organic Syntheses*; Wiley: New York, 1973; Collect. Vol. V, p. 8.

For a review of oxidation, see:

Fuson, R. C.; Bull, B. A. *Chem. Rev.* **1934**, *15*, 275.

Kurti, L.; Czako, B. *Application of Named Reactions in Organic Synthesis*; Elsevier: Amsterdam, 2005.

Smith, M. B. *March's Advanced Organic Chemistry*, 8th ed.; Wiley: New York, 2019, Chap. 19,

Wiberg, K. B. In *Oxidation in Organic Chemistry*; Trahanovsky, W. S., Ed.; Academic Press: New York, 1978, Vol. V, *Part A*, Chap. 2.

EXPERIMENT 35

Conversion of Cyclohexyl Bromide to Cyclohexene—An E2 Elimination Reaction: Factors Affecting the Rate of a Chemical Reaction

Common name: cyclohexene
CA number: [110-83-8]
CA name as indexed: cyclohexene

Purpose. A dehydrohalogenation reaction is designed and carried out for conversion of cyclohexyl bromide (bromocyclohexane) to cyclohexene using a strong base. The relative rate of this reaction is followed with gas chromatography. You will explore the effect of varying several parameters that alter the rate of the reaction such as temperature, concentration of the base, and nature of the leaving group.

Prior Reading

Technique 1: Gas Chromatography (pp. 59–65)
Technique 4: Solvent Extraction (p. 71)
Chapter 3: Reflux and reflux apparatus (pp. 25–26)
　　　　　　Automatic delivery pipets: (pp. 39–40)
　　　　　　Theoretical yield calculations: (p. 44)
Chapter 6: Experiment [10]: (pp. 226–233)
Lecture text on kinetic theory and mechanisms of E2 reactions

REACTION

Cyclohexyl bromide + NaOCH$_2$CH$_3$ $\xrightarrow{\text{CH}_3\text{CH}_2\text{OH}}$ Cyclohexene + NaBr + CH$_3$CH$_2$OH

Sodium ethoxide

DISCUSSION

The rates of chemical reactions and the factors that affect them are of great importance in chemistry. This area of chemistry is termed *kinetics*. Most of the ionic reactions observed in inorganic chemistry take place so fast that their rates cannot be easily measured, but in many organic reactions the study of chemical kinetics is particularly important. This is because measurement of reaction rates is a powerful guide for probing reaction mechanisms.

These rates can vary greatly. Five factors are involved in establishing the rate at which a reaction may proceed: (1) the nature of the reacting substances, (2) the states of the reactants, (3) the temperature of the reactants, (4) the concentration of the reactants, and (5) the presence of a catalyst. The rate of a chemical reaction is measured by the decrease in concentration of a reactant or the increase in concentration of a product in a unit period of time.

The relationship between the rate of a chemical reaction and the change in concentration of the reactants can be described quantitatively based on experimental measurements. Each reaction has its own *rate equation*. In general, a rate equation has the form

$$\text{rate} = k[A]^a[B]^b[C]^c \qquad\qquad 6\text{--}1$$

in which [A], [B], and [C] represent molar concentrations of reactants (or products or other substances), k is the rate constant for a certain reaction at a particular temperature, and the exponents a, b, and c are positive and usually integers. Both k and the exponents must be determined experimentally by observing the variation in the rate of the reaction as the concentrations of the reactants are varied. It is important to recognize that the rate equation depends on the mechanism of the reaction and on the individual steps therein.

In the general rate equation, the exponent a is called the order of the reaction with respect to the reactant A. Likewise, the exponent b is the order of the reaction with respect to B, and so on. The sum of the exponents is the overall order of the reaction. In the E2 reaction, investigated in the present experiment, we have (see above) a dehydrohalogenation following this mechanism.

Experimentation shows us that if the concentration of the reactant (cyclohexyl bromide) is doubled, the reaction rate doubles. The rate is also doubled if the concentration of the reagent, sodium ethoxide ($NaOC_2H_5$), is doubled. Thus, the concentration of both species present affects the rate of the reaction. The rate expression for this type of reaction (E2) can be quantitatively expressed as

$$\text{rate} = k[C_6H_{11}Br]^1[NaOC_2H_5]^1 \qquad\qquad 6\text{--}2$$

This expression indicates that the rate equation is *first-order* in both the substrate (cyclohexyl bromide) and the basic reagent (sodium ethoxide). The overall rate expression is *second-order* (the sum of $1 + 1 = 2$).

NOTE. The powers of 1 are not usually expressed, but they are shown here for clarity.

Kinetics and Thermodynamics. Chemical reactions are governed by two basic relationships.

1. Chemical Thermodynamics: These phenomena deal with the changes in energy that occur when molecules react. Thus, thermodynamics ultimately controls the *extent* to which a reaction will go to completion.

2. Chemical Kinetics: As outlined above, kinetics concern the *speed* at which a reaction will go to completion.

All chemical reactions are reversible. When the reactants and products reach the stage where changes in concentration can no longer be measured, the reaction is considered to be in a **state of equilibrium**. If the equilibrium lies far to the side of product formation, the reaction is said to have essentially *gone to completion*. Thus, when the equilibrium constant, K, is large, the reaction is said to have a large *driving force*. When a reaction satisfies this latter condition, a certain amount of *energy* is necessarily released when equilibrium is reached.

In terms of thermodynamics, for a reaction to take place spontaneously, the *free energy*, ΔG, of the products must be lower than the free energy of the reactants; that is, ΔG must be negative. Free energy is made up of two components: enthalpy H (the difference in bond energies between reactants and products) and entropy S (energy related to the randomness of the system). These quantities are related by the expression

$$\Delta G = \Delta H - T\Delta S \qquad\qquad 6\text{--}3$$

In many reactions, however, it is the enthalpy that mainly determines whether the reaction can take place spontaneously.

A negative ΔG value it does not necessarily mean that the reaction will take place in a reasonable length of time (that is reach equilibrium). A negative ΔG value is necessary, but not sufficient, to guarantee a spontaneous reaction. The *free energy of activation*, $\Delta G^{\ddagger}$, must also be considered. This condition is illustrated in the potential energy diagram (Fig. 6.49) for a one-step reaction of molecule A with molecule B to form molecules C and D (with no reaction intermediate assumed). This energy profile is typical of the base-catalyzed dehydrohalogenation reaction being studied in the current experiment, the E2 elimination reaction.

The generalized reaction profile is as shown in Figure 6.49

$$A + B \rightleftharpoons C + D \qquad\qquad 6\text{--}4$$

In the diagram the horizontal axis, called the reaction coordinate, indicates the progression of the reaction. $\Delta G^{\ddagger}$, is the free energy of activation for the forward reaction. The point at the top of the curve is called the *transition state*. Bond breaking and bond formation are involved over the range of this transition. Energy must be supplied through collision of molecules A and B (this involves both velocity and orientation in the collision) to move the system into and through the transition state and allow the products to form. As the products form, the activation energy is recovered plus additional energy (mainly from bond formation) so that the total energy evolved, $\Delta G^{\ddagger}$; is lower than that of the initial mixture.

The interpretation of the rate constant in terms of experimentally derived energy data is facilitated by the relationship known as the **Arrhenius equation**, which rather successfully predicts the temperature dependence of the rate constants:

$$k = A \exp\left(-\frac{E_a}{RT}\right) \qquad\qquad 6\text{--}5$$

where

R is the gas constant with the value 8.314 J mol^{-1} K^{-1}
T is the temperature on the Kelvin scale
E_a is the Arrhenius activation energy in joules/mole
A is a constant (for a given reaction) called the *frequency factor*

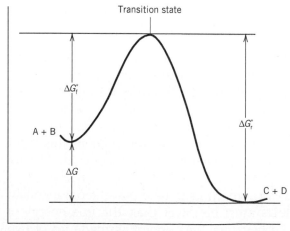

Reaction coordinate

Figure 6.49 Activation energy.

The *frequency factor* is related to the frequency of collisions and the orientation of the reacting molecules. *A* indicates how many collisions have the correct orientation to lead to products. The rest of the equation, $e^{-E_a/RT}$, gives the fraction of collisions in which the energy of the reacting species is greater than the activation energy for the reaction.

Also note that we can rewrite equation 6–5 as $-E_a = RT \ln(k/A)$, which closely resembles the expression between the equilibrium constant and the Gibbs free energy change, $-\Delta G° = RT \ln k$, and underlines the treatment of E_a as an energy term.

Thus, the Arrhenius equation describes quantitatively the discussion given here. For two reactions *at the same temperature*, the reaction with the higher activation energy has the lower rate constant and thus the slower rate. In other words, there is a smaller fraction of molecules possessing sufficient energy to react. An *increase in temperature* results in more molecules having enough energy to react and thus a faster rate. A change in conditions in which the frequency factor *A* is increased (by an increase in collisions in which the orientation of the molecules is right for reaction) also results in an increase in rate.

In the present experiment, the temperature is held relatively constant at the boiling point of the reaction mixture. Remember that a boiling solvent has a constant temperature at its boiling point. This is critical in reaction rate investigations for, as noted above, temperature change can exhibit a profound effect on rate. This phenomenon will be investigated in the second half of the experiment.

Geometrical Considerations in the E2 Elimination Reaction.

Base-induced elimination of alkyl halides (dehydrohalogenation) is a general reaction and is an excellent method for preparing alkenes. This process is often referred to as *β-elimination*, since a hydrogen atom is always removed β to the halide (leaving group):

A high concentration of a strong base in a relatively nonpolar solvent is used to carry out the dehydrohalogenation reaction (see Experiment [10] for further examples).

Experimental evidence indicates that the five atoms involved in the E2 elimination reaction must lie in the same plane; the *anti-periplanar* conformation is preferred. This conformation is necessary for the orbital overlap that must occur for the π bond to be generated in the alkene. The sp^3-hybridized atomic orbitals on carbon that comprise the C—H and C—X σ bonds broken in the reaction develop into the p orbitals comprising the π bond of the alkene being formed. The reaction is termed *stereospecific*.

The *anti-periplanar* conformation is particularly important in cyclohexane rings. This configuration can be met only if the hydrogen and the halide groups are *trans diaxial* to one another (see Fig. 6.50). If either the hydrogen or halide group is *equatorial*, the reaction does not occur. Furthermore, while there are exceptions to the relative disposition of the β-hydrogen and leaving group (Cope Elimination, for example), the number of systems which do not rely upon an anti-periplanar conformation are few and far between.

There is a smooth transition between reactant and product. Analogous to the S_N2 reaction, *no intermediate has been isolated or detected*. The reaction

Anti-periplanar conformation

H and Br are anti-periplanar

Figure 6.50 **Geometric requirements for E2 elimination reaction in cyclohexanes.**

takes place in a single step through a high-energy *transition state* in which the double bond begins to form at the same time the hydrogen and halide groups leave. Figure 6.49 illustrates the energy profile for the reaction.

Evidence in support this mechanism is obtained from the measurement of the reaction kinetics of these reactions. Since both the alkyl halide and base concentrations appear to enter into the rate determining step, E2 reactions follow the second-order rate law:

$$\text{Rate} = k[\text{RX}]\,[\text{Base}] \qquad\qquad 6\text{–}6$$

As discussed above, a change in concentration of the base or of the alkyl halide should, therefore, affect the overall rate of the reaction. This dependence on concentration can be demonstrated by changing the base concentration. The current reaction, as it proceeds, will be sampled at 30-min intervals to track the generation of the alkene product, cyclohexene. Monitoring the formation of the product is carried out by removing a small aliquot sample from the reaction vessel and immediately quenching (stopping) the reaction with 1 M HCl. Rapid neutralization of the reacting medium (in the aliquot) with acid blocks the reaction. Each quenched sample is then extracted with methylene chloride to isolate the product and any unreacted starting material (cyclohexyl bromide). The rapid analysis of the small quantities of isolated reactants and products is accomplished by GC. Methylene chloride is effective as the extracting solvent, since it elutes from the GC column well before the compounds of interest (cyclohexene and cyclohexyl halide). From the chromatographic data obtained it is possible to determine if the amount of cyclohexyl halide has decreased and if the amount of cyclohexene has begun to increase. The results are reported as the percent cyclohexyl halide reacted.

EXPERIMENTAL PROCEDURE

Estimated time to complete the experiment: two laboratory periods.

NOTE. It is recommended that the experiment be done in two parts. Use the first period to work out the procedure (equipment setup, sampling, GC analysis, etc.) and the second period to study the effect of several parameters on the rate of the reaction.

All drops are measured using a Pasteur pipet.

Physical Properties of Reactants						
Compound	MW	Amount	mmol	bp (°C)	d	n_D
Cyclohexyl bromide	163.06	0.6 mL	4.87	166.2	1.3359	1.4957
Sodium ethoxide (21% in ethanol)		2 mL				

Reagents and Equipment. Using a 10-mL graduated cylinder, measure 2 mL of a 21 wt% solution of sodium ethoxide in ethanol and add it to a 5-mL conical vial containing a magnetic spin vane (use a precalibrated Pasteur pipet for the transfer). Then equip the reaction vial with a Claisen head to which is connected a reflux condenser (➡).

Cyclohexyl bromide, 0.6 mL Sodium ethoxide (21% in EtOH), 2.0 mL

Reaction Conditions. Heat the reagent, with stirring, to reflux using a sand bath temperature of approximately 100 °C. Remove the cap from the Claisen head and use an automatic delivery pipet to add, in one portion, 0.6 mL (4.87 mmol) of cyclohexyl bromide. *Note the time of addition in your laboratory notebook.* Allow a few seconds for intimate mixing of the reactants to take place and then **immediately** remove 10 drops (about 0.4–0.5 mL) of the reaction solution using a Pasteur pipet, *recap the Claisen head*, and place the aliquot in a small sample vial containing 4–5 drops of 1 M HCl.

NOTE. You should notice that after a short time solid NaBr begins to form in the reaction vial.

To the sample vial containing the initial aliquot, add 10 drops of methylene chloride. Cap the vial and shake gently (or mix on a Vortex mixer). Loosen the cap carefully to vent the two-phase mixture. Remove the bottom, brown organic layer using a Pasteur filter pipet and transfer it to a second vial containing a small amount of anhydrous sodium sulfate. Cap the vial and label it as the *time-zero* sample.

*NOTE. For separation of the layers, it is recommended that **both** layers be drawn into the pipet. The bottom layer is then transferred to the second vial.*

Sampling Procedure. Two further samples for analysis are removed, *as outlined above*, one at 30 min into the reaction and the second after 1 h.

Each time the aliquot is treated,

1. Remove 10 drops (0.4–0.5 mL) of reaction solution (recap the Claisen head)

2. Quench the reaction with acid (4–5 drops of 1 M HCl)

3. Extract the product and unreacted cyclohexyl bromide with methylene chloride, followed by drying the methylene chloride solution over anhydrous sodium sulfate

NOTE. Having the 5 drops of 1 M HCl in each of three sample vials and sodium sulfate in each of three second sample vials (a total of six vials) before you add the cyclohexyl bromide to initiate the reaction will help you facilitate this sampling procedure. Carefully label all the vials!

At the end of the sampling period you should have three samples for analysis by GC: time-zero, 30 min, and 60 min.

Gas Chromatographic Analysis.   The composition of the aliquot samples may be determined by gas chromatographic analysis.

Parameters for the GOW-MAC 350 Column

Packed Column:	DC 710 ($\frac{1}{4}$ in. × 8 ft)
Injection:	10 μL
Temperature:	132 °C
Flow rate (He):	50 mL/min
Chart speed:	1 cm/min

Gas chromatograph data may be used to determine the change in concentrations of cyclohexene and cyclohexyl bromide in the reacting medium during the 1-h reaction period. The samples can be injected after the extracted solution has had a 15- to 20-min period to dry over anhydrous sodium sulfate. *Each sample must elute from the GC before another sample is injected.* Each sample takes approximately 20 min to elute from the column, so start sample injections as soon as possible. For a typical chromatogram see Figure 6.51.

INSTRUCTOR PREPARATION. For each chromatograph used, a mixture of cyclohexene and cyclohexyl bromide (approximately 80:20 by volume) should be injected once the parameters of the column are set. This is to establish the retention times of the components in the mixture.

NOTE. If the laboratory is done in two parts, it is advisable to stop at this point.

Figure 6.51   Example cyclohexene–cyclohexyl bromide GC.

DATA ANALYSIS

The percentage of each component (cyclohexene and cyclohexyl bromide) in the sample can be determined by measuring the area under the curves. How would you expect the areas under the cyclohexene peak and the cyclohexyl bromide peak to vary with time?

The boiling point of cyclohexene (82.9 °C) is close to that of the refluxing ethanol (78.2 °C). Therefore, when samples are removed from the reaction mixture some of the cyclohexene may be lost, since some of it will be in the vapor phase in the reaction vial. How would this loss affect the observed concentrations over the reaction period?

The *time-zero* fraction should be analyzed, since the reaction starts immediately upon addition of the cyclohexyl bromide to the basic solution.

NOTE. Several alternative techniques may be used to quantitatively determine the composition of the sample mixture from the GC data, if an integrating recorder is not available.

Two methods are described here:

1. Determination of the areas under the peaks gives reproducible results of ±3–4% when these areas are assumed equal to the peak heights (mm) × the peak widths at half-height (mm), measured from the baseline of the curve.

2. An alternative method for determining these areas is to cut out the peaks from the chromatogram and weigh them on an analytical balance (sensitivity to 0.1 mg). The weights of the peaks are directly proportional to the relative amount of each compound in the sample (assuming that the detector is similar sensitivity to the two components).

The percentage of the individual components is then calculated by dividing each area by the total sum of the two areas and multiplying by 100.

Attach a copy of your chromatogram to your Data Analysis Sheet. Label the cyclohexene and cyclohexyl bromide peaks. Show the area measurements on the chromatogram and all the calculations in your report. What conclusions can you draw based on your results? Include an explanation of the conclusions drawn.

VARIATION OF PARAMETERS

The effect of varying several parameters that influence the *rate* of the reaction can also be investigated. Of particular interest for study are the effect of changes in the concentration of the base, the reaction temperature, and the nature of the leaving group.

INSTRUCTOR NOTE. It is suggested that the class be divided into groups, with each group assigned to investigate a different parameter. The results of each group are then shared by the entire class and each student then submits a report.

The parameters to be investigated are

1. A control (see conditions below)
2. Halving the base concentration

3. Room temperature reaction
4. Replacing cyclohexyl bromide with cyclohexyl chloride
5. Replacing cyclohexyl bromide with cyclohexyl iodide

Determine the percentage of unreacted cyclohexyl bromide in the sample mixture from the chromatogram for parameters 1, 2, and 3. Determine the percentage of cyclohexyl chloride and iodide in parameters 4 and 5, respectively. A comparison of unreacted starting halide with the control reaction shows the effect on the relative rate of dehydrohalogenation of changing the conditions under which the reaction is carried out. Tabulate your results.

In this set of experiments track only the disappearance of cyclohexyl bromide, and not the appearance of cyclohexene. As mentioned previously, the amount of cyclohexene observed in successive aliquots tends to drop below the predicted values because of the sampling techniques employed. As the reaction progresses, however, the amount of cyclohexyl bromide continues to decrease at the predicted rate (see below).

Use of an Internal Standard.   How can we quantitatively track the disappearance of cyclohexyl bromide? As we have seen, the ratio of cyclohexyl bromide to cyclohexene cannot be determined accurately because of the inaccuracies in measuring the cyclohexene formed in the reaction. Also, there are small inaccuracies in the amount of the reaction mixture removed from the system for any given sample. These problems are solved by *standardizing* the samples. An *internal standard* is added to the reaction mixture. Ideally, this standard should not interfere with the reaction being studied and the substance must be easy to track by GC. In these experiments we will use *toluene*. Toluene is not very volatile (bp 110.6 °C) and has a retention time on the GC that does not interfere with the resolution of any other peaks on the chromatogram. It will be added *in a known amount* to the reaction mixture before you begin. For each gas chromatogram we can then measure *the area of the standard's peak* and use it to determine the amount of cyclohexyl bromide in the reaction mixture at that point.

Experimental Conditions

NOTE. Use the same experimental setup and workup procedure as in the previous experiment. A sample for analysis will be taken only at the 30-min mark. You may wish to take two samples so as to determine whether you can duplicate your results. The GC is setup at the same operating conditions. Drops are measured using a Pasteur pipet.

1. **Control**

    Reaction conditions

    2 mL of 21 wt% solution of sodium ethoxide in ethanol

    300 µL toluene

    Heat to reflux and stir.

    Add 0.6 mL cyclohexyl bromide.

    Time the reaction

    At the 30-min mark, remove 20 drops of reaction solution and add it to 10 drops of 1 M HCl. Extract the mixture with 20 drops of methylene chloride

and dry the methylene chloride layer over anhydrous sodium sulfate (15 min).
Analyze the sample by GC.

2. **Room Temperature Conditions**

Reaction conditions

2 mL of 21 wt% solution of sodium ethoxide in ethanol

300 μL of toluene

Stir at room temperature **(do not heat the hot plate)**.

Add 0.6 mL of cyclohexyl bromide.

Time the reaction

At the 30-min mark, remove 20 drops of reaction solution and add it to
10 drops of 1 M HCl. Extract the mixture with 20 drops of methylene chloride
and dry the methylene chloride layer over anhydrous sodium sulfate (15 min).
Analyze the sample by GC.

3. **Variation of the Concentration of Base**

Reaction conditions

1 mL of 21 wt% solution of sodium ethoxide in ethanol

1 mL of absolute (anhydrous) ethanol

300 μL of toluene

Heat to reflux and stir.

Add 0.6 mL cyclohexyl bromide.

Time the reaction

At the 30-min mark, remove 20 drops of reaction solution and add it to
10 drops of 1 M HCl. Extract the mixture with 20 drops of methylene chloride
and dry the methylene chloride layer over anhydrous sodium sulfate (15 min).
Analyze the sample by GC.

4. **Variation of Substrate—Cyclohexyl Chloride**

Reaction conditions

2 mL of 21 wt% solution of sodium ethoxide in ethanol

300 μL of toluene

Heat to reflux and stir.

Add 0.58 mL cyclohexyl chloride.

Time the reaction

At the 30-min mark, remove 20 drops of reaction solution and add it to
10 drops of 1 M HCl. Extract the mixture with 20 drops of methylene chloride
and dry the methylene chloride layer over anhydrous sodium sulfate (15 min).
Analyze the sample by GC.

5. **Variation of Substrate—Cyclohexyl Iodide**

Reaction conditions

2 mL of 21 wt% solution of sodium ethoxide in ethanol

300 μL of toluene

Heat to reflux and stir.

Add 0.63 mL of cyclohexyl iodide.

Time the reaction

At the 30-min mark, remove 20 drops of reaction solution and add it to 10 drops of 1 M HCl. Extract the mixture with 20 drops of methylene chloride and dry the methylene chloride layer over anhydrous sodium sulfate. Analyze the sample by GC.

Data Analysis

To calculate the % of cyclohexyl bromide (chloride or iodide) reacted, follow the outline given here.

Given:

a. The amount of toluene added as the internal standard:

$$\underline{\text{259 mg}}$$

Calculation (show here):

$$\frac{\rule{4cm}{0.4pt}}{\text{Volume of toluene used} \times \text{its density}}$$

b. The amount of cyclohexyl bromide used in the reaction:

$$\underline{\text{794 mg}}$$

Calculation (show here):

$$\frac{\rule{5cm}{0.4pt}}{\text{Volume of cyclohexyl bromide used} \times \text{its density}}$$

If cyclohexyl chloride or iodide is used in place of the bromide:

$$
\begin{array}{ll}
\text{Chloride:} & \text{581 mg} \\
\text{Iodide:} & \text{1029 mg}
\end{array}
$$

NOTE. The amount of cyclohexyl chloride and iodide used is the same molar amount as the cyclohexyl bromide.

The Calculations Using an Internal Standard. The calculations allow us to compare the amount (%) of cyclohexyl bromide (or chloride or iodide) reacted at the 30-min time point for each of the different reaction parameters.

Step 1. The amount of cyclohexyl bromide (chloride or iodide) **left** (unreacted) at the 30-min mark:

$$\frac{\text{Area of } C_6H_5\text{—Br, —Cl, or — I peaks}}{\text{Area of toluene peak}} \times$$

$$[259 \text{ mg of toluene standard}] = \underline{\hspace{1cm}} \text{ mg unreacted}$$

Step 2. The amount of cyclohexyl bromide (chloride or iodide) **reacted** at the 30-min mark:

$$[\text{Amount of } C_6H_5\text{—Br, —Cl, or —I used in the reaction (in mg)}] -$$
$$[\text{the amount left after 30 min (in mg), see step 1}] = \underline{\hspace{1cm}} \text{ mg reacted}$$

Step 3. The percent of cyclohexyl bromide (chloride or iodide) **reacted** at the 30-min mark:

$$\frac{\text{Amount reacted (in mg)}}{\text{Amount used (in mg)}} \times 100 = \underline{\hspace{1.5cm}} \% \text{ reacted}$$

Attach a copy of your chromatogram to your Data Analysis Sheet. Indicate which parameter you worked on. Label the toluene and cyclohexyl bromide, chloride, or iodide peaks. Show the area measurements on the chromatogram. Show all the above calculations in your report. Draw conclusions regarding the effect on the rate of the reaction for each of the parameters investigated in the laboratory. Include an explanation of the conclusions drawn.

QUESTIONS

6-222. What reaction occurs between $NaOCH_2CH_3$ and HCl that stops the dehydrohalogenation reaction?

$$NaOCH_2CH_3 + HCl \rightarrow$$

6-223. Would you expect the replacement of cyclohexyl bromide with cyclohexyl iodide to result in a decrease, an increase, or no change in the rate of the reaction? Explain.

6-224. In the second part of the experiment, why was it necessary to use toluene as an internal standard? What are the requirements for an internal standard? How would the loss of some toluene in the sampling procedure (from overheating the reaction system) affect the data?

6-225. A negative ΔG value does not necessarily mean that a reaction will take place in a reasonable length of time. A negative ΔG value is necessary, but not sufficient to guarantee a spontaneous reaction. The *free energy of activation, $\Delta G^{\ddagger}$,* must be considered. Explain this statement.

6-226. In E2 elimination reactions involving cyclohexane rings, why is it critical that the β-hydrogen and the leaving group be *trans diaxial* to one another?

6-227. Please explain why *trans*-1-bromo-4-methylcyclohexane is more likely to undergo an E2 reaction when compared to *trans*-1-bromo-4-*tert*-butylcyclohexane even when working with warm toluene. Conversely, explain why *cis*-1-bromo-4-methylcyclohexane is less likely to undergo an E2 reaction as quickly when compared to *cis*-1-bromo-4-*tert*-butylcyclohexane.

6-228. What effect on the rate of reaction would be observed if some of the ethanol solvent escaped from the reaction flask (as volatile vapor) while samples are being removed? Explain.

BIBLIOGRAPHY

For a discussion of transition states, see

Laidler, K. J. *J. Chem. Educ.* **1988**, *65*, 540.

For texts on Chemical Kinetics see, for example, (a) Laidler, K. J. *Chemical Kinetics*, 3rd ed.; Benjamin/Cummings: Menlo Park, CA, 1987; (b) Logan, S. R. *Fundamentals of Chemical Kinetics*; Addison-Wesley Longman: Menlo Park, CA, 1996; and (c) House, J. *Principles of Chemical Kinetics*, 2nd ed.; WCB/McGraw-Hill: Burr Ridge, IL, 2007.

Physical Chemistry texts generally present a kinetic analysis of equilibria. For example, see:

(a) Atkins, P. W.; DePaula, J. *Physical Chemistry*; Vol. I, 8th ed.; Freeman: New York, 2006. (b) Laidler, K. J.; Meiser, J. H.;

Sanebuary, B. C. *Physical Chemistry*, 4th ed.; Houghton Mifflin: Boston, 2003; (c) Silbey, R. J.; Alberty, R. A.; Bawendi, M. G. *Physical Chemistry*, 4th ed.; Wiley: New York, 1997; and (d) Engle, T.; Reid, P. *Physical Chemistry*; Pearson Benjamin Cummings: San Francisco, 2006.

For an introduction to the E2 mechanism, see your lecture text or Smith, M. B. *March's Advanced Organic Chemistry*, 8th ed.; Wiley: New York, 2019, Chap. 17 or Carey, F. A.; Sundberg, R. J. *Advanced Organic Chemistry*, 4th ed.; Plenum: New York, 2008, Chap. 6.

EXPERIMENT 36

Aqueous Suzuki Synthesis of 4-Phenylphenol

Common name: 4-Phenylphenol
CA number: [92-69-3]
CA name as indexed: [1,1'-Biphenyl]-4-ol

B
G

Purpose. This experiment illustrates a Suzuki reaction in water to synthesize 4-phenylphenol ([1,1'-biphenyl]-4-ol), a non-steroidal anti-inflammatory drug analog. The solid product is isolated by vacuum filtration and purified by recrystallization. Aspects of green chemistry are considered, and 4-phenylphenol is characterized by IR spectroscopy, proton NMR spectroscopy (see Chapter 8), and melting-point analysis.

Prior Reading

Standard Experimental Apparatus (pp. 23–29)
Technique 5: Filtration Techniques (pp. 92–95)
Technique 5: Crystallization (pp. 89–91)

REACTION

Phenylboronic acid   4-Iodophenol   4-Phenylphenol

DISCUSSION

The Suzuki Reaction. Suzuki coupling reactions are very popular, mild approaches for synthesis of carbon–carbon σ-bonds. The first examples were reported by Norio Miyaura and Akira Suzuki in 1979, and generally involve reaction of a boronic acid or ester with an aryl or vinyl halide under basic conditions in the presence of a palladium (0) catalyst.

This experiment is reproduced from "Greening Up" the Suzuki Reaction: *J. Chem. Educ.* **2008,** *85,* 555. People who are co-authors on this paper and contributed to the experimental design are as follows:

Evangelos Aktoudianakis
Elton Chan
Amanda R. Edward
Isabel Jarosz
Vicki Lee
Leo Mui
Sonya S. Thatipamala
Andrew P. Dicks

Adapted with permission from the American Chemical Society, Aktoudianakis, E.; Chan, E.; Edward, A. R.; Jarosz, I.; Lee, V.; Mui, L.; Thatipamala, S. S.; Dicks, A. P.; *Journal of Chemical Education,* 2008, *85,* 555. Copyright 2008, American Chemical Society.

$$R'-B{\overset{Y}{\underset{Y}{|}}} + R''-X \xrightarrow[\text{organic solvent}]{\text{Pd (0), base}} R'-R''$$

$$Y = -OH, -OR$$
$$X = Br, I$$
$$R', R'' = \text{aryl, vinyl}$$

Akira Suzuki (1930–) was born in Mukawa, Japan, and studied at Hokkaido University in Sapporo, where, as a freshman, he was inspired by reading *"Textbook of Organic Chemistry"* by Louis and Mary Fieser. The title of his doctoral thesis was *"Synthesis of the Model Compounds of Diterpene Alkaloids."* At Hokkaido Suzuki became very familiar with use of Grignard (organomagnesium) and organozinc reagents as important compounds in the context of synthesis. He subsequently undertook post-doctoral work at Purdue University in the United States with the Nobel Laureate Herbert Brown, who advised him to "do research that will be in the textbooks." Suzuki became a full professor on returning to Hokkaido and eventually worked at the Okayama University of Science and the Kurashiki University of Science and the Arts, before retiring in 2002. Since its discovery, the Suzuki reaction has become so significant that he shared the 2010 Nobel Prize in Chemistry with Richard Heck (University of Delaware) and Ei-ichi Negishi (Purdue University) "for palladium-catalyzed cross couplings in organic synthesis."

Nowadays, many fundamental reactions involved in the manufacture of industrially significant substances are achieved by Suzuki couplings. A recent example is seen during the preparation of carbazole derivatives, which are of interest as parts of dye-sensitized solar cells and building blocks for polymer synthesis. A variety of *N*-heteroaryl halides (e.g., 2-bromopyridine) have been coupled with 4-(9*H*-carbazol-9-yl)phenylboronic acid to form a range of *N*-heteroaryl substituted 9-arylcarbazolyl compounds. This methodology uses various catalytic palladium sources (Pd/C, PdCl₂ and Pd(OAc)₂ amongst others) with aqueous ethanol as solvent and potassium carbonate as base.

$$\text{(carbazolylphenylboronic acid)} + \text{(2-bromopyridine)} \xrightarrow[\text{aq. EtOH, K}_2\text{CO}_3]{5\% \text{ Pd/C}} \text{(product)} \quad 71\%$$

Green Chemistry Considerations. The previously described reaction illustrates that Suzuki couplings are often undertaken in an organic solvent, with water as a co-solvent. An inorganic base (such as sodium, potassium or cesium carbonate) is also commonly used. There is currently much active research in the area of "green chemistry" from a synthetic perspective, employing reaction conditions that are less deleterious to the environment than previous approaches. Many traditional organometallic transformations (e.g., Grignard and Gilman reactions) do not lend themselves well to aqueous conditions, but it is possible to use pure water as the solvent in other cases. Water is nonflammable, has no related toxicity issues, and is the cheapest solvent for reactions on an industrial scale. In addition, boronic acids and esters are generally stable and environmentally benign compounds. The aqueous Suzuki synthesis of 4-phenylphenol illustrates advances made in improving the "greenness" of an industrially important synthetic process. Phenylboronic acid is coupled with 4-iodophenol using potassium carbonate and a catalytic amount of 10% palladium on carbon *under purely aqueous conditions*. Water is then consequently used with methanol as the recrystallization solvent mixture. The reaction affords facile preparation of a biaryl product of the type currently sold as nonsteroidal anti-inflammatory drugs (NSAIDs) including diflunisal and felbinac. Diflunisal is marketed in the United States by Merck under the brand name Dolobid, and is a structural derivative of 4-phenylphenol. Two over-the-counter NSAIDs that you are likely familiar with are acetylsalicylic acid (aspirin) and ibuprofen.

B

Diflunisal (Dolobid)   Felbinac

An interesting green chemistry conflict here is that *the Suzuki reaction is not especially atom-economical*. The term "atom economy" was coined in 1991 by Barry Trost of Stanford University as a way of considering the efficiency of a chemical transformation in terms of the atoms concerned. It is an important measure of reaction "greenness" and is defined as follows:

$$\% \text{ Atom Economy} = (\text{Mol. Wt. of Product}/\Sigma(\text{Mol. Wt. of Reactants})) * 100$$

Based on this equation, if all reactant atoms are incorporated into a desired product (for example, during the Diels–Alder reaction of ethene and 1,3-butadiene to form cyclohexene) then the atom economy will be 100% (see Experiment 14 Discussion, p. 266). Lower atom economies reflect reactions that are less efficient, as some of the reactant atoms are converted to unwanted side-products or byproducts.

Reaction Mechanism. The mechanism of the Suzuki reaction is quite different to most classical polar reactions, due to the activity of the metal catalyst (palladium on carbon). Because of this, the mechanism is best represented as a *catalytic cycle* where Pd (0) from Pd/C is oxidized to Pd (II), which becomes reduced back to Pd (0). In the first step (*oxidative addition*), Pd metal becomes inserted into the C—I bond of 4-iodophenol to form an organopalladium (II) species. *Ligand exchange* replaces an iodide ion with a carbonate ion at the Pd (II) center. Phenylboronic acid is activated by a second equivalent of carbonate, which enhances the polarization of the organic ligand (a phenyl group). This facilitates the *transmetalation* step where the phenyl group is transferred from boron to palladium. Finally, *reductive elimination* couples the two aromatic ligands together to generate 4-phenylphenol and reform Pd (0), which can participate in another catalytic cycle.

EXPERIMENTAL PROCEDURE

Estimated time to complete the experiment: 2.5 h.

reflux
thermometer

Reflux conditions
100-mL RB flask

Physical Properties of Reactants					
Compound	MW	Amount	mmol	mp (°C)	bp (°C)
Phenylboronic acid	121.93	122 mg	1.00	216–219	
4-Iodophenol	220.01	220 mg	1.00	92–94	
Potassium carbonate	138.21	415 mg	3.00	891	
10% Palladium on carbon	106.42	3 mg			
water	18.01	11 mL		0	100
2 M HCl	36.46				
Methanol	32.04			−98	64

Reaction Setup. Place the following in a 50-mL round-bottom flask: phenylboronic acid (122 mg); potassium carbonate (415 mg); 4-iodophenol (220 mg) and distilled water (10 mL). Add a magnetic stir bar and proceed to stir the reaction mixture rapidly. Add water (~1 mL) to a supplied vial containing 10% palladium on carbon (3 mg) to create a suspension, and then transfer the suspension to the reaction mixture by Pasteur pipet.

Reaction Conditions. Heat the mixture vigorously under reflux using a water condenser for 30 min., using a sand bath as the heat source and maintaining rapid stirring. Some solid may precipitate during this time. Remove the flask from the sand bath and allow cooling to room temperature.

Suzuki product
collected here

H$_2$O, 25 mL

Isolation of Crude Product. Once cooled to room temperature, add by pipet aqueous HCl (2 M) dropwise to the crude reaction mixture until strongly acidic to litmus paper. Collect the crude solid (which still contains the Pd/C catalyst) by vacuum filtration, using a Hirsch funnel. Wash the solid with ice-cold water (~10 mL).

Pd/C

Crude reaction
mixture in
methanol/water

Purification and Characterization. Dissolve the crude solid in methanol (~10 mL) using a 25-mL Erlenmeyer flask. Remove the Pd/C catalyst by gravity filtration, collecting the reaction mixture in a 50-mL Erlenmeyer flask. Add water (~10 mL) to the methanolic solution, causing a solid to precipitate. Heat until the entire product has gone into solution (add 1–2 mL portions of methanol if necessary). Once complete dissolution has occurred, allow the solution to cool slowly to room temperature and then cool in an ice bath for 15 min. Using a Hirsch funnel, collect the recrystallized product by vacuum filtration and dry very thoroughly. Remove the solid from the funnel, weigh, and calculate the percentage yield. Take appropriate physical measurements (mp, IR, proton NMR spectra) to ascertain the purity of your compound, comparing with literature data (SciFinder Scholar) or that of an authentic sample.

Adapted with permission from the American Chemical Society,Aktoudianakis, E.; Chan, E.; Edward, A. R.; Jarosz, I.; Lee,V.; Mui, L.; Thatipamala, S. S.; Dicks, A. P. *J. Chem. Educ.*, **2008**, *85*, 555. Copyright 2008, American Chemical Society.

QUESTIONS

6-229. A fellow student misread the label on a reagent bottle and attempted the synthesis of 4-phenylphenol from 4-fluorophenol (rather than 4-iodophenol) via a Suzuki reaction. Unfortunately, only unreacted starting materials were recovered. Provide an explanation for this result.

6-230. In many reports of the Suzuki synthesis, two equivalents of base are added to the reaction. Explain why, in the preparation of 4-phenylphenol from phenylboronic acid and 4-iodophenol, three equivalents of base (potassium carbonate) are used.

6-231. Calculate the atom economy of the reaction between phenylboronic acid and 4-phenylphenol. How might the atom economy be improved by changing the structure of one of the starting materials? What is a potential drawback to doing this?

6-232. Although water seems like the ideal solvent for conducting organic reactions, it cannot be utilized universally. Discuss some problems and drawbacks associated with its use.

6-233. A journal article from 2007 (*Chem. Eur. J.* **2007,** *13*, 5937) describes the Pd/C catalysis of Suzuki reactions. Outline advantages and disadvantages of this approach compared to the preparation of 4-phenylphenol in terms of green chemistry.

6-234. Two more well-known metal-catalyzed cross-coupling processes are known as the *Stille reaction* and the *Negishi reaction*. Explain why the Stille reaction has a poor environmental profile, and why the Negishi reaction is not commonly undertaken under aqueous conditions.

BIBLIOGRAPHY

This experiment is adapted from that described by:

Sakurai, H.; Tsukuda, T.; Hirao, T. *J. Org. Chem.* **2002**, *67*, 2721.

Background on the Suzuki reaction, including examples using water as a solvent:

Franzén, R.; Xu,Y. *Can. J. Chem.* **2005**, *83*, 266.
Miyaura, N.; Suzuki, A. *Chem. Rev.* **1995**, *95*, 2457.
Miyaura, N.; Suzuki, A. *Chem. Commun.* **1979**, 866.
Mondal, M.; Bora, U. *Green Chem.* **2012**, 14, 1873.
Rao, X.; Liu, C.; Qiu, J.; Jin, Z. *Org. Biomol. Chem.* **2012**, *10*, 7875.
Polshettiwar,V.; Decottignies, A.; Len, C.; Fihri, A. *ChemSusChem* **2010**, *3*, 502.

The 2010 Chemistry Nobel Prize:

Kauffman, G. B.; Adloff, J-P. *Chem. Educ.* **2011**, *16*, 51.

Aqueous organic reactivity in undergraduate laboratories:

Dicks, A. P. *Green Chem. Lett. Rev.* **2009**, *2*, 9.
Sauer, E. L. O. Organic Reactions under Aqueous Conditions. In *Green Organic Chemistry in Lecture and Laboratory*; Dicks, A. P., Ed.; CRC Press: Boca Raton, FL, 2012, pp. 103–130.

Atom economy and green chemistry:

Abhyankar, S. B. Introduction to Teaching Green Organic Chemistry. In *Green Organic Chemistry in Lecture and Laboratory*; Dicks, A. P., Ed.; CRC Press: Boca Raton, FL, 2012, pp. 1–28.
Trost, B. M. *Science* **1991**, *254*, 1471.

C₇H₆apter 7

SEQUENTIAL SYNTHESES: THE TRANSITION FROM MACRO TO MICRO

The synthesis of a vast array—now numbering in the millions—of new organic molecules in academic and industrial laboratories over the past 100 years is one of the great achievements of modern science. Many of these new compounds have had profound effects on our way of life, both good and bad. A great challenge in the next century will be how society applies these powerful materials, and the new molecules yet to be born, to the common good.

Our ability to synthesize highly complex organic substances has taken a number of dramatic jumps during this century, and has resulted in a bewildering collection of substances that have been devised, synthesized, and applied to practically every facet of our lives. Many of these materials are now vital to our daily life (consider penicillin), and we all too often take them for granted. In just the last 30–40 years, new advances in pharmaceutical compounds have saved, extended, and made more comfortable the lives of hundreds of millions of people. The list could go on and on, including textiles, surfactants, plastics, and synthetic oils, to name only a few.

Historically, the synthesis of complex organic substances was primarily driven by the need to obtain large quantities of biologically active material that occurred as the product of plant or animal metabolism, but that could be obtained only in very small quantities from nature. For example, the synthesis of the adrenal cortex hormone cortisone was a major breakthrough for hormone therapy. The synthesis of this material initially required 33 steps. That is, the research chemist carried out a sequence of 33 reactions in which stable isolable intermediates were formed sequentially, leading ultimately to the desired cortisone molecule. Industrial sequences of this length are now rare, but those requiring three to six steps are common.

The strategy of the synthetic chemist is to devise a route whereby the desired compound can be prepared efficiently, using inexpensive, readily available starting materials in the fewest steps, and have a minimal impact on the environment when considering the waste(s) generated. For each individual step, the yield of intermediate should be as high as possible with a minimum of side reactions. In industry the overall cost of the proposed synthetic sequence must be considered, including the time involved, type of equipment required, and safety factors. Today, with the worldwide demand for organic materials in vast quantities (e.g., petroleum products), it is becoming increasingly important to assess the impact on our environment of these synthetic materials prior to large-scale production.

Chapter 7: C₇H₆, [1,2]Spirene
Simmons and Fukunaga (1967).

Because organic transformations almost always take place with some loss of material (similar bond energies lead to alternative reaction pathways and easy byproduct formation), the yield of intermediate from each individual step can have a significant impact on the overall yield of the final product. *In a multistep synthesis, the overall yield is the mathematical product of the yields of the individual steps.* For example, if we assume that in a five-step sequence for the preparation of a new dye, each step takes place in 85% yield, the overall yield would be $(0.85)^5 \times 100$ or 44% (in a 33-step synthesis with an 85% yield for every step, the overall yield would be $\sim 0.5\%$). This property of organic synthesis emphasizes why a synthetic route devised to produce a particular molecular structure must be carefully planned to minimize losses at each stage of the chosen pathway. This problem also illustrates why the initial steps of a sequence usually use larger quantities of reactants (macro or semi-micro quantities), and why in the last stages, experience at running reactions at the microscale level can be invaluable.

In this chapter we describe sequential experimentation. These experiments vary in the number of intermediates that are required from seven to three, and they vary in the complexity of the chemistry, from straightforward extensions of Chapter 6 to relatively demanding experimentation similar to that described in Chapter 10. Three of the six target molecules below are part of this chapter. Located in the appendix to this chapter (Advanced Laboratory Experimentation: Sequences D, E, and F), are the remaining three examples illustrating sequential experimentation.

The target molecules include

- The drug **sulfanilamide** (the first of the antibiotics), which is obtained in a novel three-step sequence not usually found in the introductory laboratory:

$$H_2N-\!\!\!\left\langle\right\rangle\!\!\!-SO_2NH_2$$

p-Aminobenzenesulfonamide
(sulfanilamide)

- The industrially important polymer, **nylon-6,6** (the first of the commercial synthetic textile fibers), formed in three steps closely paralleling the original synthesis:

$$\left[\begin{array}{c} \ddot{\text{O}} \quad\quad \ddot{\text{O}} \quad\quad\quad \text{H} \\ \| \quad\quad\quad \| \quad\quad\quad\quad | \\ \text{C(CH}_2)_4\text{C}\!-\!\ddot{\text{N}}\text{H(CH}_2)_6\ddot{\text{N}} \end{array}\right]_n$$

Nylon-6,6

- The synthesis of **2'-bromostyrene**, which requires three steps. This compound is an interesting substance because of its commercial use as a fragrance in soap products:

$$\text{CH}\!\!=\!\!\text{CHBr}$$

2'-Bromostyrene

- The synthesis of **piperonylonitrile** is an example of a novel conversion, in three steps, of an aromatic aldehyde to an aromatic nitrile:

Piperonylonitrile

- The synthesis of **hexaphenylbenzene** requires two parallel three-step sequences to obtain two intermediates that then react with each other to give the target compound after a seven-step synthesis. The final product is a most unusual organic material with one of the highest melting points observed for an organic substance:

Hexaphenylbenzene

- The synthesis of a **photochromic imine** in four steps yields perhaps the most intriguing substance in the entire chapter. The ability of this material to turn a deep-blue color when exposed to light, and then to lose its color when placed back in the dark makes this structure one of the most interesting of the sequential products. It also involves the most challenging chemistry of the multistep syntheses:

2-exo-6-exo-2,4-Diphenyl-
6-(4-nitrophenyl)-
1,3-diazabicyclo[3.1.0]hex-3-ene

Blue azomethine ylide
(syn configuration)

In a number of the sequences, the stereochemistry of the reactions is vitally important and controlled by the mechanisms that are operating under the prescribed conditions. This aspect of the transformations is discussed in detail. By performing one or all six of these multistep sequences you will have a chance to challenge your experimental technique under conditions essentially identical to those found in the modern synthetic organic research laboratory. You will quickly recognize the reason why laboratory technique is so vital to the success of multistep syntheses.

The Synthesis of Hexaphenylbenzene

The molecule to be prepared in the Sequence A synthesis is hexaphenylbenzene (**I**):

I
Hexaphenylbenzene

This system, which contains seven aromatic rings, was first made by Dilthey at the University of Bonn in 1933 by using a classic Diels–Alder reaction with exactly the same two reactants as you will generate in Sequence A. The Bonn group showed that an earlier claim, by Durand, to have prepared this compound via a massive Grignard attack by phenylmagnesium bromide on hexachlorobenzene, had not actually yielded hexaphenylbenzene. The compound isolated by Durand melted at 266 °C, while Dilthey's material melted at 421–422 °C. Hexaphenylbenzene was later synthesized photochemically by Büchi at the Massachusetts Institute of Technology (MIT) in 1962. The MIT group improved the purity of the isolated material, and reported a melting point of 439–441 °C. Fieser, at Harvard University, refined Dilthey's synthetic route, and published the definitive preparation in *Organic Syntheses* in 1966. Fieser obtained melting points *without decomposition* in evacuated melting point capillaries (see Chapter 4) in the range 454–456 °C.

This hydrocarbon system possesses a number of interesting structural and physical properties. First, we should note that it has a relatively high molecular weight, near 534, and a molecular formula of $C_{42}H_{30}$. Hexaphenylbenzene exhibits an extremely high melting point for a nonionic organic material. For example, of the 15,000 plus substances listed in the Table of Physical Constants for Organic Compounds in the *CRC Handbook of Physics and Chemistry*, only two materials melt above 450 °C (and both of these compounds decompose at their melting point), and only 10 melt above 400 °C. Indeed, hexaphenylbenzene melts above hexabenzobenzene (**II**), the completely delocalized seven-ring fused system, mp = 438–440 °C.

II
Hexabenzobenzene
(coronene)

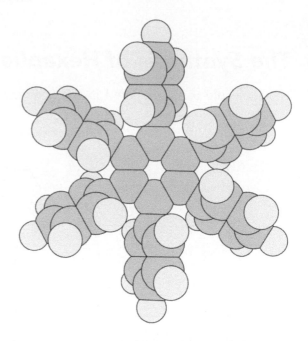

Figure 7.1 Molecular model of one of the chiral rotamers of hexaphenylbenzene.

From molecular modeling studies of hexaphenylbenzene, it appears that the six substituent rings surrounding the central system will be sterically restricted from lying in the plane of that ring by ortho-position interaction (**III**), and Figure 7.1 shows a molecular model of one of the chiral rotamers of hexaphenylbenzene.

III
Hexaphenylbenzene with steric interactions
between ortho positions

The twisted structure presents a particularly interesting problem in stereoisomerism. It is clear that when all the rings are coplanar (dihedral angle of 0°) we have maximum overlap of the π system and delocalization energy, but we also have a maximum of steric repulsion energies. On the other hand, when the dihedral angle approaches 90°, all delocalization is blocked, though steric repulsion between rings is at a minimum. It would seem reasonable to expect the system to reach some energetic compromise between these two extreme orientations. If this is the case, is it possible to establish the angle at which the external rings are positioned? An X-ray crystallographic study carried out by Bart in 1968 on solid crystalline hexaphenylbenzene did, in fact, show that in the crystal lattice the phenyl groups are twisted 65° out of the plane of the central ring. In the crystal lattice, the molecule adopts a conformation, similar to a six-bladed propeller, which is chiral. That is, in the solid state, hexaphenylbenzene can exist in two enantiomeric forms. Indeed, if this molecule happened to undergo resolution of the optical isomers during crystallization, in much the same fashion as Pasteur's tartaric acid salts, it should be possible to mechanically separate the racemate into crystals in which all the rings are tilted only in one possible conformation or in the other. These enantiomers, however, should possess a relatively low barrier to rotation, so that when dissolved in solution rapid racemization, via rotation of the rings (propeller blades) to the opposite pitch, would be expected.

This does seem to be the case. In 1977, Gust at Arizona State University showed that if a derivative of hexaphenylbenzene were synthesized in which two different groups were substituted on adjacent rings in the ortho positions (e.g., a methyl group and a methoxy group, **IV**), two possible sets of diastereomers would result:

IV
Substituted hexaphenylbenzene with steric
interactions between ortho position

It is presumed that it would be impossible for the rings to rotate the two ortho substituents past one another, but that other rotations may or may not be facile. If a large rotational barrier were present, the external rings would remain tilted at 65° with the same pitch. If this were the case, we would expect four diastereomers, and thus four different C—CH_3 resonances in the ^{1}H NMR spectrum. If rapid interconversion of the tilted conformers occurred, we would expect that the two bulky ortho groups would restrict full rotation of those two rings, even though the barrier to pitch inversion is low. Thus, in this latter case we would expect two diastereomeric pairs of enantiomers (one with the two ortho groups up and one with one up and one down), and two different C—CH_3 resonances in the NMR. When the compound was synthesized, and its ^{1}H NMR spectrum obtained, two resonances for methyl groups attached to aromatic rings and two O—CH_3 resonances were observed. These two diastereomers, **V**, were separated by column chromatography, and it was found that they slowly interconverted upon being heated to 215 °C. Thus, on the NMR time scale, it would appear that in hexaphenylbenzene, a reasonably rapid interconversion of the propeller conformations is taking place in solution at room temperature:

V

ortho-Disubstituted hexaphenylbenzene
exists as two diastereomers

The Synthesis of Hexaphenylbenzene from Benzaldehyde

As shown on the flow charts, the total synthesis of hexaphenylbenzene involves two parallel sets of three reactions each, the **a** series and the **b** series. The two series culminate in synthesizing tetraphenylcyclopentadienone (**a** series) and diphenylacetylene (**b** series), which are then reacted together (Experiment [A4$_{ab}$]) to produce the final product, hexaphenylbenzene.

The **a** series uses benzaldehyde as a starting material, which is first converted to the α-hydroxyketone benzoin in Experiment [A1$_a$]. Benzoin is then oxidized to benzil (Experiment [A2$_a$]). Benzil (along with diphenylacetone) is used in the construction of tetraphenylcyclopentadienone in Experiment [A3$_a$], the third and last of the Sequence A$_a$ intermediates.

The **b** series of synthetic experiments also begins with benzaldehyde, which is converted in Experiment [A1$_b$] into (*E*)-stilbene. (*E*)-Stilbene is the precursor to *meso*-1,2-dibromo-1,2-diphenylethane (*meso*-stilbene dibromide) prepared in Experiment [A2$_b$]. This dibromide is in turn converted

Benzaldehyde

  Benzoin [**A1$_a$**]

    Benzil [**A2$_a$**]

      Tetraphenylcyclopentadienone [**A3$_a$**]

        ⟶ Hexaphenylbenzene [**A4$_{ab}$**]

      Diphenylacetylene [**A3$_b$**]

    meso-Stilbene dibromide [**A2$_b$**]

  (*E*)-Stilbene [**A1$_b$**]

Benzaldehyde

Benzaldehyde → $A1_a$ → Benzoin → $A2_a$

Benzil → $A3_a$ → Tetraphenylcyclopentadienone

Sequence A_a

Hexaphenylbenzene

← $A4_{ab}$ ←

meso-Stilbene dibromide → $A3_b$ → Diphenylacetylene

↑ $A2_b$

(E)-Stilbene ← $A1_b$ ← Benzaldehyde

Sequence A_b

by the double dehydrohalogenation reaction in Experiment [$A3_b$] into diphenylacetylene.

The diphenylacetylene is then reacted with the tetraphenylcyclopentadiene, synthesized in the last step of the **a** series, in a Diels–Alder reaction to produce the final product of the sequences, hexaphenylbenzene, in Experiment [$A4_{ab}$].

This preparation of hexaphenylbenzene demonstrates the manner in which a variety of basic organic reactions can be integrated to prepare a desired end product.

BIBLIOGRAPHY

Bart, J. C. J. *Acta. Crystallogr. Sect. B* **1968,** 24, 1277.

Büchi, G.; Perry, C. W.; Robb, E. W. *J. Org. Chem.* **1962,** *27,* 4106.

Dilthy, W.; Schommer, W.; Trosken, O. *Berichte* **1933,** *66B,* 1627.

Durand, J. F.; Hsun, L. W. C. R. *Hebd. Seances Acad. Sci.* **1931,** *191,* 1460.

Fieser, L. F. *Organic Syntheses*; Wiley: New York, 1973; Collect. Vol. V, p. 604.

Gust, D. *J. Am. Chem. Soc.* **1977,** *99,* 6980.

Experiment A1$_a$

The Benzoin Condensation of Benzaldehyde: Benzoin

Common name: benzoin
CA number: [579-44-2]
CA name as indexed: ethanone, 2-hydroxy-1,2-diphenyl-

Purpose. One of the classic reactions of organic chemistry, the *benzoin condensation,* is carried out. You will examine the properties of the α-hydroxyketone product of this well-known reaction. The particular α-hydroxyketone generated in this experiment is the compound from which the reaction gains its name, *benzoin.*

This reaction provides quantities of benzoin for use in the multistep synthesis of hexaphenylbenzene (see Experiment [A4$_{ab}$]). Benzoin is synthesized in this first step of the **a** series of the Sequential Experiments. In this sequence of reactions, benzoin is converted by oxidation (Experiment [A2$_a$]) to benzil and then to tetraphenylcyclopentadienone (Experiment [A3$_a$]). The latter compound undergoes a Diels–Alder addition with diphenylacetylene (Experiment [A3$_b$]) to give hexaphenylbenzene (Experiment [A4$_{ab}$]).

NOTE. If the benzoin product is to be used in the reaction sequence, it is recommended that one of the semi-micro procedures be followed so that sufficient material will be available for the subsequent steps. The conditions for a one-step microscale experiment are listed following the two semi-microscale procedures.

Prior Reading
 Technique 5: Crystallization
  Use of the Hirsch Funnel (pp. 92–93)
  Craig Tube Crystallizations (pp. 93–95)

REACTION

Benzaldehyde   Benzoin

DISCUSSION

Aromatic aldehydes, in the presence of catalytic cyanide ion, dimerize to form the corresponding α-hydroxyketone (acyloin). This reaction, which is reversible, is known as the benzoin condensation, even though we now know that it is not actually a condensation reaction, because no water or alcohol is produced. Cyanide ion is a specific catalyst for the reaction with aromatic aldehydes, and can function in this capacity because it is a good nucleophile, it stabilizes the intermediate carbanion, and it is a good leaving group. In the mechanism outlined below, the nucleophilic cyanide ion attacks a molecule of the aromatic aldehyde to form the conjugate base of a cyanohydrin. The effect of the —CN group is to increase the acidity of the α-hydrogen atom, thus allowing the formation of the anion (**I**):

$$C_6H_5-\overset{\overset{\displaystyle \ddot{O}:}{\|}}{\underset{\underset{\displaystyle H}{|}}{C}} + {}^-CN \;\rightleftharpoons\; C_6H_5-\overset{\overset{\displaystyle :\ddot{O}:^-}{|}}{\underset{\underset{\displaystyle CN}{|}}{C}}-H \;\rightleftharpoons\; C_6H_5-\overset{\overset{\displaystyle :\ddot{O}H}{|}}{\underset{\underset{\displaystyle CN}{|}}{\overset{..}{C}:^-} \tag{1}$$

I

$$C_6H_5-\overset{\overset{\displaystyle :\ddot{O}H}{|}}{\underset{\underset{\displaystyle CN}{|}}{\overset{..}{C}:}} + \overset{\overset{\displaystyle \cdot\ddot{O}\cdot}{\|}}{\underset{\underset{\displaystyle H}{}}{C}}-C_6H_5 \;\rightleftharpoons\; C_6H_5-\overset{\overset{\displaystyle :\ddot{O}H}{|}}{\underset{\underset{\displaystyle CN}{|}}{C}}-\overset{\overset{\displaystyle :\ddot{O}:^-}{|}}{\underset{\underset{\displaystyle H}{|}}{C}}-C_6H_5 \tag{2}$$

I

$$C_6H_5-\overset{\overset{\displaystyle \cdot\ddot{O}\cdot}{\|}}{\underset{\underset{\displaystyle H}{|}}{C}}-\overset{\overset{\displaystyle :\ddot{O}H}{|}}{\underset{}{C}}-C_6H_5 + {}^-CN \;\rightleftharpoons\; C_6H_5-\overset{\overset{\displaystyle :\ddot{O}:^-}{|}}{\underset{\underset{\displaystyle CN}{|}}{C}}-\overset{\overset{\displaystyle :\ddot{O}H}{|}}{\underset{\underset{\displaystyle H}{|}}{C}}-C_6H_5$$

(an acyloin)

Once generated, the nucleophilic carbanion (**I**) attacks a second molecule of the aromatic aldehyde to yield a substituted cyanohydrin. This species can then be stabilized by loss of cyanide ion to form the α-hydroxyketone product.

The electronic effects of various substituents on the aromatic ring have been investigated. Because, in this reaction, the same aldehyde functions as both the nucleophile and the electrophile, electronic effects that enhance one of these functions are likely to inhibit the other. If a strongly electron-donating group is in the para position of the ring, the reaction fails due to the increase in electron density at the carbonyl carbon, brought about by the presence of the electron donor, which renders the carbonyl carbon less electrophilic. The benzoin condensation is also hindered by strong electron-with drawing groups on the ring. The presence of a para-nitro group decreases the electron density on the carbonyl carbon atom in the cyanohydrin anion, making its carbanion less nucleophilic, which drastically retards the rate of addition of the anion to the second molecule of aldehyde.

The cyanide-ion catalysis works only for aromatic aldehydes. Aliphatic aldehydes can, however, be condensed to α-hydroxyketone in the presence of thiazolium salts, such as N-dodecylthiazolium bromide:

N-Dodecylthiazolium bromide

SEMI-MICROSCALE EXPERIMENTAL PROCEDURE

(The microscale reaction quantities are increased by a factor of 2.)

Estimated time to complete the experiment: 2.0 h.

Physical Properties of Reactants						
Compound	MW	Amount	mmol	bp (°C)	d	n_D
Benzaldehyde	106.13	400 μL	3.92	178	1.04	1.5450
Sodium cyanide (0.54 M) in ethanol (95%)		2 mL				

HOOD

Reagents and Equipment. Using an automatic delivery pipet in the **hood**, place 400 μL (416 mg, 3.92 mmol) of fresh, acid-free, benzaldehyde in a weighed 10-mL round-bottom flask containing a magnetic spin bar. Reweigh the flask and record the weight. Now add 2 mL of a 0.54 M solution of sodium cyanide in ethanol (in the **hood**). Remember to use a fresh tip on the automatic delivery pipet.

NOTE. The 0.54 M NaCN solution should be prepared by the instructor.

CAUTION: Sodium cyanide (NaCN) is extremely toxic.

Attach the flask to a reflux condenser and mount the assembly in a sand bath on a magnetic stirring hot plate (◄).

Reaction Conditions. Heat the mixture with stirring in a sand bath at 90–95 °C. Maintain this temperature for 30 min. The reaction solution turns yellow, and may then become cloudy within approximately 5 min.

NOTE. Do not overheat the reaction mixture. If the solution begins to darken, immediately remove the vial from the sand bath.

Isolation of Product. At the end of the reflux period, cool the solution to room temperature and then place it in an ice bath for 10 min. Collect the benzoin product by filtration under reduced pressure using a Hirsch funnel (◄). Wash the filter cake on the Hirsch filter bed with two 1-mL portions of cold water, and air-dry the material under suction using plastic food wrap (see Prior Reading) for 5 min. The crude material is further dried on a porous clay plate or on filter paper.

90–95 °C
Thermometer

Speed   Heat

10–mL RB flask
C_6H_5CHO, 400 μL
+ NaCN solution, 2mL

Benzoin product collected here

–CN,
H_2O ~4 mL

Purification and Characterization.   This crude material may be purified by recrystallization from methanol or ethanol (95%) in a Craig tube.

Weigh the benzoin product and calculate the percent yield. Determine the melting point and compare your value to that found in the literature.

Obtain the IR spectrum of the compound. Compare your spectrum to that recorded in the literature (*The Aldrich Library of IR Spectra* and/or SciFinder Scholar).

This compound also has a characteristic ultraviolet spectrum showing a peak of maximum absorption (λ_{max}) at 247 nm ($\varepsilon_{max} = 13{,}200$, at a concentration of 6.0×10^{-5} M in ethanol) characteristic of the benzoyl group, $C_6H_5C\!\!=\!\!O$.

Chemical Tests.   Benzoin contains an aromatic ring. Confirm this fact by performing the ignition test (Chapter 9). To confirm the presence of the alcohol and ketone functions in benzoin, carry out the chromic anhydride test for the —OH group and the 2,4-dinitrophenylhydrazine test for the C=O group. Isolate the solid 2,4-dinitrophenylhydrazone derivative and compare its melting point to the literature value.

There is a specific test for the presence of benzoin. Place a few crystals of your material in 800 μL of 95% ethanol. The addition of a few drops of 10% sodium hydroxide solution produces a purple coloration. The color fades when the solution is shaken in air but reappears if the solution is allowed to stand.

OPTIONAL SCALES

These procedures are identical to that given above with the following exceptions:

Fivefold Scaleup

1. Increase the scale of the reaction by a factor of 5 compared to the microscale procedure.

Physical Properties of Reactants						
Compound	MW	Amount	mmol	bp (°C)	d	n_D
Benzaldehyde	106.13	1.0 mL	9.8	178	1.04	1.5450
Sodium cyanide (0.54 M) in ethanol (95%)		5.0 mL				

2. Use a 10-mL round-bottom flask.

3. The product is washed with two 2-mL portions of cold water.

MICROSCALE REACTION PROCEDURE

Physical Properties of Reactants						
Compound	MW	Amount	mmol	bp (°C)	d	n_D
Benzaldehyde	106.13	200 μL	1.96	178	1.04	1.5450
Sodium cyanide (0.54 M) in ethanol (95%)		1.0 mL				

1. The product is washed with two 0.5-mL portions of cold water.

2. Use a 5-mL conical vial (➡).

90–95°C
Thermometer

C_6H_5CHO, 200 μL
+ NaCN solution, 1 mL

Speed   Heat

QUESTIONS

7-1. The benzoin produced in this experiment contains a chiral carbon atom (a stereocenter), but the product itself is not optically active. Explain.

7-2. The cyanide ion is a highly specific catalyst for the benzoin condensation. Can you list three functions this ion performs in this catalytic role?

7-3. Can you suggest a reason why *p*-cyanobenzaldehyde does not undergo the benzoin condensation to yield a symmetrical benzoin product?

7-4. Tollens' reagent is used as a qualitative test for the presence of the aldehyde functional group (see Chapter 9, Classification Tests). However, benzoin, which does not contain an aldehyde, gives a positive test with this reagent. Explain.

7-5. Show how one might accomplish the conversion of *p*-methylbenzaldehyde to each of the following compounds.

BIBLIOGRAPHY

Adams, R.; Marvel, C. S. *Organic Syntheses; Wiley*: New York, 1941; Collect. Vol. I, p. 88.

Hassner, A.; Rai, K. M. L. *Comp. Org. Syn.* **1991,** *1,* 541.

Ide, W. S.; Buck, J. S. *Org. React.* **1948,** *4,* 269.

Kuebrich, I. P.; Schowen, R. L.; Wang, M.; Lupes, M. E. *J. Am. Chem. Soc.* **1971,** *93,* 1214.

Smith, M. B. *March's Advanced Organic Chemistry,* 8th ed., Wiley: New York, 2019, Chap. 16.

Stetter, H.; Kuhlmann, H. *Organic Syntheses; Wiley*: New York, 1990; Collect. Vol. VII, p. 95.

For the use of thiamine hydrochloride as a catalyst in the reaction, see:

Pasto, D. J.; Johnson, C. R.; Miller, M. J. *Experiments and Techniques in Organic Chemistry*; Prentice-Hall: New Jersey, 1992, p. 500.

Experiment A2ₐ

Copper(II) Ion Oxidation of Benzoin: Benzil

Common name: benzil

CA number: [134-81-6]

CA name as indexed: ethanedione, diphenyl-

Purpose. Benzil is the second of three synthetic intermediates in the **a** series of Sequential Reactions, which lead to the synthesis of hexaphenylbenzene. Benzoin, the starting material for this step, is prepared in Experiment [A1ₐ]. Benzil, the product formed in the present reaction, is used in the synthesis of tetraphenylcyclopentadienone in Experiment [A3ₐ]. Tetraphenylcyclopentadienone is then converted to hexaphenylbenzene in Experiment [A4ₐᵦ].

This experiment also affords an excellent opportunity for you to investigate the use of a soluble, metal-ion catalyst as an oxidizing agent. In this case, a secondary alcohol is oxidized to a ketone. Because nitrogen gas is formed as a byproduct, the progress and rate of the reaction can be followed by measuring the evolution of nitrogen.

NOTE. If you plan to continue the synthetic sequence to hexaphenylbenzene, the semi-microscale procedure described below should be used. If you wish to study this reaction as an individual microscale experiment, those conditions follow the semi-micro discussion.

Prior Reading

REACTION

Benzoin   Benzil

DISCUSSION

Benzil, a diketone, is obtained by the catalytic oxidation of benzoin using Cu^{2+} ion as the catalytic oxidant. The reaction is general for α-hydroxyketones, and is the basis of Fehling's qualitative test for certain sugars. The mechanism of the oxidation shows the catalytic effect of the Cu^{2+} ion as it is continuously reduced and reoxidized in the sequence outlined below. A key ingredient is the nitrate ion, which oxidizes the Cu^+ to the Cu^{2+} oxidation state, and is in turn reduced to nitrite ion. The nitrite ion in the presence of acid and an ammonium ion decomposes to yield nitrogen gas and water.

and

$$2\,Cu^+ + 2\,H^+ + NO_3^- \rightarrow 2\,Cu^{2+} + NO_2^- + H_2O$$

$$NH_4NO_2 \xrightarrow{\ H^+\ } N_2 + 2\,H_2O$$

SEMI-MICROSCALE EXPERIMENTAL PROCEDURE

(The microscale reaction is increased fourfold.)

Estimated time of the experiment: 2.5 h.

Physical Properties of Reactants				
Compound	MW	Amount	mmol	mp (°C)
Benzoin	212.25	400 mg	1.88	137
Cupric acetate solution		1.4 mL		

140°C thermometer

Benzoin, 400 mg
+ cupric acetate catalyst, 1.4 mL,
5–mL conical vial

HOOD

*INSTRUCTOR PREPARATION. Prepare the catalyst solution by dissolving 0.1 g of cupric acetate and 5 g of ammonium nitrate in 7.0 mL of deionized water (may require warming), followed by addition of 28 mL of glacial acetic acid. Place the container in the **hood** and dispense the solution by use of an automatic delivery pipet.*

Reagents and Equipment. Equip a 5.0-mL conical vial, containing a magnetic spin vane, with a reflux condenser, to which is attached a gas exit delivery tube (◄). Weigh and add to the vial 400 mg (1.88 mmol) of benzoin, followed by 1.4 mL of cupric acetate catalyst solution (using a calibrated Pasteur pipet).

100-mL graduated cylinder

Arrange the gas delivery tube so that it fits into an inverted 100-mL graduated cylinder that is filled with, and immersed in, a beaker of water (◄). This arrangement facilitates the measurement of the nitrogen gas evolved during the course of the reaction.

NOTE. It is absolutely necessary that all connections be tight to prevent leakage of the gas evolved. Lightly grease the ground-glass joint on the gas delivery tube.

Reaction Conditions. Heat the reaction mixture with stirring at a sand bath temperature of 140–145 °C for about 1 h or until the collected gas volume remains constant. *As the benzoin dissolves, the reaction mixture turns green and evolution of nitrogen gas commences. The theoretical volume of gas from the oxidation of 400 mg of benzoin is 42.4 mL at standard temperature and pressure (STP).*

Isolation of Product. If the gas delivery tube is used, *disconnect* it from the top of the condenser *before* removing the reaction vial from the heat source.

Benzil collected here

Aqueous acetic acid,
~7.5 mL and other
reaction byproducts

Cool the reaction mixture to room temperature, add 2 mL of cold water (using a calibrated Pasteur pipet), and then place the reaction vial in an ice bath for 10 min. Collect the yellow crystals of benzil by vacuum filtration using a Hirsch funnel (◄). Rinse the reaction vial and crystals with two additional 2-mL portions of cold water.

Purification and Characterization. Purify the crude product by recrystallization from methanol, or from 95% ethanol. Dry the recrystallized yellow benzil on a porous clay plate, or on filter paper. *The benzil obtained after recrystallization often contains a small amount of benzoin impurity. It may be purified by* *chromatography on silica gel using the procedure outlined below.* Before carrying out the column chromatography purification of benzil, the purity of the initial product may be assessed using thin-layer chromatography (see Prior Reading). Use methylene chloride as the elution solvent, silica gel as the stationary phase, and UV light for visualization. Typical R_f values are benzil, 0.62; benzoin, 0.33.

Pack a 1.0-cm-diameter column to a height of 5 cm with a slurry of activated silica gel in methylene chloride (➡). Introduce the sample of benzil to the column, followed by 100 mg of sand. Use approximately 10–15 mL of methylene chloride to elute the benzil, which is easily identified on the column because of its yellow color. Concentrate the eluate collected in a 25-mL filter flask to obtain the pure benzil (see Prior Reading). A rotary evaporator is an effective alternative.

Weigh the dried product and calculate the percent yield. Determine the melting point and compare your value to that reported in the literature.

Obtain an IR spectrum of the product and compare it with that of the starting material and to that reported in the literature (*The Aldrich Library of IR Spectra* and/or *SciFinder Scholar*).

Benzil also has a characteristic UV spectrum (see Fig. 7.2). It exhibits a wavelength maximum (λ_{max}) at 259 nm (ε_{max} = 16,329 methanol). It is of interest to compare this absorption spectrum of benzil with that of benzoin (Experiment [A1$_a$]). If the melting point and infrared spectrum compare reasonably closely to the literature values, this material may be used in the synthesis of tetraphenylcyclopentadienone (Experiment [A3$_a$]). If the melting point is low, check the product's purity by thin-layer chromatography.

Chemical Tests.   Ketones and aldehydes are often characterized by the preparation of a solid derivative. To assist in the characterization of benzil, prepare its 2,4-dinitrophenylhydrazone or semicarbazone (Chapter 9). The melting points of these derivatives are listed in Appendix A, Table A.5.

Sand (0.1 g)
Benzil

SiO₂, (~3 g)

Sand, 50 mg

Benzil in
~15 mL CH₂Cl₂

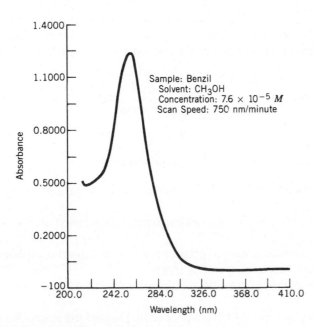

Sample: Benzil
Solvent: CH₃OH
Concentration: 7.6 × 10⁻⁵ *M*
Scan Speed: 750 nm/minute

Figure 7.2   UV-visible spectrum: benzil.

OPTIONAL MICROSCALE PREPARATION

140°C thermometer

Speed   Heat

Benzoin, 100 mg + oxidation catalyst, 350 µL, 3–mL conical vial

The microscale procedure is similar to that outlined above for the semi-microscale preparation, with the following exceptions:

1. Use a 3.0-mL conical vial containing a magnetic spin vane fitted with a reflux condenser to which is attached a gas delivery tube (➡).
2. Collect the gas in an inverted calibrated collection tube (➡).
3. Decrease the amount of the reagents and solvents fourfold.

Physical Properties of Reactants				
Compound	MW	Amount	mmol	mp (°C)
Benzoin	212.25	100 mg	0.47	137
Cupric acetate solution		350 µL		

4. Heat the reaction mixture at about 140 °C until the evolution of gas has ceased.
5. Add 0.5 mL of cold water to the cooled reaction product and, after filtration, wash the material with two 0.5-mL portions of cold water.
6. Purify and characterize the benzil product as described in the semi-microscale procedure, but using a Craig tube for the recrystallization.

QUESTIONS

7-6. In the directions given for the experiment, it is emphasized that the gas delivery tube must be disconnected from the top of the condenser before removing the reaction vial from the heat source. Why is this necessary?

7-7. What qualitative chemical tests would you perform to distinguish between benzoin and benzil? (See Chapter 9.)

7-8. 1,2-Dicarbonyl compounds, such as benzil, can be characterized by reaction with *o*-phenylenediamine to form a substituted quinoxaline:

o-Phenylenediamine   (a quinoxaline)

(a) Write the structure for the quinoxaline derivative obtained when benzil is the reactant. Do you think this compound would be colored? If so, why?

(b) Suggest a suitable mechanism for the formation of the quinoxaline compounds based on the reaction scheme shown above.

(c) What reagent would you react with *o*-phenylenediamine to prepare the unsubstituted compound quinoxaline? Show a reaction scheme giving the structure of reactants and products.

7-9. For the conversion of benzoin to benzil, what spectroscopic tool(s) would you select to monitor/confirm product formation? Explain..

7-10. Suggest a method for the synthesis of $C_6H_5CH(OH)CH(OH)C_6H_5$ from benzil. Discuss the stereochemical aspects of this 1,2 diol.

7-11. Based on the ultraviolet data given for benzil in the Purification and Characterization section of this experiment, what concentration of the benzil must have been used if a 1-cm cell was used, and a maximum absorption of 0.5 was observed?

BIBLIOGRAPHY

The synthesis of benzil is reported in Organic Syntheses:

Adams, R.; Marvel, C. S. *Org. Syn.* **1921,** *1*, 25.

Clarke, H.T.; Dreger, E. E. *Organic Syntheses; Wiley*: New York, 1941; Collect. Vol. I, p. 87.

For further information on the oxidation, see:

Depreux, P.; Betheginès, G.; Marcinal-Lefebure, A. *J. Chem. Educ.* **1988,** *65*, 553.

Haines, A. H. *Methods for the Oxidation of Organic Compounds*; Academic Press: New York, Vol. I, 1985, Vol. II, 1988.

Hudlicky, M. *Oxidations in Organic Chemistry*; American Chemical Society: Washington, DC, 1990.

Mijs, W. J.; de Jonge, C. R. H. I., Eds. *Organic Synthesis by Oxidation with Metal Compounds*; Plenum: New York, 1987.

Nigh, W. G. In *Oxidation in Organic Chemistry*; Trahanovsky, W. S. Ed.; Academic Press: New York, 1973, Vol. V, *Part B*, Chap. 1.

Weiss, M.; Appel, M. *J. Am. Chem. Soc.* **1948,** *70*, 3666.

Tetraphenylcyclopentadienone

Common name: tetraphenylcyclopentadienone

CA number: [479-33-4]

CA name as indexed: 2,4-cyclopentadien-1-one, 2,3,4,5-tetraphenyl-

Purpose.   The cyclic dienone product of this aldol condensation is the *third* intermediate in the **a** series of Sequential Reactions, which lead to the target molecule, hexaphenylbenzene. It is the last intermediate in the **a** series, and when reacted with the *last* intermediate in the **b** series (Experiment [A4ₐᵦ]) will give the target molecule.

In addition to supplying a key intermediate in the **a** series synthetic sequence, the experiment illustrates the use of the aldol condensation for the synthesis of a five-membered carbocyclic ring. The product also is a good demonstration of the impact of extended conjugation on the absorption of visible light. Starting with bright-yellow benzil, we form an even more extended π system in this experiment, and as a result, the tetraphenyldienone product absorbs strongly over a significant portion of visible spectrum and is thus deeply colored.

NOTE. *If you plan to continue the synthetic sequence to hexaphenylbenzene, the first microscale procedure described on page 439 should be used. If you wish to study this reaction as an individual microscale experiment, follow the second set of conditions.*

Prior Reading

Experiment [20] (pp. 320–322)

Technique 5: Crystallization

Use of the Hirsch Funnel (pp. 92–93)

Craig Tube Crystallizations (pp. 93–95)

REACTION

Benzil

+

1,3-Diphenylacetone

Tetraphenylcyclopentadienone

DISCUSSION

This experiment is a further example of the aldol condensation reaction (see Experiment [20] for discussion and Experiment [F1] for another example).[1] The reaction carried out here differs from the earlier example in that in this case two ketones, one of which has no α-hydrogen atoms, are the reactants. It is also different because the reagents selected lead to the formation of a carbocyclic ring system. The initially formed aldol product undergoes an elimination reaction to yield a material that has a highly conjugated system of double bonds. In general, the more extended the conjugation in a molecule, the less energy is required to promote the π electrons to a higher energy level. In this case, energy in the visible region of the spectrum is absorbed, which results in a product possessing a deep purple color.

The mechanism involves a sequence of two aldol condensations. The first is intermolecular; the second is intramolecular. The mechanism is similar to that outlined in Experiment [20].

The product of this reaction is the dienone intermediate used in the final step of the synthesis of hexaphenylbenzene (see Experiment [A4$_{ab}$]).

MICROSCALE REACTION PROCEDURE (1)

(The first microscale reaction is increased by a factor of 2.)

Estimated time to complete the experiment: 1.5 h.

Physical Properties of Reactants					
Compound	MW	Amount	mmol	mp (°C)	bp (°C)
1,3-Diphenylacetone	210.28	100 mg	0.48	35	
Benzil	210.23	100 mg	0.48	95	
Triethylene glycol	150.18	0.5 mL			278
Benzyltrimethylammonium hydroxide (40% solution in MeOH)		100 μL			

150–160 °C thermometer

Benzil, 100 mg + 1,3–diphenylacetone, 100 mg + triethylene glycol, 0.5 mL + [Immediately upon removal from the sand bath 100 μL of 40% benzyltrimethyl-ammonium hydroxide in methanol is added.]

Speed   Heat

Reagents and Equipment. Measure and place 100 mg (0.48 mmol) of 1,3-diphenylacetone and 100 mg (0.48 mmol) of benzil, followed by 0.5 mL of triethylene glycol, in a 3.0-mL conical vial containing a magnetic spin vane. Equip the vial with an air condenser (◄).

NOTE. *The benzil used in this reaction must be free of benzoin impurity. If benzil is prepared according to Experiment [A2$_a$], it should be purified by the chromatographic procedure cited therein if benzoin is detected by thin-layer chromatography.*

Reaction Conditions. Heat the mixture with stirring in a sand bath at 150–160 °C for 5–10 min. The benzil should dissolve during this time.

Next, remove the apparatus containing the homogeneous reaction solution from the heat source, and immediately add 100 μL of a 40% benzyltrimethylammonium hydroxide–methanol solution to the hot reactants, with gentle shaking. As cooling occurs, dark purple crystals of tetraphenylcyclopentadienone appear. *Cooling may be accelerated by placing the vial under a stream of cold water.*

---

[1]For references to the aldol condensation reaction, refer to Experiment [20].

Figure 7.3   UV-visible spectrum: tetraphenylcyclopentadienone.

Isolation of Product.   Add 1.4 mL of cold methanol with stirring (glass rod), and then cool the mixture in an ice bath for 5–10 min. Collect the dark crystals by filtration under reduced pressure using a Hirsch funnel. Rinse the reaction vial and crystals with a few drops of cold methanol. Continue dropwise addition of cold methanol to the crystals on the filter bed until the product appears purple and is no longer brownish in color. Finally, allow the crystalline product to air-dry on a porous clay plate (or in an oven at about 80 °C for 1 h) (➡).

Purification and Characterization.   The product is often of sufficient purity for direct use in the preparation of hexaphenylbenzene (Experiment [A4$_{ab}$]). If further purification is required, the intermediate dienone may be recrystallized from triethylene glycol using a Craig tube.

Weigh the tetraphenylcyclopentadienone product and calculate the percent yield. Determine the evacuated melting point and compare it with the literature value. Obtain an IR spectrum of the material and compare it with that of an authentic sample.

The comparison of the UV-visible spectra of benzil (see data given in Experiment [A2$_a$]) and the product may be used to demonstrate the shift of absorption bands with increased conjugation in the molecule.

The UV-visible data for tetraphenylcyclopentadienone are summarized as follows and as shown in Figure 7.3 (see Chapter 8).

$$\lambda_{max}\ 510\ nm\ (\varepsilon_{max} = 1080,\ chloroform)$$
$$\lambda_{max}\ 345\ nm\ (\varepsilon_{max} = 6380,\ chloroform)$$

NOTE. If you have synthesized the tetraphenylcyclopentadienone from benzaldehyde, calculate the overall yield to this point in the synthesis of hexaphenylbenzene. Base these calculations on the amount of benzaldehyde you started with.

MICROSCALE REACTION PROCEDURE (2)

When the microscale procedure is used as a single-step experiment, and not as part of the synthesis of hexaphenylbenzene, the scale is conveniently reduced to one-half that outlined above, with the following experimental modifications.

Physical Properties of Reactants						
Compound	MW	Amount	mmol	mp (°C)	bp (°C)	d
1,3-Diphenylacetone	210.28	50 mg	0.24	35		
Benzil	210.23	50 mg	0.24	95		
Triethylene glycol	150.18	0.25 mL			278	1.124
Benzyltrimethylammonium hydroxide (40% solution in MeOH)		50 μL				

1. The reaction is carried out in a 3-mL conical vial.
2. The reaction mixture is maintained at 155–165 °C for 10 min.
3. Following addition of the base, the vial is reheated to 150–160 °C (2–3 min) and then allowed to cool.

QUESTIONS

7-12. The above reaction carried out to construct dienone tetraphenylcyclopentadienone parallels an earlier example of an aldol condensation in Experiment [20B] in which another dienone was synthesized. In both cases all the reactants possessed carbonyl groups. What further structural similarities between the key reactants were required so that both pathways would lead to aldol condensations?

7-13. Outline a complete mechanistic sequence to account for the formation of the tetraphenylcyclopentadienone compound.

7-14. Cyclopentadienone is unstable and rapidly undergoes the Diels–Alder reaction with itself. Write the structure for this Diels–Alder addition product.

7-15. The Diels–Alder addition product of Question 7-14 undergoes a fragmentation reaction on heating to produce a bicyclotrienone compound plus carbon monoxide. Suggest a structure for this product.

7-16. Using the Hückel [4n + 2] rule for aromaticity, predict which of the following species might be expected to show aromatic properties:

7-17. Based on Questions 7-14 and 7-15, why is tetraphenylcyclopentadienone such a stable compound?

BIBLIOGRAPHY

An *Organic Syntheses* preparation of tetraphenylcyclopentadienone is available:

Johnson, J. R.; Grummitt, O. *Organic Syntheses*; Wiley, New York, 1955; Collect. Vol. III, p. 80.

For preparing this material under microwave conditions, see:
Elder, J. W. *J. Chem. Educ.* **1994**, *71*, A142.

Experiment A1ᵦ

(E)-Stilbene

Common names: (*E*)-stilbene; *trans*-1,2-diphenylethene
CA number: [103-30-0]
CA name as indexed: benzene, 1,1'-(1,2-ethenediyl)bis-, (*E*)-

Purpose.   The purpose of this experiment is to prepare a sufficient quantity of (*E*)-stilbene to complete a multistep sequence of synthetic reactions to obtain the target compound, hexaphenylbenzene (Experiment [A4$_{ab}$]).

A further purpose of experiment [A1$_b$] is to investigate the use of the Horner–Wadsworth–Emmons modified Wittig reaction to complete the synthetic objective. You will also study a reaction that involves the condensation of an aldehyde and a phosphonate ester, using a highly effective *phase-transfer catalyst*.

*NOTE. (E)-stilbene is the first intermediate to be synthesized in the **b** series of Sequential Reactions en route to hexaphenylbenzene. If you plan to carry out the entire sequence to hexaphenylbenzene, the semi-microscale procedure presented here should be used. Conditions to run the reaction as an individual microscale experiment are given in detail in Experiment [19B].*

Prior Reading

> *Technique 4:* Solvent Extraction
>> Liquid–Liquid Extraction (pp. 76–79)
>> Drying of a Wet Organic Layer (pp. 84–87)
>
> *Technique 5:* Crystallization
>> Use of the Hirsch Funnel (pp. 92–93)
>> Craig Tube Crystallization (pp. 93–95)
>
> *Technique 6:* Chromatography
>> Packing the Column (pp. 97–98)
>> Elution of the Column (pp. 98–99)
>> Thin-Layer Chromatography (pp. 101–104)
>> Concentration of Solutions (pp. 105–106)

REACTION

$$C_6H_5CHO + (C_6H_5O)_2\overset{\overset{\displaystyle :O:}{\|}}{P}-CH_2C_6H_5 \xrightarrow[\text{Hexane}]{\substack{\text{40\% NaOH} \\ \\ \text{Aliquat 336}}} C_6H_5CH{=}CHC_6H_5 + (C_2H_5O)_2\overset{\overset{\displaystyle :O:}{\|}}{P}-O^-, Na^+$$

Benzaldehyde   Diethylbenzyl phosphonate   *E*-Stilbene   Sodium diethyl phosphate

DISCUSSION

For a discussion of the Wittig reaction and a list of references, including the mechanism and modifications, see Experiment [19]. The role of the phase-transfer catalyst in the Horner–Wadsworth–Emmons modification of the Wittig reaction is also discussed in some detail in that experiment.

EXPERIMENTAL PROCEDURE

Estimated time to complete the experiment: 2.5 h.

Physical Properties of Reactants						
Compound	MW	Amount	mmol	bp (°C)	d	n_D
Benzaldehyde	106.12	600 μL	5.88	178	1.04	1.5463
Diethyl benzylphosphonate	228.23	1.20 mL	5.57	106–108 (@1 mm)	1.095	1.4970
Aliquat 336 (tricaprylmethylammonium chloride)	404.17	528 mg	1.30			
Hexane	86.18	12.0 mL				
40% Sodium hydroxide		12.0 mL				

90–100°C
thermometer

HOOD

C₆H₅CHO, 600 μL + diethyl
benzylphosphonate, 1.2 mL +
Aliquat 336, 528 mg +
40% NaOH, 12 mL, hexane, 12 mL,
50–mL RB flask

Reagents and Equipment.   Weigh and place 528 mg (600 μL of tricaprylmethylammonium chloride (Aliquat 336) in a 50-mL round-bottom flask containing a magnetic stirrer. Add 600 μL (5.88 mmol) of *freshly distilled* benzaldehyde, 1.20 mL (5.57 mmol) of diethyl benzylphosphonate, 12.0 mL of hexane, and 12 mL of 40% sodium hydroxide solution. Attach the flask to a reflux condenser (◄).

NOTE. The benzaldehyde, (automatic delivery pipet), diethyl benzylphosphonate (2-mL glass pipet), hexane, and NaOH solution are dispensed in the **hood**.
Aliquat 336 is very viscous and is best measured by weighing. A medicine dropper is used to dispense this material.
It is advisable to lightly grease the bottom joint of the condenser because strong base is being used in the reacting medium. At the end of the reaction, it is also important to loosen the Cap-seal and twist the joint to make sure it is free to rotate as the apparatus is cooling.

Reaction Conditions.   Heat the two-phase mixture at reflux on a sand bath at about 90–100 °C for 1 h. Stir the reaction mixture vigorously during this period. Allow the resulting orange solution to cool to nearly room temperature. Crystals of product may appear as cooling occurs.

Isolation of Product.   Add 4.0 mL of methylene chloride, which will dissolve any crystalline material that may have formed. Now transfer the contents of the round-bottom flask to a 125-mL separatory funnel. Rinse the flask with an additional 2.0–3.0 mL of methylene chloride, and transfer this rinse to the separatory funnel using a Pasteur filter pipet. Remove the aqueous layer carefully (*because the densities of the aqueous and organic layers are rather close, it is wise to test the solubility of a few drops of the bottom layer in water to ascertain which phase is the aqueous one in the separatory funnel*), and then wash the remaining organic layer with two 6.0-mL portions of water. Save the combined aqueous extracts in a 50-mL Erlenmeyer flask until you have successfully isolated and characterized the product. Now transfer the remaining wet methylene chloride solution to a 50-mL Erlenmeyer flask. Dry the solution by addition of anhydrous sodium sulfate (~1–2 g). Use a 50-mL glass pipet to transfer the dried solution to a second 50-mL Erlenmeyer flask. Wash the sodium sulfate remaining in the first Erlenmeyer flask with two 3-mL portions of methylene chloride. Remove these washings, using a Pasteur filter pipet, and transfer them to the same Erlenmeyer flask.

You may concentrate the solution, which contains the desired product, to dryness on a warm sand bath under a gentle stream of nitrogen gas, but a more efficient and rapid procedure is to remove the solvent by rotary evaporation (see Concentration of Solutions, Technique 6B).

Purification and Characterization. Recrystallize the crude (*E*)-stilbene from 95% ethanol in a small Erlenmeyer flask or a test tube. Collect the re-crystallized material by vacuum filtration using a Hirsch funnel (➡). Maintain the vacuum for an additional 10 min to partially dry the crystalline product. Now place the material on a clay plate, or on filter paper, and allow it to air-dry thoroughly. As an alternative procedure, the final traces of water may be removed by placing the sample (use an open test tube with the mouth covered by filter paper retained by a rubber band) in a vacuum drying oven (or pistol) for 10–15 min at 30 °C (1–2 mm).

The recrystallized compound, generally, is sufficiently pure to use in the next step of the **b** series of Sequential Reactions, Experiment [A2$_b$].

Weigh the (*E*)-stilbene and calculate the percent yield. Obtain a melting point and IR spectrum of the material, and compare your results with those reported in the literature (e.g., SciFinder Scholar).

Further characterization of the (*E*)-stilbene, including thin-layer chromatography, gas chromatography, and ultraviolet–visible spectroscopy may be carried out as outlined in Experiment [19A].

Chemical Tests. Further characterization may be accomplished by performing the Br$_2$/CH$_2$Cl$_2$ test for unsaturation. Note that the dibromo compound is prepared in Experiment [A2$_b$]. It may be used here as a derivative to characterize the (*E*)-stilbene. The ignition test (see Chapter 9) may be used to confirm the presence of the aromatic portion of the molecule.

QUESTIONS

7-18. Give the structure of the phosphorus ylide and carbonyl compound you might use to prepare the following alkenes:
   (a) Ethylenecyclohexane   **(b)** 3-ethyl-2-pentane   **(c)** C$_6$H$_5$CH=C(CH$_3$)$_2$

7-19. When an alkene is made using a phosphorus ylide and carbonyl compound, a retrosynthetic analysis will reveal two possibilities to consider. Review your selection of phosphorus ylide and carbonyl compound in Question 7-18 and propose an alternative phosphorus ylide and carbonyl compound which yields the same three systems..

7-20. How, starting from triphenylphosphine, (C$_6$H$_5$)$_3$P:, can you prepare the following ylide:
   (C$_6$H$_5$)$_3$P=C(CH$_3$)CH$_2$CH$_2$CH$_3$

7-21. Consult *The Aldrich Library of IR Spectra* (and/or SciFinder Scholar) for IR spectra of (*E*)- and (*Z*)-stilbene. Which absorption bands are most useful in determining the difference between the two compounds?

7-22. Compare the mechanisms of the aldol condensations (Experiments [20A] and [20B]) with that of the Wittig reaction. Point out any similarities and/or differences.

Bromination of (*E*)-Stilbene: *meso*-Stilbene Dibromide

Experiment A2$_b$

Common names: *meso*-stilbene dibromide, *meso*-1,2-dibromo-1,
2-diphenylethane
CA number: [13440-24-9]
CA name as indexed: benzene, 1,1'-(1,2-dibromo-1,2-ethanediyl)bis-,
(*R**,*S**)-

Purpose. You will synthesize the *second* intermediate in the **b** series of Sequential Reactions by carrying out the bromination of (*E*)-stilbene to obtain *meso*-stilbene dibromide. This product is the precursor to diphenylacetylene, the next synthetic intermediate in the **b** series. A further purpose of this experiment is to demonstrate the *stereospecific* addition of bromine to alkenes.

NOTE. If you plan to continue the synthetic sequence to hexaphenylbenzene, the semi-microscale procedure described below should be used. If you wish to study this reaction as an individual microscale experiment, those conditions and other scaleup options follow the semi-micro discussion.

Prior Reading

    Technique 5: Crystallization
                    Use of the Hirsch Funnel (pp. 92-93)

REACTION

| (*E*)-Stilbene | Pyridinium bromide perbromide | *meso*-Stilbene dibromide |

DISCUSSION

The bromination of alkenes is an example of an electrophilic addition reaction (also see Experiments [D2] and [F2]).

In the present reaction, bromination of (*E*)-stilbene yields *meso*-stilbene dibromide. Thus, this reaction is classed as *stereospecific*, because the other possible diastereomers are not formed.

The reaction proceeds in two stages. The first step involves the formation of an intermediate cyclic *bromonium ion*. The concept of a three-membered cyclic intermediate was first proposed as early as 1937. Subsequent studies have provided solid evidence that cyclic halonium ions do, indeed, exist. For example, stable solutions of cyclic bromonium ions in liquid SO_2 (−60 °C) have been prepared as SbF_6^- salts. Two examples are given here:

Ethylene bromonium ion salt   Tetramethylethylene bromonium ion salt

Nuclear magnetic resonance spectroscopic measurements have provided powerful evidence that these and other selected alkenes form stable *bridged* bromonium ion salts. A solid bromonium ion tribromide salt of adamantylidene adamantane has been isolated, and its structure determined by X-ray crystallography:

$$Br^+ \ Br_3^-$$

Tribromide salt of adamantylidene adamantane

The bromine molecule (Br_2) is normally symmetrical. However, as it approaches the nucleophilic and electron-rich π bond of the alkene, it becomes polarized by induction and can then function as the electrophile in an addition reaction. The result is the generation of a cyclic bromonium ion:

Induced polarization of Br_2 as it approaches the alkene

Bromonium ion

In the present reaction, both the bromine and the (E)-stilbene are achiral. However, the bromonium ion that is produced is chiral. In this ion, the bromine atom bridges both carbon atoms of the original carbon–carbon double bond to form a three-membered ring intermediate. The generation of a cyclic species has a profound effect on the *stereochemistry* of the second step of the bromine addition.

The second stage of the bromination involves nucleophilic attack by bromide ion on the intermediate bromonium ion. Since the nucleophile must approach from the face opposite the leaving group, bond formation involves inversion of configuration at the carbon center under attack in the second stage of the bromination reaction.

Note that *either* carbon can be approached by the nucleophile (one attack is shown). This second step is a classic backside S_N2 sequence. The bromination of cyclic alkenes provides further evidence that this type of halogenation is an anti addition, with the bromine atoms introduced trans to one another:

Bromonium ion

Dibromo product

It is important to realize that if two different groups are present on one or both of the sp^2 carbon atoms of the alkene linkage, chiral carbon centers are generated when bromine is added, though if a chiral product were formed from achiral reagents, one would expect it to be racemic. In the case of (E)-stilbene, two chiral centers are generated. However, due to the symmetry of the reactants and the stereoselectivity of the reaction, only the meso diastereomer is formed:

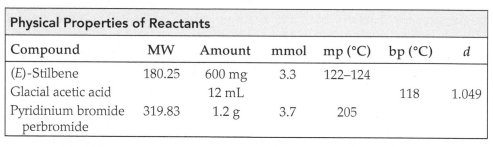

(E)-Stilbene   Identical
                 meso-Dibromostilbene

Refer to the Discussion section of Experiment [D2] for further information on the stereochemistry of bromination reactions.

Bromination of alkenes using a Br_2—CCl_4 solution (a red-brown color) is frequently used as a qualitative test for the presence of unsaturation in a compound. Rapid loss of color from the reagent solution is a positive test (see Chapter 9). Pyridinium bromide perbromide, a solid brominating agent, is used as a source of bromine in this experiment. The material is more convenient to handle than liquid bromine (see Experiment [D2]).

SEMI-MICROSCALE EXPERIMENTAL PROCEDURE

(The microscale reaction is increased by a factor of 2.6.)

Estimated time to complete the reaction: 1.0 h.

Physical Properties of Reactants						
Compound	MW	Amount	mmol	mp (°C)	bp (°C)	d
(E)-Stilbene	180.25	600 mg	3.3	122–124		
Glacial acetic acid		12 mL			118	1.049
Pyridinium bromide perbromide	319.83	1.2 g	3.7	205		

130–140°C thermometer

Air condenser

50 mL RB flask and stir bar

Sand

Speed   Heat

(E)-Stilbene, 600 mg + glacial acetic acid, 12 mL + pyridinium bromide perbromide, 1.2 g

Reagents and Equipment. In a 50.0-mL round-bottom flask containing a magnetic spin bar and equipped with an air condenser, weigh and place 600 mg (3.3 mmol) of (E)-stilbene. Next add 6 mL of glacial acetic acid (using a graduated cylinder), and warm the resulting mixture in a sand bath at 130–140 °C with stirring until the solid dissolves (~5 min) (◄).

HOOD   CAUTION: Glacial acetic acid is corrosive and toxic. It is dispensed in the **hood** using an automatic delivery pipet.

HOOD   Remove the condenser from the flask, and to the warm solution, in the **hood**, add 1.2 g (3.7 mmol) of pyridinium bromide perbromide in one portion. Wash down any perbromide adhering to the sides of the flask with an additional 6 mL of acetic acid using a Pasteur pipet. Reattach the air condenser.

HOOD   CAUTION: The brominating agent is a mild lachrymator. It should be dispensed in the **hood**. An alternative solid brominating agent is tetra-N-butylammonium tribromide.

Reaction Conditions.   With stirring, heat the reaction mixture at a sand bath temperature of 130–140 °C for an additional 5–6 min. (The product often begins to precipitate during this period.)

Isolation of Product.   Remove the reaction flask from the heat source and allow it to cool to approximately 40–50 °C (water bath). Add 12 mL of water, with swirling, and then place the flask in an ice bath for 5–8 min. Collect the crystalline solid by vacuum filtration using a Hirsch funnel (➡).

Collect *meso*-stilbene dibromide here

Aqueous acid solution and other byproducts retained in flask

Purification and Characterization.   Wash the material with three 2-mL portions of cold water to obtain white crystals, and then with two 2-mL portions of acetone. Air-dry the product on a clay plate or on filter paper.

Weigh the *meso*-stilbene dibromide and calculate the percent yield. Determine the evacuated melting point, and compare your result with the literature value. Obtain IR and NMR spectra and compare them with those reported in the literature (*The Aldrich Library of IR Spectra, The Aldrich Library of NMR Spectra*, and/or the corresponding spectral data available online [e.g., SciFinder Scholar]).

Generally, the material is sufficiently pure to be used in the next stage of the **b** series of Sequential Reactions, the preparation of diphenylacetylene. If desired, a small portion (~10–20 mg) may be recrystallized from hot xylene using the Craig tube.

Chemical Tests.   You may wish to perform several classification tests on the product (see Chapter 9). Carry out the ignition test to confirm the presence of an aromatic group. The Beilstein test can be used to detect the presence of bromine. The silver nitrate test for alkyl halides should also give a positive result.

OPTIONAL MACROSCALE AND MICROSCALE PREPARATIONS

Macroscale Reaction Procedure.   The procedure is similar to that for the 2.6-fold scaleup preparation with the following exceptions:

1. The reagent and solvent amounts are increased approximately 4.3-fold over the microscale preparation.

Physical Properties of Reactants						
Compound	MW	Amount	mmol	mp (°C)	bp (°C)	*d*
(*E*)-Stilbene	180.25	1.0 g	5.5	122–124		
Glacial acetic acid		12 mL			118	1.049
Pyridinium bromide perbromide	319.83	2.0 g	6.2	205		

2. After cooling the reaction mixture, add 20 mL of water to assist in precipitating the product. Wash the collected crystals with three 3-mL portions of cold water followed by two 3-mL portions of acetone.

Microscale Reaction Procedure. The procedure is similar to that for the 2.6-fold scaleup preparation with the following modifications:

1. Use a 10-mL round-bottom flask.

2. The reagent and solvent amounts are *decreased* by a factor of approximately 2.6.

Physical Properties of Reactants						
Compound	MW	Amount	mmol	mp (°C)	bp (°C)	d
(E)-Stilbene	180.25	230 mg	1.28	122–124		
Glacial acetic acid		4.2 mL			118	1.049
Pyridinium bromide perbromide	319.83	450 mg	1.4	205		

3. Add 2.2 mL of glacial acetic acid at the same time as the addition of the (E)-stilbene.

4. An additional 2 mL of glacial acetic acid is added with the brominating reagent.

5. The reaction mixture is diluted with 4.5 mL of water, swirled, and placed in an ice bath for 5–8 min.

6. The filter cake is washed with three 2-mL portions of cold water, followed by two 2-mL portions of acetone.

QUESTIONS

7-23. Using suitable structures, draw the sequence for the addition of bromine to (Z)-stilbene.

7-24. Are the results different for the answer in Question 7-23 than for the result in this experiment? If so, how? What is the stereochemical relationship between the products formed in the two reactions?

7-25. Bromine undergoes addition to ethylene in the presence of a high concentration of Cl⁻ ion to give 1-bromo-2-chloroethane, as well as 1,2-dibromoethane. Chloride ion does not add to the C=C unless bromine is present. Suggest a suitable mechanism to explain these results. Is the rate of bromination significantly affected by the presence of the Cl⁻ ion?

7-26. Offer an explanation for the fact that bromine adds to 2,3-dimethyl-2-butene 920,000 times faster than to ethylene, to produce the respective dibromides.

7-27. Why does the CA name of the product in the bromination of (E)-stilbene as indexed utilize "*" next to the stereochemical descriptors R and S?

BIBLIOGRAPHY

A large number of examples of the bromination of alkenes appear in *Organic Syntheses*. Selected references are given below:

Allen, C. F. H.; Abell, R. D.; Normington, J. B. *Organic Syntheses; Wiley*: New York, 1941; Collect. Vol. I, p. 205.

Cromwell, N. H.; Benson, R. *Organic Syntheses; Wiley*: New York, 1955; Collect. Vol. III, p. 105.

Fieser, L. F. *Organic Syntheses; Wiley*: New York, 1963; Collect. Vol. IV, p. 195.

Khan, N. A. *Organic Syntheses; Wiley*: New York, 1963; Collect. Vol. IV, p. 969.

McElvain, S. M.; Kundiger, D. *Organic Syntheses; Wiley*: New York, 1955; Collect. Vol. III, p. 123.

Paquette, L. A.; Barrett, J. H. *Organic Syntheses; Wiley*: New York, 1973; Collect. Vol. V, p. 467.

Rhinesmith, H. S. *Organic Syntheses; Wiley*: New York, 1943; Collect. Vol. II, p. 177.

Snyder, H. R.; Brooks, L. A. *Organic Syntheses; Wiley*: New York, 1943; Collect. Vol. II, p. 171.

Also see:

Smith, M. B. *March's Advanced Organic Chemistry*, 8th ed.; Wiley: New York, 2019.

Dehydrohalogenation of *meso*-Stilbene Dibromide: Diphenylacetylene

Common names: diphenylacetylene, diphenylethyne
CA number: [501-65-5]
CA name as indexed: benzene, 1,1′-(1,2-ethynediyl)bis-

Purpose. The product formed in this multiple elimination reaction is the third intermediate in the **b** series of Sequence A, and is one of the immediate precursors to our target molecule, hexaphenylbenzene. You will investigate the synthesis and properties of alkynes and become familiar with E2 elimination reactions.

NOTE. If you plan to continue the synthetic sequence to hexaphenylbenzene, the semi-microscale procedure described below should be used. If you wish to study this reaction as an individual microscale experiment, those conditions and other scaleup options follow the semi-micro discussion.

Prior Reading

Technique 5: Crystallization
 Use of the Hirsch Funnel (pp. 92–93)

REACTION

meso-Stilbene didromide   Diphenylacetylene

DISCUSSION

This reaction illustrates the double dehydrohalogenation of a *vicinal* dibromo compound to form an alkyne. It is a useful reaction for the synthesis of alkynes, because the starting dibromides are readily available from alkenes (see e.g., Experiment [A2ᵦ]).

The double dehydrohalogenation reaction is usually run in the presence of a strong base, such as KOH or $NaNH_2$, and proceeds in two stages. In the first, an intermediate bromoalkene is formed, which can be isolated under more mildly basic conditions. In fact, this reaction is a valuable route to vinyl halides. The mechanism of elimination involves the abstraction of the proton on the carbon atom β to the halogen. The E2 mechanism, which operates under these strongly basic conditions, is fastest when it involves removal of a proton, H^+, anti-periplanar to the leaving group, Br^-. The E2 sequence of bond breakage and formation involves a smooth transition from reactant to product without the formation of an intermediate (concerted mechanism). The general mechanism is shown here:

This type of elimination reaction is stereospecific, because the geometry of the transition state requires that the H, both Cs, and the Br all lie in the same plane.

If *meso*-stilbene dibromide is treated with KOH in ethanol solvent, it is possible to isolate the monodehydrohalogenation product, the bromoalkene.

meso-Stilbene dibromide   (*E*)-1-Bromo-1,2-diphenylethylene

The second stage of the reaction involves a higher activation energy, and therefore it requires higher temperatures to proceed. In the presence of a strong base near 200 °C, the bromoalkene undergoes an E2 elimination to form the triple bond. Part of the reluctance to eliminate, in this particular case, results from the fact that the elimination proceeds by a syn pathway:

$$H_5C_6-C\equiv C-C_6H_5 + H_2O + :\ddot{B}r:^-$$

Thus, the stereochemistry of the reactant used necessitates somewhat higher temperatures for the second elimination reaction.

SEMI-MICROSCALE EXPERIMENTAL PROCEDURE

(The microscale reaction is increased by a factor of 5.)

Estimated time to complete the reaction: 1.0 h.

Physical Properties of Reactants						
Compound	MW	Amount	mmol	mp (°C)	bp (°C)	*d*
meso-Stilbene dibromide	340.07	400 mg	1.2	241 (decomp.)		
Potassium hydroxide	56.11	387 mg	6.9	360		
Triethylene glycol	150.18	2 mL			278	1.124

Reagents and Equipment. Weigh and place 400 mg (1.2 mmol) of *meso*-stilbene dibromide and 387 mg (6.9 mmol) of KOH flakes in a 10-mL Erlenmeyer flask containing a magnetic stir bar. Using a graduated cylinder, measure and add 2 mL of triethylene glycol to the flask.

Diphenylacetylene collected here

Aqueous glycol solution plus reaction byproducts

Reaction Conditions. Place the reaction flask in a ***preheated*** sand bath set at a temperature of 190–195 °C, and stir the reaction for 7–8 min.

Isolation of Product. Allow the resulting dark-colored reaction mixture to cool to approximately 40–50 °C (water bath), and then add 5.0 mL of water. Now place the flask in an ice bath for 15 min. Collect the solid product by filtration under reduced pressure using a Hirsch funnel (◀).

Purification and Characterization.   Rinse the product crystals with two 1-mL portions of cold 70% ethanol and air-dry them on a porous clay plate or on filter paper. These crystals can be recrystallized from 95% ethanol (~2.0 mL). If desired, the product can be further purified by a second crystallization from 95% ethanol using the Craig tube.

Weigh the recrystallized product and calculate the percent yield. Determine the melting point and compare your result with the literature value. Obtain IR and NMR spectra of the material and compare them with those recorded in the literature (*The Aldrich Library of IR Spectra, The Aldrich Library of NMR Spectra*, and/or the corresponding spectral data available online [e.g., SciFinder Scholar]).

NOTE. If you have synthesized the diphenylacetylene from benzaldehyde, calculate the overall yield to this point in the synthesis of hexaphenylbenzene. Base these calculations on the starting amount of benzaldehyde.

Chemical Test.   The ignition test for aromatic groups indicates the presence of the phenyl groups. Decolorization of a Br_2–CH_2Cl_2 solution should give a positive test for unsaturation (see Chapter 9).

OPTIONAL MACROSCALE AND MICROSCALE PREPARATIONS

Macroscale Reaction Procedure.   (This reaction is scaled up by a factor of 10 over the microscale procedure.)

The procedure is similar to that outlined above with the following exceptions:

1. Carry out the reaction in a 25-mL Erlenmeyer flask containing a boiling stone. Run the reaction in the **hood**.   HOOD

2. Increase the reagent and solvent amounts approximately twofold over the semi-microscale procedure, as indicated here.

Physical Properties of Reactants						
Compound	MW	Amount	mmol	mp (°C)	bp (°C)	d
meso-Stilbene dibromide	340.07	800 mg	2.4	241 (decomp.)		
Potassium hydroxide	56.11	756 mg	13	360		
Triethylene glycol	150.18	4 mL			278	1.124

3. After cooling the reaction mixture, add 10 mL of water.

4. Rinse the product crystals with two 1-mL portions of cold 70% ethanol. They can be recrystallized from 95% ethanol (~5.0 mL).

Microscale Reaction Procedure.   The procedure is similar to that outlined above with the following exceptions:

1. Carry out the reaction in a 3-mL conical vial containing a boiling stone.

2. The reaction is heated in a sand bath at 190 °C for 5 min.

3. Decrease the amounts of reagents and solvents as given here.

Physical Properties of Reactants

Compound	MW	Amount	mmol	mp (°C)	bp (°C)	d
meso-Stilbene dibromide	340.07	80 mg	0.24	241 (decomp.)		
Potassium hydroxide	56.11	75 mg	1.3	360		
Triethylene glycol	150.18	0.4 mL			278	1.124

4. After cooling of the reaction mixture to 40–50 °C (in a water bath), add 1.0 mL of water, and place the reaction vessel in an ice bath for 15 min.

5. Rinse the product crystals with one 0.25-mL portion of cold 70% ethanol. The alkyne can be recrystallized from 95% ethanol (~0.5 mL).

QUESTIONS

7-28. Both (*E*)- and (*Z*)-2-chlorobutenedioic acids dehydrochlorinate to give acetylene dicarboxylic acid:

$$HO_2C—C(Cl)\!=\!CH—CO_2H \rightarrow HO_2C—C\!\equiv\!C—CO_2H$$

The *Z* acid reacts about 50 times faster than the *E* acid. Explain.

7-29. Compounds containing a carbon–carbon triple bond undergo the Diels–Alder reaction. Formulate the product formed by the reaction of (*E, E*)-1,4-diphenyl-1,3-butadiene with diethyl acetylenedicarboxylate.

7-30 Alkynes can be hydrated in the presence of acid and $HgSO_4$ by electrophilic addition of a molecule of water to the triple bond. The reaction proceeds by way of a carbocation intermediate. Hydration of acetylene (ethyne) produces acetaldehyde (ethanal). Outline the steps that occur in this transformation.

7-31. Use the IR tables (online correlation tables and/or Chapter 8) to locate the absorption bands of the stretching frequencies of the alkyne C—H bond, the alkyne C≡C bond, the alkene C—H bond, and the alkane C—H bond. Using these data, is it possible to distinguish between 1-butyne, 2-butyne, 2-butene, and butane?

BIBLIOGRAPHY

For a review on the preparation of alkynes, see:

Jacobs, T. L. *Org. React.* **1949,** *5,* 1.

Stang, P. J.; Diederich, F., Eds. *Modern Acetylene Chemistry;* VCH: New York, 1995.

A large number of elimination reactions leading to the formation of acetylenes appear in *Organic Syntheses.* Selected references are given below:

Abbott, W. T. *Organic Syntheses; Wiley:* New York, 1943; Collect. Vol. II, p. 515.

Campbell, K. N.; Campbell, B. K. *Organic Syntheses; Wiley:* New York, 1963; Collect. Vol. IV, p. 763.

Guha, P. C.; Sankaren, D. K. *Organic Syntheses; Wiley:* New York, 1955; Collect. Vol. III, p. 623.

Hessler, J. C. *Organic Syntheses; Wiley:* New York, 1941; Collect. Vol. I, p. 438.

Khan, N. A. *Organic Syntheses; Wiley:* New York, 1963; Collect. Vol. IV, p. 967.

Le Coq, A.; Gorgues, A. *Org. Syn.* **1980,** *59,* 10.

The synthesis of diphenylacetylene has been reported:

Smith, L. I.; Falkof, M. M. *Organic Syntheses; Wiley:* New York, 1955; Collect. Vol. III, p. 350.

Experiment A4ab

Hexaphenylbenzene

Common name: hexaphenylbenzene

CA number: [992-04-1]

CA name as indexed: 1,1′:2′,1″-terphenyl, 3′,4′,5′,6′-tetraphenyl-

Purpose. This reaction completes the Sequence A experiments. The Diels–Alder reaction is used to form six-membered aromatic rings. You will carry out the decarbonylation and aromatization of an intermediate bicyclic Diels–Alder adduct. You will examine the properties of our synthetic target molecule, hexaphenylbenzene.

NOTE. If you plan to continue the synthetic sequence to hexaphenylbenzene, you should have enough of the two starting reactants to carry out the first true microscale reaction used in Sequence A. The details of this interesting Diels–Alder addition, first carried out in 1933, are described below. You may, of course, wish to study this reaction as an individual microscale experiment, in which case use the same conditions given here.

Prior Reading

    Technique 5: Crystallization

                    Use of the Hirsch Funnel (pp. 92–93)

                    Craig Tube Crystallizations (pp. 93–95)

    Experiment [14]: (pp. 266–278)

    Experiment [15]: (pp. 278–284)

REACTION

Tetraphenylcyclopentadienone   Diphenylacetylene   Hexaphenylbenzene

DISCUSSION

This experiment (Experiment [A4$_{ab}$]) completes the Sequence A set of seven experiments that lead to the synthesis of hexaphenylbenzene. As we have noted earlier (see introduction to Sequence A), this compound is a rather unique and interesting organic system possessing a number of unusual properties. For example, it has one of the highest known melting points for a nonionic organic molecule, 465 °C, and it is perhaps even more intriguing that it melts without decomposition. Indeed, its melting point exceeds that of all 15,000+ organic compounds listed in the *CRC Handbook*. Hexaphenylbenzene also contains particularly novel stereochemistry as discussed in the introduction.

    The Diels–Alder reaction is one of the most useful synthetic tools in organic chemistry. It is an example of a cycloaddition reaction between a conjugated diene and a dienophile, which leads to the formation of six-membered rings. Here, the initial bicyclic Diels–Alder adduct can undergo a reaction that is the reverse of a concerted cycloaddition reaction between a benzene ring and the lone electron pair on the carbon of carbon monoxide. This retro cycloaddition is thermodynamically favored here, because the retro reaction generates an aromatic system, along with the quite stable carbon monoxide molecule. Under the high-temperature conditions used in this experiment, the initial bicyclic Diels–Alder adduct is quickly decarbonylated to yield hexaphenylbenzene and is not itself isolated.

    By varying the nature of the diene and dienophile, a very large number of structures can be prepared using the Diels–Alder reaction. In the majority of cases, carbocyclic rings are generated, but ring closure can also occur with reactants containing heteroatoms. This leads to the synthesis of compounds containing heterocyclic rings. For further and more detailed discussion of the Diels–Alder reaction see Experiments [14] and [15].

In the present reaction, an excess of diphenylacetylene is used to ensure that all the tetraphenylcyclopentadienone is consumed in the reaction, since diphenylacetylene is far easier to separate from hexaphenylbenzene in the purification steps.

EXPERIMENTAL PROCEDURE

Estimated time to complete the experiment: 1.0 h.

Physical Properties of Reactants and Product				
Compound	MW	Amount	mmol	mp (°C)
Tetraphenylcyclopentadienone	384.48	100 mg	0.26	220–221
Diphenylacetylene	178.23	100 mg	0.56	61
Hexaphenylbenzene	534.66			465

Reagents and Equipment.   In a 13 × 100-mm Pyrex test tube, place 100 mg (0.26 mmol) of tetraphenylcyclopentadienone and 100 mg (0.56 mmol) of diphenylacetylene. Then using a calibrated Pasteur pipet, transfer to the test tube about 1 mL of high-boiling silicone oil (temperature range [−40–230 °C]). Clamp the test tube at a slight angle, facing it away from both yourself and your laboratory neighbors.

Reaction Conditions.   Bring the mixture gently to a boil over a 3- to 5-min period by heating the test tube with the moving flame of a microburner. On melting, the reagents dissolve in the hot silicone oil to yield a dark red-purple solution. Continue to gently boil the solution for an additional 10 min. During this latter period, the deeply colored solution fades and hexaphenylbenzene begins to separate from solution as tan crystals.

Isolation of Product.   After heating for 15 min, cool the test tube to room temperature and add 4 mL of hexane, with stirring, to dilute the silicone oil and any unreacted starting material. The crude, precipitated hexaphenylbenzene is then collected by filtration on a Hirsch funnel (◄).

Hexaphenylbenzene collected here

~1 mL Silicone oil + 6 mL hexane + 4 mL toluene

Purification and Characterization.   Wash the filter cake with 2 mL of hexane to yield tan crystals of the addition product. Then wash it twice with 2-mL portions of cold toluene to yield white crystalline hexaphenylbenzene. Air-dry the product on filter paper or a porous clay plate. Weigh the Diels–Alder adduct and calculate the percent yield. Obtain an IR spectrum and compare it to that of an authentic sample. The melting point of this material is well over 400 °C, therefore, melting point determinations with apparatus that use

CAUTION   oil baths should **not** be attempted. The best approach, if a melting point is required, is to carry out an evacuated melting point determination with one of the metal heating block systems that accept the normal capillaries, *but remember to first check the maximum temperature reading on the thermometer used in the apparatus* (see evacuated melting points, Chapter 4, pp. 53–54).

CAUTION   If necessary, the product can be recrystallized in a Craig tube from diphenyl ether (5–10 mg maximum, since this very high-melting material is very insoluble even in this high-boiling aromatic ether; recrystallization is rather difficult).

This experiment completes the seven-step Sequence A synthesis of hexa-phenylbenzene from benzaldehyde. Calculate the overall yield based on both the earlier calculations for each pathway (for the diene and the dienophile used in the final Diels–Alder reaction), see Experiments [A3ₐ] and [A3_b], and the actual yield for this step in the synthesis.

QUESTIONS

7-32. What starting materials would you use to prepare each of the following compounds by the Diels–Alder reaction?

7-33. Diels–Alder reactions with benzene are rare and require a very reactive (electron-deficient) dienophile, because benzene is a rather unreactive diene. Two are shown below. Give the structures of the product produced in each reaction.

7-34. Shown below are two heteroatom compounds that undergo the Diels–Alder reaction. Formulate the product obtained in each reaction.

BIBLIOGRAPHY

Review articles:

Bastide, J.; Henri-Rousseau, O. In *The Chemistry of the Carbon–Carbon Triple Bond*; Patai, S., Ed.; Wiley: New York, 1978, Part 1, p. 447.

Butz, L. W.; Rytina, A. W. *Org. React.* **1949,** 5, 136.

Carruthers, W. *Cycloaddition Reactions in Organic Synthesis*; Pergamon Press: New York, 1990.

Holm, H. L. *Org. React.* **1948,** 4, 60.

Kloetzel, M. C. *Org. React.* **1948,** 4, 1.

Norton, J. A. *Chem. Rev.* **1942,** 31, 319.

Sauer, J. *Angew. Chem. Int. Ed. Engl.* **1966,** 5, 211; ibid., **1967,** 6, 16.

Stang, P. J.; Diederich, F., Eds. *Modern Acetylene Chemistry*; VCH: New York, 1995.

An *Organic Syntheses* preparation using tetraphenylcyclopentadienone in a Diels–Alder reaction to obtain tetraphenylphthalic anhydride has been recorded:

Grummitt, O. *Organic Syntheses*; Wiley: New York, 1955; Collect. Vol. III, p. 807.

The Stepwise Synthesis of Nylon-6,6

Purpose. The important industrial polymer, nylon-6,6, is prepared by the technique of step-growth polymerization. The physical properties of the polymer are examined. The two monomers used in the polymerization are synthesized.

Background of an Industrial Polymer. The type of polymerization used in the nylon preparation described in this series of experiments is called "step-growth" polymerization. The technique uses two different difunctional monomers that undergo ordinary organic reactions. In the present case an acid chloride is treated with an amine to produce an amide linkage.

Nylon is a polyamide. In industry it is produced by reaction of two difunctional monomers (or comonomers): a dicarboxylic acid and a diamine. The polymer that you are going to study is of great historical significance in polymer chemistry, because it was the first of the polyamides to be recognized as possessing excellent physical properties for forming very strong fibers. Nylon-6,6 was, in fact, the first commercially produced synthetic polyamide. The "6,6" nomenclature refers to the number of carbon atoms in each of the two comonomers. Industrially, nylon-6,6 is prepared from 1,6-hexanediamine (hexamethylenediamine) and hexanedioic acid (adipic acid):

$$n \; \text{HOCCH}_2\text{CH}_2\text{CH}_2\text{CH}_2\text{COH} \;+\; n \; \text{H}_2\text{N(CH}_2)_6\text{NH}_2 \;\xrightarrow{\text{heat}}\; \left[\text{C(CH}_2)_4\text{C}-\text{NH(CH}_2)_6\text{N} \right]_n$$

Adipic acid   1,6-Hexanediamine   Nylon-6,6

In the industrial process, the diacid and diamine are mixed to form the corresponding amine salt (hexamethylene diammonium adipate), which is then heated under steam pressure (250 psig) at 275 °C to form the amide bonds. The resulting polymer has an average molecular weight of about 10,000, with an average of over 400 repeating monomer units in each molecule of polymer and a melting point of about 150 °C. Fibers can be drawn from the melted polymer by a "cold-drawing" technique. This method of drawing fibers physically orients the polymer molecules into linear chains that are stabilized by the presence of hydrogen bonds between C=O and the N—H groups of adjacent chains, and the strength of the fiber is thereby increased. The synthetic polyamide linkages in the various forms of nylon are very similar (identical in some cases) to those found in proteins. For example, silk fibers gain their great strength from this type of interaction.

Numerous combinations of diacids and diamines have been evaluated as fiber materials. However, only a few have reached commercial production, which depends on low-cost, easy-to-access intermediates, and satisfactory general and physical properties of the polymer. One such group of materials are the "Aramid" class of fibers, which are prepared from aromatic monomers (see Experiment [B3], Question 7-47). One trade name for a material prepared from these type of fibers is Nomex. It has a high degree of heat and flame resistance. Race car driving suits are made from it, and it is also used as an insulator in the space shuttles.

Nylon-6,6 was first synthesized in 1899 by Gabriel and Maas in Germany. It was not until 1929, however, that the substance was shown to possess practical commercial properties. It was Carothers' research program on polyamides at DuPont that made the major discoveries that initiated the world's polymer industry. DuPont began production of nylon in October 1939, and the first nylon stockings were manufactured in May 1940. By 1950, 14 chemical plants in 10 countries produced 55,000 metric tons of polyamide fiber. By 1980 worldwide production had expanded to 3.05×10^6 tons, with about one-third of the polymer synthesized in the United States. Fast forward nearly half a century, projections for 2025 range from 5 to 10×106 tons of the polymer synthesized globally!

Thus, you should appreciate that the chemical industry carries out organic reactions on massive quantities of material for use in today's highly technological society. The discovery and characterization of these materials all starts in the research laboratory, with many of them initially prepared in microscale quantities. One of the great triumphs of our technology has been the successful scaleup of synthetic organic reactions, but that is a story for another day.

Oxidation of Cyclohexanol: Adipic Acid

Experiment B1

Common name: adipic acid
CA number: [124-04-9]
CA name as indexed: hexanedioic acid

Purpose. You will carry out the nitric acid oxidation of cyclohexanol to obtain adipic acid, an intermediate in the route to prepare nylon, a polyamide.

Prior Reading

Technique 5: Crystallization
Use of the Hirsch Funnel (pp. 92–93)
Standard Experimental Apparatus: Reflux Apparatus (pp. 25–26)

REACTION

$$\text{Cyclohexanol} \xrightarrow{\text{concd HNO}_3} \text{Adipic acid}$$

DISCUSSION

Industrially, the production of adipic acid is a two-step sequence. The main route involves the oxidation of cyclohexane with air to form a mixture of cyclohexanol and cyclohexanone. This mixture is then further oxidized using nitric acid, oxygen, and a Cu–V catalyst to yield the acid.

Ninety percent of all the synthetically produced adipic acid is used in the manufacture of nylon-6,6. Estimates now place the total amount of adipic acid generated annually at greater than 2.5 billion kilograms. In 2004, DuPont alone produced 1.1×10^6 tons of adipic acid accounting for 38% of the world's total. In the early years of nylon production, adipic acid was also used to prepare the 1,6-hexanediamine (more commonly known as hexamethylenediamine) comonomer. Treatment of the adipic acid with ammonia gave hexanedinitrile, which, on catalytic hydrogenation, produced the diamine. This monomer is now generally obtained from 1,3-butadiene. A recent DuPont industrial process involves direct *regioselective* addition of two molecules of HCN to the diene, in the presence of a transition metal catalyst, to produce the dinitrile intermediate.

The oxidation of cyclohexanol by concentrated nitric acid is mechanistically complex. A reasonable mechanistic route to the dicarboxylic acid is given here. The first stage of the oxidation is considered to proceed by a mechanism similar to that found in chromic acid oxidations of alcohols (see Experiment [33]). The reaction here involves the initial formation of a nitrate ester intermediate, which, under the reaction conditions, cleaves by proton abstraction to form the ketone. This reaction is accompanied by reduction of the nitrate to nitrite. The proton transfer may involve a cyclic intramolecular rearrangement during the oxidation–reduction cleavage step. A likely mechanism is outlined below:

| Cyclohexanol | Cyclohexyl nitrate | | Cyclohexanone | Nitrous acid |

The next stage of the reaction can be viewed as a further oxidation to yield a diketone. This stage is initiated by nucleophilic attack on a nitronium ion (NO^+) derived from either the nitric or nitrous acid. The nucleophile is the enol tautomer of the ketone, and the reaction forms an α-nitrosoketone, which is in tautomeric equilibrium with a mono-oxime. This species rapidly hydrolyzes under acidic conditions to yield an α-diketone intermediate. This sequence is shown here:

$$H^+ + HNO_2 \rightleftharpoons NO^+ + H_2O$$

Under strongly acidic conditions, the diketone (these highly electrophilic systems are reactive toward weak nucleophiles) likely undergoes nitrate addition, which is followed by attack of water, ring opening, and reduction of nitrate to nitrite. All this activity ultimately leads to the formation of the desired compound, the open-chain dicarboxylic acid, adipic acid:

| 1,2-Cyclohexandione | | Adipic acid |

EXPERIMENTAL PROCEDURE

Estimated time to complete the experiment: 2.0 h.

Physical Properties of Reactants						
Compound	MW	Amount	mmol	mp (°C)	bp (°C)	*d*
Cyclohexanol	100.16	1.0 mL	9.6	25.1	161.1	0.96
Concd HNO₃	31	2.0 mL				

Reagents and Equipment. Using a graduated glass pipet, measure and add 2.0 mL of concentrated nitric acid to a 10-mL round-bottom flask containing a boiling stone.

CAUTION: Nitric acid is very corrosive. Dispense the material in the **hood**.

HOOD

100 °C, 5 min ⟶
60-100 °C, 10 min ⟶
55-60 °C, 30 min ⟶

2 mL concd Nitric acid,
1 mL cyclohexanol

Attach the flask to a water-cooled reflux condenser, place the assembly in a sand bath, and heat the acid solution to 55–60 °C. Now add dropwise, using a calibrated 9-in. Pasteur pipet inserted down the throat of the reflux condenser, 1.0 mL of cyclohexanol at a rate of one drop every 30 s. (*Gently swirl the reaction mixture in the bath after each addition*.) The slow addition is necessary to control the reaction temperature (➡).

Reaction Conditions. Maintain the sand bath temperature at 55–60 °C for 30 min after the addition of cyclohexanol is complete. Now gradually raise the sand bath temperature to 100 °C over 10 min, and then maintain this temperature for an additional 5 min.

Isolation of Product. Remove the assembly from the hot sand bath and allow the solution to cool to room temperature. Detach the round-bottom flask, and clamp it in an ice bath for 5–10 min. Collect the light yellow crystals by vacuum filtration using a Hirsch funnel.

CAUTION: This solution is still strongly acidic.

Wash the product crystals with 200 µL portions of *ice-cold* water until the crystals turn white.

Purification and Characterization. Dry the adipic acid crystals in an oven at 110–125 °C for 10 min. Weigh the product and calculate the percent yield. Determine the melting point and obtain an IR spectrum. Compare your results to those recorded in the literature (*The Aldrich Library of IR Spectra* and/ or SciFinder Scholar).

Chemical Test. Add several crystals (~5 mg) of the adipic acid to 1 mL of a 10% aqueous solution of sodium bicarbonate. Does evolution of CO_2 indicate the presence of a carboxylic acid?

QUESTIONS

7-35. Which of the following compounds is the stronger acid: $CF_3CH_2CO_2H$ or $CH_3CH_2CO_2H$? Explain.

7-36. What spectroscopic method would you use to unambiguously distinguish between the following isomeric acids? Give an explanation of how the spectra are interpreted to give you an assignment for each compound.

    (a) $CH_3(CH_2)_3CO_2H$

    (b) $(CH_3)_2CHCH_2CO_2H$

    (c) $(CH_3)_3CCO_2H$

7-37. Indicate how you would use both ^1H and ^{13}C NMR spectroscopy to tell the difference between the following isomeric carboxylic acids:

    (a) $HO_2CCH_2CH_2CO_2H$

    (b) $CH_3CH(CO_2H)_2$

7-38. The two carboxyl groups in 3-chlorohexanedioic acid are not equivalent and thus have different dissociation constants. Which carboxylic group is more acidic? Explain.

7-39. Research the first and second dissociation constants of the following dioic acids: oxalic acid, malonic acid, and succinic acid. Note the trend line when correlated to the carbon count between the carboxylic acid moieties. Be it inductive and/or electrostatic effects between the first and second dissociation constant, provide reasoning for the trend line observed.

BIBLIOGRAPHY

For detailed information on the production and use of adipic acid, see:

Kirk-Othmer Encyclopedia of Chemical Technology, 5th ed.; Vol. I, Wiley-VCH: New York, 2004.

Ullmann's Encyclopedia of Industrial Chemistry; Wiley-VCH: Weinheim, 2005.

For oxidation methods in organic chemistry, see:

Haines, A. H. *Methods for the Oxidation of Organic Compounds*; Academic Press: New York, Vol. I, 1985, Vol. II, 1988.

Hudlicky, M. *Oxidations in Organic Chemistry*; American Chemical Society: Washington, DC, 1990.

Mijs, W. J.; de Jonge, C. R. H. I. *Organic Synthesis by Oxidation with Metal Compounds*; Plenum: New York, 1987.

Taber, D. F. *Organic Synthesis: State of the Art 2003–2005*; Wiley: New York, 2006.

————. *Organic Synthesis: State of the Art 2006–2007*; Wiley: New York 2008.

The synthesis of adipic acid is given in Organic Syntheses:

Ellis, B. A. *Organic Syntheses*; Wiley: New York, 1941; Collect. Vol. I, p. 18.

Experiment B2

Preparation of an Acid Chloride: Adipoyl Chloride

Common name: adipoyl chloride

CA number: [111-50-2]

CA name as indexed: hexanedioyl dichloride

Purpose. Adipic acid is converted to its corresponding acid chloride by reaction with thionyl chloride. The experiment will help you further understand the nucleophilic substitution reaction pathway by which carbonyl-containing compounds undergo reaction.

Prior Reading

    Standard Experimental Apparatus: Reflux Apparatus (pp. 25–26)

    Collection or Control of Gaseous Products: (pp. 109–111)

REACTION

    Adipic acid   Thionyl chloride   Adipoyl chloride

$$\text{Adipic acid} + 2\,SOCl_2 \longrightarrow ClCCH_2CH_2CH_2CH_2CCl + 2\,SO_2 + 2\,HCl$$

DISCUSSION

Carboxylic acids react with thionyl chloride ($SOCl_2$) to produce the corresponding acid chlorides, as shown in the above reaction. Thionyl chloride is an attractive reagent due to its low cost, and the fact that both byproducts produced in the reaction are gases. Thus, the reaction is driven to completion by the evolution of HCl and SO_2, and a nearly pure acid chloride is obtained. The major drawback to the reaction is that it produces a strong acid (HCl) and thus cannot be used with compounds that are acid sensitive. Oxalyl chloride is often used as an alternative reagent.

The reaction proceeds by a nucleophilic acyl substitution pathway. With thionyl chloride, a chlorosulfite intermediate is generated. Thus, the —OH group is converted into a relatively good leaving group. The chlorosulfite intermediate then undergoes attack by the chloride ion at the carbonyl carbon to yield the final product. The sequence is shown here:

Acid halides are important intermediates and they are used extensively for the conversion of carboxylic acids into other derivatives. For example, acid halides can be used to prepare (in addition to amides): anhydrides, esters, aldehydes, and ketones. Acid halides readily undergo reaction with water (hydrolysis) to form the corresponding carboxylic acid. For this reason, the reaction system must be protected from atmospheric moisture when acid halides are formed and/or used.

In the present sequence leading to the formation of nylon, the adipoyl chloride provides a reactive species, which, when treated with a diamine, forms the desired amide linkage inherent to nylon.

EXPERIMENTAL PROCEDURE

Estimated time to complete the experiment: 2.0 h.

Physical Properties of Reactants						
Compound	MW	Amount	mmol	mp (°C)	bp (°C)	d
Adipic acid	146.14	500 mg	3.4	153	265	1.35
Thionyl chloride	118.97	0.75 mL	10.4	−105	78.8	1.66

Reagents and Equipment.   Place and dry in an oven at about 125 °C for 30 min, a 10-mL round-bottom flask, a water-cooled reflux condenser, and a drying tube packed with glass wool and alumina. After removing them from the oven, place these items in a desiccator and allow them to cool to room temperature. Measure and place 500 mg (3.4 mmol) of adipic acid and 0.75 mL (10.4 mmol) of thionyl chloride into the 10-mL flask (➡).

Adipic acid, 500 mg + thionyl chloride, 0.75 mL reflux conditions, 10-mL RB flask

HOOD

*NOTE. Store the dry adipic acid and the container of thionyl chloride in a desiccator. Dispense the thionyl chloride in the **hood** using an automatic delivery pipet.*

CAUTION: Thionyl chloride is both corrosive and a lachrymator.

Add a magnetic stir bar to the flask, quickly attach the flask to the reflux condenser protected by the drying tube, and mount the assembly in a sand bath.

HOOD

*NOTE. If a larger scale reaction is to be run outside of the **hood**, it will be necessary to construct a gas trap to control the evolution of the SO₂ and HCl gases (see Prior Reading).*

Reaction Conditions.   With stirring, heat the reaction mixture to 90–95 °C within 5 min. Continue to heat the system within this temperature range for an additional 1 h.

NOTE. Heating the solution above 95 °C causes decomposition of the product. The presence of unreacted adipic acid in the flask indicates incomplete reaction.

Isolation and Characterization.   The progress of the reaction may be followed by IR analysis. With a glass capillary, remove a small sample from the flask and obtain the spectrum of the material using the capillary film technique. Remove the sample from the instrument sampling compartment immediately following the spectral scan to prevent the HCl gas buildup from damaging the instrument. The reaction is considered incomplete if the IR spectrum displays a weak band on the low-wavenumber side of the acid halide carbonyl peak (see Chapter 8).

If the reaction is determined to be incomplete after 1 h, add an additional 200 µL of SOCl₂ and continue to heat the mixture until IR analysis (the disappearance of the weak side band) indicates that completion has occurred. The adipoyl chloride is quite labile, and thus is not purified further, but it is used directly in the preparation of nylon as described in Experiment [B3].

QUESTIONS

7-40. Diagram a complete mechanistic sequence showing the hydrolysis reaction of butanoyl chloride with water to form butanoic acid.

7-41. Give an explanation of why acid chlorides are more reactive toward nucleophilic substitution than are the corresponding ethyl esters. *Hint:* Consider the nature of the leaving group and the rate-determining step in an addition–elimination sequence.

7-42. What major organic product would you expect to be formed when acetyl chloride reacts with each of the following reagents?
(a) Methanol and pyridine   **(c)** Excess methylamine
(b) Water   **(d)** Acetic acid and pyridine

7-43. Acid chlorides are used extensively as electrophiles in the Friedel–Crafts reaction to prepare aromatic ketones. The reaction involves the treatment of an aromatic hydrocarbon with an acyl chloride in the presence of a Lewis acid, such as aluminum chloride (see Experiment [27]). Using this reaction, outline the reaction sequence you would use to prepare
(a) Acetophenone
(b) 1-Indanone

7-44. Formulate a suitable mechanism to account for the reaction of thionyl chloride with carboxylic acids to yield the corresponding chlorosulfite (see the reaction diagrammed in the Discussion section). *Hint:* The first step is the nucleophilic attack of the oxygen of the —OH group of the acid on the sulfur atom of the S═O group of the thionyl chloride.

BIBLIOGRAPHY

Adipoyl chloride is prepared as an intermediate in several preparations reported in Organic Syntheses:

Fuson, R. C.; Walker, J. T. *Organic Syntheses; Wiley*: New York, 1943; Collect. Vol. II, p. 169.

Guha, P. C.; Sankaran, O. K. *Organic Syntheses; Wiley*: New York, 1955; Collect. Vol. III, p. 623.

See your organic textbook for an introduction to this reaction and its scope in synthesis. For example:

Solomons, T. W. G.; Fryhle, C. B.; Snyder, S. A. *Organic Chemistry,* 12th ed.; Wiley: New York, 2016, Chap. 17, p. 761.

For a review on the preparation and reactions of acyl halides, see:

Ansell, M. F. in *The Chemistry of Acyl Halides*; Patai, S., Ed.; Wiley: New York, 1972, p. 35.

Preparation of a Polyamide: Nylon-6,6

Experiment B3

Common names: nylon-6,6, polyhexamethylene adipamide

CA number: [32131-17-2]

CA name as indexed: poly[imino(1,6-dioxo-1,6-hexanediyl)imino-1, 6-hexanediyl]

Purpose.   The polyamide, nylon, is prepared by the step-growth condensation polymerization of adipoyl chloride with 1,6-hexanediamine. An interfacial (emulsion) polymerization technique is used to generate nylon fibers.

REACTION

DISCUSSION

The preparation of nylon outlined in this experiment is not the industrial method (see initial discussion). The use of the reactive diacid chloride reagent allows one to carry out the step-growth polymerization reaction under very mild conditions more convenient to the instructional laboratory. The interfacial (emulsion) polymerization technique used consists of dissolving the adipoyl chloride reagent in a water-immiscible solvent (cyclohexane) and bringing this solution into contact with an aqueous solution of the diamine. A thin film forms at the *interface* of the two solvents as the condensation reaction proceeds. A "rope" of nylon polymer can be pulled from the interface of the two solvents because the film is continuously generated as the reaction occurs. This polymer has an average molecular weight of ~10,000! In this particular experiment, about 5–7 meter lengths of nylon polymer can be obtained. This particular synthesis of nylon is often used in lecture demonstrations and chemical magic shows.

EXPERIMENTAL PROCEDURE

Estimated time to complete the experiment: 0.5 h.

Physical Properties of Reactants					
Compound	MW	Amount	mmol	mp (°C)	d
Adipoyl chloride	183.05	~622 mg	3.4		
1,6-Hexanediamine (5% aq)	116.21	8 mL		41–42	1.259
Cyclohexane	84.16	8 mL			

Reagents and Equipment.   Transfer the clear solution of adipoyl chloride prepared in Experiment [B2] to a 50-mL beaker using a Pasteur pipet. Rinse the flask with 2 mL of cyclohexane and transfer this rinse to the same beaker. Add an additional 8.0 mL of cyclohexane. Now slowly add 8.0 mL of a 5% aqueous solution of 1,6-hexanediamine containing eight drops of 25% NaOH solution.

NOTE. Add the solution using a Pasteur pipet, taking care to run it down the side of the beaker.

Isolation and Characterization.   Using a copper wire bent into a small hook, hook the film in the center of the beaker and draw up the nylon fiber from the solution interface. A slow, steady pull will result in long strands of the polymer. Wash the fibers in a beaker of water before handling them.

QUESTIONS

7-45. Explain why the 25% NaOH solution is added to the reaction mixture.

7-46. Draw a structure to illustrate the hydrogen bonding that may occur when two polymer molecules of this polyamide are cold-drawn together.

7-47. Predict the structure of the polymer that would result in the condensation of the following reactants. These monomers are used to produce the polyamide Nomex, a high-melting material used as an insulator in space shuttles and as the fire-resistant fabric in clothing worn in race cars.

7-48. Amides undergo hydrolysis to carboxylic acids on treatment with alkali. Diagram a suitable mechanism for the conversion of benzamide to benzoic acid using sodium hydroxide as the base.

BIBLIOGRAPHY

For detailed information on the production and use of nylon, see:

Heckert, W. W. *J. Chem. Educ.* **1953,** *30,* 166.

Kirk-Othmer Encyclopedia of Chemical Technology, 4th ed.; Wiley: New York, 1996, Vol. XIX, pp. 454, 470, 485.

For information on the interfacial polymerization technique, see:

Sprague, B. S.; Singleton, R. W. *Text. Res. J.* **1965,** *35,* 999.

Morgan, P. W.; Kwolek, S. L. *J. Chem. Educ.* **1959,** *35,*182.

Nikonov, V. Z.; Savinov, V. M. In *Interfacial Synthesis*; Millich, F.; Carraher, C. E., Jr., Eds.; Marcel Dekker: New York, 1977, Vol. II, Chap. 15.

Odian, G. *Principles of Polymerization,* 4th ed.; Wiley-Interscience: New York, 2004.

The Synthesis of Sulfanilamide

Purpose. You will carry out the multistep synthesis of the epoch-making antibacterial drug, sulfanilamide. You will learn the techniques for handling moisture sensitive materials and investigate the strategies involved in solving a synthetic organic problem.

THE SULFA DRUGS

B

The sulfanilamides were the first effective systemic bactericides. The discovery, in the early 1930s, of these substances transformed medical care. Indeed, a major fraction of the world's population is alive today as a result of these compounds and their microbiological successors. Early recognition of the importance of these materials was underlined when their discoverer, Gerhard Johannes Paul Domagk, was honored by receiving the 1938 Nobel Prize in Medicine. Even 60 years later, "sulfa" drugs are still some of the most effective antibacterial substances available to physicians.

These compounds, to which human existence owes a considerable debt of gratitude, are all derivatives of the parent material, *p*-aminobenzenesulfonamide (sulfanilamide):

$$H_2N-\!\!\left\langle\bigcirc\right\rangle\!\!-SO_2NH_2$$

p-Aminobenzenesulfonamide
(sulfanilamide)

This weakly basic compound contains two —NH_2 groups located in quite different environments, so that each one possesses quite different chemical properties and reactivities. One is substituted directly on the aromatic ring; the other is contained in a sulfonamide group. A wide variety of these derivatives can be prepared by introducing different substituents on the nitrogen atom of the sulfonamide functional group. Any attempt at modifying the ring —NH_2 group, however, was found to completely destroy the biological activity.

Prontosil, *p*-[(2,4-diaminophenyl)azo]benzenesulfonamide, was first synthesized in Germany in 1932 as a product of azo-dye research at I. G. Farbenindustrie. It was soon tested for chemotherapeutic activity, because it bound very strongly to protein fibers. In 1933, it became the first drug to effectively cure blood stream infections. The early investigations of the biological activity of Prontosil required direct testing on animal infections, since the drug appeared to be inactive in vitro.

Prontosil

In 1935, Trefouels (Pasteur Institute) established that Prontosil breaks down in the body to yield, as one of the metabolites, sulfanilamide. It was shown later that year, by Fourneau, that the biological activity of Prontosil was entirely contained in the sulfanilamide section of the molecule, and that no biological activity was lost during the cleavage. This later discovery, and the fact that sulfanilamide itself had been earlier synthesized by Gelmo in 1908 (unfortunately, by not exploring the biological activity of the substance, he missed the opportunity at a Nobel Prize), voided any patent protection and led to a worldwide effort to synthesize a wide variety of these materials. By 1944 over 5000 sulfanilamide derivatives had been prepared and tested.

Currently the sulfa drugs are still very important therapeutic agents, but have to a large extent been replaced by antibiotics, such as the penicillins. Their low cost and general effectiveness for urinary tract infections, however, still make sulfanilamide derivatives attractive alternatives. These compounds have also found a valuable role in veterinary medicine.

Production of sulfanilamide reached a peak of 9000 metric tons/year in 1943; present production is about one-half that amount.

The following experiments in this sequence outline a novel and efficient procedure for the preparation of this important substance. The chemistry involves the trifluoroacetylation of aniline (Experiment [C1]), the chlorosulfonation of the resulting acetanilide (Experiment [C2]), which is followed by concomitant aminolysis and hydrolysis of the sulfonyl chloride intermediate (Experiment [C3]) to produce the final product, sulfanilamide:

Experiment C1

Acetylation of Aniline: 2,2,2-Trifluoroacetanilide

Common name: 2,2,2-trifluoroacetanilide

CA number: [404-24-0]

CA name as indexed: acetamide, 2,2,2-trifluoro-N-phenyl-

Purpose. The goal of this experiment is to protect the easily oxidized amine group of aniline from electrophilic attack during the sulfonation of the benzene ring in the next step of the synthesis. This deactivation is accomplished by reducing the electron density on the amine nitrogen (and thus its nucleophilicity) by acetylating aniline with the strongly electron-withdrawing trifluoroacetyl group. You will become familiar with techniques for handling moisture-sensitive materials.

Prior Reading

    Standard Experimental Apparatus: Moisture-protected Claisen head with
                                    3- or 5-mL conical vial, arranged for
                                      syringe addition (pp. 27–28)

    Technique 6B: Concentration of Solutions:
                      Evaporation with Nitrogen Gas (p. 106)

REACTION

Aniline   Trifluoroacetic anhydride

DISCUSSION

Acid anhydrides react with amines, or ammonia, to yield amides. The reaction of anhydrides is similar to that of acid chlorides (see Experiment [B3]), although the anhydrides normally react at slower rate.

    Acetic anhydride, $CH_3C(O)OC(O)CH_3$, is an important industrial reagent used to prepare acetamides, $CH_3C(O)NR_2$, from a variety of amines. For example, acetaminophen, an over-the-counter analgesic, is synthesized by the reaction of *p*-hydroxyaniline (4-aminophenol) with acetic anhydride:

p-Hydroxyaniline   Acetic anhydride   Acetaminophen

    In the present reaction, *trifluoro*acetic anhydride is used in place of acetic anhydride. The halogenated anhydride is considerably more reactive because of the presence of the three fluorine atoms on the α-carbon atom of the anhydride. The electron-with drawing power of the fluorine substituents helps to enhance nucleophilic attack on the anhydride by the weak aromatic nucleophile. Once the amide is formed, the trifluoromethyl group makes the amide a better protecting group than the acetyl group, as the trifluoroacetyl group renders the amine lone electron pair less nucleophilic. Furthermore, the byproduct of the reaction, trifluoroacetic acid, is fairly volatile (bp 72 °C) and is therefore easily removed from the reaction mixture by evaporation. This enhanced reactivity helps avoid the use of an added base, as is required in the preparation of acetaminophen. Chemically, the procedure is simplified and the intermediate trifluoroacetanilide is usually obtained in a relatively pure condition. *This highly reactive anhydride, however, requires considerable care in its use.*

CAUTION: No moisture can remain on the surface of the clean glassware, or hydrolysis may occur with explosive force!

EXPERIMENTAL PROCEDURE

Estimated time to complete the experiment: 0.75 h.

Physical Properties of Reactants							
Compound	MW	Mass	Volume	mmol	mp (°C)	bp (°C)	d
Aniline	93.13	235 mg	230 μL	2.5	−6.2	184.3	1.02
Trifluoroacetic anhydride	201.04	744 mg	500 μL	3.54	−6.5	39.5	1.49
Methylene chloride	84.93		500 μL			40	1.33

Reagents and Equipment. Place and dry in an oven at about 125 °C for 30 min, a 5-mL conical vial, a Claisen head, a drying tube packed with glass wool and calcium chloride, a 1-mL glass syringe, and a $\frac{1}{2}$-dram screw-cap vial. After removal from the oven, place these items in a desiccator and allow them to cool to room temperature. Measure and place 230 μL (2.5 mmol) of aniline and 500 μL of methylene chloride in the 5-mL conical vial. Immediately attach the vial to the Claisen head equipped with a septum cap and protected by the drying tube (see Prior Reading). Cool the vial in a cold water bath.

HOOD   *NOTE. Dispense the aniline and methylene chloride in the **hood** using automatic delivery pipets.*

CAUTION: Aniline is a highly toxic substance and a cancer suspect agent.

HOOD   In the **hood**, place 500 μL of trifluoroacetic anhydride and 500 μL of methylene chloride in the dried $\frac{1}{2}$-dram vial.

NOTE. The anhydride hydrolyzes rapidly in moist air, so make the transfer quickly.

Now draw this solution into the dried syringe, and insert the needle
SLOWLY   through the septum cap of the Claisen head. Dropwise, **slowly** add this solution to the cold aniline-methylene chloride solution.

NOTE. Heat is produced in the reaction. If the addition is too rapid, the methylene chloride starts to reflux and fumes are emitted through the drying tube.

Reaction Conditions. After the addition is complete, allow the reaction mixture to stand at room temperature for 10 min.

Isolation of Product. Remove the reaction vial and, in the **hood**, concentrate the solution on a warm sand bath under a slow stream of nitrogen gas. The crude 2,2,2-trifluoroacetanilide intermediate is obtained as a white powder. Cap the vial.

HOOD

Purification and Characterization. The crude material is not purified further, but rather is used directly in the next reaction of the sequence without further characterization.

QUESTIONS

7-49. Outline a complete mechanistic sequence for the reaction to prepare acetaminophen as presented in the discussion section of this experiment.

7-50. In the preparation of amides, acid chlorides or anhydrides may be used to react with the selected amine. It is known that acid chlorides are more reactive than the corresponding anhydrides in this type of nucleophilic acyl substitution reaction. Offer a reasonable explanation for this observation.

7-51. Offer an explanation of why 2,2,2-trifluoroacetic anhydride is more reactive toward aniline than acetic anhydride in the nucleophilic acyl substitution reaction presented in this experiment.

7-52. Offer a reasonable explanation of why acetamide has a higher melting and boiling point then *N,N*-dimethylacetamide even though it has a lower molecular weight:

$$CH_3\overset{\overset{\displaystyle \cdot \ddot{O} \cdot}{\|}}{C}-\ddot{N}H_2 \qquad CH_3\overset{\overset{\displaystyle \cdot \ddot{O} \cdot}{\|}}{C}-\overset{\displaystyle CH_3}{\underset{\displaystyle CH_3}{N\colon}}$$

Acetamide
mp 82 °C
bp 221 °C

N,N-Dimethylacetamide
mp −20 °C
bp 166 °C

7-53. Outline a simple chemical test that would distinguish between ethyl benzoate and benzamide (*Hint:* See Chapter 9).

BIBLIOGRAPHY

Smith, M. B. *March's Advanced Organic Chemistry*, 8th ed.; Wiley: New York, 2019.

Selected acylation reactions in *Organic Syntheses* between anhydrides and amines:

Herbst, R. M.; Shemin, D. *Organic Syntheses*; Wiley: New York, 1943; Collect. Vol. II, p. 11.

Noyes, W. A.; Porter, P. K. *Organic Syntheses*; Wiley: New York, 1941; Collect. Vol. I, p. 457.

Wiley, R. H.; Borum, O. H. *Organic Syntheses*; Wiley: New York, 1963; Collect. Vol. IV, p. 5.

The use of the perfluoroacetic anhydride reagent in this synthesis is reported in:

Hurdis, E. C.; Yang, J. W. *J. Chem. Educ.* **1969,** *46*, 697.

Chlorosulfonation of 2,2,2-Trifluoroacetanilide: p-(Trifluoroacetamido)benzenesulfonyl Chloride

Common name: p-(trifluoroacetamido)benzenesulfonyl chloride
CA number: [31143-71-2]
CA name as indexed: benzenesulfonyl chloride,
4- [(trifluoroacetyl)amino]-

Purpose. The substitution of the aniline ring is carried out by introduction of the sulfonyl group. This substitution reaction is the second stage on the route to preparing sulfanilamide. p-(Trifluoroacetamido)benzenesulfonyl chloride is prepared by treatment of the intermediate, 2,2,2-trifluoroacetanilide, with chlorosulfonic acid. You will gain experience at handling highly reactive moisture-sensitive reagents.

Prior Reading

Technique 5: Crystallization
Use of the Hirsch Funnel (pp. 92–93)

REACTION

2,2,2-Trifluoroacetanilide   Chlorosulfonic acid   p-(Trifluoroacetamido)-
benzenesulfonyl chloride

DISCUSSION

As shown here, the sulfonyl chloride group (—SO$_2$Cl) can be conveniently introduced to an aromatic ring via an electrophilic aromatic substitution reaction using chlorosulfonic acid. The reaction is usually referred to as *chlorosulfonation*. It has been determined that two equivalents of the acid are required per equivalent of the aromatic compound. In the initial attack the system first forms the corresponding sulfonic acid, which in turn is converted to the sulfonyl chloride. It is believed that the initial stage of the reaction involves SO$_3$ as the electrophile. It is likely that this reagent results from the establishment of the equilibrium reaction shown here:

$$ClSO_3H \rightleftharpoons SO_3 + HCl$$

The intermediate sulfonic acid is converted to its sulfonyl chloride derivative by reaction with a second equivalent of the chlorosulfonic acid:

Product isolation is achieved by transferring the solution of the sulfonyl chloride onto crushed ice to precipitate the insoluble acid chloride. This isolation procedure can be used because sulfonyl chlorides are much less susceptible to hydrolysis than their carboxylic acid chloride counterparts (see Experiment [B2]). Sulfonyl chlorides *do* undergo this hydrolysis reaction, albeit rather slowly, resulting in the formation of the corresponding sulfonic acid.

Notice that the para-substituted product is formed in the reaction. Why is this observed given that two other isomers are possible (the meta- and/or ortho-products)? (See Question 7-57.)

EXPERIMENTAL PROCEDURE

Estimated time to complete the experiment: 0.75 h.

Physical Properties of Reactants							
Compound	MW	Amount	mmol	mp (°C)	bp (°C)	*d*	
2,2,2-Trifluoroacetanilide	189.06	478 mg (theoretical value)	2.5	87.6			
Chlorosulfonic acid	116.52	900 μL	13.7	−80	158	1.77	

NOTE. All reagent transfers must be made and the reaction must be conducted in the hood.   HOOD

Reagents and Equipment.   Place and dry in an oven at about 125 °C for 30 min, an air condenser and a 9-in. Pasteur pipet calibrated to deliver 0.9 mL. After removal from the oven, place these items in a desiccator and allow them to cool to room temperature.

Attach the vial containing the 2,2,2-trifluoroacetanilide (prepared in Experiment [C1]) to the *dry* air condenser. Measure, using the 9-in. Pasteur pipet, 0.9 mL of chlorosulfonic acid, and **slowly** add this reagent to the flask by   SLOWLY inserting the pipet down the neck of the air condenser.

HOOD

CAUTION: Chlorosulfonic acid is a very corrosive substance. It reacts violently with water and causes serious burns on contact with the skin. Dispense the reagent in the **hood**. Leave all equipment used to transfer this material in the **hood** until it is obvious that the residual reagent has reacted with the moist air (white fumes subside), and leave these materials in contact only with *glass* surfaces.

NOTE. An alternative procedure for making the transfer is to measure the amount of this reagent using an oven-dried 1-mL glass pipet and then transfer it to an oven-dried screw-cap vial. The material is then transferred to the reaction vial as described above, using the dry 9-in. pipet.

Reaction Conditions.   Place the reaction assembly on a sand bath in the **hood** and heat the mixture at 60–70 °C for 10 min.

Isolation and Characterization.   Allow the reaction mixture to cool to room temperature, and then place the flask in an ice bath. Using a second, dry 9-in. Pasteur pipet, transfer the cold reaction solution **slowly** to a 10-mL beaker containing ice ($-\frac{1}{2}$ full) in the **hood**.

SLOWLY
HOOD

CAUTION
HOOD

CAUTION: This is a very vigorous reaction. Use **caution**. Leave all equipment in the **hood** until it is obvious the residual reagent has reacted with the moist air (white fumes subside). Make sure that only glass surfaces are in contact with the reagent residues.

A precipitate of the product, *p*-(trifluoroacetamido)benzenesulfonyl chloride, forms at this stage. Collect the tannish-white precipitate by vacuum filtration and wash the filter cake with three 1-mL portions of ice cold water.

Allow the material to air-dry. Determine the weight of the crude product, and calculate the percent yield from aniline. Determine its melting point, and compare it to the reported value of 142–145 °C.

Chemical Test.   Carry out a sodium fusion test to confirm the presence of nitrogen, sulfur, and chlorine in the product (see Chapter 9).

QUESTIONS

7-54. Outline a suitable mechanism to illustrate the reaction of benzenesulfonyl chloride with excess methylamine to form the corresponding sulfonamide.

7-55. In reference to Question 7-54, why is an excess of methylamine used?

7-56. Account for the fact that the amide group (—NHCOCF₃) is ortho and para directing. Use resonance structures to illustrate your written explanation.

7-57. As mentioned in the discussion, offer an explanation of why only the para isomer is obtained in the chlorosulfonation reaction carried out in this experiment.

BIBLIOGRAPHY

For a summary of electrophilic aromatic substitution reactions, see:

Hapworth, J. D.; Waring, D. R.; Waring, M. *J. Aromatic Chemistry; Wiley: New York*, 2003, Chap. 2.

For the use of chlorosulfonic acid in aromatic substitution reactions, see:

Gilbert, E. E. *Sulfonation and Related Reactions*; Wiley: New York, 1965, p. 84.

Rathke, M. E.; Millard, A. A. *Org. Syn.* **1978,** *58*, 32.

Reid, J. R.; Dufresne, R. F.; Chapman, J. J. *Org. Syn.* **1997,** *74*, 217.

Scheifele, H. J., Jr.; De Tar, D. F. *Organic Syntheses*; Wiley: New York, 1963; Collect. Vol. IV, p. 34.

Smiles, S.; Stewart, J. *Organic Syntheses*; Wiley: New York, 1941; Collect. Vol. I, p. 8.

Webb, C. N. *Organic Syntheses*; Wiley: New York, 1941; Collect. Vol. I, p. 85.

Also see:

Smith, M. B. *March's Advanced Organic Chemistry*, 8th ed.; Wiley: New York, 2019.

Preparation of an Arene Sulfonamide: Sulfanilamide

Experiment C3

Common names: sulfanilamide, *p*-aminobenzenesulfonamide

CA number: [63-74-1]

CA name as indexed: benzenesulfonamide, 4-amino-

Purpose. The sequential synthesis is completed and the sulfa drug, *sulfanilamide*, is obtained by treatment of the sulfonyl chloride intermediate prepared in Experiment [C2], with an aqueous ammonia solution.[2] This interesting product is fully characterized.

Prior Reading

  Technique 5: Crystallization

Use of the Hirsch Funnel (pp. 92–93)

Craig Tube Crystallization (pp. 93–95)

DISCUSSION

The final step in the synthesis of the target molecule, sulfanilamide, is the conversion of the sulfonyl chloride intermediate to a sulfonamide and the removal of the 2,2,2-trifluoroacetyl protecting group. These transformations are accomplished in one step, because both reactions take place upon heating the protected sulfonyl chloride with aqueous ammonia. The reaction sequence is shown here:

p-(Trifluoroacetamido)-benzenesulfonyl chloride

Sulfanilamide

[2]This experiment is adapted from the work of S. Danishefsky, Yale University (personal communication).

We can now see several reasons for first protecting the reactive amino group on the aniline molecule. First, to attempt the introduction of the sulfonyl group directly on aniline would very likely lead to a species (*p*-aminobenzenesulfonyl chloride) that would react with itself (act as a difunctional monomer) to produce a sulfonamide polymer:

Second, because the sulfonyl group is introduced using chlorosulfonic acid, an acidic medium is present. Under these conditions, because free HCl is available (see discussion in Experiment [C2]), the amino group would be protonated and become a meta directing group. This situation would lead to the formation of the wrong isomer or, at best, a mixture of isomers.

Third, as mentioned above, the amino group is highly prone to oxidation and the unavoidable presence of traces of the strong oxidant, SO_3, would be expected to lead to considerable oxidation of the aromatic amine.

The 2,2,2-trifluoroacetyl group meets all the requirements of a good protecting group. First, the acetylation reaction goes rapidly in nonaqueous media to give a clean, dry product. Second, the protecting group is easily and rapidly removed in the final stage without disturbing the other functional groups in the molecule.

EXPERIMENTAL PROCEDURE

Estimated time to complete the experiment: 0.75 h.

Physical Properties of Reactants				
Compound	MW	Amount	mmol	mp (°C)
(2,2,2-Trifluoroacetamido)-benzenesulfonyl chloride	287	400 mg	1.39	142–145
Aqueous ammonia		600 μL		

HOOD   CAUTION: All reagent transfers and the *reaction* must be conducted in the **hood**.

Reagents and Equipment.   Weigh and place in a 10-mL Erlenmeyer flask 400 mg (1.39 mmol) of the *p*-(2,2,2-trifluoroacetamido)benzenesulfonyl chloride, which was prepared in Experiment [C2]. In a 10 × 75-mm test tube, prepare a solution of 0.6 mL of fresh, concentrated aqueous ammonia (ammonium hydroxide) and 0.4 mL of deionized water. Add this solution to the solid sulfonyl chloride in the Erlenmeyer flask. Now add a boiling stone. Use a glass rod to break up any lumps of the solid that may form.

HOOD   CAUTION: Dispense the ammonia solution in the **hood**.

CAUTION: If the sulfonyl chloride reagent contains acidic impurities, a vigorous reaction may occur when the reagents are mixed.

Reaction Conditions. Place the Erlenmeyer flask on a hot (100–110 °C) sand bath in the **hood** and heat the mixture until the solid material dissolves. HOOD Agitate with a glass rod, if necessary, to assist the dissolution process. Now heat the solution to boiling for an additional minute.

Isolation and Characterization. Remove the flask from the sand bath and allow the reaction mixture to cool to room temperature. Place the flask in an ice bath for 15–20 min. During this time light-yellow crystals of product precipitate.

Collect the crystalline product by vacuum filtration, and wash the crystals with three 0.5-mL portions of ice-cold water. Air-dry the sulfanilamide, weigh, and calculate a crude yield, both for this step and for the overall sequence based on the amount of aniline used in the first step. Determine the melting point of this material.

Recrystallize the crude material from water using the Craig tube. Air-dry the white crystals overnight, or in an oven (110 °C) for 10–15 min. Determine the melting point and compare your value to that reported in the literature. To further characterize the material obtain an IR and NMR spectrum and compare them to those found in the literature (*The Aldrich Library of Infrared Spectra, The Aldrich Library of NMR Spectra*, and/or the corresponding spectral data available online (e.g., SciFinder Scholar)).

Chemical Test. Carry out a sodium fusion test to confirm the presence of nitrogen and sulfur in the product (see Chapter 9).

QUESTIONS

7-58. Based on the results of this experiment, which is more nucleophilic: hydroxide ion (HO^-) or ammonia (NH_3)? Explain.

7-59. Outline a suitable mechanism to account for the conversion of the sulfonyl chloride group to the sulfonamide group by reaction with ammonia.

7-60. Outline a synthesis for the sulfa drug, sulfathiazole, starting from benzene. The primary aminothiazole ring system is available:

Sulfathiazole

7-61. Sulfonyl chlorides also react readily with alcohols to yield sulfonate esters:

$$ArSO_2Cl + HOR \xrightarrow[\substack{or \\ 3° \text{ amine}}]{\text{pyridine}} ArSO_2OR + HCl$$

Propose a suitable mechanism to account for this reaction.

7-62. In reference to the Hinsberg test used to distinguish between primary, secondary, and tertiary amines (see Chapter 9), primary amines yield a sulfonamide that is soluble in aqueous sodium hydroxide, whereas secondary amines give a sulfonamide that is not soluble in this reagent. Offer an explanation for this observation.

BIBLIOGRAPHY

For an overall view of the sulfa drug story, see:

Kirk–Othmer Encyclopedia of Chemical Technology, 4th ed.; Wiley: New York, 1992, Vol. II, p. 876.

Examples of treatment of sulfonyl chlorides with amines:

De Boer, Th. J.; Backer, H. J. *Organic Syntheses; Wiley*: New York, 1963; Collect. Vol. IV, p. 943.

Reid, J. R.; Dufresne, R. F.; Chapman, J. J. *Org. Syn.* **1997,** *74,* 217.

Scheifele, H. J., Jr.; De Tar, D. F. *Organic Syntheses; Wiley*: New York, 1963; Collect. Vol. IV, p. 34.

The use of the perfluoroacetic anhydride reagent in this synthesis is reported in:

Hurdis, E. C.; Yang, J. W. *J. Chem. Educ.* **1969,** *46,* 697.

ADVANCED LABORATORY EXPERIMENTATION: SEQUENCES D, E, AND F

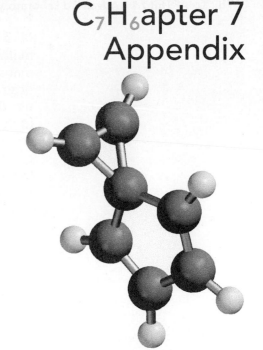

The Synthesis of 2'-Bromostyrene

Purpose. A three-step synthesis of 2'-bromostyrene is carried out starting with benzaldehyde. You will gain experience working with semimicroscale quantities of organic materials. (See Sequence A, where benzaldehyde is also used as the starting material, in that case for the formation of two different intermediates in the seven-step synthesis of hexaphenylbenzene.) You will become familiar with the stereo-chemistry involved in the addition of molecular halogen to an alkene (see also Experiment [A2ᵦ]). You will observe the elimination of a hydrogen halide promoted by a concerted decarboxylation reaction (see also Experiment [A3ᵦ]). As an option, ¹H NMR spectroscopy will be used to determine the cis/trans ratio of isomers in the product (see also Experiment [5B]).

THE SYNTHESIS OF A FRAGRANCE

Numerous classes of organic compounds have characteristic odors. For example, volatile esters have pleasant odors (see Experiment [8B]) and are often used in perfumes and artificial flavorings where they contribute to the fragrance of the material; acid chlorides have sharp penetrating odors (see Experiment [B2]); the alkyl amines have a "fishy" smell (see Experiment [B3]); benzaldehyde has the odor of bitter almond oil (see Experiment [20]); and cinnamaldehyde, the odor of cinnamon (see Experiment [11C]). Experiment [11C] also contains an expanded discussion of naturally occurring fragrant materials, called *essential oils*. As chain length increases, the odors of the short-chain alkyl carboxylic acids progress from the sharp, irritating odors of formic and acetic acids to the very rank, disagreeable odors of butyric (rancid butter), valeric, and caproic (dirty socks and goats) acids. Low molecular weight thiols (mercaptans), sulfides, and disulfides have intensely disagreeable odors. Examples of these materials are the active principals in the chemical defense spray used by the skunk. Via evolution, this animal has developed a highly effective mixture of 3-methylbutane-1-thiol, *trans*-2-butene-1-thiol, and *trans*-2-buten-1-yl methyl disulfide, which is able to deter most skunk predators, not just humans. Our sense of smell can detect one part of ethanethiol in 50 billion parts of air. This sounds spectacular, but humans have developed other highly sensitive detectors (the retinas of our eyes) to a band of electromagnetic radiation in the region 400–750 nm (the visible region of the spectrum). We have, as a consequence, become less dependent on our sense of chemical communication (our sense of smell). When we compare our detection limit of mercaptans to the male silkworm's detection of the female sex *pheromone*, we find that the *Bombyx mori* respond to threshold levels close to 1 million times lower (in the concentration range of 100 molecules per milliliter)!

Clearly, chemical communication (odor) plays a critical role in mediating animal and plant behavior within and between species. These metabolites are often stored and released when the animal (or plant) responds to certain stimuli. Historically, these substances served as the major communication link between living systems during the evolutionary development of single cell organisms. Thus, they predated hormones, which function in a similar fashion, but within a host collection of cells. While these substances may be used by a given insect species as a sex attractant, pheromones can also mediate other intraspecies functions, such as acting as trail guides and alarm signals. Thus, pheromones are substances that are used to communicate information between individuals of the same species to their adaptive advantage. The activity of a pheromone frequently depends on the configuration around a double bond (*E* or *Z*) as well as the absolute configuration around an *R* or *S* chiral center, just as humans select for these structural features (see discussion in Experiment [11D]). In a number of cases, the response has been shown to depend on the ratio of the isomers. There are several other classes of chemical communication substances used by plants and animals. The *allomones* are interspecies materials, which, on release, are used to the adaptive advantage of the host but to the disadvantage of the receiver. A typical function of an allomone is as a chemical defense agent; the thiols and disulfides used by the skunk are excellent examples. Three typical chemical communication substances are shown here:

Termite defense allomone

and

Sex pheromones of the European corn borer (a mixture of cis + trans acetates)

The synthetic route undertaken in the Sequence D experiments leads ultimately to a mixture of the diastereomeric cis–trans isomers of 2'-bromostyrene:

cis-2'-Bromostyrene   trans-2'-Bromostyrene

This material is often used as an additive to impart a pleasant fragrance to soaps, since it has a very pleasant aroma, similar to that of hyacinth. It is not, however, a naturally occurring material. As a nice illustration of the fact that stereoisomeric compounds can have markedly different physiological effects, it has been demonstrated that a single diastereomer, *trans*-2'-bromostyrene, is responsible for the hyacinth-like odor.

The chemistry involved in this particular synthesis of 2'-bromostyrene contains a number of fundamental organic reactions and is rather interesting to study. The first stage (Experiment [D1]) involves the condensation of benzaldehyde with malonic acid to produce the intermediate, *trans*-cinnamic acid. This reaction is mechanistically similar to the well-known *aldol* reaction, and is often referred to as the *Knoevenagel* condensation:

Benzaldehyde   Malonic acid   trans-Cinnamic acid

The second stage (Experiment [D2]) involves the addition of bromine to the intermediate formed in the first step, *trans*-cinnamic acid. The product of this halogenation is *erythro*-2,3-dibromo-3-phenylpropanoic acid. The stereochemistry involved in the formation of this second intermediate is a result of the nature of the anti addition of molecular bromine to a trans alkene. The details are given in the discussion in Experiment [D2].

trans-Cinnamic acid   erythro-2,3-Dibromo-3-phenylpropanoic acid

In Experiment [D3], the third step in Sequence D, the target molecule, 2'-bromostyrene, is generated by a novel elimination reaction that involves the loss of both CO_2 and HBr from the dibromophenylpropanoic acid. This stage of the synthesis offers an excellent opportunity to study the effect of solvent on the stereochemical course of a reaction by spectral analysis. In this case NMR (or infrared) data can be used to establish whether a cis–trans mixture is obtained, or just one pure isomer (cis):

erythro-2,3-Dibromo-3-
phenylpropanoic acid

cis- + trans-2'-Bromostyrene

Experiment D1

The Verley–Doebner Modification of the Knoevenagel Reaction: *trans*-Cinnamic Acid

Common name: *trans*-cinnamic acid
CA number: [140-10-3]
CA name as indexed: 2-propenoic acid, 3-phenyl-, (*E*)-

Purpose.   *trans*-Cinnamic acid is prepared as the first intermediate in the Sequence D set of reactions that ultimately lead to 2'-bromostyrene. You will review the chemistry associated with an important variety of aldol-type condensation, the modified Knoevenagel reaction.

Prior Reading

Technique 5: Crystallization
Use of the Hirsch Funnel (pp. 92–93)

REACTION

Benzaldehyde   Malonic acid   *trans*-Cinnamic acid

DISCUSSION

Those reactions that are called aldols derive their name from the early nineteenth-century organic literature. The term was first applied to a self-addition product of acetaldehyde that forms under basic conditions. The product, 3-hydroxybutanal, can form in yields as high as 50% in the presence of sodium hydroxide. This substance eventually became referred to as *aldol*, because it was both an aldehyde and an alcohol.

The aldol reaction can be defined in the broad sense as a reaction in which a nucleophile-generated alpha to an electron-withdrawing functional group (in the large majority of cases, carbonyl groups are responsible for the α-hydrogen acidity, although they are not required) adds to a carbonyl group (it may be self-condensation, as in the example of acetaldehyde, but it may also involve attack on another carbonyl-containing species, if it is present). Several experiments in this text illustrate this type of reactivity. Experiments [20] and [A3$_a$] are classic aldol reactions in which conditions are vigorous enough to result in elimination of the β-hydroxyl group to yield α,β-unsaturated ketones as products. In the first step of Sequence D, we now add the Knoevenagel reaction as another condensation possessing a mechanism similar to that of the aldol.

Emil Knoevenagel (1865–1921) Knoevenagel was born in Hanover, Germany. He was the son of a chemist and started his studies at the Technical Institute at Hanover. Later he studied with both Victor Meyer and Gattermann at Göttingen where he received a Ph.D. in 1889. When Victor Meyer moved to Heidelberg, Knoevenagel accompanied him. In 1896 he was appointed assistant professor of organic chemistry at Heidelberg and, in the same year, published his studies on the reaction that now bears his name. He eventually became professor of organic chemistry in 1900.

Knoevenagel had a variety of interests, including stereoisomerism. He worked extensively with aldol-type reactions, and pioneered the use of amine bases to promote these condensations. He was particularly interested in pyridine chemistry, and was the first to synthesize the pyridine ring by heating hydroxylamine with 1,5-diketones.

Following World War I, in which he saw active service, he continued his studies, only to die suddenly, at the age of 56, during an appendectomy.

Interestingly, although Knoevenagel demonstrated the effectiveness of amine bases in promoting aldol-type reactions and though, as noted above, he was particularly interested in pyridine, he overlooked this material's potential application to aldol condensations. It was left to Verley (1899) and Doebner (1900) to introduce successful modifications of the condensation in which pyridine appears to play a number of roles: as a solvent, as a base, and assisting in the decarboxylation.

The Knoevenagel reaction in its simplest form is the condensation of malonic esters (or their analogues) with aldehydes or ketones in the presence of an amine base catalyst plus a small amount of a carboxylic acid (or amino acid) cocatalyst. The condensation products are often α,β-unsaturated carbonyl compounds. For example,

| Benzaldehyde | Diethyl malonate | | Diethyl benzalmalonate |

Other substances possessing an acidic methylene group have been incorporated in aldol-type reactions. These materials include ethyl cyanoacetate and ethyl acetoacetate, as well as phenylacetonitrile, benzyl ketones, and aliphatic nitro compounds. The reaction is often run in benzene or toluene solvent so that the water formed can be continuously removed. Other cocatalysts, besides carboxylic acids, are various ammonium salts, such as ammonium acetate and piperidinium acetate. The Knoevenagel reaction gives highest yields when aldehydes are used as the electrophile, although selected ketones can sometimes give acceptable yields. One of the more important properties of these reactions from a synthetic perspective is that they offer a route to the formation of C—C bonds.

In the formation of the first synthetic intermediate in Sequence D, the very effective Verley–Doebner modification of the fundamental Knoevenagel condensation is used. This modification uses malonic acid in place of the conventional ester to promote enolization. In addition, the heterocyclic amine, pyridine, functions as both the base catalyst and the solvent. A cocatalyst, β-alanine (an amino acid), is also introduced. Mechanistically, the reaction closely resembles the aldol condensation in that in both cases a carbanion is generated by abstraction, by base, of a proton *alpha* to a carbonyl group. The resulting carbanion is stabilized as an enolate anion (see below).

The enolate anion then acts as a nucleophile and attacks the carbonyl carbon of a second carbonyl-containing molecule in the reaction. (An example would be the aldol reactions in Experiment [20].) The intermediate aldol product, the β-hydroxyacid, undergoes rapid dehydration under these reaction conditions, to give the α,β-unsaturated diacid shown here:

(acidic hydrogen)   (carbanion)   (enolate anion)
                                                    (one resonance structure shown)

Conjugate addition (1,4 addition) of pyridine and ionization of a carboxylic acid are followed by decarboxylation and concomitant elimination of pyridine to yield the α,β-unsaturated carboxylic acid as shown here:

EXPERIMENTAL PROCEDURE

Estimated time to complete the experiment: 2.5 h.

Physical Properties of Reactants						
Compound	MW	Amount	mmol	mp (°C)	bp (°C)	d
Benzaldehyde	106.12	580 μL (603 mg)	5.7	−26	179	1.04
Malonic acid	104.06	1.5 g	14.4	135 (decomp.)		
Pyridine	79.10	3 mL			115	0.978
β-Alanine	89.09	120 mg	1.35	205 (decomp.)		

Reagents and Equipment. In a 10-mL round-bottom flask containing a magnetic stir bar, weigh and place 1.5 g (14.4 mmol) of malonic acid followed by 120 mg of β-alanine. Now, in the **hood**, add 3 mL of pyridine and 580 μL of freshly distilled benzaldehyde. HOOD

NOTE. Dispense the benzaldehyde using an automatic delivery pipet. Measure the pyridine using a 10-mL graduated cylinder.

CAUTION: Pyridine is a toxic amine with a strong unpleasant odor. Both reagents should be stored and dispensed in the **hood**. HOOD

Attach the flask to a water-jacketed reflux condenser, and place the assembly in a sand bath on a magnetic stirring hot plate.

Reaction Conditions. Heat the mixture with stirring at a sand bath temperature of about 130 °C for 1.5 h.

Isolation of Product. Allow the reaction mixture to cool to room temperature and, in the **hood,** remove the flask. Transfer the reaction solution using a Pasteur pipet to a 50-mL Erlenmeyer flask containing 12 mL of ice-cold water. Rinse the round-bottom flask with an additional 3 mL of ice-cold water, and transfer the rinse to the same Erlenmeyer flask in like manner. Now add, in small portions, 6 M HCl (~8 mL) until a white precipitate forms and the solution tests weakly acidic to pH paper. Collect the precipitate of *trans*-cinnamic acid by vacuum filtration using a Hirsch funnel. Wash the white crystals with three 2-mL portions of cold water. To partially dry the material, continue the suction for an additional 15 min (remember to cover the Hirsch funnel with filter paper to protect the filter cake from contamination with dust from the laboratory atmosphere during the extended air-drying period) (➡).

trans-cinnamic acid collected here

aqueous rinse

Purification and Characterization. Oven-dry the partially dried product at 100 °C overnight. Weigh the crude acid and calculate the percent yield. Determine the melting point. Compare your result with the literature value. This material is likely to be of sufficient purity to be used in the next step of Sequence [D], Experiment [D2]. If your material melts below the literature value, for characterization recrystallize approximately 30 mg from hot water using the Craig tube. Dry as before and redetermine the melting point. To further characterize the material, obtain IR and ¹H NMR spectra. Compare your spectra with those recorded in the literature (*The Aldrich Library of IR Spectra, The Aldrich Library of NMR Spectra,* and/or the corresponding spectral data available online (e.g., SciFinder Scholar)).

NOTE. Approximately 500 mg of purified product with a melting point within 2–3 °C of the literature value is suggested for continuing the sequence on to Experiment [D2].

Chemical Tests. Add several crystals (~5 mg) of the *trans*-cinnamic acid to 1 mL of 5% sodium bicarbonate on a watch glass. Does evolution of CO_2 indicate the presence of a carboxylic acid? Does the material give a positive bromine test for unsaturation? (See Chapter 9.)

QUESTIONS

7A-1. Write a complete mechanism for the addition of diethyl malonate to ethanal in the presence of base to form a β-hydroxy ester.

7A-2. Outline a synthesis that forms at least one C—C bond for each of the following compounds:
   (a) $CH_3CH_2CH_2CH_2CH_2CH{=}CHCO_2H$
   (b) $CH_3CH_2CH_2CH{=}C(CN)(CO_2CH_2CH_3)$

7A-3. As mentioned in the discussion, ketones generally give poor yields in the Knoevenagel reaction with diethyl malonate. However, the reaction with ketones gives good yields with ethyl cyanoacetate and malononitrile. Explain.

7A-4. Give the structure of the products of the Knoevenagel reaction for the following pairs of reactants:
   (a) Cyclopentanone + malononitrile (dicyanomethane)
   (b) Benzaldehyde + ethyl acetoacetate
   (c) Propanal + nitromethane

BIBLIOGRAPHY

Knoevenagel, E. *Chem. Ber.* **1898**, *27*, 2345.

Descriptions of the Knoevenagel reaction:

Cope, A. C. *J. Am. Chem. Soc.* **1937**, *59*, 2327.

Cope, A. C.; Hofmann, C. M.; Wyckoff, C.; Hardenberg, E. *J. Am. Chem. Soc.* **1941**, *63*, 3452.

House, H. O. *Modern Synthetic Reactions*; Benjamin: Menlo Park, CA, 1972, pp. 649–653.

Johnson, J. R. *Org. React.* **1942**, *1*, 210.

Jones, G. *Org. React.* **1967**, *15*, 204.

Smith, M. B.; March, J. *Advanced Organic Chemistry*, 7th ed.; Wiley: New York, 2013, Chap. 16, p. 1142 and references therein.

Tietze, L. F.; Beifuss, U. *Comp. Org. Syn.* **1991**, *2*, 341.

Selected references of the use of the Knoevenagel reaction recorded in Organic Syntheses:

Allen, C. F. H.; Spanger, F. W. *Organic Syntheses*; Wiley: New York, 1955; Collect. Vol. III, p. 377.

Cope, A. C.; Hancock, E. M. *Organic Syntheses*; Wiley: New York, 1955; Collect. Vol. III, p. 399.

Horning, E. C.; Koo, J.; Fish, M. S.; Walker, G. N. *Organic Syntheses*; Wiley: New York, 1963; Collect. Vol. IV, p. 408.

McElvain, S. M.; Clemens, D. H. *Organic Syntheses*; Wiley: New York, 1963; Collect. Vol. IV, p. 463.

This step in the sequence was adapted from the work of:

Kolb, K. E.; Field, K. W.; Schatz, P. F. *J. Chem. Educ.* **1990**, *67*, A304.

Experiment D2

Bromination of *trans*-Cinnamic Acid: *erythro*-2,3-Dibromo-3-phenylpropanoic Acid

> Common name: *erythro*-2,3-dibromo-3-phenylpropanoic acid
> CA number: [31357-31-0]
> CA name as indexed: Benzenepropanoic acid, α,β-dibromo-, (*R**,*S**)-

Purpose. The bromination of *trans*-cinnamic acid is carried out to obtain *erythro*-2,3-dibromo-3-phenylpropanoic acid, the direct precursor to 2'-bromostyrene, which is the synthetic target of Sequence D. You will review the stereospecificity of the addition of molecular bromine to an alkene.

Prior Reading

> *Technique 5:* Crystallization
> > Use of the Hirsch Funnel (pp. 92–93)

REACTION

trans-Cinnamic acid   + Br₂ →(acetic acid)→   *erythro*-2,3-Dibromo-3-phenylpropanoic acid

DISCUSSION

The bromination of alkenes is known to be a **stereospecific** reaction. An example and detailed discussion of this addition to *trans*-stilbene is given in Experiment [A2₆], and another example is illustrated in Experiment [F2]. In the present case, bromine undergoes a similar addition, and the halogenated product obtained is the erythro diastereomer of 2,3-dibromo-3-phenylpropanoic acid, which is produced as a racemic mixture of the two enantiomers.

The term *erythro* is derived from carbohydrate chemistry, and is used to describe the *relative* configurations of the two adjacent chiral centers (stereocenters). Specifically, the term erythro describes structures whose Fischer projections place identical substituents (on adjacent carbon atoms) on the *same side* of the (otherwise identical) Fischer projection. The opposite of erythro is *threo*, which would describe structures with identical substituents (on adjacent carbons) on opposite sides of the (otherwise identical) Fischer projection. Because erythro and threo describe only the relative configurations of the two chiral centers with respect to one another, there are two enantiomers of each erythro diastereomer, and of each threo diastereomer, as shown here both in Fischer projections and more familiar illustrations:

Enantiomers of the erythro diastereomer
of 2,3-dibromo-3-phenylpropanoic acid

Enantiomers of the threo diastereomer
of 2,3-dibromo-3-phenylpropanoic acid

The result of the halogenation of *trans*-cinnamic acid, as in the case of (*E*)-stilbene in Experiment [A2$_b$], is an anti addition of molecular bromine. The enantiomeric products are shown here:

trans-Cinnamic acid

erythro-(2*R*,3*S*- and 2*S*,3*R*)-2,3-dibromo-3-phenylpropanoic acid

What mechanism best explains this observed stereochemistry? The starting unsaturated acid and bromine are both *achiral*, and thus the product, if a chiral molecule, must be racemic. The outer-shell electrons of molecular bromine, however, are highly polarizable and molecular bromine reacts as an electrophile with the nucleophilic π electrons of the alkene group of cinnamic acid, in a reaction that is effectively a nucleophilic substitution reaction on a bromine atom. The product is a carbocation, which is, however, stabilized by the ability of the bromine atom to donate a lone pair to the carbocation. This interaction results in the formation of a *chiral, cyclic bromonium ion* intermediate. This intermediate ion is a rigid structure in which the ring section is only open to further attack by a nucleophile (S$_N$2 attack by the Br$^-$ ion generated in the first stage) from the side opposite the bromine atom.

Because both cinnamic acid and molecular bromine are achiral, two enantiomeric bromonium ions are formed at equal rates. Ring opening of each bromonium ion may preferentially occur as shown here at the carbon bearing the phenyl group, since this carbon bears more fractional positive charge than the carbon adjacent to the carboxyl group. Because the two reaction pathways shown here are enantiomeric, they proceed at equal rates to produce racemic *erythro*-2,3-dibromo-3-phenylpropanoic acid:

trans-Cinnamic acid

erythro-2,3-Dibromo-3-phenylpropanoic acid

Recent investigations have shed further light on the nature of the cyclic intermediate. Stable solutions of cyclic bromonium ions in liquid SO_2 have been prepared as SbF_6^- salts. For example,

See the discussion in Experiment [A2_b] for further evidence relating to these intermediate species.

The *erythro*-2,3-dibromo-3-phenylpropanoic acid product prepared in this experiment has a melting point of 203–204 °C. The corresponding threo diastereomer has been synthesized and its racemate melts at 91–93 °C. Thus, the experimental results support the stereospecific nature of the bromine addition that is rationalized by the proposed mechanism.

NOTE. Pyridinium bromide perbromide, a solid reagent, is used as a source of bromine in this experiment. This material is much easier to handle than liquid bromine (also see Experiment [A2_b]).

EXPERIMENTAL PROCEDURE

Estimated time to complete the experiment: 2.0 h.

Physical Properties of Reactants and Product						
Compound	MW	Amount	mmol	mp (°C)	bp (°C)	d
trans-Cinnamic acid	148.2	500 mg	3.37	133		
Pyridinium bromide perbromide (tech)	319.84	1.2 g				
Acetic acid	60.05	12 mL	16.2		116–118	1.05
erythro-2,3-Dibromo-3-phenylpropanoic acid	308.0			203–204		

Reagents and Equipment.   Weigh and place 500 mg (3.37 mmol) of *trans*-cinnamic acid in a 25-mL round-bottom flask containing a magnetic stir bar. Now weigh and place (in the **hood**) 1.2 g of pyridinium bromide perbromide in the same flask.   HOOD

CAUTION: Pyridinium bromide perbromide is a corrosive reagent and a lachrymator. Dispense this material in the **hood**.   HOOD

Using a graduated cylinder, measure 12 mL of glacial acetic acid and add this to the solid reagents. Attach the flask to a water-cooled reflux condenser and place the assembly in a sand bath.

Reaction Conditions.   Heat the mixture with stirring in a sand bath at 120–125 °C for 45 min.

Isolation of Product.   Cool the orange solution to room temperature. With a Pasteur pipet, transfer the solution in small portions to a 50-mL Erlenmeyer flask. Rinse the reaction flask with an additional 2 mL of acetic acid, and transfer this rinse to the same Erlenmeyer flask.

Cool the solution in an ice bath for 5 min, and then slowly add 15 mL of cold water in three 5-mL portions. Swirl the contents of the flask between additions. A pale yellow precipitate should form. Collect the crude product by vacuum filtration using a Hirsch funnel. Wash the filter cake with three 2-mL portions of cold water, during which time the solid product should become white.

Purification and Characterization.   Dry the product in an oven at 110 °C overnight. Weigh the product and calculate the crude yield. Determine the melting point and compare it to that listed in the Reactant and Product table on page 507.

HOOD
A 20- to 30-mg sample of the *erythro*-2,3-dibromo-3-phenylpropanoic acid may be recrystallized (Craig tube) from chloroform (in the **hood**), if desired.

Chemical Test.   Perform the Beilstein test to detect the presence of a halogen (see Chapter 9).

NOTE. *Approximately 300 mg of purified (washed) product, with a melting point within 3–5 °C of the literature value, is suggested for continuing the sequence on to Experiment [D3].*

QUESTIONS

7A-5. The product prepared in this experiment has 2 chiral centers (stereocenters), which give rise to 4 stereoisomers. Draw each of these stereoisomers, and indicate the relationships between each of the stereoisomers.

7A-6. The addition of bromine to ethylene in the presence of high concentrations of chloride ion in an inert solvent results in the formation of 1,2-dibromoethane and 1-bromo-2-chloroethane. No 1,2-dichloroethane is obtained. Account for these results using a suitable mechanistic sequence.

7A-7. 2,3-Dibromobutane contains 2 chiral centers. Therefore, the possibility of 4 stereoisomers exists. However, the addition of bromine to an equimolar mixture of *cis*- and *trans*-2-butene generates only three stereoisomers. Explain.

7A-8. Cyclohexene undergoes bromination in methanol solvent to give *trans*-1-bromo-2-methoxycyclohexane. Draw a suitable mechanism to account for this.

BIBLIOGRAPHY

erythro-2,3-Dibromo-3-phenylpropanoic acid has been previously prepared:

Corvari, L.; McKee, J. R.; Zanger, M. *J. Chem. Educ.* **1991,** *68,* 161.

Murahashi, S.; Naota, T.; Tanigawa, Y. *Organic Syntheses;* Wiley: New York, 1990; Collect. Vol. VII, p. 172.

Sudborough, J. J.; Thompson, K. J. *J. Chem. Soc.* **1902,** *83,* 666.

The bromination of the ethyl ester of *trans*-cinnamic acid has been reported:

Abbott, T. W.; Althousen, D. *Organic Syntheses;* Wiley: New York, 1943; Collect. Vol. II, p. 270.

McElvain, S. M.; Kundiger, D. *Organic Syntheses;* Wiley: New York, 1955; Collect. Vol. III, p. 123.

Additional selected references of bromination of alkenes from *Organic Syntheses*:

Allen, C. F. H.; Edens, C. O., Jr. *Organic Syntheses;* Wiley: New York, 1955; Collect. Vol. III, p. 731.

Cromwell, N. H.; Benson, R. *Organic Syntheses;* Wiley: New York, 1955; Collect. Vol. III, p. 105.

Fieser, L. F. *Organic Syntheses;* Wiley: New York, 1963; Collect. Vol. IV, p. 195.

Also see:

Smith, M. B. *March's Advanced Organic Chemistry,* 8th ed.; Wiley: New York, 2019.

An Elimination Reaction with *erythro*-2,3-Dibromo-3-phenylpropanoic Acid: 2'-Bromostyrene

Common names: (Z)-β-bromostyrene, *cis*-2'-bromostyrene
CA number: [588-73-8]
CA name as indexed: benzene, (2-bromoethenyl)-, (Z)-
Common names: (E)-β-bromostyrene, *trans*-2'-bromostyrene
CA number: [588-72-7]
CA name as indexed: benzene, (2-bromoethenyl)-, (E)-

Purpose. An elimination reaction is carried out using *erythro*-2,3-dibromo-3-phenylpropanoic acid, the direct precursor to 2'-bromostyrene. The influence of solvents and base on the course of the elimination reaction is illustrated. You will consider factors that control the stereospecificity of a reaction. You will explore the option of using NMR spectroscopy to establish the cis/trans isomer ratio in the 2'-bromostyrene product.

Prior Reading

> *Technique 4:* Solvent Extraction
> > Liquid–Liquid Extraction (p. 76)
> *Technique 6B:* Concentration of Solutions (pp. 105–108)

REACTION

erythro-2,3-Dibromo-3-phenylpropanoic acid   →[base / solvent]   *cis*- + *trans*-2'-Bromostyrene $+ CO_2 + Br^- + B:H$

DISCUSSION

The course of the elimination depicted in the above reaction is both base and solvent dependent. In this experiment you will have the opportunity to investigate this solvent dependence. In Part A, you will study the effect of carbonate base in aqueous solvent. Under these conditions, the reaction sequence proceeds via an E1 mechanism in which the first step involves the ionization of one of the C—Br bonds. Clearly, of the two possibilities, halide formation that results in a resonance stabilized benzylic carbocation will be highly favored over development of a positive charge on a carbon alpha to a carboxyl group (see Question 7-W1). Formation of the carbocation intermediate is then rapidly followed by decarboxylation. Although the conjugated section of the positively charged intermediate is planar, free rotation or partially hindered rotation remains possible about the C—C bond, leading to the remaining tetrahedral carbon. Thus, both *cis*- and *trans*-2'-bromostyrene may be obtained during the subsequent decarboxylation. The proposed sequence is shown here:

erythro-2,3-Dibromo-3-phenylpropanoic acid

(benzylic carbocation)

$+ HCO_3^- + Br^- \longrightarrow$

$CH{=}CHBr$

cis- + *trans-*2'-Bromostyrene

$+ CO_2$

In Part B of the experiment, the solvent is changed from water (used in Part A) to a much less polar solvent, acetone. Under these nonaqueous conditions, elimination also occurs, but in this case the experimental data demand a different mechanism, one involving a smooth concerted electron flow without short-lived intermediate formation. Furthermore, when carried out in acetone, the reaction yields exclusively *cis*-2'-bromostyrene. Details of the mechanism follow:

erythro-2,3-Dibromo-3-phenylpropanoic acid anion

$\xrightarrow[\text{acetone}]{\text{K}_2\text{CO}_3}$

cis-2'-Bromostyrene

$+ CO_2 + Br^-$

Under these conditions, therefore, the reaction becomes stereospecific. The mechanism is classified as an E2 elimination. Note that when the two leaving groups are anti to one another, as is thermodynamically preferred, elimination yields the cis alkene.

In the two sets of reaction conditions to be studied in Parts A and B of this experiment, potassium carbonate is used as the base in both cases and it is the solvent that is varied, as mentioned above. In another experimental procedure, sodium azide is used as the base and *N,N*-dimethylformamide (DMF) is used as the solvent, and higher selectivity for the formation of the cis isomer is reported (see **Bibliography**, Corvari et al.).

EXPERIMENTAL PROCEDURE

Estimated time to complete the reaction: 1.5 h, Part A; 2.0 h, Part B.

Physical Properties of Reactants				
Compound	MW	Amount	mmol	mp (°C)
erythro-2,3-Dibromo-3-phenylpropanoic acid	308.0	300 mg	0.97	203–204
Potassium carbonate	38.21	300 mg	2.1	891

PART A Conditions: Water Solvent

Reagents and Equipment. Weigh and place 300 mg (0.97 mmol) of *erythro*-2,3-dibromo-3-phenylpropanoic acid (prepared in Experiment [D2]) and 300 mg (2.1 mmol) of potassium carbonate in a 25-mL Erlenmeyer flask. Add 5 mL of distilled water.

Reaction Conditions. Warm the mixture in a 120 °C sand bath until the solids dissolve (5–10 min) and then heat for an additional 10 min. A yellow oil should separate from the aqueous phase during this period.

Isolation of Product. Cool the two-phase system to room temperature and, using a Pasteur pipet, transfer the contents of the flask to a 12-mL centrifuge tube. Rinse the Erlenmeyer flask with a 2-mL portion of methylene chloride, which is then transferred to the centrifuge tube. Extract the aqueous phase with the CH_2Cl_2 rinse and follow this extraction with two more extractions of the aqueous phase with 2-mL portions of methylene chloride. Transfer the methylene chloride extracts (the organic phase should be the lower phase, but check solubility in water to be sure) to a 25-mL Erlenmeyer flask containing enough anhydrous sodium sulfate to dry the solution. Allow the extracts to dry for about 10 min.

Transfer a portion of the dried solution, using a Pasteur filter pipet, to a *tared* 5-mL conical vial containing a boiling stone. Concentrate this portion on a warm sand bath under a gentle stream of nitrogen gas in the **hood**.

HOOD

Now transfer the remaining dried solution to the same tared vial and concentrate as before. Finally, rinse the sodium sulfate drying agent with an additional 1-mL portion of methylene chloride, transfer it to the vial and concentrate again to yield the *cis/trans*-2'-bromostyrene as a yellow oil.

NOTE. Exercise care during the concentration steps. Product yield will be greatly reduced by overheating or leaving the solution on the sand bath for too long a period.

Purification and Characterization. No further purification of the product is usually necessary. Note the hyacinth-like odor of the liquid. Determine the refractive index and compare it to the literature value (lit. value = 1.6070 at 20 °C, 78% trans isomer). Obtain IR and ^1H NMR spectra of the oil. In the infrared, the cis isomer shows a phenyl C—H bend at 770 cm^{-1}, while the trans isomer shows this bending mode at 731 and the alkene C—H out-of-plane mode is found at 941 cm^{-1} (see **Bibliography**, Strom et al.). Compare your results with the spectrum recorded in the literature (*The Aldrich Library of IR Spectra* and/or SciFinder Scholar).

NMR Analysis. The ^1H NMR spectrum of a mixture of *cis*- and *trans*-2'-bromostyrene is shown in Figure 7A.1. The doublets corresponding to the two olefinic protons of the major isomer can be clearly seen, centered at 6.75 and 7.1 ppm. One of the corresponding doublets from the minor isomer is evident at 6.42 ppm. This doublet has a coupling constant of 8.0 Hz. The two larger doublets from the major isomer have a coupling constant of 14.0 Hz, and thus, in this sample, the major isomer must be *trans*-2'-bromostyrene.

Figure 7A.1 ¹H NMR spectrum: Mixture of *cis*- and *trans*-2′-bromostyrene.

We can readily locate the second doublet from the minor, cis isomer by examining the ¹H–¹H COSY two-dimensional NMR spectrum (see Chapter 8, NMR discussions and Figure 7A.2). The spectrum here is shown at a large vertical scale in order to discern the minor (cis) isomer. By looking below the doublet at 6.42 ppm, a cross-peak can be seen at about 7.1 ppm. Careful inspection reveals that these signals must be slightly upfield from the doublet at 7.1 ppm from the major (trans) isomer. Referring back to the ¹H NMR spectrum (Figure 7A.1), one can see one peak of the hidden doublet just emerging from the upfield end of the doublet from the major isomer. Thus, the second doublet from the minor (cis) isomer must coincidentally be obscured by the NMR signal from the major isomer.

PART B Conditions: Acetone Solvent

Reagents and Equipment. Weigh and place 300 mg (0.97 mmol) of *erythro*-2,3-dibromo-3-phenylpropanoic acid (prepared in Experiment [D2]) and 300 mg (2.1 mmol) of potassium carbonate in a 10-mL round-bottom flask containing a magnetic stir bar. Add 5 mL of acetone that has been dried over sodium sulfate. Attach the flask to a reflux condenser and assemble the apparatus on an 80 °C sand bath.

Figure 7A.2 ¹H–¹H COSY NMR spectrum: Mixture of *cis*- and *trans*-2′-bromostyrene.

Reaction Conditions. Heat (bath temperature ~80 °C) the mixture, with stirring, at reflux for 1 h.

Isolation of Product. Allow the solution to cool to room temperature and then concentrate it to dryness in the **hood** on a warm sand bath under a gentle stream of nitrogen gas. **Do not overheat!**   HOOD

Now add 5 mL of distilled water to dissolve the solid cake. A yellow oil appears at this point, forming a two-phase system. Using a Pasteur pipet, transfer this aqueous–oily mixture to a 12-mL centrifuge tube. Rinse the round-bottom flask with a 2-mL portion of methylene chloride, which is then transferred to the centrifuge tube. Extract the aqueous phase with the CH₂Cl₂ rinse and follow this extraction with two more extractions of the aqueous phase with 2-mL portions of methylene chloride. Transfer the organic extracts to a second 25-mL Erlenmeyer flask using a Pasteur filter pipet and dry the combined extracts over anhydrous sodium sulfate for at least 10 min.

Transfer the dried solution in 2- to 3-mL portions to a *tared* 5-mL conical vial containing a boiling stone. Concentrate the solution in the **hood** on   HOOD
a *warm* sand bath under a gentle stream of nitrogen gas. Rinse the sodium sulfate with an additional 1 mL of methylene chloride and combine this rinse with the dried solution. Concentrate to yield a yellow oil, *cis*-2′-bromostyrene.

NOTE. Exercise care during the concentration steps. Product yield will be greatly reduced by overheating or leaving the solution on the sand bath for too long a period.

Purification and Characterization.   No further purification of the product is usually necessary. Weigh the product and calculate the crude percent yield. Obtain IR and ^{1}H NMR spectra of the oil. In the NMR, the peaks due to both isomers, if present, can be discerned (see Discussion, Part A). Compare your IR and NMR results with the spectra obtained for the cis–trans mixture formed in Part A and to those recorded in the Corvari et al. reference (see **Bibliography**) for pure *cis*-2′-bromostyrene. This reference also gives conditions under which the purity of the material may be more carefully determined by gas chromatography.

Chemical Tests.   Several tests can be run to determine the presence of unsaturation, aromatic character, and the presence of bromine. Select the appropriate tests from Chapter 9. Do your results confirm the presence of these groups?

QUESTIONS

7A-9. 1-Chloro-1-phenylethane ionizes easily under E1 conditions to form a benzylic carbocation intermediate. The ion is stabilized due to the delocalization of the positive charge to the aromatic ring. Draw resonance structures that indicate the stability of this ion.

7A-10. Comment on the fact that *erythro*-2,3-dibromo-3-phenylpropanoic acid undergoes elimination by an E1 pathway in water solvent (Part A conditions), but by an E2 pathway (Part B conditions) when acetone is used as the solvent.

7A-11. Explain in terms of the mechanism why conditions in Part A lead to a mixture of cis–trans isomers, but that those used in Part B give only the cis isomer.

7A-12. A benzylic carbocation is generated under the conditions of Part A. Would the presence of a para CH_3O— group on the benzene ring increase or decrease the stability of the benzylic carbocation? Explain.

7A-13. Consider the stereochemistry of the carbocation intermediate formed under the conditions of Part A. Because the starting material used in this experiment, *erythro*-2,3-dibromo-3-phenylpropanoic acid, is racemic, the enantiomer of the carbocation shown must also be generated. Does the other enantiomer lead to the same diastereomer (cis) of the product or does it lead to the trans diastereomer?

BIBLIOGRAPHY

The synthesis of 2′-bromostyrene has been reported:

Corvari, L.; McKee, J. R.; Zanger, M. *J. Chem. Educ.* **1991**, *68*, 161.
Cristol, S. J.; Norris, W. P. *J. Am. Chem. Soc.* **1953**, *75*, 2645.
Grovenstein, E., Jr.; Lee, D. E. *J. Am. Chem. Soc.* **1953**, *75*, 2639.
L'abbé, G.; Miller, M. J.; Hassner, A. *Chem. Ind. (London)* **1970**, 1321.
Mestdagh, H.; Puechberty, A. *J. Chem. Educ.* **1991**, *68*, 515.
Murahashi, S.; Naota, T.; Tanigawa, Y. *Organic Syntheses*; Wiley: New York, 1990; Collect. Vol. VII, p. 172.

Strom, L. A.; Anderson, J. R.; Gandler, J. R. *J. Chem. Educ.* **1992**, *69*, 588.

For an overview of elimination reactions, see:

Smith, M. B. *March's Advanced Organic Chemistry*, 8th ed.; Wiley: New York, 2019.

SEQUENCE E

The Synthesis of Piperonylonitrile from Piperonyl Alcohol

Purpose.   The aromatic nitrile, piperonylonitrile, is obtained via a multistep synthesis that does not depend on potentially dangerous diazonium intermediates. The selective oxidation of a primary alcohol to an aldehyde is investigated. You will explore, at the same time, the use of resin-bound reagents that simplify the isolation of products. You will carry out a novel elimination reaction using an oxime to obtain the target nitrile.

Synthetic Perfumes

The first step in this sequence of reactions is the oxidation of piperonyl alcohol to obtain piperonal, an aromatic aldehyde. The reaction is selective, and only the aldehyde is obtained; no oxidation to the carboxylic acid is observed:

Piperonyl alcohol   Piperonal

Piperonal has an agreeable odor and is marketed to the perfume industry as *heliotropine*. This commercial term comes from *heliotrope*, which was a name given by early herbalists to any plant that turned to the sun (from the Greek *helios*, the sun, and *trepein*, to turn), such as the sunflower. The name now is more narrowly defined as applying to those plants of the genus *Heliotropium* and specifically to *H. peruvianum*, a common species that is widely cultivated. The fragrance of this particular species also is referred to as *heliotrope*. Since piperonal possesses a very similar fragrance, the origin of the commercial terminology is clear. The commercial source of piperonal, however, is safrole, the formaldehyde ketal (acetal) of 3,4-dihydroxyallylbenzene. Safrole is the chief constituent of the oil of sassafras and is itself obtained commercially from oil of camphor. When safrole is heated in the presence of a strong base, isomerization of the side-chain double bond yields isosafrole. Subsequent oxidation of the isomerized double bond yields piperonal. For more detailed discussions of odor–fragrance molecules and their role in living systems see Sequence D and Experiment [11C].

Safrole   Isosafrole (cis and trans)   Piperonal

The second step of the sequence is the conversion of the intermediate aldehyde to a substituted oxime. By design, the oxime intermediate is constructed with a good leaving group as the oxygen substituent on the nitrogen atom. The second stage, therefore, creates a molecular system prone to undergo the subsequent desired elimination reaction:

Piperonal   O-(2,4-Dinitrophenyl)-hydroxylamine   Piperonal O-(2,4-dinitrophenyl)oxime (an aldoxime)

Treatment of the oxime intermediate with alcoholic KOH causes an elimination reaction that yields the target nitrile product, piperonylonitrile:

Piperonal O-(2,4-dinitrophenyl)oxime   Piperonylonitrile   2,4-Dinitrophenol
(an aldoxime)

The present synthetic sequence offers an alternative approach to the preparation of aromatic nitriles. The classic route to this class of compounds is to use the Sandmeyer reaction. The key steps in this latter pathway involve the diazotization of an aromatic amine followed by reaction of the diazonium salt, which is not isolated (it is explosive!), with CuCN to give the aromatic nitrile directly. Aromatic halides (Br or Cl) may also be prepared in this manner using the corresponding copper(I) halides, as shown here:

Piperonal

Common names: piperonal, 3,4-(methylenedioxy)benzaldehyde
CA number: [120-57-0]
CA name as indexed: 1,3-benzodioxole-5-carboxaldehyde

Purpose. The selective oxidation of a primary (1°) alcohol to an aldehyde is investigated. The use of polymer-bound chromium trioxide as an oxidizing agent is explored. You will prepare the intermediate aldehyde, piperonal, as the first step in the sequence to synthesize piperonylonitrile.

Prior Reading

    Technique 5: Crystallization

               Use of the Hirsch Funnel (pp. 92–93)

    Technique 6: Chromatography

               Column Chromatography (pp. 96–99)

               Concentration of Solutions (pp. 105–108)

REACTION

Piperonyl alcohol   Piperonal

Experiment E1

DISCUSSION

Many of the preferred reagents for the oxidation of primary alcohols to aldehydes (secondary alcohols to ketones) contain the transition metal chromium in its highest oxidation state, VI. Upon reaction with an alcohol, the yellow-orange chromium(VI) species is reduced to the blue-green chromium(III) state. Normally the reaction is carried out in aqueous acid solution using the sodium dichromate salt, $Na_2Cr_2O_7$, or the oxide, CrO_3. A typical reaction is shown here:

$$3\ CH_3CH_2CH_2OH + 4\ H_2SO_4 + Na_2Cr_2O_7 \longrightarrow 3\ CH_3CH_2CHO + Cr_2(SO_4)_3 + Na_2SO_4 + 7\ H_2O$$

In this case, the aldehyde, propanal, can be isolated in moderate yield because it is relatively volatile and can be removed from the reaction mixture by distillation as it is formed. If not removed, the aldehyde, which is in equilibrium with the corresponding hydrate (geminal diol), is oxidized further to the carboxylic acid:

$$CH_3CH_2CH_2OH \xrightarrow[H_2SO_4]{Na_2Cr_2O_7} CH_3CH_2CHO \underset{}{\overset{H_2O}{\rightleftharpoons}} CH_3CH_2CH(OH)_2 \xrightarrow[H_2SO_4]{Na_2Cr_2O_7} CH_3CH_2COOH$$

1-Propanol   Propanal   1,1-Dihydroxypropane   Propanoic acid

Because the hydrate is the species oxidized to the acid, oxidation of the aldehyde can largely be prevented by running the oxidation in a nonaqueous medium.

A number of selective oxidizing agents, such as pyridinium chlorochromate, are frequently used for this purpose. With this reagent, the oxidation stops at the aldehyde stage, because the oxidation, as pointed out earlier, is conducted in a nonaqueous solution. Thus, decanal can be obtained from 1-decanol using this reagent (methylene chloride solvent) in 92% yield. Pyridinium chlorochromate (PCC) is a solid, yellow salt prepared from chromium trioxide as shown here:

$$CrO_3 + HCl + \text{(pyridine)} \longrightarrow \text{(pyridinium)} , CrO_3Cl^-$$

Pyridine   Pyridinium chlorochromate

An alternate reagent is the chromium trioxide(pyridine)$_2$ complex, $CrO_3 \cdot (py)_2$ (where py = pyridine), often referred to as the Collins reagent:

$$RCH_2CH_2\ddot{O}H + CrO_3(py)_2 \xrightarrow[\underset{Na^+,\ ^-:\ddot{O}CCH_3}{\ddot{O}}]{HCl,\ CH_2Cl_2} RCH_2\overset{\ddot{O}}{\overset{\|}{C}}H$$

(80–90% average yield)

The chief difficulty with this reagent is that the complex is highly hygroscopic. However, it can be prepared in situ, thus avoiding this major drawback. Pyridinium dichromate and chromium trioxide · 3,5-dimethylpyrazole are also effective as selective oxidizing agents for these reactions.

In the present experiment, the selective oxidation of a 1° alcohol to an aldehyde is carried out using a chromium trioxide polymer-bound oxidizing agent. This reagent is also used in Experiment [33A] to oxidize a 2° alcohol to a ketone. As noted there, the reagent is easy to prepare and has the advantage that the toxic reduced chromium species is bound to a polymer resin and can thus be easily separated from the reaction mixture by simple filtration.

As shown here, the oxidation proceeds through the formation of a chromate ester. Note that the oxidation state of chromium, Cr(VI), at this stage remains the same as in the starting reagent. The second stage is equivalent to an E2 elimination reaction, with the water molecule acting as a base. The donation of an electron pair to the metal atom changes its oxidation state to Cr(IV):

G

(chromate ester)

(ketone or aldehyde)

NOTE. If you plan to carry out the total sequence of reactions, the scaled-up procedure should be followed. If an individual oxidation reaction is to be studied, the microscale procedure may be used.

EXPERIMENTAL PROCEDURE

Semimicroscale Preparation. (This procedure is scaled up to 10 times the amounts of the microscale preparation.)

Estimated time to complete the experiment: 3.5 h.

Physical Properties of Reactants						
Compound	MW	Amount	mmol	mp (°C)	bp (°C)	d
Piperonyl alcohol	152.16	1.0 g	6.57	58		
Chromic oxide resin		7.5 g				
Dioxane	88.12	10 mL			101	1.034

Reagents and Equipment. Weigh and place 1.0 g (6.57 mmol) of piperonyl alcohol and 7.5 g of freshly prepared chromic oxide resin in a 25-mL round-bottom flask containing a magnetic stirring bar. Using a graduated cylinder in the **hood**, measure and add to the flask 10 mL of dioxane. Attach the flask to a reflux condenser (◂).

CAUTION: Dioxane is a suspected carcinogen. It should be dispensed and handled in the **hood**. That is, the entire experiment should be set up and run in the **hood**, at least until the solvent is removed.

INSTRUCTOR PREPARATION. The CrO₃ resin is prepared by adding 35 g of Amberlyst A-26 resin to a solution of 15 g of CrO₃ in 100 mL of water. Stir the mixture

125–130°C
thermometer

HOOD

Speed   Heat

HOOD

25–mL RB flask
piperonyl alcohol, 1.0 g
+ chromic oxide oxidizing resin, 7.5 g
+ dioxane, 10 mL

for 30 min at room temperature. Collect the resin by filtration under reduced pressure and successively rinse the material with water and acetone. Partially dry the resin on a Büchner funnel by drawing air through it for 1 h, then air-dry the material overnight.

Reaction Conditions.   With stirring, heat the reaction mixture at reflux in the **hood** for a period of 1.0 h using a sand bath temperature of 125–130 °C.   HOOD

Isolation of Product.   Cool the reaction product to room temperature. Transfer the solution by Pasteur filter pipet to a funnel fitted with fast-grade filter paper. Collect the filtrate in a 25-mL Erlenmeyer flask containing approximately 1 g of anhydrous sodium sulfate. Rinse the reaction flask and resin (include the walls of the condenser) with three 1.0-mL portions of methylene chloride solvent (calibrated Pasteur pipet). Transfer each rinse by Pasteur filter pipet to the funnel. Combine each rinse filtrate with the original filtrate.

Now transfer the solution in three portions by Pasteur filter pipet to a 10-mL round-bottom flask and concentrate the solution following each addition using a rotary evaporator. If a rotary evaporator is not available, transfer the solution to a 25-mL Erlenmeyer flask containing a boiling stone and remove the solvent in the **hood** by warming the flask on a sand bath at 125 °C. A gentle stream of nitrogen aids this concentration process.   HOOD

The crude piperonal is obtained as an oil and is purified by column chromatography.

Purification and Characterization.   Preweigh (and number 1–10) ten 25-mL Erlenmeyer flasks containing a boiling stone. Wet-pack a 25-mL buret using hexane and silica gel in the following manner. Place a cotton plug in the buret followed by a 1-in. layer of sand. Add 15 mL of hexane to the column, and then slowly add 8 g of silica gel while tapping the sides of the buret to release any air bubbles. Finally, add an additional 1 in. of sand to the top of the column.

Allow the hexane to drain from the buret until it reaches the top of the sand level in the column. Collect the eluted hexane in a waste container.

Dilute the crude piperonal oil with five drops of methylene chloride and then, using a Pasteur filter pipet, add this solution to the top of the column. Again, drain a portion of the elution solvent until the solvent head is level with the sand at the top of the column. Rinse the sides of the column with 0.5 mL of 4:1 methylene chloride/hexane solution, and again drain the column to the same sand level.

Now add, in order, the following amounts of elution solvents: three 10-mL portions of 4:1 methylene chloride/hexane; three 10-mL portions of methylene chloride; and three 10-mL portions of 9:1 methylene chloride/ethyl acetate.

Collect 10-mL fractions in each of the preweighed Erlenmeyer flasks. After collecting fraction No. 9, use flask No. 10 to collect a final fraction as the column completely drains. (Fraction No. 1 can be added directly to the waste container.)

Concentrate each of the collected fractions to dryness on a warm sand bath under a gentle stream of nitrogen gas in the **hood**. Weigh the flasks to   HOOD
determine the total amount of piperonal product. Now preweigh an additional 25-mL Erlenmeyer flask containing a boiling stone. To each of the flasks containing white or light yellow fanlike crystals, add sufficient methylene chloride to just dissolve them and then transfer the resulting solutions to the tared Erlenmeyer flask using a Pasteur filter pipet. Concentrate this final solution as before, and weigh the flask to obtain the total amount of piperonal isolated.

NOTE. If a rotary evaporator is available, the eluate can be concentrated in a 10-mL round-bottom flask. Transfer of each fraction to a single evaporation flask shortens the above procedure.

Determine the melting point of your product and compare your value to that found in the literature. Obtain IR and/or ¹H NMR spectra and compare your results to those reported in the literature (*The Aldrich Library of IR Spectra, The Aldrich Library of NMR Spectra,* and/or the corresponding spectral data available online (e.g., SciFinder Scholar)).

Chemical Tests. Chemical classification data may also be useful. The ignition test, the Tollens test, and the 2,4-dinitrophenylhydrazine test should all give a positive result for the piperonal compound.

NOTE. Approximately 300 mg of purified product with a melting point within 2–4 °C of the literature value is suggested for continuing the sequence on to Experiment [E2].

OPTIONAL MICROSCALE PREPARATION

The microscale preparation is similar to that of the scaled-up synthesis outlined above, with the following exceptions.

HOOD

*NOTE. Even in the case of the microscale preparation this reaction should be **entirely** carried out in the **hood** at all times until the dioxane solvent has been removed.*

1. The reagent and solvent amounts are one-tenth of those used above.

Physical Properties of Reactants					
Compound	MW	Amount	mmol	mp (°C)	bp (°C)
Piperonyl alcohol	152.16	100 mg	0.66	58	
Chromic oxide resin		750 mg			
Dioxane	88.12	1.0 mL			101

2. A 5-mL round-bottom flask is used.

3. The reaction mixture is heated for 1 h.

4. After the reaction product is cooled to room temperature, transfer the solution by Pasteur filter pipet to a funnel containing a loose cotton plug covered with 500 mg of anhydrous sodium sulfate. Collect the filtrate in a 10-mL Erlenmeyer flask. Rinse the reaction flask and resin (include the walls of the condenser) with three 0.5-mL portions of methylene chloride solvent (calibrated Pasteur pipet). Transfer each rinse by Pasteur filter pipet to the funnel and, finally, rinse the sodium sulfate with an additional 0.1 mL of methylene chloride. Combine each rinse filtrate with the original filtrate. Concentrate the solution as described.

5. Purify the crude oil using chromatography as follows:

(a) Pack a short buret column with approximately 1.75 g of silica gel, and transfer the crude product by Pasteur filter pipet to the column. Rinse the flask with 1.0 mL of methylene chloride (calibrated Pasteur pipet) and also transfer the rinse to the column.

(b) Now add methylene chloride to the column, 2.0 mL at a time (calibrated Pasteur pipet), and collect the eluate in three tared 10-mL Erlenmeyer flasks. Set aside the first 4.0 mL of eluate collected in the first flask. Collect the next 6.0 mL in one flask; this fraction contains the bulk of the product. Also collect one additional fraction of 3.0 mL in the third flask.

(c) Concentrate the second fraction (6.0 mL) in the **hood** using a warm sand bath with a slow stream of nitrogen to assist solvent removal. *Do not forget to add a boiling stone to the flask.*

HOOD

(d) Weigh the flask and contents. Reheat for 1 min, cool, and weigh again. If the two weights are within 2.0 mg, the product is quite pure. If not, repeat the evaporation process until a constant weight is obtained.

(e) Cool the product. If it does not solidify, place it in an ice bath and scratch the sides and bottom with a glass rod to induce crystallization.

A small amount of additional product can be isolated from the third fraction in like manner.

QUESTIONS

7A-14. Primary alcohols can be oxidized to aldehydes and carboxylic acids. Often it is difficult to stop at the aldehyde oxidation state. One method frequently used to accomplish this, for those that boil below about 100 °C is to remove the product from the reaction mixture as it is formed. This method is based on the fact that aldehydes have a lower boiling point than their corresponding alcohols. Explain this difference in boiling point.

7A-15. A specific oxidizing agent for the conversion of primary alcohols to aldehydes is pyridinium chlorochromate, abbreviated as $^+py \cdot CrO_3Cl^-$. Generally, the oxidation is run in methylene chloride solvent. For example,

Oxidation of this alcohol with the conventional $Na_2Cr_2O_7$–H_2SO_4–water system produces the carboxylic acid. Offer an explanation for the difference in these results.

7A-16. In reference to Question 7A-15, can you see another advantage of the pyridine chlorochromate reagent over that of the conventional conditions?

7A-17. An excellent way to identify the presence of an aldehyde group is by ^1H NMR. The chemical shift of the aldehydic proton occurs in the 9- to 10-ppm region, where few other proton signals occur. Why is this chemical shift so far downfield?

7A-18. A series of compounds with increasing boiling point is listed below. Offer an explanation for this trend.
   (a) Butane (8 °C)
   (b) Propanal (56 °C)
   (c) 1-Propanol (97 °C)

BIBLIOGRAPHY

The present procedure is based on the work reported by:

Cainelli, G.; Cardillo, G.; Orena, M.; Sandri, S. *J. Am. Chem. Soc.* **1978**, *98*, 6737.

Several preparations involving the oxidation of a 1° alcohol to an aldehyde are given in *Organic Syntheses*:

Collins, J. C.; Hess, W. W. *Organic Syntheses*; Wiley: New York, 1988; Collect. Vol. VI, p. 644.

Hurd, C. D.; Meinert, R. N. *Organic Syntheses*; Wiley: New York, 1943; Collect. Vol. II, p. 541.

Ratcliffe, R. W. *Organic Syntheses*; Wiley: New York, 1988; Collect. Vol. VI, p. 373.

Also see:

Smith, M. B. *March's Advanced Organic Chemistry*, 8th ed.; Wiley: New York, 2019.

Piperonal *O*-(2,4-Dinitrophenyl)oxime

Common name: piperonal *O*-(2,4-dinitrophenyl)oxime
CA number: [17188-61-3]
CA name as indexed: 1,3-benzodioxole-5-carboxaldehyde, *O*-(2,4-dinitro-phenyl) oxime

Purpose. You will carry out the second step in the sequence of synthetic reactions leading to an aromatic nitrile, piperonylonitrile. You will prepare a specific oxime derivative of the aldehyde obtained in Experiment [E1]. This oxime derivative is derivatized on oxygen so as to allow the oxygen to function as a good leaving group, which will allow an elimination reaction to form a nitrile in the final step of the synthetic sequence leading to piperonylonitrile.

Prior Reading

Technique 5: Crystallization
Use of the Hirsch Funnel (pp. 92–93)

REACTION

| Piperonal | *O*-(2,4-Dinitrophenyl)-hydroxylamine | Piperonal *O*-(2,4-dinitrophenyl)oxime |

DISCUSSION

The preparation of this oxime is the second step in this sequence to obtain the aromatic nitrile target molecule. In Experiment [E2] you are going to convert the aromatic aldehyde formed in Experiment [E1] into an *O-phenylated oxime*, which on treatment with base in Experiment [E3] yields the desired nitrile via an elimination reaction. Oxime formation is also involved in the well-known Beckmann rearrangement (see Experiment [6$_{adv}$]), which is used for the synthesis of amides.

The generation of the oxime intermediate involves a nucleophilic addition of the amine group of the hydroxylamine to the carbonyl carbon, followed by a dehydration to form the carbon–nitrogen double bond (and the oxime group). A general mechanism for the reaction is given here.

The first step is a nucleophilic addition to the carbonyl group. Rapid proton transfer gives an intermediate that generally is not isolated:

The next stage is an acid-catalyzed dehydration reaction in which water is eliminated to produce the oxime. This stage has been shown to be the rate-determining step in oxime formation.

$$R-\underset{\underset{H}{|}}{\overset{\overset{\ddot{O}:}{|}}{C}}-\underset{\underset{H}{|}}{\overset{H}{N}}-\ddot{O}R' + H^+ \rightleftharpoons R-\underset{\underset{H}{|}}{\overset{\overset{\overset{H}{|}}{\overset{+}{\ddot{O}H}}}{C}}-\underset{\underset{H}{|}}{N}-\ddot{O}R' \rightleftharpoons R-\underset{\underset{H}{|}}{C}=\overset{+}{N}-\ddot{O}R' + H_2O \rightleftharpoons R-\underset{\underset{H}{|}}{C}=\ddot{N}-\ddot{O}R' + H^+$$

Oxime
(an aldoxime)

NOTE. *If you are conducting the total sequence of reactions to obtain piperonyloni-trile, follow the scaled-up procedure. If the individual reaction is to be studied, follow the microscale procedure.*

EXPERIMENTAL PROCEDURE

Semimicroscale Preparation.   (This procedure is scaled up to 10 times the amounts of the microscale preparation.)

Estimated time to complete the experiment: 1 h.

Room Temp

Speed   Heat

100–mL RB three-necked flask containing
O–(2,4–dinitrophenyl)hydroxylamine, 400 mg +
absolute ethanol, 30 mL + piperonal, 300 mg +
12 *M* HCl, 2 mL

Physical Properties of Reactants and Product						
Compound	MW	Amount	mmol	mp (°C)	bp (°C)	*d*
Piperonal	150.14	300 mg	2.0	37	263	
O-(2,4-Dinitrophenyl) hydroxylamine	199.12	400 mg	2.0	111–112		
Ethanol	46.07	30 mL			78	0.789
HCl (12 M)		2 mL				
Piperonal *O*-(2,4-dinitrophenyl) oxime	330.24			194–195		

Reagents and Equipment.   In a 100-mL three-necked round-bottom flask containing a magnetic stir bar and equipped with a reflux condenser protected by a calcium chloride drying tube and two caps or glass stoppers, weigh and place 400 mg (2.0 mmol) of *O*-(2,4-dinitrophenyl)hydroxylamine, followed by 30 mL of absolute ethanol (➡).

NOTE. *The three-necked flask may be replaced with a conventional round-bottom 100-mL flask. In this case, addition of reagents can be carried out in the first instance by removing the condenser and in the second case by addition directly down the condenser with a 9-in. Pasteur pipet.*

NOTE. *The preparation of O-(2,4-dinitrophenyl)hydroxylamine is described in the Instructor's Manual, which is available from the publisher.*

Attach the flask to the condenser and warm (60–65 °C) the contents, with stirring, in a sand bath until a homogeneous solution is obtained. Now add 300 mg (0.2 mmol) of piperonal via one of the unused flask necks. After the aldehyde has dissolved, add slowly, with stirring, 2 mL of 12 M HCl through an unused flask neck using a Pasteur pipet. Stir the reaction mixture for an additional 1 min after the addition is complete.

Reaction Conditions.
The oxime forms immediately. Complete the precipitation of the product by cooling the reaction flask in an ice bath for 10 min.

Isolation of Product.
Collect the solid product by filtration under reduced pressure using a Hirsch funnel. Rinse the flask with two 3-mL portions of cold absolute ethanol, using the rinse to wash the crystals on the filter funnel. Now rinse the crystals with three additional 3-mL portions of cold absolute ethanol (◄).

NOTE. *Approximately 50–70 mg of purified product with a melting point within 2–4 °C of the literature value is the minimum quantity and quality suggested for continuing the sequence in Experiment [E3].*

Air-dry the product on a clay plate or on filter paper. Collect and refrigerate the filtrate for at least 24 h. This procedure generally produces another crop of oxime crystals. This second crop, collected by the same technique, may be combined with the initial product, if its melting point is above 180 °C.

Purification and Characterization.
Determine the melting point of the oxime. It is of sufficient purity to proceed with the next step of the sequence if its mp is 187 °C or greater. If necessary, recrystallize the material from acetic acid or hot acetone.

Obtain an IR spectrum of the oxime and compare it with that of the reference standard shown in Figure 7A.3.

Oxime product

125-mL Filter flask

35 mL ethanol +
20 mL aqueous HCl

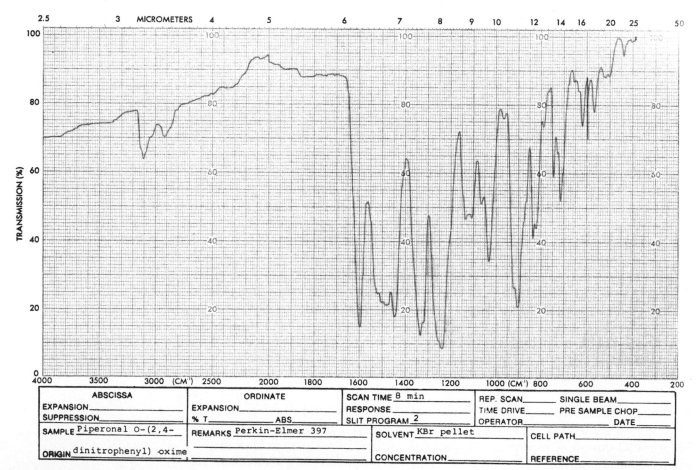

Figure 7A.3   IR spectrum: piperonal *O*-(2,4-dinitrophenyl)oxime.

OPTIONAL MICROSCALE PREPARATION

The microscale preparation is similar to that of the scaled-up synthesis outlined above, with the following exceptions:

1. The reagent and solvent amounts are one-tenth of those used above.

Physical Properties of Reactants and Product						
Compound	MW	Amount	mmol	mp (°C)	bp (°C)	d
Piperonal	150.14	30 mg	0.2	37	263	
O-(2,4-Dinitrophenyl) hydroxylamine	199.12	40 mg	0.2	111–112		
Ethanol	46.07	3.0 mL			78	0.789
HCl (12 M)		2 drops				
Piperonal O-(2,4-dinitrophenyl)oxime	330.24			194–195		

2. A 5-mL conical vial is used.

3. The hydrochloric acid is added dropwise through the top of the condenser using a 9-in. Pasteur pipet, after removing the drying tube. As the acid is delivered, the tip of the pipet should be held just above the surface of the solution. The drying tube is then reattached, and the contents are mixed by swirling the flask.

QUESTIONS

7A-19. Referring to the first step in the mechanism of oxime formation outlined in the discussion section, offer an explanation of why a high acid concentration would hinder the formation of the intermediate generated in this first stage.

7A-20. Oximes prepared from aldehydes or ketones by reaction with hydroxylamine can be reduced to primary amines in high yields. One can use various reagents for the reduction step, including Ni–H_2 in methanol or $LiAlH_4$ in ether. Using a reduction reaction, carry out the following transformations:
(a) 3-Pentanone to 3-aminopentane
(b) Propanal to 1-aminopropane
(c) Benzaldehyde to benzylamine

7A-21. Oximes prepared from unsymmetrical ketones are likely to exist as mixtures of geometrical isomers. For example,

(E)- and (Z)-isomeric oximes

Which isomer is designated as *E*? How do you account for both of these isomers being formed?

BIBLIOGRAPHY

For the preparation of oximes using hydroxylamine hydrochloride as the reagent, see:

Bousquet, E. W. *Organic Syntheses;* Wiley: New York, 1943, Collect. Vol. II, p. 313.

Fuson, R. C.; Curtin, D.Y.; Morrill, T. C.; Hermann, C. K. F.; Shriner, R. L. *The Systematic Identification of Organic Compounds,* 7th ed.; Wiley: New York, 1998.

Hach, C. C.; Banks, C.V.; Diehl, H. *Organic Syntheses;* Wiley: New York, 1963, Collect. Vol. IV, p. 229.

Lachman, A. *Organic Syntheses;* Wiley: New York, 1943, Collect. Vol. II, p. 70.

Pasto, D. J.; Johnson, C. R.; Miller, M. J. *Experiments and Techniques in Organic Chemistry;* Prentice Hall: Englewood Cliffs, NJ, 1992, p. 332.

Semon, W. W.; Damerell, V. R. *Organic Syntheses;* Wiley: New York, 1943, Collect. Vol. II, p. 204.

Smith, M. B. *March's Advanced Organic Chemistry,* 8th ed.; Wiley: New York, 2019.

Piperonylonitrile

Common names: piperonylonitrile, 3,4-methylenedioxybenzonitrile
CA number: [4421-09-4]
CA name as indexed: 1,3-benzodioxole-5-carbonitrile

Purpose. The piperonal O-(2,4-dinitrophenyl)oxime intermediate, prepared in the previous experiment (Experiment [E2]), is converted into the target molecule, piperonylonitrile. This completes the set of Sequence E reactions for the conversion of a substituted benzyl alcohol into an aromatic nitrile. You will investigate the use of a novel elimination reaction to convert an oxime derivative to a nitrile.

Prior Reading

    Technique 4: Solvent Extraction

              Liquid–Liquid Extraction (p. 76)

    Technique 6: Chromatography

              Column Chromatography (pp. 96–99)

              Concentration of Solutions (pp. 105–108)

REACTION

Piperonal O-(2,4-dinitrophenyl)oxime   Piperonylonitrile   2,4-Dinitrophenol

DISCUSSION

The Sequence E reactions illustrate the oxidation of a benzylic alcohol to its corresponding aldehyde and the subsequent two-step conversion of this aldehyde via an intermediate aldoxime to an aromatic nitrile. This synthetic route is an attractive alternative to the preparation of aromatic nitriles via the Sandmeyer reaction.

Aromatic nitriles are usually prepared by the diazotization of the corresponding aromatic amine followed by treatment of the diazonium salt with copper(I) cyanide. This sequence is an example of the Sandmeyer reaction, a specific example of which is shown here:

(64–70%)

This particular variation of the Sandmeyer reaction is useful, because it allows the conversion of an aromatic amine, readily available by reduction of the corresponding nitro compound, to a reactive carbon substituent. This substitution involves replacement of the C—N bond with a C—C bond.

In the present reaction, treatment of an *O*-phenylated oxime with base yields the nitrile by an elimination reaction. The role of the dinitrophenyl group is to enhance the oxime oxygen to function as a good leaving group. Thus, the phenoxy system departs as the conjugate base of the relatively acidic 2,4-dinitrophenol. The mechanistic sequence is outlined below:

$$\text{:ÖH} + R\text{—}C\overset{\overset{\displaystyle \ddot{\text{Ö}}R'}{|}}{=}N \longrightarrow H\ddot{\text{Ö}}H + R\text{—}C\equiv N\text{:} + \text{:}\ddot{\text{Ö}}R' \rightleftharpoons R\text{—}C\equiv N\text{:} + H\ddot{\text{Ö}}R' + \text{:}\ddot{\text{Ö}}H$$

Oximes derived from aldehydes (aldoximes) can be dehydrated to nitriles by many different dehydrating reagents; acetic anhydride is one of the most common reagents used:

$$\underset{H}{\overset{R}{\diagdown}}C=N\overset{\ddot{\text{Ö}}H}{\diagup} \xrightarrow{\text{Ac}_2\text{O}} R\text{—}C\equiv N\text{:}$$

The reaction proceeds more rapidly when the hydrogen and the hydroxyl group are trans to one another (see also Experiments [10], [D3], and [A3$_b$]). Various derivatives of the hydroxylamine other than the ethers, RCH=NOR, illustrated in the present reaction, also undergo the conversion to nitriles. Among these are the RCH=NOCOR and RCH=NOSO$_2$Ar compounds. In some cases it is also possible to convert aldehydes to nitriles in one step by refluxing the reagents in concentrated hydrochloric acid (or by reaction with sodium formate in formic acid or sodium acetate in acetic acid) as follows:

$$R\text{—}\overset{\overset{\displaystyle H}{|}}{C}=\ddot{\text{O}}\text{:} + H_2\ddot{N}\ddot{\text{O}}H\cdot HCl \xrightarrow[\Delta]{\text{concd HCl}} R\text{—}C\equiv N\text{:}$$

When oximes are treated with strong acid they are converted to amides by a rearrangement sequence known as the Beckmann rearrangement. This reaction is illustrated in Experiment [6$_{adv}$].

Nitriles are synthetically versatile functional groups, because they are readily converted to carboxylic acids by hydrolysis under acidic or basic conditions, reduced with LiAlH$_4$ to form primary amines, and reaction with Grignard reagents leads to the formation of ketones. These reactions are illustrated here:

$$R\text{—}C\equiv N\text{:} \xrightarrow[\text{H}_2\text{O}]{\text{H}^+ \text{ or NaOH}} R\text{—}\overset{\overset{\displaystyle \ddot{\text{O}}}{\|}}{C}\text{—}\ddot{\text{O}}H$$

$$R\text{—}C\equiv N\text{:} \xrightarrow[\text{2. H}_2\text{O}]{\text{1. LiAlH}_4\text{, ether}} R\text{—}CH_2\ddot{N}H_2$$

$$R\text{—}C\equiv N\text{:} + R\text{—}MgX \xrightarrow[\text{2. H}_3\text{O}^+]{\text{1. ether}} R\text{—}\overset{\overset{\displaystyle \ddot{\text{O}}}{\|}}{C}\text{—}R$$

EXPERIMENTAL PROCEDURE

Estimated time to complete the experiment: 2.5 h.

Physical Properties of Reactants and Product						
Compound	MW	Amount	mmol	mp (°C)	bp (°C)	d
Piperonal O-(2,4-dinitrophenyl)oxime	330.24	50 mg	0.15	194–195		
Ethanol		5 mL			78.5	0.789
Potassium hydroxide (0.2 M)		2 mL				
Piperonylonitrile	147.13			92–93		

90–100 °C
thermometer

Speed   Heat

Piperonal O–(2,4–dinitrophenyl)oxime, 50 mg +
95% ethanol, 5.0 mL +
0.2 N ethanolic KOH, 2.0 mL
10 mL RB flask

Reagents and Equipment.   Weigh and place 50 mg (0.15 mmol) of piperonal O-(2,4-dinitrophenyl)oxime in a 10-mL round-bottom flask containing a magnetic stir bar. Now add 5.0 mL of 95% ethanol and 2.0 mL of 0.2 M ethanolic KOH. Attach the flask to a reflux condenser (◄).

NOTE. The oxime is prepared in Experiment [E2]. The 0.2 M ethanolic KOH is prepared using 95% ethanol.

Reaction Conditions.   Slowly heat the reaction mixture, while stirring, to reflux by use of a sand bath (100–110 °C) and maintain the mixture at this temperature (gentle reflux) for 1 h. During the initial warming period, the solution turns a deep yellow.

NOTE. If the laboratory time is not sufficient to continue the isolation and purification of the product after the heating is terminated, cool the solution and remove the reaction vial. Cap the vial and store it until the next laboratory period.

Isolation of Product.   Remove the reaction flask and concentrate the reaction mixture to a volume of 0.5 mL or less with a gentle stream of nitrogen gas

HOOD   and/or warming in a sand bath in the **hood**. *This concentration process takes a considerable length of time.*

Now prepare an alkaline solution by diluting 1 mL of 5% aqueous NaOH with 5 mL of distilled water. Use this solution to transfer the concentrated reaction mixture to a 12-mL centrifuge tube in the following manner.

Add a 2-mL portion of the alkaline solution to the reaction flask, mix by swirling, and then transfer the resulting suspension to the centrifuge tube using a Pasteur pipet. Repeat this operation twice, using 2 mL of the alkaline solution each time.

Extract the resulting suspension with four 2-mL portions of methylene chloride (calibrated Pasteur pipet). Remove the methylene chloride extract (bottom layer) using a Pasteur filter pipet, and place the combined fractions in a 25-mL Erlenmeyer flask. Dry the solution over granular anhydrous sodium sulfate (0.5 g).

By use of a Pasteur filter pipet, transfer the dried solution to a second 25-mL Erlenmeyer flask containing a boiling stone. Rinse the drying agent with two 1-mL portions of methylene chloride. Combine the rinses with the origi-

HOOD   nal solution. Remove the solvent in the **hood** under a gentle stream of nitrogen and/or by warming in a sand bath to obtain the crude piperonylonitrile.

Purification and Characterization.   Purify the crude product by column chromatography using a Pasteur filter pipet filled with 300 mg of alumina (neutral, activity 1, see Glossary). Wet the column with 1.0 mL of 1:1 methylene chloride/hexane solution.

Alumina, 300 mg

Nitrile product in ~3 mL of CH$_2$Cl$_2$/hexane solvent

HOOD

Dissolve the residue of crude nitrile in a minimum amount of 1:1 methylene chloride/hexane solvent, and transfer the resulting solution by Pasteur pipet to the column. Elute the nitrile from the column with 2.0 mL of the 1:1 CH$_2$Cl$_2$/hexane solvent and collect the eluate in a 10-mL Erlenmeyer flask containing a boiling stone (➡).

Evaporate the solvent under a gentle stream of nitrogen while warming in a sand bath in the **hood**. Dry the white needles of piperonylonitrile on a porous clay plate or on filter paper.

Weigh the product and calculate the percent yield. Determine the melting point and compare it with the literature value.

Obtain an IR spectrum and ^{1}H NMR spectra of the sample and compare your results with those in the literature (*The Aldrich Library of IR Spectra, The Aldrich Library of NMR Spectra,* and/or the corresponding spectral data available online (e.g., SciFinder Scholar)).

QUESTIONS

7A-22. The Sandmeyer reaction is based on the replacement of the diazonium group in aryldiazonium salts by chloro, bromo, or cyano groups. Copper salt reagents are used:

CH$_3$ $-N_2^+$, Cl$^-$ $\xrightarrow[\Delta]{\text{CuCl}}$ CH$_3$ Cl $+ N_2$

Show how one could carry out the following transformations using the Sandmeyer reaction:

NO$_2$ / CH$_3$ → CN / CH$_3$

NO$_2$ / Cl → Br / Cl

7A-23. When CuCN is used in the Sandmeyer reaction, the preparation is generally carried out in a neutral medium. Can you offer an explanation for why this is done?

7A-24. Outline a synthetic route for the preparation of nitriles using a carboxylic acid as the starting material.

7A-25. A yellow azo dye once used to color margarine has been outlawed because it is carcinogenic. Outline a synthesis of this dye, butter yellow, starting from benzene and *N,N*-dimethylaniline:

Butter Yellow

BIBLIOGRAPHY

The procedure outlined above for the preparation of piperonylonitrile is based on the work of:

Miller, M. J.; Loudon, G. M. *J. Org. Chem.* **1975**, *40*, 126.

Several examples of the conversion of an aromatic aldehyde to the corresponding nitrile are given in *Organic Syntheses*:

Back, J. S.; Ide, W. S. *Organic Syntheses*; Wiley: New York, 1943; Collect. Vol. II, p. 622.

Clarke, H. T.; Nagy, S. M. *Organic Syntheses;* Wiley: New York, 1955; Collect. Vol. III, p. 690.

For a one-step conversion of aromatic aldehydes to the corresponding nitrile, see:

DeMott, J. M. Jr.; Kelley, C. J. *J. Chem. Educ.* **2001,** *78,* 780. This procedure works for the conversion of the present experiment as verified by DeMott, J. M., Jr. (Massachusetts College of Pharmacy and Allied Health, Boston), private communication.

There are numerous references to the use of the Sandmeyer reaction in Organic Syntheses. Several are cited here:

Bigelow, L. A. *Syntheses;* Wiley: New York, 1941; Collect. Vol. I, pp. 135, 136.

Cleland, G. H. *Organic Syntheses;* Wiley: New York, 1988; Collect. Vol. VI, p. 21.

Grundstone, F. D.; Tucker, S. H. *Organic Syntheses;* Wiley: New York, 1963; Collect. Vol. IV, p. 160.

Hartwell, J. L. *Organic Syntheses;* Wiley: New York, 1955; Collect. Vol. III, p. 185.

Marvel, C. S. *Organic Syntheses;* Wiley: New York, 1941; Collect. Vol. I, p. 170.

Rutherford, K. G.; Redmond, W. *Organic Syntheses;* Wiley: New York, 1963; Collect. Vol. IV, p. 133.

Smith, M. B. *March's Advanced Organic Chemistry,* 8th ed.; Wiley: New York, 2019.

SEQUENCE F

Introduction to Photochromism: The Synthesis of a Photochromic Imine

The photochromic effect is a property of relative rarity in both organic and inorganic molecular structures. When it is present, a material is found to exhibit reversible color change upon exposure to radiation:

$$\text{Photochromic Substance A} \underset{\Delta \text{ or } h\nu_B}{\overset{h\nu_A}{\rightleftharpoons}} \text{Photochromic Substance B}$$

The photoproduct generally reverts to the initial system via thermal pathways, but there are examples where the reverse reaction is induced by radiation of a different wavelength from that driving the forward reaction, or by both thermal and photochemical processes. Generally, sensitivity to thermal effects controls the concentrations obtained from the forward reactions and, therefore, their effectiveness in producing the product.

CLASSES OF PHOTOCHROMIC REACTIONS

Cis–Trans Isomerizations. Experiment [6] studies the photochromic properties of *trans*-dibenzoylethylene. In this case, the highly conjugated bright-yellow trans diastereomer is rapidly isomerized under intense sunlamp visible radiation, via excited electronic states, to the colorless cis alkene. A π electron is promoted to an anti-bonding π* molecular orbital, which destroys the π bond, and thus permits facile rotation about the remaining σ bond and formation of the cis alkene:

trans-1,2-Dibenzoylethylene
(bright yellow)

cis-1,2-Dibenzoylethylene
(colorless)

This is an endothermic reaction that does not reach equilibrium for but goes to completion because the cis isomer's electronic transitions are shifted to the ultraviolet and do not absorb visible radiation. The cis alkene is a structurally shorter chromophore than its trans isomer, and it is also likely to experience steric crowding with resultant distortion of the π system. The cis isomer, therefore, absorbs at shorter wavelengths (higher energies) and has a lower molecular extinction coefficient (weaker) than the trans isomer. Thus, once formed, the cis isomer is trapped. Upon heating, however, the cis isomer undergoes exothermic isomerization back to the original, more stable, trans alkene under equilibrium conditions.

A good example of photochromic behavior is the highly colored *cis-* and *trans*-azobenzene. In this case the π → π* transition is promoted by ultraviolet light so that nonequilibrium isomerization to the cis isomer requires UV irradiation. The cis isomer is considerably less stable, however, and it undergoes relatively easy reversion back to the trans isomer by other mechanisms (the thermal conversion of visible radiation absorbed by the colored cis compound is an additional isomerization route apparently open to the cis isomer):

trans-Azobenzene   *cis*-Azobenzene

Tautomerism. A number of proton and valence tautomers are subject to photo-chemical induction. One example is 2-methylbenzophenone (**I**), a colorless compound that can be photochemically induced to tautomerize to a system with extended conjugation. The tautomer (**II**) is a yellow material that will revert to the colorless form under thermal conditions:

I   **II**
2-Methylbenzophenone   (enol, yellow)
(colorless)

Homolysis and Heterolysis of Bonds. Photochromic homolysis has been observed with materials such as 2,3,4,4-tetrachloro-1-(4*H*)-naphthalenone (**III**):

III   (violet)
(colorless)

Excitation leading to heterolysis and zwitterion formation has been observed in numerous spiropyrans, as shown here for **IV**. These compounds undergo ring opening to yield a zwitterion (**V**):

IV	V
(colorless)	(violet)

The spiropyran example in which the photochromism develops following the transformation of a colorless isomer to a violet system is closely related to the structural isomerism observed in the target photochromic substance, a diazabicyclo[3.1.0]hex-3-ene (**VI**) synthesized in Sequence F. In both instances, the photoisomerization involves heterocyclic ring opening with formation of zwitterion **VII**:

VI	**VII**
2-*exo*-6-*exo*-2,4-Diphenyl-6-(4-nitrophenyl)-1,3-diazabicyclo[3.1.0]hex-3-ene	Blue azomethine–ylide (syn configuration)

The diazabicyclo[3.1.0]hexenes form a series of compounds of which many exhibit photochromic properties.

APPLICATIONS OF PHOTOCHROMISM

One successful application of inorganic photochromic systems has been in the manufacture of sunglasses. When a particular silver salt is incorporated in the lenses, the glass will darken on exposure to sunlight in order to protect the eyes, but then bleach quickly when the light intensity drops, so that the same glasses can be used at night or indoors.

The cis–trans isomerism of azobenzene has been incorporated in a novel chemical method for information storage at ultrahigh densities with nondestructive readout. A device based on this chemical information storage system would have, it is estimated, the capacity of 100 million bits per square centimeter.

There also have been some investigations by the military directed toward developing camouflage "photochromic paints."

In most applications the ability of the system to undergo essentially endless recycling is an important factor. Thus, the degradation response time impacts heavily on the effectiveness of the system. In this regard, the valence bond tautomeric isomerizations would appear to possess the most promising properties, while those mechanisms that involve fragmentation open the system up to the possibility of irreversible byproduct formation. The heterolytic processes described in the photochromism of the target compound (internal ring opening) would appear to fall in between the two boundaries.

THE REACTIONS OF SEQUENCE F[3]

The synthesis begins with an aldol reaction between 4-nitrobenzaldehyde and acetophenone in Experiment [F1] to yield 4-nitrochalcone:

p-Nitrobenzaldehyde   Acetophenone   trans-4-Nitrochalcone

The chalcone is brominated in the second step (Experiment [F2]) to yield erythro-2,3-dibromo-3-(4-nitrophenyl)propiophenone:

trans-4-Nitrochalcone   erythro-2,3-Dibromo-3-(4-nitrophenyl)propiophenone

When this halogenated compound is treated with ethanolic ammonium hydroxide for several days, as in Experiment [F3], the system undergoes several reactions (dehydrohalogenation, amination, and ring closure) to ultimately yield a trans-substituted aziridine product:

erythro-2,3-Dibromo-3-(4-nitrophenyl)propiophenone   trans-2-(4-Nitrophenyl)-3-benzoylaziridine

The photochromic substance, a diazabicyclo[3.1.0]hex-3-ene, is obtained in the fourth step (Experiment [F4]) when the aziridine is treated with benzaldehyde, anhydrous ammonia, and ammonium bromide. The reaction requires several days to go to completion.

[3]This synthetic sequence is based on a set of experiments first developed for the undergraduate instructional laboratory by Professor R. Marshall Wilson and Laboratory Director D. L. Lieberman of the University of Cincinnati. We are grateful to Paulette Messier of Bowdoin College and Dr. Joanne M. Holland of Sepracor, Inc., for further development and optimization work.

trans-2-(4-Nitrophenyl)-3-benzoylaziridine

2-exo-6-exo-2,4-Diphenyl-6-(4-nitrophenyl)-1,3-diazabicyclo[3.1.0]-hex-3-ene

The product is a colorless crystalline substance that possesses the property of turning a deep blue when exposed to indoor light:

2-exo-6-exo-2,4-Diphenyl-6-(4-nitrophenyl)-1,3-diazabicyclo[3.1.0.]hex-3-ene

syn-Azomethine-ylide

The mechanism involved in the photochromic isomerization reaction is relatively rare, which makes these substances all the more interesting to study.

BIBLIOGRAPHY

Coyle, J. D.; Hill, R. R.; Roberts, D. R. *Light, Chemical Change and Life: A Source Book in Photochemistry*; The Open University: Milton Keynes, England, 1982, pp. 306–309.

de la Mare, P. D. H.; Suzuki, H. *J. Chem. Soc.* **1968**, 648.

Coxon, J. M.; Halton, B. *Organic Photochemistry*, 2nd ed.; Cambridge University Press: New York, 1987.

Kopecky, J. *Organic Photochemistry*; VCH: New York, 1992.

Also see:

Horspool, W. M.; Lenci, F., Eds., *CRC Handbook of Photochemistry and Photobiology*, 2nd ed.; CRC Press: Boca Raton, FL, 2003, Vols. 1, 2.

Smith, M. B. *March's Advanced Organic Chemistry*, 8th ed.; Wiley: New York, 2019.

Experiment F1

An Aldol Reaction: *trans*-4-Nitrochalcone

Common name: 4-nitrochalcone

CA number: [2960-55-6]

CA name as indexed: 2-propen-1-one, 3-(4-nitrophenyl)-1-phenyl-, (*E*)-

Purpose. We prepare the first of three intermediates on the synthetic pathway to our target molecule, a photochromic imine. A base-catalyzed aldol reaction is carried out in which an aromatic aldehyde is condensed with an aryl alkyl ketone. This addition reaction is followed by dehydration to form an α,β-unsaturated ketone; this particular product is commonly called a chalcone. This intermediate is isolated and purified for use as the starting material in the next stage of the synthesis. You will carry out a semimicroscale reaction to gain experience at conducting larger-scale organic reactions.

Prior Reading

Technique 4: Solvent Extraction

Liquid–Liquid Extraction (p. 76)

Technique 5: Crystallization

Use of the Hirsch Funnel (pp. 92–93)

REACTION

4-Nitrobenzaldehyde   Acetophenone   *trans*-4-Nitrochalcone

DISCUSSION

The aldol reaction (aldol condensation) is one of the fundamental reactions of organic chemistry because it leads to the formation of a new carbon–carbon bond (see Experiment [20]). In this version, Experiment [20A] details the condensation of 4-nitrobenzaldehyde (an aldehyde without an α-hydrogen atom) with acetophenone (a ketone) gives *trans*-4-nitrochalcone. The aldol condensation of the unsubstituted aromatic aldehyde, benzaldehyde with acetophenone, yields *trans*-1,3-diphenyl-2-propenone (PhCH=CHCOPh), which has the common name, chalcone. Thus, the substituted derivatives of this system are known collectively as chalcones.

The extended conjugation in the product favors the formation of the chalcone product. Furthermore, this product is insoluble in the aqueous ethanol solvent and rapidly precipitates from the reaction medium as it is formed, whereas the starting materials are all soluble in aqueous ethanol. Thus, the experimental conditions assist in driving this equilibrium reaction to completion.

The classic aldol condensation involves generation of an *enolate* by removal of an acidic proton from a carbon alpha to the carbonyl group of an aldehyde or ketone, and subsequent nucleophilic addition of this enolate to the carbonyl carbon of an aldehyde or ketone. This reaction is base catalyzed and involves the following mechanistic steps:

An enolate

Nucleophilic attack   Protonation

The reaction involves (a) base-catalyzed generation of the enolate, (b) nucleophilic attack of this anion on a carbonyl carbon, and (c) proton transfer to the resulting anion to yield the initial aldol product, a β-hydroxycarbonyl compound. The β-hydroxycarbonyl product may be isolated in many cases, if desired, since the subsequent dehydration is generally much slower than the addition reaction that precedes it. The final stage of the aldol reaction, as in the present reaction, is a hydroxide-catalyzed dehydration of the initial product by way of the enolate. Though hydroxide ion (HO⁻) is generally not a good leaving group, the H alpha to the carbonyl in the β-hydroxyketone is quite acidic. In addition, the elimination produces a highly conjugated α,β-unsaturated ketone. Under these strongly basic conditions, the hydroxide ion becomes an adequate leaving group. In these systems, both during the loss of the proton in the formation of the enolate anion and during the loss of hydroxide to yield the α,β-unsaturated ketone, the molecular conformations involved favor development of the more-stable trans product:

Enolate anion

Trans α,β-unsaturated ketone

In the present experiment both aldol condensation routes yield a chalcone derivative. In Experiment [20B], a nearly identical double aldol reaction yields dibenzalacetone.

EXPERIMENTAL PROCEDURE

Estimated time to complete the experiment: 3.0 h.

Physical Properties of Reactants						
Compound	MW	Amount	mmol	mp (°C)	bp (°C)	d
Acetophenone	120.16	488 μL	4.16	20.5	202.6	
Ethanol (95%)		20 mL			78.5	0.789
4-Nitrobenzaldehyde	151.12	500 mg	3.31	106		
Sodium hydroxide (aq, 10%)		2 mL				

Reagents and Equipment.   Weigh and place 500 mg (3.31 mmol) of 4-nitrobenzaldehyde in a 50-mL Erlenmeyer flask containing a magnetic stir bar. Now add 20 mL of 95% ethanol and 488 μL of acetophenone.

Reaction Conditions.   Warm the reaction mixture, while stirring in a sand bath at 65–70 °C, until the aldehyde dissolves to yield a clear light yellow solution.

At this point, cool the Erlenmeyer flask in an ice bath for at least 5 min and then, while continuing to cool the system, *carefully* add, dropwise with stirring, 2 mL of 10% (aq) sodium hydroxide over a second 5-min period. During this time, the reaction mixture often turns a dark orange color and considerable precipitation may occur. Cool the Erlenmeyer flask for an additional 15 min following the last addition of base.

Nitrochalcone product

20 mL ethanol + ~2 mL 10% NaOH

Isolation of Product. Collect the solid tan precipitate, which formed during the reaction, on the filter bed of a Hirsch funnel under reduced pressure (➡).

If solid continues to form in the filtrate, refilter the reaction solution and combine the second collection of crystals with the first batch. You want to maximize your total yield of the aldol product, because this is the first step of a four-step synthesis. Thus, you will need efficient recovery of product at each intermediate stage of the synthesis to successfully obtain a reasonable quantity of the photochromic target molecule. Rinse the Erlenmeyer flask once or twice with ice-cold water to effect as closely as possible a quantitative transfer of the chalcone to the Hirsch funnel.

Purification and Characterization. Wash the tan filter cake containing the reaction product, dropwise with an ice-cold 80:20 ethanol/water solution until the product appears as pale yellow crystals. The wash dissolves and removes a red-brown amorphous material that contaminates the crude product in many instances. Transfer the purified and partially dried chalcone to a watch glass, and then place it in a desiccator for final drying. Characterize the anhydrous intermediate accurately by an evacuated melting point and infrared spectrum; the latter can be compared to that recorded in the literature (*The Aldrich Library of IR Spectra* and/or SciFinder Scholar). The chalcone normally is of sufficient purity to be carried directly on to Experiment [F2].

NOTE. *Approximately 400 mg of purified product with a melting point within 2–4 °C of the literature value is the minimum quantity of intermediate suggested for continuing the sequence on to Experiment [F2].*

QUESTIONS

7A-26. In a number of cases it is possible to successfully isolate the β-hydroxyketone intermediate prior to the dehydration that forms the α,β-unsaturated ketone.
    (a) Does the isolation of the β-hydroxyketone suggest which step in the aldol condensation is the rate-determining step in this case?
    (b) If so, which one *is* the rate-determining step?
    (c) Suggest what reaction conditions may have a significant impact on determining which step becomes rate determining.
    (d) What structural changes might lead to a change in the rate-determining step?

7A-27. Ketones also undergo the aldol condensation, although a successful reaction often requires "enhanced" conditions, since the addition involves an unfavorable equilibrium constant. This is the situation in the reaction in which 4-nitrochalcone is synthesized. With the odds against it, why is the reaction successful in this case?

7A-28. If you had obtained both the *cis-* and *trans-*chalcone products, and had purified them by recrystallization, how could you instantly know which one was cis and which one was trans without any further characterization?

Trans   Cis

7A-29. The mixed aldol reaction between propionaldehyde and acetone gives an 85% yield of 4-hydroxy-2-hexanone when run in THF at -78 °C with lithium diisopropylamide (LDA, a powerful base). The reaction is carried out by first adding the ketone to the base, cooling the solution, and then adding the aldehyde.
   (a) Why does this mixed system give essentially a single product?
   (b) Why is there no self-condensation of the acetone?
   (c) Why does the system not rapidly go on to dehydrate?
   (d) Why is the ketone added to the base rather than vice versa?

7A-30. Give the aldol product or products from the following reaction:

$$CH_3CH_2CH_2CH_2CH_2CHO + CH_3CH_2CH_2CH_2CHO \xrightarrow{\text{base}}$$

7A-31. The chalcone structure is particularly interesting. It was known in nature long before it was synthesized in the laboratory. This structure is incorporated biosynthetically into a large class of over 300 natural pigments called flavonoids. These substances heavily contribute to the spectacular New England autumn colors and many flower pigments. Flavonoids arise from chain extension of shikimic acid–derived cinnamic acids (see Experiment [10C] for a more detailed discussion of biological origin of these materials). A typical example of a flavone would be luteolin (5,7,3′,4′-tetrahydroxyflavone, **I**), the orange-yellow pigment of the snapdragon:

I
Luteolin

Flavonoids may be synthesized by using reactions similar to those used in the chalcone synthesis. For example, the basic flavone structure (**II**) can be simply derived from a Claisen condensation between ethyl benzoate and 2-methoxyacetophenone, followed by treatment with HI:

II
Flavone (skeleton)

Show the mechanistic route leading to this flavone from the ester and the ketone.

BIBLIOGRAPHY

Selected examples of the Claisen–Schmidt reaction from *Organic Syntheses* are given here:

Conrod, C. R.; Dolliver, M. A. *Organic Syntheses*; Wiley: New York, 1943, Collect. Vol. II, p. 413.

Kohler, E. P.; Chadwell, H. M. *Organic Syntheses*; Wiley: New York, 1941, Collect. Vol. I, p. 78.

Leuck, G. J.; Cejka, L. *Organic Syntheses*; Wiley: New York, 1941, Collect. Vol. I, p. 283.

Wawzonek, S.; Smolin, E. M. *Organic Syntheses*; Wiley: New York, 1955, Collect. Vol. III, p. 715.

Also see:

Heathcock, C. H. In *Asymmetric Synthesis*; Morrison, J. D., Ed.; Academic Press: 1984, Vol. III.

Heathcock, C, H. In *Comprehensive Carbanion Chemistry*; Durst, T.; Buncel, E., Eds.; Elsevier: New York, 1984, Vol. II.

Nielson, A. T.; Houlihan, W. J. *Org. React.* **1969,** *16,* 1.

Smith, M. B. *March's Advanced Organic Chemistry*, 8th ed.; Wiley: New York, 2019.

erythro-2,3-Dibromo-3-(4-nitrophenyl)propiophenone

> ### Experiment F2

Common name: *erythro*-2,3-dibromo-3-(4-nitrophenyl)propiophenone

CA number: [24213-17-0]

CA name as indexed: 1-propanone, 2,3-dibromo-3-(4-nitrophenyl)-1-phenyl-, (*R*,S**)-

Purpose. You will prepare the appropriate dibromide, an intermediate in your synthetic sequence, to act as the precursor to the aziridine ring system. You will carry out a semimicroscale halogenation with bromine as the active reagent. A further purpose of this experiment is to demonstrate the stereospecific addition of bromine to alkenes.

Prior Reading

   Technique 4: Solvent Extraction

           Liquid–Liquid Extraction (p. 76)

           Concentration of Solutions (pp. 105–108)

   Technique 5: Crystallization

           Use of the Hirsch Funnel (pp. 92–93)

REACTION

trans-4-Nitrochalcone   →(Br$_2$/CHCl$_3$)   *erythro*-2,3-Dibromo-3-(4-nitrophenyl)propiophenone

DISCUSSION

The bromination of alkenes is an example of an electrophilic addition reaction. In the present reaction, bromination of 4-nitrochalcone yields *erythro*-2,3-dibromo-3-(4-nitrophenyl)propiophenone.

This reaction is *stereospecific*, because the other possible diastereomer [threo or (R*, R*)] is not formed:

trans-4-Nitrochalcone

erythro-2,3-Dibromo-
3-(4-nitrophenyl)propiophenone

The reaction proceeds in two steps. The first involves the formation (from either side of the plane of the double bond; attack from below is shown here) of an intermediate cyclic *bromonium ion*:

The bromine molecule (Br_2) is normally symmetrical. However, as it approaches the electron-rich, and nucleophilic, π bond of the alkene, it becomes polarized by induction and then functions as the electrophile in an addition reaction. The result is the generation of a cyclic bromonium ion:

Induced polarization
of Br_2 as it approaches
the alkene

Bromonium ion

Nucleophilic
attack by bromide

In the present reaction, both the bromine and the 4-nitrochalcone are achiral, though the bromonium ion that is produced is chiral. Because it results from the reaction of achiral molecules in an achiral environment, the bromonium ion must be racemic. In this ion, the bromine atom bridges both carbon atoms of the original carbon–carbon double bond to form a three-membered ring intermediate. The generation of this high-energy cyclic species has a profound effect on the *stereochemistry* of the second step of the bromine addition: the ring restricts rotation about the C—C single bond in the carbocation.

The second stage of the bromination involves nucleophilic attack by bromide ion on the intermediate bromonium ion. Since the nucleophile must approach from the face opposite the leaving group, bond formation involves inversion of configuration at the carbon center under attack in the second stage of the bromination reaction.

Note that *either* carbon can be approached by the nucleophile (one attack is shown). This second step is a classic backside S_N2 type displacement:

| Bromonium ion | Dibromo product |

In the case with 4-nitrochalcone, two chiral centers are generated in the bromonium ion and we might, therefore, expect that two diastereomeric pairs would be formed. However, due to the stereoselectivity of the reaction, only a single diastereomer is generated, as a racemic pair of enantiomers (refer to Experiment [D2] for a further discussion of the stereochemistry of this halogenation).

EXPERIMENTAL PROCEDURE

Estimated time to complete the experiment: 2.0 h.

Physical Properties of Reactants and Product

Compound	MW	Amount	mmol	mp (°C)	bp (°C)	d
trans-4-Nitrochalcone	253	400 mg	1.58	161–164		
Chloroform		~8 mL			61.7	1.492
Bromine/chloroform (2.5%)	179.8	(200 μL Br₂)	3.91		58.8	
erythro-2,3-Dibromo-3-(4-nitrophenyl)propiophenone	413			151–153		

Reagents and Equipment.   Weigh and place 400 mg (1.58 mmol) of 4-nitrochalcone in a tared 50-mL round-bottom flask containing a magnetic stir bar. Now add about 8 mL of chloroform (dispensed in the **hood**). Connect the flask to a water-jacketed reflux condenser fitted with a drying tube.   HOOD

INSTRUCTOR PREPARATION. The active brominating reagent in this reaction is liquid Br₂ dissolved in chloroform. Prepare a solution of 200 μL (624 mg) of bromine dissolved in 8 mL of chloroform multiplied by the number of students carrying out the experiment. The reagent should be prepared, dispensed, and added to the reaction in the hood.   HOOD

CAUTION: Bromine is a highly reactive substance. Even in chloroform solution you must handle it with care. Be very careful not to get this reagent on your skin. All transfers of the reagent should be made in the **hood**.   HOOD

Chloroform itself is highly toxic and a cancer suspect agent. Handle it with respect.

Reaction Conditions.   Warm the round-bottom flask in a sand bath between 60 and 70 °C, with stirring, until the 4-nitrochalcone dissolves to yield a clear light yellow solution. Once dissolution has occurred, continue to maintain the bath temperature at 60–70 °C and add the bromine reagent. The addition is carried out dropwise using a 9-in. Pasteur pipet inserted down the condenser

40–50°C
thermometer

Speed   Heat

400 mg, 4–nitrochalcone +
16 mL, chloroform +
200 µL, bromine +
50–mL RB flask

(briefly remove the drying tube during this operation), with stirring, over a 10-min period. By the end of the Br$_2$ addition the solution turns a dark orange. Continue to heat the stirred reaction mixture for an additional 20 min ($\leftarrow$).

After cooling the reaction mixture to room temperature, remove a small aliquot and spot it on a silica gel TLC plate next to a reference spot of the 4-nitrochalcone starting material. Elute the plate with 50:50 methylene chloride/hexane and visualize the spots by UV. (A number of unidentified by-products are formed in small quantities during the reaction that are often observed on the TLC plates.) There also will probably be a trace of unreacted 4-nitrochalcone left in the reaction solution. The major (and highest R_f) spot ($R_f = 0.3$) is the brominated product. Because an excess of bromine was used in the reaction, however, only small quantities of 4-nitrochalcone are normally detected at this stage. If significant unreacted substrate remains (i.e., more than a faint or weak spot on TLC), add an additional small amount of the bromine reagent and then reanalyze the reaction mixture again by TLC.

Isolation of Product. Once it is established that the reaction is largely complete, remove the solvent and excess reagent from the reaction mixture by rotary evaporation. Weigh the crude residue.

*NOTE. At this point the crystalline residue that remains following rotary evaporation may be stored until the next laboratory period by first flushing the round-bottom flask with dry nitrogen (or argon) and then quickly sealing the flask with a ground-glass standard-taper stopper that is sealed with Parafilm. The flask should be labeled and given to your instructor for storage in the freezer. While it is possible to safely interrupt the workup at this stage, most organic materials are much more stable when they are stored in as pure a state as possible. **If time permits, you are urged to finish the workup of the bromination**.*

Purification and Characterization. The crude material, a yellow-orange solid is now partially purified by column chromatography.

The chromatographic column (short buret) is packed with silica gel (10 g) after first positioning a plug of cotton and about 1 cm of sand at the bottom. Then a portion of 50:50 methylene chloride/hexane is added to the column (~15 mL), followed by 10 g of silica gel. The solid substrate is slowly added while the column is tapped to promote even settling of the packing material. During this process the column stopcock is slightly opened to create a slow drip rate of the packing solvent out of the column. As a result of this drainage more solvent may be required to keep the solvent level above that of the silica gel during the settling operation. Finally, carefully drain any excess solvent to the top of the column and close the stopcock after the packing procedure is complete.

The crude *erythro*-2,3-dibromo-3-(4-nitrophenyl)propiophenone is then dissolved in a minimum amount of methylene chloride (~10 mL) and applied to the top of the column by slowly pipetting the solution down the side of the column without disturbing the silica gel (as this solution is added, it is also slowly drained onto the column by cracking open the stopcock). As the final quantity of crude product drains to the top of the column, elution is started with 50:50 methylene chloride/hexane solution (again by careful addition so as not to disturb the upper layers of silica gel containing the adsorbed reaction products). Collect 3 × 30-mL fractions in 50-mL Erlenmeyer flasks (labeling each flask with the fraction number). You may observe some yellow zones of material slowly moving down the column during the elution. This colored material usually does not begin to elute with this chromatographic scheme. Once the fractions have been collected, use TLC analysis (silica gel plates,

50:50 methylene chloride/hexane) to determine the composition of the fractions. The purest fractions are combined to give sufficient material to continue on to the next step (Experiment [F3]) in the sequence. Separate and remove the solvent (N₂ and warm sand bath). Determine the weight and melting point (evacuated) of your brominated product. The *erythro*-2,3-dibromo-3-(4-nitrophenyl)-propiophenone should appear as white to light yellow needles. It may be further recrystallized from 95% ethanol if desired.

If the melting point is only a few degrees low (~142–147 °C), increased purity often is quickly obtained by simply adding a few milliliters of ice cold chloroform to the product residue that is cooled in an ice bath. Triturate the residue for a few seconds with the cold solvent. Withdraw the solvent by Pasteur filter pipet leaving the washed crystals behind (this treatment may be repeated if necessary). Remove traces of the solvent remaining on the residue by a short rotary evaporation. Recheck the melting point (evacuated) to determine if product purity is improved enough to continue on to the third step (➡).

Weigh your purified *erythro*-2,3-dibromo-3-(4-nitrophenyl)propiophenone intermediate and calculate the percent yield based on both the starting 4-nitrochalcone and the 4-nitrobenzaldehyde (the starting material used in Experiment [F1]). Compare your spectrum to that of a reference standard shown in Figure 7A.4.

Figure 7A.4 IR spectrum: *erythro*-2,3-dibromo-3-(4-nitrophenyl)propiophenone.

OPTIONAL. Obtain the ^{1}H NMR spectrum of the product in CDCl$_3$. Identify the resonances for the two protons attached to the halogenated carbon atoms, and ascertain that the product is a single diastereomer (see Chapter 8 for discussion). This information will be useful in Experiment [F3].

NOTE. The erythro-2,3-dibromo-3-(4-nitrophenyl)propiophenone is a particularly sensitive substance and will decompose in contact with air at room temperature over several days. If you do not have time to continue on to Experiment [F3] during this laboratory period, you should store your purified material as described above for the crude product.

If you have time to start the reaction, you are urged to continue on to the next step. *Once you have the pure dibromide in hand, it takes only a relatively short time to set up and get the next reaction running. Since this latter reaction will be left to run for a week, you have a lot to gain by getting the third step started during the period when the second intermediate is worked up.*

NOTE. Approximately 450 mg of purified product with a melting point within 2–4 °C of the literature value is the minimum quantity of intermediate suggested for continuing the sequence on to Experiment [F3].

QUESTIONS

7A-32. A considerable excess of Br$_2$ in chloroform is required to successfully drive the halogenation of 4-nitrochalcone to completion. Offer a suggestion as to the role of the excess reagent.

7A-33. What product(s) would you expect to obtain from the bromination (Br$_2$ in CCl$_4$) of cyclobutene?

7A-34. In the bromination of 4-nitrochalcone, a racemic dibromide is formed. A second diastereomeric dibromide is structurally possible, but it is not formed in the reaction.

   (a)   Give stereochemically detailed drawings of the stereoisomer(s) isolated from the reaction mixture.

   (b)   Give stereochemically detailed drawings of the stereoisomer(s) that is/are not formed.

   (c)   Why is this reaction stereoselective?

   (d)   How could the second diastereomer be synthesized if its preparation was required?

   (e)   Assign, by the R and S convention, the stereocenters in each of the diastereomers.

7A-35. Show the stereoisomer(s) generated by bromination of each of the enantiomers of *cis*- and *trans*-4-bromo-2-pentene. Show the relationship (enantiomer, diastereomer, etc.) and assign R and S centers in all product(s) and starting materials.

7A-36. Show, with the correct absolute configuration, the stereoisomer(s) formed on bromination of (*S*)–4-*tert*-butyl-1-cyclohexene. Would you expect them to be formed in equal amounts?

BIBLIOGRAPHY

A large number of examples of the bromination of alkenes appear in *Organic Syntheses*. Selected references are given below:

Allen, C. F. H.; Abell, R. D.; Normington, J. B. *Organic Syntheses*; Wiley: New York, 1941; Collect. Vol. I, p. 205.

Cromwell, N. H.; Benson, R. *Organic Syntheses*; Wiley: New York, 1943; Collect. Vol. II, p. 105.

Fieser, L. F. *Organic Syntheses*; Wiley: New York, 1963; Collect. Vol. IV, p. 195.

Khan, N. A. *Organic Syntheses*; Wiley: New York, 1963; Collect. Vol. IV, p. 969.

McElvain, S. M.; Kundiger, D. *Organic Syntheses*; Wiley: New York, 1955; Collect. Vol. III, p. 123.

Paquette, L. A.; Barrett, J. H. *Organic Syntheses*; Wiley: New York, 1973; Collect. Vol. V, p. 467.

Rhinesmith, H. S. *Organic Syntheses*; Wiley: New York, 1943; Collect. Vol. II, p. 177.

Snyder, H. R.; Brooks, L. A. *Organic Syntheses*; Wiley: New York, 1943; Collect. Vol. II, p. 171.

Also see:

Smith, M. B. *March's Advanced Organic Chemistry*, 8th ed.; Wiley: New York, 2019.

trans-2-(4-Nitrophenyl)-3-benzoylaziridine

Common names: *trans*-2-(4-nitrophenyl)-3-benzoylaziridine

CA number: [76336-95-3]

CA name as indexed: methanone, [3-(4-nitrophenyl)-2-aziridinyl]phenyl-, *trans*-

Purpose.   The third intermediate on the pathway to the target photochromic imine is synthesized. A heterocyclic three-membered ring, an aziridine derivative, is formed. This is the first ring of the diazabicyclohexene system that you will ultimately convert into the photochromic imine. You will study a process that involves three reactions and the formation of two intermediates en route to the final product. You will study a number of interesting stereoselective reactions. You will work with organic reactions that require several days to come to completion.

Prior Reading

   Technique 4: Solvent Extraction

          Liquid–Liquid Extraction (p. 76)

REACTION

erythro-2,3-Dibromo-
3-(4-nitrophenyl)propiophenone

trans-2-(4-Nitrophenyl)-3-benzoylaziridine

DISCUSSION

The conversion of *erythro*-2,3-dibromo-3-(4-nitrophenyl)propiophenone to a substituted aziridine involves a number of interesting steps and intermediates. The *first stage* of the reaction involves attack on the halogenated intermediate by base (concentrated ammonium hydroxide, $NH_4OH \rightleftharpoons NH_3 + H_2O$). Under the highly polar conditions, the reaction likely proceeds via an E1cb mechanism involving the initial attack by ammonia, acting as a base, on the **2** proton to yield the resonance stabilized anion:

(Z)-2-Bromo-4'-nitrochalcone

You should refer to the detailed discussions in Experiments [9] and [10], which describe in detail the chemistry associated with the E1 and E2 elimination mechanisms. The E1cb mechanism, like the E1 mechanism, involves a two-step process, but in this case the order of charge development is the reverse of the E1 mechanism. Here proton abstraction precedes loss of the leaving group. A generalized scheme is shown here:

E1cb Mechanism

$$RCH_2CHR' + B^- \longrightarrow R\overset{-}{C}HCHR' + BH \longrightarrow RCH{=}CHR' + X^-$$

with X below the first CHR' and X below the second CHR'.

The mechanisms of a large majority of elimination reactions can be explained by invoking various positions along the continuum between the three elimination mechanisms mentioned here; E1cb and E1 are at the extremes of the continuum, and the E2 mechanism lies exactly halfway between the two.

In the present case, carbanion (or near-carbanion) formation following α-proton abstraction appears to be favored, and formation of the anionic intermediate has the attractive feature that it then allows rotation about the incipient π bond. Stereoelectronic requirements of the elimination mechanism require the carbon–bromine bond to be parallel to the p orbital of the adjacent enolate (or α-keto carbanion, depending on which resonance structure is being discussed), just as the proton being removed and the leaving group prefer to be anti in an E2 elimination. Two possible conformations (or rotamers) meet this requirement. The most stable one will be that with the carbonyl and nitrophenyl groups anti to one another, and elimination of bromide ion from this conformation leads to the alkene with the carbonyl and nitrophenyl groups trans, which is the Z alkene, as illustrated here:

(Z)-2-Bromo-4'-nitrochalcone

Thus, the first stage of the reaction primarily leads to formation of an α-bromo-α,β-unsaturated ketone, (Z)-2-bromo-4'-nitrochalcone. This alkene could be easily isolated from the reaction medium, if required, since the second step in the reaction occurs at a considerably slower rate than the first step. Thus, appreciable concentrations of this unsaturated intermediate are obtained.

The *second stage* of the reaction involves 1,4 addition of the amine (in this case ammonia in equilibrium with the ammonium hydroxide) to the unsaturated ketone.

There are many examples of nucleophilic reagents that add to α,β-unsaturated aldehydes and ketones in a manner in which the addition is formally 1,4. This result is called conjugate addition. Under basic conditions, these transformations involve initial attack by the nucleophile to the β-carbon atom, followed by electrophilic addition (normally of a proton) on the carbonyl oxygen; the nucleophile and electrophile add at the 1 and 4 positions relative to one another. The enolate formed in the early stages of the reaction is generally quickly protonated to give an enol. The enol will subsequently tautomerize to the ketone. A general mechanistic scheme is shown below for the 1,4 addition of water to an α,β-unsaturated carbonyl system.

In the present case, conjugate addition has the potential to result in two diastereomers. As in the case of the starting dibromide, these stereoisomers can be either erythro or threo. The distribution between these two products is important because it ultimately determines the ratio of products in the next step (third) of the reaction.

From the geometry of the aziridine final product, and the fact that the internal nucleophilic substitution reaction will proceed with inversion of configuration, we can reliably postulate that conjugate addition must eventually result in the erythro diastereomer. In the conjugate addition of ammonia to 2-bromo-4′-nitrochalcone, the stereochemistry of the product is determined by which face of the resulting enolate receives the proton. The proton source may be the $—NH_3^+$ group resulting from the initial step of the conjugate addition reaction (as shown here), or it may be an ammonium ion or a molecule of ethanol or water hydrogen bonded to the amino group. A further consideration may be the geometry of the resulting enolate, since there is the potential to generate either the Z or the E enolate; the E enolate is illustrated here because it would appear to be the least hindered of the two diastereomeric enolates:

Threo   Erythro

It is also possible that both the threo and erythro diastereomers are produced at comparable rates and that the erythro is thermodynamically preferred by an equilibrium process between the two.

A comparison of the most-favored conformations, based on relative group sizes, for the erythro and threo diastereomers, indicates that the conformation **A** for the erythro isomer would be less sterically crowded than conformation **B** for the threo isomer, as shown here.

A
(erythro)

B
(threo)

L = large, m = medium, s = small
Ph = phenyl

The *final stage* on the route to the aziridine product involves an internal S_N2 ring closure in which the primary amine group attacks the α-carbon atom holding the remaining bromine from the backside to close an aziridine ring and displace bromide with inversion of configuration at the α carbon. This displacement has been shown to go exclusively by this mechanism in the case of the β-amino-α-bromoketones; an S_N1 reaction is less likely because of the instability of an α-keto carbocation. For a detailed discussion of the mechanism of the classic S_N2 substitution reaction, refer to Experiment [22].

The stereochemistry of aziridine ring substitution, as pointed out above, is controlled by the product distribution in the β-amino-α-bromoketone intermediate, which in this case favors the erythro configuration. Thus, inversion of configuration during ring formation leads to the trans-substituted aziridine ring in the present example:

erythro-3-Amino-2-bromo-3-
(4-nitrophenyl)propiophenone

trans-2-(4-Nitrophenyl)-3-
benzoylaziridine

EXPERIMENTAL PROCEDURE

Estimated time to complete the experiment: 3.5 h.

Estimated time to complete Part A of the experiment: 1.5 h.

Estimated time to complete Part B of the experiment: 2.0 h.

Physical Properties of Reactants and Product						
Compound	MW	Amount	mmol	mp (°C)	bp (°C)	d
erythro-2,3-Dibromo-3-(4-nitrophenyl)propiophenone	411	450 mg	1.10	151–153		
Ethanol (95%)		4.4 mL			78.5	0.789
Ammonium hydroxide (concd)		1.3 mL				
trans-2-(4-Nitrophenyl)-3-benzoylaziridine	268.3			142–143		

PART A

Reagents and Equipment. Weigh and place 450 mg (1.10 mmol) of the *erytho*-2,3-dibromo-3-(4-nitrophenyl)propiophenone intermediate synthesized in Experiment [F2] in a labeled 25-mL round-bottom flask containing a magnetic stir bar. In the **hood**, add 4.4 mL (9.8 mL/g of chalcone) of 95% ethanol (graduated cylinder) and 1.3 mL (2.8 mL/g of chalcone) of concentrated ammonium hydroxide (automatic delivery pipet or 2-mL glass pipet) to the flask. Stopper the flask, swirl to mix the contents, and then seal it with Parafilm. (A polypropylene standard taper stopper is preferred for sealing the vessel for long periods in the presence of base.)

HOOD

CAUTION: Concentrated ammonium hydroxide is a strongly caustic reagent. You must handle it with care. Be particularly alert not to get this reagent on your skin or breathe the vapors. All transfers of the reagent should be made in the **hood**.

HOOD

Reaction Conditions. Stir the reaction mixture for 15 min. Then wrap the flask with aluminum foil and continue to stir the system for 24 h. The reaction may be worked up at this point or stored in the dark for 1 week if necessary. Do not expect the chalcone to immediately dissolve in the reaction medium; it will do so over the course of several hours as the aziridine product begins to precipitate. If you wish, you can occasionally magnetically stir the reaction mixture for 15–20 min during this intervening period.

This completes Part A of the experiment.

PART B

Isolation of Product. Add 15 mL of ice-cold water to the reaction mixture and swirl for 5 min. Collect the solid residue formed in the reaction by vacuum filtration on a Hirsch funnel. The product will appear as pale orange, very fine needles. There may also be a small amount of a more powdery yellow material that should dissolve in three 1.0-mL washes with ice-cold water. Use the first wash to rinse the Erlenmeyer flask and transfer the rinse to the Hirsch funnel. Fine orange needles of the aziridine product will be deposited on the filter bed (➡).

Aziridine as orange needles collected here

~4 mL 95% ethanol + ~3 mL dil NH₄OH

Purification and Characterization. Collect the orange needles, dry them to constant weight (under reduced pressure), and determine an evacuated melting point. Recrystallize the crude product from hot methanol (~10–20 mL). After crystallization has begun, cool the system further in an ice bath for 10–15 min to complete the collection. The purified aziridine is obtained as long, shiny, pale orange needles via Hirsch filtration.

Weigh the *trans*-2-(4-nitrophenyl)-3-benzoylaziridine and calculate the percentage yield based on both the starting materials: *erythro*-2,3-dibromo-3-(4-nitrophenyl)propiophenone (Experiment [F3]), 4-nitrochalcone (Experiment [F2]), and 4-nitrobenzaldehyde (Experiment [F1]). Recheck the evacuated melting point, to see if further purification is required, and obtain an IR spectrum. Compare the spectrum to that of a reference standard shown in Figure 7A.5.

OPTIONAL. Obtain the ¹H NMR spectrum of the aziridine. Establish from this spectrum and the NMR data obtained in Experiment [F2] if any unreacted dibromochalcone still contaminates the aziridine sample that has been purified for use in the preparation of the photochromic target molecule. Determine the diastereomeric purity of your product, and determine that the aziridine product is indeed trans substituted.

NOTE. Approximately 50 mg of purified product with a melting point within 2–4 °C of the literature value is the minimum quantity of intermediate suggested for continuing the sequence on to Experiment [F4].

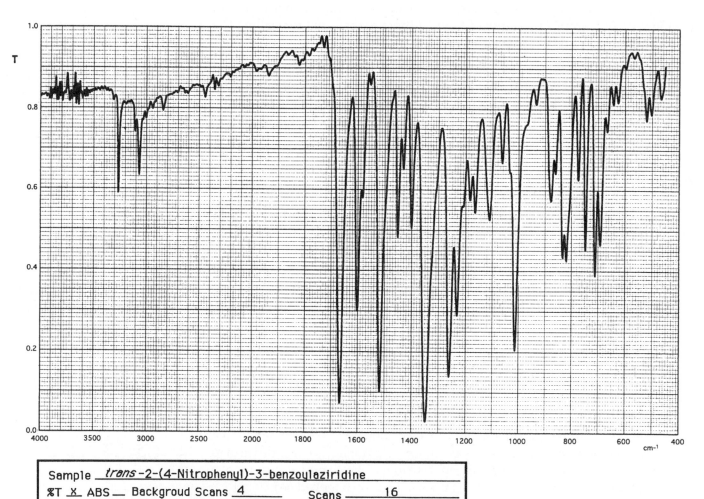

Sample *trans*-2-(4-Nitrophenyl)-3-benzoylaziridine

%T X ABS __ Backgroud Scans 4   Scans 16
Acquisition & Calculation Time 42 sec   Resolution 4.0 cm⁻¹
Sample Condition solid   Cell Window _____
Cell Path Length _____   Matrix Material KBr

Figure 7A.5 IR spectrum: *trans*-2-(4-nitrophenyl)-3-benzoylaziridine.

7A-37. The synthesis of 1-cyclohexyl-2-phenyl-3-(4-phenylbenzoyl)aziridine from cyclohexylamine and 4'-phenylchalcone dibromide, gives a mixture of 47% of the cis isomer and 44% of the trans isomer. Chromatography on activated alumina yielded, in the first eluates, a crystalline material, mp = 118–119 °C while the final eluates produced a higher melting material, mp = 144–146 °C. The IR spectrum of lower melting substance exhibited a $\tilde{v}_{C=O} = 1656$ cm^{-1} while the higher melting compound had $\tilde{v}_{C=O} = 1686$ cm^{-1}. Which product is the transsubstituted aziridine, and which is the cis compound?

Cis   Trans

7A-38. *N*-Haloaziridines have been found to possess an extremely high nitrogen inversion barrier. The isolation of the cis (**I**) and trans (**II**) isomers of 1-chloro-2-methylaziridine has been accomplished, and they represent the first isolated inversion isomers of trivalent nitrogen. These inversion isomers appear to have remarkable stability. For example, 1-chloro-2,2-dimethylaziridine (**III**) retains configurational stability at temperatures as high as about 135 °C! Make a list of the potential stereoisomerism available to structures **I**, **II**, and **III**. Use perspective drawings to illustrate each isomer, and label the stereocenters by the *R* and *S* convention:

I   II   III

7A-39. Another route to the effective synthesis of aziridines (**III**) is through the ring closure of the β-amino alcohol (**I**). The alcohol must first be converted to the β-amino hydrogen sulfate (**II**), which is the actual species that undergoes cyclization with strong base. Why is it necessary to convert the hydroxyl group into a hydrogen sulfate group prior to base treatment?

$$C_6H_5\overset{OH}{\underset{}{CHCH_2NH_2}} \xrightarrow{H_2SO_4} C_6H_5\overset{OSO_3H}{\underset{}{CHCH_2\overset{+}{NH_3}}} \xrightarrow{NaOH} C_6H_5\text{—}\underset{III}{\triangle}\overset{H}{\underset{}{N}}$$

I   II   III

7A-40. Aziridines are relatively reactive systems and undergo nucleophilic ring opening with the accompanying release of ring strain. Give a Fischer projection drawing of the expected product(s) of the reaction of *N*-ethyl-(2*S*,3*S*)-*trans*-2,3-dimethylaziridine (**I**) in aqueous ethyl amine:

$$\xrightarrow{\text{70\% CH}_3\text{CH}_2\text{NH}_2,\ \text{H}_2\text{O},\ 120\ °C,\ 16\ \text{days}}$$

I

BIBLIOGRAPHY

Cromwell, N. H.; Cram, D. J. *J. Am. Chem. Soc.* **1943**, *65*, 301.

Cromwell, N. H.; Mercer, G. D. *J. Am. Chem. Soc.* **1957**, *79*, 3819.

Cromwell, N. H.; Barker, N. G.; Wankel, R. A.; Vanderhorst, P. J.; Olson, F.W.; Anglin, J. H. *J. Am. Chem. Soc.* **1951**, *73*, 1044.

Cromwell, N. H.; Hudson, G.V.; Wankel, R. A.; Vanderhorst, P. J. *J. Am. Chem. Soc.* **1953**, *75*, 5384.

Cromwell, N. H.; Cahoy, R. P.; Franklin, W. E.; Mercer, G. D. *J. Am. Chem. Soc.* **1957**, *79*, 922.

Do Minh, T.; Trozzolo, A.M. *J. Am. Chem. Soc.* **1972**, *94*, 4046.

Heine, H.W.; Hanzel, R. P. *J. Org. Chem.* **1969**, *34*, 171.

Heine, H.W.; Weese, R. H.; Cooper, R. A.; Durbetaki, A. J. *J. Org. Chem.* **1967**, *32*, 2708.

Heine, H.W.; Smith, III, A. B.; Bower, J.D. *J.Org.Chem.* **1968**, *33*, 1097.

Padwa, A.; Clough, S.; Glazer, E. *J. Am. Chem. Soc.* **1970**, *92*, 1778.

Smith, M. B. *March's Advanced Organic Chemistry*, 8th ed.; Wiley: New York, 2019.

Experiment F4

A Photochromic Imine: 2-*exo*-6-*exo*-2,4-Diphenyl-6-(4-nitrophenyl)-1,3-diazabicyclo[3.1.0]hex-3-ene

Common name: *exo*-2,4-diphenyl-6-(*trans*-4-nitrophenyl)-1,3-diazabicyclo [3.1.0]hex-3-ene

CA number: [36799-57-2]

CA name as indexed: 1,3-diazabicyclo[3.1.0]hex-3-ene, 6-(4-nitrophenyl)-2,4-diphenyl-, (2α,5β,6β)-

Purpose. This experiment completes the synthesis of the photochromic imine, which is incorporated into the diazabicyclo[3.1.0]hex-3-ene skeleton. A rare molecular system is obtained in which an aziridine ring is fused to another heterocyclic ring. You will explore the exceedingly interesting photochromic properties of the target molecule. Microscale techniques are used during the conversion and isolation of this light-sensitive material.

Prior Reading

> *Technique 4:* Solvent Extraction
>> Liquid–Liquid Extraction (p. 76)
> *Technique 5:* Crystallization
>> Craig Tube Crystallization (pp. 93–95)

REACTIONS

The Photosensitive Compound

trans-2-(4-Nitrophenyl)-3-benzoylaziridine

2-*exo*-6-*exo*-2,4-Diphenyl-6-(4-nitrophenyl)-1,3-diazabicyclo[3.1.0]-hex-3-ene

Photoproduct

2-*exo*-6-*exo*-2,4-Diphenyl-6-(4-nitrophenyl)-
1,3-diazabicyclo[3.1.0]hex-3-ene

syn-Azomethine-ylide

DISCUSSION

This reaction completes the synthesis of the photochromic target molecule, which possesses the 1,3-diazabicyclo[3.1.0]hex-3-ene ring system. The substituted aziridine ring system formed in Experiment [F3] is condensed with benzaldehyde and ammonia to yield this bicyclic system. The reaction may be viewed as proceeding under anhydrous conditions via an initial reaction between the aromatic aldehyde and the base, aided by the ammonium bromide catalyst, to generate an imine as indicated in the following scheme:

Aldehyde   Hemiaminal   Imine

The imine is subsequently attacked by the aziridine nucleophile to yield an aminal, again catalyzed by the ammonium bromide as shown in the following scheme:

Aminal

Finally, the aminal undergoes an internal condensation involving ring closure by nucleophilic attack of the primary amine group on the carbon of the carbonyl followed by dehydration of the hemiaminal to yield the diazabicyclo[3.1.0]hex-3-ene ring system, again catalyzed by ammonium bromide. A reasonable scheme is shown below:

Aminal

$+ NH_3 + Br^-$

$+ H_2O$
$+ H_4N^+, Br^-$

2-*exo*-6-*exo*-2,4-Diphenyl-6-(4-nitrophenyl)-1,3-diazabicyclo[3.1.0]-hex-3-ene

The photochromic compound appears to interconvert in the solid state to a bright blue azomethine–ylide (**I**) by an electrocyclic ring cleavage. An *ylide* is a neutral species whose Lewis structure contains opposite charges on adjacent atoms. The atoms involved are carbon, and an element from either group 5A or 6A of the periodic table, such as N, P, or S. Of considerable interest is the fact that the photochemical process has been shown to give exclusively the syn isomer.

Conrotatory
ring opening

I
Blue azomethine-ylide
(syn configuration)

This result is consistent with a *conrotatory ring opening* involving a symmetry-allowed concerted transformation in the *ground state*. Since orbital symmetry requires a *disrotatory ring opening* from an *excited state* in these aziridines, the photochemically induced formation of the *syn*-azomethine–ylide (which is isoelectronic with the allyl anion) has been proposed to proceed via a dark reaction in which *electronically excited* states internally convert to *vibrationally excited* ground states. Indeed, evidence supporting this mechanism comes from the thermochromic behavior (a *ground-state* process) of the close relative 6-*exo*-2-dimethyl-4-diphenyl-6-(4-nitrophenyl)-1,3-diazabicyclo [3.1.0]-hex-3-ene (**II**), which when heated to 150 °C turns the same bright blue color (**III**).

II

III
Blue azomethine-ylide
(syn configuration)

Exposure to light from tungsten lamps will not photochemically induce the ring-opening step. Most laboratories, however, are illuminated by fluorescent lights that emit small amounts of short wavelength radiation at the edge of the ultraviolet region (a wavelength long enough to not be absorbed by Pyrex glass), which will initiate the photochemical reaction. It is important to recognize that the azomethine-ylide photoproduct is a highly reactive species. Thus, if it is formed in solution where it can easily interact with other species, it rapidly decays to various byproducts, generally turning the solution yellow (not blue). It appears that if the ylide is produced in an environment that isolates it from other molecules, the system is stable and, given enough time, it will slowly revert to the diazabicyclic starting material.

The crystalline state is the ideal solution to this problem. Thus, when the nearly colorless, solid imine is irradiated, it turns bright blue. The neighbors to any ylide in the crystal can be either the ylide itself or the diazabicyclo starting material. If colored material is then placed in the dark it will slowly revert to the colorless form once again. These molecular systems have been successfully cycled between the colorless and colored states many hundreds of times with little degradation of the crystalline material. Obviously, it is exceedingly important that (a) you do not expose the imine to light when it is in solution during recrystallization (or at least that contact with light is kept to a minimum; working in a red-light darkroom would be ideal), and (b) recrystallizations be carried out as quickly as possible, but obviously with care. Remember, you have a lot of time invested in this product, so work as quickly as is consistent with avoiding a costly spill.

EXPERIMENTAL PROCEDURE

Estimated time to complete the experiment: 2.5 h.

Physical Properties of Reactants						
Compound	MW	Amount	mmol	mp (°C)	bp (°C)	d
trans-2-(4-Nitrophenyl)-3-benzoylaziridine	268.3	40 mg	0.15	142–143		
Ethanol, absolute		800 μL			78.5	0.789
Benzaldehyde	106.13	140 μL	1.37		178	1.045
Ammonium bromide	97.95	14 mg	0.14	452 sub		
Ammonia	17.03	Excess			−33	

Reagents and Equipment. Weigh and place 40 mg (0.15 mmol) of the trans-2-(4-nitrophenyl)-3-benzoylaziridine synthesized in Experiment [F3] in a 15-mL screw-capped centrifuge tube containing a magnetic spin vane. Dissolve the aziridine (20 μL of EtOH/mg of aziridine) and benzaldehyde (140 μL, 1.37 mmol) in 800 μL of absolute ethanol. To this solution add 14 mg (0.14 mmol) of ammonium bromide (0.35 mg of NH_4Br/mg of aziridine). Stir this mixture for 1 min and saturate the system with anhydrous ammonia (NH_3) in the **hood** (see your instructor for directions on this addition).

HOOD

HOOD

CAUTION: Anhydrous ammonia is a dangerous substance, particularly under pressure. The addition of this material to the reaction must be carried out in the **hood** under the *direct* supervision of the laboratory instructor.

Reaction Conditions. Gently bubble the ammonia gas through the reaction mixture until the system is saturated (until the reaction mixture cools off, ~5 min). Tightly cap the tube (Teflon liner), wrap it in aluminum foil, and stir the mixture for a minimum of 24 h. As in Experiment [F3], the reaction takes place at room temperature over a period of several hours. You can safely store the sealed reaction tube in your locker, protected from light with aluminum foil, for up to 1 week if necessary.

Isolation of Product. After the 24-h period, remove the supernatant liquid by centrifuging the tube and then transferring the liquid by means of a 9-in. Pasteur filter pipet to a Craig tube. Dry the remaining crystals in a stream of nitrogen gas. Remove a small sample of the solid material from the centrifuge tube on a glass rod or spatula and expose this material to direct sunlight or fluorescent light. If the solid material slowly turns blue, you have successfully synthesized the photochromic product. Concentration of the solution in the Craig tube may yield further quantities of the azomethine product. Weigh the tube and determine your percentage crude yield.

Purification and Characterization. Recrystallize the crude product in the dark or red light (best) or with the laboratory lights off from hot 95% ethanol. Dissolution may require as much as 10 mL of solvent and is best carried out in the centrifuge tube. After cooling the centrifuge tube in ice and scratching the sides with a glass rod to induce crystallization, centrifuge the system, and remove the mother liquors with a Pasteur filter pipet. Dry the white (or near white) crystals using a stream of nitrogen gas (crystals should be white if the recrystallization is done without fluorescent lights or sunlight).

NOTE. The photoproduct is an azomethine–ylide that is highly reactive. It is wise, therefore, to reduce the time that the material is in solution to a minimum, and to protect, while in solution, the contents of the capped centrifuge tube from light as much as possible. Once in the crystalline state, the large majority of the photochromic reactions take place in an environment in which the azomethine–ylide is protected from further reaction and thus, given the opportunity (in the dark), the ylide will slowly recycle back to the diazabicyclic precursor with little loss.

After removing the mother liquors and drying the purified photochromic product, determine an evacuated melting point. (The diazabicyclic intermediate substance is oxidatively sensitive and will decompose during atmospheric melting point measurements.) Be alert to color changes during the melting point determination.

Compare your results with the literature value of 169–172 °C. Obtain an infrared spectrum (if using a KBr disk, prepare in the dark or red light) of the white form of the product. Compare your spectral data to that of the reference standard shown in Figure 7A.6. Then expose the disk to a bright fluorescent

Figure 7A.6   IR spectrum: photochromic imine.

light for 1 min and redetermine the infrared spectrum. What does a comparison of the two experimentally derived spectra tell you about this photochemically induced reaction?

Store the remainder of your azomethine–ylide in a clean, sealed vial, flushed with N_2, and protected from light. By your next laboratory period this material should be colorless.

QUESTIONS

7A-41. The heavily halogenated napthalenones were synthesized well over 100 years ago, but the structural details were sorted out less than 20 years ago. The compound, 2,3,4,4-tetrachloro-1-(4*H*)-naphthalenone (**I**), was discussed in the introduction to this sequence as an example of a substance that gains its photochromic activity by *bond homolysis*. In the synthesis of **I** two tetrachloro isomers were originally isolated. Compound **I** has photochromic activity while the second isomer (**II**) is simply a yellow-colored material. The structure of II has been determined to be 2,2,3,4-tetrachloro-1-(2*H*)-naphthalenone. A key piece of physical evidence that allowed the assignment of the structures was the infrared frequencies of the carbonyl groups. These frequencies were found to be 1701 and 1675 cm^{-1}. Which wavenumber value belongs with which structure?

Explain your reasoning.

7A-42. Explain why the neutral aziridine ring system is isoelectronic with the cyclopropyl carbanion.

7A-43. As shown in Question 7A-42 , the aziridine ring system is isoelectronic with the cyclopropyl anion. Based on the theory of electrocyclic reactions these *even-number* π-electron-pair systems would be expected to undergo photochemical ring-cleavage in *disrotatory* fashion. Huisgen et al. have shown that in dimethyl 1-(4-methoxyphenyl) aziridine-*trans*-2,3-dicarboxylate (**I**), upon photochemical excitation the ring opens to a 1,3-dipolar azomethine–ylide in a *disrotatory* cleavage that can be trapped by the addition of dimethyl acetylenedicarboxylate, which acts as a dipolarophile:

I

(a) Give a perspective drawing of the expected adduct.
(b) Give a perspective drawing of the expected adduct that would be formed from the blue azomethine ylide formed from our photochromic imine.
(c) Is the relative stereochemistry the same in the two adducts in *a* and *b*? Explain.

7A-44. The dramatic change in color that occurs when the diazabicyclo[3.1.0]hex-3-ene ring system isomerizes on exposure to long wavelength UV radiation is related to what structural or electronic changes (or both) in the photochemically induced system?

7A-45. Discuss the following:
(a) What role does ammonium bromide play in the conversion of the trans-substituted aziridine to the diazabicyclo[3.1.0] hex-3-ene derivative?
(b) Why is this particular bromide salt used?

BIBLIOGRAPHY

DoMinh, T.; Trozzolo, A. M. *J. Am. Chem. Soc.* **1972,** *94,* 4046.

Heine, H. W.; Weese, R. H.; Cooper, R. A.; Durbetaki, A. J. *J. Org. Chem.* 1967**,** *32,* 2708.

Hermann, H.; Huisgen, R.; Mäder, H. J. *Am. Chem.* **1971,** *93,* 1779.

Huisgen, R.; Scheer, S.; Huber, H. *J. Am. Chem. Soc.* **1967,** *89,* 1753.

Marckwald, W. Z. *Phys. Chem. (Leipzig)* **1899,** *30,* 140.

Padwa, A; Glazer, E. *J. Am. Chem. Soc.* **1972,** *94,* 7788.

Padwa, A.; Clough, S; Glazer, E. *J. Am. Chem. Soc.* **1970,** *92,* 1778. *ibid.,* **1970,** *92,* 6997.

Turner, A. B.; Heine, H.W.; Irving, J.; Bush, J. B. *J. Am. Chem. Soc.* **1965,** *87,* 1050.

Ullman, E. F.; Henderson, W. A. *J. Am. Chem. Soc.* **1964,** *86,* 5050.

Woodward, R. B.; Hoffmann, R. *Angew. Chem. Int. Ed. Engl.* **1969,** *8,* 781.

Zincke, T.; and Kegel, O. *Berichte* **1888,** *21,* 1030.

For an overview of photochemistry, see:

Smith, M. B. *March's Advanced Organic Chemistry*, 8th ed.; Wiley: New York, 2019.

Also see references cited in Experiment [6].

C₈H₈apter 8

SPECTROSCOPIC IDENTIFICATION OF ORGANIC COMPOUNDS

INFRARED SPECTROSCOPY

The wavelike character of electromagnetic radiation can be expressed in terms of velocity v, frequency ν, and wavelength λ of sinusoidally oscillating electric and magnetic vectors traveling through space (Fig. 8.1). Frequency is defined as the number of waves passing a reference point per unit time, usually expressed as cycles per second (s^{-1}) or hertz (Hz). The velocity of the wave, therefore, equals the product of frequency and wavelength:

$$v = \nu\lambda$$

If the wavelength (the distance between the wave maxima or alternate nodes) is measured in centimeters, v is expressed in centimeters per second (cm/s). For radiation traveling in a vacuum, v becomes a constant, c ($c \sim 3 \times 10^{10}$ cm/s), for all wavelengths. When electromagnetic radiation traverses other media, however, the velocity changes. The ratio of the speed in a vacuum, c, to the matrix velocity, v, is termed the *refractive index, n,* of the material:

$$n = \frac{c}{v}$$

Since n is frequency-dependent, the frequency at which the refractive index is measured must be specified. Frequency, however, has been shown to be independent of the medium and, therefore, remains constant. Wavelength thus varies inversely with n.

$$\lambda = \frac{c}{n\nu}$$

Since the velocity of electromagnetic radiation in a vacuum is normally greater than that in any other medium, n will generally be greater than 1 at all frequencies. Thus the wavelength must become shorter for a particular frequency when measured in any matrix.

Frequency can be considered to be a more fundamental property of radiation because it is independent of the medium. This property also requires that the energy E associated with the radiation be matrix independent, because E is directly proportional to frequency by

$$E = h\nu$$

Chapter 8: C₈H₈, Cubane
Eaton and Cole (1964).

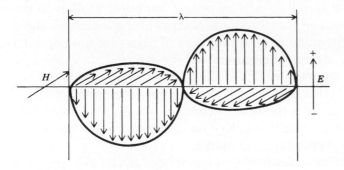

Figure 8.1   Electromagnetic wave. *H*, magnetic field; *E*, electric field; λ wavelength.

where *E* equals the energy of a photon, which is related to frequency *v* by Planck's constant (*h*) (6.6×10^{-27} erg s or 6.6×10^{-34} J s).

The vibrational states present in molecules can be excited by absorption of photons. The nuclear masses and bond force constants determine the separation of these states and, therefore, the energies of the photons involved in the absorption process. The corresponding radiation frequencies fall predominantly in the IR region (10^{14}–10^{12} Hz) of the electromagnetic spectrum.

The IR spectrum is currently measured in *wavenumbers*, $\tilde{v}$, which are units proportional to frequency and energy. The wavenumber is defined as

$$\tilde{v} = \frac{v}{c} = \frac{E}{hc}$$

and as

$$v = \frac{c}{n\lambda} \qquad \text{then in air} \qquad \tilde{v} = \frac{\tilde{1}}{\lambda}$$

The wavenumber, as expressed in units of reciprocal centimeters (cm^{-1}) (the number of waves per centimeter), offers several advantages:

1. Wavenumbers are directly proportional to frequency and are expressed in much more convenient numbers (in this region of the spectrum), 5000–500 cm^{-1}.

2. As shown above, wavenumbers are easily converted to wavelength values. The reciprocal of $\tilde{v}$ and conversion of centimeters to wavelength units are all that is required (this is particularly handy because much early IR data were recorded linearly in wavelength). The wavelength unit employed in most of these spectra was the micron, μ (1×10^{-4} cm). The micron has been replaced by a unit expressed in meters, the micrometer, μm (1 μm = 1×10^{-6} m).

3. Because the wavenumber is directly proportional to frequency and energy, the use of wavenumbers allows spectra to be displayed linearly in energy, which is a distinct aid in sorting out related vibrational transitions.

INTRODUCTION TO GROUP FREQUENCIES: INTERPRETATION OF INFRARED SPECTRA

Studies of the vibrational spectra of thousands of molecules have revealed that many of the normal modes associated with particular atomic arrangements may be transferred from one molecule to another. A *normal mode* of vibration is one of the residue *fundamental* vibrations of a molecular system in which the atomic displacements are all related by simple harmonic motion to the overall total vibrational motion (or vibrational energy) of the molecule. There are $3N - 6$ (where N = the number of atoms) *normal modes* (or fundamental vibrations or vibrational degrees of freedom—all these terms are essentially synonymous) present in all nonlinear molecules. Linear molecules have only $3N - 5$ normal modes—in this case there is one more normal mode of vibrational energy present, because a rotational degree of freedom has been lost. Rotation around the molecular axis involves no energy because the atomic nuclei are assumed to be point sources of matter.

Operating under selection rules these normal modes of vibration give rise to absorption bands in the infrared region of the spectrum (see, for example, the infrared spectrum of *n*-hexane, Fig. 8.2, p. 565). In the analysis these modes are often assigned numbers. For example, the 30 modes of benzene (where $N = 12$ in the $3N - 6$ expression) can be assigned 1 through 30 or the numbering can be done using any one of a number of different criteria. Subscripts *a* and *b* are often used to indicate doubly degenerate modes, that is, modes that have identical energies (and thus required to have the same frequency. One of the numbering systems for the benzene ring is used here when the aromatic ring stretching vibrations are identified (see Table 8.6 for a more detailed discussion of normal modes).

Many of these vibrational frequencies are associated with small groups of atoms that are essentially uncoupled from the rest of the molecule. The absorption bands that result from these modes, therefore, are characteristic of the small group of atoms regardless of the composition of other parts of the molecule. These vibrations are known as the *group frequencies*. Interpretation of infrared spectra of complex molecules based on group frequency assignments is an extremely powerful aid in the elucidation of molecular structure.

The following four factors make significant contributions to the development of a good group frequency from a molecular vibration:

1. The group has a large dipole-moment change during vibrational displacement. This change in moment is formally related to the efficiency of absorption of radiation during the molecular displacement by the expression $I\alpha(\delta\mu/\delta Q)^2$ (where I = intensity, μ = electric dipole moment, and Q = the normal coordinate [a mathematical description of the vibration]). Thus, if $\delta\mu/\delta Q$ is large, there is a large absorption of infrared radiation, which gives rise to very intense bands (the intensity is dependent on the square of the moment change at that vibrational frequency).

2. The presence of a large force constant, so that for many of these groups the stretching frequency occurs at high values above the fingerprint region.

3. The fundamental mode occurs in a frequency range that is reasonably narrow (little coupling), but sensitive enough to the local environment to allow for considerable interpretation of the surrounding structure.

4. The range of frequencies is determined by a number of factors that are now well understood in terms of the mass, geometric, electronic, intramolecular, and intermolecular effects.

Strategies for Interpreting Infrared Spectra

1. Divide the spectrum into two parts at 1350 cm^{-1}.

2. Above 1350 cm^{-1}, absorption bands have a high probability of being good group frequencies. The interpretation is usually reliable and free from ambiguities. We can be much more confident of our assignments in this region even with rather weak bands.

3. Because of the reliability of the high-wavenumber region, we always begin the interpretation of a spectrum at this end.

4. Bands below 1350 cm^{-1} may be either group frequencies or fingerprint frequencies.

5. Below 1350 cm^{-1}, group frequencies are less easily assigned. In addition, even if a reliable group frequency occurs in this region, absorption at that frequency is not necessarily a result of that mode. That is, fingerprint bands can also randomly occur in the same location as reliable group frequency bands and the observer cannot usually distinguish which type of band is present.

6. To make more confident assignments below 1350 cm^{-1}, it is helpful to be able to associate a secondary property, such as band shape, with the particular mode. For example, it helps to know whether the band is very intense, broad, sharp, occurs as a characteristic doublet, gives the correct frequency shift on isotopic substitution, or the like.

7. A good rule to remember is that in the fingerprint region the **absence** of a band is more important than the presence of a band. If a band is absent, you can conclude with confidence that a reliable group frequency assigned to this region is absent and therefore the group must be absent from the sample. At the same time you also know that no interfering fingerprint bands occur in the region.

8. Before beginning the interpretation, note the sampling conditions and determine as much other information about the sample as possible—such as molecular weight, melting point, boiling point, color, odor, elemental analysis, solubility, and refractive index.

9. In the interpretation try to assign the most intense bands first. These bands very often will be associated with a polar functional group.

10. Do not try to assign all the bands in the spectrum. Fingerprint bands are unique to a particular system. Occasionally, intense bands will be fingerprint-type absorptions; these bands, generally, will be ignored in the interpretation. Fingerprint bands do, however, play an important role when infrared data are employed for identification purposes.

11. The correlation chart (back endpaper) can act as a helpful quick aid for checking potential assignments. It is not a substitute for understanding the theory and operation of group frequency logic. *The use of the correlation chart without a good knowledge of group frequencies is the shortest path to disaster!*[1]

12. Try to utilize the so-called *macro group frequency* approach. That is, if the functionality or molecular structural group requires the presence of more

[1]Bellamy, L. J. *The Infrared Spectra of Complex Molecules*, 3rd ed.; Chapman & Hall: London, 1975, p. 3.

than a single group frequency vibrational mode, make sure that all modes are correctly represented. The *macro frequency train* represents a very powerful approach to the interpretation of relatively complex spectral data. This technique is at the core of current work on the automatic computer interpretation of infrared spectra. Contained in the product characterization section of the experiments given in Chapters 6 and 10 are 12 detailed discussions (in Chapter 6, Experiments [5A], [5B], [6], [7], [8], [11], [14], [19], and [20B]; in Chapter 10, Experiments [1adv], [4adv], and [6adv]) that demonstrate the operational use of the *macros*. Careful reference to these discussions will be very helpful in the initial stages of learning these interpretation techniques. It should become relatively easy to extend this interpretive approach to other reactions by reference to common infrared library files. Practice using *macro group frequencies* will pay big dividends in the laboratory. This last suggestion is perhaps the most important strategy to master in learning to interpret infrared spectra.

A SURVEY OF GROUP FREQUENCIES IDENTIFIED IN ORGANIC MOLECULES

The useful group frequencies are listed in the following sets of tables.

In the following tables the vibration motion of the localized sections of the molecules assigned to a particular group's frequencies is often described using the following terms:

1. *Symmetric stretch* or *symmetric bend (deformation):* Here the local group retains its symmetry during displacement. The symmetric bend of the methylene group, CH_2, is often termed the *scissoring* bend, while the symmetric bend of a methyl group, CH_3, is termed an *umbrella* mode—both descriptions imply the type of displacements that are taking place in the vibration.

2. *Antisymmetric stretch* or *bend* (deformation): Here the vibrating system loses its symmetry during the vibration. The displacements involve a reflective (mirror image) displacement during the opposite phase of the simple harmonic vibration and the motion is termed antisymmetric rather than asymmetric. Antisymmetric bends (deformations) are often classified as *twisting, rocking,* and *wagging* vibrations.

3. Some vibrations involving planar sections of the molecules are referred to as *in-plane* or *out-of-plane*. They can be either symmetric or antisymmetric in nature and if they involve the bending of all the displaced bonds of a set of atoms moving together in the same direction they will be termed *all-in-phase*.

4. *Degenerate* vibrations are defined as the case where two or more molecular vibrations are required to occur at the same frequency.

5. *Overtones* (integral multiples of the fundamental mode frequency) and *sum tones* (the sum of two different fundamental modes) are forbidden bands that are almost always very weak. Occasionally these bands are good group frequencies.

Group Frequencies of the Hydrocarbons

Alkanes   Alkynes

Alkenes   Arenes

Alkanes. The C—H vibrational modes of the alkanes (or mixed compounds containing alkyl groups) that are characteristic and reliable group frequencies are summarized in Table 8.1.

These modes give rise to characteristic bands found in the infrared spectrum of alkanes, such as in the spectrum of *n*-hexane shown in Figure 8.2.

Alkenes C═C Stretching. It is possible to classify open-chain unsaturated systems into two groups, those with C═C stretching modes falling above 1660 cm^{-1} and those with modes falling below 1660 cm^{-1} as shown in Table 8.2.

Alkenes C—H. Several fundamental modes associated with the alkene C—H groups are group frequencies and are summarized in Table 8.3.

Alkynes. The group frequencies of the alkynes are summarized in Table 8.4.

Table 8.1 Alkane Vibrational Normal Modes

C—H Vibrational Modes	$\tilde{v} \pm 10$ (cm^{-1})
Methyl groups	
Antisymmetric (degenerate) stretch	2960
Symmetric stretch	2870
Antisymmetric (degenerate) deformation	1460
Symmetric (umbrella) deformation	1375
Methylene groups	
Antisymmetric stretch	2925
Symmetric stretch	2850
Symmetric deformation (scissors)	1450
Rocking mode (all-in-phase)	720

Table 8.2 Substitution Classification of C═C Stretching Frequencies

C═C Normal Modes	$\tilde{v}$ (cm^{-1})
Trans-, tri-, tetrasubstituted	1680–1665
Cis-, vinylidene- (terminal 1,1-disubstituted), vinyl-substituted	1660–1620

Figure 8.2   IR spectrum: *n*-hexane.

Table 8.3 Alkene Vibrational Normal Modes	
C—H Vibrational Modes	$\tilde{\nu} \pm 10$ (cm^{-1})
Stretching modes	
Antisymmetric stretch (=CH$_2$)	3080
Symmetric stretch (=CH$_2$)	3020
Uncoupled stretch (=CH$_2$)	3030
Out-of-plane bending modes	
Vinyl group	
Trans hydrogen atoms (in-phase)	990
Terminal hydrogen atoms (wag)	910
Vinylidene group (=CH$_2$)	
Terminal (wag)	890
Trans alkene	
Trans hydrogen atoms (in-phase)	965
Cis alkene	
Cis hydrogen atoms (in-phase)	~700
Trisubstituted alkene	
Uncoupled hydrogen atom	820
Tetrasubstituted alkene: no vibrational modes seen in IR	

Table 8.4 Alkyne Vibrational Normal Modes	
C≡C, C—H Vibrational Modes	$\tilde{\nu} \pm 10$ (cm^{-1})
Triple-bond stretch (monosubstituted)	2120
Triple-bond stretch (disubstituted)	2225
R—C≡C—H bond stretch (monosubstituted)	3300

Arenes. The group frequencies of the *phenyl* group can be classified as carbon-hydrogen vibrations consisting of stretching and out-of-plane bending modes, plus carbon–carbon ring stretching and out-of-plane bending modes. The in-plane bending modes in both cases are not effective group frequencies.

The wavenumber values for the all-in-phase C—H bending vibrations are presented in Table 8.5.

The generalized group frequencies of the arenes are summarized in Table 8.6.

Group Frequencies of Carbonyl Groups: C=O

The carbonyl group is perhaps the single most important functional group in organic chemistry. It is certainly the most commonly occurring functionality. Infrared spectroscopy can play a powerful role in the characterization of the carbonyl because this group possesses all of the properties that give rise to an excellent group frequency (Table 8.7).

Table 8.5 Arene Out-of-Ring-Plane C—H Deformation Modes	
Arene Fundamentals (C—H bend) (Number of C—H groups directly adjacent)	$\tilde{\nu}$ Range (cm^{-1})
5	770–730
4	770–735
3	810–750
2	860–800
1	900–845

Table 8.6 Arene Group Frequencies

Arene Fundamentals	$\tilde{v}$ Range (cm^{-1})
C—H stretch	3100–3000
C=C ring stretch (v_{8a})	1600 ± 10
C=C ring stretch (v_{8b})	1580 ± 10
C=C ring stretch (v_{19a})	1500 ± 10
C=C ring stretch (v_{19b})	1450 ± 10
C—H out-of-plane bend (1H)	900–860
C—H out-of-plane bend (2H)	860–800
C—H out-of-plane bend (3H)	810–750
C—H out-of-plane bend (4H)	770–735
C—H out-of-plane bend (5H)	770–730
C—C ring out-of-plane bend (1; 1,3; 1,3,5-substituted)	690 ± 10
C—H out-of-plane bend sum tones	2000–1650

The major factors perturbing carbonyl frequencies can be summarized as follows:

Factors That Raise the C=O Frequency

1. Substitution with electronegative atoms
2. Decrease in C—CO—C internal bond angle

Factors That Lower the C=O Frequency

1. Conjugation
2. Hydrogen bonding

Several of these factors may be operating simultaneously, so careful judgment as to the contribution of each individual effect must be exercised in predicting

Table 8.7 Carbonyl Group Vibrational Frequencies

Compound	$\tilde{v}$ (cm^{-1})
Ketones, aliphatic, open-chain (R$_2$CO)	1725–1700
Ketones, conjugated	1700–1675
Ketones, cyclic (listed by ring size (3MR to 7MR))	1906, 1782, 1744, 1710, 1699
Acyl halides	>1800
Esters, aliphatic	1755–1735
Esters, conjugated	1735–1720
Esters (conjugated to oxygen)	1780–1760
Lactones (listed by ring size (4MR to 7MR))	1832, 1775, 1740, 1727
Anhydrides: aliphatic, open-chain	1840–1810 and 1770–1740
Carboxylic acids, aliphatic	1725–1710
Amides	(see Tables 8.22–8.24)
Lactams (listed by ring size (4MR to 6MR))	1750, 1695, 1670
Aldehydes	1735–1720

Table 8.8 Vibrational Normal Modes of the Hydroxyl Group

$\tilde{v}$ (cm^{-1})	Intensity	Mode Description
3500–3200	Very strong	O—H stretch (only strong when hydrogen bonded)
1500–1300	Medium strong	O—H in-plane bend (overlap CH$_2$, CH$_3$ bend)
1260–1000	Strong	C—C—O antisymmetric stretch
650	Medium	O—H out-of-plane bend

Table 8.9 Substitution Effects on C—O Stretch of Aliphatic Alcohols

Type of Alcohol	$\tilde{v}_{c-o}$ (cm^{-1})
RCH$_2$—OH (primary)	1075–1000
R$_2$CH—OH (secondary)	1150–1075
R$_3$C—OH (tertiary)	1200–1100
C$_6$H$_5$—OH (phenol)	1260–1180

carbonyl frequencies. This judgment develops rapidly with practice at interpretation.

Group Frequencies of the Heteroatom Functional Groups

(Alkanes)

Alcohols	Aldehydes	Ketones	Esters
Acyl halides	Carboxylic acids	Anhydrides	Ethers
Amines, primary	Nitriles	Amides, primary	Amides, secondary
Isocyanates	Thiols	Halogens	Phenyl

Hexane. Refer to Table 8.1.

Alcohols. A very intense band appears at ~3350 cm^{-1}, which is assigned to the stretching mode of the O—H group (Table 8.8).

Of particular importance is a strong band in the spectrum of aliphatic alcohols usually located near 1060 cm^{-1}. This absorption has been identified as the C—O stretching mode. The vibrational displacements of this fundamental are similar to the antisymmetric stretch of water. Since the vibration involves significant displacement of the adjacent C—C oscillator, the vibration will be substitution-sensitive. These latter shifts can be of value in determining the nature of the alcohol (primary, secondary, or tertiary—see Table 8.9).

Aldehydes. The aldehyde functional group gives rise to several good group frequencies (Table 8.10).

Ketones. The only group frequency mode associated directly with aliphatic ketones is the stretching frequency ($\tilde{v}_{c=o}$ ~1720 cm^{-1}), which occurs within the expected region as discussed above. There are, however, several other related bands (Table 8.11).

Table 8.10 Vibrational Normal Modes of the Aliphatic Aldehyde Group

$\tilde{v}$ (cm^{-1})	Intensity	Mode Description
2750–2720	Weak to medium	C(O)—H stretch
1735–1720	Very strong	C=O stretch
1420–1405	Medium	CH$_2$ symmetric bend, —CH$_2$—α to —CHO
1405–1385	Medium	C—H in-plane bend

Table 8.11 Normal Vibrational Modes of Aliphatic Ketones

$\tilde{\nu}$ (cm^{-1})	Intensity	Mode Description
3430–3410	Very weak	Overtone of carbonyl stretch
1725–1700	Very strong	C=O
1430–1415	Medium	—CH$_2$— symmetric bend, —CH$_2$—α to ketone C=O

Table 8.12 Vibrational Normal Modes of the Aliphatic Ester Group

$\tilde{\nu}$ (cm^{-1})	Intensity	Mode Description
1755–1735	Very strong	C=O stretch
1370–1360	Medium	CH$_3$ symmetric bend α to ester C=O
1260–1230	Very strong	C—CO—O antisymmetric stretch —acetates
1220–1160	Very strong	C—CO—O antisymmetric stretch —higher esters
1060–1030	Very strong	O—CH$_2$—C antisymmetric stretch —1° acetates
1100–980	Very strong	O—CH$_2$—C antisymmetric stretch —higher esters (may overlap with upper band)

Table 8.13 Vibrational Normal Modes of the Acyl Halide Group

$\tilde{\nu}$ (cm^{-1})	Intensity	Mode Description
1810–1800	Very strong	C=O stretch, acyl chlorides
1415–1405	Strong	—CH$_2$—symmetric bend, α to —COCl carbonyl

Esters

The very strong band found at ~1745 cm^{-1} is typical of the carbonyl frequency of an aliphatic ester, particularly aliphatic acetate esters (Table 8.12).

Acyl Halides. The carbonyl stretching mode dominates the spectrum in aliphatic acyl halides. In acyl chlorides it is an extremely intense band occurring near 1800 cm^{-1} (Table 8.13).

Carboxylic Acids. Acids, observed in the solid or pure liquid states, often possess a very intense band with a width at one-half peak height of about 1000 cm^{-1}, which covers the region 3500–2200 cm^{-1}. This absorption is characteristic of very strongly hydrogen-bonded carboxylic acid groups (Table 8.14).

Anhydrides. The coupling of the anhydride carbonyls through the ether oxygen splits the carbonyls (in the aliphatic case $\tilde{\nu}_{c=o}$ = ~ 1830, 1760 cm^{-1}) by about 70 cm^{-1} (Table 8.15).

Ethers. The large intensity associated with antisymmetric C—O—C stretching mode relative to the other bands occurring in this part of the fingerprint

Table 8.14 Vibrational Normal Modes of the Carboxylic Acid Group

$\tilde{v}$ (cm^{-1})	Intensity	Mode Description
3500–2500	Very very strong	O—H stretch intensified by hydrogen bonding
2800–2200	Very weak	Overtone and sum tones
1725–1710	Very strong	C=O antisymmetric hydrogen-bonded dimer stretch
1450–1400	Strong	CH$_2$—CO—O antisymmetric stretch mixed with O—H bend
1300–1200	Strong	CH$_2$—CO—O antisymmetric stretch mixed with O—H bend
950–920	Medium	Out-of-plane O—H bend, acid dimer

Table 8.15 Vibrational Normal Modes of the Anhydride Group

$\tilde{v}$ (cm^{-1})	Intensity	Mode Description
1840–1810	Very strong	C=O in-phase stretch
1770–1740	Very strong	C=O out-of-phase stretch
1420–1410	Strong	—CH$_2$— symmetric bend α to C=O
1100–1000	Very strong	C=O stretch, mixed modes

Table 8.16 Vibrational Normal Modes of the Ether Group

$\tilde{v}$ (cm^{-1})	Intensity	Mode Description
1150–1050	Strong	C—O—C antisymmetric stretch, mixed mode

region, particularly in aliphatic compounds, makes it possible, in most cases, to assign with confidence the observed strong band (Table 8.16).

Primary Amines. The spectra of these bases usually possess two bands ($\tilde{v}_{N—H}$ = ~ 3380, ~3300 cm^{-1}) of medium-to-weak intensity. These bands are assigned to the antisymmetric and symmetric N—H stretching modes, respectively, of the primary amino group (Table 8.17).

Nitriles. The very strong triple bond present in the nitrile group (as in the case of the alkynes) contributes to an unusually high stretching frequency, and the polar character of the group gives rise to very intense bands (Table 8.18).

Primary Amides. The highly polar amide group leads to very strong hydrogen bonding, which in turn leads to greatly intensified N—H antisymmetric and symmetric stretching modes ($\tilde{v}_{N—H}$ = ~ 3375, ~ 3200 cm^{-1} see Table 8.19).

Secondary Amides. The single N—H group present in secondary amides gives rise to a very strong band near about 3300 cm^{-1}, which is indicative of strong hydrogen bonding. The drop in frequency from that of the primary —NH$_2$ scissoring mode near 1600 cm^{-1} allows for confident assignment of substitution on secondary amide groups (Table 8.20).

Table 8.17   Vibrational Normal Modes of the Primary Amine Group

$\tilde{v}$ (cm^{-1})	Intensity	Mode Description
3400–3200	Weak to medium	NH$_2$ stretch doublet, (antisymmetric and symmetric modes)
1630–1600	Medium	NH$_2$ symmetric bend
820–780	Medium	NH$_2$ wag

Table 8.18   Vibrational Normal Modes of the Nitrile Group

$\tilde{v}$ (cm^{-1})	Intensity	Mode Description
2260–2240	Strong	C≡N stretch, aliphatic
2240–2210	Strong	C≡N stretch, conjugated

Table 8.19   Vibrational Normal Modes of the Primary Amide Group

$\tilde{v}$ (cm^{-1})	Intensity	Mode Description
3400–3150	Very strong	—NH$_2$ antisymmetric and symmetric stretching modes, hydrogen bonded
1680–1650	Very strong	C=O stretch, hydrogen bonded
1660–1620	Strong	—NH$_2$ symmetric bend (overlap with C=O stretch)
1430–1410	Strong	—CH$_2$— symmetric bend α to amide carbonyl
750–650	Medium	—NH$_2$ wag

Table 8.20   Vibrational Normal Modes of the Secondary Amide Group

$\tilde{v}$ (cm^{-1})	Intensity	Mode Description
3350–3250	Strong	—NH stretch, intensified by hydrogen bonding
3125–3075	Medium	Overtone N—H bend (see also)
1670–1645	Very strong	C=O stretch, hydrogen bonded
1580–1550	Strong	N—H in-plane bend (see also)
1415–1405	Strong	—CH$_2$—symmetric bend α to amide C=O
1325–1275	Medium	C—N stretch mixed with N—H in-plane bend
725–680	Medium	N—H out-of-plane bend

Table 8.21   Vibrational Normal Modes of the Amide Carbonyl: Solution and Solid-Phase Data

Amide	Dilute Solution (cm^{-1})	Solid (cm^{-1})
R—CO—NH$_2$ (primary)	~1730	~1690–1650
R—CO—NHR (secondary)	~1700	~1670–1630
R—CO—NR$_2$ (tertiary)	~1650	~1650

Studies of amide carbonyl frequencies in dilute nonpolar solution indicate that hydrogen-bonding effects are largely responsible for the low frequencies observed with primary and secondary amides, but play no role in tertiary amides (see Table 8.21).

Table 8.22	Vibrational Normal Mode of the Isocyanate Group	
$\tilde{v}$ (cm^{-1})	Intensity	Mode Description
2280–2260	Very strong	—N=C=O antisymmetric stretch

Table 8.23	Vibrational Normal Mode of the Thiol Group	
$\tilde{v}$ (cm^{-1})	Intensity	Mode Description
2580–2560	Weak	S—H stretch

Table 8.24	Vibrational Normal Mode of the Alkyl Chloro Group	
$\tilde{v}$ (cm^{-1})	Intensity	Mode Description
750–650	Strong	C—Cl stretch

Table 8.25	Vibrational Normal Modes of the Aryl Chloro Group	
$\tilde{v}$ (cm^{-1})	Intensity	Mode Description
3080	Medium	C—H stretch, bonded to ring carbon
1585	Strong	v_{8a} ring stretching
1575	Weak	v_{8b} ring stretching
1475	Strong	v_{19a} ring stretching
1450	Strong	v_{19b} ring stretching
747	Strong	C—H all-in-phase, out-of-plane bend
700	Strong	C—Cl stretch
688	Strong	Ring deformation
1945, 1865, 1788, 1733	All weak	Sum tones, out-of-plane C—H bends, pattern matches monosubstitution of ring

Isocyanates. The range of stretching frequencies observed for alkyl-substituted isocyanates is very narrow, $\tilde{v} = 2280 - 2260 \text{ cm}^{-1}$, which implies little coupling to the rest of the system (Table 8.22).

Thiols. Although weak absorption is associated with the S—H stretching fundamental the band is generally found in a very open region of the infrared spectrum (Table 8.23).

Alkyl Halides. The massive halogen atom is connected to the alkyl section by a fairly weak but highly polarized bond, which dictates that the C—X stretching frequency appears as an intense band at low frequencies (Table 8.24).

Aryl Halides (Chlorobenzene). The final system to be considered in this section is the aryl halide, chlorobenzene. Based on the above assignments the group frequencies of the complete hydrocarbon portion and the heteroatom functional group can be assigned as in Table 8.25.

INFRARED SPECTROSCOPY INSTRUMENTATION AND SAMPLE HANDLING

Instrumentation

The workhorse infrared instrument used for routine characterization of materials in the undergraduate organic laboratory is the optical-null double-beam grating spectrometer (Fig. 8.3a). For a discussion of double-beam spectrometers, see the UV–vis instrumentation discussion (p. 680). Although many undergraduate instructional laboratories still utilize this type of instrumentation, the winds of change are blowing. ATR (attenuated total reflectance)

Figure 8.3a   The Perkin–Elmer model 710B IR spectrometer. From Zubrick, J. W. *The Organic Chem Lab Survival Manual,* 9th ed.; Wiley: New York, 2013. (Reprinted by permission of John Wiley & Sons, Inc., New York.)

Figure 8.3b   Attenuated Total Reflectance (ATR).

FT-IRs and lower-cost FT-IR spectrometers, which both depend on computer manipulation of the spectral data, are becoming the infrared instructional instrumentation of choice. Two of the many benefits when using the more expensive ATR FT-IR spectrometer are faster sampling and spectral reproducibility. Even though ATR plates are prone to contamination, one key advantage when working with an ATR crystal (Fig. 8.3b) is that no sample preparation is required. This avoids the need to use KBr or mineral oil (Nujol) when working with solids! Many of the spectra utilized in the interpretive discussions were generated on a prototype of this kind of infrared instrumentation, the Perkin–Elmer model 1600.

This instrument can acquire 16 scans and carry out the required calculations in 42 s. While the spectrum is being printed out (~40 s) the data on a second sample can be acquired. The 42-s acquisition data are significantly superior to those recorded by dispersive instruments that take from 5 to 8 min to scan a sample from 4000 to 600 cm^{-1}.

Sample Handling in the Infrared

If not working with an ATR FT-IR, the standard techniques of sample preparation employed to obtain infrared spectra of microscale laboratory products are the use of capillary films with liquids on NaCl (or AgCl) plates and the use of KBr disks and melts in solids. This, of course, assumes that for a spectrum to be obtained in the infrared region, the sample must be mounted in a cell that is transparent to the radiation. With an ATR FT-IR spectrometer, liquids and solids are placed in direct contact with the ATR crystal. Since glass and quartz absorb in this spectral region, cells constructed of these materials cannot

Figure 8.4 **IR salt plates and holder.** From Zubrick, J. W. *The Organic Chem Lab Survival Manual*, 9th ed.; Wiley: New York, 2013. (Reprinted by permission of John Wiley & Sons, Inc., New York.)

be used when working with spectrometers not utilizing ATR technology. Accordingly, alkali metal halides have large spectral regions of transmission in the infrared, as do silver halides. Sodium chloride is the most commonly used material in cell windows in infrared sampling.

Liquid Samples. For materials boiling above 100 °C, the procedure is very simple. Using a syringe or Pasteur pipet, place 3–5 µL of sample on a polished plate of sodium chloride or silver chloride or directly on the ATR crystal. If working with a NaCl or AgCl plate, cover it with a second plate of the same material and clamp it in a holder that can be mounted vertically in the instrument. Be sure that the plates are clean when you start and when you are through! Obviously, the sodium chloride plates cannot be cleaned with water. Silver chloride is very soft and scratches easily; it also must be kept in the dark when not in use, because it darkens quickly in direct light. Spectra obtained in this fashion are referred to as *capillary film spectra* (Fig. 8.4).

Solution Spectra and the Spectra of Materials Boiling Below 100 °C. These samples generally require a sealed cell constructed of either sodium chloride or potassium bromide windows. Such cells are expensive and need careful handling and maintenance. They are assembled as shown in Figure 8.5.

Solid Samples Using Non-ATR Spectrometers. Solid powders could be mounted on horizontal sodium chloride plates, and the beam diverted through

Figure 8.5 **Sealed demountable cell or demountable cell with ports.** (Courtesy of the Perkin-Elmer Corp., Norwalk, CT.)

the sample by mirrors. This procedure would make sample preparation very easy for solids. Unfortunately, powders tend to scatter the entering radiation very efficiently by reflection, refraction, and molecular scattering. Some of these effects become rapidly magnified at higher frequencies, since they vary with the fourth power of the frequency. Thus, in solid-sample scattering a lot of energy is scattered away from the sample beam. This results in poor absorption spectra, as the instrument is forced to operate at very low energies. The detector cannot differentiate between a drop in energy from absorption or one derived from scattering.

For materials melting below 80 °C, the simplest technique is to mount the sample between two salt plates and **gently** warm with a heat lamp until melting occurs. With the fast acquisition times of interferometers, the melting point range is now as high as 100 °C and the spectrum can be obtained so rapidly that the sample does not have time to cool and crystallize. (Heated cells are used in research laboratories, but they are rather expensive and difficult to maintain.) For substances melting above 100 °C, the sampling routine most often employed to avoid scattering problems is the potassium bromide (KBr) disk. Potassium bromide is transparent to infrared radiation in the region of interest. Most important, however, the KBr makes a much better match of the refractive indexes between the sample and its matrix than does air. Thus, reflection and refraction effects at the crystal faces of the sample are greatly suppressed.

In the KBr method the sample (2–3 mg) is finely ground in a mortar, the finer the better for lower reflection or refraction losses. Then, 150 mg of previously ground and dried KBr is added to the mortar and quickly mixed by stirring, not grinding, it with the sample. (Potassium bromide is very hygroscopic and will rapidly pick up water while being ground in an open mortar.) When mixing is complete the mixture is transferred to a die and pressed into a solid disk. Potassium bromide will flow under high pressure and seal the solid sample in a glasslike matrix. Several styles of dies are commercially available. For routine use a die consisting of two stainless steel bolts and a barrel is the simplest to operate (see Fig. 8.6). The ends of the bolts are polished flat to form the die faces. The first bolt is seated to within a turn or two of the head. Then the sample mixture is added (avoid breathing over the die while adding the sample). The second bolt is firmly seated in the barrel, and then the clamped assembly is tightened by a torque wrench to 240 in./lb. After standing for 1.5 min, the two bolts are removed, leaving the KBr disk mounted in the center of the barrel, which can then be mounted in the instrument. After the spectrum of the sample is run, the disk can be retrieved and the sample recovered if necessary (Fig. 8.6). *Always clean the die immediately after use. KBr is highly corrosive to steel.*

When infrared spectra are obtained, it is important to establish that the wavenumber values have been accurately recorded. Successful interpretation of the data often depends on very small shifts in these values. Calibration of the frequency scale is usually accomplished by obtaining the spectrum of a reference compound, such as polystyrene film. To save time, record absorption peaks only in the region of particular interest (this applies only to dispersive instrument derived data).

Turn "top" bolt down to form pellet

Barrel

Sample on this bolt

"Bottom" bolt halfway in

Figure 8.6   The KBr pellet minipress. From Zubrick, J. W. *The Organic Chem Lab Survival Manual,* 9th ed.; Wiley: New York, 2013. (Reprinted by permission of John Wiley & Sons, Inc., New York.)

NOTE. Most of the infrared spectra found in the experimental sections of the text, which are Fourier-transform-derived (Perkin–Elmer 1600), have been plotted on a slightly different scale than the other spectra presented in the text and on the website. The former spectra utilize a 12.5-cm^{-1}/mm format below 2000 cm^{-1} and undergo a 2:1 compression above 2000 cm^{-1} (25 cm^{-1}/mm).

QUESTIONS

NOTE. Some of the following questions assume that the student is familiar with the infrared material contained on the website available to refer to if needed.

8-1. The form of the C—H out-of-plane bending vibrations of the vinyl group are shown below:

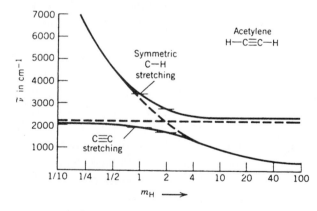

The first two vibrational modes give rise to excellent group frequencies, while the third fundamental does not lend itself to these correlations.

(a) Explain the factors that lead to the third vibrational mode being such a poor group frequency.
(b) Predict the location in the spectrum of the third fundamental vibration.

8-2. In the figure below, the mass of the terminal hydrogen atoms on the acetylene is hypothetically varied from zero to infinity. The response of the C—H symmetric stretching (3374 cm^{-1}) and triple-bond stretching (1974 cm^{-1}) modes to the change in mass is shown.

(a) Calculate the expected deuterium isotope shift for the C—H symmetric stretching mode. Is the hypothetical value close to the calculated value? Explain.
(b) Explain why the triple-bond stretching frequency is approximately 100 cm^{-1} higher for high-mass terminal isotopes (>100) than for the low-mass terminal isotopes (<3).

8-3. Acetylene has two C—H groups. It will have two C—H stretching frequencies, the in-phase and out-of-phase stretching modes. The in-phase (symmetric) stretch occurs at 3374 cm^{-1} and the out-of-phase stretch at 3333 cm^{-1}. Explain why the in-phase vibration is located at a higher frequency than the out-of-phase stretch.

8-4. The carbonyl stretching frequencies of a series of benzoyl derivatives are listed below:

x	$\tilde{v}_{C=O}$ (CCl$_4$)
2	1677
3	1686
4	1687
5	1686

Consider the $\tilde{v}_{C=O}$ of acetone at 1715 cm^{-1} as a reference frequency and identify the factors affecting $\tilde{v}_{C=O}$ in the series of compounds listed.

8-5. Explain how mass effects act to lower the carbonyl frequency, as well as how inductive and hyperconjugation effects act to raise the carbonyl frequency of aldehydes relative to ketones.

8-6. The carbonyl stretching frequency of aliphatic carboxylic acids in dilute solution is located near 1770 cm^{-1}. This frequency is much higher than the carbonyl frequency of these substances when measured neat (~1720 cm^{-1}). Also, it is considerably higher than the corresponding simple aliphatic ester value (1745 cm^{-1}). Explain.

8-7. In a number of cases, dipolar interactions control the frequency shifts found in carbonyl stretching vibrations. The table lists wavenumber shifts in going from neat to dilute nonpolar solutions. Explain the observed values.

Carbonyl Dipolar Interactions[a]	
Compound	$\Delta \tilde{v}$ (cm^{-1})
Acetyl chloride	15
Phosgene	13
Acetone	21
Acetaldehyde	23
N,N-Dimethylformamide	50
[a]Shift measured between dilute nonpolar solution and neat sample.	

8-8. The antisymmetric —CH$_2$—CO—O— stretching vibration in carboxylic acids is heavily mixed with the in-plane bending mode of the O—H group. In alcohols these two vibrations seldom show evidence of mechanical coupling. Explain.

8-9. Conjugation of the functional group in alkyl isocyanates has little impact on the antisymmetric —N=C—O stretching vibration located near 2770 cm^{-1} Explain.

8-10. In the infrared spectrum of 2-aminoanthraquinone (**I**) two carbonyl stretching frequencies are observed at 1673.5 and 1625 cm^{-1}:

2-Aminoanthraquinone
I

1-Hydroxyanthraquinone
II

(a) Assign carbonyl bands in the infrared spectrum to the carbonyl groups in structure **I** and explain your reasoning.

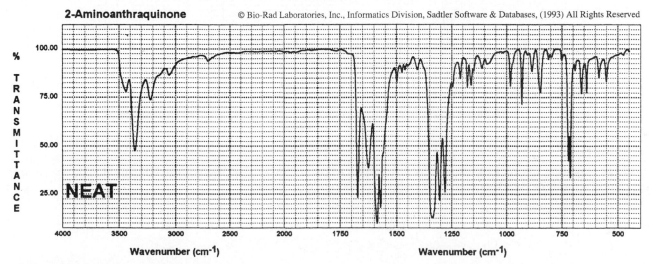

2-Aminoanthraquinone   © Bio-Rad Laboratories, Inc., Informatics Division, Sadtler Software & Databases, (1993) All Rights Reserved

IR spectrum: 2-Aminoanthraquinone.

(b) The infrared spectrum of 1-hydroxyanthraquinone (**II**) also exhibits two carbonyl frequencies, which are located at 1675 and 1637 cm^{-1}. Assign the carbonyl groups to the related absorption bands. Explain your reasoning.

IR spectrum: 1-Hydroxyanthraquinone.

(c) The spectrum of 2-hydroxyanthraquinone exhibits a single carbonyl stretching frequency near 1673 cm^{-1}. Explain why a single carbonyl band would be expected in the system and why this vibration is located at 1673 cm^{-1}.

8-11. Suggest a possible structure for the hydrocarbon C_6H_{14}, which has the infrared spectrum shown here: Is there more than one correct structure?

IR unknown spectrum: C_6H_{14}.

8-12. The hydroxylamine **I** can be oxidized by MnO_2 to the amide oxohaemanthidine (**II**). In dilute solution, the carbonyl absorption band of **II** occurs at 1702 cm^{-1}. Explain this observation.

8-13 The infrared spectra of the three xylene (dimethylbenzene) isomers, and an additional aromatic hydrocarbon, are given below. Assign the spectra to the isomers and suggest a potential structure for the remaining unknown substance.

Infrared unknown spectrum a.

Infrared unknown spectrum b.

Infrared unknown spectrum c.

Infrared unknown spectrum d.

8-14. The C—H stretching mode of chloroform ($CHCl_3$), which occurs at 3022 cm^{-1}, is one of the rare exceptions to the 3000-cm^{-1} rule. What is the rule? Suggest an explanation for this exception.

BIBLIOGRAPHY

Aldrich Library of Infrared Spectra, Aldrich Chemical Co., Inc., 940 West Saint Paul Avenue, Milwaukee, WI 53233. 3rd ed.; 1981, 12,000 spectra, 8 per page, in one volume arranged by chemical type.

Aldrich Library of FT-IR Spectra, Aldrich Chemical Co., Inc., 940 West Saint Paul Avenue, Milwaukee, WI 53233. 2nd ed.; 1997, 18,000 spectra, 4 per page, in three volumes arranged by chemical type.

API collection (American Petroleum Institute). *About 4500 spectra. M.C.A. collection (Manufacturing Chemists' Association). About 3000 spectra.* Chemical Thermodynamics Property Center, Texas A&M University Department of Chemistry, College Station, TX 77843.

Coblentz Society Infrared Spectra Collections for ACS/Labs. (http://www.coblentz.org). An extensive collection of critically evaluated spectra.

Grasselli, J. G.; Ritchey, W. M., Eds., *Atlas of Spectral Data and Physical Constants for Organic Compounds,* 2nd ed.; CRC Press: 2000 Corporate Blvd. NW, Boca Raton, FL 33431. 1975.

Bio-Rad—Sadtler Collection. Over 1.3 million spectra of pure organic and commercial products. Bio-Rad Laboratories, Inc., Informatics Division, Two Penn Center Plaza, Suite 800, 1500 John F. Kennedy Blvd., Philadelphia PA 19102.

For reviews on ATR FT-IR spectroscopy, see:

Grdadolnik, J. *Acta Chimica Slovenica.* **2002**, 49, 631.

Hind, A. R.; Bhargava, S. K.: McKinnon, *A. Adv. Colloid Interface Sci.* **2001**, 93, 91.

Milosevic, M. *Applied Spectroscopy Reviews* **2004**, 39, 365.

For reports on the use of ATR FT-IR in the undergraduate laboratory, see:

Schuttlefield, J. D.; Grassian, V. H. *J. Chem. Educ.* **2008**, 85, 279.

Schuttlefield, J. D.; Larsen, S. C.; Grassian, V. H. *J. Chem. Educ.* **2008**, 85, 282.

NUCLEAR MAGNETIC RESONANCE SPECTROSCOPY

Nuclear spin

Nuclear spin is an energy property intrinsic to a nucleus and analogous to the electron spin that plays such an important role in determining electron configurations. Nuclear spin values are quantized, as are electron spins, and are represented by I, the nuclear spin quantum number. Nuclear spin quantum numbers range from 0 through $\frac{7}{2}$ in increments of $\frac{1}{2}$. The nuclei of greatest interest to organic chemists, the ^1H, ^{13}C, ^{19}F, and ^{31}P nuclei, have spins of $\frac{1}{2}$; the

^{12}C, ^{16}O, and ^{32}S nuclei have spins of 0 (and thus cannot be observed by nuclear magnetic resonance spectroscopy, NMR); the ^{2}H (deuterium, D) and ^{14}N nuclei have spins of 1. Since any spinning charged particle (or body) produces a magnetic moment, a nucleus with a nonzero spin quantum number has a magnetic moment, μ.

Nuclear spin values are quantized because the nuclear angular momentum, and thus the nuclear magnetic moment, is quantized. When placed in an external magnetic field, nuclei orient their magnetic moments in certain ways with respect to the magnetic field, which is assumed to be aligned with the z axis of a Cartesian coordinate system. These orientations are referred to as the z components of the nuclear magnetic moment, μ_z. For a nucleus with a spin of $\frac{1}{2}$, μ_z may be $+\frac{1}{2}$ or $-\frac{1}{2}$. In general, for a nucleus of spin I, the μ_z takes quantized values from $[-I, -I+1, ..., I-1, I]$; or $(2I + 1)$ different values in all. For this discussion we will limit ourselves to nuclei with spin $\frac{1}{2}$, since this is easier to describe, and since most nuclei of interest in organic chemistry are of spin $\frac{1}{2}$.

When placed in a static magnetic field of strength H_0, the magnetic moment, μ_z, of the spinning nucleus precesses about the magnetic field at a frequency, v, such that $v = \gamma H_0/2\pi$, where H_0 is the strength of the applied magnetic field, and γ is a characteristic property of the nucleus known as the gyromagnetic ratio. When a nucleus of spin $\frac{1}{2}$ is placed in a magnetic field, the energies of the $\mu_z = +\frac{1}{2}$ and $-\frac{1}{2}$ states are separated, since in one spin state the nuclear magnetic moment is aligned with the applied magnetic field, and in the other spin state the nuclear magnetic moment is opposed to the applied magnetic field.

The amount of separation of the two energy states, ΔE, is proportional to the magnetic field, and is given by the following expression:

$$\Delta E = \frac{h\gamma H_0}{2\pi} = hv$$

When nuclei in the magnetic field are exposed to radiation of the proper frequency, transitions between the two energy states are stimulated, and the nucleus is said to be in *resonance*, or to *resonate*. This transition occurs when the frequency and the energy difference are related by the Planck relation, $\Delta E = hv$, and thus the sample will absorb energy of frequency v. The study of these energy changes is known as nuclear magnetic resonance, or NMR, spectroscopy.

INSTRUMENTATION

In an NMR spectrometer, the magnetic field is provided by a large permanent magnet, electromagnet, or superconducting electromagnet. Commercially available NMR spectrometers have magnets with field strengths that range from 1.4 to 20 tesla (the earth's magnetic field at its surface is roughly 5×10^{-5} tesla), and thus operate at frequencies from 60 to 850 MHz for protons. In general, most spectrometers with an operating frequency above 100 MHz use a superconducting electromagnet.

Traditionally, NMR spectra were acquired either by holding the applied magnetic field constant and sweeping the radio frequency (rf), or by holding the rf constant and sweeping the applied magnetic field. Energy absorption by the sample was detected, and the result was the NMR spectrum, a plot of intensity (of energy absorption) versus frequency (or field). This instrumental technique is referred to as continuous wave, or CW, spectroscopy (Fig. 8.7).

Figure 8.7   Schematic of NMR spectrometer.

Over the last 20 years, however, it has been commonly replaced by pulsed, or Fourier transform (FT), NMR spectroscopy. Among many other benefits, FT-NMR spectroscopy allows very rapid acquisition of spectral data, which permits analysis of small samples and rare nuclei, such as ^{13}C.

The basic principles of FT-NMR spectroscopy can be qualitatively explained as follows. Take, for example, an NMR spectrum that contains a single peak at a given frequency. The graph of this spectrum (Fig. 8.8) is a plot of intensity versus frequency. The same information can be conveyed by a plot of intensity versus time that shows a cosine wave at the frequency described by the graph of the usual NMR spectrum. This is shown in Figure 8.9, and for a spectrum with a single frequency, this plot of intensity versus time is almost as easy to interpret as the usual NMR spectrum shown in Figure 8.8.

Of course, it would be very difficult to determine the frequencies of many superimposed cosine waves from this kind of plot, and it would be at best awkward to interpret a complex NMR spectrum presented in such a fashion (Fig. 8.10). The use of the Fourier transform allows us mathematically to interconvert these time domain (Fig. 8.10) and frequency domain spectra (Fig. 8.11).

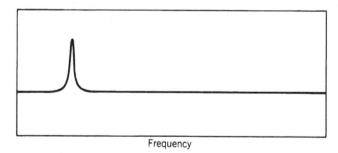

Frequency

Figure 8.8   Intensity versus frequency (usual NMR spectrum).

Figure 8.9   Intensity versus time.

Figure 8.10 Three-signal NMR spectrum: intensity versus time.

Figure 8.11 Three-signal NMR spectrum: intensity versus frequency.

Figure 8.12 Basic pulsed NMR experiment.

Fourier transform of the apparently complex spectrum in Figure 8.10 gives the spectrum in Figure 8.11. It is then easy to see that there are actually only three different resonance signals contained in the time domain of the data of Figure 8.10.

Fourier transform NMR spectra are obtained by applying a short (~1–10 ms), high-powered pulse of rf energy to the sample (Fig. 8.12). This pulse affects all the nuclei to be observed. Before the pulse is applied, the equilibrium net nuclear magnetization is aligned with the applied magnetic field, along the z axis. The coordinate system is presumed to be rotating about the z axis at the frequency of the rf pulse. The pulse, applied down the x axis, applies a torque to the nuclear magnetic moments and rotates them into the xy plane. At this point the pulse is turned off and the nuclear magnetic moments return to their equilibrium alignments. In the process, they precess about the z axis (applied magnetic field) in the xy plane and induce a current in a detector coil, which can be thought of as being aligned with the y axis. This current varies in a sinusoidal manner, and the observed frequency will be the difference between the resonance frequency of the nuclei and the frequency of the rf pulse.

The detected signal, which is called the free induction decay (FID), is digitized and stored. For small organic molecules in a nonviscous solution, the FID will disappear after a few seconds, which corresponds to the time it takes the nuclei to regain equilibrium alignments after the rf pulse. Thus, an entire ^{1}H NMR spectrum can be obtained in approximately 2 s, in contrast to the 10–15 min usually needed to obtain a CW spectrum. A major advantage of FT-NMR is that many spectra of a sample can be rapidly obtained and added together to increase the signal-to-noise (S/N) ratio. Noise is presumably random about some zero level, so when many spectra are added together, the noise level is reduced, while real signals are reinforced when added. The S/N ratio is proportional to the square root of the number of spectra added together. Thus, one can obtain the ^{1}H spectrum of a 10-μmol sample in a minute or two. With FT-NMR it is possible to obtain spectra of isotopes that are insensitive and/or of low natural abundance, as well as spectra of large biological molecules in dilute solution. By adding a few hundred spectra together, an adequate ^{13}C NMR spectrum of a 100-μmol sample can be obtained in about 20–30 min.

CHEMICAL SHIFT

In a molecule, the magnetic field at a nucleus depends not only on H_0, the field generated by the instrument (the external field), but also on the magnetic fields associated with the electron density near the nucleus. Electrons are influenced by the external field in such a way that their motion generates a small magnetic field that opposes the applied field, and reduces the actual field experienced at the nucleus. This reduction is very small (relative to the external field) and is on the order of 0.001%, 10 ppm, for most protons, and about 200 ppm for ^{13}C nuclei. Reduction of the external field is known as *shielding*, and it gives rise to differences in the energy separation for nuclei in different electronic environments in a molecule. The differences in the energy separation are known as *chemical* shifts.

The magnitude of the chemical shift depends on the nature of the valence and inner electrons of the nucleus and even on electrons that are not directly associated with the nucleus. Chemical shifts are influenced by inductive effects, which reduce the electron density near the nucleus and reduce the shielding. The orientation of the nucleus relative to π electrons also plays an important role in determining the chemical shift. A proton located immediately outside a π-electron system (as in the case of the protons on benzene rings) will be significantly deshielded. In most molecules the chemical shift is determined by a combination of these factors. Chemical shifts are difficult to predict using theoretical principles, but have been well studied and can usually be easily predicted empirically upon comparison to reference data.

In an NMR spectrum, the absorption of rf energy is detected, as in Figure 8.13, where the energy absorption is shown for increasing frequency. In this example we illustrate the case with two different nuclei, A and X. Since A and X are different, they absorb energy at different frequencies while in the same applied magnetic field.

The spectrum would be displayed as in Figure 8.14. The difference in the resonances is known as a *chemical shift* and is expressed in parts per million (ppm). The use of frequency units is cumbersome and is complicated because NMR spectrometers of different magnetic field strengths (and thus operating frequencies) are used. The use of ppm units allows direct comparisons of

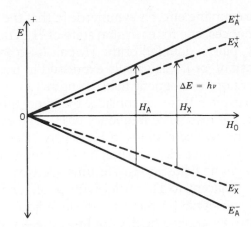

Figure 8.13   The energy splitting for two chemically different protons. The differences between the A energy levels (solid lines) and the X levels (dashed lines) have been amplified for illustrative purposes. At 60 MHz, nucleus A absorbs energy at field H_A and nucleus X absorbs energy at field H_X. Nucleus X is said to be more strongly shielded than A. The resonance for X is said to occur upfield of that for A.

Figure 8.14   The spectrum for the system in Figure 8.13 as it would be displayed. It is conventional to display the spectrum with magnetic field strength increasing to the right so that upfield (and more strongly shielded) is toward the right and downfield (and deshielded) is toward the left.

spectroscopic data obtained on different instruments, when chemical shifts are referenced relative to a reference compound whose chemical shift is arbitrarily defined as 0 ppm. The accepted reference standard for ^{1}H and ^{13}C NMR in organic solutions is tetramethylsilane (TMS), $(CH_3)_4Si$. The chemical shift relative to TMS is symbolized by δ.

Tetramethylsilane is used as a reference substance for a number of reasons. It is more strongly shielded (Si is more electropositive than C) than most other protons and carbon atoms, and its resonance is thus well removed from other areas of interest in the NMR spectrum. Tetramethylsilane is inert and thus unlikely to react with the compound being analyzed, it is volatile (bp 26 °C) and thus easily removed after a sample has been analyzed, and its 12 identical protons per molecule provide a strong signal per molecule of TMS.

SPIN–SPIN COUPLING

In a molecule with several protons, the exact frequency at which a proton resonates depends not only on the chemical shift of that proton, but also on the spin states of nearby protons. This occurs because the magnetic moments of the nearby protons can either shield or deshield the proton in question from the applied magnetic field, depending on the orientation of the nearby magnetic moments relative to the applied magnetic field. The extent of this perturbation is independent of the applied magnetic field strength. The effect of the spin state of one nucleus on the resonance of another is known as *coupling* or *splitting*.

The spectra resulting from spin–spin coupling depend on the types of nuclei, the distance and geometry between the nuclei, the nature of the bonding, the electronic environment, and the total number of spin states possible. The latter may be illustrated by looking at the spectrum of an imaginary compound that has protons H_A and H_X on adjacent carbons, connected by three bonds: H_A—C—C—H_X (Fig. 8.15). In the first approximation we would expect one resonance for H_A and one resonance for H_X, and the spectrum would resemble that shown in Figure 8.14. In the presence of coupling, the resonance for H_A splits into two signals, one of which corresponds to H_X having $\mu_z = +\frac{1}{2}$ and the

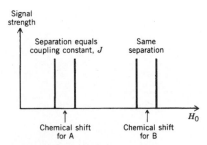

Figure 8.15   Spectrum of two chemically different protons that are coupled.

other to $\mu_z = -\frac{1}{2}$. The coupling effect is symmetric in that the H_X resonance also splits into two resonances, one for each spin state of H_A. The magnitude of the separation of the H_A pair (a doublet) or the H_X pair (also a doublet) is known as the coupling constant, or J. It is usually expressed in frequency units (Hz), since J is independent of the magnetic field strength.

A simple way to explain this is to consider the effect the two possible spin states of H_X have on the resonance frequency of H_A. The equilibrium population distribution of the two spin states in H_X is very close to 1:1, since ΔE is only about 10^{-3} cal/mol. Since H_X has a magnetic moment, there are then two slightly different magnetic fields at H_A. We thus see two signals for H_A, one for those H_A nuclei adjacent to H_X nuclei with μ_z aligned with the applied magnetic field, and one signal for those H_A nuclei adjacent to H_X nuclei with μ_z aligned opposed to the applied field. Coupling between protons connected by more than three bonds does occur, but its magnitude, J, is usually small and often not directly observed in a usual NMR spectrum.

The splitting becomes more interesting when there are several nuclei of one type. 1,1-Dibromoethane (CH_3CHBr_2) has three equivalent protons in the methyl group and one proton on the C-1 atom. The methyl group exhibits rapid internal rotation so that its three protons are equivalent. The chemical shift for the C-1 proton is 5.86 ppm, and that for the methyl protons is 2.47 ppm. Here we can see an example of decreased shielding resulting from the presence of electronegative substituents. Equivalent protons do not couple with one another (this is an important rule in interpreting spectra), but the methyl protons will affect the proton on C-1, and vice versa.

To analyze the splitting pattern, we need to consider the orientations of the nuclear magnetic moments, with respect to the applied magnetic field, for all three methyl protons. Since each of the three protons may have two spin states that are of nearly equal probability, there are $2^3 = 8$ possible combinations of spin states in all for the methyl protons. The net sums of these may have only four different values, as shown in Figure 8.16. The symbol (+) is used to represent $\mu_z = +\frac{1}{2}$ for a single proton and (−) is used to represent $\mu_z = -\frac{1}{2}$. Thus (+)(+)(−) means that protons 1 and 2 have $\mu_z = +\frac{1}{2}$, while proton 3 has $\mu_z = -\frac{1}{2}$.

The number of different μ_z states is $(2N + 1)$, where N is the number of equivalent nuclei (of spin ½) Thus the three methyl protons can generate four slightly different magnetic fields, and the proton on the C-1 of CH_3CHBr_2 sees (in different molecules) four different magnetic fields. Since these different magnetic fields are not of equal probability, but rather are populated in a ratio of 1:3:3:1, the four signals we see for the proton on C-1 when coupled to the

Individual μ_z	SUM
(+)(+)(+)	$+\frac{3}{2}$
(+)(+)(−) or (+)(−)(+) or (−)(+)(+)	$+\frac{1}{2}$
(+)(−)(−) or (−)(+)(−) or (−)(−)(+)	$-\frac{1}{2}$
(−)(−)(−)	$-\frac{3}{2}$

Figure 8.16   Possible combinations of spin states for a methyl group.

methy group are of intensities 1:3:3:1, and are referred to as a *quartet*. This is shown schematically in Figure 8.17b (downfield set of peaks).

Since the proton on C-1 has two possible spin states of nearly equal probability, the protons of the methyl group experience two slightly different magnetic fields and are observed in the spectrum as two slightly separated signals of equal intensity, or a *doublet*. The separation between each of the C-1 proton signals is the coupling constant, J, and will equal the J of the methyl signal. The proposed spectrum is shown in Figure 8.17b. The coupling constant in this case is about 7 Hz. The 60-MHz NMR spectrum of 1,1-dibromoethane is shown in Figure 8.17c.

Figure 8.17a   The spectrum without any spin–spin coupling.

Figure 8.17b   A "stick figure" spectrum indicating the expected intensities.

Figure 8.17c   The actual 60-MHz spectrum of 1,1-dibromoethane. The TMS signal at 0 ppm is seen along with a weak signal at 7.3 ppm, which is not from this molecule.

Figure 8.18   The 60-MHz spectrum of 1-nitropropane. (©Agilent Technologies, Inc. 2011 Reproduced with permission, courtesy of Agilent Technologies, Inc.) **Starting from the right, the TMS signal at 0 ppm is seen. Next is a 1:2:1 triplet at 1.03 ppm. This triplet results from the protons on C-3 and their coupling with the two protons on C-2. Next is the sextet centered at 2.07 ppm. This multiplet is from the protons on C-2 and their coupling with the protons on C-1 and C-3. Finally, we have the signal from the protons closest to the nitro group centered at 4.38 ppm. These protons appear as a 1:2:1 triplet due to their coupling with the protons on C-2.**

NOTE. The net effect of spin–spin coupling is that a proton (or group of equivalent protons) adjacent to N other protons will be observed as a multiplet with (N + 1) lines.

A proton, or group of equivalent protons, may be coupled to more than one group of nuclei. The spectrum of 1-nitropropane ($CH_3CH_2CH_2NO_2$) is shown in Figure 8.18. The signal from the central methylene (CH_2) group is seen at about 2.0 ppm. Because the methylene group is adjacent to (and thus coupled to) five protons, its signal is a (5 + 1) or six-line multiplet—a *sextet*.

Nuclei with spins of 1 or greater exhibit more complex spin–spin coupling, since they can exist in more than two different spin states. For a nucleus coupled to N nuclei of spin I, a multiplet of $2IN$ lines will be observed. Nuclei of spin zero do not couple.

INTENSITIES

The area under an NMR peak is proportional to the number of nuclei giving rise to that signal. The intensity of a resonance is thus best determined by the *integral* of the NMR spectrum over a resonance, or group of resonances. Nuclear magnetic resonance spectrometers can measure the integral, though integration data from an FT spectrometer are less reliable than those from a CW spectrometer. In more complex spectra, the intensities are useful as a measure of the number of protons of a given type. For instance, in the above case the integral over both peaks of the methyl group doublet will be three times the integral over the quartet of the proton on C-1. Integration can thus often provide useful information for determining the identity of a compound.

SECOND-ORDER EFFECTS

So far, all of our examples have consisted of first-order spectra. First-order spectra are those multiplets interpretable through elementary coupling analysis, such as that above; second-order spectra are those that are not interpretable in this manner. These highly symmetric and fairly simple first-order spectra are generally observed when the chemical-shift differences (expressed as a frequency) are much greater than the coupling constant. Second-order effects occur when the coupling constants become comparable to or greater than the chemical-shift differences. Thus, spectra obtained on instruments with higher magnetic fields are more likely to be first-order, since the frequency differences between given signals increase with increasing magnetic fields. However, the chemical-shift differences (in ppm) remain the same regardless of magnetic field strength.

Second-order effects may be understood in qualitative terms by considering the limiting cases. Let us consider the hypothetical disubstituted ethylene shown in Figure 8.19, where R and M are substituents that might be identical or may have very different effects on the alkenyl protons. In Figure 8.19a the spectrum is shown for the situation in which R and M have very different effects. In this case we will observe a first-order spectrum consisting of two doublets. The coupling constant is the separation in the doublets, and the chemical shift of each nucleus is the geometric midpoint of each doublet.

In Figure 8.19b, groups R and M are identical. H_A and H_B are identical in this case and only a single resonance is observed (coupling between equivalent nuclei is not observed).

In Figure 8.19c the difference in the chemical environment of H_A and H_B is very slight. The spectrum shown may be seen as intermediate between the

Figure 8.19   Second-order effects. (a) The chemical-shift difference is much larger than the coupling constant and a first-order spectrum is observed. (b) Protons A and B are equivalent and a single resonance is observed. (c) The chemical-shift difference is of the same order of magnitude or less than the coupling constant. Note the "leaning in" of the peak intensities in this spectrum relative to that in part a.

limiting cases in Figure 8.19*a* and Figure 8.19*b*. Note that there is a "leaning in" of the doublets as the central members increase in intensity at the expense of the outer members. A full continuum of behavior may be expected with cases observed in which the outer members are lost in the noise and the central members take the appearance of a doublet. This would be one example of a class of spectra known as "deceptively simple spectra."

The second-order spectra of systems with more than two protons are difficult to describe even in qualitative terms. Second-order spectra may well display more lines than one would predict from simple coupling theory. Also, in second-order spectra, the coupling constants and the chemical-shift differences may not be obtainable as simple differences in the positions of spectral lines. Thus, spectra obtained at high frequencies (and magnetic fields) are often more useful. As the operating frequency of the instrument is increased, the chemical-shift differences (in frequency terms) increase while the spin–spin coupling remains constant. Thus, the complicating second-order effects are likely to be less noticeable in high-field spectra. The reader is referred to more extensive treatments of NMR for a discussion of second-order cases.

INTERPRETATION OF ¹H NMR SPECTRA

The first issues that must be addressed are molecular symmetry and the magnetic equivalence or nonequivalence of protons or other functional groups. Even if two protons, or groups, are chemically equivalent, they may or may not be magnetically equivalent. Although molecular symmetry can often simplify NMR spectra, one must be able to discern which protons or groups are equivalent by symmetry. The two most useful symmetry properties (or symmetry operators) are the plane of symmetry and the axis of symmetry.

A plane of symmetry is simply a mirror plane such that one half of the molecule is the mirror image of the other half, as in *meso*-pentane-2,4-diol:

meso-Pentane-2,4-diol

The methyl groups are identical by symmetry, and one would expect this stereoisomer to show one methyl doublet in its ¹H NMR spectrum and one signal for a methyl group in its ¹³C spectrum.

Consider the other diastereomer of pentane-2,4-diol, the chiral *d,l* isomer. This isomer has an axis of symmetry. If the molecule is rotated 180° about an axis in the plane of the paper passing through the central carbon, the molecule can be converted into itself:

d,l-Pentane-2,4-diol

Here, too, the methyl groups are identical by symmetry, and one would expect this stereoisomer to show one methyl doublet in its ¹H NMR spectrum and one signal for a methyl group in its ¹³C spectrum.

It is, however, relatively simple to use NMR spectroscopy to distinguish between these stereoisomers. To do this, look at the two methylene protons on the central carbon, C-3, of each isomer:

The plane of symmetry in the meso isomer bisects each of the two protons. In the *d,l* isomer, the axis of symmetry interconverts the two protons. Thus, in the *d,l* isomer, the two methylene protons are equivalent by symmetry, but they are not equivalent in the meso isomer. This can also be seen by inspecting the molecule. On the left, one H is syn to both —OH groups and the other is anti to both —OH groups. On the right, each H atom is syn to one —OH group and anti to the other.

The more rigorous way to determine equivalence or nonequivalence is to determine whether the two protons (or groups) are homotopic (identical), diastereotopic, or enantiotopic. To compare two protons, we use the usual Cahn–Ingold–Prelog system for the nomenclature of stereoisomers. We artificially distinguish the relative priority of two protons by a method such as drawing them in different colors or pretending that one is deuterium (as long as the molecule does not contain D). We draw the two possibilities (i.e., the first H as D and then the second H as D) and then determine the stereochemical relationship between the two:

If the two are identical, the two protons are identical, or *homotopic*. If the two structures are diastereomers, the two protons are *diastereotopic*, and if the two structures are enantiomers, the two protons are *enantiotopic*. Diastereotopic protons, or groups, will be magnetically nonequivalent. Enantiotopic protons, or groups, will be magnetically equivalent only in an achiral environment and may appear nonequivalent in a chiral environment, such as a chiral solvent or in a biological sample. Homotopic protons may or may not be magnetically equivalent. Of course, it is possible for magnetically nonequivalent signals to be so close to one another in the NMR spectrum as to overlap (accidentally degenerate).

Homotopic protons may be magnetically nonequivalent if the two protons have different coupling constants to the same third proton. The most common example of this occurs in para-substituted benzenes:

By symmetry, H_{1A} and H_{1B} are equivalent. These protons are not, however, magnetically equivalent because H_{1A} and H_{1B} have different coupling constants to, for example, H_{2A}, and the spectrum of this molecule may well be more complex than one would at first expect.

The equivalence or nonequivalence of functional groups, as well as protons, can easily be determined. The ^1H NMR spectrum of menthol shows three methyl doublets, since the two methyls in the isopropyl group are diastereotopic. The ^{13}C spectrum of menthol shows three distinct resonances for the three different methyl groups:

Menthol

Figure 8.20   NMR ¹H chemical shifts. (From Zubrick, J. W. *The Organic Lab Survival Manual,* 9th ed.; Wiley: New York, 2013. Reprinted by permission of John Wiley & Sons, New York.)

¹H CHEMICAL SHIFTS

Figure 8.20 summarizes the chemical shifts of protons in a large range of chemical environments. It is, however, a bit dangerous to use figures such as this one without understanding some of the factors that underlie shielding and the chemical shift. To give some flavor of the factors that determine chemical shifts and the range of values observed, we will briefly examine chemical shifts in methyl groups and chemical shifts for protons on sp² carbon atoms.

Methyl groups bonded to an sp³ carbon generally have chemical shifts in the range 0.8–2.1 ppm as long as there is no more than one electron-withdrawing group attached to the carbon. The shifts generally increase as the strength of the electron withdrawing group increases, or as more electron-withdrawing groups are added. Groups that inductively withdraw electrons reduce the electron density near the methyl group protons. This results in less shielding and a downfield shift of the methyl resonance. This effect is clearly seen in the spectra of 1-nitropropane and 1,1-dibromoethane (Fig. 8.17c and Fig. 8.18), respectively. The chemical shifts for methyl groups bonded to sp² carbon atoms fall in the range 1.6–2.7 ppm.

In the case of a proton bonded to an sp² carbon, the location of the proton relative to the π cloud plays an important role in determining the chemical shift. In unconjugated alkenes the chemical shifts fall in the range 5–6 ppm. Where more than one proton is bonded to an alkene, complex second-order spectra can be expected at low operating frequencies since the coupling constants are usually fairly large relative to the difference in resonance frequencies.

In aldehydes, RCHO, the increased electronegativity of the oxygen increases the deshielding and the chemical shift falls in the range 9–10.5 ppm.

The chemical shift in an aromatic system is generally greater than that for alkenes. For example, the chemical shift of benzene is 7.37 ppm, which is substantially greater than the 5.6 ppm for the alkenyl protons of cyclohexene. Much of this difference results from the "ring current" effect and the orientation of the proton relative to the aromatic π electrons. If the ring substituents are not strongly electron withdrawing or electron donating, such as alkyl groups, the chemical shift for ring protons will not be shifted greatly from that of benzene itself. Furthermore, these substituents generate only small chemical-shift differences among the ring protons. Thus, the 60-MHz spectra for toluene (methylbenzene) appears to have a single resonance in the aromatic region at about 7.1 ppm. If, on the other hand, the substituents are electron-withdrawing, the ortho and para ring protons will be somewhat deshielded relative to benzene. π-electron-donating substituents, such as a methoxy group, will increase the shielding of groups ortho and para to it.

SPIN–SPIN COUPLING

Coupling information is the primary reason that ^{1}H NMR is such a powerful tool for organic structure determination. Since coupling information is transmitted through bonds, coupling provides information about nearby protons and can often be used to deduce stereochemistry.

The sign of the coupling constant (usually symbolized as J) may be positive or negative. However, first-order spectra are not sensitive to the sign of the coupling constant. In second-order cases, the sign of J may be determined by a detailed analysis of the spectrum, though the sign of J is generally of little value for organic structure determination.

Geminal Coupling

Nonequivalent protons attached to the same carbon (geminal protons) will couple with one another. These coupling constants tend to be large (>10 Hz) for sp^3 carbon atoms and small (<4 Hz) for sp^2 carbon atoms. Geminal coupling constants tend to decrease with decreasing ring size, because of hybridization changes at carbon, and with the increasing electronegativity of the substituents on a given methylene group.

Vicinal Coupling

Vicinal coupling describes the coupling over three bonds observed between protons attached to two bonded carbon atoms, H—C—C—H. Vicinal coupling constants (J values) can range from near 0 to greater than 15 Hz, depending on the stereochemical relationship (dihedral angle) between the coupled protons, the hybridization of the carbon atoms, and the electronegativity of other substituents. For vicinal protons on sp^3 carbon atoms, the coupling constant is related to the dihedral angle and is expressed graphically by the Karplus curve (Fig. 8.21).

Though the magnitude of vicinal coupling is very sensitive to the angle of rotation about the central bond, in many simple cases nearly all coupling constants are equal. This situation is often the case if internal rotation about a C—C single bond can occur on a time scale that is very short relative to the NMR time scale, such as in acyclic systems. In these cases, the effect of internal

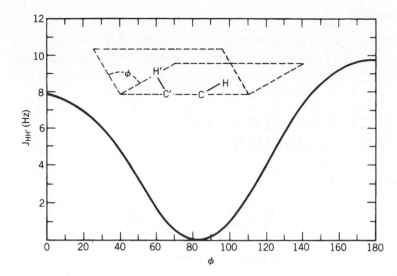

Figure 8.21 The vicinal Karplus correlation showing the relationship between dihedral angle and coupling constants for vicinal protons.

rotation is completely blurred as far as NMR is concerned, and only an average coupling constant is observed. Vicinal coupling constants in freely rotating alkyl groups are usually observed in the 6.5- to 8-Hz range.

When the central C—C bond between two coupled protons is a double bond, rotation is restricted and separate coupling constants for cis and trans protons may be observed. Cis coupling constants fall in the range 5–12 Hz, whereas trans coupling constants range from 12 to 20 Hz. As a result of these large coupling constants, second-order effects are often observed in substituted alkenes in instruments of lower field strengths.

When rotation about carbon–carbon single bonds is restricted, or when stereochemistry dictates significant conformational preferences, nonaveraged coupling constants may be observed that can complicate the appearance of the NMR spectrum. For example, two diastereotopic protons of a methylene (CH₂) group may well each couple to a given third proton with different coupling constants. The familiar coupling explained at the elementary level suggests that if one proton is adjacent to, for example, two others, the NMR signal of the first proton will be a triplet. This simplification will be true only if the two coupling constants are identical. A triplet is merely a doublet of doublets with equal coupling constants, which gives rise to the familiar triplet with intensities of 1:2:1 (Fig. 8.22*a*). A doublet of doublets, on the other hand, gives rise to a four-line multiplet with peaks of roughly equal intensity (Fig. 8.22*b*).

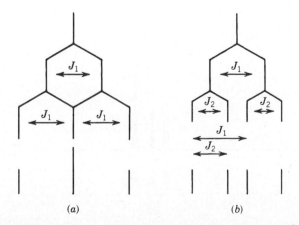

Figure 8.22   (a) Triplet equals doublet of doublets with equal *J* values; (b) doublet of doublets.

Long-Range Coupling

Longer range coupling involving four or more bonds is common in allylic systems and in aromatic rings and other conjugated π systems. These coupling constants are generally smaller than the values considered above (i.e., <3 Hz).

EXAMPLES OF COMPLEX, YET FIRST-ORDER, COUPLING

Ethyl Vinyl Ether

The coupling constants of even a seemingly complex multiplet can be discerned in a relatively simple manner. First, the total width (outside peak to outside peak) of a first-order multiplet is equal to the sum of all the coupling constants, keeping in mind that, for example, a triplet of $J = 7$ Hz is really a doublet of doublets with both J values equal to 7 Hz. The expansion of the proton spectrum of ethyl vinyl ether is presented as an example in Figure 8.23. Integration data are displayed between the spectrum and the horizontal axis in Figure 8.23a.

Ethyl vinyl ether

Consider the multiplet centered at 6.45 ppm (Fig. 8.23b). By measuring the distance (in Hz) from either outside line to the next inner line, which is 6.8 Hz, the first coupling constant is determined. Then, by measuring from

Figure 8.23a   Ethyl vinyl ether in CDCl₃.

Figure 8.23*b*   Ethyl vinyl ether in CDCl₃ (expansion).

Figure 8.23c   Ethyl vinyl ether in CDCl₃ (expansion).

the outside line to the second line in, the second coupling constant is found to be 14.4 Hz.

We know that this is the last coupling constant to be found for several reasons. First, if we measure from the outside line to the third line in, we get a value, 21.2 Hz, which is equal to the sum of the previously determined coupling constants. Second, the width of the multiplet (the same measurement in this simple case) is equal to our two coupling constants. Thus, the NMR signal at 6.45 ppm is a doublet of doublets with $J = 14.4$ and 6.8 Hz.

The two doublets of doublets at 4.15 and 3.96 ppm (Fig. 8.23c) must be coupled to one another because they both have the coupling constant of 1.9 Hz in common. This geminal coupling constant is typical of the terminal methylene of an alkene.

Since the proton at 3.96 is coupled to the proton at 6.45 ppm by $J = 6.8$ Hz, and the proton at 4.15 ppm is coupled to the one at 6.45 ppm by $J = 14.4$ Hz, the proton at 3.96 ppm must be cis, and the proton at 4.15 ppm must be trans, to the alkene proton at 6.45 ppm.

The simple coupling observed for the ethyl group in ethyl vinyl ether can be readily assigned. The triplet at about 1.25 ppm, which integrates for three protons, is due to the methyl group; it is a triplet because the equivalent protons of the methyl group are coupled to the two protons on the adjacent carbon with equal coupling constants. The O—CH₂ protons are observed in the NMR spectrum as the quartet at about 3.75 ppm; they are a quartet because they are coupled equally to the three equivalent protons of the methyl group:

1.25 ppm

4.15 ⟶ H$_A$

H$_B$

H$_C$

3.96   3.73 ppm

6.45

Chemical shifts

Ethyl vinyl ether

1.9 Hz   H$_A$ 14.4 Hz

H$_B$

H$_C$

6.8 Hz

Coupling constants

Allyl Acetate

For a more complex example, refer to the ^{1}H NMR spectrum of allyl acetate (the NMR signal for the methyl group has been omitted) in Figure 8.24*a*.

H$_C$

H$_B$ O   CH$_3$

H$_A$ H$_D$ H$_D$ O

Protons A, B, and C are all chemically distinct, and the two protons labeled D are equivalent to one another by symmetry (the plane of the paper). The multiplet at 4.58 ppm (Fig. 8.24*d*) corresponds to H$_D$ and is a doublet of triplets. The coupling constant for the triplet is 1.4 Hz, and the coupling constant for the doublet is 5.8 Hz, which can be measured between any two corresponding peaks in the two triplets.

The four quartets around 5.3 ppm (Fig. 8.24*c*) are actually two doublets of quartets at 5.32 and 5.24 ppm and correspond to H$_A$ and H$_B$ in the structure above. At 5.32 ppm, the multiplet is a doublet of quartets, J = 17.2, 1.5 Hz. At 5.24 ppm, we have another doublet of quartets, J = 10.4, 1.3 Hz. We see quartets because the long-range allylic coupling to the two H$_D$ signals gives a triplet that has a coupling constant J that is approximately equal to the geminal coupling constant (~1.4 Hz) between H$_A$ and H$_B$. Since NMR line widths are naturally several tenths of a hertz, it is not possible to distinguish between coupling constants such as these that differ only by 0.2 Hz. We can unambiguously

Figure 8.24*a* Allyl acetate in CDCl$_3$.

HERTZ

1798.40 1792.63 1788.01 1786.92 1782.25 1781.20 1776.51 1775.43 1770.80 1769.69 1765.04 1759.30

6.08 6.06 6.04 6.02 6.00 5.98 5.96 5.94 5.92 5.90 5.88 5.86 5.84 5.82 5.80
PPM

Figure 8.24*b*   Allyl acetate in CDCl$_3$ (expansion).

1608.77 1607.23 1605.69 1604.15   1591.58 1590.03 1588.49 1586.95   1581.26 1579.98 1578.64 1577.33   1570.88 1569.59 1568.24 1566.94

5.40 5.38 5.36 5.34 5.32 5.30 5.28 5.26 5.24 5.22
PPM

Figure 8.24*c*   Allyl acetate in CDCl$_3$ (expansion).

HERTZ

1377.41 1376.01 1374.61   1371.66 1370.26 1368.86

4.64 4.62 4.60 4.58 4.56 4.54 4.52 4.50
PPM

Figure 8.24*d*   Allyl acetate in CDCl$_3$ (expansion).

distinguish H_A and H_B by the magnitudes of their coupling constants to H_C, which are 17.2 and 10.4 Hz. Since trans coupling constants are larger than cis coupling constants, H_A must have the 17.2-Hz coupling constant to H_C and is thus assigned to the signal centered at 5.32 ppm. Since H_B is coupled to H_C by $J = 10.4$ Hz, it is assigned to the signal centered at 5.24 ppm.

Finally, we already know what the multiplet for H_C should look like, since we know all of its coupling constants. It is coupled to the two H_D protons with a coupling constant of 5.8 Hz, to H_A with $J = 17.2$, Hz, and to H_B with $J = 10.4$ Hz. The multiplet for H_C at 5.93 ppm should be, therefore, a doublet of doublets of triplets with $J = 17.2$, 10.4, and 5.8 Hz, respectively. There should be $2 \times 2 \times 3 = 12$ lines and the width should be $17.2 + 10.4 + (2 \times 5.8) = 39.2$ Hz. There are indeed 12 lines (Fig. 8.24b), and the distance between the outside peaks is 39.1 Hz, which is a perfectly reasonable deviation from the ideal:

Allyl acetate, chemical shift assignments

Allyl acetate, proton–proton coupling constants

Since cyclohexane rings are often held in at most two potential conformations (both chairs), coupling constants may allow the determination of relative stereochemistry. On cyclohexane rings in chair conformations, the axial–axial coupling constants for vicinal protons (180° dihedral angle) are on the order of 9–12 Hz. Equatorial–equatorial and equatorial–axial coupling constants (60° dihedral angles) are on the order of 2–4 Hz. Thus it is often a relatively simple matter to determine stereochemical relationships on a six-membered ring using NMR spectroscopy.

Take, for example, the two diastereomers of 4-*tert*-butylcyclohexanol:

Cis

Trans

We know that the very large *tert*-butyl group will effectively always be equatorial. By examining the coupling to the methine proton of the alcohol, it is simple to determine whether that proton is axial or equatorial. It is also possible to distinguish between these stereoisomers by using chemical-shift information (in one isomer the alcohol methine is seen at ~3.52 ppm and in the other at ~4.04 ppm), but use of coupling information provides a far more definitive and unambiguous determination of stereochemistry.

The alcohol methine proton in the cis isomer is equatorial and thus has a 60° dihedral angle to all four adjacent protons that give rise to a pentet (which is really a doublet of doublets of doublets of doublets with equal coupling constants) with a coupling constant J = ~3 Hz, which is seen in Figure 8.25.

The alcohol methine in the trans isomer is axial and thus has a 180° dihedral angle to each of the two adjacent axial protons and a 60° dihedral angle to each of the two adjacent equatorial protons. This arrangement gives rise to a triplet of triplets with J = ~13 and 3 Hz, which is shown in Figure 8.26.

Figure 8.25 *cis-4-tert-Butylcyclohexanol.*

¹³C NMR SPECTROSCOPY

With the advent of Fourier transform (FT) NMR spectrometers, ¹³C NMR spectroscopy is now available as a simple and routine tool for the structure determination of organic molecules. Since ¹³C is of low natural abundance (1.1%), addition of many spectra is required to obtain acceptable signal-to-noise (S/N) levels. With modern spectrometers, ¹³C spectra can often be acquired simply by issuing software commands; in some instruments a different probe is inserted into the magnet. Since ¹³C resonates at roughly 25% of the proton operating frequency of a spectrometer system, an instrument that acquires ¹H spectra at 300 MHz will be reset to about 75 MHz for ¹³C work.

Generally, ¹³C NMR spectra are acquired while the entire ¹H frequency range is irradiated by a second rf coil inside the probe assembly. These spectra are referred to as broadband-decoupled ¹³C spectra and they do not show the effect of spin–spin coupling to ¹H nuclei. Such decoupling is done because ¹H–¹³C coupling constants can be quite large (a few hundred Hz) relative to chemical-shift differences, which leads to multiplets split over a large portion of the spectrum and subsequent confusion (Fig. 8.27). It is often simpler to see a single line for each distinct carbon atom in a molecule. Furthermore, irradiation of the ¹H spectrum results in signal enhancement of the ¹³C signals of the attached carbon atoms. This enhancement is the nuclear Overhauser effect (NOE).

Figure 8.26 *trans-4-tert-Butylcyclohexanol.*

The ¹³C NMR chemical shifts follow the same rough trends as seen in ¹H chemical shifts. ¹³C chemical shifts, however, are not nearly as amenable to prediction based on the electronegativity of substituents as are ¹H chemical shifts. The ¹³C chemical shifts are, in general, less sensitive to substituent electronegativities, and are far more sensitive to steric effects than are ¹H chemical shifts. A brief listing of approximate ¹³C chemical shifts is provided in Table 8.26; a more extensive and thorough listing is available in the Silverstein et al. reference (Bibliography). As in ¹H NMR spectroscopy, TMS (Si(CH₃)₄) is used as the internal reference and the chemical shift of TMS is defined as zero. Except for functional groups such as acetals and ketals, sp³-hybridized carbon atoms appear upfield (to the right) of 100 ppm, and sp²-hybridized carbon atoms appear downfield of 100 ppm. Common carbonyl-containing functional groups appear downfield of 160 ppm. Aldehydes and ketones appear at 195–220 ppm; esters, amides, anhydrides, and carboxylic acids appear at 165–180 ppm.

Figure 8.27 Fully ¹H-coupled ¹³C NMR spectrum of 5-(1-acetoxy-1-methylethyl)-2-methyl-2-cyclohexenone in CDCl₃.

Table 8.26 Approximate ¹³C NMR Chemical Shifts		
Functional Group	**Carbon**^{a}	**Chemical Shift/δ (ppm)**
Alkyl carbon atoms		~5–45
	1° R—CH₃	~5–30
	2° R—CH₂—R′	~15–35
	3° R—CHR′R″	~20–40
	4° RCR′R″R‴	~25–45
Alkenyl carbon atoms		~110–150
	H₂C=C	~100–125
	HRC=C	~125–145
	RR′C=C	~130–150
Aromatic carbon atoms		~120–160
Alkynyl carbon atoms	C≡C	~65–90
Nitriles	R—C≡N	~115–125
Alcohols and ethers	C—OH(R)	~50–75
	C—O (epoxides)	~35–55
Amines	C—N	~30–55
Alkyl halides	C—X	~0–75
Carbonyl groups	C=O	~165–220
Ketones, aldehydes	RCOR′, RCHO	195–220
Carboxylic acids, esters	RCO₂H, RCO₂R′	165–180
Amides, anhydrides	RCON, RCO₂OCR′	160–175
^{a} R = alkyl group.		

Typical ^{13}C NMR spectroscopy provides an NMR spectrum that is not amenable to integration because of the NOE and insufficient relaxation delays. Therefore, the number of carbon atoms giving rise to a given signal cannot generally be determined by these techniques. It is possible to obtain ^{13}C NMR spectra that can be accurately integrated (inverse-gated decoupling), but this experiment requires a great deal of acquisition time to achieve adequate signal-to-noise levels.

Information about C—H coupling can be readily obtained, however. Fully coupled ^{13}C NMR spectra are not very useful for structure determination because C—H couplings are large (~120–270 Hz, depending mainly on the hybridization at carbon) and multiplets tend to overlap. Furthermore, when the hydrogen atoms are not irradiated, there is no NOE, and the signal-to-noise ratio suffers significantly. The most common use for coupling information is to determine the number of protons attached to a given carbon atom. This can be done in a variety of ways, some of which do not actually display the carbon signals as multiplets due to coupling to attached protons.

Single-frequency off-resonance decoupling (SFORD) is a useful technique for determining the number of hydrogen atoms attached to a given carbon. The decoupler is tuned off to one side of the proton spectrum and the sample is irradiated at a single frequency giving rise to ^{13}C spectra that show C—H couplings as a fraction of their actual values and that show a partial NOE. The apparent C—H coupling is dependent on both the actual coupling constant and the difference between the decoupler frequency and the resonance frequency of the hydrogen in question. The major disadvantage of SFORD is its low signal-to-noise ratio, which is due to two factors. First, there is only a partial NOE. Second, when NMR signals are split into multiplets, the signal intensity becomes distributed among several peaks. Thus, SFORD spectra require significantly more spectral acquisitions than do fully decoupled ^{13}C NMR spectra, and to some extent have been replaced with distortionless enhancement by polarization transfer (DEPT) spectra.

DEPT ^{13}C NMR spectroscopy provides a rapid way of determining the number of hydrogen atoms attached to a given carbon atom. DEPT spectra result from a multiple-pulse sequence that terminates in a "read pulse," which can be varied according to the spectrum desired. In DEPT spectra, all peaks are singlets; quaternary carbon atoms (without attached hydrogen atoms) are not seen in any DEPT spectra. In DEPT-135° spectra, CH and CH$_3$ groups appear as singlets of positive intensity, and CH$_2$ groups appear as negative peaks. The DEPT-90° spectra show only CH groups, and thus allow CH$_3$ and CH groups to be distinguished. In combination with a routine fully decoupled spectrum, DEPT spectra allow unambiguous assignment of the number of hydrogen atoms attached to each carbon. In practice, such spectral editing techniques are not perfect, and small residual peaks are often seen where, in principle, there should be none; these are usually small enough to be readily distinguished from the "real" peaks.

The fully coupled ^{13}C NMR spectrum of the acetoxy-enone (**I**) is shown in Figure 8.27, the broadband decoupled spectrum in Figure 8.28, and the SFORD spectrum in Figure 8.29. The DEPT-135° spectrum is shown in Figure 8.30, and the DEPT-90° spectrum in Figure 8.31. The 1:1:1 triplet centered at 77 ppm is due to the solvent, CDCl$_3$.

Figure 8.28   Broadband ¹H-decoupled ¹³C NMR spectrum of 5-(1-acetoxy-1-methylethyl)-2-methyl-2-cyclohexenone in CDCl₃.

Interpretation and assignment of the ¹³C NMR spectrum are much easier when we unambiguously know how many protons are attached to each carbon.

Figure 8.29   Single-frequency off-resonance decoupled (SFORD) ¹³C NMR spectrum of 5-(1-acetoxy-1-methylethyl)-2-methyl-2-cyclohexenone in CDCl₃.

Figure 8.30   The DEPT-135° spectrum of 5-(1-acetoxy-1-methylethyl)-2-methyl-2-cyclohexenone in CDCl₃.

The ¹³C NMR is often better than ¹H NMR for distinguishing functional groups, because typical ¹³C chemical shifts are in the range 0–200 ppm relative to TMS, as compared to 0–10 ppm for proton chemical shifts. Coupling between adjacent ¹³C nuclei is not observed (except in isotopically enriched samples) because the probability of having two rare isotopes adjacent to one another is very small. Because of the absence of decoupling, ¹³C spectra are less complex than ¹H spectra, and ¹³C spectra are often better suited for the detection and identification

Figure 8.31   The DEPT-90° spectrum of 5-(1-acetoxy-1-methylethyl)-2-methyl-2-cyclohexenone in CDCl₃.

Figure 8.32 The ¹H NMR spectrum of 4-cyclohexene-*cis*-1,2-dicarboxylic acid anhydride in CDCl₃.

of isomeric or other impurities in a sample; it is easy for small peaks to be concealed underneath a complex second-order multiplet in the ¹H NMR spectrum.

The 300-MHz ¹H spectrum of 4-cyclohexene-*cis*-1,2-dicarboxylic acid anhydride is shown in Figure 8.32. Owing in part to the presence of two stereocenters, as well as to long-range coupling through the π system of the alkene, the entire ¹H spectrum is second order at this field strength, and no information is available from the coupling constants because the spectrum is too complex. Limited assignments to peaks could be made on the basis of chemical shift, but it would be difficult to make any statements regarding purity of our sample based on the ¹H NMR spectrum, because an impurity could easily be hidden underneath any of the complex signals.

The ¹³C spectrum of 4-cyclohexene-*cis*-1,2-dicarboxylic acid anhydride (Fig. 8.33) is much less complex. Because of the mirror plane of symmetry in the compound, there are only four different carbon atoms, and thus only four lines are seen in the fully decoupled ¹³C NMR spectrum. This simplicity makes it easy to detect isomeric or other impurities in this sample. These impurities

Figure 8.33 Broadband ¹H-decoupled ¹³C NMR spectrum of 4-cyclohexene-*cis*-1,2-dicarboxylic acid anhydride in CDCl₃.

were not as easy to detect in the ^{1}H NMR spectrum. The 1:1:1 triplet centered at 77 ppm is due to the solvent, $CDCl_3$. Although no ^{1}H–^{13}C coupling is observed because of the ^{1}H broadband decoupling, ^{2}H–^{13}C coupling is observed, because ^{1}H and ^{2}H resonate at different frequencies.

TWO-DIMENSIONAL NMR SPECTROSCOPY

Two significant developments in NMR spectroscopy are the use of Fourier transform techniques, and the development of two-dimensional (2D) NMR spectroscopy. Two-dimensional spectra are obtained using a sequence of rf pulses that includes a variable delay or delays. A set of FIDs is acquired and stored. The variable delay is incremented by a small amount of time and a new set of FIDs are obtained and stored, and so on. At the end, the resulting matrix of FID data is Fourier transformed twice: first with respect to the acquisition time (as in normal FT-NMR) and second with respect to the time of the variable delay in the pulse sequence. The resulting data represent a surface and are presented as a contour plot of that surface.

The most useful 2D spectra for organic compound identification are called correlation spectra. Correlation Spectroscopy (COSY) spectra are presented as a contour plot with routine proton spectra along both of the axes, as shown in the COSY spectrum of ethyl vinyl ether in Figure 8.34. The spectra along the axes are low-digital-resolution spectra and appear to be a bit different from

Figure 8.34   The COSY NMR spectrum of ethyl vinyl ether.

those generated as usual NMR spectra. Note that the 2D spectrum is symmetric about the diagonal that runs from the lower left corner to the upper right corner. Every peak is represented by a peak along the diagonal and, in fact, the diagonal *is* the normal proton spectrum. Where the contour plot indicates a peak other than on the diagonal, the interpretation is that the corresponding peaks on the two axes represent protons coupled to one another.

For example, draw a line down from the signal at 1.2 ppm on the horizontal axis. This line encounters the diagonal at 1.2 ppm and then there is a peak at 3.7 ppm. By drawing a horizontal line over to the spectrum on the vertical axis, we can see that the peaks at 1.2 and 3.7 ppm are coupled to one another; these are the signals from the methyl triplet and methylene quartet of the ethyl group. We can also see that neither of these is coupled to the remainder of the spectrum, which of course is the vinyl group. Each proton in the vinyl group is coupled to each of the others, as we can see in the COSY spectrum. For example, the signal at 4.2 ppm is coupled to both the signal at 4.0 ppm and the one at 6.4 ppm.

It is possible to obtain 2D spectra that correlate the spectra of different nuclei, such as ^1H and ^{13}C. The heteronuclear correlation spectrum of ethyl vinyl ether is shown in Figure 8.35. The ^{13}C spectrum is along the horizontal axis and the ^1H spectrum is along the vertical axis. The peaks of the 2D spectrum indicate that the corresponding peaks on the axes represent a carbon and a proton (or protons) that are directly bonded. By using this spectrum, we can easily verify the ^{13}C and ^1H chemical-shift assignments below:

There are many other powerful 2D NMR spectroscopic techniques that can provide a wealth of information about molecular structure, even in organic molecules as large as proteins and nucleic acids. A few of the good texts that provide further information about these powerful tools are listed in the Bibliography.

NUCLEAR MAGNETIC RESONANCE SAMPLING

It is usually simple to prepare a sample of a small organic molecule (MW 500) for NMR analysis. For ^1H NMR spectroscopy, the sample size compatible with CW spectrometers is in the range 30–50 mg dissolved in about 0.5 mL of solvent. Fourier transform spectrometers require only 2–3 mg of sample in the same volume of solvent. Most samples are measured in solution in thin-walled tubes 5 mm in diameter and 18–20 cm long. NMR sample tubes are expensive and delicate; they must be as perfectly straight, and have as perfectly concentric walls, as possible. The sample tube is filled to a depth of about 3 cm. Filling the tube to this depth maximizes sample concentration in the active part of

Figure 8.35   Ethyl vinyl ether ^{1}H–^{13}C correlation spectrum.

the NMR probe, and thus the strength of the signal. Adding more solvent just wastes sample by dilution. The sample tube is spun about its axis in the instrument to average out small changes in the tube walls and in the magnetic field strength over the sample volume. There must be enough solvent in the tube to ensure that the vortex, or whirlpool, created when the tube is spun, does not extend down into the portion of the tube where the rf coils in the NMR probe are active. Many instruments use a depth gauge that shows exactly where this area is. Glass microcells are now commercially available. An inexpensive microcell technique for CW-NMR spectrometers using ordinary 5-mm NMR tubes has been described.[2]

The most practical NMR solvent is deuterochloroform ($CDCl_3$); it is relatively cheap and dissolves many different compounds. Handle this solvent with care, in the **hood**, because it is **toxic**! Many other deuterated solvents are commercially available, including acetone, methanol, and water. The universally accepted internal reference compound employed in making these measurements is tetramethylsilane, $Si(CH_3)_4$ (TMS). The most convenient source of TMS is commercially available $CDCl_3$, which contains about 1% TMS for use with CW spectrometers (commercially available 0.03% solutions are more appropriate for FT spectrometers).

HOOD

The most significant problem in sample preparation is the exclusion of small pieces of dust and dirt, because they may contain magnetic material,

---

[2]Yu, S. J. *J. Chem. Educ.* **1987,** *64,* 812.

which will result in poor spectra. Scrupulously clean samples of liquids can often simply be added to the NMR tube, followed by the solvent. Liquids containing visible impurities, and solids, are best prepared by dissolving the sample in about 0.3–0.4 mL of solvent in a small vial. The solution can then be filtered into the NMR tube through a Pasteur pipet plugged with a small piece of cotton, and the pipet can be rinsed with the NMR solvent to achieve the appropriate volume in the NMR tube. Since TMS is volatile (bp = 26.5 °C), the tube should be capped following addition, or the TMS will evaporate. If several people in a laboratory section are going to be obtaining NMR spectra, you will find that the spectrometer will be easier to tune with each new sample if all the sample tubes are filled to exactly the same level.

At this point, all specific instructions are dependent on the NMR spectrometer available to you. In general, the sample is inserted into the magnet and the magnetic field is adjusted very slightly (called shimming or tuning) to obtain a magnetic field that is as homogeneous as possible throughout the sample volume observed by the spectrometer. These adjustments are accomplished by energizing a collection of small electromagnetic coils around the sample, a process that is often done by the spectrometer's computer. Symptoms of a poorly shimmed spectrometer include broad and/or asymmetric peaks. The best place to check for this is on the TMS peak, because it is the one peak in the spectrum you *know* should be narrow and symmetric. Once the spectrometer is tuned, the spectrum is obtained and plotted, and the sample is removed from the magnet.

HOOD

Your NMR sample can be easily recovered by emptying the NMR tube into a small vial, rinsing the tube once or twice with (a nondeuterated) solvent, and then evaporating the solvent in a **hood** under a gentle stream of dry nitrogen.

QUESTIONS

Several 60-MHz ^{1}H NMR spectra are given below (Figs. 8.36–8.40) along with the molecular formula of the compound.[3] You should be able to account for at least one acceptable structure and for all of the observed resonances.

8-15. C_4H_8O. Spectrum *a*:

HSP-03-535

ppm

Figure 8.36   NMR unknown spectrum *a*. (Courtesy of SDBS, National Institute of Advanced Industrial Science and Technology.)

[3]From Pouchert, C. J. *The Aldrich Library of NMR Spectra*, 2nd ed.; Aldrich Chemical Co.: Milwaukee, WI, 1983.

8-16. $C_3H_6O_2$. Spectra *b* and *c*. Two compounds with the same empirical formula:

Figure 8.37a   NMR unknown spectrum *b*.
(Courtesy of SDBS, National Institute of Advanced
Industrial Science and Technology.)

Figure 8.37b   NMR unknown spectrum *c*.
(Courtesy of SDBS, National Institute of Advanced
Industrial Science and Technology.)

8-17. C_4H_8O. Spectrum *d*. Also give some thought to the weak resonances at 0.5 and 1.8 ppm:

Figure 8.38   NMR unknown spectrum *d*.
(Courtesy of SDBS, National Institute of
Advanced Industrial Science and Technology.)

8-18. C_7H_7Cl. Spectrum *e*:

HSP-00-390

ppm

Figure 8.39   NMR unknown spectrum *e*. (Courtesy of SDBS, National Institute of Advanced Industrial Science and Technology.)

8-19. $C_8H_{10}O$. Spectrum *f*:

HSP-05-935

ppm

Figure 8.40   NMR unknown spectrum *f*. (Courtesy of SDBS, National Institute of Advanced Industrial Science and Technology.)

BIBLIOGRAPHY

Nuclear magnetic resonance theory and principles of interpretation:

Abraham, R. J.; Fisher, J.; Loftus, P. *Introduction to NMR Spectroscopy*; Wiley: London, 1988.

Cooper, J. W. Spectroscopic *Techniques for Organic Chemists*; Wiley: New York, 1980.

Field, L. D.; Sternhell, S.; Kalman, J. R. *Organic Structures from Spectra*, 3rd ed.; Wiley: London, 2002.

Richards, S. A. Laboratory *Guide to Proton NMR Spectroscopy*; Blackwell Scientific Publications: London, 1988.

Silverstein, R. M.; Webster, F. X.; Kiemle, D. J. *Spectrometric Identification of Organic Compounds*, 7th ed.; Wiley: New York, 2005.

Sorrell, T. N. *Interpreting Spectra of Organic Molecules*; University Science Books: Mill Valley, CA, 1988.

Advanced theory and spectroscopic techniques:

Atta-ur-Rahman. *Nuclear Magnetic Resonance. Basic Principles*; Springer-Verlag: New York, 1986.

Claridge, T. D. W. *High-Resolution NMR Techniques in Organic Chemistry. Tetrahedron Organic Chemistry Series Volume 19*; Pergamon: Oxford, UK, 1999.

Derome, A. E. *Modern NMR Techniques for Chemistry Research*; Pergamon Press: Oxford, UK, 1987.

Duddeck, H.; Dietrich, W.; Toth, G. *Structure Elucidation by Modern NMR. A Workbook*, 3rd ed.; Springer-Verlag: New York, 1998.

Sanders, J. K. M.; Hunter, B. K. *Modern NMR Spectroscopy. A Guide for Chemists*; 2nd ed.; Oxford University Press: Oxford, UK, 1993.

Sanders, J. K. M.; Constable, E. C.; Hunter, B. K. *Modern NMR Spectroscopy. A Workbook of Chemical Problems*; 2nd ed.; Oxford University Press: Oxford, UK, 1993.

Libraries of NMR spectra:

Bhacca, N. S.; Hollis, D. P.; Johnson, L. F.; Pier, E. A.; Shoolery, J. N. *NMR Spectra Catalog*; Varian Associates: Palo Alto, CA, 1963.

Pouchert, C. J. *The Aldrich Library of NMR Spectra*, 2nd ed.; Aldrich Chemical Co.: Milwaukee, WI, 1983.

Pouchert, C. J.; Behnke, J. *The Aldrich Library of ^{13}C and ^{1}H FT-NMR Spectra*; Aldrich Chemical Co.: Milwaukee, WI, 1992.

Online libraries of NMR spectra:

SciFinder Scholar (https://scifinder.cas.org/)

ULTRAVIOLET–VISIBLE SPECTROSCOPY: INTRODUCTION TO ABSORPTION SPECTROSCOPY

In an atom, molecule, or ion, a limited number of electronic energy states are available to the system because of the quantized nature of the energies involved. The absorption of a photon by the system can be interpreted as corresponding to the occupation of a new energy state by an electron. The difference in energy between these two states may be expressed as ΔE:

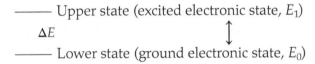

where the energy of the photon, E, is related to the frequency of the radiation by the Planck equation,

$$E = h\nu_i$$

where h is Planck's constant, 6.626×10^{-34} J s, and ν_i is the frequency in hertz. In the case above, $\Delta E = E_1 - E_0 = h(\nu_1 - \nu_0) = h\nu_i$

Thus, when a frequency match between the radiation and an energy gap (ΔE) in the substance occurs, a transition between the two states involved may be induced. The system can either absorb or emit a photon corresponding to ΔE, depending on the state currently occupied (emission would occur if the system relaxed from an upper-level excited state to a lower state). All organic molecules absorb photons with energies corresponding to the visible or ultraviolet regions of the electromagnetic spectrum, but to be absorbed, the incident energy in this frequency range must correspond to an available energy gap between an electronic ground state and an upper-level electronic excited state. The electronic transitions of principal interest to the organic chemist are those that correspond to the excitation of a single electron from the highest occupied molecular orbital (HOMO) to the lowest unoccupied molecular orbital (LUMO). As we will see, this will be the molecule's absorption occurring at the longest wavelength in the electronic absorption spectrum; it is, therefore, the most easily observed.

Electromagnetic radiation can be defined in terms of a frequency v, which is inversely proportional to a wavelength λ times a velocity c ($v = c/\lambda$, where c is the velocity of light in a vacuum, 2.998×10^8 m/s, and $c = v\lambda$ is the wave velocity). Thus,

$$\Delta E = hv = \frac{hc}{\lambda} = hc\tilde{v}$$

where $\tilde{v}$ is the wavenumber, defined as the reciprocal of the wavelength ($1/\lambda$) $\times$ the velocity of light.

Most ultraviolet and visible (UV and vis) spectra are recorded linearly in wavelength, rather than linearly in frequency or in units proportional to frequency (the wavenumber) or in energy values. Wavelength in this spectral region is currently expressed in nanometers (nm, where 1 nm = 10^{-9} m) or angstrom units (Å, where 1 Å = 10^{-10} m). The older literature is full of UV–vis spectra in which wavelength is plotted in millimicrons (mμ), which are also equivalent to 10^{-9} m.

It is unfortunate that because of instrumentation advantages this region of the spectrum is most often plotted in units that are nonlinear in energy (note the inverse relationship of E to λ) A convenient formula for expressing the relationship of wavelength and energy in useful values is

$$E = 28,635/\lambda \text{ kcal/mol} \quad (\lambda \text{ in mm})$$

or in terms of wavenumbers

$$E = (28.635 \times 10^{-4})\tilde{v} \quad (\tilde{v} \text{ in cm}^{-1})$$

The electromagnetic spectrum and the wavelength ranges corresponding to a variety of energy-state transitions are listed in Table 8.27. Infrared, UV–vis, and rf are of particular interest to the organic chemist because the excitation of organic substances by radiation from these regions of the spectrum can yield significant structural information about the molecular system being studied.

The absorption of rf energy by organic molecules immersed in strong magnetic fields involves exceedingly small energy transitions (~0.05 cal/mol), which correspond to nuclear spin excitations and result in NMR spectra. When a molecule absorbs microwave radiation, the energy states available for excitation correspond to molecular rotations and involve energies of roughly

Table 8.27 Spectroscopic Wavelength Ranges

Region	Wavelength (m)	Energy (kJ/mol)	Change Excited
Gamma ray	Less than 10^{-10}	$> 10^6$	Nuclear transformation
X-ray	10^{-8}–10^{-10}	10^4–10^6	Inner shell electron transitions
Ultraviolet (UV)	4×10^{-7}–1×10^{-8}	10^3–10^4	Valence shell electrons
Visible (vis)	8×10^{-7}–4×10^{-7}	10^2–10^3	Electronic transitions
Infrared (IR)	10^{-4}–2.5×10^{-6}	1–50	Bond vibrations
Microwave	10^{-2}–10^{-4}	10–1000	Molecular rotations
ESR	10^{-2}	10	Electron spin transitions
NMR	0.5–5	0.02–0.2	Nuclear spin transitions

1 cal/mol. With relatively simple molecules (in the gas phase) possessing a dipole moment (required for the absorption process) the analysis of the microwave spectrum can yield highly precise measurements of the molecular dimensions (bond lengths and angles). Unfortunately, relatively few organic systems exhibit pure rotational spectra that can be rigorously interpreted.

Absorption of radiation in the infrared region of the spectrum involves the excitation of vibrational energy levels and corresponds to energies in the range of about 1–12 kcal/mol. The excitation of electronic states requires considerably higher energies, from a little below 40 to nearly 300 kcal/mol. The corresponding radiation wavelengths would fall across the visible (400–800 nm), the near-UV (200–400 nm), and the far- (or vacuum) UV (100–200 nm) regions. The long-wavelength visible and near-UV regions of the spectrum hold information of particular value to the organic chemist. Here the energies correspond to the excitation of loosely held bonding (π) or lone-pair electrons. The far-UV region, however, involves high-energy transitions associated with the inner-shell and σ-bond electronic energy transitions. This region is difficult to access, because atmospheric oxygen begins to absorb UV radiation below 190 nm, which requires working in evacuated or purged instruments (which is why this region is often referred to as the vacuum UV).

UV–VIS SPECTROSCOPY

As we have seen, the application of electronic absorption spectroscopy in organic chemistry is restricted largely to excitation of ground-state electronic levels in the near-UV and vis regions. When photons of these energies are absorbed, the excited electronic states that result have bond strengths appreciably less than their ground-state values, and the internuclear distances and bond angles will be altered within the region of the molecules where the electronic excitation occurs (see Fig. 8.41). It is normally reasonable to assume that nearly all of the molecules are present in the ground vibrational state within the ground electronic state. The upper electronic state also contains a set of vibrational levels and any of these may be open to occupation by the excited electron (see Fig. 8.41). Thus, an electronic transition from a particular ground-state level can be to any number of upper-level vibrational states on the excited electronic state.

The shape of an electronic absorption band will be determined to a large extent by the spacing of the vibrational levels and the distribution of band intensity over the vibrational sublevels. In most cases these effects lead to broad absorption bands in the UV–vis region.

The wavelength maximum at which an absorption band occurs in the UV–vis region is generally referred to as the λ_{max} of the sample (where wavelength is determined by the band maximum).

The quantitative relationship of absorbance (the intensity of a band) to concentration is expressed by the Beer–Lambert equation:

$$A = \log\frac{I_0}{I} = \varepsilon c I$$

where

A = absorbance, expressed as I_0/I
I_0 = the intensity of the incident light
I = the intensity of the light transmitted through the sample

Figure 8.41   **Two electronic energy levels in a diatomic molecule.**

ε = molar absorptivity, or the extinction coefficient (a, constant characteristic of the specific molecule being observed); values for conjugated dienes typically range from 10,000 to 25,000

c = concentration (mol/L)

λ = length of sample path (cm)

The calculated extinction coefficient and solvent are usually listed with the wavelength at the band maximum. For example, data for methyl vinyl ketone (3-buten-2-one) would be reported as follows:

$$\lambda_{max}\, 219 \text{ nm } (\varepsilon = 3600, \text{ ethanol})$$
$$\lambda_{max}\, 324 \text{ nm } (\varepsilon = 24, \text{ ethanol})$$

Typical UV–vis spectra are shown in Experiments [6], [19D], [A2$_a$], and [A3$_a$]. As part of the characterization data, UV–vis information is also given in Experiments[16], [A1$_a$], [19A], and [33A].

APPLICATION TO ORGANIC MOLECULES

In organic compounds containing *conjugated* systems of π electrons, a particular *chromophore* present can often be identified by the use of UV–vis spectroscopy. A *chromophore* is, in this case, a group of atoms able to absorb light in the UV–vis region of the spectrum. Since the electronic transitions involved are limited primarily to π-electron (and lone-pair) systems, this type of spectroscopy is less commonly used than the other modern spectroscopic techniques—which, in fact, it predates by several decades. Ultraviolet–visible spectroscopy, however, can play a valuable role in certain situations. For example, if a research problem involves synthesizing a series of derivatives of a complex organic molecule that possesses a strong chromophore, the UV–vis spectrum will be highly sensitive to structural changes involving the arrangement of the π-electron system (see e.g., Experiment [6]).

In a conjugated alkene, such as 1,3-butadiene, the long-wavelength photon absorbed corresponds to the energy required for the excitation of a π electron from the HOMO, π_2 to the LUMO, π^*_3. For these alkenes, this transition is represented as $\pi \rightarrow \pi^*$ that is, an electron is promoted from a π

(bonding) molecular orbital to a π^* (antibonding) orbital. This type of excitation is depicted below for both ethylene and 1,3-butadiene. Note that as a consequence of extending the chromophore and raising the energy of the highest occupied level in butadiene, the energy gap between the HOMO and LUMO levels of ethylene is larger than that in the conjugated system. Thus, the photon required for excitation of ethylene has a higher energy (higher frequency = shorter wavelength, $\lambda_{max} = 171$ nm) than the photon absorbed by 1,3-butadiene($\lambda_{max} = 217$ nm):

If we continue to extend the chromophore, the decrease of the energy gap between the HOMO and LUMO levels also continues. This drop in ΔE is then reflected in a drop in energy of the photon required to excite the $\pi \rightarrow \pi^*$ transition. This effect is illustrated in Table 8.28.

As the extension of the chromophore continues, the λ_{max} of the $\pi \rightarrow \pi^*$ transition will eventually shift into the visible region. At this point the substance exhibits color. Because the absorbed wavelength is coming from the blue end of the visible spectrum, these compounds will appear yellow. The color will deepen and become red as the energy of the photon required for electronic excitation continues to drop. For example, tetraphenylcyclopentadienone is purple (Experiment [A3a]); the dye, methyl red, is deep red (Experiment [26]); and *trans*-9-(2-phenylethenyl)anthracene is golden yellow (Experiment [19D]).

Compounds that contain a carbonyl chromophore $\diagdown C{=}\ddot{O}\!:$ also absorb radiation in the UV region. A π electron in this unsaturated system undergoes a $\pi \rightarrow \pi^*$ transition. However, unless the carbonyl is part of a more extended

Table 8.28 Absorption Maxima of Conjugated Alkene		
Name	**Structure**	**λ_{max} (nm)**
Ethylene	$CH_2{=}CH_2$	165
1,3-Butadiene	$CH_2{=}CH{-}CH{=}CH_2$	217
1,3,5-Hexatriene	$CH_2{=}CH{-}CH{=}CH{-}CH{=}CH_2$	268
1,3,5,7-Octatetraene	$CH_2{=}CH{-}CH{=}CH{-}CH{=}CH{-}CH{=}CH_2$	290

chromophore, such as an α,β-unsaturated ketone system, the $\pi \rightarrow \pi^*$ transition requires a fairly high-energy photon for excitation, usually below 190 nm in the far-UV and similar to the energy required for excitation of a carbon–carbon double bond. The edge of the $\pi \rightarrow \pi^*$ absorption band may just barely be observed on instrumentation designed for near-UV studies. This partially observed absorption band is generally referred to as *end absorption*. In the case of carbonyls, however, the heteroatom also loosely holds two pairs of nonbonding electrons that are often termed *lone-pair* electrons. These nonbonding electrons reside in orbitals (n) that are higher in energy than the bonding π orbital, but lower in energy than the antibonding π^* orbital. Thus, while a transition from an n level to a π^* level is formally forbidden, in fact, weak bands are observed at λ_{max} in the near-UV that have their origin in the excitation of a lone-pair electron by an $n \rightarrow \pi^*$ transition. An energy diagram of a typical carbonyl system follows:

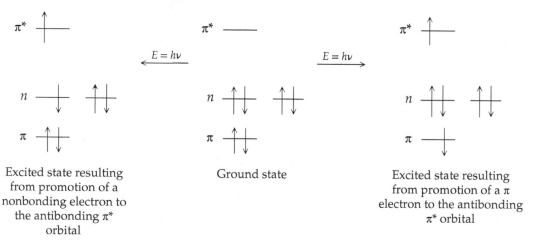

Excited state resulting from promotion of a nonbonding electron to the antibonding π^* orbital

Ground state

Excited state resulting from promotion of a π electron to the antibonding π^* orbital

Thus, those substances that contain the carbonyl chromophore absorb radiation of wavelengths that corresponds to both the $n \rightarrow \pi^*$ and the $\pi \rightarrow \pi^*$ transitions. For a simple ketone, such as acetone (CH_3COCH_3), the $\pi \rightarrow \pi^*$ transition is found in the far-UV and the $n \rightarrow \pi^*$ in the near-UV. When the carbonyl becomes part of an extended chromophore, such as in methyl vinyl ketone (3-buten-2-one), the spectra reveal that these two transitions have shifted to longer wavelengths—a bathochromic shift (see Fig. 8.42 for the

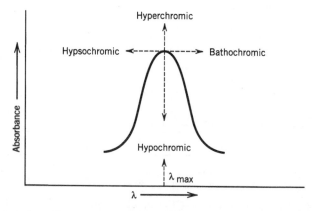

Figure 8.42 Terms describing direction of wavelengths and intensity shifts.

definition of terms used in UV–vis spectra to indicate the direction of wavelength and intensity shifts):

		$CH_3-\overset{\overset{\displaystyle \cdot\cdot}{O}}{\underset{\|}{C}}-CH_3$	$CH_3-\overset{\overset{\displaystyle \cdot\cdot}{O}}{\underset{\|}{C}}-CH=CH_2$
$n \Rightarrow \pi^*$	λ_{max} 270 nm ε_{max} 16	λ_{max} 324 nm ε_{max} 24
$\pi \Rightarrow \pi^*$	λ_{max} 187 nm ε_{max} 900	λ_{max} 219 nm ε_{max} 3600

Saturated systems containing heteroatoms with nonbonded electrons also exhibit weak absorption bands, often as end absorptions, which have their origin in forbidden $n \rightarrow \sigma^*$ transitions. When these heteroatomic groups are attached to chromophores, both the wavelength and the intensity of the absorption can be altered. These are often referred to as *auxochromes* and *auxochromic shifts*.

Often, model compounds containing a chromophore of interest are referred to as an aid in the interpretation of the UV–vis spectrum of a new structure. Substantial collections of data have been developed for a wide variety of chromophores as an aid to this type of correlation. A number of empirical correlations, such as the Woodward–Fieser rules, of substituent effects on λ_{max} values are available. The Woodward–Fieser rules are a set of empirical correlations derived from studies of UV–vis spectral data. Using these rules it is possible to predict with reasonable accuracy the λ_{max} for *new* systems containing various substituents on *known* chromophores. The rules are summarized in Table 8.29.

Examples of homoannular and heteroannular dienes are shown below:

A heteroannular diene   A homoannular diene

An example illustrating the use of these rules follows. Calculate the wavelength at which the following steroidal methyl sulfide will absorb:

The base value for the diene is 214 nm, because the system is heteroannular (if a homoannular diene were present it would take precedence over the heteroannular diene; see the following example). There are three-ring residues (or alkyl substituents) attached to the chromophore. Through hyperconjugation, the π system is slightly extended by this type of substitution. The residues are labeled *a*, *b*, and *c*. Each of these substituents is assumed to add 5 nm to the λ_{max} of the parent heteroannular diene, for a total of 15 nm. The 5,6-double bond in the B ring marked *z* is exocyclic to the A ring, so empirically we add an additional 5 nm. Finally, for the thiomethyl substituent at the 3 position we add 30 nm. The total is 214 + 15 + 5 + 30 = 264 nm.

Thus we have

$$\text{Predicted value} \quad \lambda_{max}\ (calcd) = 264\ nm$$
$$\text{Observed value} \quad \lambda_{max}\ (obsd) = 268\ nm \quad (\varepsilon = 22{,}600)$$

Table 8.29 Woodward–Fieser Rules for Conjugated Dienes	
Functionality	Increment (nm)
Base value for homoannular diene	253
Base value for heteroannular diene	214
Add:	
For each double bond extending conjugation	+30
For double bond outside of ring (exocyclic)	+5
For alkoxy groups	+ 6
For S-alkyl groups	+30
For Cl, Br groups	+5
For dialkylamino groups	+ 60
For parts of rings attached to butadiene fragment	+5

As another example, consider ergosta-3,5,7,9-tetraene-3-acetate (**I**):

I

Prediction of λ_{max} for a homoannular diene

Parent homoannular diene in ring **B**		253 nm
Increments for		
Double bond extending conjugation	c [2 × 30]	60
Alkyl substituent or ring residue	a [5 × 5]	25
Exocyclic double bond	b [3 × 5]	15
Polar substituents	d [0]	0
	λ_{calcd}	353

Predicted value λ_{max} (calcd) = 353 nm

Observed value λ_{max} (obsc) = 355 nm (ε = 19,700)

There are additional rules for carbonyl-containing compounds, such as ketones, aldehydes, carboxylic acids, and so on, and for aromatic compounds. Table 8.30 lists the parameters for conjugated carbonyl systems. Note that in contrast to the conjugated diene compounds, in which we are observing $\pi \to \pi^*$ transitions, the $n \to \pi^*$ transitions of the carbonyl λ_{max} chromophore are often solvent dependent. Thus, solvent effects will have to be considered when predicting λ_{max} values in these systems.

An example of λ_{max} (ethanol) calculation for a carbonyl system is presented here:

Table 8.30 Conjugated Carbonyl Systems

α,β-Unsaturated Functionality	Base Value (nm)
Acyclic or six-membered or higher cyclic ketone	215
Five-membered ring ketone	202
Aldehydes	210
Carboxylic acids and esters	195

	Increment (nm)
Extended conjugation	+30
Homoannular diene	+ 39
Exocyclic double bond	+5

	Substituent Increment (nm)		
Substituent	α	β	δ
Alkyl	+ 10	+ 12	+ 18 (γ and higher)
Hydroxyl	+35	+30	+50
Alkoxy	+35	+30	+31 (γ + 17)
Acetoxy	+ 6	+6	+6
Dialkylamino		+ 95	
Chloro	+ 15	+ 12	
Bromo	+25	+30	
Alkylthio		+ 85	

Solvent	Solvent Increment (nm)
Water	−8
Ethanol	0
Methanol	0
Chloroform	+1
Dioxane	+5
Ether	+7
Hexane	+11
Cyclohexane	+11

The base value for the α,β-unsaturated six-membered ring ketone system is 215 nm. Extended conjugation adds an additional 30 nm. The presence of an exocyclic double bond, marked *a*, extends the λ_{max} another +5 nm. There is a substituent on the β-carbon atom (+12 nm) and on the δ-carbon atom (+18 nm). There is no solvent effect, because the spectrum was obtained in ethanol (0 shift). The total is 215 + 30 + 5 + 12 + 18 = 280 nm:

$$\text{Predicted value} \quad \lambda_{1max} \text{ (calcd)} = 280 \text{ nm}$$
$$\text{Observed value} \quad \lambda_{1max} \text{ (obsd)} = 284 \text{ nm}$$

The Woodward–Fieser rules work well for systems with four or fewer double bonds. For more extensively conjugated systems, λ_{max} values are more accurately predicted using the Fieser–Kuhn equation:

$$\text{Wavelength} = 114 + 5\,M + n(48.0 - 1.7n) - 16.5\,R_{endo} - 10\,R_{exo}$$

where

n = number of conjugated double bonds

M = number of alkyl substituents in the conjugated system

R_{endo} = number of rings with endocyclic double bonds in the system

R_{exo} = number of rings with exocyclic double bonds in the system

B   *Sample calculation:* Find the UV λ_{max} of β-carotene:

β-Carotene

In the structure there are 11 conjugated double bonds, $n = 11$. There are 6 alkyl groups and 4 ring residues on the conjugated system, $M = 10$. Both rings have an endocyclic double bond, $R_{endo} = 2$. Neither ring has any exocyclic double bonds, therefore $R_{exo} = 0$. Substituting in the equation gives

$$\text{Wavelength} = 114 + 5(10) + 11[48 - 1.7(11)] - 16.5(2) - 10(0)$$

$$= 114 + 50 + 322.3 - 33 - 0 = 453 \text{ nm}$$

$$\text{Predicted value} \qquad \lambda_{lmax}(\text{calcd}) = 453 \text{ nm}$$

$$\text{Observed value} \qquad \lambda_{max}(\text{obsd}) = 455 \text{ nm}$$

Two examples of this correlation scheme (see Table 8.31) are

Table 8.31 The Benzoyl Chromophore

Parent Chromophore C_6H_5—CO—R

Function	Wavelength (nm)
R = alkyl or ring residue	246
R = H	250
R = OH, O-alkyl	230

Substituent	Substituent Increment (nm)		
	$o-$	$m-$	$p-$
Alkyl or ring residue	3	3	10
—OH, —OCH$_3$, —O-alkyl	7	7	25
—O$^-$ (p-sensitive to steric effects)	11	20	78
—Cl	0	0	10
—Br	2	2	15
—NH$_2$	13	13	58
—NHAc	20	20	45
—NHCH$_3$			73
—N(CH$_3$)$_2$	20	20	85

Note. Spectra obtained in alcohol solvents.

1. 6-Methoxytetralone

Predicted λ_{max} is calculated by taking

Parent value, 246 nm + one *o*-ring residue, 3 + one *p*-OMe, 25 = 274 nm

Predicted value $\quad \lambda_{max}(calcd) = 274$ nm
Observed value $\quad \lambda_{max}(obsd) = 276$ nm $(\varepsilon = 16,500)$

2. 3-Carboethoxy-4-methyl-5-chloro-8-hydroxytetralone

Parent value, 246 nm + one *o*-ring residue, 3 + one *o*-OH, 7 + 0 *m*-Cl = 256 nm

Predicted value $\quad \lambda_{max}(calcd) = 256$ nm
Observed value $\quad \lambda_{max}(obsd) = 257$ nm $(\varepsilon = 8000)$

In summary, UV–vis spectra can make substantial contributions to understanding the molecular structure of organic substances that possess *chromophores*:

1. Interpretation of ultraviolet–visible spectra often can be a powerful approach for identifying the molecular structure of that section of a new substance that contains the chromophore.

2. The λ_{max} increases within a series of compounds that contain a common chromophore that is lengthened (increased conjugation) over the series. The intensity of the absorption (ε_{max}) also generally becomes greater as conjugation increases, but can be very sensitive to steric effects (see Experiment [6]).

3. The λ_{max} is sensitive to hyperconjugation by alkyl substituents, conformational changes that restrict π-system overlap, configurational, or geometric isomerization in which π systems are perturbed, and structural changes, such as the isomerization of a double bond from an *exocyclic* to an *endocyclic* position and changes in ring size.

4. In many instances, accurate prediction of the λ_{max} of a new molecular system can be made based on empirical correlations of the parent chromophore giving rise to the absorption.

Table 8.32 lists the λ_{max} values of a number of common organic molecules.

INSTRUMENTATION

The acquisition of UV–vis absorption spectra for use in the elucidation of organic molecular structure is now carried out with instrumentation that is typically an automatic-recording photoelectric spectrophotometer. The optical

Table 8.32 Absorption Maxima of Several Unsaturated Molecules

Compound	Structure	λ_{max} (nm)	ε_{max}
Ethylene	$CH_2{=}CH_2$	171	15,530
1,3-Butadiene	$CH_2{=}CH{-}CH{=}CH_2$	217	21,000
Cyclopentadiene		239	3,400
1-Octene	$CH_3(CH_2)_5CH{=}CH_2$	177	12,600
trans-Stilbene		295	27,000
cis-Stilbene		280	13,500
Toluene		189	55,000
		208	7,900
		262	260
4-Nitrophenol	$HO{-}\langle\ \rangle{-}NO_2$	320	9,000
3-Penten-2-one	$CH_3CH{=}CHCCH_3$	220	13,000
		311	35

components of one of the classic spectrophotometers is given in Figure 8.43. This system is typical of a high-quality double-beam double-monochromator instrument. The instrument consists of a number of components: the radiation source, monochromator, sample compartment, detector, amplifier, and recorder.

The Source of Radiation

Radiant energy may be generated by either a deuterium discharge lamp or a tungsten–halogen lamp depending on the spectral region to be observed. Deuterium is generally preferred over hydrogen since the intense radiating ball of plasma is slightly larger in the case of deuterium, and therefore source brightness is enhanced by a factor of about 4. Below 360 nm, deuterium gas emits an intense continuum band that covers a major portion of the UV region. With special windows, the short wavelength cutoff can be extended down to about 160 nm well out into the vacuum-UV. Emission line spectra limit the long wavelength use of these lamps to about 380 nm. The lamps of choice for the region above 350 nm (the visible) are incandescent filament lamps, because they emit a broad band of radiation from 350 nm on the short wavelength end all the way to about 2.5 μm (the near-IR) on the long wavelength side. Most of the radiation emitted falls outside the visible, peaking at about 1 μm in the near-IR. Nevertheless, tungsten lamps are *the* choice for measurements in the visible region, because they are extremely stable light sources.

Thus, radiation sources must possess two basic characteristics: (1) they must emit a sufficient level of radiant energy over the region to be studied so that the instrument detection system can function, and (2) they must maintain constant power during the measurement period. Source power fluctuations can result in spectral distortion.

The Monochromator

As the name implies, a monochromator (making a single color or hue) functions to isolate a single frequency from the source band of radiation. In practice we settle for isolating a small collection of overlapping frequencies surrounding the monochrome radiation we wish to observe. Thus, the monochromator section of the instrument takes all the source radiation in at one end and releases a very narrow set of bands of radiation at the other end. This function is accomplished, as shown in Figure 8.43, by focusing the entering radiation on an entrance slit that forms a narrow image of the source. After passing through the entrance slit, the spreading radiation is collimated by being reflected off a parabolic mirror, and is converted into parallel light rays (just as in a search light). The collimated radiation is then directed to the dispersing agent, which is usually a quartz prism (quartz is transparent to UV, glass is not) or diffraction grating. The dispersing device spreads the different wavelengths of collimated light out in space. After emerging from the prism, the dispersed radiation is redirected to either the same or a new collimator mirror and refocused as

Figure 8.43 Schematic optical diagram of a double beam-in-time spectrophotometer with double monochromation (Cary Model 17D). (Courtesy of Varian Associates, Inc.)

an image of the source on the exit slit of the monochromator. The exit slit has only a small fraction of the original radiation focused on it, and allows it to pass through in the image of the source. The remaining frequencies lie at different angles on either side of the exit slit. By mechanically turning the prism or grating, and thus changing the angle of the dispersing device with respect to the exit slit, all of the narrowly dispersed bands of radiation can be passed out of the monochromator in sequential fashion.

Instruments that are designed to reduce unwanted radiation to an absolute minimum will place two monochromators in tandem with an intermediate slit connecting the dispersing systems. In the case illustrated in Figure 8.43 the first monochromator uses a prism, while the second uses a grating. The two monochromators, however, must be in perfect synchronization or no light at all will be transmitted.

Sample Compartment

After leaving the monochromator the radiation is directed to the sample compartment by a rotating sector mirror, where it is alternately focused on the substance to be examined (which is contained in a cell with quartz windows) and a reference cell (which holds the pure solvent used to dissolve the sample). The system now has two beams, hence the name *double-beam spectrophotometer*. After passing through the sample where the absorption of radiation may occur, the beams are recombined.

The sampling position could be placed either before or after the monochromator. In infrared instruments (such as the PE Model 710B, Fig. 8.3*a*, p. 507) it was generally found before the monochromator until the introduction of interferometers. In UV systems, the sampling area is most often placed after the monochromator, and for good reasons. If the sample were placed before the monochromator, it would be exposed to the entire band of high-energy UV radiation being emitted by the source over the entire sampling period. By positioning the sample after the monochromator, at any one time the sample sees only the very small fraction of the dispersed radiation passed by the exit slit. Thus, sample stability is greatly protected by this arrangement. Remember that near-UV radiation carries photons with energies that approach those of the bond energies of organic molecules.

The Detector

The recombined beams are then focused on the detector. Detectors function as transducers, because they convert electromagnetic radiation into electrical current. There are a number of radiation-sensitive transducers available as detectors for these instruments. One is the photomultiplier tube. These detectors operate with photocathodes that emit electrons in direct proportion to the number of photons striking the photosensitive surface and possess very large internal amplification. Thus, they operate at low power levels. One particular advantage of the photomultiplier is that you can adjust their sensitivity over a wide range simply by adjusting the supply voltage.

The Electronics: The Amplifier and Recorder

In double-beam instruments, the two signals generated by the sample and reference beams (each referenced against a dark current) in the detector are amplified and the ratio of the sample signal to the reference signal is plotted on

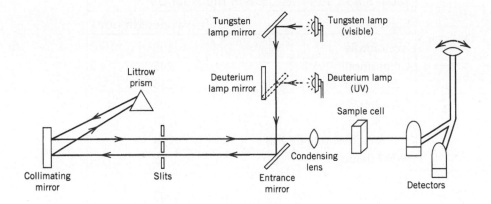

Figure 8.44   UV–visible single-beam spectrometer.

a recorder. The simplest of the absorption spectrometers are the single-beam instruments (see Fig. 8.44). These spectrometers are generally employed for problems involving simple one-component analyses. The photometric accuracy of scanned spectra should not be of paramount importance with these systems. Single-beam spectrometers require extremely stable sources and detectors.

SAMPLE PREPARATION

Ultraviolet spectra are usually obtained on samples in solution using quartz cells. Quartz is used, because it is transparent to both UV and visible light. For spectra restricted to the visible region, Pyrex cells are satisfactory (and a good deal less expensive), but because Pyrex absorbs UV radiation, these cells cannot be employed for measurements in this region.

Solution cells usually have a horizontal cross section of 1 cm^2 and require about 3 mL of sample solution. Cells must be absolutely clean, and it is advisable to rinse the cell several times with the solvent used to dissolve the sample. A background spectrum of the solvent-filled cell (*without* a reference sample) can easily be obtained at this time and used as a check against contamination of either the cell or the solvent or both.

Because the intensities of electronic transitions vary over a very wide range, the preparation of samples for UV–vis spectra determination is highly concentration dependent. Intense absorption can result from the high molecular extinction coefficients found in many organic chromophores. The sampling of these materials requires very dilute solutions (on the order of 10^{-6}–10^{-4} M). These solutions can be conveniently obtained by the technique of *serial dilution*. In this method a sample of the material to be analyzed is accurately weighed, dissolved in the chosen solvent, and diluted to volume in a volumetric flask. Sample weights of 4–5 mg in 10-mL volumetric flasks are typical. An aliquot is then taken from this original solution, transferred to a second volumetric flask, and diluted as before. This sequence is repeated until the desired concentration is obtained.

Numerous choices of solvent are available (a list is given in Table 8.33) and most of them are available in "spectral grade." The most commonly used solvents are water, 95% ethanol, methanol, and cyclohexane.

Table 8.33 Solvents Used in the Near-UV	
Solvent	Cutoff Wavelength (nm)
Acetonitrile	190
Chloroform	245
(toxic, substitute CH_2Cl_2)	235
Cyclohexane	205
1,4-Dioxane	215
(toxic, substitute EtOEt)	218
95% Ethanol	205
n-Hexane	195
Methanol	205
Isooctane	195
Water	190

Note. Since these solvents have no color, they are transparent in the visible.

Criteria for Choosing a Solvent

- The most important factor is solubility of the sample. UV–vis spectra can be very intense, so even low solubility may be quite acceptable in sample preparation.

- The wavelength cutoff for the solvent may be important if the sample absorbs below about 250 nm.

- Sample–solvent molecular interactions must be considered. An example of these effects would be hydrogen bonding of protic solvents with carbonyl systems. Hydrocarbon chromophores are less influenced by solvent character than are the more polar chromophores.

BIBLIOGRAPHY

American Petroleum Research Institute Project 44 *Selected Ultraviolet Spectral Data*, Vols. I–IV; Thermodynamics Research Center, Texas A&M University: College Station, TX, 1945–1977 (1178 compounds).

Feinstein, K. *Guide to Spectroscopic Identification of Organic Compounds*; CRC Press: Boca Raton, FL, 1995.

Field, L. D.; Sternhell, S.; Kalman, J. R. *Organic Structures from Spectra*, 4th ed.; Wiley: New York, 2008.

Grasselli, J. G.; Ritchey, W. M. *Atlas of Spectral Data and Physical Constants for Organic Compounds*, 2nd ed.; CRC Press: Cleveland, OH, 1975.

Harwood, L. M.; Claridge, T. D. W. *Introduction to Organic Spectroscopy*; Oxford University Press: New York, 1997.

Hesse, M; Meier, H; Zeeh, B. *Spectroscopic Methods in Organic Chemistry*, 2nd ed.; Thieme Medical Publishing: New York, 2008.

Lang, L., Ed.; *Absorption Spectra in the Ultraviolet and Visible Region*, Vols. 1–20; Academic Press: New York, 1961–1975; Vols. 21–24; Krieger: New York, 1977–1984.

Pavia, D. L.; Lampman, G. M.; Kriz, G. S. *Introduction to Spectroscopy: A Guide for Students of Organic Chemistry*, 3rd ed.; Saunders College: Philadelphia, 2000.

Pretsch, E.; Clerc, J. T. *Spectra Interpretation of Organic Compounds*; VCH/Wiley: New York, 1997.

Rouessac, F.; Rouessac, A. *Chemical Analysis: Modern Instrumental Methods and Techniques*, 2nd ed.; John Wiley & Sons: New York, 2007.

Standard Ultraviolet Spectra; Sadtler Research Laboratories: Philadelphia.

UV Atlas of Organic Compounds, Vols. I–IV; Butterworths: London, 1966–1971.

Williams, D. H.; Fleming, I. *Spectroscopic Methods in Organic Chemistry*, 6th ed.; WCB/McGraw-Hill: New York, 2007.

MASS SPECTROMETRY*

In comparison with other forms of spectroscopy, such as NMR, IR, or UV–vis, mass spectrometry is unique in terms of how we generate and interpret the spectrum. Instead of monitoring the absorption of electromagnetic radiation in terms of frequency or wavelength, a mass spectrum can be thought of as a snapshot of a rather unconventional organic reaction involving one energetic reactant that decomposes to give a variety of reaction products. By characterizing the composition of this "reaction mixture" we are able to learn the identity of the starting reactant.

The reaction gets started when a molecule in the gas phase is converted to a radical cation by an energetic collision with an electron, as shown below for N_2O:

$$N_2O_{(g)} + e^- \longrightarrow N_2O^{+\cdot} + 2e^-$$

The fact that we have formed a positive ion (cation) with an unpaired electron (radical) becomes important for understanding the decomposition reactions. The process of forming the radical cation yields a collection of energized *molecular ions* that contain a range of internal energies. The molecular ion is produced in a low-pressure environment where it is unable to bump into other molecules. Fragmentation (bond breaking) results in the formation of charged and neutral products. Ideally, intramolecular rearrangements, which could complicate determination of the molecules original structure, do not occur. Depending on characteristics of the molecule and the amount of energy deposited, a variety of fragmentation reactions can take place. For example, the molecular ion (N_2O^+) may fragment to give the following products through one- or two-step reactions:

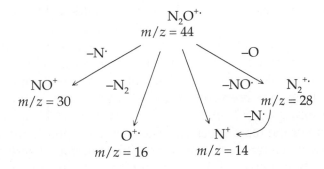

Mass spectrometry derives its name from its ability to distinguish the molecular ion and the different charged reaction products based on the ratio of the ion mass to its charge (m/z ratio). In most cases, the charge, z, is equal to 1 and we can easily tell the difference between the molecular ion ($m/z = 44$) and products, such as N_2^+ ($m/z = 28$). In addition, a mass spectrometer can be used to determine the relative amounts of molecular ions and fragment ions present after the reaction has had a little (very little!) time to proceed. A mass spectrum, typically shown in a bar-graph format, is a display of the relative number of each type of ion plotted as a function of the m/z ratio (see Fig. 8.45).

*This section has been written by Elizabeth A. Stemmler, Professor of Chemistry, Bowdoin College.

Figure 8.45 Electron ionization mass spectrum of nitrous oxide, N$_2$O.

Instead of reporting the actual number of each type of ion, we normalize the data and give the most abundant ion a value of 100%. Note that only charged species are detected by a mass spectrometer. The neutral products are not observed and their identity must be inferred.

Mass spectrometry is useful to organic chemists because of the information it provides about molecular structure. For example, if the molecular ion is present, that peak can be used to determine the molecular weight (MW) of the neutral molecule. With precise measurement of the m/z ratio (to ± 0.0001, for example), the elemental formula of the molecular ion can be determined. For example, N$_2$O and CO$_2$ both have a molecular weight of 44; however, measurement of their exact masses (44.0011 and 43.9898, respectively) can be used to assign their elemental formula. Mass spectrometry was used to originally determine the exact mass of each element (see Table 8.34), and these exact masses, *not* the atomic weights, are used to calculate mass.

Even when precise mass measurements are not available, the products in the mass spectrum may provide enough information to determine the structure of the neutral molecule. Interpretation of a mass spectrum involves working backward from the observed charged fragments to a proposed molecular structure. For example, CO$_2$ or N$_2$O (same MW) could be distinguished by an examination of the mass spectrum. The fragment ions at m/z 14 and 30 in Figure 8.45 clearly eliminate CO$_2$ as a possible structure. There are no combinations of carbon (mass = 12) and oxygen (mass = 16) that could produce these ions. In addition, the mass spectrum allows us to distinguish between the isomers NNO and NON. What would the mass spectrum of NON show? We would expect to see only a peak at m/z 30 (NO$^+$) and no signal at m/z 28 (N$_2^+$). The mass spectrum would thus support your chemical intuition that NON is an unlikely and unstable molecular structure.

As you will see below, mass spectral interpretation is not always as straightforward as the case given above. Like the outcome of an organic reaction, a mass spectrum will reflect the outcome of competing sequential and simultaneous reaction pathways. For some molecules, very little fragmentation will take place and only the molecular ion is observed. For other less stable molecules, we may have complete conversion of the molecular ion to products, although, because of the high energy required for their formation, we will rarely see complete fragmentation down to products at the atomic level.

Table 8.34 Exact Masses and the Atomic Weights for Isotopes of Some Common Elements

Element	Nuclide	Mass	Atomic Weight[a]
Hydrogen	^{1}H	1.0078	1.0079
	D(^{2}H)	2.0141	
Carbon	^{12}C	12.00000 (std)	12.011
	^{13}C	13.0034	
Nitrogen	^{14}N	14.0031	14.0067
	^{15}N	15.0001	
Oxygen	^{16}O	15.9949	15.9994
	^{17}O	16.9991	
	^{18}O	17.9992	
Fluorine	^{19}F	18.9984	18.9984
Silicon	^{28}Si	27.9769	28.0855
	^{29}Si	28.9765	
	^{30}Si	29.9738	
Phosphorus	^{31}P	30.9738	30.9738
Sulfur	^{32}S	31.9721	32.066
	^{33}S	32.9715	
	^{34}S	33.9679	
Chlorine	^{35}Cl	34.9689	35.4527
	^{37}Cl	36.9659	
Bromine	^{79}Br	78.9183	79.904
	^{81}Br	80.9163	
Iodine	^{127}I	126.9045	126.904

[a]Average mass of the naturally occurring isotopes of the element; *not* used for mass calculations in mass spectrometry.

The interpretation of a mass spectrum requires developing an understanding of important, characteristic reaction pathways and an appreciation of factors influencing ion stability. In many ways, the interpretation of mass spectra provides a place to apply principles of organic chemistry to a unique kind of chemical reaction.

INSTRUMENTATION

All mass spectrometric instruments contain regions where ionization, mass analysis, and ion detection take place. Mass spectrometry takes place at low pressure; all of the mass spectrometric components are contained in a vacuum system at pressures of 10^{-7} to 10^{-5} torr. Because the instrument must be sealed from the atmosphere to maintain the low pressure, and because samples must be converted to the gas phase prior to ionization, all mass spectrometers have a region devoted to sample introduction. In this region the sample—in the form of a solid, liquid, or gas—is transferred to the low pressure of the mass spectrometer, while preventing the introduction of air. A block diagram of a basic mass spectrometer is shown in Figure 8.46. Ions are generated and

| Sample inlet (GC or DIP) | | Ion source | Mass analyzer | | Detector | | Computer |

Vacuum system

Figure 8.46   Block diagram of components of a mass spectrometer.

fragment in the ion source; the molecular ion and fragments are separated, based upon m/z ratios, in the mass analyzer; and the ion signals are converted by the detector into a signal that may be input to a computer.

Ion Source

The ion source is the region where ions are generated. Mass spectrometrists have many ways of creating ions from different types of samples, including biological materials or the surface of a particle. In our discussions, we will focus only on the most common ionization method, electron ionization (EI). In an EI ion source (Fig. 8.47), we send current through a wire, called a filament. As the filament gets hot, electrons are emitted from the surface. The electrons are produced in an electric field, which results in electron acceleration through the ion source region where the sample vapor is found. If you work with an electron energy that is too low (below the ionization potential for the molecule), no ions will be produced. As the electron energy increases, the molecular ion ($M^{+\cdot}$) will appear. With further increases in electron energy, fragment ions are

Figure 8.47   Schematic diagram of an electron ionization source. (From Watson, J. T. *Introduction to Mass Spectrometry*, 3rd ed.; Lippincott-Raven Publishers: Philadelphia, 1997, p. 140.)

observed. Formation of doubly charged ions (M^{2+}) occurs, but the ion intensities are very low:

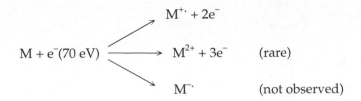

$$M^{+\cdot} + 2e^-$$
$$M + e^-(70\ eV) \longrightarrow M^{2+} + 3e^- \quad \text{(rare)}$$
$$M^{-\cdot} \quad \text{(not observed)}$$

Mass spectra, by convention, are measured with 70-eV electrons. At this energy, the ion intensity is high and the distribution of products remains relatively constant with small changes in electron energy. Formation of negative molecular ions with 70-eV electrons does not occur.

Another important role for the ion source is directing the ions toward the mass analyzer. The ions are pushed and pulled as they pass through one or more metal plates that have a hole in the center for ion transmission. These plates accelerate the ions and keep them directed at the mass analyzer. Depending on the mass analyzer in use, the ions are accelerated toward the analyzer with high or low energy.

Mass Analyzer

Ion formation and fragmentation in the source is followed by mass analysis. Mass analyzers are used to separate ions based on their mass-to-charge ratios. Organic chemists commonly use two types of mass analyzers: magnetic sector instruments (low- and high-resolution) and quadrupole instruments. Magnetic sectors separate ions based on dispersion of the ions into beams with different m/z ratios; quadrupoles are mass filtering devices.

In a magnetic sector instrument, ions are accelerated out of the ion source into a magnetic field with high (kilovolt) kinetic energies. The magnet field, applied perpendicular to the path of the ions, exerts a force that causes the ions to follow a curved path through the magnet (Fig. 8.48). The extent to which the path is bent depends on the mass-to-charge ratio (more specifically, the momentum) of the ions. Light ions are bent more than heavier ions. If the path that the ions must travel is fixed, ions that are too light or too heavy will run into the walls of the mass analyzer, where they are neutralized, and will then be pumped away by the vacuum system. Only the ions with the correct radius (correct m/z ratio) will make it to the detector. To measure a complete mass spectrum, the magnetic field strength is varied to bring ions of different m/z ratio to focus on the detector.

High-resolution magnetic sector instruments incorporate an additional energy analyzer prior to mass analysis by the magnetic sector. This more precisely defines the kinetic energies of ions entering the magnetic sector, which improves the mass resolution. High-resolution instruments require more expertise to operate and are less common because of their expense, but they can provide the precise and accurate mass measurements needed to determine elemental composition.

A quadrupole is composed of a set of four rods to which (electric) potentials are applied (Fig. 8.49). To allow ions of a particular m/z ratio through the rods, a constant positive potential is applied to two opposing rods (the x-rods), while the remaining two rods experience a constant negative potential (the

Figure 8.48 **Schematic diagram of a magnetic sector mass analyzer.** (From McLafferty, F. W.; Turevek, F. *Interpretation of Mass Spectra*, 4th ed.; University Science Books: Sausalito, CA, 1993, p. 8.)

y-rods). In addition, each set of rods experiences a time-varying potential that causes the rod potentials to vary between positive and negative potentials, with the signals 180° out-of-phase. When the x-rods are positive, the y-rods are negative. Mass filtering occurs when a group of ions enters the analyzer. Ions that have m/z ratios that are too low or too high will experience unstable

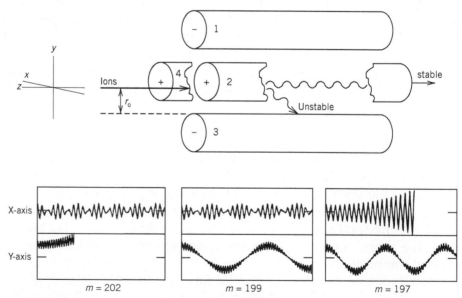

Figure 8.49 **Schematic diagram of a quadrupole mass analyzer.** X and Y axis trajectories for m/z 202, 199, and 197. (From Steel, C.; Henchman, M. J. *Chem. Educ.* **1998**, 75, 1049–1054.)

trajectories through the rods and will strike a rod, become neutralized, and be pumped out of the system. Only ions with an appropriate m/z ratio will have a stable trajectory and will make it through the rods to the detector. Figure 8.49 shows the trajectories of three ions with respect to the x- and y-rods. Only the ion with $m/z = 199$ makes it through the quadrupole. To change the m/z ratio of the ions that are transmitted, the magnitude of the constant and time-varying potentials are changed. Mass spectrometers that fit on a laboratory benchtop have a mass range of $m/z = 10$ to 650. With the quadrupole mass analyzer a mass spectrum can be measured rapidly (roughly 1 scan per second), which is important when capillary GC columns are used for sample introduction.

Detector

Ions can be detected directly through the current produced when they strike a plate; however, we usually make this signal larger through the use of electron multiplier detectors.

Tuning the Mass Spectrometer

Before the mass spectrometer can be used to collect mass spectra, the instrument must be tuned and calibrated. The tuning procedure involves setting voltages associated with the ion source, lenses, and detector (to optimize sensitivity), and selecting values for potentials applied to the quadrupole (to set the instrument resolution). These tasks are accomplished while a calibration standard is continuously added to the instrument. A common calibration standard is perfluorotributylamine (PFTBA), $(CF_3CF_2CF_2CF_2)_3N$. Usually ions at m/z 69, 219, and 502 are monitored (Fig. 8.50).

Figure 8.50 Electron ionization mass spectrum of perfluorotributylamine (PFTBA). Insets show the peak profiles for m/z 69, 219, and 502.

Sample Introduction

Samples analyzed by EI mass spectrometry must be converted to gas phase. For pure gases or volatile liquids the samples may be introduced directly through a small orifice that allows an appropriate amount of material into the vacuum chamber. A small amount of a solid sample can be placed in a melting point capillary tube and inserted into the mass spectrometer at the end of a metal rod, called a *direct insertion probe* (DIP). The temperature at the tip of the probe can be varied to promote sublimation of the sample. Another common method of sample introduction is gas chromatography, which is the ideal choice for samples that are impure.

Gas Chromatography/Mass Spectrometry (GC/MS)

While the goal of a synthetic organic reaction is the production of one pure product in high yield, organic reactions often produce a mixture of reaction products. Chromatographic separation of those products is a useful complement to the mass spectral analysis. The components of the mixture elute from the chromatographic column, ideally, as pure peaks. The mass spectrometer, which is scanning rapidly, is then able to collect a few spectra for each eluting peak. Both the chromatographic retention time and the mass spectrum can be used to help identify components of the mixture. Because the compounds are detected with little bias for one type of compound over another, GC/MS has provided organic chemists with a powerful tool to characterize reaction mixtures and assess product purity.

Sensitivity is another distinguishing feature of mass spectrometry. This sensitivity has allowed mass spectrometers to act as detectors for capillary columns, which can separate mixture containing hundreds of compounds, when less than a nanogram (10^{-9} g) of each compound is injected.

Capillary Columns

Most GC/MS instruments use capillary columns for chromatographic separation. Capillary columns are very long (15- to 30-m), open tubes of fused silica that are coated with a thin coating of the stationary phase (Fig. 8.51). A carrier gas, typically helium, is used as the mobile phase. Capillary column diameters are commonly in the range of 0.25 to 0.53 mm, and the coating of the stationary phase is in the range of 0.25 to 1 μm. Thicker coatings are used for the separation of low-boiling compounds.

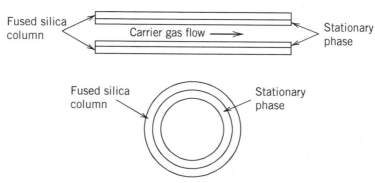

Figure 8.51   Longitudinal and radial cross sections of a capillary column.

Chromatographic resolution increases as a function of the square root of the column length, and the extraordinary length of capillary columns means that most simple mixtures are easily resolved on just a few types of stationary phases. One common nonpolar stationary phase is poly(dimethylsiloxane) ($R = CH_3$) which can be made slightly more polar by the incorporation of phenyl groups (typically 5% phenyl) in place of methyl groups:

$$
R-\underset{\underset{R}{|}}{\overset{\overset{R}{|}}{Si}}-O-\left[\underset{\underset{R}{|}}{\overset{\overset{R}{|}}{Si}}-O-\right]_n\underset{\underset{R}{|}}{\overset{\overset{R}{|}}{Si}}-R
$$

These two stationary phases interact with solutes primarily through dispersion interactions, and compounds elute as a function of boiling point. More polar stationary phases are also available. Because capillary columns are used to separate compounds with a wide range of boiling points, we often make use of a technique called *temperature programming*. This techniques allows you to start with a low oven temperature, to optimize the elution of low boiling components, and then increase the oven temperature at a controlled rate, to decrease stationary phase interactions for high-boiling compounds. The increase in temperature decreases retention times and produces narrower peaks for compounds that would otherwise require a long time to elute as a very broad peak.

While samples may be directly injected onto a packed column, the small diameter of the capillary column presents a problem. In addition, it is easy to overload the capillary column with sample (Table 8.35). Two techniques for getting the sample into the column are split and split/splitless injection.

Split Injection

In the split injector the sample is injected into the heated injection port and the evaporated sample is mixed with the carrier gas. The sample/carrier gas mixture is then split between the column and a vent, and a fraction of the sample (determined from the column and vent flow) is introduced to the column (Fig. 8.52a). This technique is used to introduce concentrated samples.

Split/Splitless Injection

Splitless injections are used to introduce dilute solutions. The sample is injected into the heated injection port, which is in the "purge off" mode. In this

Table 8.35 Sample Capacity as a Function of Column Diameter and Stationary Phase Thickness		
Column Diameter (mm)	Stationary Phase Thickness (µm)	Approximate Capacity (ng/component)
0.25	0.10 (thin)	25
	0.25 (most common)	80
	1.0 (thick)	250
0.53	0.10 (thin)	53
	1.0 (most common)	530
	5.0 (thick)	2600

Source: Alltech Capillary Instruction Manual, Bulletin No. 242; Alltech Associates, Inc.: Deerfield, IL, 1991, p. 9.

Figure 8.52 Schematic diagram showing the operation of a split/splitless injection port.

mode, carrier gas flows through the injector directly to the column (Fig. 8.52b). This flow rate is very low (0.5–3 mL/min). Of critical importance to splitless injection is the "solvent effect." The oven temperature is maintained below the solvent boiling point, and the vaporized solvent condenses in the column inlet. This condensed solvent acts like a thick layer of stationary phase and traps sample components. After this concentration period (typically 1 min), the injector is changed to the "purge on" mode. This purge sweeps excess solvent (and other volatile components) out of the injector. Purging too early risks venting volatile components, while purging too late increases interference from the solvent tail.

FEATURES OF THE MASS SPECTRUM

A low-resolution mass spectrum can provide many pieces of information that help an organic chemist determine the structure of a molecule. One of the most useful pieces of information is the compound's nominal molecular weight, MW, as determined by identification of the molecular ion, $M^{+\cdot}$. In addition, by careful examination of the region around the molecular ion for the presence of isotopes, we can learn more about the elemental formula for the molecule. The mass spectrum also reveals information about the molecular structure through the appearance of groups of ions characteristic of certain compound types. With more experience and an understanding of mass spectral fragmentation pathways, a molecular structure can be proposed by gathering all information from the spectrum and determining if a proposed structure is consistent with the observed ions. Here we will provide a limited introduction to this process with an emphasis on identification of the molecular ion. We will present one case study to show how mass spectrometry, coupled with gas chromatography, can be used to characterize the products of a synthetic organic reaction.

Terms

The *molecular ion*, represented by $M^{+\cdot}$, is the intact molecule with one electron missing. This should be the peak in the spectrum with the largest mass, but it is not always observed. All spectra have an ion that we call the *base peak*. This is the most abundant peak in the spectrum (m/z 44 in the spectrum of N_2O or m/z 69 in the spectrum of PFTBA). In the next section you will find that more than one peak may correspond to the $M^{+\cdot}$ or fragment ion when that ion contains elements with different isotopes. We use the term *nominal mass* to describe the mass of the molecule in terms of the most abundant (and, generally, the lightest) isotopes of the element. Relative isotopic abundances for common elements are summarized in Table 8.36.

Isotope Peaks

The mass spectrum for $N(C_4F_9)_3$ (PFTBA; MW = 671) is shown in Figure 8.50. The peaks at m/z 69, 219, and 502 are shown above the spectrum as they were measured by the instrument; the spectrum shows their bar-graph representation. These peaks correspond to CF_3^+, $C_4F_9^+$, and $NC_9F_{20}^+$. If you look carefully at the enlarged peaks, you will notice smaller peaks that appear one mass unit above that of the ion. We call these $[A + 1]^+$ peaks. Where do these peaks come from and why does the abundance increase with the mass of the fragment?

If you look at Table 8.36, you will find that fluorine is an isotopically pure element; however, 1.1% of carbon is the ^{13}C isotope. You may recall that it is this low abundance of ^{13}C that you measure with ^{13}C NMR. For nitrogen there is a small amount of ^{15}N. When a molecule contains more than one atom of an isotopically impure element, you increase the chance of finding the higher mass isotope in the molecule. For example, the ^{13}C peaks for m/z 69, 219, and 502 of PFTBA (Fig. 8.50) are 1.1, 4.4, and 10.3% of the ^{12}C peaks. The relative intensity increases, because there is a higher statistical probability of finding one ^{13}C when the ion has nine versus one carbon atom. The $[A + 1]^+$ peak intensity from ^{13}C is equal to n times 1.1% the height of the peak A^+, where n is the number of carbon atoms. In addition, we need to add contributions from other A + 1 elements, like nitrogen (0.4%). With precise ion intensity measurements, the relative abundance of the [A + 1] peak can be used to determine the number of carbons present in an organic molecule.

Table 8.36	Relative Isotope Abundances of Common Elements					
Element	Mass	%	Mass	%	Mass	%
Carbon	12	100	13	1.1		
Hydrogen	1	100	2	0.015		
Nitrogen	14	100	15	0.37		
Oxygen	16	100	17	0.04	18	0.2
Fluorine	19	100				
Silicon	28	100	29	5.10	30	3.4
Phosphorus	31	100				
Sulfur	32	100	33	0.79	34	4.4
Chlorine	35	100			37	32.0
Bromine	79	100			81	97.3
Iodine	127	100				

Mass spectra get very interesting when chlorine or bromine is present. Both elements exist as mixtures of the A and A + 2 isotopes (Table 8.36). The characteristic isotope distribution for bromine, with nearly equal abundances for ^{79}Br and ^{81}Br, is apparent in the spectrum of bromobenzene (Fig. 8.53a). The M^{+} is observed as a cluster of peaks, with m/z 156 containing the lightest isotopes (^{12}C and ^{79}Br). Note that the fragment at m/z 77 results from loss of Br, and consequently no Br isotope peaks are observed. When more atoms of A + 2 elements are present, characteristic peak distributions are produced (Fig. 8.54). For example, an ion that contains two chlorine atoms will show three peaks, A, [A + 2], and [A + 4], with a 100:65:10.6 intensity distribution. Working from Figure 8.54, you can use a pattern recognition approach to determine the number of chlorine or bromine atoms present in an ion. For example, the spectrum in Figure 8.53b shows two ion clusters that suggest the presence of chlorine. A close examination of the distributions indicates that two chlorines are found in the m/z 84 cluster, while the m/z 49 cluster contains one chlorine. When looking at the mass difference between these ions we work with the *nominal* mass (mass of the ion that has the lightest isotopes). The mass difference, defined by the nominal mass of each cluster, is 35, which corresponds to the mass of one ^{35}Cl (Fig. 8.53b).

Figure 8.53 Electron ionization mass spectrum of (a) bromobenzene and (b) dichloromethane.

Figure 8.54   Isotope peak distributions for ions containing chlorine and bromine atoms.

Recognizing the Molecular Ion

Compound molecular weight is a valuable piece of information that is not always available from the NMR or IR spectrum. In this section we discuss some things to consider as you examine a mass spectrum and attempt to identify the $M^{+\cdot}$ ion. If you look back at the mass spectra that have appeared above, you will find examples of spectra where the molecular ion is the base peak in the spectrum. In some other cases the molecular ion may be weak or not observed at all! For example, PFTBA fragments extensively and the molecular ion does not appear in the spectrum. The following are some things to consider as you attempt to identify the molecular ion.

The molecular ion should be the highest mass peak in the spectrum. When you have tentatively identified a peak as the molecular ion, you should then determine the masses lost from the molecular ion to give high-mass fragments. For example, in Figure 8.53b we found a mass difference of 35 between $M^{+\cdot}$ and the first fragment. A listing of common losses from $M^{+\cdot}$ can be found in Table 8.37. If an observed fragment corresponds to an unreasonable loss, such as M − 12, this strongly suggests that your tentative identification of the molecular ion is incorrect. Remember that it is always possible that no molecular ion is present.

Another useful feature to consider is the *nitrogen rule*. For most elements found in organic molecules, the compound molecular weight will be *even* if the compound has an even number of nitrogen atoms (remember, zero is an even number). In contrast, the mass will be *odd* if the compound contains an odd number of nitrogen atoms. If you are sure that your product could not contain nitrogen, then an odd mass ion could not correspond to the molecular ion.

Mass spectrometrists also use "softer" ionization techniques to obtain MW information. These techniques included measuring EI mass spectra at

Table 8.37 Some Reasonable[a] Losses From $M^{+\cdot}$

Fragment[b]	Radical Lost	Neutral Loss
M-1	H	
M-2		H_2
M-15	CH_3	
M-18		H_2O
M-28		CO or C_2H_4
M-29	C_2H_5	
M-31	OCH_3	
M-32		CH_3OH
M-43	C_3H_7	

[a]Unreasonable losses include [M-4] to [M-14]; [M-21] to [M-25].
[b]For a more complete listing see McLafferty and Turecek (1993).

lower ionization energies and using a higher-pressure ionization technique called *chemical ionization.*

Mass Spectral Interpretation

The following list contains some factors to consider when interpreting mass spectra. To make best use of this summary, the interested reader should consult the text by McLafferty and Turecek, which is considered by many to be the best resource to learn mass spectral interpretation.

1. Using the considerations described above, identify the molecular ion.
2. If possible, determine the elemental composition for $M^{+\cdot}$ and other important peaks using isotopic abundances. In particular, look for isotope peaks from "M + 2" elements like Cl, Br, S, and Si (Table 8.36 and Fig. 8.54). If you are able to establish a molecular formula, calculate "rings + π bonds":

$$\text{For } C_xH_YN_zO_N \text{ (rings} + \pi \text{ bonds)} = X - \frac{Y}{2} + \frac{Z}{2} + 1$$

NOTE. An even-electron ion, with no unpaired electrons, will have a fractional value. If halogens are present, they are counted as hydrogens.

3. Is the molecular weight odd? If so, this indicates an odd number of nitrogen atoms (for organic molecules).
4. Consider the general appearance of the EI mass spectrum: Is it "aliphatic" (lots of fragmentation) or "aromatic" character (minimal fragmentation)?
5. Look for important low-mass ions (Table 8.38).
6. In the region near M^+, identify fragments lost from the molecular ion (neutral losses) (Table 8.37). Look for intense high-mass ions that may indicate a characteristic, stable fragment ion.
7. Postulate a structure by assembling the various mass fragments/neutral losses. Do the observed fragment ions make sense in terms of fragment/neutral loss stability considerations? Does the structure make sense

Table 8.38 Some Common Ion Series

Ion Series	Compound Type
m/z 15, 29, 43, 57, 73	Aliphatic hydrocarbons
m/z 38, 39, 50–52, 63–65, 75–78	Aromatic hydrocarbons (not all peaks in ranges will be observed)
m/z 30, 44, 58	Amines
m/z 31, 45, 59	Alcohols

Note. For a more complete listing, see McLafferty and Turecek (1993).

in terms of other information, such as the reaction conditions, NMR or IR spectra?

8. Verify a postulated structure by comparing the spectrum with a reference spectrum. The reference spectrum may be found in the literature or it may be measured by purchasing or synthesizing a standard of the postulated structure.

CASE STUDY: SYNTHESIS OF METHYL BENZOATE*

To illustrate how gas chromatography and mass spectrometry can be used to characterize the products of an organic reaction, we consider the synthesis of methyl benzoate using a base-catalyzed esterification of benzoic acid. The reaction proposed for this synthesis involved deprotonation of benzoic acid by n-butyllithium in dry tetrahydrofuran (THF), followed by the addition of methyl iodide, with dimethylformamide (DMF) added to promote the S_N2 displacement of iodide by the benzoate anion:

We isolated the reaction products from the reaction mixture by quenching the reaction with water, adding saturated $NaHCO_3$, and extracting the neutral products with diethyl ether. The ethereal solution containing the reaction products was then analyzed by capillary column GC/MS, and the chromatogram shown in Figure 8.55 was produced. The chromatogram displays total ionization as a function of time. The total ionization is a summation of all the ions detected in one scan of the mass spectrometer (one spectrum) plotted as a point as a function of time. The display is often called the TIC (total ionization chromatogram), and the peak areas should reflect the relative amounts of each compound detected by the mass spectrometer.

The chromatogram shown in Figure 8.55 is not quite what we would hope to see. Instead of detecting a single chromatographic peak for our product, we see three peaks. To determine if we made *any* methyl benzoate, and to determine what other components are present in our mixture, we examine the

*The synthetic work presented below was carried out by Joshua Pacheco, Bowdoin College Class of 1999.

Figure 8.55 **Total ionization chromatogram of a reaction mixture.**

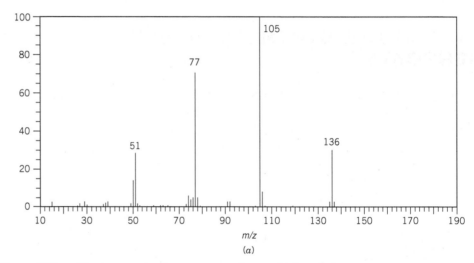

Figure 8.56*a* **Electron ionization mass spectrum of (***a***) peak 1.**

mass spectrum for each peak. Because we are eager to determine if we made any methyl benzoate, we start by trying to locate a chromatographic peak that has a mass spectrum that corresponds to methyl benzoate. Even if we are not sure what the mass spectrum will look like, we can try to find a spectrum that shows a molecular ion, $M^{+\cdot}$, that corresponds to the molecular weight of methyl benzoate ($C_8H_8O_2$, MW = 136). The mass spectrum for peak **1** (Fig. 8.56*a*) shows an ion at *m/z* 136 that appears as the highest mass peak in the spectrum. Let's now take a closer look at the mass spectrum to see if the fragment ions are consistent with the structure of methyl benzoate.

The base peak in the spectrum appears at *m/z* 105. This intense peak results from a loss of 31 from the molecular ion, which corresponds to loss of OCH_3. This is a predicted loss. Upon ionization, we expect one of the nonbonding electrons on the carbonyl oxygen to be lost, and can consider the charge and unpaired electron to be localized on that oxygen. The unpaired electron initiates a cleavage that results in loss of OCH_3 radical. The ions at *m/z* 77 and 51, which are characteristic of aromatic rings, may form by cleavage on the other side of the carbonyl group. The ions at *m/z* 105 and 77 may undergo another fragmentation, but the loss of another radical species is not generally observed from ions of this type, where all electrons are now paired up. Instead, even-electron neutrals, such as CO or C_2H_2, are lost to give fragments m/z 77 and 51, respectively:

Thus, the mass spectrum for peak **1** is consistent with the structure of methyl benzoate. We could further confirm our identification by consulting a library of mass spectra. If we had any pure methyl benzoate around, we could also prepare a standard and use both the GC retention time and the mass spectrum of the standard as a means of confirming the compound identification.

We can now move on to some other peaks in the chromatogram. The mass spectra for peaks **2** and **3** (Fig. 8.56b, c) have many similar features to those of methyl benzoate. Both **2** and **3** show ions at m/z 51, 77, and 105. We can conclude that both compounds have a carbonyl group attached to an aromatic ring. We can next consider identification of the molecular ion. For peak **2**, the highest mass ion is m/z 162. To determine if this is a reasonable assignment for M⁺, we determine losses from m/z 162. The ions at m/z 133, 120, and 105 could result from losses of 29, 42, and 57, respectively. None of these losses is unreasonable. We next want to consider two important pieces of information. First, if we assume that the MW is 162, we must add 57 to the carbonyl

Figure 8.56b   Electron ionization mass spectrum of peak 2, and (c).

Figure 8.56c Electron ionization mass spectrum of peak 3.

substituted aromatic ring to make our molecule. Addition of a butyl group is a logical choice:

$$\text{MW} = 162$$

Next we take note of the peak at *m/z* 120. The even mass of this ion, resulting from an even mass molecular ion, indicates that it is a special ion! Even-mass fragments generally result from rearrangement reactions, and rearrangements involving hydrogen transfers to carbonyl groups can produce particularly informative product ions. If a butyl group is attached to the carbonyl, the following fragmentation pathway will occur, which nicely explains the *m/z* 120 peak:

m/z 162   *m/z* 120

To determine if this assignment makes sense, we go back to consideration of our reaction. How could this product be generated? If we assume that there is some unreacted *n*-butyllithium around after the methyl benzoate has been formed, then the following reaction is possible. Thus, we can feel quite confident in our assignment for peak **2**:

$$\text{MW} = 162$$

Moving on to peak **3**, we detect an ion at m/z 134. Let's start by assuming that this is our molecular ion. The m/z 105 peak would result from a loss of 29 (C_2H_5) from m/z 134, which is a reasonable loss from $M^{+\cdot}$. This *suggests,* erroneously, that ethyl benzoate is our product:

MW = 134

Why is this identification incorrect? If we look back at the chromatogram in Figure 8.55 the chromatographic retention times tell us that something is amiss. Remember that compounds elute from the column in approximate order of increasing boiling point. We would expect that the butyl phenyl ketone would elute *after,* not before, the ethyl phenyl ketone! In addition, you would be hard pressed to propose a mechanism for formation of ethyl benzoate in the context of this reaction. Let's go back and take another look at the mass spectrum. A careful examination shows a small peak at m/z 176. We may have incorrectly identified the molecular ion! If $M^{+\cdot}$ is m/z 176, this gives losses of 42 and 71 to form m/z 134 and 105, respectively. Now it looks like we have a pentyl group attached to the carbonyl, which agrees nicely with the chromatographic retention times. The m/z 134 ion becomes one of our special, even mass ions. What does this ion reveal about the pentyl group? The fact that the ion results from loss of 42, and not 58, clearly indicates that this is not an *n*-pentyl group. What makes the most sense is the branching shown below:

m/z 176 m/z 134

In terms of the chemistry of the reaction, this product also makes sense if deprotonation by *n*-butyllithium occurs α to the carbonyl, followed by reaction with methyl iodide:

MW = 176

BIBLIOGRAPHY

Mass Spectral Interpretation:

Beynon, J. H. *The Mass Spectra of Organic Molecules*; Elsevier: Amsterdam, 1968.

Biemann, K. *Mass Spectrometry. Organic Chemical Applications*; McGraw-Hill: New York, 1962.

Budzikiewicz, H.; Djerassi, C.; Williams, D. H. *Mass Spectrometry of Organic Compounds*; Holden-Day: San Francisco, 1967.

McLafferty, F. W.; Turecek, F. *Interpretation of Mass Spectra*, 4th ed.; University Science Books: Sausalito, CA, 1993.

Silverstein, R. M.; Webster, F. X.; Kiemle, D. *Spectrometric Identification of Organic Compounds*, 7th ed.; Wiley: New York, 2005.

Watson, J. T.; Sparkman, O. D. *Introduction to Mass Spectrometry: Instrumentation, Applications, and Strategies for Data Interpretation*, 4th ed.; Wiley: New York, 2007 (also a good introduction to instrumentation).

Theory and Instrumentation:

Dass, C. *Fundamentals of Contemporary Mass Spectrometry*; Wiley: New York, 2007.

De Hoffmann, E.; Stroobant, V. *Mass Spectrometry: Principles and Applications*, 3rd ed.; Wiley: New York, 2007.

Kitson, F. G.; Larsen, B. S.; McEwen, C. N. *Gas Chromatography and Mass Spectrometry: A Practical Guide*; Academic Press: San Diego, 1996.

Message, G. M. Practical *Aspects of Gas Chromatography/Mass Spectrometry*; Wiley: New York, 1984.

Quadrupole Mass Spectrometry and Its Applications; Dawson, P. H., Ed., Elsevier: New York, 1976.

Libraries:

McLafferty, F. W.; Stauff, D. B.; *The Important Peak Index of the Registry of Mass Spectral Data*, Wiley: New York, 1991.

McLafferty, F. W.; Stauffer, D. B. *Wiley/NBS Registry of Mass Spectral Data*; Wiley: New York, 1989.

NIST/EPA/MSDC Mass Spectral Library; National Institute of Standards and Technology. You may also access some mass spectral data through the internet: *WebBook.nist.gov*

QUALITATIVE IDENTIFICATION OF ORGANIC COMPOUNDS

ORGANIC QUALITATIVE ANALYSIS

One of the exciting challenges that a chemist faces on a regular basis is identifying organic compounds. This challenge is an excellent way for a student to be initiated into the arena of chemical research. Millions of organic compounds are recorded in the chemical literature. At first glance it may seem a bewildering task to attempt to identify one certain compound from this vast array, but most of these substances can be grouped, generally by functional groups, into a comparatively small number of classes. In addition, chemists have an enormous database of chemical and spectroscopic information, which has been correlated and organized over the years, at their disposal. Determination of the physical properties of a molecule, the functional groups present, and the reactions the molecule undergoes has allowed the chemist to establish a systematic, logical identification scheme.

Forensic chemistry, the detection of species causing environmental pollution, the development of new pharmaceuticals, progress in industrial research, and development of polymers all depend to a large extent on the ability of the chemist to isolate, purify, and identify specific chemicals. The objective of organic qualitative analysis is to place a given compound, through screening tests, into one of a number of specific classes, which in turn greatly simplifies the *identification* of the compound. This screening is usually done by using a series of preliminary observations and chemical tests, in conjunction with the instrumental data that developments in spectroscopy have made available to the analyst. The advent of infrared (IR) and nuclear magnetic resonance (NMR) spectroscopy and mass spectrometry (MS) have had a profound effect on the approach taken to identify a specific organic compound. Ultraviolet (UV) spectra may also be utilized to advantage with certain classes of materials.

The systematic approach taken in this text for the identification of an unknown organic compound is as follows:

1. Preliminary tests are performed to determine the physical nature of the compound.

2. The solubility characteristics of the unknown species are determined. This identification can often lead to valuable information related to the structural composition of an unknown organic compound.

Chapter 9: C$_9$H$_{12}$, Triasterene (*L. aster*, star)
Musso and Biethan (1964).

3. Chemical tests, mainly to assist in identifying elements other than C, H, or O, may also be performed.

4. Classification tests are carried out to detect *common functional groups* present in the molecule. Most of these tests may be done using a few drops of a liquid or a few milligrams of a solid. An added benefit to the student, especially in relation to the chemical detection of functional groups, is that a vast amount of chemistry can be *observed* and *learned* in performing these tests. The successful application of these tests requires that you develop the ability to think in a logical manner and to interpret the significance of each result based on your observation. Later, as the *spectroscopic techniques* are introduced, *the number of chemical tests performed are usually curtailed.*

5. The spectroscopic method of analysis is utilized. As your knowledge of chemistry develops, you will appreciate more and more the revolution that has taken place in chemical analysis over the past 25–30 years and the powerful tools now at your disposal for the identification of organic compounds. In the introductory laboratory, the techniques of IR, NMR, and UV–vis spectroscopy, and mass spectrometry are generally explored.

It is important to realize that *negative* findings are often as important as *positive* results in identifying a given compound. Cultivate the habit of following a *systematic pathway or sequence* so that no clue or bit of information is lost or overlooked along the way. It is important also to develop the *attitude* and *habit* of planning ahead. Outline a logical plan of attack, depending on the nature of the unknown, and follow it. As you gain more experience in this type of investigative endeavor, the planning stage will become easier.

At the *initial* phase of your training, the unknowns to be identified will be relatively pure materials and will all be known compounds. The properties of these materials are recorded in the literature. Later, perhaps, mixtures of compounds or samples of commercial products will be assigned for separation, analysis, and identification of the component compounds.

Record all observations and results of the tests in your laboratory notebook. Review these data as you execute the sequential phases of your plan. This method serves to keep you on the path to success.

A large number of texts have been published on organic qualitative analysis. Several references are cited here.

BIBLIOGRAPHY

Cheronis, N. D.; Entriken, J. B. *Identification of Organic Compounds: A Student's Text Using Semimicro Techniques*; New York: Interscience, 1963.

Cheronis, N. D.; Ma, T. S. *Organic Functional Group Analysis by Micro and Semimicro Methods*; Interscience: New York, 1964.

Cheronis, N. D.; Entrikin, J. B.; Hodnett, E. M. *Semimicro Qualitative Organic Analysis*, 3rd ed.; Interscience: New York, 1965.

Feigl, F.; Anger, V. *Spot Tests in Organic Analysis*, 7th ed.; Elsevier: New York, 1966.

Pasto, D. J.; Johnson, C. R.; Miller, M. J. *Experiments and Techniques in Organic Chemistry*; Prentice Hall: Englewood Cliffs, NJ, 1992.

Schneider, F. L. In *Qualitative Organic Microanalysis*, Vol. II of *Monographien aus dem Gebiete der qualitativen Mikroanalyse*; Benedetti-Pichler, A. A., Ed.; Springer-Verlag: Vienna, 1964.

Shriner, R. L.; Hermann, C. K. F.; Morrill, T. C.; Curtin, D. Y.; Fuson, R. C. *The Systematic Identification of Organic Compounds*, 8th ed.; Wiley: New York, 2003.

Vogel, A. I. *Qualitative Organic Analysis*, Part 2 of *Elementary Practical Organic Analysis*; Wiley: New York, 1966.

Vogel's Textbook of *Practical Organic Chemistry, Including Qualitative Organic Analysis*, 5th ed.; London: Longman Group, 1989.

PRELIMINARY TESTS

Preliminary tests help you select a route to follow to ultimately identify the unknown material at hand. These tests frequently consume material, so, given the amounts of material generally available at the micro- or semi-microscale level, judicious selection of the tests to perform must be made (in some tests, the material analyzed may be recovered). Each preliminary test that can be conducted with *little expenditure of time and material* can offer valuable clues as to which class a given compound belongs.

Nonchemical Tests

Physical State. If the material is a *solid*, a few milligrams of the sample may be viewed under a magnifying glass or microscope, which may give some indication as to the homogeneity of the material. Crystalline shape is often an aid in classifying the compound.

Determine the melting point, using a small amount of the solid material. A narrow melting point range (1–2 °C) is a good indication that the material is quite pure. If a broad range is observed, the compound must be recrystallized from a suitable solvent before proceeding. If the material undergoes decomposition on heating, try an evacuated (sealed-tube) melting point. If any evidence indicates that sublimation is occurring, an evacuated melting point should be run. Furthermore, this result indicates that sublimation might be used to purify the compound, if necessary.

If the material is a *liquid*, the boiling point is determined by the ultramicro method. If sufficient material is on hand and the boiling point reveals that the material is relatively pure (narrow boiling point range), the *density* and the *refractive index* can provide valuable information for identification purposes.

Color. Since the majority of organic compounds are colorless, examination of the color can occasionally provide a clue as to the nature of the sample. Use caution, however, since tiny amounts of some impurities can color a substance. Aniline is a classic example. When freshly distilled it is colorless, but on standing a small fraction oxidizes and turns the entire sample a reddish-brown color.

Colored organic compounds contain a *chromophore*, usually extended conjugation in the molecule. For example, 1,2-dibenzoylethylene (Experiments [3A] and [6]) is yellow; 5-nitrosalicylic acid (Experiment [29C]) is light yellow; tetraphenylcyclopentadienone (Experiment [A3$_a$]) is purple.

Can you identify the chromophores that cause these compounds to be colored? Note that a colorless liquid or white solid would not contain these units. Thus, compounds containing these groupings would be excluded from consideration as possible candidates in identification of a given substance.

Odor. Detection of a compound's odor can occasionally be of assistance, since the vast majority of organic compounds have no definitive odor. You should become familiar with the odors of the common compounds or classes. For example, aliphatic amines have a fishy smell; benzaldehyde (like nitrobenzene and benzonitrile) has an almond odor; esters have fruity odors (Experiments [8A–C]). Common solvents, such as acetone, diethyl ether, and toluene, all have distinctive odors. Butyric and caproic acids have rancid odors. Low molecular weight mercaptans (—SH) have an intense smell of rotten eggs. In many cases, extremely small quantities of certain relatively high molecular weight

Figure 9.1 Heating on the microspatula. (*Courtesy of Springer-Verlag, Vienna.*)

compounds can be detected by their odor. For example, a C_{16} unsaturated alcohol released by the female silkworm moth elicits a response from male moths of the same species at concentrations of 100 molecules/cm³. Odors are an important facet of chemical communication between plants and animals and often result in a spectacular behavioral response (see also Experiment [8]).

Odor detection in humans involves your olfactory capabilities and thus can be a helpful lead, *but very rarely can this property be used to strictly classify or identify a substance*. As mentioned above, contamination by a small amount of an odorous substance is always a possibility.

CAUTION | CAUTION: You should be very cautious when detecting odors. Any odor of significance can be detected several inches from the nose. Do not place the container closer than this to your eyes, nose, or mouth. Open the container of the sample and gently waft the vapors toward you.

Ignition Test

CAUTION | CAUTION: Make sure you are wearing safety glasses.

Valuable information can be obtained by carefully noting the manner in which a given compound burns. The ignition test[1] is carried out by placing 1–2 mg of the sample on a spatula, followed by heating and ignition with a microburner flame. Do not hold the sample directly in the flame; heat the spatula about 1 cm from the flat end and move the sample slowly into the flame (see Fig. 9.1).

Important observations to be made concerning the ignition test are summarized in Table 9.1.

As the sample is heated, you should make the following observations:

1. Any melting or evidence of sublimation: This observation gives an approximate idea of the melting point by the temperature necessary to cause melting.

2. Color of the flame as the substance begins to burn (see Table 9.1).

3. Nature of the combustion (flash, quiet, or an explosion). Rapid, almost instantaneous combustion indicates high hydrogen content. Explosion indicates

---

[1]For an extensive discussion on examination of ignition residues, see Feigl, F.; Anger, V. *Spot Tests in Organic Analysis*, 7th ed.; Elsevier: New York; 1966, p. 51.

Table 9.1 **Ignition Test Observations**

Type of Compound	Example	Observation
Aromatic compounds, unsaturated, or higher aliphatic compounds	Toluene	Yellow, sooty flame
Lower aliphatic compounds	Hexane	Yellow, almost nonsmoky flame
Compounds containing oxygen	Ethanol	Clear bluish flame
Polyhalogen compounds	Chloroform	Generally do not ignite until burner flame applied directly to the substance
Sugars and proteins	Sucrose	Characteristic odor
Acid salts or organometallic compounds	Ferrocene	Residue

Source. Cheronis, N. D.; Entrikin, J. B. *Semimicro Qualitative Analysis*; Interscience: New York, 1947, p. 85.

the presence of nitrogen- or nitrogen–oxygen-containing groups, for example, nitro groups (Experiment [29]).

4. Nature of the residue, if present, after ignition.

   a. If a black residue remains and disappears on further heating at higher temperature, the residue is carbon.

   b. If the residue undergoes swelling during formation, the presence of a carbohydrate or similar compound is indicated.

   c. If the residue is black initially but still remains after heating, an oxide of a heavy metal is indicated.

   d. If the residue is white, the presence of an alkali or alkaline earth carbonate or SiO_2 from a silane or silicone is indicated.

SEPARATION OF IMPURITIES

If the preliminary tests outlined above indicate that the unknown in question contains impurities, it may be necessary to carry out one of several purification steps. These techniques are discussed in earlier chapters and are summarized below for correlation purposes:

1. For a liquid, distillation is generally used (see Techniques 2 and 3).

2. For a solid, recrystallization is generally used (see Technique 5).

3. Extraction is used if the impurity is insoluble in a solvent in which the compound itself is soluble (see Technique 4).

4. Sublimation is a very efficient technique if the compound sublimes (see Technique 9).

5. Chromatography (gas, column, and thin-layer) is often used (see Techniques 1 and 6A).

These techniques may be applied to the separation of mixtures as well.

DETECTION OF ELEMENTS OTHER THAN CARBON, HYDROGEN, OR OXYGEN

Other than C, H, and O, the elements that are most often present in organic compounds are nitrogen, sulfur, and the halogens (F, Cl, Br, or I). To detect the presence of these elements, the organic compound is generally fused with metallic sodium. This reaction converts these heteroatoms to the water-soluble inorganic compounds, $NaCN$, Na_2S, and NaX. Inorganic qualitative analysis tests enable the investigator to determine the presence of the corresponding anions:

$$\text{Organic compound containing} \begin{Bmatrix} C \\ H \\ O \\ N \\ S \\ X \end{Bmatrix} \xrightarrow[\Delta]{Na} \begin{Bmatrix} NaCN \\ Na_2S \\ NaX \end{Bmatrix}$$

Sodium Fusion[2]

HOOD

*NOTE. The fusion reaction is carried out in the **hood**. Make sure you are wearing safety glasses. All reagents must be of analytical grade, and deionized water must be used.*

CAUTION

CAUTION: Sodium metal can cause serious burns, and it reacts violently with water.

In a small (10 × 75-mm) test tube (soft glass preferred), supported in a transite board (see Fig. 9.2), place about 25–30 mg of clean sodium metal (about one-half the size of a pea).

CAUTION

CAUTION: Use forceps to make this transfer; never touch sodium metal with your fingers.

Heat the tube with a flame until the sodium melts and sodium vapor is observed rising in the tube (see Fig. 9.2).

Mix a small sample of your unknown compound (1–2 drops of a liquid; 6–10 mg if a solid) with about 15–25 mg of *powdered* sucrose.[3] Gentle mixing of solids may be done on filter paper or glassine weighing paper; liquids can be mixed on a watch glass. Add this mixture to the tube, being careful not to get any material on the sides of the test tube.

NOTE. The addition of sucrose to the sample helps reduce various nitrogen or sulfur compounds. It also absorbs volatile materials so that they may undergo the desired reaction before significant vaporization can occur.

Now heat the tube gently to initiate the reaction with sodium. Remove the flame until the reaction subsides, and then heat to redness for 1–2 min. Allow the

---

[2]See Campbell, K. N.; Campbell, B. K. *J. Chem. Educ.* **1950**, *27*, 261 for a discussion of the procedure.
[3]Ordinary confectioner's sugar purchased at the supermarket can be used.

Figure 9.2   Apparatus for sodium fusion.

tube and contents to cool to room temperature. Then, *and only then*, **cautiously** add several drops of methanol (using a Pasteur pipet) to decompose any unreacted metallic sodium. Gently warm the mixture to drive off the excess methanol.

Reheat the tube to a bright red. While the tube is still red hot, lift the transite board and test tube from the iron ring and place the tube in a small beaker (30 mL) containing about 15 mL of deionized water (the transite board acts as a cover on the beaker).

CAUTION: The soft-glass tube usually cracks and breaks during this operation.   CAUTION

Break up the tube with a glass rod, heat the solution to boiling and filter it by gravity into a clean 50-mL Erlenmeyer flask. Wash the filter paper with an additional 2.0 mL of distilled water and combine this wash with the original filtrate.

NOTE. If a Pyrex test tube is used, after the unreacted sodium metal is completely destroyed by adding methanol, add 2 mL of deionized water directly to the tube and contents. Place a glass stirring rod in the tube and heat the solution to boiling with stirring and then filter as described above. Dilute the filtrate with deionized water to about 5 mL.

Using the Fusion Solution. The clear, colorless fusion solution is used to test for the presence of CN^- (nitrogen), S^{2-} (sulfur), and X^- (halogens, except F^-) as described in the following sections.

Sulfur

1. Place 2–3 drops (Pasteur pipet) of the fusion solution on a white spot plate, followed by 2 drops of water. Now add 1 drop of dilute (2%) aqueous sodium nitroprusside solution. The formation of a deep blue-violet color is a positive test for sulfur:

$$Na_2S \ + \ Na_2Fe(CN)_5NO \ \longrightarrow \ Na_4[Fe(CN)_5NOS]$$

Sodium nitroprusside   Blue-violet complex

2. Place 3–4 drops (Pasteur pipet) of the fusion solution on a white spot plate followed by 1–2 drops of acetic acid. Now add 1 drop of 1% lead(II) acetate solution. The formation of a black precipitate (lead sulfide) indicates the presence of sulfur.

Nitrogen[4]

Reagents

1. A 1.5% solution of *p*-nitrobenzaldehyde in 2-methoxyethanol

2. A 1.7% solution of *o*-dinitrobenzene in 2-methoxyethanol

3. A 2.0% solution of NaOH in distilled water

NOTE. All reagent drops are dispensed using Pasteur pipets.

On a white spot plate, place together: 5 drops of reagent **1**, 5 drops of reagent **2**, and 2 drops of reagent **3**. Stir this mixture gently with a glass rod.

Now add 1 drop of the fusion solution. The formation of a deep-purple color is a positive test for the presence of CN⁻ ion; a yellow or tan coloration is negative. If a positive result is obtained, nitrogen is present in the sample.

The test is valid in the presence of halogens (NaX) or sulfur (Na₂S). It is much more sensitive than the traditional Prussian Blue test.[5]

The Soda Lime Test. In a 10 × 75-mm test tube, mix about 50 mg of soda lime and 50 mg of MnO_2. Add 1 drop of a liquid unknown or about 10 mg of a solid unknown. Place over the mouth of the tube a moist strip of Brilliant Yellow paper (moist, red litmus paper is an alternative). Using a test tube holder, hold the tube at an incline (*pointing away from you and others*) and heat the contents gently at first and then quite strongly. Nitrogen-containing compounds will usually evolve ammonia.

A positive test for nitrogen is the deep red coloration of the Brilliant Yellow paper (or blue color of the litmus paper).

The Halogens (Except Fluorine)

Using the Fusion Solution. In a 10 × 75-mm test tube containing a boiling stone, place 0.5 mL (calibrated Pasteur pipet) of the fusion solution. Carefully acidify this solution by the dropwise addition of dilute HNO_3 (nitric acid), delivered from a Pasteur pipet (test acidity with litmus paper). If a positive test for nitrogen or sulfur was obtained, heat the resulting solution to a gentle boil (stir with a microspatula to prevent boilover) for 1 min over a microburner in the **hood** to expel any HCN or H_2S that might be present. Then cool the tube to room temperature.

HOOD

To the resulting cooled solution, add 2 drops (Pasteur pipet) of aqueous 0.1 M $AgNO_3$ solution.

---

[4]Adapted from Guilbault, G. G.; Kramer, D. N. *Anal. Chem.* **1966**, *39*, 834. *Idem. J. Org. Chem.* **1966**, *31*, 1103. See also Shriner, R. L.; Fuson, R. C.; Morrill, T. C. *The Systematic Identification of Organic Compounds*, 6th ed.; Wiley: New York, 1980, p. 80.
[5]See Vogel, A. I. *Elementary Practical Organic Chemistry, Part 2*, 2nd ed.; Wiley: New York, 1966, p. 37.

A heavy curdy-type precipitate is a positive test for the presence of Cl^-, Br^-, or I^- ion. A faint turbidity is a negative test.

AgCl precipitate is white.

AgBr precipitate is pale yellow.

AgI precipitate is yellow.

AgF is not detected by this test, since it is relatively soluble in water.

The silver halides have different solubilities in dilute ammonium hydroxide solution.

Centrifuge the test tube and contents and remove the supernatant liquid using a Pasteur filter pipet. Add 0.5 mL (calibrated Pasteur pipet) of dilute ammonium hydroxide solution to the precipitate and stir with a glass rod to determine whether the solid is soluble.

AgCl is soluble in ammonium hydroxide due to the formation of the complex ion, $[Ag(NH_3)_2]^+$.

AgBr is slightly soluble in this solution.

AgI is insoluble in this solution.

A Further Test. Once the presence of a halide ion has been established, a further test is available to help you distinguish between Cl^-, Br^-, and I^- ions.[6]

As described above, acidify 0.5 mL of the fusion solution with dilute HNO_3. To this solution, add 5 drops (Pasteur pipet) of a 1.0% aqueous $KMnO_4$ solution and shake the test tube for about 1 min.

Now add 10–15 mg of oxalic acid, enough to decolorize the excess purple permanganate, followed by 0.5 mL of methylene chloride solvent. Stopper, shake, and vent the test tube and allow the layers to separate. Observe the color of the CH_2Cl_2 (lower) layer.

A clear methylene chloride layer indicates Cl^- ion.

A brown methylene chloride layer indicates Br^- ion.

A purple methylene chloride layer indicates I^- ion.

The colors may be faint and should be observed against a white background.

The Beilstein Test.[7] In the Beilstein test organic compounds that contain chlorine, bromine, or iodine, and hydrogen decompose on ignition in the presence of copper oxide, to yield the corresponding hydrogen halides. These hydrogen halides react to form the volatile cupric halides that impart a green or blue-green color to a nonluminous flame. It is a very sensitive test, but some nitrogen-containing compounds and some carboxylic acids also give positive results.

Pound one end of a 4-in.-long copper wire to form a flat surface that can act as a spatula. The other end of the wire is stuck in a cork stopper to serve as an insulated handle.

Heat the flat tip of the wire in a flame until coloration of the flame is negligible.

---

[6]For a further test to distinguish between the three halide ions, see Shriner, R. L.; Fuson, R. C.; Morrill, T. C. *The Systematic Identification of Organic Compounds,* 6th ed.; Wiley: New York, 1980, p. 81. Also see this reference (p. 85) for a specific test for the F^- ion.

[7]Beilstein, F. *Berichte* **1872**, *5*, 620.

On the **cooled** flat surface of the wire, place a drop (Pasteur pipet) of liquid unknown or a few milligrams of solid unknown. Gently heat the material in the flame. The carbon present in the compound will burn first, so the flame will be luminous, but then the characteristic green or blue-green color may be evident. It may be fleeting, so watch carefully.

It is recommended that a known compound containing a halogen be tested so that you become familiar with the appearance of the expected color.

Fluoride ion is not detected by this test, since copper fluoride is not volatile.

SOLUBILITY CHARACTERISTICS

Determination of the solubility characteristics of an organic compound can often give valuable information as to its structural composition. It is especially useful when correlated with spectral analysis. Several schemes have been proposed that place a substance in a definite group according to its solubility in various solvents. The scheme presented below is similar to that outlined in Shriner et al.[8]

There is no sharp dividing line between soluble and insoluble, and an arbitrary ratio of solute to solvent must be selected. We suggest that a compound be classified as soluble if its solubility is greater than 15 mg/500 μL of solvent.

Carry out the solubility determinations, at ambient temperature, in 10 × 75-mm test tubes. Place the sample (~15 mg) in the test tube and add a total of 0.5 mL of solvent in three portions from a graduated or calibrated Pasteur pipet. Between addition of each portion, stir the sample vigorously with a glass stirring rod for 1.5–2 min. If the sample is water soluble, test the solution with litmus paper to assist in classification according to the solubility scheme that follows.

NOTE. *To test with litmus paper, dip the end of a small glass rod into the solution and then gently touch the litmus paper with the rod.* **Do not dip the litmus paper into the test solution.**

In doing the solubility tests follow the scheme in the order given. *Keep a record of your observations.*

Step I Test for water solubility. If soluble, test the solution with litmus paper.

Step II If water soluble, determine the solubility in diethyl ether. This test further classifies water-soluble materials.

Step III Water-insoluble compounds are tested for solubility in a 5% aqueous NaOH solution. If soluble, determine the solubility in 5% aqueous NaHCO$_3$. The use of the NaHCO$_3$ solution aids in distinguishing between strong (soluble) and weak (insoluble) acids.

Step IV Compounds insoluble in 5% aqueous NaOH are tested for solubility in a 5% HCl solution.

---

[8]Shriner, R. L.; Fuson, R. C.; Morrill, T. C. *The Systematic Identification of Organic Compounds,* 6th ed.; Wiley: New York, 1980.

Step V Compounds insoluble in 5% aqueous HCl are tested with concentrated H_2SO_4. If soluble, further differentiation is made using 85% H_3PO_4, as shown in the scheme.

Step VI Miscellaneous neutral compounds containing oxygen, sulfur, or nitrogen are normally soluble in strong acid solution:

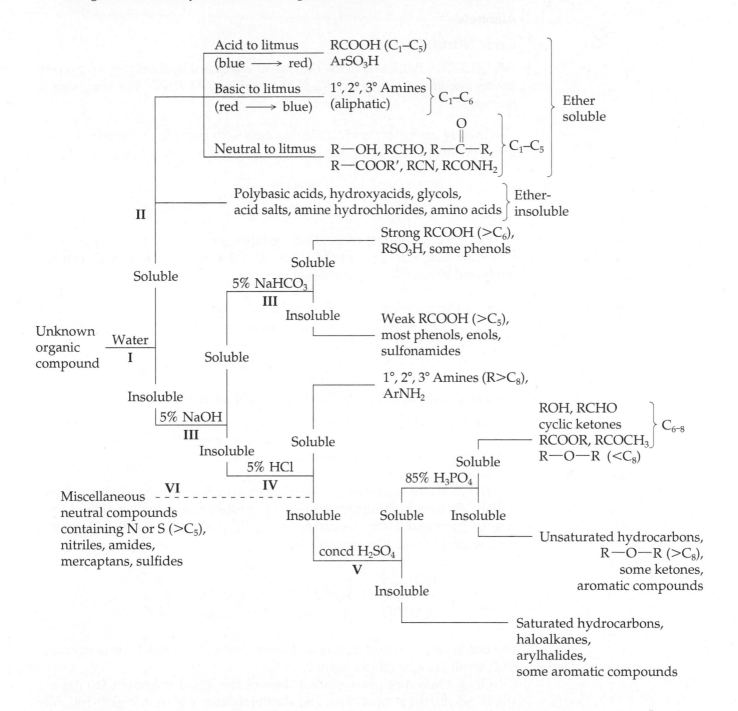

To classify a given compound, it may not be necessary to test its solubility in every solvent. *Do only those tests that are required to place the compound in one of the solubility groups.* Make your observations with care, and proceed in a logical sequence as you make the tests.

THE CLASSIFICATION TESTS[9]

NOTE. For all tests given in this section, drops of reagents are measured out using Pasteur pipets.

Alcohols

Ceric Nitrate Test

INSTRUCTOR PREPARATION. The reagent is prepared by dissolving 4.0 g of ceric ammonium nitrate $[(NH_4)_2Ce(NO_3)_6]$ in 10 mL of 2 M HNO_3. Warming may be necessary.

Primary, secondary, and tertiary alcohols with fewer than 10 carbon atoms give a positive test as indicated by a change in color from *yellow* to *red*:

$$(NH_4)_2Ce(NO_3)_6 \ + \ RCH_2OH \ \longrightarrow \ [alcohol + reagent]$$

Yellow   (Red complex)

Place 5 drops of test reagent on a white spot plate. Add 1–2 drops of the unknown sample (5 mg if a solid). Stir with a thin glass rod to mix the components and observe any color change.

1. If the alcohol is water insoluble, 3–5 drops of dioxane may be added, but run a blank to make sure the dioxane is pure. Efficient stirring gives positive results with most alcohols.

2. Phenols, if present, give a brown color or precipitate.

Chromic Anhydride Test: The Jones Oxidation

INSTRUCTOR PREPARATION. The reagent is prepared by slowly adding a suspension of 1.0 g of CrO_3 in 1.0 mL of concentrated H_2SO_4 to 3 mL of water. Allow the solution to cool to room temperature before using.

The Jones oxidation test is a rapid method to distinguish primary and secondary alcohols from tertiary alcohols. A positive test is indicated by a color change from *orange* (the oxidizing agent, Cr^{6+}) while the oxidizing agent is itself reduced to the *blue green* (Cr^{3+}):

$$\left.\begin{array}{c} RCH_2OH \\ or \\ R_2CHOH \end{array}\right\} + H_2Cr_2O_7 \xrightarrow{H_2SO_4} Cr_2(SO_4)_3 + \begin{array}{c} RCO_2H \\ or \\ R_2C{=}O \end{array}$$

Orange   Green

The test is based on oxidation of a primary alcohol to an aldehyde or acid, and of a secondary alcohol to a ketone.

On a white spot plate, place 1 drop of the liquid unknown (10 mg if a solid). Add 10 drops of acetone and stir the mixture with a thin glass rod. Add

---

[9]For a detailed discussion of classification tests, see (a) Shriner, R. L.; Fuson, R. C.; Morrill, T. C. *The Systematic Identification of Organic Compounds,* 6th ed.; Wiley: New York, 1980, p. 138; (b) Pasto, D. J.; Johnson, C. R.; Miller, M. J. *Experiments and Techniques in Organic Chemistry;* Prentice Hall: Englewood Cliffs, NJ, 1992.

1 drop of the test reagent to the resulting solution. Stir and observe any color change within a 2-second time period.

1. Run a blank to make sure the acetone is pure.

2. Tertiary alcohols, unsaturated hydrocarbons, amines, ethers, and ketones give a negative test within the 2-s time frame for observing the color change. Aldehydes, however, give a positive test, since they are oxidized to the corresponding carboxylic acids.

The HCl/ZnCl$_2$ Test: The Lucas Test

INSTRUCTOR PREPARATION. The Lucas reagent is prepared by dissolving 16 g of anhydrous ZnCl$_2$ in 10 mL of concd HCl while it is cooling in an ice bath.

The Lucas test is used to distinguish between primary, secondary, and tertiary monofunctional alcohols having fewer than six carbon atoms:

$$R—OH + H^+ \xrightarrow{ZnCl_2} R^+ + H_2O$$

Soluble

$$\xrightarrow{Cl^-} RCl$$

Insoluble

The test requires that the alcohol initially be soluble in the Lucas test reagent solution. As the reaction proceeds, the corresponding alkyl chloride is formed, which is insoluble in the reaction mixture. As a result, the solution becomes cloudy. In some cases a separate layer may be observed.

1. Tertiary, allyl, and benzyl alcohols react to give an immediate cloudiness to the solution. You may be able to see a separate layer of the alkyl chloride after a short time.

2. Secondary alcohols generally produce a cloudiness within 3–10 min. The solution may have to be heated to obtain a positive test.

3. Primary alcohols having less than six carbon atoms dissolve in the reagent but react very, very slowly. Those having more than six carbon atoms do not dissolve to any significant extent, no reaction occurs, and the aqueous phase remains clear.

4. A further test to aid in distinguishing between tertiary and secondary alcohols is to run the test using concentrated hydrochloric acid. Tertiary alcohols react immediately to give the corresponding alkyl halide, whereas secondary alcohols do not react under these conditions.

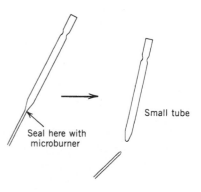

Seal here with microburner

Small tube

In a small test tube prepared by sealing a Pasteur pipet off at the shoulder (➡), place 2 drops of the unknown (10 mg if a solid) followed by 10 drops of the Lucas reagent.

Shake or stir the mixture with a thin glass rod and allow the solution to stand. Observe the results. Based on the times given above, classify the alcohol.

Additional points to consider:

1. Certain polyfunctional alcohols also give a positive test.

2. If an alcohol having three or fewer carbons is expected, a 1-mL conical vial equipped with an air condenser should be used to prevent low molecular weight alkyl chlorides (volatile) from escaping and thus remaining undetected.

The Iodoform Test. This test is positive for compounds that on oxidation generate methyl ketones (or acetaldehyde) under the reaction conditions. For example, methyl carbinols (secondary alcohols having at least one methyl group attached to the carbon atom to which the (—OH) is attached), acetaldehyde, and ethanol give positive results.

For the test, see Methyl Ketones and Methyl Carbinols (p. 671).

Periodic Acid: Vicinal Diols

INSTRUCTOR PREPARATION. This reagent solution is prepared by dissolving 250 mg of periodic acid (H_5IO_6) in 50 mL of deionized water.

Vicinal diols (1,2 diols) are differentiated from the simple alcohols by the characteristic reaction below. Metaperiodic acid (HIO_4) selectively oxidizes 1,2-diols to give carbonyl compounds:

1,2-Diol

The test is based on the *instantaneous* formation of a white precipitate of silver iodate ($AgIO_3$) following addition of silver nitrate:

$$HIO_3 + AgNO_3 \longrightarrow HNO_3 + AgIO_3 \downarrow$$

Place 2 mL of the periodic acid reagent solution in a small test tube.

Add 2 drops of concentrated nitric acid and mix the solution thoroughly. Add 2 drops of a liquid unknown (~2–5 mg of a solid) and mix again. Now add 2–3 drops of 5% aqueous silver nitrate solution. An *instantaneous white precipitate* constitutes a positive test.

α-Hydroxyaldehydes, α-hydroxyketones, α-hydroxyacids, 1,2-diketones, and α-aminoalcohols also give a positive test.

Aldehydes and Ketones

The 2,4-Dinitrophenylhydrazine Test

INSTRUCTOR PREPARATION. The reagent solution is prepared by dissolving 1.0 g of 2,4-dinitrophenylhydrazine in 5.0 mL of concentrated sulfuric acid. This solution is slowly added, with stirring, to a mixture of 10 mL of water and 35 mL of 95% ethanol. After mixing, filter the solution.

Aldehydes and ketones react rapidly with 2,4-dinitrophenylhydrazine to form 2,4-dinitrophenylhydrazones. These derivatives range in color from *yellow* to *red*, depending on the degree of conjugation in the carbonyl compound:

2,4-Dinitrophenylhydrazine   Yellow-to-red precipitate

On a white spot plate, place 7–8 drops of 2,4-dinitrophenylhydrazine reagent solution.

Then add 1 drop of a liquid unknown. If the unknown is a solid, add 1 drop of a solution prepared by dissolving 10 mg of the material in 10 drops of ethanol. The mixture is stirred with a thin glass rod. The formation of a red-to-yellow precipitate is a positive test.

NOTE. The reagent, 2,4-dinitrophenylhydrazine, is orange-red and melts at 198 °C (dec). Do not mistake it for a derivative!

Reactive esters or anhydrides react with the reagent to give a positive test. Allylic or benzylic alcohols may be oxidized to aldehydes or ketones, which in turn give a positive result. Amides do not interfere with the test. Be sure that your unknown is pure and does not contain aldehyde or ketone impurities.

Phenylhydrazine and *p*-nitrophenylhydrazine are often used to prepare the corresponding hydrazones. These reagents also yield solid derivatives of aldehydes and ketones.

Silver Mirror Test for Aldehydes: Tollens Reagent. This reaction involves the oxidation of aldehydes to the corresponding carboxylic acid, using an alcoholic solution of silver ammonium hydroxide. A positive test is the formation of a *silver* mirror, or a black precipitate of finely divided silver:

$$\underset{\overset{|}{H}}{R\ddot{C}}=\ddot{O}\colon + 2\,Ag(NH_3)_2\ddot{O}H \longrightarrow 2\,Ag\downarrow + R-\overset{\ddot{O}\colon}{\underset{\ddot{O}\colon^-,\,NH_4^+}{C}} + H_2\ddot{O} + 3\,\ddot{N}H_3$$

The test should be run only after the presence of an aldehyde or ketone has been established.

In a small test tube prepared from a Pasteur pipet (see the Lucas test) place 1.0 mL of a 5% aqueous solution of $AgNO_3$, followed by 1 drop of aqueous 10% NaOH solution. Now add concentrated aqueous ammonia, drop by drop (2–4 drops) with shaking, until the precipitate of silver oxide just dissolves. Add 1 drop of the unknown (10 mg if a solid), with shaking, and allow the reaction mixture to stand for 10 min at room temperature. If no reaction has occurred, place the test tube in a sand bath at 40 °C for 5 min. Observe the result.

Additional points to consider:

1. Avoid a large excess of ammonia.

2. Reagents must be well mixed. Stirring with a thin glass rod is recommended.

3. *This reagent is freshly prepared for each test. It should not be stored since decomposition occurs with the formation of AgN_3, which is explosive.*   CAUTION

4. This oxidizing agent is very mild, and thus alcohols are not oxidized under these conditions. Ketones do not react. Some sugars, acyloins, hydroxylamines, and substituted phenols do give a positive test.

Chromic Acid Test

INSTRUCTOR PREPARATION. The reagent is prepared by dissolving 1 g of chromium trioxide in 1 mL of concd H_2SO_4, followed by 3 mL of H_2O.

Chromic acid in acetone rapidly oxidizes aldehydes to carboxylic acids. Ketones react very slowly, or not at all.

In a 3-mL vial or small test tube, place 2 drops of a liquid unknown (~10 mg if a solid) and 1 mL of spectral-grade acetone. Now add several drops of the chromic acid reagent.

A green precipitate of chromous salts is a positive test. Aliphatic aldehydes give a precipitate within 30 s; aromatic aldehydes take 30–90 s.

The reagent also reacts with primary and secondary alcohols (see Chromic Anhydride Test: Jones Oxidation, p. 660).

Bisulfite Addition Complexes

INSTRUCTOR PREPARATION. The reagent is prepared by mixing 1.5 mL of ethanol and 6 mL of a 40% aqueous solution of sodium bisulfite. Filter the reagent before use, if a small amount of the salt does not dissolve.

Most aldehydes react with a saturated sodium bisulfite solution to yield a crystalline bisulfite addition complex:

$$\underset{/}{\overset{\backslash}{C}}=\overset{..}{\underset{..}{O}} \underset{H^+ \text{ or } HO^-}{\overset{NaHSO_3}{\rightleftarrows}} \underset{/}{\overset{\backslash}{C}}\overset{SO_3^-,\,Na^+}{\underset{\overset{..}{\underset{..}{O}}H}{}}$$

The reaction is reversible, and thus the carbonyl compound can be recovered by treatment of the complex with aqueous 10% $NaHCO_3$ or dilute HCl solution.

Place 50–75 μL of the liquid unknown (~15 mg if a solid) in a small test tube and add 150 μL of the sulfite reagent and mix thoroughly.

A crystalline precipitate is a positive test.

Alkyl methyl ketones and unhindered cyclic ketones also give a positive test.

Alkanes and Cycloalkanes: Saturated Hydrocarbons

Iodine Charge-Transfer Complex. Alkanes exhibit a *negative* iodine charge-transfer complex test. Species containing π electrons or nonbonded electron pairs produce a brown solution. This color formation is due to the charge-transfer complex between iodine and the available electrons:

$$\underset{/\backslash}{\overset{\backslash/}{\underset{C}{\overset{C}{\|}}}}\cdots I_2 \quad \text{or} \quad \overset{\backslash}{\underset{/}{\,:\!O\!:}}\cdots I_2$$

Solutions of iodine and nonparticipating compounds are violet in color.

On a white spot plate, place a small crystal of iodine. Now add 2–3 drops of a liquid unknown. Alkanes give a *negative* test (violet color).

The test is run only on liquid unknowns. Saturated hydrocarbons, fluorinated and chlorinated saturated hydrocarbons, and aromatic hydrocarbons and their halogenated derivatives all give violet solutions. All other species give a positive test (brown solution).

Concentrated Sulfuric Acid. Saturated hydrocarbons, halogenated saturated hydrocarbons, simple aromatic hydrocarbons, and their halogenated derivatives are insoluble in *cold* concentrated sulfuric acid.

CAUTION   In a small test tube, using **caution**, place 100 μL of *cold* concentrated sulfuric acid. Now add 50 μL of an unknown. A resulting heterogeneous solution (the unknown does *not* dissolve) is a positive test for a saturated hydrocarbon.

Alkenes, and compounds having a functional group containing a nitrogen or oxygen atom are soluble in cold, concentrated acid.

Alkenes and Alkynes: Unsaturated Hydrocarbons

Bromine in Methylene Chloride. Unsaturated hydrocarbons readily add bromine (Br_2). An example of this reaction is given in Experiment [F2]:

$$\underset{}{>}C=C\underset{}{<} + Br_2 \longrightarrow \underset{}{C}\overset{Br^+}{\underset{}{\diagup\diagdown}}\underset{}{C} \xrightarrow{Br^-} \overset{Br}{\underset{Br}{C-C}}$$

The test is based on the decolorization of a red-brown bromine–methylene chloride solution.

CAUTION: Bromine is highly toxic and can cause burns.   CAUTION

In a 10 × 75-mm test tube, or in a small tube prepared from a Pasteur pipet (see Lucas test), place 2 drops of a liquid unknown (~15 mg if a solid) followed by 0.5 mL of methylene chloride. Add dropwise, in the **hood** with   HOOD
shaking, a 2% solution of bromine in methylene chloride solvent. The presence of an unsaturated hydrocarbon will require 2–3 drops of the reagent before the reddish-brown color of bromine persists in the solution.

Additional points to consider:

1. Methylene chloride is used in place of the usual carbon tetrachloride (CCl_4) because it is less toxic.

2. Phenols, enols, amines, aldehydes, and ketones interfere with this test.

Permanganate Test: Baeyer Test for Unsaturation. Unsaturation in an organic compound can be detected by the decolorization of permanganate solution. The reaction involves the cis hydroxylation of the alkene to give a 1,2 diol (glycol):

$$\underset{}{>}C=C\underset{}{<} + 2\,MnO_4^- + 4\,H_2O \longrightarrow \underset{:\underset{..}{O}H \;\; :\underset{..}{O}H}{C-C} + 2\,MnO_2 + 2\,OH^-$$

On a white spot plate, place 0.5 mL of *alcohol-free* acetone, followed by 2 drops of the unknown compound (~15 mg if a solid). Now add dropwise (2–3 drops), with stirring, a 1% aqueous solution of potassium permanganate ($KMnO_4$). A positive test for unsaturation is the discharge of purple permanganate color from the reagent and the precipitation of brown manganese oxides.

Any functional group that undergoes oxidation with permanganate interferes with the test (phenols, aryl amines, most aldehydes, primary and secondary alcohols, etc.).

Alkyl Halides

Silver Nitrate Test. Alkyl halides that undergo the S_N1 substitution reaction react with alcoholic silver nitrate ($AgNO_3$) to form a precipitate of the corresponding silver halide.

Secondary and primary halides react slowly or not at all at room temperature. However, they do react at elevated temperatures. Tertiary halides react immediately at room temperature.

In a 1.0-mL conical vial place 0.5 mL of 2% ethanolic $AgNO_3$ solution and 1 drop of unknown (~10 mg if a solid). A positive test is indicated by the appearance of a precipitate within 5 min. If no reaction occurs, add a boiling stone and equip the vial with an air condenser. Heat the solution at *gentle* reflux for an additional 5 min using a sand bath. Cool the solution.

If a precipitate is formed, add 2 drops of dilute HNO_3. Silver halides will not dissolve in nitric acid solution.

Additional points to consider:

1. The order of reactivity for R groups is allyl $\cong$ benzyl > tertiary > secondary > > > primary. For the halide leaving groups the order is I > Br > Cl.

2. Acid halides, α-haloethers, and 1,2-dibromo compounds also give a positive test at room temperature. Only activated aryl halides give a positive test at elevated temperatures.

Sodium Iodide in Acetone

INSTRUCTOR PREPARATION. The reagent is prepared by dissolving 3 g of sodium iodide (NaI) in 25 mL of acetone. Store in a dark bottle.

Primary alkyl chlorides and bromides can be distinguished from aryl and alkenyl halides by reaction with sodium iodide in acetone (Finkelstein reaction):

$$R{-}X + NaI \xrightarrow{\text{acetone}} R{-}I + NaX \downarrow$$
$$X = Cl, Br$$

Primary alkyl bromides undergo an S_N2 displacement reaction within 5 min at room temperature, and primary alkyl chlorides only at 50 °C.

In a 1.0-mL conical vial, place 1 drop of a liquid unknown (~10 mg if a solid) and 3 drops of acetone. To this solution add 0.5 mL of sodium iodide–acetone reagent.

A positive test is the appearance of a precipitate of NaX within 5 min. If no precipitate is observed, add a boiling stone and equip the vial with an air condenser. Warm the reaction mixture in a sand bath at about 50 °C for 5 min. Cool to room temperature and determine whether a reaction has occurred.

Additional points to consider:

1. Benzylic and allylic chlorides and bromides, acid chlorides and bromides, and α-haloketones, α-haloesters, α-haloamides, and α-halonitriles also give a positive test at room temperature.

2. Primary and secondary alkyl chlorides, and secondary and tertiary alkyl bromides, react at 50 °C under these conditions.

3. If the solution turns red brown in color, iodine is being liberated.

Amides, Ammonium Salts, and Nitriles

Hydroxamate Test for Amides. Unsubstituted (on nitrogen) amides, and the majority of substituted amides, will give a positive hydroxamate test:

where Q = NH$_2$, NHR', or NR'R''

The hydroxamic acid is identified by formation of a red-to-purple color in the presence of the Fe^{3+} ion, as for the test with esters (see page 670).

In a 3.0-mL conical vial containing a boiling stone and equipped with an air condenser place 1 drop of a liquid unknown (~10 mg if a solid), followed by 0.5 mL of 1 M hydroxylamine hydrochloride–propylene glycol solution. Heat the resulting mixture to reflux temperature (~190 °C) using a sand bath, and reflux for 3–5 min. Cool the solution to room temperature, and add 2 drops of 5% aqueous FeCl$_3$ solution. The formation of a red-to-purple color is a positive test.

Alkaline Hydrolysis. Ammonium salts, amides, and nitriles undergo hydrolysis in alkaline solution to form ammonia gas, or an amine:

Detection of ammonia from ammonium salts, primary amides, and nitriles, by use of a color test using copper sulfate solution, constitutes a positive test for these functional groups. The same test may also be used for secondary and tertiary amides that can generate low molecular weight (volatile) amines upon hydrolysis.

In a 1.0-mL conical vial containing a boiling stone, and equipped with an air condenser, place 1–2 drops of the unknown liquid (~10 mg if a solid) and 0.5 mL of 20% aqueous NaOH solution. Heat this mixture to *gentle* reflux on a sand bath. Moisten a strip of filter paper with 2 drops of 10% aqueous copper sulfate solution and place it over the top of the condenser. Formation of a *blue* color (copper ammonia [or amine] complex) is a positive test.

The filter paper may be held in place using a small test tube holder or other suitable device.

Amines

Copper Ion Test. Amines will give a blue-green coloration or precipitate when added to a copper sulfate solution. In a small test tube, place 0.5 mL of a 10% copper sulfate solution. Now add 1 drop of an unknown (~10 mg if a solid). The blue-green coloration or precipitate is a positive test. Ammonia will also give a positive test.

Hinsberg Test. The Hinsberg test is useful for distinguishing between primary, secondary, and tertiary amines. The reagent used is *p*-toluenesulfonyl chloride in alkaline solution.

Primary amines with fewer than seven carbon atoms form a sulfonamide that is soluble in the alkaline solution. Acidification of the solution results in the precipitation of the insoluble sulfonamide:

$$H_3C-\text{⟨⟩}-SO_2Cl + R-NH_2 \xrightarrow{NaOH} H_3C-\text{⟨⟩}-SO_2\bar{N}R, Na^+$$

$$H_3C-\text{⟨⟩}-SO_2\bar{N}R, Na^+ \underset{\substack{\text{excess} \\ \text{base}}}{\overset{\substack{\text{excess} \\ \text{acid}}}{\rightleftharpoons}} H_3C-\text{⟨⟩}-SO_2NHR + NaCl + H_2O$$

(soluble)   (insoluble)

Secondary amines form an insoluble sulfonamide in the alkaline solution:

$$H_3C-\text{⟨⟩}-SO_2Cl + R_2NH \xrightarrow{NaOH} H_3C-\text{⟨⟩}-SO_2NR_2 + NaCl + H_2O \xrightarrow[\text{base}]{\text{excess}} \text{no change}$$

(insoluble)

Tertiary amines normally give no reaction under these conditions:

$$H_3C-\text{⟨⟩}-SO_2Cl + R_3N \xrightarrow{NaOH} H_3C-\text{⟨⟩}-SO_3^- + NR_3 + 2\,Na^+ + Cl^- + H_2O$$

(soluble)   (oil)

In a 1.0-mL conical vial containing a boiling stone, and equipped with an air condenser, place 0.5 mL of 10% aqueous sodium hydroxide solution, 1 drop of the sample unknown (~10 mg if a solid), followed by 30 mg of

HOOD   *p*-toluenesulfonyl chloride (in the **hood**). Heat the mixture to reflux for 2–3 min on a sand bath, and then cool it in an ice bath. Test the alkalinity of the solution using litmus paper. If it is not alkaline, add additional 10% aqueous sodium hydroxide dropwise.

Using a Pasteur filter pipet, separate the solution from any solid that may be present. Transfer the solution to a clean 1.0-mL conical vial and **save**.   SAVE

*NOTE. If an oily upper layer is obtained at this stage, remove the lower alkaline phase using a Pasteur filter pipe and **save**. To the remaining oil add 0.5 mL of cold*   SAVE
water and stir vigorously to obtain a solid material.

If a solid is obtained, it may be (1) the sulfonamide of a secondary amine; (2) recovered tertiary amine, if the original amine was a solid; or (3) the insoluble salt of a primary sulfonamide derivative, if the original amine had more than six carbon atoms.

Additional points to consider:

1. If the solid is a tertiary amine, it is soluble in aqueous 10% HCl.
2. If the solid is a secondary sulfonamide, it is insoluble in aqueous 10% NaOH.
3. If no solid is present, acidify the alkaline solution by addition of 10% aqueous HCl. If the unknown amine is primary, the sulfonamide will precipitate.

Bromine Water. Aromatic amines, since they possess an electron-rich aromatic ring, can undergo electrophilic aromatic substitution with bromine, to yield the corresponding arylamino halide(s). Therefore, if elemental tests indicate that an aromatic group is present in an amine, treatment with the bromine water reagent may indicate that the amine is attached to an aromatic ring.

For the test, see Phenols and Enols (p. 673).

Aromatic Hydrocarbons with NO Functional Groups

Fuming Sulfuric Acid. Simple aromatic hydrocarbons are insoluble in sulfuric acid (H_2SO_4) but are soluble in fuming sulfuric acid. If these hydrocarbons contain more than two alkyl substituents, they may be sulfonated under these conditions.

In a small test tube place 100 µL of fuming sulfuric acid, using **caution**.   CAUTION
Now add 50 µL of the unknown (~15 mg if a solid) suspected to be aromatic. A resulting homogeneous solution is a positive test.

Azoxybenzene and Aluminum Chloride. This color test is run only on those aromatic compounds that are insoluble in sulfuric acid (see previous test). The color produced in this test results from the formation of a complex of $AlCl_3$ and a *p*-arylazobenzene derivative:

In a small dry test tube, place 250 µL of the aromatic unknown (~15 mg if a solid). Add a small crystal of azoxybenzene and about 12 mg of anhydrous aluminum chloride. If a color is not produced immediately, warm the mixture for a few minutes.

Aryl halides and other simple aromatic hydrocarbons give a deep-orange to dark-red color or precipitate. Polynuclear aromatic hydrocarbons, such as naphthalenes and anthracenes, give brown colors. Aliphatic hydrocarbons give no color, or at most a light yellow tint.

Carboxylic Acids

The presence of a carboxylic acid is detected by its solubility behavior. An aqueous solution of the acid will be acidic to litmus paper (or pH paper may be used). Since a sulfonic acid would also give a positive test, the test for sulfur (sodium fusion) is used to distinguish between the two types of acids. A water-soluble phenol is acidic toward litmus paper but also would give a positive ferric chloride test.

Carboxylic acids also react with a 5% solution of sodium bicarbonate (see Experiment [4B]).

Place 1–2 mL of the bicarbonate solution on a watch glass and add 1–2 drops of the acid (~10 mg if a solid). Gas bubbles of CO_2 constitute a positive test.

Esters

Hydroxamate Test. Carboxylic esters can be identified by conversion to hydroxamic acid salts. Acidification of this salt produces the corresponding hydroxamic acid (RCONHOH), which is identified by formation of a red-to-purple color in the presence of Fe^{3+} ion:

In a 3.0-mL conical vial containing a boiling stone, and equipped with an air condenser, place 1 drop of the liquid unknown (~10 mg if a solid) followed by 0.5 mL of 1.0 M ethanolic hydroxylamine hydrochloride solution. Add 10% methanolic KOH to this solution (dropwise) until the resulting solution has pH ~10 (pH paper). Heat this mixture to reflux temperature using a sand bath for 5 min, cool to room temperature, and acidify to pH = 3–4 by dropwise addition of 5% aqueous HCl solution. Now add 2 drops of 5% aqueous $FeCl_3$ solution. The formation of a red-to-purple color is a positive test.

Additional points to consider:

1. It is suggested that a blank be run for comparison purposes.
2. Acid chlorides, anhydrides, lactones, and imides also give a positive test.

Saponification. This well-known reaction of esters can often be used to classify these compounds. It also may lead to a useful derivative if the corresponding carboxylic acid is isolated.

In a 3.0-mL conical vial containing a magnetic spin vane, place 100 μL of the liquid unknown (~150 mg if a solid) and add 1 mL of 6 M NaOH solution. Attach the vial to a reflux condenser. Now place the vial in a sand bath on a magnetic stirring hot plate and, with stirring, heat the mixture at reflux for 0.5 h, or until the solution becomes homogeneous.

A positive test is the disappearance of the organic layer (if the original unknown was water insoluble) or the lack of the usually pleasant aroma of the unknown ester.

High-boiling esters (bp > 200 °C) are usually not saponified under these conditions due to their low solubility in the aqueous solvent.

Ethers

CAUTION: Upon standing, ethers may form peroxides. Peroxides are very explosive. To test for the presence of these substances, use starch iodide paper that has been moistened with 6 M HCl. Peroxides cause the paper to turn blue. To remove peroxides from ethers, pass the material through a short column of highly activated alumina (Woelm basic alumina, activity grade 1).[10] Always retest for peroxides before using the ether.

Ferrox Test. The ferrox test is a color test sensitive to oxygen, which may be used to distinguish ethers from hydrocarbons that, like most ethers, are soluble in sulfuric acid.

In a dry 10 × 75-mm test tube using a glass stirring rod, grind a crystal of ferric ammonium sulfate and a crystal of potassium thiocyanate. The ferric hexathiocyanatoferrate that is formed adheres to the rod.

In a second clean 10 × 75-mm test tube, place 2–3 drops of a liquid unknown. If dealing with a solid, use about 10 mg and add toluene until a saturated solution is obtained. Now, using the rod with the ferric hexathiocyanatoferrate attached, stir the unknown. *If the unknown contains oxygen, the ferrate compound dissolves and a reddish-purple color is observed.*

Some high molecular weight ethers do not give a positive test.

Bromine Water. Since the aromatic ring is electron rich, aromatic ethers can undergo electrophilic aromatic substitution with bromine to yield the corresponding aryl ether–halide(s). Therefore, if elemental tests indicate that an aromatic group is present in an ether, treatment with the bromine water reagent may substantiate the presence of an aryl ether.

For the test, see Phenols and Enols (p. 673).

Methyl Ketones and Methyl Carbinols

Iodoform Test

INSTRUCTOR PREPARATION. Dissolve 3 g of KI and 1 g I$_2$ in 20 mL of water.

---

[10]Pasto, D. J.; Johnson, C. R.; Miller, M. J. *Experiments and Techniques in Organic Chemistry;* Prentice Hall: Englewood Cliffs, NJ, 1992, p. 33.

The iodoform test involves hydrolysis and cleavage of methyl ketones to form a yellow precipitate of iodoform (CHI_3):

$$R-\overset{\ddot{O}}{\overset{\|}{C}}-CH_3 + 3\,I_2 + 3\,KOH \longrightarrow R-\overset{\ddot{O}}{\overset{\|}{C}}-CI_3 + 3\,KI + 3\,H_2O$$

$$\downarrow KOH$$

$$R-\overset{\ddot{O}}{\overset{\|}{C}}-O^-, K^+ + CHI_3 \downarrow$$
$$\text{Yellow}$$

It is also a positive test for compounds that, upon oxidation, generate methyl ketones (or acetaldehyde) under these reaction conditions. For example, methyl carbinols (secondary alcohols having at least one methyl group attached to the carbon atom to which the (—OH) unit is linked), acetaldehyde, and ethanol give positive results.

In a 3.0-mL conical vial equipped with an air condenser, place 2 drops of the unknown liquid (10 mg if a solid), followed by 5 drops of 10% aqueous KOH solution.

HOOD

*NOTE. If the sample is insoluble in the aqueous phase, either mix vigorously or add dioxane (in the **hood**) or bis(2-methoxyethyl) ether to obtain a homogeneous solution.*

Warm the mixture on a sand bath to 50–60 °C and add the $KI-I_2$ reagent dropwise until the solution becomes dark brown in color (~1.0 mL). Additional 10% aqueous KOH is now added (dropwise) until the solution is again colorless.

CAUTION

CAUTION: Iodine is highly toxic and can cause burns.

After warming for 2 min, cool the solution and determine whether a yellow precipitate (CHI_3, iodoform) has formed. If a precipitate is not observed, reheat as before for another 2 min. Cool and check again for the appearance of iodoform.

Additional points to consider:

1. The iodoform test is reviewed elsewhere.[11]
2. An example of the general haloform reaction, using bleach to oxidize a methyl ketone, is given in Experiment [34].

Nitro Compounds

Ferrous Hydroxide Test. Many nitro compounds give a positive test based on the following reaction:

$$R-NO_2 + 4\,H_2O + 6\,Fe(OH)_2 \longrightarrow R-NH_2 + 6\,Fe(OH)_3 \downarrow$$
$$\text{Red-brown}$$

The nitro derivative oxidizes the iron(II) hydroxide to iron(III) hydroxide; the latter is a red-brown solid.

In a 1.0-mL conical vial place 5–10 mg of the unknown compound (~2 drops if a liquid), followed by 0.4 mL of freshly prepared 5% aqueous ferrous

---

[11]Fuson, R. C.; Bull, B. A. *Chem. Rev.* **1934**, *15*, 275.

ammonium sulfate solution. After mixing, add 1 drop of 3 M sulfuric acid followed by 10 drops of 2 M methanolic KOH. Cap the vial, shake vigorously, vent, and then allow it to stand over a 5-min period. The formation of a red-brown precipitate, usually within 1 min, is a positive test for a nitro group.

Sodium Hydroxide Color Test. Treatment of an aromatic nitro compound with 10% sodium hydroxide solution may often be used to determine the number of nitro groups present on the aromatic ring system.

> Mononitro compounds produce no color (a light yellow may be observed).
> Dinitro compounds produce a bluish-purple color.
> Trinitro compounds produce a blood-red color.

The color formation is due to formation of Meisenheimer complexes (for a discussion, see Pasto et al.[12]).

To run the test, dissolve 10 mg of the unknown (1–2 drops if a liquid) in 1 mL of acetone in a small test tube. Now add about 200 µL of 10% NaOH solution and shake. Observe any color formation.

If amino, substituted amino, or hydroxyl groups are present in the molecule, a positive color test is not obtained.

Phenols and Enols

Ferric Ion Test. Most phenols and enols form colored complexes in the presence of ferric ion, Fe^{3+}:

Phenols give red, blue, purple, or green colors. Sterically hindered phenols may give a negative test. Enols generally give a tan, red, or red-violet color.

On a white spot plate place 2 drops of water, or 1 drop of water plus 1 drop of ethanol, or 2 drops of ethanol, depending on the solubility characteristics of the unknown. To this solvent system add 1 drop (10 mg if a solid) of the substance to be tested. Stir the mixture with a thin glass rod to complete dissolution. Add 1 drop of 2.5% aqueous ferric chloride ($FeCl_3$) solution (light yellow in color). Stir and observe any color formation. If necessary, a second drop of the $FeCl_3$ solution may be added.

Additional points to consider:

1. The color developed may be fleeting or it may last for many hours. A slight excess of the ferric chloride solution may or may not destroy the color.

2. An alternative procedure using $FeCl_3–CCl_4$ solution in the presence of pyridine is available.[13]

---

[12]Pasto, D. J.; Johnson, C. R.; Miller, M. J. *Experiments and Techniques in Organic Chemistry;* Prentice Hall: Englewood Cliffs, NJ, 1992, p. 321.
[13]Soloway, S.; Wilen, S. H. *Anal. Chem.* **1952**, *4*, 979.

Bromine Water. Phenols, substituted phenols, aromatic ethers, and aromatic amines, since the aromatic rings are electron rich, undergo aromatic electrophilic substitution with bromine to yield substituted aryl halides. For example,

HOOD

CAUTION: The test should be run in the **hood.**

HOOD

In a small test tube, place 1–2 drops of the unknown (~20 mg if a solid) and add 1–2 mL of water. Check the pH of the solution with pH paper. In the **hood**, add saturated bromine water dropwise until the bromine color persists. A precipitate generally forms.

A positive test is the decolorization of the bromine solution, and often the formation of an off-white precipitate. If the unknown is a phenol, this should cause the pH of the original solution to be less than 7.

PREPARATION OF DERIVATIVES

Based on the preliminary and classification tests carried out to this point, you should have established the type of functional group (or groups) present (or lack of one) in the unknown organic sample. The next step in qualitative organic analysis is to consult a set of tables containing a listing of known organic compounds sorted by functional group and/or by physical properties or by both. Using the physical properties data for your compound, you can select a few possible candidates that appear to "fit" the data you have collected. On a chemical basis, the final step in the qualitative identification sequence is to prepare one or two *crystalline derivatives* of your compound. Selection of the specific compound, and thus final confirmation of its identity, can then be made from the extensive derivative tables that have been accumulated. With the advent of spectral analysis, the preparation of derivatives is often not necessary, but the wealth of chemistry that can be learned by the beginning student in carrying out these procedures is extensive and important. The preparation of selected derivatives for the most common functional groups are given below. For extensive tables and alternative derivatives that can be utilized, see the following Bibliography.

BIBLIOGRAPHY

Pasto, D. J.; Johnson, C. R.; Miller, M. J. *Experiments and Techniques in Organic Chemistry*; Prentice Hall: Englewood Cliffs, NJ, 1992.

Rappoport, Z. *Handbook of Tables for Organic Compound Identification*, 3rd ed.; CRC Press: Boca Raton, FL, 1967.

Shriner, R. L.; Hermann, C. K. F.; Morrill, T. C.; Fuson, R. C. *The Systematic Identification of Organic Compounds*, 8th ed.; Wiley: New York, 2003.

NOTE. In each of the procedures outlined below, drops of reagents are measured using Pasteur pipets.

CARBOXYLIC ACIDS

Preparation of Acid Chlorides

$$R\overset{\overset{\cdot\cdot}{O}}{\underset{\|}{C}}\text{—}\ddot{O}H + Cl\text{—}\overset{\overset{\cdot\cdot}{O}}{\underset{\|}{S}}\text{—}Cl \xrightarrow{\text{DMF}} R\overset{\overset{\cdot\cdot}{O}}{\underset{\|}{C}}\text{—}Cl + HCl{\uparrow} + SO_2{\uparrow}$$

Weigh and place 20 mg of the unknown acid in a dry 3.0-mL conical vial containing a boiling stone and fitted with a cap. Now, in the **hood**, add 4 drops of thionyl chloride and 1 drop of *N,N*-dimethylformamide (DMF). Immediately attach the vial to a reflux condenser that is protected by a calcium chloride drying tube.   **HOOD**

CAUTION: This reaction is run in the **hood**, since hydrogen chloride and sulfur dioxide are evolved. Thionyl chloride is an irritant and is harmful to breathe. Immediately recap the vial after each addition until the vial is attached to the reflux condenser.   **HOOD**

Allow the mixture to stand at room temperature for 10 min, heat it at gentle reflux on a sand bath for 15 min, and then cool it to room temperature. Dilute the reaction mixture with 5 drops of methylene chloride solvent.

The acid chloride is not isolated, but is used directly in the preparations that follow.

Amides

$$R\overset{\overset{\cdot\cdot}{O}}{\underset{\|}{C}}\text{—}Cl + 2\,\ddot{N}H_3 \longrightarrow R\overset{\overset{\cdot\cdot}{O}}{\underset{\|}{C}}\text{—}\ddot{N}H_2 + NH_4Cl$$

Cool the vial in an ice bath and add 10 drops of concentrated aqueous ammonia, in the **hood** via Pasteur pipet, *dropwise,* with stirring. *It is convenient to make this addition down the neck of the air condenser.* The amide may precipitate during this operation. After the addition is complete, remove the ice bath and stir the mixture for an additional 5 min. Now add methylene chloride (10 drops) and stir the resulting mixture to dissolve any precipitate. Separate the methylene chloride layer from the aqueous layer using a Pasteur filter pipet and transfer it to another Pasteur filter pipet containing 200 mg of anhydrous sodium sulfate. Collect the eluate in a Craig tube containing a boiling stone. Extract the aqueous phase with an additional 0.5 mL of methylene chloride. Separate the methylene chloride layer as before and transfer it to the same column. Collect this eluate in the same Craig tube. Evaporate the methylene chloride solution using a warm sand bath in the **hood** under a gentle stream of nitrogen gas. Recrystallize the solid amide product using the Craig tube. Dissolve the material in about 0.5 mL of ethanol, add water (dropwise) until the solution becomes cloudy, cool the Craig tube in an ice bath, and collect the crystals in the usual manner. Dry the crystalline amide on a porous clay plate and determine the melting point.   **HOOD**

Anilides

$$R-\overset{\overset{\ddot{O}}{\|}}{C}-Cl + 2\,H_2\ddot{N}-\!\!\!\langle\bigcirc\rangle\!\!\!-CH_3 \longrightarrow R-\overset{\overset{\ddot{O}}{\|}}{C}-\underset{\underset{H}{|}}{\ddot{N}}-\!\!\!\langle\bigcirc\rangle\!\!\!-CH_3 + CH_3-\!\!\!\langle\bigcirc\rangle\!\!\!-NH_3{}^+,\,Cl^-$$

HOOD

In a 3.0-mL conical vial containing a magnetic spin vane, and equipped with an air condenser, place 5 drops of aniline and 10 drops of methylene chloride. Cool the solution in an ice bath and transfer the acid chloride solution (prepared above) via Pasteur pipet, *dropwise*, with stirring, to the aniline solution in the **hood**. *It is convenient to make this addition down the neck of the condenser.* After the addition is complete, remove the ice bath and stir the mixture for an additional 10 min.

Transfer the methylene chloride layer to a 10 × 75-mm test tube, and wash it with 0.5 mL of H_2O, 0.5 mL of 5% aqueous HCl, 0.5 mL of 5% aqueous NaOH, and, finally, 0.5 mL of H_2O. For each washing, shake the test tube and remove the top aqueous layer by Pasteur filter pipet. Transfer the resulting wet methylene chloride layer to a Pasteur filter pipet containing 200 mg of anhydrous sodium sulfate. Collect the eluate in a Craig tube containing a boiling stone. Rinse the original test tube with an additional 10 drops of methylene chloride. Collect this rinse and pass it through the same column. Both eluates are combined.

HOOD

Evaporate the methylene chloride solvent on a warm sand bath under a gentle stream of nitrogen gas in the **hood**. Recrystallize the crude anilide from an ethanol–water mixture using the Craig tube. Dissolve the material in about 0.5 mL of ethanol, add water (dropwise) to the cloud point, cool in an ice bath, and collect the crystals in the usual manner. Dry the purified derivative product on a porous clay plate, and determine its melting point.

Toluidides

$$R-\overset{\overset{\ddot{O}}{\|}}{C}-Cl + 2\,H_2\ddot{N}-\!\!\!\langle\bigcirc\rangle\!\!\!-CH_3 \longrightarrow R-\overset{\overset{\ddot{O}}{\|}}{C}-\underset{\underset{H}{|}}{\ddot{N}}-\!\!\!\langle\bigcirc\rangle\!\!\!-CH_3 + CH_3-\!\!\!\langle\bigcirc\rangle\!\!\!-NH_3{}^+,\,Cl^-$$

The same procedure described for the preparation of anilides is used, except that *p*-toluidine replaces the aniline.

ALCOHOLS

Phenyl- and α-Naphthylurethanes (Phenyl- and α-Naphthylcarbamates)

$$Ar-\ddot{N}=C=\ddot{O}: + R-\ddot{O}-H \longrightarrow Ar-\underset{\underset{H}{|}}{\ddot{N}}-\overset{\overset{\ddot{O}:}{\|}}{C}-\ddot{O}-R$$

Isocyanate   Urethane

NOTE. For the preparation of these derivatives, the alcohols must be anhydrous. Water hydrolyzes the isocyanates to produce arylamines that react with the isocyanate reagent to produce high-melting, disubstituted ureas.

In a 3.0-mL conical vial containing a boiling stone and equipped with an air condenser protected by a calcium chloride drying tube place 15 mg of an anhydrous alcohol or phenol. Remove the air condenser from the vial and add 2 drops of phenyl isocyanate or α-naphthyl isocyanate. Replace the air condenser immediately. If the unknown is a phenol, add 1 drop of pyridine in a similar manner.

CAUTION: This addition must be done in the **hood**. The isocyanates are lachrymators! Pyridine has the characteristic strong odor of an amine.

If a spontaneous reaction does not take place, heat the vial at about 80–90 °C using a sand bath, for a period of 5 min. Then cool the reaction mixture in an ice bath. It may be necessary to scratch the sides of the vial to induce crystallization. Collect the solid product by vacuum filtration, using a Hirsch funnel, and purify it by recrystallization from ligroin. For this procedure, place the solid in a 10 × 75-mm test tube and dissolve it in 1.0 mL of warm (60–80 °C) ligroin. If diphenyl (or dinaphthyl) urea is present (formed by reaction of the isocyanate with water), it is insoluble in this solvent. Transfer the warm ligroin solution to a Craig tube using a Pasteur filter pipet. Cool the solution in an ice bath and collect the resulting crystals in the usual manner. After drying the product on a porous clay plate, determine the melting point.

3,5-Dinitrobenzoates

3,5-Dinitrobenzoyl
chloride

NOTE. The dinitrobenzoyl chloride reagent tends to hydrolyze on storage to form the corresponding carboxylic acid. Check its melting point before use (3,5-dinitrobenzoyl chloride, mp = 74 °C; 3,5-dinitrobenzoic acid, mp = 202 °C).

In a 3.0-mL conical vial containing a boiling stone, and equipped with an air condenser protected by a calcium chloride drying tube, place 25 mg of pure 3,5-dinitrobenzoyl chloride and two drops of the unknown alcohol. Heat the mixture to about 10 °C below the boiling point of the alcohol (but not over 100 °C) on a sand bath for a period of 5 min. Cool the reaction mixture, add 0.3 mL of water, and then place the vial in an ice bath to cool. Collect the solid ester by vacuum filtration, using a Hirsch funnel, and wash the filter cake with three 0.5-mL portions of 2% aqueous sodium carbonate (Na$_2$CO$_3$) solution, followed by 0.5 mL of water. Recrystallize the solid product from an ethanol–water mixture using a Craig tube. Dissolve the material in about 0.5 mL of ethanol, add water (dropwise) until the solution is just cloudy, cool in an ice bath, and collect the crystals in the usual manner. After drying the product on a porous clay plate, determine the melting point.

ALDEHYDES AND KETONES

2,4-Dinitrophenylhydrazones

$$O_2N \overset{\cdot\cdot}{}\!\!-\!\!\ddot{N}H\!-\!\ddot{N}H_2 + R\!-\!\overset{\overset{\cdot\cdot}{\underset{\|}{O}}}{C}\!-\!R' \longrightarrow O_2N\overset{\cdot\cdot}{}\!\!-\!\!\ddot{N}H\!-\!\ddot{N}\!\!=\!\!C\overset{R}{\underset{R'}{}}$$

2,4-Dinitrophenylhydrazine   A 2,4-Dinitrophenylhydrazone

The procedure outlined in the Classification Test Section for aldehydes and ketones (p. 662) is used. Since the derivative to be isolated is a solid, it may be convenient to run the reaction in a 3-mL vial or in a small test tube. Double the amount of the reagents used. If necessary, the derivative can be recrystallized from 95% ethanol.

The procedure is generally suitable for the preparation of phenylhydrazone and p-nitrophenylhydrazone derivatives of aldehydes and ketones.

Semicarbazones

$$H_2\ddot{N}\!-\!\overset{\overset{\cdot\cdot}{\underset{\|}{O}}}{C}\!-\!\ddot{N}H\!-\!\ddot{N}H_2 + R\!-\!\overset{\overset{\cdot\cdot}{\underset{\|}{O}}}{C}\!-\!R' \longrightarrow H_2\ddot{N}\!-\!\overset{\overset{\cdot\cdot}{\underset{\|}{O}}}{C}\!-\!\ddot{N}H\!-\!\ddot{N}\!\!=\!\!C\overset{R}{\underset{R'}{}}$$

Semicarbazide   A semicarbazone

In a 3.0-mL conical vial place 12 mg of semicarbazide hydrochloride, 20 mg of sodium acetate, 10 drops of water, and 12 mg of the unknown carbonyl compound (~2 drops if a liquid). Cap the vial, shake vigorously, vent, and allow the vial to stand at room temperature until crystallization is complete (varies from a few minutes to several hours). Cool the vial in an ice bath if necessary. Collect the crystals by vacuum filtration, using a Hirsch funnel, and wash the filter cake with 0.2 mL of cold water. Dry the crystals on a porous clay plate. Determine the melting point.

AMINES

Primary and Secondary Amines: Acetamides

$$R\!-\!\ddot{N}H_2 + \begin{matrix} CH_3\!-\!\overset{\overset{\cdot\cdot}{\underset{\|}{O}\!:}}{C} \\ \overset{\cdot\cdot}{\underset{\cdot\cdot}{\underset{\|}{:O:}}} \\ CH_3\!-\!\overset{}{\underset{\underset{\cdot\cdot}{\overset{\|}{O}\!:}}{C}} \end{matrix} \xrightarrow{\text{NaOAc}} CH_3\!-\!\overset{\overset{\cdot\cdot}{\underset{\|}{O}\!:}}{C}\!-\!\overset{}{\underset{H}{\ddot{N}}}\!-\!R + CH_3\!-\!\overset{\overset{\cdot\cdot}{\underset{\|}{O}\!:}}{C}\overset{}{\underset{\ddot{O}H}{}}$$

In a 3.0-mL conical vial equipped with an air condenser, place 20 mg of the unknown amine (~2 drops if a liquid), 5 drops of water, and 1 drop of concentrated hydrochloric acid.

In a small test tube, prepare a solution of 40 mg of sodium acetate trihydrate dissolved in 5 drops of water. Stopper the solution and set it aside for use in the next step.

Warm the solution of amine hydrochloride to about 50 °C on a sand bath. Then cool it, and add 40 μL of acetic anhydride in one portion (in the **hood**) through the condenser by aid of a 9-in. Pasteur pipet. In like manner, *immediately* add the sodium acetate solution (prepared previously). Swirl the contents of the vial to ensure complete mixing. **HOOD**

Allow the reaction mixture to stand at room temperature for about 5 min, and then place it in an ice bath for an additional 5–10 min. Collect the white crystals by vacuum filtration, using a Hirsch funnel, and wash the filter cake with two 0.1-mL portions of water. The product may be recrystallized from ethanol–water using the Craig tube, if desired. Dry the crystals on a porous clay plate and determine the melting point.

Primary and Secondary Amines: Benzamides

In a 3.0-mL conical vial in the **hood** place 0.4 mL of 10% aqueous NaOH solution, 25 mg of the amine, and 2–3 drops of benzoyl chloride. Cap and shake the vial over a period of about 10 min. Vent the vial periodically to release any pressure buildup. **HOOD**

Collect the crystalline precipitate by vacuum filtration, using a Hirsch funnel, and wash the filter cake with 0.1 mL of dilute HCl followed by 0.1 mL of water. It is generally necessary to recrystallize the material from methanol or aqueous ethanol using the Craig tube. Dry the product on a porous clay plate and determine the melting point.

Primary, Secondary, and Tertiary Amines: Picrates

Picric acid   Picrate salt

In a 3.0-mL conical vial containing a boiling stone and equipped with an air condenser, place 15 mg of the unknown amine (~2 drops if a liquid) and 0.3 mL of 95% ethanol.

NOTE. If the amine is not soluble in the ethanol, shake the mixture to obtain a saturated solution and then transfer this solution, using a Pasteur filter pipet, to another vial.

Now add 0.3 mL of a saturated solution of picric acid in 95% ethanol.

CAUTION: Picric acid explodes by percussion or when rapidly heated.   **CAUTION**

Heat the mixture at reflux, using a sand bath, for about 1 min and then allow it to cool slowly to room temperature. Collect the yellow crystals of the picrate by vacuum filtration, using a Hirsch funnel. Dry the material on a porous clay plate and determine the melting point.

ACID CHLORIDES AND ANHYDRIDES

Amides

$$R-\overset{\overset{\ddot{O}}{\|}}{C}-Cl + 2\,NH_3 \longrightarrow R-\overset{\overset{\ddot{O}}{\|}}{C}-NH_2 + NH_4Cl$$

HOOD

In a 10 × 75-mm test tube, place 0.4 mL of ice cold, concentrated ammonium hydroxide solution. To this solution, in the **hood**, slowly add, with shaking, about 15 mg of the unknown (~3 drops if a liquid) acid chloride or anhydride. Stopper the test tube and allow the reaction mixture to stand at room temperature for about 5 min. Collect the crystals by vacuum filtration, using a Hirsch funnel, and wash the filter cake with 0.2 mL of ice-cold water. Recrystallize the material, using a Craig tube, from water or an ethanol–water mixture. Dry the purified crystals on a porous clay plate and determine the melting point.

AROMATIC HYDROCARBONS

Picrates

Ar—H + [Picric acid] ⟶ Ar—H · [Picrate complex]

Picric acid / Picrate complex

The procedure outlined on page 679 is used to prepare these derivatives.

NITRILES

Hydrolysis to Amides

$$\text{Ar—CN:} \xrightarrow[\text{NaOH}]{H_2O_2} \text{Ar—}\overset{\overset{\ddot{O}}{\|}}{C}-NH_2 + O_2 + H_2O$$

Conversion of nitriles to water-insoluble amides, by hydrolysis with alkaline hydrogen peroxide, is a possible method of characterization for these compounds. It is especially useful for aromatic nitriles.

In a 5-mL conical vial containing a magnetic spin vane, weigh and place about 50 mg of the nitrile and 500 μL of a 1 M NaOH solution. Cool the mixture in a water bath and, with stirring, add dropwise 500 μL of 12% H_2O_2 solution. Attach the vial to an air condenser and warm the solution on a sand bath while stirring at 50–60 °C for approximately 45 min. To the cooled reaction mixture add 1–2 mL of cold water, and then collect the solid amide by vacuum filtration. Wash the product with two 1-mL portions of cold water, and recrystallize the amide from aqueous ethanol using the Craig tube. Dry the solid and determine the melting point.

PHENOLS

α-Naphthylurethanes (α-Naphthylcarbamates)

The procedure outlined under Alcohols: Phenyl-, and α-Naphthylurethanes is used to prepare these derivatives (p. 676).

Bromo Derivatives

INSTRUCTOR PREPARATION *The brominating reagent is prepared by adding 1.0 mL (3 g) of bromine in the **hood** to a solution of 4.0 g of KBr in 25 mL of water.*   HOOD

In a 1.0-mL conical vial, place 10 mg of the unknown phenol (~2 drops if a liquid) followed by 2 drops of methanol and 2 drops of water. To this solution, in the **hood**, add 3 drops of brominating agent from a Pasteur pipet.

Continue the addition (dropwise) until the reddish-brown color of bromine persists. Now add water (4 drops), cap the vial, shake, vent, and then allow it to stand at room temperature for 10 min. Collect the crystalline precipitate by vacuum filtration using a Hirsch funnel and wash the filter cake with 0.5 mL of 5% aqueous sodium bisulfite solution. Recrystallize the solid derivative from ethanol, or from an ethanol–water mixture, using a Craig tube. Dissolve the material in about 0.5 mL of ethanol, add water until it becomes cloudy, cool in an ice bath, and collect the crystals in the usual manner. Dry the purified product on a porous clay plate and determine the melting point.

ALIPHATIC HYDROCARBONS, HALOGENATED HYDROCARBONS, AMIDES, NITRO COMPOUNDS, ETHERS, AND ESTERS

These compounds do not give derivatives directly, but are usually converted into another material that can then be derivatized. The procedures are, for the most part, lengthy, and frequently give mixtures of products. It is recommended that compounds belonging to these classes be primarily identified using spectroscopic methods. Measurement of their physical properties is also of utmost importance.

QUESTIONS

9-1. The following six substances have approximately the same boiling point and are all colorless liquids. Suppose you were given six unlabeled bottles, each of which contained one of these compounds.

Explain how you would use simple chemical tests to determine which bottle contained which compound.

Ethanoic acid	Toluene
Propyl butanoate	Diisobutylamine
1-Butanol	Styrene

9-2. A colorless liquid (C_4H_6O) with a boiling point of 81 °C was found to be soluble in water and also in ether. It gave a negative test for the presence of halogens, sulfur, and nitrogen. It did, however, give a positive test with the Baeyer reagent and also gave a positive test with the 2,4-dinitrophenylhydrazine reagent. It gave negative results when treated with ceric nitrate solution and with Tollens reagent. Treatment with ozone followed by hydrolysis in the presence of zinc gave formaldehyde as one of the products.

What is the structure and name of the colorless liquid?

9-3. A colorless liquid, compound A (C_3H_6O), was soluble in water and ether, and had a boiling point of 94–96 °C. It decolorized a Br_2–CH_2Cl_2 solution and gave a positive ceric nitrate test. On catalytic hydrogenation it formed compound B (C_3H_8O), which did not decolorize the above bromine solution, but did give a positive ceric nitrate test. Treatment of compound A with ozone, followed by hydrolysis in the presence of zinc, gave formaldehyde as one of the products. Compound A formed an α-naphthylurethane with a melting point of 109 °C.

What are the names and structures of compounds A and B?

9-4. A compound of formula $C_{14}H_{12}$ gave a positive Baeyer test and burned with a yellow, sooty flame. Treatment with ozone followed by hydrolysis in the presence of zinc gave formaldehyde as one of the products. Also isolated from the ozonolysis reaction was a second compound, $C_{13}H_{10}O$, which burned with a yellow, sooty flame, and readily formed a semicarbazone with a melting point of 164 °C. The ^1H NMR spectrum of this compound ($C_{13}H_{10}O$) showed only complex multiplets that were near 7.5 ppm; the fully ^1H-decoupled ^{13}C NMR spectrum showed only five peaks.

What are the structures and names of the two compounds?

9-5. Compound A ($C_7H_{14}O$) burned with a yellow, nonsooty flame and did not decolorize a bromine–methylene chloride solution. It did give a positive 2,4-dinitrophenylhydrazine test, but a negative Tollens test. Treatment of the compound with lithium aluminum hydride followed by neutralization with acid, produced compound B, which gave a positive Lucas test in about 5 min. Compound B also gave a positive ceric nitrate test. The ^1H NMR spectrum for compound A gave the following data:

1.02 ppm	9H, singlet
2.11 ppm	3H, singlet
2.31 ppm	2H, singlet

Give suitable structures for compounds A and B.

9-6. A friend of yours, who is a graduate student attempting to establish the structure of a chemical species from field clover, isolated an alcohol that was found to have an optical rotation of +49.5°. Chemical analysis gave a molecular formula of $C_5H_{10}O$. It was also observed that this alcohol readily decolorized Br_2–CH_2Cl_2 solution. On this basis, the alcohol was subjected to catalytic hydrogenation and it was found to absorb 1 mol equivalent of hydrogen gas. The product of the reduction gave a positive ceric nitrate test, indicating that it, too, was an alcohol. However, the reduced compound was optically inactive.

Your friend has come to you for assistance in determining the structures of the two alcohols. What do you believe the structures are?

9-7. An unknown compound burned with a yellow, nonsmoky flame and was found to be insoluble in 5% sodium hydroxide solution but soluble in concentrated sulfuric acid. Measurement of its boiling point gave a range of 130–131 °C. Combustion analysis gave a molecular formula of C_5H_8O. It was found to give a semicarbazone with a melting point of 204–206 °C. However, it gave a negative result when treated with Tollens reagent, and it did not decolorize the Baeyer reagent. It also gave a negative iodoform test.

Identify the unknown compound.

9-8. An unknown organic carboxylic acid, mp = 139–141 °C, burned with a yellow, sooty flame. The sodium fusion test showed that nitrogen was present. It did not react with *p*-toluenesulfonyl chloride, but did give a positive test when treated with 5% aqueous ferrous ammonium sulfate solution, acidified with 3 M H_2SO_4, and then followed by methanolic KOH solution. A 200-mg sample of the acid neutralized 12.4 mL of 0.098 M sodium hydroxide solution.

Identify the acid.

Does your structure agree with the calculated equivalent weight?

9-9. An unknown organic liquid, compound A, was found to burn with a yellow, sooty flame and give a positive Lucas test (~5 min). Upon treatment with sodium dichromate–sulfuric acid solution it produced compound B, which also burned with a yellow, sooty flame. Compound B gave a positive 2,4-dinitrophenylhydrazine test, but a negative result when treated with the Tollens reagent. However, compound B did give a positive iodoform test.

The 1H NMR spectrum for compound A showed the following:

1.4 ppm	3H (doublet)	4.8 ppm	1H (quartet)
1.9 ppm	1H (singlet)	7.2 ppm	5H (complex multiplet)

Give the structures and suitable names for compounds A and B.

9-10. A hydrocarbon, compound A (C_6H_{10}), burned with a yellow, almost nonsmoky flame. On catalytic hydrogenation over platinum catalyst, it absorbed 1 mol of hydrogen to form compound B. It also decolorized a Br_2–CH_2Cl_2 solution to yield a dibromo derivative, compound C. Ozonolysis of the hydrocarbon gave only one compound, D. Compound D gave a positive iodoform test when treated with iodine–sodium hydroxide solution. On treatment of compound D with an alcoholic solution of silver ammonium hydroxide, a silver mirror was formed within a few minutes.

Identify the hydrocarbon A and compounds B–D.

9-11. A high-boiling liquid, bp = 202–204 °C burns with a yellow, sooty flame. Sodium fusion indicates that halogens, nitrogen, and sulfur are not present. The compound is not soluble in water, dilute sodium bicarbonate solution, or dilute hydrochloric acid. However, it proved to be soluble in 5% aqueous sodium hydroxide solution. The compound gives a purple color with ferric chloride solution and a precipitate when reacted with bromine–water. Treatment with hydroxylamine reagent did not give a reaction, but a white precipitate was obtained when the compound was treated with α-naphthylisocyanate. On drying, this white, solid derivative had an mp = 127–129 °C. Identify the original liquid and write a structure for the solid derivative.

After identifying the unknown liquid, can you indicate what the structure of the precipitate obtained on reaction with bromine might be?

9-12. A colorless liquid, bp = 199–201 °C, burns with a yellow, sooty flame. The sodium fusion test proved negative for the presence of halogens, nitrogen, and sulfur. The compound was not soluble in water, 5% aqueous sodium hydroxide, or 5% hydrochloric acid. However, it dissolved in sulfuric acid with evolution of heat. It did not give a precipitate with 2,4-dinitrophenylhydrazine solution, and it did not decolorize bromine–methylene chloride solution. The unknown liquid did give a positive hydroxamate test and was found to have a saponification equivalent of 136. Identify the unknown liquid.

9-13. Your friend of Question 9-6 still needs your help. A week later a low-melting solid, compound A, was isolated, which combustion analysis showed had composition $C_9H_{10}O$. The substance gave a precipitate when treated with 2,4-dinitrophenylhydrazine solution. Furthermore, when reacted with iodoform reagent, a yellow precipitate of CHI_3 was observed. Acidification of the alkaline solution from the iodoform test produced a solid material, compound B.

Reduction of compound A with $LiAlH_4$ gave compound C ($C_9H_{12}O$). Compound C also gave compound B when treated with iodoform reagent. Vigorous oxidation of compound A, B, or C with sodium dichromate–sulfuric acid solution gave an acid having an mp = 121–122 °C.

Your friend needs your assistance in determining the structures for compounds A, B, and C. Can you identify the three compounds?

9-14. An organic compound ($C_9H_{10}O$) showed strong absorption in the IR spectrum at 1735 cm^{-1} and gave a semicarbazone having a melting point of 198 °C. It burned with a yellow, sooty flame and also gave a positive iodoform test. The ^1H NMR spectrum of the compound provided the following information:

2.11 ppm	3H (singlet)
3.65 ppm	2H (singlet)
7.20 ppm	5H (complex multiplet)

Identify the unknown organic compound.

9-15. An unknown compound (A) was soluble in ether but only slightly soluble in water. It burned with a clear blue flame and combustion analysis showed it to have the molecular formula of $C_5H_{12}O$. It gave a positive test with the Jones reagent producing a new compound (B) with a formula of $C_5H_{10}O$. Compound B gave a positive iodoform test and formed a semicarbazone. Compound A on treatment with sulfuric acid produced a hydrocarbon (C) of formula C_5H_{10}. Hydrocarbon C readily decolorized a Br_2–CH_2Cl_2 solution, and on ozonolysis, produced acetone as one of the products.

Identify the structure of each of the lettered compounds.

9-16. Compound A (C_7H_{14}) decolorized a Br_2–CH_2Cl_2 chloride solution. It reacted with $BH_3 \cdot THF$ reagent, followed by alkaline peroxide solution, to produce compound B. Compound B, on treatment with chromic acid–sulfuric acid solution, gave carboxylic acid C, which could be separated into two enantiomers. Compound A, on treatment with ozone, followed by addition of hydrogen peroxide, produced compound D. Compound D was identical to that material isolated from the oxidation of 3-hexanol with chromic acid–sulfuric acid reagent.

Identify the structures of compounds A, B, C, and D.

9-17. Compound A (C_8H_{16}) decolorized a bromine–methylene chloride solution. Ozonolysis produced two compounds, B and C, which could be separated easily by gas chromatography. Both B and C gave a positive 2,4 dinitrophenylhydrazine test. Carbon–hydrogen analysis and molecular weight determination of B gave a molecular formula of $C_5H_{10}O$. The ^1H NMR spectrum revealed the following information for B:

0.92 ppm	3H, triplet	2.17 ppm	3H, singlet
1.6 ppm	2H, pentet	2.45 ppm	2H, triplet

Compound C was a low-boiling liquid (bp 56°C). The ^1H NMR of this material showed only one singlet.

Identify compounds A, B, and C.

C₁₀H₁₀apter 10

ADVANCED MICROSCALE ORGANIC LABORATORY EXPERIMENTS

From a theoretical perspective, the chemistry described in this chapter is more demanding of the student–investigator than that described in Chapter 6.

1. The organic reactions performed are less familiar and are not as likely to be found in most introductory texts.
2. The mechanisms proposed for these systems are more involved and not generally developed at the introductory level.
3. Many of the reagents used are rarely used in the introductory organic laboratory.

Thus, the experiments contained in Chapter 10 are specifically tailored to challenge the more advanced undergraduate students, those who are already able to access the chemical literature. This chapter can also offer a special laboratory experience for those few beginning students who are particularly interested in the subject and wish to spend extra preparation time. The techniques involved are not, in most instances, any different from those used in the introductory microscale laboratory reactions described in Chapter 6, and therefore the manipulations involved should not be considered a barrier to undertaking any of the advanced experiments. The reaction conditions, however, are less forgiving to slight deviations from the suggested ones, and the ultimate success of the transformations is less secure.

Hence, the experimentation contained in Chapter 10 can be adapted to several levels of undergraduate laboratory programs. For example, the study of these reactions can potentially make significant contributions to advanced undergraduate programs where microscale techniques are being introduced to research-oriented students for the first time.

The formats for the discussions and experimental procedures are similar to those used in Chapter 6. The reactions selected for study in this chapter include

- An unusual borane reduction of a carbonyl directly to a methylene group.
- The trapping of an α,β-unsaturated ketone as its enol acetate, by acylium ion formation with chlorotrimethylsilane and acetic anhydride.

Chapter 10: C₁₀H₁₀, Housane
Eaton, Or, Branca, and Shankar (1968).

- The synthesis of a heterocyclic ring using diethyl carbonate and sodium hydride. The discovery of the medicinal properties of this class of heterocycles led directly to modern anticoagulation therapy.

- The synthesis of an isotopically labeled molecule through the use of a unique Grignard cross-coupling reaction in the presence of dichloro[1,3-bis(diphenylphosphino)propane]nickel(II).

- An oxidative coupling of a naphthol by ferric chloride in a reaction that mimics nature's method of coupling phenolic substances into important pigments, such as hypericin (**I**):

I

- An important molecular rearrangement of oximes to amides discovered by Ernst Otto Beckmann in 1886 and so named in his honor. The modern version uses one of the most powerful acid reagents used in organic chemistry, triflic acid.

The study of the reactions outlined in Chapter 10 should help to facilitate the student's smooth transition into the organic research laboratory. We hope you find the collection as exciting as we did.

EXPERIMENT
[1 adv]

Diborane Reductions: Thioxanthene and Xanthene

Common name: thioxanthene
CA number: [261-31-4]
CA name as indexed: 9*H*-thioxanthene
Common name: xanthene
CA number: [92-83-1]
CA name as indexed: 9*H*-xanthene

Purpose.   This experiment investigates an unusual example of a hydroboration in which a carbonyl group is directly and fully reduced to a methylene group by borane (BH$_3$). This example is an atypical reduction of an aldehyde or ketone, since the use of this reagent usually leads to the formation of the corresponding alcohol. You will explore the mechanistic rationale for this unexpected product.

Prior Reading

    Standard Experimental Apparatus: Moisture Protected Reaction Apparatus
                           (pp. 27–28)
    Technique 6: Chromatography
          Packing the Column (p. 97)
          Elution of the Column (p. 98)
    Technique 6B: Concentration of Solutions
          Removal of Solvent under Reduced Pressure (pp. 106–108)

NOTE. See Experiment [13] for a biography of Herbert C. Brown, Nobel Laureate, who discovered and developed the boron hydride reagents. This experiment also contains further details about the use of these powerful reagents, which have revolutionized reduction reactions in organic chemistry.

DISCUSSION

Borane is a useful and selective reducing agent. It is prepared by the reaction of boron trifluoride etherate with sodium borohydride. The borane produced, as the etherate, may be distilled as the dimer, which is a colorless, toxic gas (B_2H_6). Collection of the dimer distillate in tetrahydrofuran (THF) again forms the monomer, in this case as the $BH_3 \cdot THF$ complex. The latter is commercially available as a 1.0 M solution.

$$3\,NaBH_4 + 4\,BF_3 \cdot O(CH_2CH_3)_2 \longrightarrow 3\,NaBF_4 + 2\,B_2H_6 \uparrow (gas)$$

    Borane complexes can also be formed with other ethers, such as diethyl ether (as just discussed) or diglyme (diethylene glycol dimethyl ether). These complexes form readily, because the ether, acting as a Lewis base (electron donor), can satisfy the electron-deficient boron atom, which acts as a Lewis acid (electron acceptor). Borane reacts rapidly with water, and therefore procedures using the $BH_3 \cdot THF$ complex must be conducted under anhydrous conditions.

    Borane is a Lewis acid that is attacked by electron-rich centers. Thus, when aldehydes or ketones are treated with the $BH_3 \cdot THF$ complex, the borate ester (H_2B—OR) is rapidly formed, which, upon hydrolysis, gives the corresponding alcohol. The reduction of the carbonyl group is believed to take place by addition of the oxygen atom to the electron-deficient boron atom, followed by irreversible transfer of hydride ion from the now anionic boron to the carbon atom of the (former) carbonyl:

    In the case of the xanthone and thioxanthone ring systems, the corresponding alcohol is not formed. The intermediate borate ester undergoes an elimination reaction, forming a borate anion and a resonance-stabilized carbocation. The second stage of the reaction is initiated by displacement of THF from a second $BH_3 \cdot THF$ complex by a lone pair from either the oxygen atom of the xanthene carbocation or the sulfur atom of the thioxanthene carbocation. This *new complex* then undergoes an internal hydride ($:H^-$) transfer from boron to the C-9 ring position (this is the second hydride attack on this position in the overall reaction) to form a stable oxonium (or sulfonium) ion intermediate. On aqueous-alcohol workup of the reaction mixture the oxonium (sulfonium) product is quickly hydrolyzed to yield xanthene (or thioxanthene) possessing a fully reduced methylene group, $—CH_2—$, at the 9 position.

Conventional methods for the reduction of a carbonyl to a methylene group that do not require the conjugative assistance of a heteroatom are the well-known Clemmensen (Zn(Hg), HCl), and Wolff–Kishner (H_2NNH_2/KOH) reductions, and the desulfurization of the corresponding thioacetal with Raney nickel.

Thioxanthene

REACTION

Thioxanthone →(BH$_3$·THF)→ Thioxanthene

EXPERIMENTAL PROCEDURE

Estimated time to complete the experiment: 1.5 h.

Physical Properties of Reactants							
Compound	MW	Amount	mmol	mp (°C)	bp (°C)	d	n_D
Thioxanthone	212.28	50 mg	0.24	209			
Tetrahydrofuran	72.12	1.7 mL			67	0.889	1.407
Borane · THF, 1 M		1.0 mL	1.0				

Reagents and Equipment. Attach a 5.0-mL conical vial containing a magnetic spin vane to a Claisen head. Then fit the Claisen head with a nitrogen inlet tube (prepared from a syringe) and a rubber septum. Weigh and add 50 mg (0.24 mmol) of thioxanthone to the conical vial.

Flush the reaction vial with a gentle stream of nitrogen gas for several minutes, add 1.7 mL of *dry* (see Note) tetrahydrofuran (THF) through the septum (syringe), and then place a small balloon over the Claisen-head outlet so as to maintain a dry atmosphere in the system (➡).

*INSTRUCTOR'S NOTE. The THF must be **absolutely** dry. It is recommended that HPLC-grade reagent be used. If you do not have a fresh bottle, distill it once from calcium hydride (or sodium benzophenone ketyl) and store it over molecular sieves. It may be stored and used safely for up to a week without adversely affecting the yield of product.*

Heat the mixture with stirring in a sand bath at 55–60 °C until the thioxanthone dissolves, yielding a yellow solution. Then, with continued stirring, add 1.0 mL of a 1.0 M solution of BH$_3$ · THF in one portion through the rubber septum with a 1.0-mL syringe.

CAUTION: The BH$_3$ · THF solution reacts *violently* with water.

Reaction Conditions. Heat the solution with stirring in a sand bath at 55–60 °C for 5 min. The solution should become colorless during this time.

Thioxanthone, 50 mg
+ THF, 1.7 mL
+ 1*M* BH$_3$·THF, 1.0 mL

Experiment [1A$_{adv}$]

Thumb
controls
pressure;
continuous
shaking

THF, 2.5 mL,
+ CH₃CH₂OH, 0.5 mL
+ H₂O, 2 mL
+ thioxanthene

Isolation of Product.   Quench the reaction by *carefully* adding dropwise approximately 10 drops of 95% ethanol (Pasteur pipet), with stirring, or until the observed foaming subsides. *The aqueous alcohol is added to decompose any unreacted BH₃ · THF reagent and to hydrolyze the sulfonium ion intermediate.*

After the solution has cooled to room temperature, transfer the mixture by Pasteur pipet to a 25-mL filter flask containing a boiling stone. Carefully remove roughly one-half of the tetrahydrofuran solvent under reduced pressure with continuous swirling of the flask (see Prior Reading) (◄). Then, use a calibrated Pasteur pipet to add two 1.0-mL portions of water to the reaction mixture. Carefully remove the remaining tetrahydrofuran and ethanol solvent under reduced pressure with continuous swirling of the flask. As the tetrahydrofuran and ethanol evaporate, white crystals of thioxanthene appear, and a slurry of these crystals in water will remain in the flask after the tetrahydrofuran and ethanol are removed. Collect the product crystals under reduced pressure by use of a Hirsch funnel, and wash the filter cake on the Hirsch filter bed with two 1.0-mL portions of water. Dry the crystals in air on a porous clay plate or on filter paper.

Purification and Characterization.   The thioxanthene product is essentially pure as isolated. It may be recrystallized from an ethanol–chloroform mixture, if necessary. Weigh the product and calculate the percent yield. Determine the melting point of the material and compare it with the value reported in the literature. Obtain IR spectra of thioxanthone and thioxanthene and compare them to each other as well as to those given in the literature (*The Aldrich Library of IR Spectra* and/or *SciFinder Scholar*).

Experiment [1B_adv]   *Xanthene*

REACTION

Xanthone $\xrightarrow{\text{BH}_3 \cdot \text{THF}}$ Xanthene

EXPERIMENTAL PROCEDURE

Estimated time to complete the experiment: 3.0 h.

Physical Properties of Reactants							
Compound	MW	Amount	mmol	mp (°C)	bp (°C)	d	n_D
Xanthone	196.22	50 mg	0.26	174			
Tetrahydrofuran	72.12	0.7 mL			67	0.889	1.407
Borane · THF, 1 M		0.75 mL	0.75				

Reagents and Equipment.   Using the experimental apparatus described in Experiment [1A$_{adv}$], weigh and place in the reaction flask 50 mg (0.26 mmol) of xanthone followed by 0.7 mL of dry THF (see Experiment [1A$_{adv}$], Reagents and Equipment). Maintain a dry nitrogen atmosphere in the system by the same procedure as used in Experiment [1A$_{adv}$]. Heat the reaction mixture, with stirring, to 55–60 °C using a sand bath. After the xanthone dissolves, use a 1.0-mL syringe to add, in one portion, 0.75 mL of a 1.0 M solution of BH$_3$ · THF through the rubber septum on the screw-cap port of the Claisen head.

CAUTION: The BH$_3$ · THF solution reacts *violently* with water.

Reaction Conditions.   Stir the reaction solution in a sand bath at 55–60 °C for a period of 1 h.

Isolation of Product.   While stirring the warm reaction mixture, quench the reaction by carefully adding (with a Pasteur pipet) 95% ethanol—approximately 10 drops or until the observed foaming subsides. *The alcohol is added to decompose any unreacted BH$_3$ · THF reagent and to hydrolyze the oxonium ion intermediate.*

Transfer the solution by Pasteur pipet to a 25-mL filter flask containing a boiling stone. Remove roughly one-half of the tetrahydrofuran solvent under reduced pressure with continuous swirling of the flask (➡). Now add two 1.0-mL portions of water (calibrated Pasteur pipet) to the solution. Carefully remove the **remaining** tetrahydrofuran–ethanol solvent under reduced pressure with continuous swirling of the flask. As the tetrahydrofuran and ethanol are removed and the solution becomes more concentrated, white crystals of xanthene appear as a slurry in the remaining water. Collect the crystals under reduced pressure using a Hirsch funnel.

Thumb controls pressure; continuous shaking

THF, 1.25 mL
+ (CH$_3$CH$_2$)$_2$O, 0.5 mL
+ H$_2$O, 2 mL
+ xanthene

Purification and Characterization.   Purify the crude xanthene by column chromatography. Place 0.5 g of activated silica gel followed by 0.1 g of anhydrous sodium sulfate in a Pasteur filter pipet (➡).

Wet the column with a small amount of hexane, and then place a solution of the crude xanthene, dissolved in 0.25 mL of methylene chloride, on the column using a Pasteur filter pipet. Elute the xanthene by adding additional hexane (~2.5 mL). Collect the eluate in a tared 5-mL conical vial containing a boiling stone. Remove the hexane solvent by evaporation in the **hood** while warming in a sand bath to yield pure xanthene. This compound may be recrystallized from ethanol if further purification is found to be necessary after characterization of the product.

Anhydrous Na$_2$SO$_4$, 100 mg

SiO$_2$, 500 mg

Sand, 50 mg

Cotton Plug

Xanthe product + 2.5 mL hexane

HOOD

After air-drying, weigh the solid and calculate the percent yield of xanthene. Determine the melting point and compare it with the value found in the literature.

Obtain an IR spectrum of your xanthene, and compare the spectrum to that shown in Figure 10.1 or SciFinder Scholar.

In this experiment, the carbonyl group of an aromatic ketone was reduced to a methylene group. Examine the infrared spectra of the starting material and of the reduction product to see what evidence is present to indicate that the desired reaction has occurred.

Infrared Analysis.   We will first consider the spectrum of xanthone (Fig. 10.2). The macro group frequency associated with six-membered carbocyclic aromatic ring systems applies in this instance (peaks at 3100–3000, 1600, 1585, 1500, and 1450 cm^{-1}). This frequency train involves the bands located

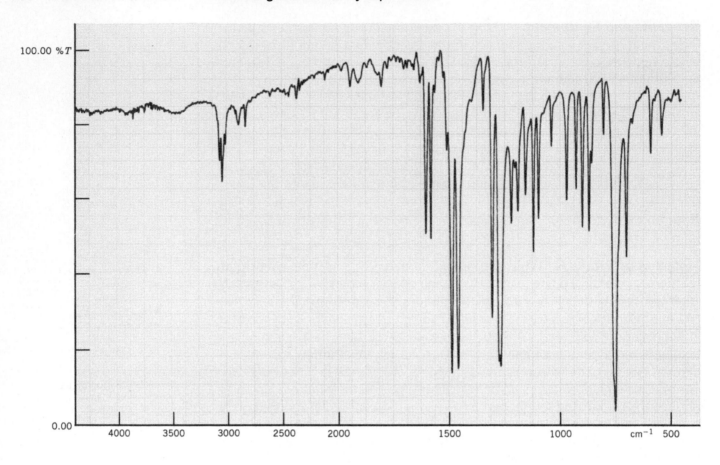

Sample	Xanthene		
%T _X_ ABS — Background Scans _4_		Scans _16_	
Acquisition & Calculation Time _42 sec_		Resolution _4.0 cm⁻¹_	
Sample Condition _solid_		Cell Window	
Cell Path Length		Matrix Material _KBr_	

Figure 10.1   IR spectrum: xanthene.

at 3085–3020, 1610–1570, 1484, and 1460 cm^{-1}. These peaks are assigned as follows:

a. **3085–3020 cm^{-1}:** C—H stretch on sp^2-hybridized carbon. The breadth and complexity of this set of absorption bands indicates the presence of a fairly complex aromatic system.

b. **1610–1570 cm^{-1}:** The peaks observed in this region are related to the two degenerate fundamental stretching motions, v_{8a} and v_{8b}, of the simple aromatic ring system. Normally, v_{8a} is found near 1600 cm^{-1} and is considerably more intense than v_{8b}, which is located near 1580 cm^{-1}. In xanthone the situation is more complicated, because the central γ-pyrone ring introduces a pseudoaromatic six-membered ring system. Thus, this ring system might be expected to possess somewhat shifted fundamental frequencies for the carbon rings. It is not unexpected, then, that we observe a band system of four major components in this region.

Sample	Xanthone

%T _X_ ABS __ Background Scans _4_   Scans _____16_____

Acquisition & Calculation Time _42 sec_   Resolution _4.0 cm⁻¹_

Sample Condition _____solid_____   Cell Window _____

Cell Path Length _____   Matrix Material _KBr_

Figure 10.2   IR spectrum: xanthone.

c. **1484 and 1460 cm⁻¹:** Aromatic ring stretch related to v_{19a} and v_{19b}. These frequencies are less disturbed by the presence of the pyrone system and occur near their expected locations of 1500 and 1450 cm⁻¹.

The very strong band at 756 cm⁻¹, in the absence of strong absorption near 700 cm⁻¹, is supporting evidence for the presence of four adjacent ring C—H groups, which implies ortho substitution.

The intense band observed at 1656 cm⁻¹ is strong evidence for the presence of a highly conjugated carbonyl group, which, of course, is consistent with the structure of the starting material.

The IR spectrum of the product (Fig. 10.1) supports the complete reduction to the methylene system. The macro group frequency train defined for six-membered carbocyclic aromatic systems still applies. The expected frequencies are very close to the observed values:

a. 3075–3025, 1603, 1589, 1490, and 1457 cm⁻¹.

b. The carbonyl band has vanished, and in its place two weak bands have arisen near 2902 and 2840 cm⁻¹. These latter peaks are assigned to the antisymmetric and symmetric stretching modes of the newly formed methylene group.

c. The most intense band in the spectrum occurs at 750 cm^{-1} (four in a row, C—H all-in-phase, out-of-plane bend) and indicates that the basic substitution pattern has not changed on the ring system during the reaction.

QUESTIONS

10-1. In the reaction performed in this experiment, assume that the first stage of the reaction is the rate-determining step. Would you then predict that the relative rate of reduction of the carbonyl group in compound A to the methylene group will be faster or slower than that of xanthone under the conditions of this experiment? Explain.

Compound A

10-2. The reduction of aldehydes or ketones to the methylene group occurs with hydride reagents only when some special feature of the substrate promotes cleavage of the C—OH linkage. Suggest a suitable mechanism by which the reductions given below might occur.

10-3. Borane also forms complexes with sulfides and amines. Draw a suitable structure to represent the complex formed between BH_3 and dimethyl sulfide, and also that formed between BH_3 and triethylamine.

10-4. Using your lecture textbook as a reference, find three different methods for the conversion of 4-methylcyclohexanone to methylcyclohexane.

10-5. The infrared spectrum of the xanthene reduction product contains evidence that demonstrates that conjugation of the rings is still maintained following removal of the carbonyl group. What is this evidence?

BIBLIOGRAPHY

The reactions outlined in Experiments [1A$_{adv}$] and [1B$_{adv}$] were adapted from the work of:

Wechter, W. J. *J. Org. Chem.* **1963**, *28*, 2935.

Borane as a reducing agent:

Carruthers, W.; Coldham, I. *Modern Methods of Organic Synthesis*, 4th ed.; Cambridge University Press: New York, 2004, Chap. 7.

Fieser, L. F.; Fieser, M. *Reagents for Organic Synthesis;* Wiley: New York, 1967, Vol. I, p. 199. Subsequent volumes of this series have further examples of diborane as a reducing agent.

Matteson, D. S. in Hartley, F. R.; Patai, S, Eds., *The Chemistry of the Metal-Carbon Bond,* Vol. IV; Wiley: New York, 1987, pp. 307–409.

Pelter, A.; Smith, K.; Brown, H. C. *Borane Reagents*; Academic Press: New York, 1988.

Borane as a hydroborating agent:

Brown, H. C. *Hydroboration;* Benjamin: New York, 1962.

Brown, H. C. *Boranes in Organic Chemistry;* Cornell University Press: New York, 1972.

Brown, H. C. *Organic Synthesis Via Boranes;* Wiley: New York, 1975.

See Zaidlewicz, M. for an instructive presentation on *Hydroboration in Kirk-Othmer Encyclopedia of Chemical Technology*, 5th ed.; Wiley-VCH: New York, 2005, Vol. XIII.

Smith, K. *Organoboron Chemistry in Organometallics in Synthesis—A Manual;* Schlosser, M. M., Ed.; Wiley: New York, 1994.

Zweifel, G.; Brown, H. C. *Org. React.* **1963**, *13*, 1.

Heterocyclic Ring Synthesis: Benzimidazole

Common name: benzimidazole
CA number: [51-17-2]
CA name as indexed: 1*H*-benzimidazole

Purpose. This experiment investigates conditions under which one of the important heterocyclic ring systems, the benzimidazole ring, may be formed. The method used involves the condensation of a 1,2-diaminobenzene with formic acid. The simplest possible benzimidazole ring system, benzimidazole itself, is prepared.

Prior Reading
Technique 5: Crystallization
Use of the Hirsch Funnel (pp. 92–93)
Craig Tube Crystallizations (pp. 93–95)

REACTION

o-Phenylenediamine   Formic acid   1-*H*-Benzimidazole

DISCUSSION

This experiment illustrates the classic method of forming the benzimidazole ring system. This heterocycle is generally prepared from 1,2-diaminobenzene (*o*-phenylenediamine) derivatives by reaction with carboxylic acids, or their derivatives, under acidic conditions. The ring system is aromatic; thus it is difficult to oxidize or reduce, and it is stable to both acids and bases. It is an important heterocyclic ring system that occurs in vitamin B_{12} and in many other biologically important compounds. Benzimidazole itself inhibits the growth of certain yeasts and bacteria.

The parent compound of this class of heterocyclic compounds is imidazole. This ring system exhibits basic properties and is protonated to give a conjugate acid with $pK_a = 6.95$. Once the imidazole ring is protonated, the two nitrogen atoms are indistinguishable, because the resonance forms of the protonated species are equivalent. As a resonance stabilized intermediate, the imidazole scaffold offers the synthetic organic chemist a multitude of opportunities when considering its role as part of an ionic liquid. As the term implies, an ionic liquid is a salt. However, when the salt is appropriately modified, it can exist as a liquid at room temperature. As a salt, the system exhibits a significantly lower vapor pressure when compared to standard organic solvents, and thus it can serve the role of a solvent for many synthetic transformations. Over the past decade, ionic liquid technologies have witnessed an unprecedented surge in interest, especially among industrial applications.

The reaction to form the ring system used in this experiment proceeds in two stages. The first involves the in situ formation of an *N*-substituted formamide, via the usual nucleophilic addition–elimination reaction. The second involves an intramolecular nucleophilic addition to a carbonyl group and subsequent elimination of water to form the unsaturated heterocyclic ring. The sequence is outlined below:

o-Phenylenediamine   Formic acid

(substituted formamide)

Benzimidazole

EXPERIMENTAL PROCEDURE

Estimated time to complete the experiment: 2.0 h.

Physical Properties of Reactants							
Compound	MW	Amount	mmol	mp (°C)	bp (°C)	*d*	n_D
o-Phenylenediamine	108.1	108 mg	1.0	102			
90% Formic acid	46.03	64 μL	1.7		101	1.22	1.3714

Reagents and Equipment. Weigh and add 108 mg (1.0 mmol) of *o*-phenylenediamine to a 3.0-mL conical vial containing a magnetic spin vane. Now add 64 µL (79 mg, 1.7 mmol) of 90% formic acid, and attach the vial to a reflux condenser protected by a calcium chloride drying tube (➡).

HOOD

150–160 °C
thermometer

CAUTION: *o*-Phenylenediamine is toxic and is a suspected carcinogen. Formic acid is very corrosive to the skin and should be dispensed in the **hood** using an automatic delivery pipet.

Speed   Heat

o-$(H_2N)_2C_6H_4$, 108 mg
+ 90% HCO_2H, 64 µL

Reaction Conditions. Heat the reaction mixture, with stirring, at a sand bath temperature of 150–160 °C for 1 h.

Isolation of Product. Allow the reaction mixture to cool to room temperature. Add 630 µL of 10% aqueous sodium hydroxide solution (automatic delivery pipet). Crude benzimidazole precipitates at this point. Collect the product by vacuum filtration using a Hirsch funnel and wash the filter cake with three 0.5-mL portions of cold water (calibrated Pasteur pipet) (➡).

Purification and Characterization. Recrystallize the crude material from water, using a Craig tube. Dry the product in a desiccator or in a vacuum drying apparatus (see Prior Reading). Weigh the crystals and calculate the percent yield. Determine the melting point and compare your result with that listed in the literature.

The UV spectrum of benzimidazole in 95% ethanol has been reported.[1]

Benzimidazole product
collected here

H_2O, ~2.0 mL

Chemical Tests. Several tests may be run to assist in the identification of this material. Does the ignition test confirm the presence of the aromatic ring system? Does the soda lime or sodium fusion test indicate that nitrogen is present? Is the material soluble in 10% hydrochloric solution?

QUESTIONS

I

II

10-6. The parent compound of the imidazole series, imidazole (**I**) itself, was first prepared in 1858: Can you account for the fact that it has a very high boiling point (256 °C), whereas 1-methyl imidazole (**II**) has the somewhat lower boiling point of 199 °C?

10-7. The imidazole ring system has a great deal of aromatic character. Can you formulate two resonance structures that account for this characteristic?

10-8. Imidazole is a weak acid, and thus reacts with strong bases to form the corresponding anion. Show this reaction, and draw resonance structures that account for the stability of the conjugate base.

10-9. Suggest a mechanism for the dehydration involved in the last step in the synthesis of benzimidazole.

10-10. Imidazole, acting as a nucleophile, catalyzes the hydrolysis of phenyl acetate by attack on the carbonyl carbon atom of the ester. The imidazole displaces the phenoxide anion and forms acetyl imidazole. In turn, the acetyl imidazole is quite unstable in water and hydrolyzes to form acetic acid, and regenerates the imidazole molecule. Write a suitable mechanism outlining these steps.

[1]Steck, E. A.; Nackod, F. C.; Ewing, G. W.; Gorman, N. H. *J. Am. Chem. Soc.* **1948,** *70*, 2406.

BIBLIOGRAPHY

The conditions of this reaction were adapted from those reported by:

Wagner, E. C.; Millett, W. H. *Organic Syntheses*; Wiley: New York, 1943; Collect. Vol. II, p. 65.

For reviews see:

Eicher, T.; Hauptmann, S.; Speicher, A. *The Chemistry of the Heterocycles*, 3rd ed.; Wiley-VCH: New York, 2012, Chap. 5.

Elderfield, R. C., Ed.; *Heterocyclic Compounds (Five-membered Heterocyclic Compounds Containing Two Hetero Atoms and Their Benzo Derivatives)*, Vol. V; Wiley: New York, 1957.

Grimmett, M. R. *Imidazole and Benzimidazole Synthesis*; Academic Press: New York, 1997.

Wright, J. B. *Chem. Revs.* **1951**, *48*, 397.

For applications involving ionic liquids, see:

Davis, J. H., Jr. *Chem. Lett.* **2004**, *33*, 1072.

Seddon, K. R. *J. Chem. Technol. Biotechnol.* **1997**, *68*, 351.

Wasserscheid, P.; Welton, T. *Ionic Liquids in Synthesis*; Wiley-VCH Verlag: Stuttgart, Germany, 2002.

Welton, T. *Chem. Rev.* **1999**, *99*, 2071.

EXPERIMENT [3 adv]

Heterocyclic Ring Synthesis: 4-Hydroxycoumarin and Dicoumarol

Common name: 4-hydroxycoumarin

CA number: [1076-38-6]

CA name as indexed: 2*H*-1-benzopyran-2-one, 4-hydroxy-

Common names: dicoumarol, dicoumarin

CA number: [66-76-2]

CA name as indexed: 2*H*-1-benzopyran-2-one, 3,3′-methylenebis [4-hydroxy]-

B

Purpose. You will synthesize a material, dicoumarol, that was the prototype for the oral anticoagulants widely used in medicine to lower blood coagulation rates. A carbon nucleophile is added to a carbonyl carbon to form a C—C bond. You will utilize a Claisen condensation reaction to prepare a β-ketoester, which, upon cyclization, forms a lactone, 4-hydroxycoumarin. Further condensation of two mole equivalents of 4-hydroxycoumarin with formaldehyde yields dicoumarol.

Prior Reading

Technique 2: Simple Distillation at the Semi-microscale Level (pp. 65–68)

Technique 4: Solvent Extraction

Liquid–Liquid Extraction (p. 76)

Technique 5: Crystallization

Use of the Hirsch Funnel (pp. 92–93)

Craig Tube Crystallizations (pp. 93–95)

REACTION

o-Hydroxyacetophenone   Diethyl carbonate   4-Hydroxycoumarin   Ethanol

DISCUSSION

This reaction sequence illustrates the well-known Claisen condensation, which is widely used to form C—C bonds. The bond formation is brought about by the nucleophilic attack of an enolate anion on the carbonyl carbon of an ester. The enolate is generated by removal of a slightly acidic hydrogen from the α-carbon atom of a ketone, nitrile, or ester, using a relatively strong base. The reaction mechanism is shown here. The methyl ketone is deprotonated by the base, sodium hydride. Sodium hydride (NaH) provides a basic and non-nucleophilic source of hydride ion (H⁻). The resulting enolate then attacks the ester, diethyl carbonate. The β-ketoester product thus formed is esterified in an *intramolecular* reaction with the phenolic —OH group to form the lactone product (**I**). Note that the methylene hydrogen atoms in this lactone (**I**) are quite acidic because they are adjacent to two carbonyl groups. In the basic medium of this reaction, the sodium enolate of the lactone (**II**) will be formed. This enolate is water-soluble, because it is a salt, which explains why the aqueous solution must be acidified to precipitate the neutral 4-hydroxycoumarin when isolating the product.

1. 4-Hydroxycoumarin

2. Dicoumarol

The aldol condensation of 4-hydroxycoumarin with formaldehyde provides an α,β-unsaturated carbonyl compound, which then undergoes a conjugate (1,4-) addition of a second molecule of 4-hydroxycoumarin. This reaction could be catalyzed by either trace base or trace acid; the acid-catalyzed reaction is shown and discussed here. The enol portion of 4-hydroxycoumarin is the nucleophile in an aldol reaction with protonated formaldehyde. The resulting product dehydrates to provide the α,β-unsaturated carbonyl compound, which, after protonation renders it a more reactive electrophile, then reacts with another nucleophilic molecule of 4-hydroxycoumarin in a conjugate addition reaction. This product, upon loss of a proton to the aqueous solvent, leads to dicoumarol. This substance is present in moldy sweet clover. It is a blood anticoagulant and its ingestion leads to the hemorrhagic sweet clover disease that kills cattle.

4-Hydroxycoumarin

The reaction is shown on page 10-15.

Experiment [3A$_{adv}$]

EXPERIMENTAL PROCEDURE

Estimated time to complete the reaction: 4.0 h.

Physical Properties of Reactants						
Compound	MW	Amount	mmol	bp (°C)	d	n_D
Sodium hydride (60% in oil dispersion)	24.0	85 mg	2.13			
Toluene	92.15	6.0 mL		111	0.86	1.4960
o-Hydroxyacetophenone	136.16	133 μL	1.1	218	1.13	1.5584
Diethyl carbonate	118.13	333 μL	2.75	126	0.96	1.3845

Reagents and Equipment.

NOTE. All equipment used in this reaction must be thoroughly dried in an oven at 110 °C for 30 min just prior to use. Upon removal from the drying oven, it should be allowed to cool to ambient temperature in a desiccator.

Weigh and place 85 mg (2.13 mmol) of sodium hydride (60% dispersion in oil) in a 10-mL round-bottom flask containing a magnetic stirring bar. Now add 3.0 mL of dry toluene. Attach the flask to a Hickman still fitted with an air condenser protected by a calcium chloride drying tube. Wrap the Hickman still 14/10 TS male joint with Teflon tape to prevent joint freeze-up (➡).

175 °C thermometer

HOOD

60% NaH dispersion, 85 mg + C$_6$H$_5$CH$_3$, 6 mL + o-HOC$_6$H$_4$COCH$_3$, 150 mg + (CH$_3$CH$_2$O)$_2$CO, 342 mg

---

CAUTION: Sodium hydride (NaH) is a flammable solid. Dispense in the **hood**. Toluene is distilled and stored over molecular sieves. Also dispense this solvent in the **hood**.

In rapid order, place 133 μL (150 mg, 1.1 mmol) of o-hydroxyacetophenone, 3.0 mL of dry toluene, and 333 μL (324 mg, 2.75 mmol) of diethyl carbonate in a stoppered 10-mL Erlenmeyer flask.

NOTE. Dispense the small volumes of the liquid reagents using an automatic delivery pipet. Be sure to dry the removable plastic tips in the oven before use. Use a 10-mL graduated cylinder to measure the toluene.

Remove the air condenser from the distilling head and add the *o*-hydroxyacetophenone solution, with stirring, to the reaction flask as rapidly as possible using a Pasteur pipet. The resulting solution turns yellow. Immediately reattach the air condenser.

Reaction Conditions. Place the reassembled apparatus in a sand bath and rapidly raise the temperature of the bath to about 175 °C.

CAUTION: To avoid excessively high temperatures, calibrate the hot plate temperature control before conducting the experiment.

Collect the ethanol and toluene distillate (1.0 mL in 2–3 fractions) in the collar of the still, and then remove the apparatus from the heat source. Allow the reaction solution to cool to room temperature, and then add 3.0 mL of water with stirring.

Cyclized product collected here

Acidified aqueous phase, ~10 mL

Isolation of Product. Transfer the resulting two-phase solution, using a Pasteur pipet, to a 15-mL centrifuge tube. Rinse the reaction flask with an additional 2.0 mL of water and add this to the centrifuge tube. Separate the toluene layer using a Pasteur filter pipet and transfer it to a second 15-mL centrifuge tube. Then extract this organic phase with 3.0 mL of water, and add the water extract to the original water phase. *This aqueous solution contains the water-soluble sodium enolate* (**II**). Cool the combined aqueous layers in an ice bath and acidify them by dropwise addition of concentrated hydrochloric acid delivered from a Pasteur pipet. The solid product precipitates from the aqueous phase. Add acid until the yellow color of the solution disappears (~ 10 drops). Collect the product by vacuum filtration using a Hirsch funnel (◄).

Purification and Characterization. Recrystallize the crude 4-hydroxycoumarin from 50% ethanol using a Craig tube. Dry the material on a porous clay plate or in a vacuum apparatus (see Prior Reading).

Weigh the product and calculate the percent yield. Determine the melting point and obtain an IR spectrum. Compare your results with those reported in the literature.

Chemical Tests. You may wish to perform the ignition test to establish that the compound is aromatic. Does the ferric chloride test (Chapter 9) for phenols give a positive result?

Experiment [3B_adv]   Dicoumarol

The reaction is shown on page 700.

REACTION

4-Hydroxycoumarin   Formaldehyde   Dicoumarol

EXPERIMENTAL PROCEDURE

Estimated time to complete the reaction: 0.5 h.

Physical Properties of Reactants					
Compound	MW	Amount	mmol	mp (°C)	bp (°C)
4-Hydroxycoumarin	162.15	50 mg	0.31	213–214	
Water		15 mL			100
Formaldehyde (37% in H$_2$O)		0.5 mL	6		

Reagents and Equipment. Weigh and place 50 mg (0.31 mmol) of 4-hydroxycoumarin, followed by 15 mL of water, in a 50-mL Erlenmeyer flask containing a boiling stone. Heat the mixture to boiling on a hot plate. Now add 0.5 mL (~500 mg, ~6 mmol) of formaldehyde (37% aqueous solution) to the resulting solution.

CAUTION: Formaldehyde is a suspected carcinogen. Dispense in the **hood.**   HOOD

Reaction Conditions. White crystals of the product form immediately on addition of the formaldehyde solution. Cool the flask in an ice bath.

Isolation of Product. Collect the solid product by vacuum filtration using a Hirsch funnel (➡). Wash the filter cake with three 1-mL portions of ice water. Remove the crystals and dry them on filter paper or on a porous clay plate.

Purification and Characterization. Recrystallize the material from a mixture of toluene/cyclohexanone (~2:1) using a Craig tube. After drying the isolated material, determine its melting point and compare your value with that in the literature. Obtain an IR spectrum and compare your spectrum with that in the literature (*The Aldrich Library of IR Spectra* and/or *Sci Finder Scholar*).

Condensation product collected here

H$_2$O, ~15 mL

QUESTIONS

10-11. Why must water be excluded from the reaction that generates 4-hydroxycoumarol?

10-12. In the first step of the above reaction, formation of 4-hydroxycoumarin, what is the purpose of removing the ethanol generated in the reaction?

10-13. The Dieckmann condensation is simply an intramolecular Claisen condensation. For example,

Suggest a suitable mechanism to account for the formation of the cyclic β-ketoester.

10-14. Predict the product formed in each of the following reactions:

(a) Acetophenone + diethyl carbonate $\xrightarrow[\text{2. H}^+]{\text{1. NaH, toluene}}$

(b) Acetophenone + ethyl formate $\xrightarrow[\text{2. H}^+]{\text{1. C}_2\text{H}_5\text{ONa, ethanol}}$

10-15. Suggest a synthesis for each of the following compounds using the Claisen condensation. Any necessary organic or inorganic reagents may be used.

10-16. Draw structures for the two possible mono enols of 1,3-cyclohexanedione. Explain which one is more stable.

BIBLIOGRAPHY

A review of the Claisen condensation is given in:

Davis, B. R.; Garratt, P. *J. Comp. Org. Syn.* **1991**, *2*, 795.
Garst, J. F. *J. Chem. Educ.* **1979**, *56*, 721.
Hauser, C. R.; Swamer, F. W.; Adams, J. T. *Org. React.* **1954**, *8*, 59.

See your organic textbook for an introduction to this reaction and its scope in synthesis. For example:

Smith, M. B. *March's Advanced Organic Chemistry*, 8th ed.; Wiley: New York, 2019.
Solomons, T. W. G.; Fryhle, C. B.; Snyder, S. A. *Organic Chemistry*, 12th ed.; Wiley: New York, 2016

Selected references of Claisen condensations from *Organic Syntheses* include:

Ainsworth, C. *Organic Syntheses*; Wiley: New York, 1963; Collect. Vol. IV, p. 536.
Floyd, D. E.; Miller. S. E. *Organic Syntheses*; Wiley: New York, 1963; Collect. Vol. IV, p. 141.

John, J. P.; Swaminathan, S.; Venkataraman, P. S. *Organic Syntheses*; Wiley: New York, 1973; Collect. Vol. V, p. 747.
Magnani, A.; McElvain, S. M. *Organic Syntheses*; Wiley: New York, 1955; Collect. Vol. III, p. 251.
Riegel, E. R.; Zwilgmeyer, F. *Organic Syntheses*; Wiley: New York, 1943; Collect. Vol. II, p. 126.
Snyder, H. R.; Brooks, L. A.; Shapiro, S. H. *Organic Syntheses*; Wiley: New York, 1943; Collect. Vol. II, p. 531.

For the synthesis of coumarin, see:

Sethna, S.; Phadke, R. *Org. React.* **1953**, *7*, 1.

It may be of interest in a broad aspect to review:

Corey, E. J.; Czakó, B; Kürti, L. *Molecules and Medicine*; Wiley: New York, 2007.

EXPERIMENT [4$_{adv}$]

Grignard and Aryl Halide Cross-Coupling Reaction: 1-Methyl-2-(methyl-d_3)-benzene

Common names: 1-methyl-2-(methyl-d_3)-benzene, α,α,α-d_3-o-xylene
CA number: [25319-53-3]
CA name as indexed: benzene, 1-methyl-2-(methyl-d_3)-

Purpose.   This experiment illustrates a selective cross-coupling reaction between an alkyl Grignard reagent[2] and an aryl halide in the presence of a nickel phosphine catalyst to form a C—C bond. The reaction is also used to demonstrate the technique of "labelling" specific hydrogen atoms with an isotope for identification purposes.

Prior Reading

    Technique 4: Solvent Extraction

                Liquid–Liquid Extraction (p. 76)

                Drying of the Wet Organic Layer (pp. 84–87)

    Technique 6: Chromatography

                Column Chromatography (pp. 96–97)

                Concentration of Solutions (pp. 105–108)

REACTIONS

D_3CMgI +   *o*-Chlorotoluene $\xrightarrow[\text{Ni[dppp]Cl}_2]{\text{ether}}$ 1-Methyl-2-(methyl-d_3)-benzene   +   MgClI   Magnesium chloroiodide

Alternatively, the use of methyl iodide in the formation of the Grignard reagent produces *o*-xylene in step 2.

DISCUSSION

Before the discovery of the nickel-phosphine catalyst, the cross-coupling of Grignard reagents with organic halides was seldom used in synthetic practice. This fact was mainly due to the formation of homocoupling products along with significant elimination side reactions. With the use of this new catalyst, the reaction now has wide application in the synthesis of unsymmetrical alkanes and alkenes.

In this experiment we will synthesize an isotopically labeled *o*-xylene. The *o*-xylene product will be labeled with a trideuteromethyl group. Deuterium, one of the hydrogen **isotopes**, can be used as a **label** in organic compounds. Its incorporation into a molecule can be detected by IR, NMR, or MS. Deuterium labeling is a particularly powerful way of investigating infrared spectra because the resultant frequency shifts (which are inversely proportional to the square root of the mass ratio; see Infrared Discussions, Chapter 8) are the largest obtainable with stable isotopes. Isotopic labeling is often used to follow hydrogen–deuterium exchange reactions, such as enolizations, to study biological reactions, and for gaining insight into organic reaction mechanisms.

---

[2]For general references on Grignard reagents, refer to Experiment [16].

The nickel catalyst used in this experiment is [1,3-bis(diphenylphosphino) propane]nickel(II) chloride, mercifully abbreviated as Ni[dppp]Cl$_2$.

[1,3-Bis(diphenylphosphino)propane]nickel(II) chloride
(Ni[dppp]Cl$_2$)

It has been proposed that the catalytic ability of the dihalodiphosphinen-ickel reagent stems from its ability to react with a Grignard reagent to form a diorgano-nickel complex (A).[3] This complex is then converted to the haloor-ganonickel complex B by reacting it with an *organic halide:*

$$L_2NiX_2 \; + \; 2\,RMgX' \; \rightarrow \; L_2Ni{\overset{R}{\underset{R}{\diagdown}}} \; + \; 2\,MgXX'$$

A

$$L_2Ni{\overset{R}{\underset{R}{\diagdown}}} \; + \; R'\!-\!X'' \; \rightarrow \; L_2Ni{\overset{X''}{\underset{R'}{\diagdown}}} \; + \; R\!-\!R$$

A   **B**

(where L$_2$ = organodiphosphine ligand;
R = alkyl or aryl; R'aryl or vinyl;
X, X' and X'' = halo

Reaction of complex B with the Grignard reagent forms a new diorgano-nickel complex C, from which the cross-coupling product is formed by attack of the organic halide. In the reaction, complex B is regenerated and thus the catalytic cycle is completed:

$$L_2Ni{\overset{X''}{\underset{R'}{\diagdown}}} \; + \; RMgX' \; \rightleftharpoons \; L_2Ni{\overset{R}{\underset{R'}{\diagdown}}} \; + \; MgX'X''$$

B   **C**

$$L_2Ni{\overset{R}{\underset{R'}{\diagdown}}} \; + \; R'\!-\!X'' \; \rightleftharpoons \; L_2Ni{\overset{X''}{\underset{R'}{\diagdown}}} \; + \; R\!-\!R'$$

C   **B**   Cross-coupling
                                                     product

---

[3]Tamao, K.; Sumitani, K.; Kumada, M. *J. Am. Chem. Soc.* **1972**, *94*, 4374.

EXPERIMENTAL PROCEDURE

Estimated time to complete the experiment: 5.0 h.

Physical Properties of Reactants and Product							
Compound	MW	Amount	mmol	mp (°C)	bp (°C)	d	n_D
Magnesium	24.31	65 mg	2.7	649			
Diethyl ether	74.12	700 μL			34.5	0.71	1.3526
Methyl-d_3-iodide	144.96	250 μL	3.93		42	2.28	1.5262
Ni[dppp]Cl$_2$	542.1	10 mg	0.02				
o-Chlorotoluene	126.59	150 μL	1.28		159	1.08	1.5268
1-Methyl-2-(methyl-d_3)-benzene	109.17				144		1.5055

NOTE. All the glassware used in the experiment should be cleaned, dried in an oven at 110 °C for at least 30 min, and then cooled in a desiccator before use.

Reagents and Equipment.   Equip a 5.0-mL conical vial containing a magnetic spin vane with a reflux condenser protected by a calcium chloride drying tube.

Step 1.   Scrape a 4- to 5-in. piece of magnesium ribbon to remove the oxide coating. Then cut it into 1-mm-long sections. *Using forceps*, weigh and add 65 mg (2.7 mmol) of the magnesium sections to the conical vial. In the **hood**, use an automatic delivery pipet to dispense 300 μL of anhydrous diethyl ether into the conical vial containing the magnesium. Reassemble the apparatus.

Using automatic delivery pipets, prepare a solution of 250 μL of methyl-d_3 iodide in 200 μL of anhydrous diethyl ether in a capped vial.

CAUTION: Methyl-d_3 iodide is a suspected carcinogen.

Draw this solution into a Pasteur filter pipet. Remove the drying tube and insert the pipet down the length of the condenser, allowing the pipet bulb to rest on the condenser lip.

Reaction Conditions.   Place the reaction vial in an ice bath and add the methyl-d_3 iodide dropwise with stirring. Withdraw the pipet, reinstall the drying tube, and stir the mixture for 20 min (➡).

Step 2.   In this step, all liquids must be dispensed in the **hood** using automatic delivery pipets. Weigh and add to a capped 0.5-dram vial, 10 mg (0.02 mmol) of Ni[dppp]Cl$_2$.

CAUTION: Ni[dppp]Cl$_2$ is a suspected carcinogen.

Now add 150 μL of o-chlorotoluene and 200 μL of anhydrous diethyl ether. Draw this solution into a Pasteur filter pipet and, in one portion, add it to the reaction vial through the condenser as described above. Reinsert the drying tube. Place this mixture in a preheated sand bath at a temperature of 55 °C. Heat with stirring for 2 h (➡).

HOOD

Step I:
Mg, 65 mg +
(CH$_3$CH$_2$)$_2$O, 0.5 mL +
CD$_3$I, 250 μL

55 °C
thermometer

HOOD

Speed   Heat

Step II:
Ni[dppp]Cl$_2$, 10 mg
+ o-CH$_3$C$_6$H$_4$Cl, 150 μL
+ (CH$_3$CH$_2$)$_2$O, 200 μL

On addition of the catalyst, the solution turns green. After approximately 30 min of heating, the mixture becomes dark brown.

Na₂SO₄, 1.0 g

SiO₂, 200 mg

Sand, 50 mg

Product +
(CH₃CH₂)₂O,
~2.0 mL

HOOD

Isolation of Product. Cool the reaction mixture in an ice bath and quench the reaction by the dropwise addition (calibrated Pasteur pipet) of 1.5 mL of 1.0 M HCl solution. *The acid should be added slowly, because frothing occurs.*

NOTE. The following extracting solutions are measured using calibrated Pasteur pipets. For each extraction operation, the vial is capped, shaken, and vented, and the layers are allowed to separate. The aqueous (lower) phase is then removed.

Add 1.0 mL of ether to the solution. Using a Pasteur filter pipet, remove the aqueous layer. Extract the remaining ether phase with 1.0 mL of water followed by 1.0 mL of saturated sodium bicarbonate solution and then 1.0 mL of deionized water. Then extract with two 1.0-mL portions of 1.0 N sodium thiosulfate (Na₂SO₃) solution and then with 1.0 mL of water.

Purification and Characterization. Isolate and purify the reaction product using column chromatography. In a modified Pasteur filter pipet, place 200 mg of activated silica gel (100 mesh) followed by 1.0 g of anhydrous sodium sulfate. Wet the column with 0.5 mL of ether using a calibrated Pasteur pipet (◄).

Transfer the ether solution of product to the column by Pasteur pipet, and elute the material from the column with two 0.5-mL portions of ether. Collect the eluate in a tared 10-mL Erlenmeyer flask and then transfer the eluate to a 5-mL conical vial for concentration. Evaporate the solvent by warming the vial in a sand bath in the **hood** to yield the"labeled" *o*-xylene.

Weigh the product and calculate the percent yield. Determine the boiling point, density, and refractive index (optional) and compare your values with those listed for *o*-xylene in the literature. Obtain an IR spectrum of the crude (dry) reaction product using the capillary film sampling technique and analyze it following the analysis given below.

Infrared Analysis. The spectrum of *o*-chlorotoluene (Fig. 10.3) nicely mimics the large macro group frequency train for an ortho-substituted dialkylbenzene system, which has the following frequency train: 3100–3000, 3000–2850, 1950, 1920, 1880, 1840, 1790, 1690, 1600, 1580, 1500, 1450, 1380, and 750 cm⁻¹.

a. **3072 and 3026 cm⁻¹:** C—H stretch on aromatic ring.

b. **3000–2850 cm⁻¹:** C—H stretch on alkyl substituents.

c. **1955, 1915, 1880, 1835, 1795, and 1690 cm⁻¹:** Combination band pattern for ortho-disubstituted rings. The fit in this case is extremely good.

d. **1596 and 1577, 1475 and 1450 cm⁻¹:** Two degenerate pairs of ring stretching modes. The 1450-wavenumber band is overlapped by the antisymmetric methyl deformation vibrations.

e. **1382 cm⁻¹:** Symmetric methyl deformation mode.

f. **749 cm⁻¹:** All in-phase out-of-plane bend of four adjacent C—H groups on the aromatic ring.

The C—Cl stretch falls below the region of measurement in these aromatic systems.

The IR spectrum of 1-methyl-2-(methyl-*d₃*)-benzene is shown in Figure 10.4. The reaction has replaced the —Cl group with a —CD₃ group. Thus, the only spectral changes expected to be observed will involve the vibrations

Sample	2-Chlorotoluene		
%T _X_ ABS __ Background Scans _4_		Scans _____16_____	
Acquisition & Calculation Time _42 sec_		Resolution _4.0 cm⁻¹_	
Sample Condition _liquid, neat_		Cell Window ___KBr___	
Cell Path Length _capillary film_		Matrix Material _____	

Figure 10.3   IR spectrum: 2-chlorotoluene.

of the labeled group. To a first approximation, the frequency should shift by a factor of $1/\sqrt{2} = 0.707$, but in practice the full shift is not observed because most vibrations are not pure modes. A factor of 0.75–0.72 is generally observed. In 1-methyl-2-(methyl-d_3)-benzene, we have two sets of doublets, with the doublet centers at 2220 and 2090 cm⁻¹. Because we would expect the C—D stretching modes to give rise to only two band systems, coupling with overtones of lower lying fundamental vibrations is likely occurring. The corresponding C—H stretching modes also evidence some coupling. If we use the major band centers observed at 2965 and 2870 cm⁻¹, however, as the C—H stretching values, the observed ratios are 2220/2965 = 0.749 and 2090/2870 = 0.728, results that are quite reasonable to expect for this type of isotopic shift. The bending modes are moved into the cluttered fingerprint region and do not easily lend themselves to analysis.

Examine the spectrum of the reaction product you have obtained as a capillary film. Discuss the similarities and differences of the experimentally derived spectral data to the reference spectra (Figs. 10.3 and 10.4).

ABSCISSA	ORDINATE	SCAN TIME 12 min	REP. SCAN_____ SINGLE BEAM_____
EXPANSION_____	EXPANSION_____	RESPONSE_____	TIME DRIVE_____ PRE SAMPLE CHOP_____
SUPPRESSION_____	% T_____ ABS_____	SLIT PROGRAM normal	OPERATOR_____ DATE_____
SAMPLE 1-methyl-2-	REMARKS Perkin-Elmer 283 B	SOLVENT neat	CELL PATH_____
[methyl-d₃]-benzene	_____		
ORIGIN_____	_____	CONCENTRATION thin film	REFERENCE_____

Figure 10.4   IR spectrum: 1-methyl-2-(methyl-d_3)-benzene.

QUESTIONS

10-17. Predict the reagents that could be used to prepare each of the compounds by the Grignard coupling reaction. Give a suitable name to each reactant.

10-18. The Wurtz coupling reaction involves the treatment of haloalkanes with an active metal, such as sodium.
   (a) Butyl bromide + Na
   (b) 1-Bromo-3-chorocyclobutane + Na
   What products would be formed in the above reactions? Name the products.
10-19. In reference to Question 10-18, explain why the coupling reaction of n-butyl bromide and n-propyl bromide under conditions of the Wurtz reaction is synthetically inefficient for the preparation of heptane.
10-20. Describe what you would expect to see in the ^1H and ^{13}C NMR spectra of 1-methyl-2-(methyl-d_3)-benzene.

10-21. The spectrum of *o*-chlorotoluene given in Figure 10.3 has two additional peaks that are not present in the library reference standard spectrum. Both these bands are weak. One occurs near 1714 cm⁻¹ and the other at 1362 cm⁻¹. Can you offer an explanation for the presence of the extra absorption bands? How successful do you think the coupling reaction would be if carried out using this sample as the starting material?

10-22. In the spectrum of *o*-chlorotoluene, the lower wavenumber band of the 1596, 1577 pair is the more intense member. In the labeled product, the 1610, 1580 pair has the higher wavenumber peak more intense. Explain this reversal of intensities.

BIBLIOGRAPHY

The present coupling reaction is based on the work reported by:

Kumada, M.; Tamao, K.; Sumitani, K. *Organic Syntheses*; Wiley: New York, 1988; Collect. Vol. VI, p. 407, and references cited therein.

Selected Grignard coupling reactions presented in *Organic Syntheses*:

Gilman, H.; Catlin, W. E. *Organic Syntheses*; Wiley: New York, 1941; Collect. Vol. I, p. 471.

Gilman, H.; Robinson, J. *Organic Syntheses*; Wiley: New York, 1943; Collect. Vol. II, p. 47.

Lespieau, R.; Bourguel, M. *Organic Syntheses*; Wiley: New York, 1941; Collect. Vol. I, p. 186.

Smith, L. I. *Organic Syntheses*; Wiley: New York, 1943; Collect. Vol. II, p. 360.

Turk, A.; Chanan, H. *Organic Syntheses*; Wiley: New York, 1955; Collect. Vol. III, p. 121.

The reaction was first described by:

Mayo, D. W.; Bellamy, L. J.; Merklin, G.T.; Hannah, R. W. *Spectrochim. Acta* **1985**, *41A*, 355.

Oxidative Coupling of 2-Naphthol: 1,1′-Bi-2-Naphthol

EXPERIMENT
[5ₐdᵥ]

Common name: 1,1′-bi-2-naphthol

CA number: [602-0-5]

CA name as indexed: [1,1′-binaphthalene]-2,2′-diol

Purpose. The coupling reaction that aromatic phenols undergo in the presence of transition metal oxidants is investigated. This reaction mimics the biogenetic process that occurs in nature.

Prior Reading

Technique 5: Crystallization

Use of the Hirsch Funnel (pp. 92–93)

Craig Tube Crystallizations (pp. 93–95)

REACTION

2-Naphthol   1,1′-Bi-2-naphthol

DISCUSSION

B This reaction illustrates oxidative coupling of phenols, which is an important biogenetic pathway in nature, leading to the formation of many different natural products.

The coupling reaction involves oxidation of 2-naphthol by electron transfer to give an aryloxy radical, which then dimerizes to yield the product. The mechanism is shown here:

This binaphthol compound, by virtue of restricted rotation about the single bond joining the two naphthalene groups, is a chiral compound. That is, the molecule cannot readily exist in a planar form because of the steric interference of the bulky —OH substituents, and, in fact, the two enantiomers have been separated. Such enantiomeric compounds contain no stereocenter, but rather have an axis of chirality, or a *chiral axis*, which in this case contains the bond joining the two naphthalene rings (see Fig. 10.5)[4]:

---

[4]Donald J. Cram (Nobel Laureate, 1987) and coworkers have incorporated this binaphthyl group into cyclic crown ethers.

Figure 10.5   Enantiomers of 1,1'-bi-2-naphthol.

An investigation of the complexation of this class of molecules with various ionic species may lead to important insights into the catalytic nature of enzymes. For example, the crown ether similar to that above containing *two* 2,2'-substituted 1,1'-binaphthyl groups as chiral barriers complexes preferentially with one enantiomer of some primary amine salts:

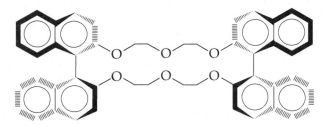

Complexation of enantiomerically pure binaphthols with metal hydride reducing agents allow enantiospecific reductions of some carbonyl compounds.

EXPERIMENTAL PROCEDURE

Estimated time to complete the experiment: 2.0. h.

Physical Properties of Reactants				
Compound	MW	Amount	mmol	mp (°C)
2-Naphthol	144.19	100 mg	0.69	123–124
Sodium hydroxide	40.00	30 mg	0.75	318.4
Ferric chloride · 6H$_2$O	270.30	297 mg	1.1	37
HCl (concd)	36.46	100 μL		
Water	18.06	3 mL		

Reagents and Equipment.   Weigh and add 100 mg (0.69 mmol) of 2-naphthol and 30 mg (0.75 mmol) of sodium hydroxide to a 10-mL round-bottom flask containing a stirring bar.

CAUTION: Sodium hydroxide is a corrosive and toxic chemical. Do not allow it to touch your skin or to come in contact with your eyes.

Add 3.0 mL of water. Attach the flask to a reflux condenser and place the assembly in a sand bath on a magnetic stirring hot plate (◂).

Heat the reaction mixture to reflux, with stirring, using a sand bath temperature of 120 °C.

In a 10-mL Erlenmeyer flask prepare a solution of 180 mg (1.1 mmol) of *anhydrous* ferric chloride (MW 162) *or* 297 mg (1.1 mmol) of ferric chloride *hexahydrate* (MW 270), 1.0 mL of water (calibrated Pasteur pipet), and 100 μL of concentrated hydrochloric acid.

~120 °C thermometer

2-Naphthol, 100 mg
+ NaOH, 30 mg
+ H₂O, 4.2 mL
+ FeCl₃· 6H₂O, 297 mg
+ concd HCl, 100 μL

CAUTION: Be careful when mixing acid with water. **Add the acid to the water.** Avoid contact with the skin. Dispense the acid with an automatic delivery pipet.

Using a Pasteur pipet, transfer the ferric chloride solution, through the top of the condenser, to the reaction flask. Rinse the Erlenmeyer flask with 200 μL of water, and add this rinse to the reaction flask as before.

Reaction Conditions.   Heat the resulting mixture at reflux, with stirring, for 45–60 min using a sand bath temperature of 120 °C. Allow the mixture to cool to room temperature, and then place the flask in an ice bath to complete crystallization of the product.

Isolation of Product.   Collect the solid product by vacuum filtration using a Hirsch funnel, and wash the filter cake with two 1.0-mL portions of cold water (calibrated Pasteur pipet) (◂).

Coupled product collected here

Purification and Characterization.   Recrystallize the crude product from 95% ethanol using a Craig tube, and dry the resulting crystals on a porous clay plate or on filter paper.

Weigh the crystals and calculate the percent yield. Determine the melting point and compare it with the literature value. Obtain the IR spectrum and compare it to that shown in the literature (*The Aldrich Library of IR Spectra* and/ or *SciFinder Scholar*).

H₂O, ~6 mL and other reaction byproducts

Chemical Tests.   Chemical classification tests (Chapter 9) may be used to assist in characterization of this material. Perform the ignition test and the ferric chloride test.

QUESTIONS

10-23. In the oxidative coupling reaction of 1-naphthol, it is *possible* to obtain three products:

Account for the formation of this mixture by suggesting a mechanistic sequence similar to that presented in the discussion section of this experiment.

10-24. Predict the structure of the diphenylquinone product, $C_{28}H_{40}O_2$, formed by the oxidative coupling of 2,6-di-*tert*-butylphenol with oxygen in the presence of base.

10-25. Substituted phenols, such as BHT, are used as antioxidants in processed foods:

$$\overset{:\ddot{O}H}{\underset{(CH_3)_3C \quad\quad C(CH_3)_3}{\bigcirc}}$$

CH₃

BHT

The role of the antioxidant is to stop spoilage caused by the free radical reactions brought about by reaction of oxygen with compounds containing $C{=}C$ bonds.

(a) Give a suitable chemical name for BHT.

(b) Can you suggest why this compound is an effective antioxidant?

10-26. Naphthalene is the simplest of the polycyclic aromatic hydrocarbons and can be represented by three resonance structures. Draw them. Indicate which of the structures are equivalent.

10-27. Some relatively simple chiral compounds contain a chiral axis. Use molecular models to convince yourself that the two molecules below are enantiomers, even though they contain no stereocenters:

BIBLIOGRAPHY

For references related to the oxidative coupling of phenolic derivatives, see:

Dewar, M. J. S.; Nakaya, T. *J. Am. Chem. Soc.* **1968**, *90*, 7134.
Scott, A. I. *Q. Rev.* **1965**, *19*, 1.

For an overview of Cram's work on the binaphthyl crown ethers, see:

Cram, D. J.; Cram, J. M. *Science* **1974**, *183*, 803.

For the resolution of 1,1′-bi-2-naphtol, see:

Dongwei, C; Hughes, D. L; Verhoeven, T. R.; Reider, P. J. *Organic Synthesis*, Collect. Vol. X, p. 93. Also see Kazlauskas, P. J. *Organic Synthesis*, Collect. Vol. IX, p. 77.

For the use of 1,1′-bi-2-naphthol as a chiral element for asymmetric catalysts, see:

Chem. Eng. News: **1995**, *9* Oct, p. 74.
Noyori, R.; Takaya, H. *Acc. Chem. Res.* **1990**, *23*, 345.

Beckmann Rearrangement: Benzanilide

EXPERIMENT [6$_{adv}$]

Common name: benzanilide

CA number: [93-98-1]

CA name as indexed: benzamide, *N*-phenyl-

Purpose. You will carry out the *Beckmann rearrangement* in which a ketone is converted, via an oxime, to an amide. The reaction is an example of a group of reactions in which migration to an electron-deficient nitrogen occurs.

Prior Reading

Technique 4: Solvent Extraction

Liquid–Liquid Extraction (p. 76)

Technique 5: Crystallization

Craig Tube Crystallizations (pp. 93–95)

REACTION

Benzophenone

Hydroxylamine
hydrochloride

Benzophenone
oxime

Benzanilide

DISCUSSION

The Beckmann rearrangement was discovered in 1886 by E. Beckmann and is the reaction of an oxime of a ketone, in the presence of acid, to yield the corresponding amide or amides. It is a very general reaction and a wide variety of reagents have been used to cause the rearrangement to take place. These include sulfuric, hydrochloric, and polyphosphoric acids, phosphorus pentachloride, and aromatic sulfonyl chlorides. The Beckmann rearrangement is the nitrogen analogue, both functionally and mechanistically, of the Baeyer–Villiger oxidation, which converts a ketone to an ester.

In the present case, only one amide is formed, because benzophenone is a symmetrical ketone. Because the oximes are prepared from ketones, the reaction was often used, before the advent of modern spectroscopic techniques, to determine the structure of the starting ketone. This structure determination was accomplished by the subsequent identification of the acid and amine obtained upon hydrolysis of the amide product of the rearrangement.

As depicted in the following mechanism, the acid catalyst converts the oxime's hydroxyl group to a good leaving group (water, a small neutral molecule). The acid used in this experiment is a very strong acid, trifluoromethanesulfonic acid, usually called *triflic acid*. The first part of the overall reaction in this experiment constitutes the formation of the oxime, which is followed by the actual migration of an aryl (or alkyl) group from the former carbonyl carbon to the nitrogen atom, the Beckmann rearrangement:

There are two significant points concerning the stereo- and regiochemistry of the reaction: (1) the group that migrates is the one *anti* to the hydroxyl group on the C=N bond and (2), the stereochemistry, if any, of the migrating group is retained.

EXPERIMENTAL PROCEDURE

Estimated time to complete the experiment: 2.5 h.

Physical Properties of Reactants				
Compound	MW	Amount	mmol	mp (°C)
Benzophenone	182.21	100 mg	0.55	48.1
Hydroxylamine hydrochloride	69.49	51 mg	0.73	155–157
Triflic acid–formic acid solution		500 μL		

Reagents and Equipment. Weigh and place 100 mg (0.55 mmol) of benzophenone and 51 mg (0.73 mmol) of hydroxylamine hydrochloride in a 5.0-mL conical vial containing a magnetic spin vane. Using an automatic delivery pipet in the **hood**, now add 500 μL of triflic acid–formic acid solution to this mixture. Attach the vial to a reflux condenser and mount the assembly in a sand bath on a magnetic stirring hot plate (➡).

125 °C
thermometer

HOOD

Speed   Heat

$(C_6H_5)_2CO$, 100 mg
+ HCl•H_2NOH, 51 mg
+ CF_3SO_3H–HCO_2H, 0.5 mL

CAUTION: Triflic acid (trifluoromethanesulfonic acid) is one of the strongest acids known. It is very corrosive and toxic!

INSTRUCTOR PREPARATION. The acid solution is prepared by adding 2 drops of triflic acid to 5.0 mL of 90% formic acid.

Reaction Conditions. With stirring, heat the reaction mixture at reflux for 1 h in a sand bath at 125 °C. Then cool the resulting solution to room temperature.

Isolation of Product. To the cooled reaction solution add 1.0 mL of water (using a calibrated Pasteur pipet), and extract the resulting mixture with three 1.0-mL portions of methylene chloride. Separate the organic phase using a Pasteur filter pipet and transfer it to a 10-mL Erlenmeyer flask. For each extraction, after addition of the methylene chloride, cap the vial, shake, vent, and then allow the layers to separate.

Dry the combined methylene chloride extracts over granular, anhydrous sodium sulfate. Using a Pasteur filter pipet, transfer the dried solution to a clean, dry 10-mL Erlenmeyer flask containing a boiling stone. Evaporate the
HOOD solvent in the **hood** under a gentle stream of nitrogen on a warm sand bath to yield the crude reaction product.

Characterization and Purification. Transfer the crude benzanilide from the Erlenmeyer flask to a Craig tube, and recrystallize the material from 95% ethanol.

Weigh the product and calculate the percent yield. Determine the melting point and compare your result to that listed in the *CRC Handbook* or found using SciFinder Scholar. Obtain an infrared spectrum.

Infrared Analysis. In this experiment, an aromatic ketone has been rearranged to a secondary amide. By examining the infrared spectra of starting material and product, we can confirm this molecular transformation.

The spectrum of benzophenone (Fig. 10.6) possesses two macro group frequency trains:

1. Conjugated aromatic ketone: This ketone is defined by the bands at 3325 (overtone of ketone carbonyl stretch), 1663 (C$=\!=$O stretch, doubly conjugated), 1601 and 1580 (ν_{8a} and ν_{8b} degenerate ring stretch), 1492 and 1450 cm^{-1} (ν_{19a} and ν_{19b} degenerate ring stretch). The intensification of the 1580-wavenumber peak confirms the conjugation of the carbonyl to the ring. The 1500-wavenumber ring stretch, which is generally a bit variable in intensity, is quite weak in this case (2 different benzene rings).

2. Monosubstituted phenyl group: This group is defined by weak bands located at 1969, 1913, 1823, 1724 cm^{-1}, and strong bands recorded at 701 and 640 cm^{-1}. For discussions of these modes see Chapter 8, and Experiment [20B]. In the case of phenyl rings conjugated to carbonyl groups, the 750- and 690-wavenumber bands often appear on the low side of their usual range and can be down as much as 40–50 cm^{-1}, as occurs here.

The rearrangement product has incorporated a heteroatom into its structure, but the carbonyl group and the ring systems have survived. The infrared spectrum of benzanilide (Fig. 10.7) must be consistent with the formation of a secondary amide. We expect to observe a macro group frequency for the presence of monosubstituted phenyl groups, plus a second macro group frequency for a secondary aromatic amide.

Sample	Benzophenone		
%T X ABS — Background Scans 4		Scans 16	
Acquisition & Calculation Time 42 sec		Resolution 4.0 cm⁻¹	
Sample Condition solid – melt		Cell Window KBr	
Cell Path Length capillary film		Matrix Material	

Figure 10.6 IR spectrum: benzophenone.

The first macro group frequency is similar to that of the starting material, with bands occurring at

a. **1950, 1912, 1840, and 1725 cm⁻¹:** All four combination bands are doubled in this case. The values given are for the centers of the doublets.

b. **752 and 718 cm⁻¹:** C—H out-of-plane bend. The lower value likely corresponds to the ring conjugated directly with the carbonyl group.

c. **693 cm⁻¹:** Ring out-of-plane bend (puckering) vibration.

The second macro group frequency for the aromatic secondary amide uses the following bands: 3350, 3060, 1659, 1602, 1587, 1539, 1322, and 680 (broad) cm⁻¹.

a. **3350 cm⁻¹:** This strong band corresponds to the highly hydrogen-bonded amide N—H stretch.

b. **3060 cm⁻¹:** C—H stretch on sp² carbon, aromatic C—H.

Sample ___Benzanilide_____

%T _X_ ABS __ Background Scans _4_____   Scans _____16_____

Acquisition & Calculation Time _42 sec____   Resolution _4.0 cm⁻¹____

Sample Condition _____solid_____   Cell Window _____

Cell Path Length _____   Matrix Material _KBr____

Figure 10.7   IR spectrum: benzanilide.

c. **1659 cm⁻¹:** C═O stretch of a conjugated and hydrogen-bonded amide. It is the most intense band in the spectrum.

d. **1602 and 1587 cm⁻¹:** Ring stretch degenerate pair, ν_{8a} and ν_{8b}.

e. **1539 and 1322 cm⁻¹:** These two bands involve both N—H bending (in-plane) and C—N stretch. The two fundamentals fall somewhere between 1450–1400 cm⁻¹. Thus, they can mechanically interact and split apart, with one component falling at 1539 cm⁻¹ and the other near 1322 cm⁻¹. Since the fundamental frequencies will vary somewhat from molecule to molecule, the interaction term that is sensitive to the frequency match (see Infrared Discussions, Chapter 8) will also vary in magnitude, and thus the wavenumber separation will be affected.

f. **680 cm⁻¹:** Broad and weak. This ill-defined band arises from the out-of-plane bend of the N—H group. It is similar to the O—H bend found in alcohols in this spectral region (see Infrared discussions in Experiments [5A], [5B], and [8B]).

Examine the IR spectrum of your reaction product. Discuss the similarities and differences of the experimentally derived spectral data to the reference spectra (Figs. 10.6 and 10.7).

Chemical Tests. Chemical classification tests (Chapter 9), such as the ignition test and the soda lime test (or sodium fusion test), may also be conducted to further establish the identity of the product. The hydroxamate test may be used to establish the presence of the amide functional group.

QUESTIONS

10-28. Draw the structure of the product expected in each of the Beckmann rearrangements presented below. Give a suitable name for each product.

(a)

(b)

(c) $CH_3—C—CH—(CH_2)_3CH_3$ $\xrightarrow{H_2SO_4}$

10-29. The picryl iminoethers (an iminoether is C=N—O—C) of oximes undergo the Beckmann rearrangement without an acid catalyst:

(picryl ester)

Explain why a catalyst is not required in this rearrangement.

10-30. Compounds A and B undergo the Beckmann rearrangement upon gentle heating of the solid compounds:

A

B

(a) Write the structure of the products expected.
(b) If a mixture of the two iminoethers, A and B, are heated in the same reaction flask, what products would be formed? Explain.

10-31. When acetophenone oxime was allowed to rearrange in ^{18}O-enriched solvent, the amide product contained the same percentage of ^{18}O as the solvent. Explain this observation.

10-32. Oximes are usually crystalline materials and have been prepared as a means of identifying liquid ketones or aldehydes. It has been found in the preparation of these derivatives that if the acid concentration is too high (low pH), then the oxime does not form. Explain.

10-33. The mechanical coupling between the N—H bending vibration and the C—N stretching vibration depends on a close frequency match. Normally the C—C, C—N, and C—O stretching vibrations are found in the 1200- to 800-wavenumber region. Why, in the case of the amides, do we find the C—N stretch approaching 1400 cm^{-1}?

BIBLIOGRAPHY

Reviews on the Beckmann rearrangement:

Donaruma, L. G.; Heldt, W. Z. *Org. React.* **1960**, *11*, 1.

Gawley, R. E. *Org. React.* **1988**, *35*, 1.

Jochims, J. C.; Hehl, S.; Herzberger, S. *Synthesis* **1990**, 1128, and references therein.

McCarthy, C. G. in *The Chemistry of the Carbon-Nitrogen Double Bond*; Patai, S., Ed.; Interscience: New York, 1970; p. 408.

Yamabe, S.; Tauchida, N.; Yamazaki, H. *J. Org. Chem.* **2005**, *70*, 10638.

An example of the Beckmann rearrangement using different reagents is given in:

Deluca, L.; Giacomelli. G.; Porcheddu, A. *J. Org. Chem.* **2002**, *67*, 6072.

Furuya, Y.; Ishihara, K.; Yanamoto, H. *J. Am. Chem. Soc.* **2005**, *127*, 11240.

Harada, T.; Ohno, T.; Kobayashi, S.; Mukaiyama, T. *Synthesis* **1991**, 1216.

Ohno, M.; Naruse, N.; Terasawa, I. *Organic Syntheses*; Wiley: New York, 1973; Collect. Vol. V, p. 266.

The present reaction was adapted from the work of:

Ganboa, I.; Palomo, C. *Synth. Commun.* **1983**, *13*, 941.

EXPERIMENT [7 adv]

Preparation of an Enol Acetate: Cholesta-3,5-dien-3-ol Acetate

Common name: cholesta-3,5-dien-3-ol acetate

CA number: [2309-32-2]

CA name as indexed: cholesta-3,5-dien-3-ol, acetate

Purpose. This experiment investigates the conditions under which the enol acetate of an α,β-unsaturated ketone is prepared. The combined use of acetic anhydride with chlorotrimethylsilane is an illustration of a technique for the generation of *acylium ions*, a reactive electrophile for acylation reactions.

Prior Reading

> *Technique 5:* Crystallization
> > Craig Tube Crystallizations (pp. 93–95)
> *Technique 6:* Chromatography
> > Column Chromatography (pp. 96–99)
> > Thin-Layer Chromatography (pp. 101–103)
> > Removal of Solvent Under Reduced Pressure (pp. 106–108)

REACTION

Cholesta-4-en-3-one   Cholesta-3,5-dien-3-ol acetate

DISCUSSION

Acylium ions are useful reactive intermediates in organic synthesis because their high reactivity allows reactions with relatively weak nucleophiles, in this case the carbonyl oxygen of an α,β-unsaturated enone. The product in this experiment, a dienol acetate of a cholesterol derivative, is itself a useful intermediate in the synthesis of steroids such as cortisone.

Acetic anhydride and chlorotrimethylsilane react, as shown here, to generate a small equilibrium concentration of acylium ion ($CH_3C{\equiv}O^+$) along with trimethylsilyl acetate, $CH_3CO_2Si(CH_3)_3$. The acylium ion is a much more reactive electrophilic acyl group than neutral sources of an electrophilic acyl group, such as acetic anhydride or acetyl chloride, CH_3COCl.

The acylium ion thus generated is a very reactive electrophile, capable of reacting with the relatively weakly nucleophilic lone pairs on the carbonyl oxygen of the cholesta-4-en-3-one. The resulting oxonium ion renders the γ-protons of the enone quite acidic, and thus they are readily removed by a very weak base, such as chloride ion, to generate the product, cholesta-3,5-dien-3-ol acetate. The proposed mechanism for this transformation is shown here:

Acylium ion   Cholesta-4-ene-3-one

Cholesta-3,5-dien-3-ol acetate

EXPERIMENTAL PROCEDURE

Estimated time to complete the experiment: 4.0 h.

Physical Properties of Reactants and Product							
Compound	MW	Amount	mmol	mp (°C)	bp (°C)	d	n_D
Cholesta-4-en-3-one	384.65	96 mg	0.25	81–82			
Acetic anhydride	102.09	1.0 mL	10.6		140	1.08	1.3901
Chlorotrimethylsilane	108.64	200 μL	1.58		57	0.86	
Cholesta-3,5-dien-3-ol acetate	425.68			79–80			

145–150 °C
Thermometer

HOOD

Speed   Heat

Cholesta-4-ene-3-one, 96 mg
+ (CH₃CO)₂O, 1.0 mL
+ (CH₃)₃SiCl, 200 μL HOOD

NOTE. The glass equipment should be dried in an oven (110 °C) and cooled to room temperature in a desiccator before use.

Reagents and Equipment. Weigh and place 96 mg (0.25 mmol) of cholesta-4-ene-3-one in a dry 5.0-mL conical vial containing a magnetic spin vane. Now, in the **hood**, add 1.0 mL of acetic anhydride and 200 μL of chlorotrimethylsilane. Immediately attach the vial to a reflux condenser protected by a calcium chloride drying tube (◄).

NOTE. The quality of the acetic anhydride has a significant influence on the reaction. For best results, the anhydride should be distilled and stored over molecular sieves before use. It is convenient to dispense the anhydride and the chlorotrimethylsilane (which hydrolyzes rapidly in moist air) using automatic delivery pipets (in the **hood***). An alternative is to place each reagent in a small bottle sealed with a septum cap. The reagents are then removed through the septum using a 1-mL syringe with a Teflon-tipped plunger while the bottle is connected to another needle providing dry* *nitrogen. The chlorotrimethylsilane (bp 57 °C) may also be purified by distillation in the **hood**.*

HOOD

Reaction Conditions. Heat the reaction mixture with stirring at reflux for 1–2 h in a sand bath (145–150 °C). Follow the course of the reaction using TLC.

TLC DIRECTIONS. Use activated silica gel plates with 1:1 methylene chloride/hexane as the elution solvent. Visualization of the separated components is achieved by placing the plate in a closed jar containing a few crystals of iodine (see Prior Reading for further details).

Enol acetate collected here

Isolation of Product. Allow the reaction mixture to cool to room temperature and then place it in an ice bath for 15–30 min. A solid product forms during this time. Collect this solid by vacuum filtration using a Hirsch funnel, and wash the filter cake with 15 mL of 5% aqueous sodium bicarbonate, followed by 5 mL of cold water (➡).

5% NaHCO₃, 15 mL + cold H₂O, 5 mL

Purification and Characterization. Purify the crude product by column chromatography. In a 1 × 10-cm buret, place 2.0 g of activated silica gel (100 mesh) packed wet (slurry) with methylene chloride (➡). Dissolve the product in about 1.0 mL of hexane and transfer the solution by Pasteur pipet to the column. Elute the material from the column using approximately 10 mL of 1:1 methylene chloride/hexane solvent. Collect the eluate in a tared 25-mL filter flask.

Remove the solvent by warming in a sand bath under reduced pressure to give the solid product, cholesta-3,5-dien-3-ol acetate (▼). A rotary evaporator, if available, is a more rapid alternative. Recrystallize this material from methanol using a Craig tube, and dry the resulting crystals on a clay plate. Weigh the product and calculate the percent yield. Determine the melting point and compare it with the literature value shown in the Reactant and Product table. Obtain an IR spectrum of the material and compare it with that of an authentic sample, as well as with the spectrum of the starting material.

Sand, 50 mg

SiO₂, 2.0 g

Sand, 50 mg

1:1 Hexane/CH₂Cl₂, 10 mL

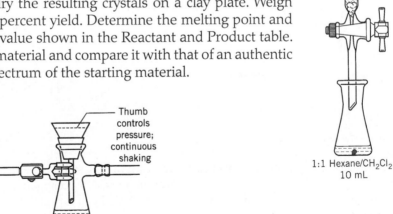

Thumb controls pressure; continuous shaking

1:1 Hexane/CH₂Cl₂, 10 mL + enol acetate

QUESTIONS

10-34. Reaction of $(CH_3)_3SiCl$ with alcohols produces a trimethylsilyl ether. For example,

$$CH_3CH_2CH_2\ddot{O}H + (CH_3)_3SiCl \xrightarrow[\text{ether}]{(C_2H_5)_3N:} CH_3CH_2CH_2\ddot{O}Si(CH_3)_3 + (C_2H_5)_3\overset{+}{N}H, Cl^-$$

The trimethylsilyl ether is more volatile than the corresponding alcohol, and is often used to facilitate GC analysis. Explain.

10-35. Chlorotrimethylsilane reacts with enolate anions to form stable silyl enol ethers. For example,

Assuming that the reaction is run under equilibrium conditions, predict which of the above silyl enol ethers is formed in the largest amount, and why.

10-36. In this experiment, the protons at *both* the γ and α′ positions relative to the intermediate oxonium ion are quite acidic. Removal of a proton from the γ position results in the formation (pathway *a* below) of the actual product, cholesta-3,5-dien-3-ol acetate (see Discussion section). Removal of a proton from the α′ position would result in the formation (pathway *b* below) of an isomeric product, cholesta-2,4-dien-3-ol acetate. Because the reaction is conducted under equilibrium conditions, the product obtained must be the thermodynamically more stable of the two possibilities. Why is cholesta-2,4-dien-3-ol acetate less stable than cholesta-3,5-dien-3-ol acetate?

Acylium ion   Cholesta-4-ene-3-one

Cholesta-2,4-dien-3-ol acetate

Cholesta-3,5-dien-3-ol acetate

10-37. There are two reasons why the enols of β-dicarbonyl compounds, such as pentane-2,4-dione are relatively stable. One is that they are conjugated and the π-electron overlap due to the conjugation provides additional stability. What is the other reason that these species are relatively stable? Would you expect the enol of cyclohexane-1,3-dione to be as stable as the enol of pentane-2,4-dione?

BIBLIOGRAPHY

The present experiment is adapted from the work reported by:

Chowdhury, P. K.; Sharma, R. P.; Barua, J. N. *Tetrahedron Lett.* **1983,** *24,* 3383.

For information on the formation and reactivity of enols, see:

House, H. O. *Modern Synthetic Reactions,* 2nd ed.; Benjamin: Reading, MA, 1972; Chapter 9, p. 492.

Rappoport, Z., Ed.; *The Chemistry of the Enols;* Wiley: New York, 1990.

Smith, M. B. *March's Advanced Organic Chemistry,* 8th ed.; Wiley: New York, 2019.

Absorb To take up matter (to dissolve), or to take up radiant energy.

Activated complex An unstable combination of reacting molecules that is intermediate between reactants and products.

Activation energy The minimum energy, $\Delta G^{\ddagger}$, necessary to form an activated complex in a reaction. Or the difference in energy levels between the ground state and transition state.

Active methylene A methylene group with hydrogen atoms rendered acidic due to the presence of an adjacent (α) electron withdrawing group, such as a carbonyl group.

Activity (of alumina) A measure of the degree to which alumina adsorbs polar molecules. The activity (absorptivity) of alumina may be reduced by the addition of small amounts of water. Thus, the amount of water present in a sample of alumina determines the activity grade. Alumina of a specific activity can be prepared by dehydrating alumina at 360 °C for about 5–6 h and then allowing the dehydrated alumina to absorb a suitable amount of water. The Brockmann scale of alumina activity is based on the amount of water (weight percent) that the alumina contains: grade I = 0%, grade II = 3%, grade III = 6%, grade IV = 10%, and grade V = 15%. For further information, see Brockmann, H.; Schodder, H. *Chem Ber.* **1941**, *74*, 73.

Adsorb The process by which molecules or atoms (either gas or liquid) adhere to the surface of a solid.

Aliphatic Term used to refer to nonaromatic species, such as alkanes, alkenes, alkynes.

Aliquot A portion.

Alkaloid A naturally occurring compound that contains a basic amine functional group. They are found particularly in plants.

Anilide A compound that contains a C_6H_5NHCO group. An amide formed by acylation of aniline (aminobenzene).

Atom economy A way of considering the efficiency of a chemical transformation using all atoms involved.

Bimolecular reaction The collision and combination of two reactants to give an activated complex in a reaction.

Capillary action The action by which the surface of a liquid, where it contacts a solid, is elevated or depressed because of the relative attractions of the molecules of the liquid for each other and for the solid. It is particularly observable in capillary tubes, where it determines the ascent (descent) of the liquid above (below) the level of the liquid in which the capillary tube is immersed.

Catalyst A substance that changes the speed of a chemical reaction without affecting the yield or undergoing permanent chemical change itself.

Characterize To conclusively identify a compound by the measurement of its physical, spectroscopic, and other properties.

Condensation reaction A condensation reaction is an addition reaction that produces water (or another small neutral molecule such as CH_3OH or NH_3) as a byproduct.

Dehydrohalogenation A reaction that involves loss of HX from a halide by treatment with strong base.

Deliquescent Liquefying by the absorption of water from the surrounding atmosphere.

Dihedral angle   The angle between two intersecting planes. In organic chemistry the term dihedral angle (or torsional angle) is used to describe the angle between two atoms (or groups) bonded to two adjacent atoms, such as H—C—C—H, and can be determined from a molecular model by looking down the axis of the bond between the two central atoms.

Dipole   The separation of charge in a bond or in a molecule with a positively and negatively charged end.

Eluant   A mobile phase in chromatography.

Eluate   The solution that is eluted from a chromatographic system.

Elute   To cause elution.

Elution   The flow, in chromatography, of the mobile phase through the stationary phase.

Emulsion   A suspension composed of immiscible drops of one liquid in another liquid (e.g., oil and vinegar in salad dressing).

Enol   A functional group composed of a hydroxyl group bonded to an alkene.

Enolate   The conjugate base of a enol, that is, a negatively charged oxygen atom bonded to an alkene. An enolate results from deprotonation α to a carbonyl group.

Enthalpy change (ΔH)   The heat lost or absorbed by a system under constant pressure during a reaction.

Entropy (S)   The randomness, or amount of disorder of a system.

Entropy change (ΔS)   The change in the amount of disorder.

Filter cake   The material that is separated from a liquid, and remains on the filter paper, after a filtration.

Free energy change (ΔG)   A predictor of the spontaneity of a chemical reaction at constant temperature. $\Delta G = \Delta H - T \Delta S$

Glacial acetic acid   Pure acetic acid containing less than 1% water.

Green Chemistry   Reaction conditions that are less deleterious to the environment than previous approaches.

Heterocycle   A cyclic molecule whose ring contains more than one kind of atom.

Heterolysis   Cleavage of a covalent bond in a manner such that both the bond's electrons end up on one of the formerly bonded atoms.

Homogeneous   Consisting of a single phase.

Homolysis   Cleavage of a covalent bond in a manner such that the bond's electrons are evenly distributed to the formerly bonded atoms.

Hydroboration   Addition of borane (BH_3) or an alkyl borane to a multiple bond.

Hydrogenation   Addition of hydrogen to a multiple bond.

Hygroscopic   Absorbs moisture.

In situ   In chemistry, the term usually refers to a reagent or other material generated directly in a reaction vessel and not isolated.

Kinetics   Referring to the rate of a reaction.

Lachrymator   A material that causes the flow of tears.

Ligroin   A solvent composed of a mixture of alkanes.

Mechanism   A complete description of how a reaction occurs.

Metabolism   The chemical processes performed by a living cellular organism.

Metabolites   The compounds consumed and produced by metabolism.

Methine   A CH group (with no other hydrogen atoms attached to the carbon atom).

Methylene   A CH_2 group (with no other hydrogen atoms attached to the carbon atom).

Mother liquor   The residual, and often impure, solution remaining from a crystallization.

Olefin   An older term for an alkene.

Optical isomers   Enantiomers. Isomers that have a mirror–image relationship.

Order of reaction   With respect to one of the reactants, the order of a reaction is equal to the power to which the concentration of that reactant is raised in the rate equation.

Oxonium ion   A trivalent oxygen cation with a full octet of electrons (e.g., H_3O^+).

Paraffins   An older name for alkanes.

Phase transfer catalysts   Agents that cause the transfer of ionic reagents between phases, thus catalyzing reactions.

Plasticizer   A substance added to a polymer to make it more flexible or to prevent embrittlement.

Polymer   A compound of high molecular mass that is built up of a large number of repeating simple molecules, or monomers.

Racemic   Consisting of an equimolar mixture of two enantiomers.

Rate equations   Equations giving the relationship between reaction rate and the concentrations of the reactants.

Reaction mechanism   The stepwise sequence of elementary reactions in an over-all reaction.

Reagent   A chemical or solution used in the laboratory to detect, measure, react with, or otherwise examine other chemicals, solutions, or substances.

Reflux   The process by which all vapor evaporated or boiled from a vessel is condensed and returned to that vessel.

Rotamers   Conformational isomers that can be interconverted by rotation about one or more single bonds (e.g., *gauche* and *anti* butane).

Spontaneous process   A physical or chemical change that occurs without the net addition of energy. $\Delta G < 0$ for a spontaneous process.

Sublimation   The passing of a solid directly into vapor state without first melting.

Tare   A tared container is one whose weight has been measured. The term may also refer to the process of zeroing a balance after a container has been placed on the weighing platform.

Thermodynamics   The chemical science that deals with the energy transfers and transformations that accompany chemical and physical changes.

Transition state   A combination of reacting molecules that is intermediate between reactants and products.

Triturate   To grind to a fine powder. (Or, washing solid organic products in a solvent in which the desired product has little solubility.)

Vapor pressure   The pressure exerted by a vapor in equilibrium with a liquid or solid at a given temperature.

Ylide   A neutral dipolar molecule in which negative and positive charges are on adjacent atoms.

Zwitterion   A neutral molecule containing separated opposite formal charges.

PERIODIC TABLE OF THE ELEMENTS

Legend:
Atomic number → 6
Symbol (IUPAC) → C
Name (IUPAC) → Carbon
Atomic mass → 12.011

IUPAC recommendations →
Chemical Abstracts Service group notation →

1 IA	2 IIA	3 IIIB	4 IVB	5 VB	6 VIB	7 VIIB	8 VIIIB	9 VIIIB	10 VIIIB	11 IB	12 IIB	13 IIIA	14 IVA	15 VA	16 VIA	17 VIIA	18 VIIIA
1 **H** Hydrogen 1.0079																	2 **He** Helium 4.0026
3 **Li** Lithium 6.941	4 **Be** Beryllium 9.0122											5 **B** Boron 10.811	6 **C** Carbon 12.011	7 **N** Nitrogen 14.007	8 **O** Oxygen 15.999	9 **F** Fluorine 18.998	10 **Ne** Neon 20.180
11 **Na** Sodium 22.990	12 **Mg** Magnesium 24.305											13 **Al** Aluminum 26.982	14 **Si** Silicon 28.086	15 **P** Phosphorus 30.974	16 **S** Sulfur 32.065	17 **Cl** Chlorine 35.453	18 **Ar** Argon 39.948
19 **K** Potassium 39.098	20 **Ca** Calcium 40.078	21 **Sc** Scandium 44.956	22 **Ti** Titanium 47.867	23 **V** Vanadium 50.942	24 **Cr** Chromium 51.996	25 **Mn** Manganese 54.938	26 **Fe** Iron 55.845	27 **Co** Cobalt 58.933	28 **Ni** Nickel 58.693	29 **Cu** Copper 63.546	30 **Zn** Zinc 65.409	31 **Ga** Gallium 69.723	32 **Ge** Germanium 72.64	33 **As** Arsenic 74.922	34 **Se** Selenium 78.96	35 **Br** Bromine 79.904	36 **Kr** Krypton 83.798
37 **Rb** Rubidium 85.468	38 **Sr** Strontium 87.62	39 **Y** Yttrium 88.906	40 **Zr** Zirconium 91.224	41 **Nb** Niobium 92.906	42 **Mo** Molybdenum 95.94	43 **Tc** Technetium (98)	44 **Ru** Ruthenium 101.07	45 **Rh** Rhodium 102.91	46 **Pd** Palladium 106.42	47 **Ag** Silver 107.87	48 **Cd** Cadmium 112.41	49 **In** Indium 114.82	50 **Sn** Tin 118.71	51 **Sb** Antimony 121.76	52 **Te** Tellurium 127.60	53 **I** Iodine 126.90	54 **Xe** Xenon 131.29
55 **Cs** Cesium 132.91	56 **Ba** Barium 137.33	57 ***La** Lanthanum 138.91	72 **Hf** Hafnium 178.49	73 **Ta** Tantalum 180.95	74 **W** Tungsten 183.84	75 **Re** Rhenium 186.21	76 **Os** Osmium 190.23	77 **Ir** Iridium 192.22	78 **Pt** Platinum 195.08	79 **Au** Gold 196.97	80 **Hg** Mercury 200.59	81 **Tl** Thallium 204.38	82 **Pb** Lead 207.2	83 **Bi** Bismuth 208.98	84 **Po** Polonium (209)	85 **At** Astatine (210)	86 **Rn** Radon (222)
87 **Fr** Francium (223)	88 **Ra** Radium (226)	89 **#Ac** Actinium (227)	104 **Rf** Rutherfordium (261)	105 **Db** Dubnium (262)	106 **Sg** Seaborgium (266)	107 **Bh** Bohrium (264)	108 **Hs** Hassium (277)	109 **Mt** Meitnerium (268)	110 **Ds** Darmstadtium (281)	111 **Rg** Roentgenium (272)	112 **Cn** Copernicium (285)	113 **Uut** (284)	114 **Fl** Flerovium (289)	115 **Uup** (288)	116 **Lv** Livermorium (293)	117 **Uus** (294)	118 **Uuo** (294)

*Lanthanide Series

58 **Ce** Cerium 140.12	59 **Pr** Praseodymium 140.91	60 **Nd** Neodymium 144.24	61 **Pm** Promethium (145)	62 **Sm** Samarium 150.36	63 **Eu** Europium 151.96	64 **Gd** Gadolinium 157.25	65 **Tb** Terbium 158.93	66 **Dy** Dysprosium 162.50	67 **Ho** Holmium 164.93	68 **Er** Erbium 167.26	69 **Tm** Thulium 168.93	70 **Yb** Ytterbium 173.04	71 **Lu** Lutetium 174.97

Actinide Series

90 **Th** Thorium 232.04	91 **Pa** Protactinium 231.04	92 **U** Uranium 238.03	93 **Np** Neptunium (237)	94 **Pu** Plutonium (244)	95 **Am** Americium (243)	96 **Cm** Curium (247)	97 **Bk** Berkelium (247)	98 **Cf** Californium (251)	99 **Es** Einsteinium (252)	100 **Fm** Fermium (257)	101 **Md** Mendelevium (258)	102 **No** Nobelium (259)	103 **Lr** Lawrencium (262)

Common Organic Solvents: Table of Properties

Solvent	Formula	MW	Boiling Point (°C)	Melting Point (°C)	Density (g/mL)	Solubility in Water (g/100g)	Dielectric Constant	Flash Point (°C)
acetic acid	$C_2H_4O_2$	60.05	118	16.6	1.049	Miscible	6.15	39
acetone	C_3H_6O	58.08	56.2	−94.3	0.786	Miscible	20.7(25)	−18
acetonitrile	C_2H_3N	41.05	81.6	−46	0.786	Miscible	37.5	6
benzene	C_6H_6	78.11	80.1	5.5	0.879	0.18	2.28	−11
1-butanol	$C_4H_{10}O$	74.12	117.6	−89.5	0.81	6.3	17.8	35
2-butanol	$C_4H_{10}O$	74.12	98	−115	0.808	15	15.8(25)	26
2-butanone	C_4H_8O	72.11	79.6	−86.3	0.805	25.6	18.5	−7
t-butyl alcohol	$C_4H_{10}O$	74.12	82.2	25.5	0.786	Miscible	12.5	11
carbon tetrachloride	CCl_4	153.82	76.7	−22.4	1.594	0.08	2.24	—
chlorobenzene	C_6H_5Cl	112.56	131.7	−45.6	1.1066	0.05	5.69	29
chloroform	$CHCl_3$	119.38	61.7	−63.7	1.498	0.795	4.81	—
cyclohexane	C_6H_{12}	84.16	80.7	6.6	0.779	<0.1	2.02	−20
1,2-dichloroethane	$C_2H_4Cl_2$	98.96	83.5	−35.3	1.245	0.861	10.42	13
diethyl ether	$C_4H_{10}O$	74.12	34.6	−116.3	0.713	7.5	4.34	−45
diethylene glycol	$C_4H_{10}O_3$	106.12	245	−10	1.118	10	31.7	143
diglyme (diethylene glycol dimethyl ether)	$C_6H_{14}O_3$	134.17	162	−68	0.943	Miscible	7.23	67
1,2-dimethoxy-ethane (glyme, DME)	$C_4H_{10}O_2$	90.12	85	−58	0.868	Miscible	7.2	−6
dimethylether	C_2H_6O	46.07	−22	−138.5	NA	NA	NA	−41
dimethyl-formamide (DMF)	C_3H_7NO	73.09	153	−61	0.944	Miscible	36.7	58
dimethyl sulfoxide (DMSO)	C_2H_6OS	78.13	189	18.4	1.092	25.3	47	95
dioxane	$C_4H_8O_2$	88.11	101.1	11.8	1.033	Miscible	2.21(25)	12
ethanol	C_2H_6O	46.07	78.5	−114.1	0.789	Miscible	24.6	13
ethyl acetate	$C_4H_8O_2$	88.11	77	−83.6	0.895	8.7	6(25)	−4
ethylene glycol	$C_2H_6O_2$	62.07	195	−13	1.115	Miscible	37.7	111
glycerin	$C_3H_8O_3$	92.09	290	17.8	1.261	Miscible	42.5	160
heptane	C_7H_{16}	100.20	98	−90.6	0.684	0.01	1.92	−4
hexamethylphosphoramide (HMPA)	$C_6H_{18}N_3OP$	179.20	232.5	7.2	1.03	Miscible	31.3	105
hexamethylphosphorous triamide (HMPT)	$C_6H_{18}N_3P$	163.20	150	−44	0.898	Miscible	??	26
hexane	C_6H_{14}	86.18	69	−95	0.659	0.014	1.89	−22
methanol	CH_4O	32.04	64.6	−98	0.791	Miscible	32.6(25)	12
methyl t-butyl ether (MTBE)	$C_5H_{12}O$	88.15	55.2	−109	0.741	5.1	??	−28
methylene chloride	CH_2Cl_2	84.93	39.8	−96.7	1.326	1.32	9.08	1.6
N-methyl-2-pyrrolidinone (NMP)	CH_5H_9NO	99.13	202	−24	1.033	10	32	91
nitromethane	CH_3NO_2	61.04	101.2	−29	1.382	9.50	35.9	35
pentane	C_5H_{12}	72.15	36.1	−129.7	0.626	0.04	1.84	−49
petroleum ether (ligroin)	—	—	30–60	−40	0.656	—	—	−30
1-propanol	C_3H_8O	88.15	97	−126	0.803	Miscible	20.1(25)	15
2-propanol	C_3H_8O	88.15	82.4	−88.5	0.785	Miscible	18.3(25)	12
pyridine	C_5H_5N	79.10	115.2	−41.6	0.982	Miscible	12.3(25)	17
tetrahydrofuran (THF)	C_4H_8O	72.11	66	−108.4	0.886	30	7.6	−21
toluene	C_7H_8	92.14	110.6	−93	0.867	0.05	2.38(25)	4
triethyl amine	$C_6H_{15}N$	101.19	88.9	−114.7	0.728	0.02	2.4	−11
water	H_2O	18.02	100.00	0.00	0.998	—	78.54	—
water, heavy	D_2O	20.03	101.3	4	1.107	Miscible	??	—
o-xylene	C_8H_{10}	106.17	144	−25.2	0.897	Insoluble	2.57	32
m-xylene	C_8H_{10}	106.17	139.1	−47.8	0.868	Insoluble	2.37	27
p-xylene	C_8H_{10}	106.17	138.4	13.3	0.861	Insoluble	2.27	27

T = 20 °C unless specified otherwise.

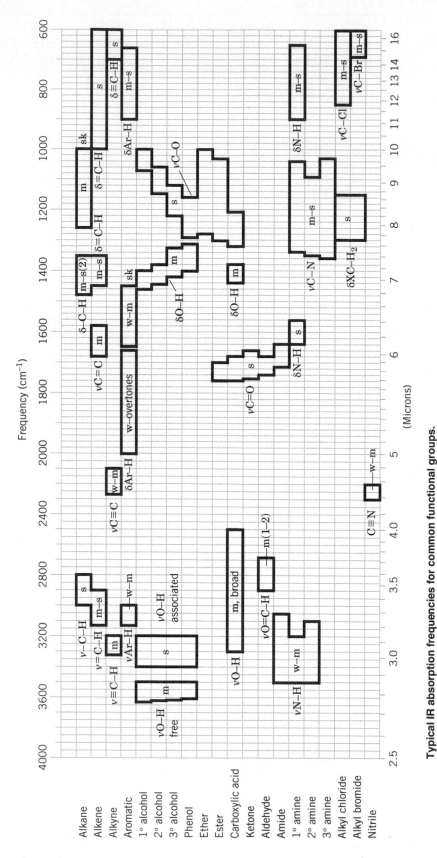

Typical IR absorption frequencies for common functional groups.
Absorptions are as follows: ν = stretching; δ = bending; w = weak; m = medium; s = strong; sk = skeletal
From *Multiscale Organic Chemistry: A Problem-Solving Approach* by John W. Lehman © 2002.
Reprinted by permission of Pearson Education, Inc., Upper Saddle River, NJ.

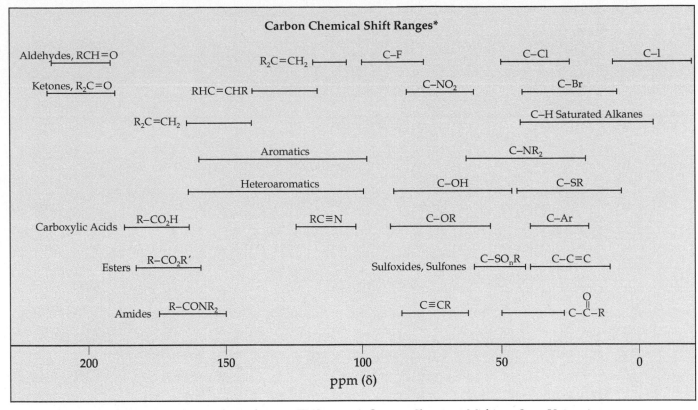

Carbon Chemical Shift Ranges*

*For samples in CDCl₃ solution. The δ scale is relative to TMS at δ = 0. Organic Chemistry Michigan State University Source, Dept. of Chemistry, Michigan State University.

¹H NMR Chemical Shift Ranges